physica status solidi

VOLUME 16 · NUMBER 2 · 1966

AKADEMIE-VERLAG · BERLIN

THEORIA CUM PRAXI

Contents

(Continued on cover three)

physica status solidi

Volume 16 · Number 2 · Pages 349 to 802 and K 105 to K 220

August 1, 1966

AKADEMIE-VERLAG · BERLIN

Subscriptions and orders for single copies should be addressed to

AKADEMIE-VERLAG GmbH, 108 Berlin, Leipziger Straße 3—4 or to Buchhandlung KUNST UND WISSEN, Erich Bieber, 7 Stuttgart 1, Wilhelmstr. 4—6 or to Deutsche Buch-Export und -Import GmbH, 701 Leipzig, Postschließfach 160

Editorial Note:

"physica status solidi" undertakes that an original paper *accepted for publication* before the 8th of any month will be published within 50 days of this date unless the author requests a postponement. In special cases there may be some delay between *receipt* and *acceptance* of a paper due to the review and, if necessary, revision of the paper.

Schriftleiter und verantwortlich für den Inhalt: Professor Dr. Dr. h. c. P. Görlich, 102 Berlin, Neue Schönhauser Str. 20 bzw. 69 Jena, Humboldtstr. 26. Redaktionskollegium: Dr. S. Oberländer, Dr. E. Gutsche, Dr. W. Borchardt. Anschrift der Schriftleitung: 102 Berlin, Neue Schönhauser Str. 20, Fernruf: 42 67 88. Verlag: Akademie-Verlag GmbH, 108 Berlin, Leipziger Str. 3—4, Fernruf: 22 04 41, Telex-Nr. 011 773, Postscheckkonto: Berlin 35 021. — Die Zeitschrift „physica status solidi" erscheint jeweils am 1. des Monats. Bezugspreis eines Bandes MDN 60,—. Bestellnummer dieses Bandes 1068/16. Jeder Band enthält zwei Hefte. Gesamtherstellung: VEB Druckerei „Thomas Müntzer" Bad Langensalza. — Veröffentlicht unter der Lizenznummer 1310 des Presseamtes beim Vorsitzenden des Ministerrates der Deutschen Demokratischen Republik.

Review Article

phys. stat. sol. **16**, 351 (1966)

Solid State Physics Division, Mullard Research Laboratories, Salfords, Redhill (Surrey)

Alloyed Semiconductor Heterojunctions

By

J. R. Dale

Contents

1. Introduction

Heterojunctions are defined as junctions between two semiconductors with different energy band structures. In principal, they may be either gradual junctions in which two bulk crystals are joined by an alloy series of continuously varying composition, or abrupt junctions in which there is a sharply defined interface between two homogeneous semiconductors.

Interest in heterojunction devices was stimulated in the late 1950's by the work of Kroemer. The concept of quasielectric and magnetic fields for variable energy gap semiconductors was initiated [1, 2] and a theory presented [3] for a wide band gap emitter transistor. Later a number of articles by other investigators were published [4, 5, 6] which dealt with similar considerations. The interest in heterojunctions has also spread to the field of solar energy convertors [7, 8]. Theoretical considerations by Emtage [9] have shown that it might be possible to incorporate a heterojunction type structure into a photovoltaic convertor which could give an efficiency as high as 40%. With recent observation of highly efficient light emission from forward biased p-n junctions in GaAs [10, 11] a number of opto-electronic transistor devices have been proposed [12 to 16] using heterojunction structures.

With the advent of epitaxial vapour growth techniques [17] considerable interest was stimulated in heterojunctions where one crystal was vapour deposited on another. Anderson [18 to 20] and Marinace [21] first reported the properties and theoretical problems associated with this type of junction. A number of workers [22 to 32] carried out a more intensive investigation into the Ge–GaAs system originally studied by Anderson and extended the work to other systems, such as Ge–Si, InP–GaAs, GaP–GaAs, Ge–GaP, InAs–GaAs, $GaAs_{1-x}P_x$–GaAs and Ga_xIn_{1-x}–GaAs. In this series of papers the general conclusion reached appears to be that there are a large number of discrepancies which make uncertain the complete validity of the dicontinuous energy gap model, which shows energy discontinuities in the conduction and valence band edges at the junction interface.

An attractive and alternative approach to the fabrication of heterojunctions by vapour transport technique is that of alloying. Three different systems have been used, these are:

a) Interface alloying where the low melting point material is placed in a temperature gradient, with the high temperature side in contact with the high melting point material.

b) A solution growth process where one of the materials is put into solution with a suitable solvent at an elevated temperature. This solution is then put in contact with the substrate and allowed to cool slowly growing a semiconductor layer which may contain some of the solvent and substrate.

c) A combination of both interface alloying and vapour deposition. Here the substrate is held at a temperature higher than the melting point of the vapour deposited material so that on deposition alloying takes place at the interface.

Interface alloying techniques have been used to make heterojunctions betweeen GaAs–Ge, GaAs–GaSb [33, 35], InSb–GaSb [34], Ge–Si [36], and InAs–GaSb [37]. Solution growth techniques have been used to grow heterojunctions between GaAs–$GaSb_xAs_{1-x}$, GaAs–$Ga_xIn_{1-x}As$, GaAs–Mn_2As [38], GaAs–GaP [28], and Ge–Ge_xSi_{1-x} [88]. The vapour transport–interface alloying technique has at present only been successfully used to make Ge–Si alloyed heterojunctions [39]. However at the present time, since little attention has so far been given to the metallurgical variables of the processes which are of considerable importance in the interpretation of junction phenomena, any interpretation of the measurements obtained from alloyed heterojunctions, in terms of simple heterojunction theory, must be viewed with considerable scepticism.

2. Heterojunction Structures

2.1 *Metallurgical considerations*

In a semiconductor heterojunction it is implied that there is an intimate contact or bond between the different semiconductor materials. The degree of perfection of fit can vary to a great extent and is dependent on the lattice match of

Table 1
Some important parameters in the selection of heterojunction pairs

Semiconductor	Crystal structure	Lattice constant [69] (Å)	Energy gap [24] (eV)	Melting point (°C)	Coefficient of linear expansion per degree	Microhardness Knoop [73]	Density [70] g/cm^3
InSb	zinc-blende	6.48	0.16	525	4.9×10^{-6} [74]	220	5.77
InAs	zinc-blende	6.05	0.36	942	4.5×10^{-6} [74]	380	5.66
GaSb	zinc-blende	6.10	0.67	712	6.9×10^{-6} [74]	450	5.32
Ge	diamond	5.68	0.66	958	6.1×10^{-6} [71]	700 to 880*)	5.33
Si	diamond	5.43	1.08	1420	2.33×10^{-6} [70]	1100 to 1400*)	2.33
InP	zinc-blende	5.87	1.29	1062	4.5×10^{-6} [74]	540	4.78
GaAs	zinc-blende	5.64	1.40	1237	5.5×10^{-6} [72]	750	5.32
GaP	zinc-blende	5.45	2.25	1465	5.3×10^{-6}	950	4.13

*) Depending on orientation.

the two materials, their orientation, contamination by impurities, and the continuity of any polar bonds which are involved in the system, for example such as exist in the Group III to V semiconductors.

Lattice constants for some diamond and zinc blende structures are listed in Table 1. From this table it can be seen that very few semiconductor pairs with close la'tice constants can be found if systems such as GaAs–$Ga_xIn_{1-x}As$ are neglected. For many applications the most desirable heterojunctions would be between two semiconductors of a three component system, here lattice mismatch of 3 to 4% or greater would be expected, except in the case of Ge–GaAs where a lattice mismatch of only 0.07% is observed. As a result of this, in most heterojunctions, defects mainly in the form of edge type dislocations must be present near the interface. These will give a number of dangling bonds or bond deficiencies which will tend to produce an interface charge.

Fig. 1a shows dislocations at the boundary of two diamond type semiconductors joined at the (111) plane and Fig. 1b shows in three dimensions, a grid consisting of three equiangularly spaced sets of edge type dislocations on the (111) plane. The minimum number of dislocations of any type which must be present has been calculated by Oldham and Milnes [40] directly from the difference in surface bond energy on the semiconductors. For semiconductors having lattice constants λ_1 and λ_2 where λ_1 is less than λ_2 the difference in surface bond energy ΔN_s is given by $\Delta N_s = N_{s1} - N_{s2}$ where N_{s1} and N_{s2} are the surface bond energies on either semiconductor which depend only on λ_1 and λ_2 and the particular interface plane. ΔN_s is the minimum dangling bond density in the interface plane and is therefore equal to the total length of dislocations per unit area times the number of dangling bonds per unit length. When there are i sets of dislocations with each set consisting of parallel dislocations with a spacing h between the dislocations in the set, then $\Delta N_s = i/hc$ Here c is the mean spacing between dangling bonds on the dislocation and i is determined by the symmetry of the interface plane. Read [41] has developed a formula for c as a function of the magnitude of the Burgers vector b and the

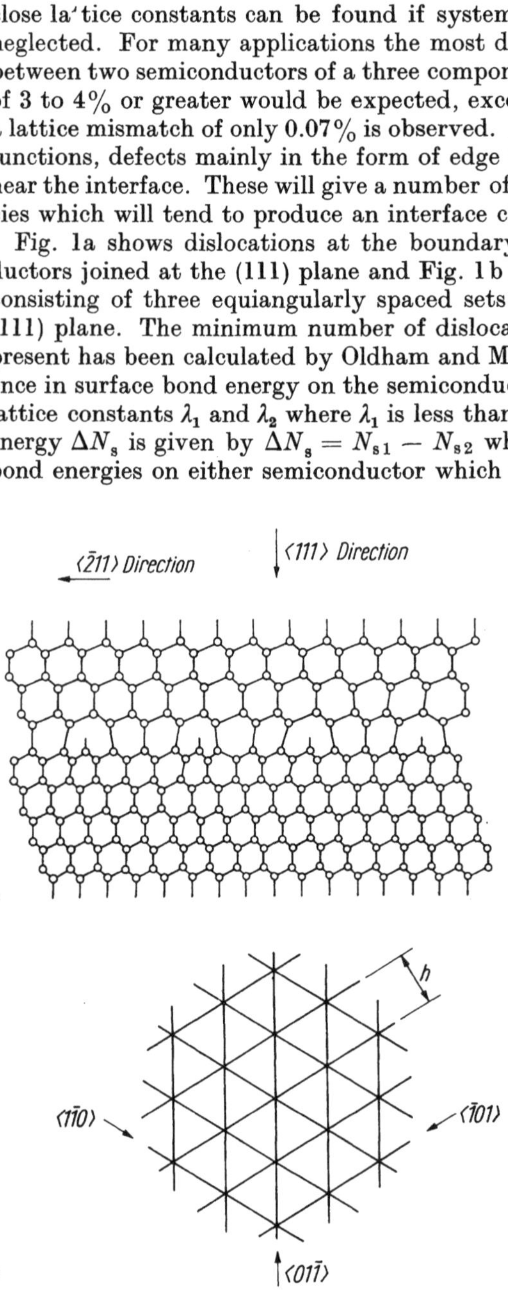

a
b

Fig. 1. a) A $\langle 01\bar{1}\rangle$ view of some partial edge dislocations lying in the (111) plane at a 29% lattice mismatch. The dislocations have burgers vectors lying in the $\langle\bar{1}10\rangle$ or $\langle 10\bar{1}\rangle$ direction.
b) A $\langle 111\rangle$ view of a possible single array of partial edge dislocations for a (111) heterojunction. The dislocation lines lie in the $\langle 01\bar{1}\rangle$, $\langle\bar{1}01\rangle$, and $\langle 1\bar{1}0\rangle$ directions and the burgers vectors in the $\langle\bar{1}10\rangle$, $\langle 0\bar{1}1\rangle$, and $\langle 10\bar{1}\rangle$ directions (not shown).
(After Oldham and Milnes [40])

angle θ between the Burgers vector and the dislocation line: $c = b\ V(3)/2 \sin \theta$. Oldham and Milnes have shown that h can be determined from these basic equations if the dislocation type and the interface plane has been selected: a typical value of h for an abrupt Ge–Si(111) junction would be 155 Å, and 8166 Å for a Ge–GaAs junction. If a structure is graded and not abrupt, the predicted number of dangling bonds is not reduced and the dislocation density not significantly changed.

Calculations by Oldham and Milnes [40] have also indicated that for systems with very large mismatch it becomes energetically favourable to produce dislocations at the interface rather than grow the layer in uniform sheet strain with the substrate lattice constant.

The overall effect of the mismatch is also to some extent dependent on the thickness of the two semiconductor materials involved. For instance in the case of a very thin layer on a thick substrate, elastic deformation of the thin layer would take place to give a better lattice match. This would prevent dislocations forming, since the elastic deformation energy can, for the very thin layers, be less than the energy needed to produce a dislocation. At present, since little is known of the interface dislocation energies it is difficult to predict the thickness of various materials at which dislocation generation occurs. However in the Ge–Si system for example, only four or five atomic layers are probably needed if they possess bulk properties. This is supported by Oldham and Milnes [40] who have predicted that dislocations in a Ge–Si junction would start to form in the overgrowth or at the interface after about 15 Å of growth and in a Ge–GaAs junction at about 2000 Å.

For heterojunctions having an interface lattice mismatch of the order of 2 to 4% Oldham and Milnes have assumed that dislocations lie in a sheet at the heterojunction interface. This would give the interface electrical properties similar to those expected from a crystal grain boundary.

Edge dislocations have been shown by many workers to act as very active recombination centres between holes and electrons. Even when two structures have essentially the same lattice constants at room temperature, a mismatch occurs at other temperatures when they have different coefficients of expansion. This means that the cooling cycle after fabrication is very critical. Dislocations will be introduced during cooling because a good lattice match is unlikely to exist over a wide range of temperatures. In brittle materials this shrinkage often causes cracking and has been observed in Ge–Si and other alloyed heterojunctions where cooling rates have been too rapid. In the Ge–Si interface the mismatch is of the order of 4×10^{-6} per °C and of the order of 5×10^{-7} per °C in the Ge–GaAs system where cracking is not normally a problem. Table 1 gives some important parameters that must be considered in the selection of materials for heterojunction pairs.

In the growth of layers of Ge on GaAs by the interface alloy technique both the lattice constants and thermal expansion coefficients (see Table 1) of the materials are close together, but other defects can be produced because of the high fabrication temperatures (approximately 950 °C) and experimental limitations encountered. Rediker et al. [33] found that Ge and GaAs single crystals when recrystallised had slight rotation or tilts relative to each other when using (111) interfaces, but that the resultant strain was relieved at a short distance from the junction. These types of defect make electrical determinations of the heterojunction properties difficult as they contribute obscuring features to the electrical parameters.

In semiconductor heterojunctions it is possible to make p-n, n-n or p-p junctions to evaluate their properties. To utilise this property fully, however, the electrical junction must coincide with the heterojunction interface and not form a homojunction on either side of the interface. In addition to this, the materials should not interact in any way that would obscure interpretation of the results. For instance in the Ge–GaAs system, Ge may diffuse into the GaAs. When incorporated into Ga sites in the GaAs, Ge acts as an n-type dopant. In a similar manner Ga and As from the GaAs may diffuse in the opposite direction and can act as p-type or n-type dopants in Ge. These types of secondary reaction can obviously lead to complete masking of the true properties of heterojunctions and must be avoided or the results treated with some caution.

2.2 Theoretical considerations

2.2.1 Energy band diagram

When materials having different work functions are brought into intimate contact, a charge redistribution occurs, and if at least one of the materials is a semiconductor, a rectifying barrier may be produced. P-n junctions in homogeneous semiconductors and metal-semiconductor contacts have been the subject of prolonged study but heterojunctions by comparison have received little attention. Impurities, interface characteristics, and the lack of crystal continuity have presented serious problems in achieving well defined models of behaviour especially in alloyed heterojunctions. The work of Anderson [18, 20, 42], Oldham and Milnes [24, 40], Hampshire and Wright [43], and Fang and Howard [44] using epitaxial vapour techniques to fabricate heterojunctions between Ge–GaAs, InP–GaAs, and Ge–Si, have brought our knowledge of this subject a considerable step forward.

Energy band diagrams for two separate uncharged n-type semiconductors are shown in Fig. 2a where Φ_1 and Φ_2 are the work functions of the two semiconductors, X_1 and X_2 are the electron affinities. The electron affinity can be defined as the energy required to remove an electron to infinity from the bottom of the conduction band of an uncharged crystal [45, 46]. From this it can be seen that the electron affinity is a fundamental crystalline property and for non-degeneracy is independent of doping. Typical values of this are 4.25 eV for Ge, 4.15 eV for Si, and 3.65 eV for GaAs [42]. Fig. 2b shows the effect of bringing these two crystals into intimate contact. An electron flow from semiconductor 1 to semiconductor 2 occurs giving rise to a diffusion voltage U_d where $U_d = (1/q) \times$

a

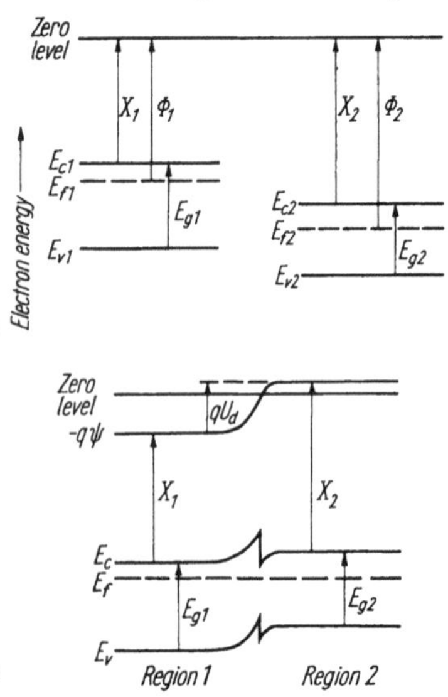

b

Fig. 2. a) Two isolated uncharged semiconductors. b) The ideal semiconductor-semiconductor junction. (After Oldham and Milnes [24])

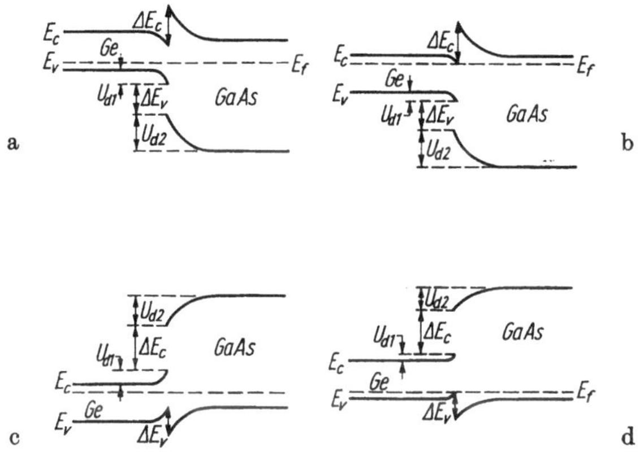

Fig. 3. Equilibrium energy band diagrams in the interface regions of p-n, n-n, n-p, and p-p Ge-GaAs heterojunctions. (After Lopez and Anderson [47])

$\times (\Phi_2 - \Phi_1)$. Since equilibrium conditions require the Fermi level to be constant, the total energy of the conduction band electrons becomes the crystal energy $-x$ plus the electrostatic energy $-q\,\psi$ from the charge redistribution. This model represents an ideal n-n rectifying heterojunction. If the two conduction band barriers are greater than a few $k\,T$, semiconductor 1 will acquire a depletion layer near the interface and semiconductor 2 an accumulation layer. Fig. 3a, b, c, and d represent the predicted equilibrium energy band diagrams for the interface regions of p-n, n-n, n-p, and p-p Ge–GaAs heterojunctions.

The Ge–GaAs system has probably received the most attention of investigators and theoreticians, mainly because vapour transport techniques have made fabrication relatively simple. The bulk properties of the two materials have also been the object of intensive study as have the properties of homojunctions in each material. In addition, since the crystals have similar crystal structures and almost the same lattice constants (Ge 5.657 Å, GaAs 5.653 Å) the strain due to lattice mismatch should be negligible. The resultant influence on the electrical characteristics due to interface states is also expected to be small.

The electrical characteristics of this type of heterojunction agree with those predicted by the energy band model where it is assumed that the energy band characteristics of either bulk material remains unchanged up to the interface. At the interface discontinuities in conduction (ΔE_c) and valence band edges (ΔE_v) occur. Regions of space charge with the resultant electrostatic fields are present in both materials because of the difference in work functions of the two materials. These heterojunctions have in general rectification properties, but since the forbidden gap in GaAs overlaps that of Ge, the direction of rectification is determined by the type of doping in the GaAs. The behaviour of these functions has been discussed by Anderson [18, 19] who has generalised the conventional homojunction theory [50] of the electrostatic charge distribution and the differential capacitance, $C \equiv dQ/dV$, (obtained by solving Poisson's equation for a space charge determined by local equilibrium) for heterojunction systems.

In p-n and n-p type heterojunctions a depletion region is found on either side of the interface. The relative potential supported in each of the semiconductors is

$$\frac{U_{d_1} - U_1}{U_{d_2} - U_2} = \frac{N_{a_2}\,\varepsilon_2}{N_{d_1}\,\varepsilon_1},$$

where U_{d_1} and U_{d_2} are the portions of the applied voltage U appearing on the Ge and GaAs sides, respectively, while N_{d_1} and ε_1 are the total donor concentration and the permitivity for the Ge side and N_{a_2} and ε_2 denote similar qualities for the GaAs side. The ratio of the transition region widths on either side of an abrupt junction is

$$\frac{W_2}{W_1} = \frac{N_{d_1}}{N_{a_2}}$$

while the transition region width on the GaAs side is

$$W_2 = \left[\frac{2}{9}\,\frac{N_{d_1}}{N_{a_2}}\,\frac{\varepsilon_1\,\varepsilon_2\,(U_d - U)}{\varepsilon_1\,N_{d_1} + \varepsilon_2\,N_{a_2}}\right]^{1/2}.$$

The transition capacitance for an abrupt n-p junction is given by a generalisation of the usual results:

$$C_{n-p} = \left[\frac{q\,N_{d_1}\,N_{a_2}\,\varepsilon_1\,\varepsilon_2}{2\,(\varepsilon_1\,N_{d_1} + \varepsilon_2\,N_{d_2})\,(U_d - U)}\right]^{1/2}.$$

In an n-n junction, however, the relative alignment of the Fermi levels with respect to the band edges is such that the space charge in the Ge region consists of mobile electrons in the conduction band. Except for very small values of $U_d - U$ the voltage drop is expected to be predominantly in the GaAs where the capacitance is determined primarily by the characteristics of the GaAs. For a constant impurity density in the GaAs the transition capacitance is

$$C_{n-n} = \left[\frac{q\,N_{d_2}\,\varepsilon_2}{2\,(U_d - U)}\right]^{1/2}.$$

However for cases where the impurity density varies with position, neither of the above equations will apply. If the capacitance can be expressed as $C = C_1 \times < (U_d - U)^{-\alpha}$ where α can be determined experimentally a plot of U vs. $(1/C)\ 1/\alpha$ extrapolated to a zero value of $(1/C)\ 1/\alpha$ gives the built in voltage. From values of C_1, α, and junction area, the nett ionised impurity concentration on the more lightly doped side of the junction can be obtained.

A junction capacitance which exhibits a cube law dependence on voltage is a characteristic of a linearly graded junction. In practice a junction approaching this is obtained when the materials used are mutually soluble or where diffusion readily occurs at the preparation temperature. In either case interpretation of junction phenomena is extremely difficult and questionable since rectification could be due to impurity distribution in either the recrystallised region of the interface alloyed junctions or at diffusion fronts in either of the two bulk materials.

2.2.2 Photocurrent response

A characteristic of both alloyed and vapour deposited heterojunctions which has aroused much interest and investigation [33, 47, 48, 52] is the photocurrent response. The general characteristics of this response can be explained by referring to Fig. 4 which shows a schematic representation of photon absorption and the resultant production of photocurrent in a Ge–GaAs heterojunction. When the doping level in the Ge is less than that in the GaAs, a considerable amount of the transition region is on the Ge side of the junction. Carriers optically generated in the GaAs, within a diffusion length of the transition region,

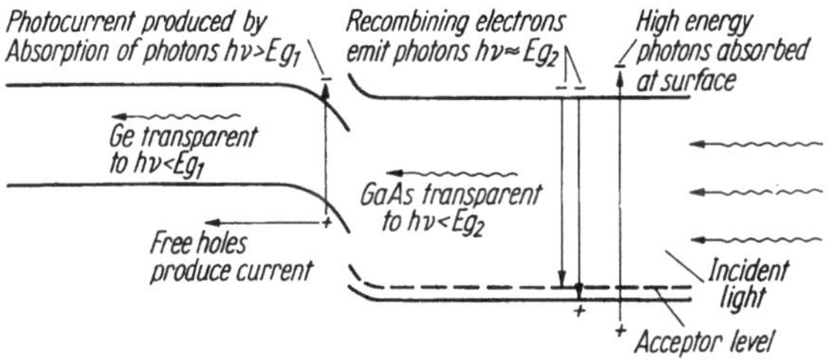

Fig. 4. A schematic representation of photon absorption and the production of photocurrent in a Ge–GaAs heterojunction.
(After Lopez and Anderson [47])

can contribute to the photocurrent. However in the Ge, only the carriers generated in the transition region itself can contribute to the photocurrent (except for higher order effects), since the discontinuity or adverse potential gradient in the band edges prevent carrier diffusion from the Ge to the GaAs. The low energy photons ($h\nu < 0.65$ eV) are unable to excite hole–electron pairs in either the Ge or the GaAs with the result that no photocurrent is generated. At photon energies close to the band gap of the Ge indirect or phonon assisted absorption occurs. Radiation is transmitted to the Ge, since the GaAs is reasonably transparent at this wavelength, and absorption therefore occurs in the Ge transition region producing a photocurrent. At the energy required for direct transition, the absorption increases substantially. At energies above the band gap for GaAs, the photons are absorbed near the GaAs surface. However, electron-hole pairs created at the surfaces cannot produce a photocurrent directly but they can contribute to the photocurrent, since on recombination lower energy photons are produced. Some of these photons will reach the transition zone since the absorption constant within the GaAs is smaller for these energies. A fraction

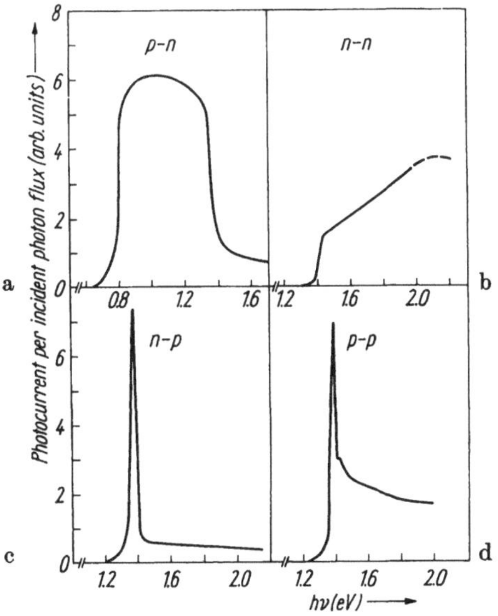

Fig. 5. Relative photocurrent response vs. photon energy for Ge–GaAs heterojunctions. n-type Ge $N_d \approx 10^{41}$ cm^{-3}, n-type GaAs $N_d \approx 5 \times 10^{16}$ cm^{-3}, p-type Ge $N_a \approx 8 \times 10^{15}$ cm^{-3}, p-type GaAs $N_a \approx 1.5 \times 10^{16}$ cm^{-3}.
(After Lopez and Anderson [47])

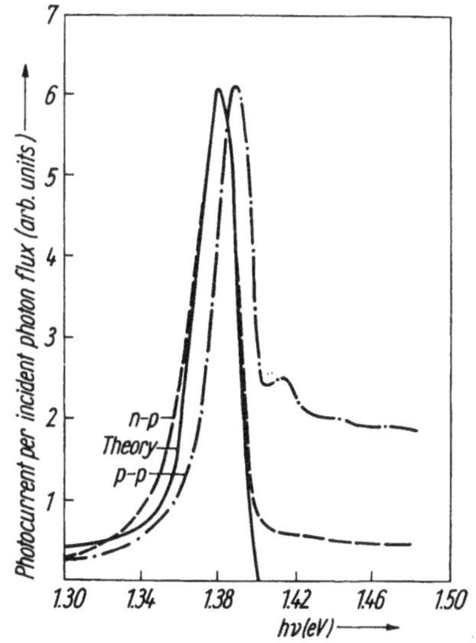

Fig. 6. Photocurrent spectra in the peak region for n-p and p-p Ge–GaAs heterojunctions compared with the theoretical response.
(After Lopez and Anderson [47])

of these photons will be absorbed on either the Ge or GaAs side which in either case will contribute to the photocurrent.

In n-n junctions, absorption in the Ge side produces a negligible photocurrent since the transition region is almost completely on the GaAs side. Below the GaAs band gap (1.4 eV) therefore, the response approaches zero. A similar response occurs for p-p junctions where again the transition region is essentially all on the GaAs side. Lopez and Anderson [47] have measured the short circuit photocurrent for p-n, n-n, n-p, and p-p Ge–GaAs heterojunctions (see Fig. 5a, b, c, and d) and have compared these with the theoretical predictions of Moss and Hawkins [49]. In general the experimental response curves for n-p and p-p junctions agree with the theoretical predictions (see Fig. 6) except for the high energy tails and these can be explained in terms of surface re-radiation.

3. Fabrication and Metallurgical Variables

The fabrication of heterojunctions is usually accomplished by one of three well-known methods: epitaxial growth from the vapour phase, growth from solution or interface alloying. However, hybrids of these techniques have also been reported by a few workers. All of these processes have growth temperatures in excess of 400 °C.

In epitaxial growth from the vapour phase the general procedure is to place the constituent elements or source intermetallic compound, together with suitable quantities of halogen or halide and a single crystal substrate of required orientation in different parts of a quartz tube. This is then sealed under high vacuum and heated in such a manner that the source elements or crystals are at a higher temperature than the substrate end of the tube. The non-volatile source materials are attacked by the reagent, transported as volatile intermediates and deposited on the cooler parts of the tube and the substrate. This cycle is repeated thus providing continuous epitaxial deposition and growth until a layer of required thickness is obtained. In practice the growth time may be from a few minutes to hours, so that where the growth temperatures are high appreciable cross diffusion of the materials on either side of the interface might occur. A typical example of such a heterojunction pair is the Ge–GaAs system, where, as discussed previously, it is possible to create homojunctions in either of the two base materials by cross diffusion. Auto-doping during growth of epitaxial layers has also been reported by many workers (e.g. [51]), this has been observed particularly when growth temperatures in excess of 600 °C have been used for the epitaxial deposition of GaAs on to Ge.

However, since numerous workers have reported and discussed the relative intricacies and merits of the growth of heterojunctions by vapour transport, only alloyed heterojunctions will be considered in this review.

In the solution growth process at least one of the materials is dissolved by a suitable solvent. The solution is then put into contact with a substrate and the solution allowed to cool slowly so that a layer of the semiconductor material recrystallises epitaxially onto a substrate. A typical example of such a heterojunction pair is $Ga_xIn_{1-x}As$–GaAs where Bi is the solvent. In this case the substrate is dissolved to some extent so that the resulting deposit contains some of the substrate material. The deposited layer and the solution are generally in contact for some time at high temperatures so that limited diffusion may take place.

In the interface alloying process the low melting point material is placed in a temperature gradient with the high temperature side in contact with the high melting point material. When the temperature is raised, in general a thin zone of the low melting point material adjacent to the high temperature material melts. When the temperature is reduced the melt recrystallises after having alloyed into the high melting point material. In this process, since the reaction temperatures must be very high, cross diffusion of the constituents is very probable especially in the liquid state. However, this process can be accomplished in a very short time.

Composite processes such as deposition by vapour transport onto a substrate held at a temperature higher than the melting point of the deposited material (with the result that alloying occurs), generally have limited advantages and some disadvantages of both vapour transport and alloying methods.

3.1 The interface alloyed heterojunction

In practice the most widely used interface alloying technique is carried out very simply by placing two wafers, which have first been prepared by careful orientation, lapping, polishing, and etching in the conventional manner, onto a carbon or similar type of heater strip (see Fig. 7). The higher melting point semiconductor 1 lies between the heater strip and the lower melting point semiconductor 2. Alloying is carried out in a hydrogen or inert atmosphere dependent on the materials used and under microscopic examination to allow easy and accurate control on the process to be maintained. The heater strip produces a temperature gradient such that semiconductor 1 is at a higher temperature than semiconductor 2. As the temperature is raised the bottom face of semiconductor 2 begins to melt. Before the complete wafer can melt, however, the temperature of the heater is reduced and the melted region recrystallises having alloyed into semiconductor 1. Recrystallised regions of the order of 5 to 20 μm thick are normally used for assessment of the heterojunction properties. During alloying, practical considerations (i.e. jigging, thermal expansion etc.) make it difficult to keep both substrate and the low melting point semiconductor perfectly aligned in the correct crystal direction, so that on solidification slight rotations or tilts may occur leading to strain at the interface. At the present time junction areas in this type of process are normally limited to a few mm^2.

Rediker, Stopek, Ward, Hinkley, and Jadus [33, 52, 68] have used and developed this process for the preparation of GaAs–Ge, GaSb–GaAs, InAs–GaSb, and GaAs–InSb interface alloyed heterojunctions. Evidence from metallographic examination and Kossel pattern techniques [54 to 56] suggests that single

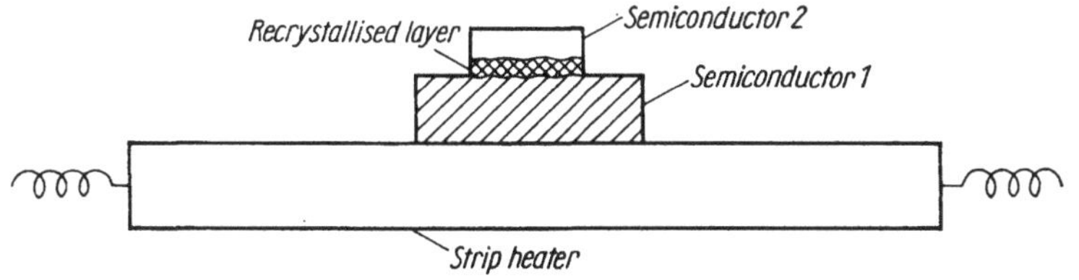

Fig. 7. Experimental arrangement for interface alloying. Semiconductor 1 has a higher melting point than semiconductor 2

crystal heterojunctions are obtained in all these systems. In the Kossel divergent beam X-ray technique, X-rays are produced by a 1 μm electron microbeam in the material being investigated. These X-rays diverge in all directions from an effective point source where the beam is absorbed, they are then diffracted by the crystal planes giving a depletion of the X-ray intensity in the directions which satisfy Bragg's law. In transmission through thin samples both absorption and diffraction conics are observed. When the X-rays pass through a single crystal portion of the material sharp Kossel lines are observed. Although Rediker and his co-workers using these examination techniques observed the heterojunctions to be single crystals, in the Ge–GaAs pair they found that the two crystals had rotated about 3° relative to each other. However, the patterns also indicated that the resultant strain was relieved at a distance of less than 1 μm from the interface. Mroczkowski, Lavine, and Gatos [53] have since reported that the transition from GaAs to Ge is not monotonic (see Fig. 10) and that a portion of the recrystallised region consists of the GaAs–Ge eutectic.

In the GaAs–GaSb heterojunction pair the faces of the GaAs and GaSb were rotated and tilted with respect to each other so that the lattices were misorientated by about 20°. This large degree of misorientation is expected to occur in single crystal GaAs–GaSb heterojunctions in order to compensate for the 8% mismatch in the lattice constants (GaAs 5.64 Å, GaSb 6.10 Å) of the two semiconductors. Since there is a relatively high X-ray absorption in antimony only simple and weak GaSb patterns are observed. Because of the weak GaSb patterns and the finite size of the electron beam an apparent broadening of the Kossel patterns was noted by Rediker in both the bulk GaSb and the transition region. Their results are almost certainly consistent with the heterojunction being a single crystal, but, owing to the diameter of the microbeam, one would not expect to detect interface polycrystallinity of much less than 1 μm. Here it must be stated that the title of "single crystal heterojunction" implies only that the lattices of the two semiconductors in the proximity of the interface are single crystals. This differs slightly from the normal definition of single crystal which states that a single crystal is an orderly periodic arrangement of atoms bounded by its own surface. However in the case of single crystal heterojunctions this is justified since in general only the materials immediately in contact at the interface need be considered, as the heterojunction utilises the backing material only as contacts.

Mroczkowski, Lavine, and Gatos [53] have investigated the metallurgical variables associated with the formation of heterojunctions using the interface alloying technique of Rediker et al. The effect of such parameters as solidification processes, composition gradients, orientation, and the accommodation of crystal structures in the Ge–GaAs system have been discussed. Before any theory of the solidification or alloy processes can be considered however, a study of the phase diagram of the Ge–GaAs pseudobinary system must be made (see Fig. 8). This has been determined experimentally by Mroczkowski et al. [53] and Takeda, Hirai, and Hirao [57] using differential thermal analysis, metallographic examination, and electron probe microanalysis. Both groups of workers treated the system as a pseudobinary pair and came to a similar conclusion in that only a simple eutectic type phase diagram with the eutectic composition at approximately 74 to 75 at% Ge was indicated. No intermediate compounds were observed. The eutectic temperature is at present unresolved, Mroczkowski et al. reporting this at 845 °C at 74 at% Ge, Takeda et al. at 865 °C at 15 mol%

Fig. 8. Phase diagram of pseudobinary GaAs-Ge system.
——— Mroczkowski et al. [53]
—·—·— Takeda et al. [57]

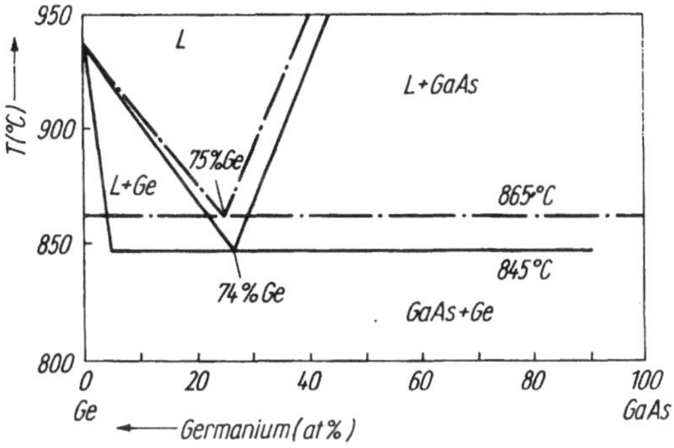

GaAs (75 at% Ge). Differences in cooling rates and experimental conditions may account for some of this disagreement. The solubility of Ge in GaAs and GaAs in Ge is more clearly defined, Takeda reports low solubilities which confirm lattice parameter measurements on Ge–GaAs alloys by Kolm et al. [58]. The values of the solubilities are less than 0.5 mol% Ge and less than 0.25 mol% GaAs in the GaAs and Ge, respectively.

Composition profiles of Ge–GaAs alloyed heterojunctions have been determined using electron probe microanalysis techniques by both Rediker et al. [33] and Mroczkowski et al. [53]. Fig. 9 shows the results of Rediker and his coworkers. The 100% lines at a distance from the interface correspond respectively to the X-ray fluorescence of unadulterated Ge and that of the Ga and As lines in unadulterated GaAs. The Ga–K_α line and the As–K_α line in this case were monitored separately and then normalised. As expected from the phase diagram little evidence of solubility of one phase in another was observed. Rediker has suggested an explanation for the maxima for both Ge and GaAs occuring in the transition zone. In this he assumes that during the interface alloying a slight amount of GaAs is dissolved by the Ge and that on resolidification the segregation coefficient for the GaAs is less than unity. As the melt recrystallises on the GaAs seed and as the recrystallised region grows the melt would become richer in GaAs so that the final portion of the material to recrystallise would be GaAs containing the equilibrium solid solubility of Ge. In this case the solubility limit was found to be below the resolution limits of the microprobe. Thus the result of interface alloying of the GaAs–Ge couple is likely to be a four layer structure of GaAs–Ge rich (thin) — GaAs rich (very thin) — Ge.

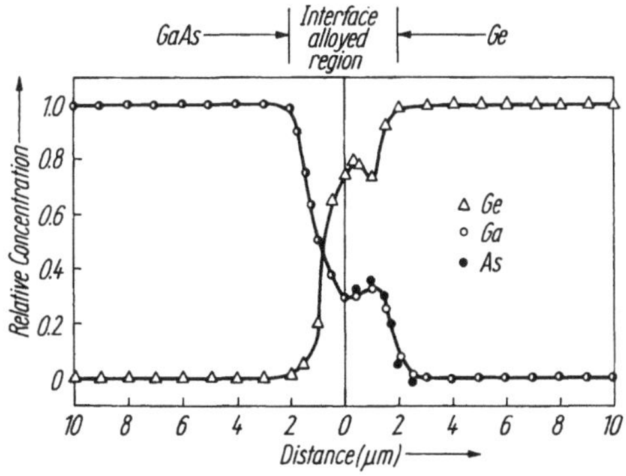

Fig. 9. Composition profile of a typical alloyed GaAs-Ge heterojunction from electron beam microprobe analysis. The 100% lines correspond to the intensity of the Ge-K_α line from the bulk Ge and the intensities of the As and Ga-K_α lines from the Bulk GaAs. (After Rediker et al. [33])

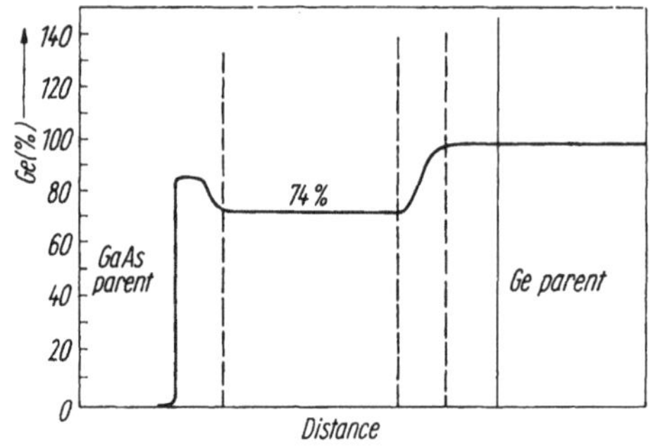

Fig. 10. Composition profile of an alloyed GaAs-Ge heterojunction from electron beam microprobe analysis and showing the distinct layer structure. (After Mroczkowski et al. [53])

Mroczkowski et al. have also reported composition profiles for this type of alloyed heterojunction (see Fig. 10). In general they support the views of Rediker et al. However, the flat portion of the profile suggests the possibility of a eutectic system which has been confirmed by the phase diagram determination. Contrary to Rediker's results, Mroczkowski has reported that from microprobe analysis the majority of the transition zone has a eutectic composition. However, this has not been confirmed by metallographic examination of the junctions. Mroczkowski has tentatively explained this phenomena by revealing what he believes to be growth cells in the junction material adjacent to a {100} GaAs substrate where excess GaAs may be accommodated in the walls of the cells. But, since uniform inner band structures have been reported the possibility of cellular growth in this region would appear to be remote. Mroczkowski has also assumed that rapid cooling rates during the formation of the junction prevent thermodynamic equilibrium being reached so that no eutectic structure is formed. This, however, is unlikely without modifications being made to the present diagram.

On the basis of metallographic examination of cross-sectioned alloyed Ge–GaAs heterojunctions, Mroczkowski has confirmed the formation of the four layer structure predicted by Rediker, good agreement was found between the optically measured widths of the layers and the measured widths of the distinct regions of the composition profile. By the location and structure of the layers these can be divided into the two outer bands which are adjacent to the parent crystals and the two inner bands which are separated by a sharp discontinuity. A characteristic of this type of heterojunction is this sharp discontinuity separating the inner bands; this represents the meeting of two growth fronts, one from the GaAs and the other from the Ge. The outer bands are always found to be free of structure and to be independent of the orientation of parent material. The boundaries between the inner and outer bands have however been found to be strongly orientation dependent, (100) oriented parent substrates giving irregular boundaries, (110) discontinuous and (111) faint straight boundaries. The structure of the inner bands was also found to be orientation dependent. At present very little study of the problems associated with the accommodation of structures and orientations across the discontinuities between the inner bands has been attempted. Until this is undertaken any interpretation of junction properties has to be made with caution.

A similar process to that described by Rediker has been reported by Shirafuji and Nakayama [35] for making GaAs–Ge heterojunctions. In this case however small chips of germanium were placed on the GaAs substrate and not oriented prepared slices. When using slow cooling rates good diodes were obtained, but the electrical measurements suggest a p-i-n structure. No metallographic data were reported but from the measurements quoted these devices may also have the four layer structure reported by Rediker and Mroczkowski.

Shewchun and Wei [36] have used an almost identical process to that described by Shirafuji and Nakayama for the preparation of Ge–Si heterojunctions. In this case the high melting point semiconductor is Si, and on to this is placed a chip of Ge, its dimensions dependent on the size of heterojunction required. Short heating, alloying (3 to 15 s), and cooling times are deliberately employed to form thin abrupt junctions and to minimise the diffusion of doping elements in the Ge and Si. Alloying is normally carried out in hydrogen to give good wetting properties. Shewchun and Wei have shown by metallographic techniques that the junction itself consists of an abrupt transition from Si to Ge through a Ge–Si regrown layer. This thin regrown layer which is normally of the order of 1 μm is thought to consist of a range of Ge–Si solid solutions of varying energy gap as would be expected from phase diagram considerations (see Fig. 11a and 11b).

Since both Ge and Si are diamond structure crystals a substitutional Ge–Si alloy can be formed where the Ge and the Si atoms are arranged at random at lattice sites of the diamond lattice. However, a great deal of work has been carried out both on a practical (e.g. [60, 61]) and a theoretical basis (e.g. [62]) by many workers on this system. Since an alloy is not a perfect crystal but is disordered, it does not have an energy band structure in the strict sense, but rather energy bands with a high density of allowed states and bands with a low density of states [63]. In Ge–Si alloys the high density of allowed states should resemble the allowed bands of perfect Ge or Si crystals according to the predominant constituent of the alloy. Theoretically therefore the transition from an allowed to a forbidden band cannot be as sharp as in a perfect crystal. However, experimental evidence [60] has shown that the trailing band edges for this system are negligible. The valence band structures have been the subject of a great deal of interest; Braunstein [64, 65] has extensively studied these to determine an accurate energy band diagram.

Since the lattice constants of Ge and Si (Ge 5.68 Å, Si 5.43 Å) differ by a few per cent the lattice mismatch can either be accomodated by dislocation generation or lattice irregularities such as voids or fine cracks. Shewchun and Wei

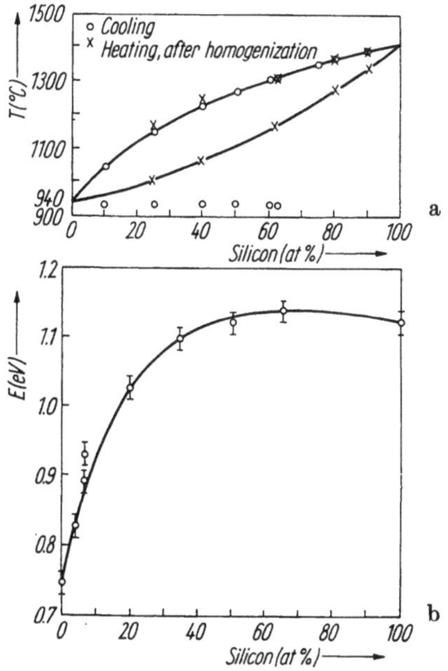

Fig. 11. a) Phase diagram of the Ge-Si system (after Hansen [67]).
b) The variation of energy gap of Ge-Si alloys with composition (after Levitas et al. [60])

using metallographic techniques showed that dislocation generation was the dominant mechanism even with extremely rapid alloying and cooling rates. This, however, is in direct contrast with results on Ge–GaAs heterojunctions reported by Shirafuji and Nakayama [35] where rapid cooling was shown to cause lattice irregularities and voids. These results are surprising, since consideration of lattice constant and thermal expansion data suggests that less strain should be associated with the Ge–GaAs system. At present, however, the limited published work tends to indicate that dislocation generation is the dominant mechanism.

A process similar in many ways to interface alloying is that reported by Kiseleva [34] who produced heterojunctions between GaSb and InSb which exhibited sharp boundaries between the two seminconductors and a sharp impurity front using a technique involving crystallisation from the melt.

In the preparation of heterojunctions by crystallisation from the melt, the conditions at the interface of the solid phase of one material and the liquid phase of the other are determined by the nature of the interaction of the materials. InSb and GaSb form a continuous range of solutions both in the liquid and solid state [66, 67] so that two possible states of the seed-melt system may occur. The non-equilibrium state where a limited amount of the seed dissolves in the melt and the equilibrium state where enough of the seed dissolves to give a saturated solution. In the former state, as the GaSb seed dissolves, the crystallisation temperature continuously increases. In the equilibrium state the concentration of GaSb in the melt is determined from the liquidus line on the phase diagram. In practice Kiseleva used both conditions to grow GaSb–InSb heterojunctions. In both cases a single crystal seed of GaSb was pulled from a melt of InSb or 4% GaSb in InSb at a controllable rate. Under non-equilibrium conditions limited dissolution of the seed material and its diffusion in the melt occurs. This results in the growth on to the seed of InSb containing a small amount of GaSb which can be controlled to a large degree by a choice of pulling rate. This will also affect the thickness of the layer to be grown, the pulling time, and the degree of mutual diffusion of the various materials. Using equilibrium conditions, pulling is commenced when dynamic equilibrium has been established between the seed and the melt. The composition of the solid growing, in this case, is determined from the equilibrium phase diagram. Since the time required for equilibrium to be established is lengthy, there is considerable cross diffusion of the semiconductor components and impurities.

Of the two processes the non-equilibrium state is preferable since the shorter period the seed is allowed to stay in the melt results in a structure more amenable to interpretation.

3.2 The solution grown heterojunction

The preparation of homojunctions in elemental and compound semiconductors using solution growth techniques has been reported by many workers (e.g. [76]). In this technique low melting point metals such as In, Bi, and Sn which may contain suitable dopants are heated in contact with a semiconductor substrate. As the temperature is raised the carrier metal dissolves a small amount of the substrate. If equilibrium is established, at the alloying temperature this amount is determined from the equilibrium phase diagram. As the temperature is reduced the dissolved material is recrystallised epitaxially as a single crystal on to the substrate. The recrystallised material in general contains a small amount of the carrier material in addition to any dopant deliberately introduced. In practice alloying temperatures greater than 600 °C are not used so that cross diffusion of the substrate and alloy components or fast diffusing impurities are kept to a minimum. (The diffusion coefficient is dependent on temperature according to the relation $D = D_0 \exp(-\Delta E/kT)$ where ΔE is the diffusion activation energy.)

Heterojunctions have been prepared using a modification of this technique by Dale and Josh [38] who produced monocrystalline epitaxial layers of $GaSb_xAs_{1-x}$, and Mn_2As on GaAs substrates using a series of Bi-based alloys (see Table 2) containing varying amounts of GaSb, InAs, and InAs + Mn. In this technique pellets of these alloys up to 1 mm diameter were pre-melted on to oriented and etched GaAs substrates at temperatures from 350 to 450 °C followed by alloying for periods of 5 to 120 min at 500 to 600 °C to enable equilibrium to be reached. Pre-melting and alloying were carried out in vacuo to facilitate wetting and dissolution of the substrates. After alloying, slow cooling over 1 to 8 h was used to prevent interface cracking due to lattice mismatch and differential thermal expansion between the recrystallised layers and the substrates.

Metallographic and microprobe analysis of the layers recrystallised from Bi–GaSb and Bi–InAs alloys showed that abrupt single crystal heterojunctions were obtained between $GaSb_xAs_{1-x}$–GaAs and $Ga_xIn_{1-x}As$–GaAs. In the former x was generally of the order of 0.9 and in the latter of the order of 0.5. That solid solutions are formed and recrystallised from the carrier material is

Table 2

Alloy systems for the solution growth of solid solution-GaAs heterojunctions (after Dale and Josh [38])

Alloy composition (part by weight)	Minimum*) pre-wetting temperature (°C)	Solution grown layer	Approximate composition of solution grown layer
90 Bi, 10 GaSb	350	$GaSb_xAs_{1-x}$	Arsenic content dependent on alloying temperature, x generally 0.9 when alloyed at 500 to 600 °C
80 Bi, 20 GaSb	350	$GaSb_xAs_{1-x}$	
90 Bi, 20 InAs	350	$Ga_xIn_{1-x}As$	x generally of the order of 0.5 for an alloying temperature of 550 to 600 °C
80 Bi, 20 InAs	350	$Ga_xIn_{1-x}As$	
90 Bi, 5 InAs, 5 Mn	400	$Ga_xIn_{1-x}As$-Mn_2As	x generally of the order of 0.5 for an alloying temperature of 550 to 600 °C
85 Bi, 10 InAs, 5 Mn	400	$Ga_xIn_{1-x}As$-Mn_2As	
75 Bi, 15 InAs, 10 Mn	450	$Ga_xIn_{1-x}As$-Mn_2As	x of the order of 0.05

*) The minimum pre-wetting temperature is the minimum temperature at which the alloy will 'wet' the surface of a GaAs substrate.

a

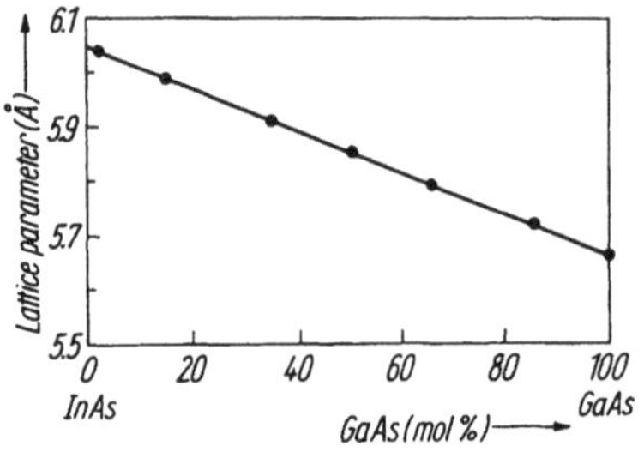

b

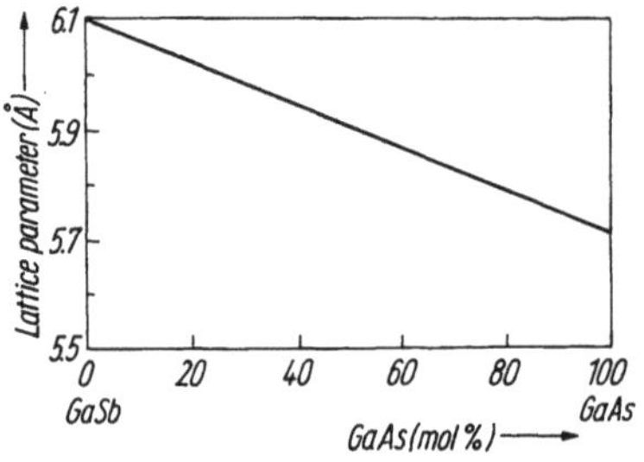

Fig. 12. Variation in lattice constant as a function of composition in the
a) InAs-GaAs system (after Woolley and Smith [78]),
b) GaSb-GaAs system (linear extrapolation)

not surprising: Dale and Josh (unpublished work) have shown that the solid solubilities of GaAs, GaSb, and InAs in Bi are very low and Hooker and Lenker [76] and Woolley and Smith [77, 78] have shown that both the GaAs–GaSb and GaAs–InAs systems exhibit a complete range of solid solutions. The high values reported for x are unexpected since from such metallurgical considerations as mutual solubilities and recrystallisation phenomena low values would be expected. Low values would also give less interface strain due to a better lattice match between the materials (see Fig. 12a and b).

Metallographic and microprobe anlaysis were also used by Dale and Josh [38] to establish the nature and composition of Mn_2As–GaAs heterojunctions and Mn_2As–$Ga_xIn_{1-x}As$–GaAs structures grown from Bi–InAs–Mn alloys. However, in the former case evidence suggests that this is not a true heterojunction since optical transmission experiments by Wright and Tansley [79] have shown that the energy-gap of bulk Mn_2As was substantially less than the value of 0.1 eV corresponding to the limit of detection of the experiment. This would in effect give a metal–semiconductor junction and not a semiconductor–semiconductor heterojunction. Measurements of the electrical and optoelectronic properties of these devices reported by Tansley and Newman [80] have, however, shown an abrupt heterojunction between two semiconductors of only slightly different band gap (1.4 and 1.3 eV). This suggests that a thin layer of $Ga_xIn_{1-x}As$ of low indium content separates the Mn_2As from the GaAs to give a $Ga_xIn_{1-x}As$–GaAs heterojunction. By consideration of the composition implied from the observed energy gap, a lattice mismatch of less than 0.3% was predicted.

From optical response curves, Tansley deduced the composition of this thin layer to be of the order of $Ga_{0.95}In_{0.05}As$. This approximate composition was derived by extrapolation of the band gap versus composition profiles of the InAs–GaAs system obtained by Abrahams, Braunstein, and Rosi [81] using infrared transmission techniques on optically polished homogeneous material (see Fig. 13). However, the negative deviation from linear dependence of the band gap on composition for InAs to GaAs is in contrast both with other group III–V systems where a linear or positive deviation from linear dependence is observed [82] and with the Ge–Si system where, as discussed previously, Johnson and Christian [61] observed a distinct discontinuity of the rate of change of band gap versus composition. (This indicates that the band edges shift to a different position in momentum space.) From the results of Abrahams et al. no break in the slope was detected or indicated, consequently the band edges in this system probably remain essentially in the same position in k-space. This was also consistent with thermo-electric power measurements [81] where

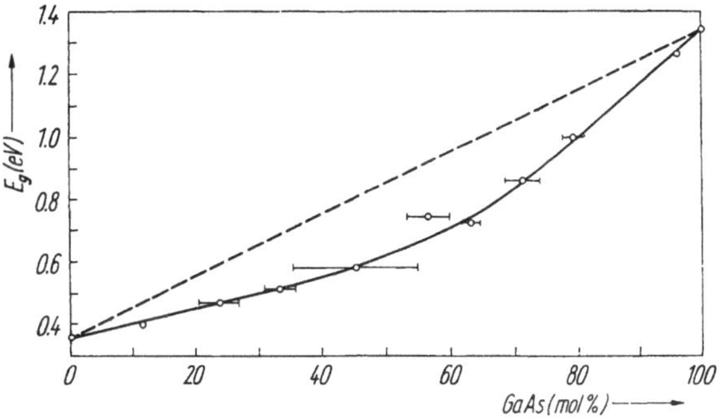

Fig. 13. Variation of the band gap with composition in the InAs-GaAs system. (After Abrahams et al. [81])

no variation in the density of effective states was observed. The effective masses of electrons in GaAs and InAs are approximately 0.07 *m* [83] and 0.03 *m* [84], respectively. Since however it appears that small effective masses are characteristic of conduction band minimal that lie at the centre of the Brillouin zone [86], the lowest conduction band minima are probably those with $k = 0$ for all compositions of InAs and GaAs.

It is interesting to note that it should be possible to produce solid solution layers by solution growth alloying techniques having any desired energy gap in the range 0.35 to 1.4 eV by a variation of such alloying conditions as temperature and alloy composition.

Recently Dale [86] has examined the growth mechanism of Mn_2As layers on GaAs from Bi–InAs–Mn alloys. This has been shown to differ from the normal solution growth process in that growth is by a precipitation mechanism and not by recrystallisation. This was suggested from observations which showed that Mn_2As layers could not be redissolved in Bi at normal alloying temperatures and from the low solubility of bulk Mn_2As in Bi (less than 0.1 mol% Mn_2As at 600 °C). In this process, therefore, when a pellet of a Bi–InAs–Mn alloy is placed in intimate contact with a GaAs substrate and the temperature raised, the Bi dissolves GaAs from the substrate together with InAs and Mn from the pellet material. Above 350 to 400 °C a complex reaction occurs which results in Mn displacing both Ga and In from their arsenides. Since Mn_2As is not soluble in Bi, precipitation of the Mn_2As onto the substrate occurs during both the heating and alloying cycles and not during cooling as is normally observed in the solution growth process.

Understanding of the mechanisms of formation of the solid solution layer separating the Mn_2As from the GaAs is still confused. This could be formed either by diffusion of indium into the GaAs from the liquid alloy or by recrystallisation. Diffusion data reported by Kendall [87] suggest that under equilibrium conditions it might be possible to form a thin layer of $Ga_{0.95}In_{0.05}As$ during the alloying cycle. However, one would expect diffusion to produce a graded junction when in fact Tansley and Newman [80] have reported that the capacitance–voltage law indicates an abrupt junction. Metallographic observations by Dale and Josh [38] have suggested that recrystallisation may take place. However, since precipitation of Mn_2As probably occurs during the heating and alloying cycles this mechanism is open to some doubt since the Mn_2As layer would be expected to prevent GaInAs from recrystallising onto the substrate. This may however tend to explain why the indium content of the intermediate $Ga_xIn_{1-x}As$ layers has been observed to decrease as the Mn content of the alloys is increased.

In heterojunctions grown from the low (5%) manganese alloys metallographic examination by Dale and Josh revealed distinct layers of GaInAs separating the Mn_2As from the GaAs substrate. In the higher Mn ones (>10%), however, these intermediate layers were not detected. This may possibily be explained by differences in indium content of the GaInAs layers. For instance in layers grown from low manganese alloys Dale [86] reported high indium content, but on layers grown from high manganese alloys Tansley and Newman [80] from optical response evidence deduced the indium content to be low. Since lattice mismatch in the low indium content layers would be much smaller than in the higher indium layers, metallographic detection by etching techniques would be more difficult.

Diedrich [88] has reported the formation of solution grown heterojunctions in experimental investigations into the formation of alloyed wide band emitters for Ge transistors. For this application Diedrich considered that suitable materials should form semiconducting solid solutions with Ge with band–band distances of over 0.65 eV at room temperature. The mixed crystal should also be capable of recrystallising as a single crystal and the lattice match at the interface should be good enough to give a lowly stressed structure. The alloying temperature should also be sufficiently low and rapid to prevent cross diffusion and evaporation of volatile components.

Of the possible materials available Diedrich investigated the Ge–Si system and some of the Group III–V compounds. Most of his experiments were however carried out on silicon alloys since plots of the energy gap versus composition in the Ge–Si system (see Fig. 11 b) show that only small concentrations of Si in Ge give a considerable increase in the energy gap. In addition to this, below about 10 at% Si single homogeneous crystals of the solid solution can easily be produced.

In these experiments Diedrich used silicon alloys with Bi, In, Sn, Al, Au, and Zn as the solvent materials and InP with In + Ga as the solvent. Binary, ternary, and quarternary alloy systems were used. Of these a quaternary In–Au–Si–Ge system with small amounts of Ga as a dopant gave the most promising results. However, since In–Ge and In–Si form pseudo eutectics,

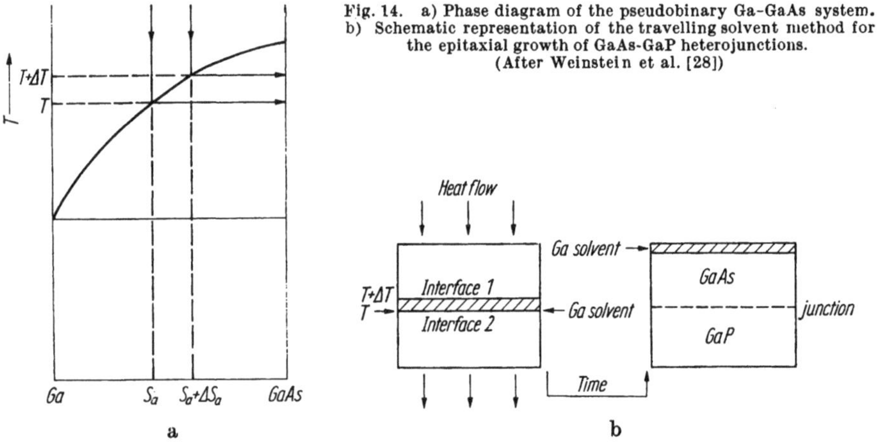

Fig. 14. a) Phase diagram of the pseudobinary Ga-GaAs system. b) Schematic representation of the travelling solvent method for the epitaxial growth of GaAs-GaP heterojunctions. (After Weinstein et al. [28])

Au–Si an eutectic and In–Au has a complicated phase diagram which includes stoichiometric compounds, interpretation of metallurgical phenomena associated with the process is very difficult. In practice the wetting properties of these alloys were found to be good at alloying temperatures of about 450 °C. By varying the composition of the alloys from 5% AuSi to 20% AuSi, Diedrich studied the metallurgical effects of the AuSi content. He found that by increasing the AuSi content, the alloying and recrystallisation properties continuously deteriorated until at 20% AuSi – 80% In, polycrystalline regrown layers were formed. The replacement of In by Sn gave similar results but metallographic examination showed severe lattice distortions in the regrown layers giving rise to poor electrical characteristics. The use of Bi, Zn, and Al in general gave poor wetting properties, inferior recrystallisation, and inadequate solubility of Si in the recrystallised layers.

Transmission and absorption measurements made on thin slices of material recrystallised from Ga–In–Au–Si alloys (after removal of the germanium substrate) suggested that the layers were composed of In and Ga doped Ge–Si solid solutions where the silicon exhibited a Gaussian distribution. The maximum amount of Si found in these layers was of the order of 2 at%.

A very different approach to the solution growth of materials was first described by Pfann [89] as a temperature gradient zone melting (TGZ M) process. This process has features which make it particularly suitable for growing single crystals of semiconducting compounds and the direct applicability of the method for the fabrication of large area junction structures during growth. With this technique Pfann [89, 90] and Wernick [91] have used thin solvent zones to grow both bulk silicon and germanium and to create large p-n junction structures. Mlavsky and Weinstein [92] have developed this technique as the "travelling solvent method" (TSM) for the growth of GaAs and homojunctions in GaAs. Using this process Weinstein, Bell, and Menna [28] have reported the preparation of single crystal heterojunction structures between GaAs and GaP. This technique is most easily described by reference to Fig. 14a and b. As can be seen from diagram b) a sandwich structure which consists of a slice of GaAs separated from a slice of GaP by a thin layer of solvent (in this case Ga) is heated from above so as to create a temperature gradient perpendicular to the gallium layer. As the temperature rises both GaAs and GaP dissolve into the gallium layer. However, since interface 1 is at a temperature of $T + \Delta T$, the solubility of GaAs in the gallium is $S_a\% + \Delta S_a\%$. The solubility of GaP at interface 2 at temperature T is $S_b\%$, however, under experimental conditions Weinstein has shown that this is extremely small relative to the rate of solution of the GaAs at the upper interface, so that the solvent can be assumed to be saturated only with GaAs. The solubility of GaAs therefore at interface 2 at temperature T is $S_a\%$. Since a liquid cannot be saturated at two different temperatures simultaneously, excess GaAs from the hotter interface diffuses across the liquid gallium zones and deposits on the cooler interface. The liquid zone therefore travels through the upper slice and finally arrives at the top surface. The rate of zone movement is dependent on T, ΔT, S_a, and ΔS_a, the liquid zone thickness, and the liquid diffusion coefficient of arsenic in gallium. In the above case if the bottom slice is a single crystal of GaP, the GaAs will grow epitaxially on to this seed as a single crystal. It is possible to achieve similar results with other materials provided the two lattices, solvents, and solubilities are compatible.

Weinstein, Bell, and Menna [28] have used this technique to prepare single crystal heterojunction structures between GaAs and GaP. They placed a temperature gradient of 170 °C across the complete sandwich structure with the GaP (solid)–Ga(liquid) interface maintained at 650 °C. Examination of these heterostructures did not reveal any transition zone and showed abrupt interfaces between the GaP seed and the single crystal epitaxial layers of GaAs which had the orientation of GaP seeds used. Capacity and current–voltage characteristics indicated abrupt junctions with a square law dependence of capacitance on the reverse bias, and in general the built in potential had a magnitude between that of the energy gaps of the constituent semiconductors.

3.3 The vapour transport-interface alloyed heterojunction

One of the problems associated the growth of heterojunctions using-open-tube vapour transport techniques is the presentation of an oxide free substrate for the deposit to grow on. Oldham [24] overcame this by using a cleaving technique after initiating a flow of the germanium halide-hydrogen mixture over a hot silicon substrate. An alternative process which combines both vapour transport and interface alloying processes to achieve this has been reported by Brownson [39, 59]. Using the Theuerer halide reduction open-tube process [31] he allowed germanium to deposit on a silicon substrate which was above the silicon self reducing temperature (i.e. the temperature above which silicon reduces SiO_2 to the volatile SiO) and also above the melting point of germanium (i.e. a temperature of about 1100 °C). Such a surface is oxide free and on this germanium will readily deposit. After deposition had been initiated at this temperature Brownson reduced the deposition temperature to the more usual value of 970 °C until the desired thickness of a thin liquid surface film of a Ge–Si alloy was formed. An advantage of this process is that the composition of these layers can be predicted from the deposition temperature and the phase diagram. Electron beam microprobe analysis carried out by Brownson showed that as expected the regrown layers had the approximate composition 75% Ge – 25% Si. Laue back-reflection patterns demonstrated that these regrown layers were single crystals but distortion of the spots suggested considerable strain. This strain in some cases, where the regrown layers were of the order of a few μm thick, resulted in cracking and exfoliation of the layers from the substrate.

Using three probe techniques on low angle lapped sections of the regrown layers Brownson [93] found that rectification occurred at the Si to Si–Ge alloy interface within the limits of resolution of ± 1 μm normal to the plane of the junction in both n-p and n-n material. The reverse characteristics of these are at present inferior to those one would expect of an alloy homojunction to a silicon substrate of the same resistivity, leakage is higher and the characteristic is generally soft.

The main advantage of this process over the conventional interface alloy technique is that large area heterojunctions can be fabricated with relative ease. It should also be possible to control the composition of the alloy layer by variation of the deposition temperature. Against this, however, there are a number of disadvantages. The equipment and technology are far more complicated, and the process temperatures are higher and are held for longer periods, so that cross diffusion and the diffusion of impurities may become a problem.

3.4 Summary

The results which have been described above show a wide variety of detailed form which is due to variations in approach of the individual authors to the formation of alloyed semiconductor heterojunctions. In attempting to compare these processes the lack of reported work on the metallurgical variables involved has limited understanding of the reactions taking place during alloying. Of the processes discussed, interface alloying has received the most attention both from a technological and metallurgical aspect. The mechanism by which these heterojunctions are formed is by no means simple and individual experimental conditions have given varied information and theories on this process. The problem of accommodation and lattice match across the interface is worthy of a more intensive study than has at present been carried out.

In general the relative merits of each process may be summarised by reference to Table 3, which gives a comparison of the advantages and disadvantages of the alloying techniques used to fabricate alloyed semiconductor heterojunctions.

Table 3
Comparison of alloying techniques for the fabrication of semiconductor heterojunctions

Process	Advantages	Disadvantages
Interface alloy	Simple equipment and technology, short process time, wide range of heterojunction pairs possible	High temperatures required, size limited (few mm ∅), generally highly strained (dependent on components), Crystal alignment may be difficult
Solution growth (from alloy system)	Simple equipment and technology, low alloying temperatures, wide range of heterojunction pairs possible including solid solutions	Complex metallurgy, long process time, high risk of contamination by fast diffusing impurities, size limited (few mm ∅)
Solution growth (T.S.M.)	Simple equipment and technology, simple metallurgy, relatively low process temperatures, lowly stressed layers, large area heterojunctions possible	Long process time, high risk of cross diffusion, limited range of compatible materials
Vapour transport — interface alloy	Large area heterojunctions possible, relatively short process time	Very high temperatures required technology and equipment more complex than by other processes, limited range of suitable materials, high risk of contamination by impurities, at present highly strained layers

4. Interface Characteristics

4.1 Electrical properties

As discussed previously, Anderson in a series of papers [18 to 20] proposed an energy diagram for abrupt heterojunctions which includes energy discontinuities in the conduction and valence band edges. These energy discontinuities, which stem from an assumed abrupt change in semiconductor energy gap and electron affinity at the junction interface, create potential barriers that impede the flow of current carriers across the junction. The potential barriers differ greatly from those found in homojunctions so that the electrical properties of a heterojunction are also expected to differ greatly from those of a homojunction.

Anderson predicted heterojunction current–voltage characteristics by combining the proposed band diagram with a diode emission theory. Saturation currents are obtained by direct application of Shockley's homojunction diffusion theory. As a means of explaining the large differences between experimental and theoretical saturation currents, Anderson defined and evaluated a transmission coefficient. Perlman and Feucht [94] later carried out a detailed analysis of the static current–voltage characteristics of abrupt p-n heterojunctions where the n-type semiconductor had a larger energy gap than the p-type. Their analysis combined a kinetic model for the evaluation of emission fluxes with a diffusion model for the determination of the extent of the minority carrier built up at the edges of the depletion region. In this case the total current was compared to the opposing emission currents to predict regions of quasi-equilibrium and to help determine the variation of the quasi-Fermi levels throughout the depletion region. The I-U characteristics derived by Perlman and Feucht differed significantly from those obtained by Anderson and are in closer agreement with experimental data reported for Ge–GaAs heterojunctions [18, 19].

From this model Perlman and Feucht have also reasoned that in most cases the transmission coefficient is equal to or less than unity for carriers crossing the interface region due to the presence of the mass barrier (i.e. the change in effective mass). Some interesting conclusions were also reached from calculations using this model. These indicate that carriers are reflected by a mass barrier only when they are travelling in the direction of decreasing effective mass. The ratios of the transmission coefficients for the forward and reverse carrier transport of the various barrier configurations were all the same and equal to the ratios of the effective masses on either side of the junction. An analogous situation also exists for the transmission coefficients of a light beam incident on a dielectric barrier where the ratios of the coefficients are equal to the square root of the dielectric constants.

An empirical approach by Newman [102] has shown that for all types of heterojunctions for which forward characteristics over a range of temperatures have been published, the empirical equation

$$J = J_0 \exp \frac{T}{T_0} \exp \frac{U}{U_0}$$

is obeyed. This has been shown to give a better correlation with published data than with the temperature dependent forms of Anderson's equation [19].

Because of the similarities in the properties of heterojunctions and homojunctions it is quite evident that when electrical characteristics are measured they must be those due to the interface between the two different semiconducting materials. At present there seems to be no single technique which clearly determines the real heterojunction properties, however, several types of measurement may be used to determine that homojunctions are not present. The proper direction of rectification can be used in some cases to indicate that the heterojunction is responsible for the electrical characteristics. The value of the diffusion potential U_d obtained either by extrapolation of the current–voltage characteristics or by voltage–capacitance characteristics can also be used as an indication of a heterojunction property. However, there are problems associated with both of these measurements. With current–voltage measurements at room temperature it is generally difficult to determine when U_d is approached due to excessive resistive drops in the bulk material. In this case extrapolation of these measurements can lead to U_d having a value more closely resembling than that of a low energy gap homojunction. Capacitance–voltage measurements are often difficult or impossible to make at high forward biases. Since dislocation and impurity pile-up at the interface and the resulting interface states do not change the capacitance-voltage characteristics at low forward and reverse biases, the extrapolated value of U_d obtained from these measurements may be in agreement with an abrupt heterojunction. However in actual practice, since the measurements are not sensitive to defects, homojunctions may exist due to impurity pile-up at the interface and not be detected.

From this discussion it will be seen that any interpretation of the electrical properties must be made with care and that other data such as metallographic, thermal, photovoltaic, photoconductive, and micro-probe results must always be taken into consideration.

Measurements on GaAs–Ge interface alloyed heterojunctions by Rediker et al. [33, 52] using all four combinations of conductivity types have demonstrated rectification properties, but low impedence was observed in all the interface alloyed junctions regardless of whether they were n-n, n-p, p-n or p-p when the germanium was biased positively. This suggests that in this case rectification cannot be a property of the impurity doping of the GaAs and Ge but is probably due to the impurity distribution in the recrystallised region. Interpretation of these results is therefore questionable since metallographic evidence has also shown variations in impurity concentration with distance in the recrystallised region (see Fig. 9 and 10). This, together with the known electrical behaviour of Ga and As in Ge and of Ge in GaAs, would make any simple heterojunction theory impossible.

In contrast to these results Shirafiji and Nakayama [35] have reported excelllent diode characteristics with breakdown voltages of up to 200 V and where the polarity of rectification agreed with that predicted by heterojunction theory. However, these results were not reproducible and from measurements of the voltage dependence of junction capacitance, p-i-n structures were suggested, this is also consistent with the observed high breakdown voltages.

Both these results suggest that to obtain an understanding of alloyed heterojunction properties more emphasis should be placed on systems between two different Group III–V compounds or Ge and Si.

Measurements of the electrical properties of alloyed GaSb–GaAs heterojunctions by Rediker et al. [33, 52] again using all four combinations of con-

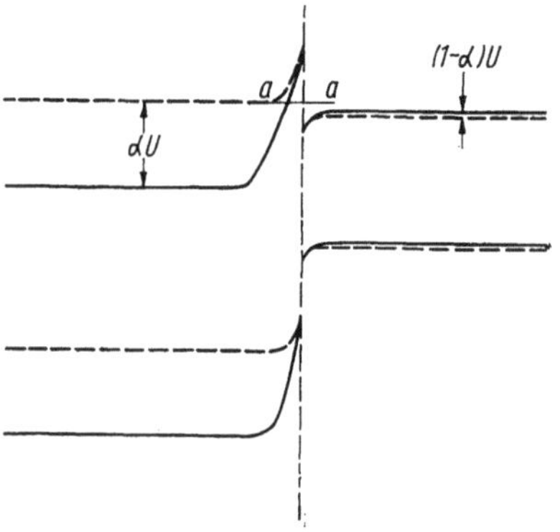

Fig. 15. Energy band diagram of a p-n heterojunction. Here a fraction α of the applied voltage U is effective in raising the conduction band of the wide band gap semiconductor with respect to the top of the barrier. (After Rediker et al. [52])

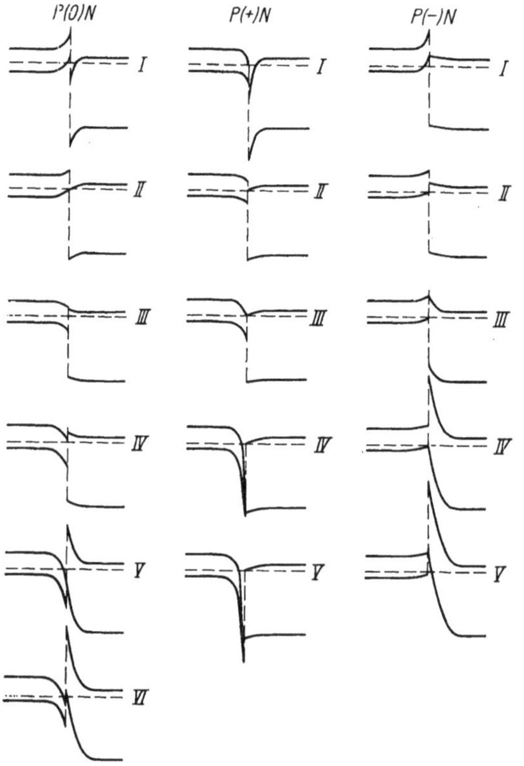

Fig. 16. Theoretical possibilites for abrupt p-n heterojunction band profiles. The wide gap semiconductor is shown on the right of each diagram. (After van Ruyven [96])

ductivity types showed that forward conduction occurred when the n-type GaAs was biased negatively or when p-type GaAs was biased positively. The rectification ratio of p-p heterodiodes however was very poor. For all the other combinations, the forward current varied as $I_f = I_0 \exp(AU)$ where A is an arbitrary numeric value which may be related to tunnelling and is relatively independent of temperature. The values of less than unity found by Rediker (for the current-voltage relationship $I_f = I_0 \exp(qU/nkT)$, indicate that the mechanism of current transport across the junction is not a thermal one and rules out explanations such as thermal injection over a barrier [18] or diffusion. The relative independence of A on temperature is consistent however with the current being due to a tunnelling mechanism.

Using the band model for an abrupt heterojunction originally proposed by Anderson [18] and later developed by Price [95], Rediker suggested tunnelling through the barrier would occur as illustrated by the line *a-a* at Fig. 15. In this model the I–U characteristics may be explained by the barrier in the conduction band at the interface. However, such a barrier could also be produced if the Fermi level is fixed at the interface by surface states [96] (see Fig. 16) or by a plane of dislocations in a non-abrupt or graded heterojunction. At the present time there is insufficient evidence to differentiate between these models. However, electron-probe microanalysis by Rediker has indicated that graded GaSb–GaAs heterojunctions are formed in practice.

Similar measurements of the electrical and strucural characteristics of InAs–GaSb alloyed heterojunctions reported by Rediker et al. [52] have also indicated that the heterojunction properties are due, at least in part, to deviations in the stoichiometry of the recrystallised region. This implies that direct application of the band model of Fig. 15 is not valid, but electrical and electro-optical results for InAs–GaSb p-n heterojunctions can still be explained by a barrier in the conduction band at the interface. Forward currents again suggest that the major transport mechanism is by tunnelling through the barrier. Work reported by Fang and Howard [44] on Ge–GaAs heterojunctions prepared by vapour transport techniques has shown that different barrier heights are obtained for different substrate crystal orientations. This effect probably also applies to alloyed heterojunctions and adds a further variable parameter to the already confused position.

Current-voltage measurements on vapour deposited Ge–Si n-n heterojunctions by Oldham and Milnes [24] and on alloyed Ge–Si heterojunctions by Shewchun and Wei [36] have shown current saturation in both directions. This characteristic has been attributed to a large number of interface states, probably of the order of 5×10^{13} cm^{-2} which are due to the 4% lattice mismatch between germanium and silicon and the large difference in thermal expansion coefficients. These interface states probably act as minority carrier recombination centres and greatly attenuate the number of minority carriers which are able to traverse the interface region. Since tunnelling can occur via these states, in thin tunnelling junctions very large excess currents (J_x) are expected. (The total current for a junction is $J = J_t + J_x + J_d$ where J_d is the diffusion current and J_t the normal tunnelling current.) In these thin junctions J_x is therefore dominant. Shewchun and Wei have confirmed that this effect occurs in n-p and p-n Si–Ge heterojunctions where the Fermi level is either above or extremely close to the conduction band edge or below or extremely close to the valence band edge for both materials. In this case the defect states by which tunnelling takes place are almost certainly regions of high dislocation density.

Shewchun and Wei have also discussed the I–U measurements on Ge–Si alloyed heterojunctions where the doping levels of the two components were of the order of 1.0 to 0.1 Ω cm (see Fig. 17). These showed characteristics having an exponential region between 10^{-4} and 10^{-2} A with a slope of $e/\eta kT$ where e is the electronic charge and $\eta = 1.22$ at room temperature and $\eta = 4.3$ at liquid air temperatures. This would result in the curves being almost parallel which is compatible with tunnelling controlled mechanisms. Tansley and Newman [80] have also reported a similar effect in $Ga_xIn_{1-x}As$–GaAs alloyed heterojunctions.

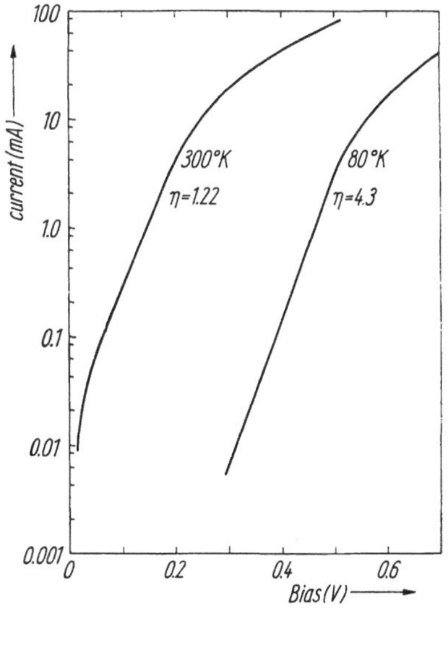

Fig. 17. Forward bias characteristics of an alloyed n (0.01 Ω cm) Si-p (0.01 Ω cm) Ge heterojunction. (After Shewchun and Wei [36])

Since the normal theory for rectification in a homogeneous diode is given by the expression

$$J_d = J_s \exp\left(\frac{e\,U_a}{\eta\,k\,T}\right) - 1\ ,$$

where J_s is the saturation current, $\eta = 1$ and U_a the applied bias, an increase in the slope of a plot of log J vs U_d with decrease in temperature would be expected. This type of behaviour has been reported by Chynoweth and McKay [97] in narrow silicon p-n junctions where field emission is prevalent, but has not been observed by Scewchun and Wei.

Since the reverse bias characteristics of heterojunctions depend strongly on the junction width which in turn depends strongly on the donor and acceptor impurity concentrations on both sides of the junction, large variations in form of the reverse bias characteristics have been reported by individual authors. In general, however, soft breakdown characteristics have been observed with few exceptions. Of these Shewchun and Wei [36] in experiments on Ge–Si alloyed heterojunctions reported "hard" characteristics which showed a soft knee prior to the sharp breakdown. As with other authors they observed a decrease in breakdown voltage with decrease in the measurement temperature. These features have been interpreted as typical of avalanche breakdown mechanisms in homojunctions by McKay [98]. Fig. 18a shows a typical reverse characteristic of this type for a Ge–Si alloyed heterojunction where the impurity levels on both sides of the junction were low, giving a junction width of the order of 5900 Å. However, Shewchun and Wei have also indicated that there is a possibility of field emission or tunnelling across the junction occurring when the junctions widths are thinner. Fig. 18b shows this type of characteristic where the junction width is of the order of 1200 Å. When the temperature is lowered in such a heterojunction the reverse current drops slightly at moderate bias but then increases more rapidly and crosses over the room temperature charac-

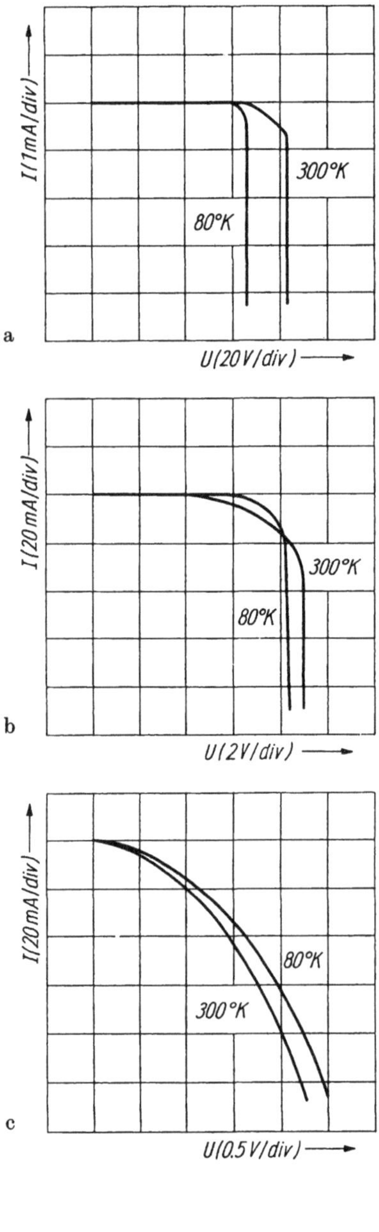

Fig. 18. Typical reverse characteristics for Ge-Si alloyed heterojunctions.
a) n (1 Ω cm) Si-p (1 Ω cm) Ge junction width approximately 5900 Å.
b) p (0.01 Ω cm) Si-n (0.01 Ω cm) Ge junction with approximately 1200 Å.
c) p (0.0001 Ω cm) Si-n (0.0001 Ω cm) Ge junction with approximately 300 Å.
(After Shewchun and Wei [36])

teristics to give a lower breakdown voltage. This lowering of the reverse current at moderate values of bias indicates field emission, while the decrease in breakdown voltage indicates avalanche breakdown. For very thin junction widths of the order of 300 Å very soft breakdowns occur as expected, with a complete reversal in breakdown voltage shift with decreasing temperature (see Fig. 18c), in this case Zener or field emission breakdown is probably predominant.

At the present time the knowledge of whether or not alloyed heterojunctions have an inherently soft reverse characteristic is still a theoretical problem since no one has yet achieved hard characteristics with these devices which compare favourably with homojunctions in similar materials.

4.2 Electro-optical properties

As previously discussed in Section 2.2.2 when heterojunctions are illuminated by a beam of light, photons are absorbed near the junction and result in the production of a photocurrent. The room temperature short circuit photocurrent spectra for heterojunctions at zero bias have been studied by Lopez and Anderson [47] who have shown that for p-n heterojunctions with approximately-equal doping levels on either side of the junction, the response peaks between the band gap energies of the two materials (see Fig. 5). Agusta and Anderson [48] have reported that by illuminating p-n heterojunctions with photons having energies between the direct band gaps of the two materials at room temperature, a photocurrent is produced which is linear with photon flux and is nearly independent of the junction voltage.

The influence of light on the I–U characteristics of a heterojunction is similar to that of a homojunction. However, by comparison with homojunctions of similar doping levels, the quantum efficiency (photocurrent per unit incident photon flux) has been generally found to be lower. (The highest reported figures of η for heterojunctions are of the order of 50% [80, 101].) Fig. 19a shows the I–U dark characteristics of a Ge–GaAs heterojunction with 1.04 eV photons applied. As the ambient temperature is reduced the response changes drastically with voltage. This is illustrated in Fig. 19b which shows the I–U characteristics of the same heterojunction at 78 °K with and without incident light where the photoresponse at zero volts is very small.

An interesting cut-off phenomenon has been reported by Agusta and Anderson [48] in Ge–GaAs heterojunctions. In the reverse direction the photocurrent increases with increasing bias, but below a critical value of forward bias the application of light tends to cut-off whatever diode current is flowing. This effect is illustrated in Fig. 19c which shows the forward I–U characteristics for different values of photon flux. It can be seen that with increasing flux the voltage range over which the current exists increases, and that in the non cut-off region the slope of the I–U characteristic also increases. The current in this region can in fact be expressed as $I \sim \exp(q\,U/\eta\,k\,T)$ where the value of η decreases with increasing photon flux. In contrast to the observations of Agusta and Anderson, van Ruyven [96] did not report cut-off phenomena in Ge–GaP heterojunctions so that at the present time any interpretation of this effect must be treated with caution.

When Agusta and Anderson [48] plotted photocurrent as a function of bias for constant photon flux, a small response was obtained near zero bias which increased with forward bias, then decreased and became slightly negative (see Fig. 19b). These effects were interpreted by Agusta and Anderson in terms of the energy band diagram for zero, reverse, and forward bias conditions (see Fig. 20a, b, and c). In the case of zero bias, photons having energies greater than the band gap of GaAs are absorbed near the surface of the GaAs. Those with energies below the band gap of Ge are transmitted through both the Ge and the GaAs. The photons with energies between the band gap of both material are transmitted through the GaAs but absorbed in the Ge. The photons which

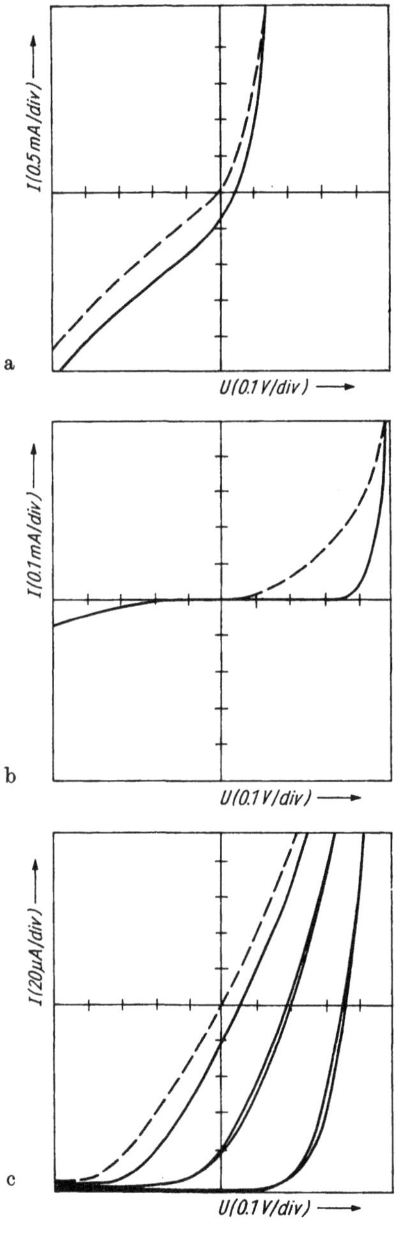

Fig. 19 a) Current — voltage characteristics of a Ge-GaAs p-n heterojunction with and without 1.04 eV energy light at room temperature.
b) As for a). But at 78 °K.
c) Forward $I-U$ characteristics for three different light intensities at 78 °K.
– – – – dark current, ——— illuminated.
(After Agusta and Anderson [48])

Fig. 20. Energy band profiles for a p-n Ge-GaAs heterojunction with or without light in
a) zero bias condition, b) reverse bias condition, c) forward bias condition.
(After Agusta and Anderson [48])

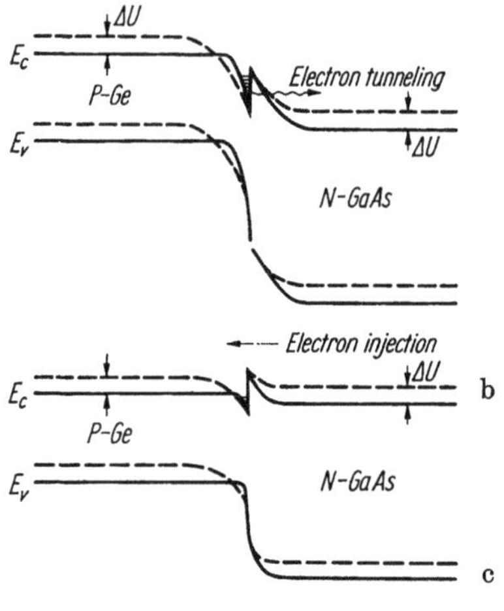

are absorbed in, or within a diffusion length of the transition region therefore excite electrons into the notch. However, since the holes are swept away by the field in the junction, few of the electrons trapped in the notch are free to recombine. To a limited extent, however, some of the electrons can tunnel into the conduction band of the GaAs to produce a positive photocurrent (i.e. the forward current is decreased or the reverse current increased), and others can be thermally excited over the barrier into the GaAs to contribute to the photocurrent. The electrons in the notch therefore constitute a negative space charge which has the same sign as ionised acceptor atoms in the Ge side of the transition region: this suggests a modified potential profile as illustrated in Fig. 20a.

On the application of light the band on the side of the charged notch is depressed and to maintain equilibrium the other band is depressed an equal amount. As a result, the electron flux emitted thermally over the barrier in either direction is reduced by the factor $\exp(-q\,\Delta U/kT)$ where ΔU is the reduction in voltage on either side of the junction relative to the interface. The electron flux at zero bias is therefore reduced by approximately the same amount in either direction. Since initially this is small, little current flows as a direct result of the modification of the electrostatic potential. The photocurrent is therefore produced by trapped electrons tunnelling through the barrier and from thermally excited electrons from the notch. Since the latter are temperature dependent a higher photocurrent is obtained at room temperature than at lower temperatures.

Agusta and Anderson [48] have also shown that this explanation is valid for reverse bias conditions (see Fig. 20b) where the modification to the potential profile has little effect on the current. In this case at low temperatures the photocurrent is mainly due to tunnelling and at room temperature is predominantly due to electrons being thermally excited over the barrier.

Agusta and Anderson have shown that incident light under forward bias conditions tends to cut off the current. In this region therefore, the nett current is slightly negative as expected from consideration of the energy band diagram (see Fig. 20c), and is proportional to the $\exp(q\,U/\eta\,kT)$ where the value of η decreases with increasing flux. For forward bias conditions without light, only that fraction of the applied voltage supported by the GaAs is effective in reducing the barrier. In a Ge–GaAs heterojunction where the Ge is the more lightly doped, Agusta and Anderson have shown that this fraction will be smaller than that supported by the Ge. This results in the variation of current with voltage being smaller than if the voltage was supported entirely by the GaAs with the result that n is expected to be large. However, under illuminated conditions electrons which are excited into the notch increase the space charge in this region and so decrease the voltage supported by the Ge and effectively increase that in the GaAs. An applied voltage therefore has a greater effect on the GaAs side and reduces n.

Agusta and Anderson [48] have suggested that at any given forward current, the difference in the applied voltage under dark conditions, and with photon flux, is a lower limit to the voltage supported on the Ge side $(U_{d_1} - U_1)$ in the dark. Thus, the maximum voltage at which cut-off occurs in the forward direction increases as the ratio of the doping levels in the Ge to the GaAs decreases. At voltages above cut-off, the increase of current with voltage is more rapid under illumination than in the dark.

Agusta and Anderson [48] have also observed that at room temperature the I–U characteristic with light applied (see Fig. 19a) was almost parallel to the dark characteristic. This implies that the band profile is influenced only slightly by light and that at room temperature very little excess charge accumulates in the notch. This means that at these temperatures the carrier recombination in this region is appreciable, or that the phonon density is high enough to excite electrons from the notch fast enough to prevent accumulation.

Tansley and Newman [80] have studied the spectral response of $Ga_xIn_{1-x}As$–GaAs alloyed heterojunctions under forward and reverse bias conditions. In general these results agree with those predicted by Agusta and Anderson [48]. However, Tansley and Newman [80] have reported that for this type of heterojunction the response curves could be divided into two distinct regions (see Fig. 21). At about 1.37 eV a predicted sharp peak having a fast response of the order of 10^{-7} s was observed. Between 1.37 and 0.97 eV an unexpected broad peak appeared which had a very slow response time and did not occur under zero bias conditions. This has been shown by Tansley and Newman [80] to be due to a bulk photoconductive effect of the GaAs. (A similar effect has been observed by Voronkova and Nasledov [99] in all GaAs having a carrier concentration of less than 10^{17} atoms/cm^3 at room temperature.) In the fast response curve at energies approaching the band gap of GaAs, incident illumination to the GaAs is absorbed within a few μm of the surface so that the generated carriers recombine without having contributed to the photocurrent. The sharp rise of the high response peak at 1.37 eV indicates the sharpness of the absorption band edge of the GaAs substrate and agrees with the results of Braunstein et al. [100]. The low energy edge falls exponentially as the probability of absorption within a diffusion length of the junction decreases, until, finally, it is reduced to zero at 1.18 eV. This is expected from band gap considerations for a solid solution composition of $Ga_{0.95}In_{0.05}As$ (see Section 3.2). Tansley and Newman [80] also reported that the shape of the high response peak was unaf-

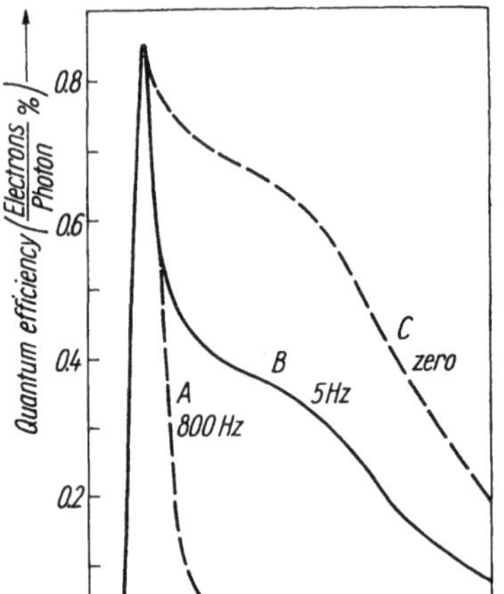

Fig. 21. Comparison of the spectral response at three different chopping rates for a GaAs-$Ga_{0.95}In_{0.05}As$ alloyd heterojunction at 5 V bias. (After Tansley and Newman [80])

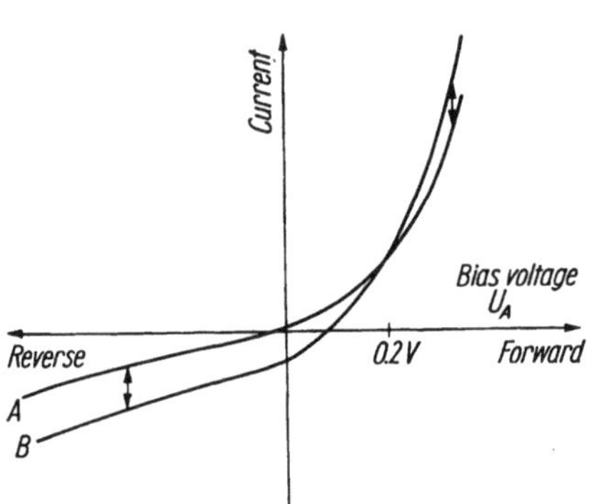

Fig. 22. The change of photocurrent polarity with respect to the dark current in an alloyed GaAs-$Ga_{0.95}In_{0.5}As$ heterojunction. (After Tansley and Newman [80])

fected by the application of high bias voltages. However, as expected, changes in the magnitude of the peak are observed with changes in bias. The change of sign of the photocurrent with respect to the dark current is illustrated in Fig. 22, this being almost certainly due to a bulk effect such as the modulation of the series resistance by light. Above a few volts reverse bias Tansley and Newman showed that the photocurrent increased as U^2 corresponding to an increase in both the transition region width and the transition field.

Rediker et al. [52] have also measured the spectral response of alloyed heterojunctions. When monochromatic incident radiation was applied to the edge of an InAs–GaSb p-n heterojunction a photovoltaic response of the form shown in Fig. 23a was observed. In the region between the energy gaps a small relatively constant response is produced. An abrupt increase in the photovoltage is then observed as the incident photon energy reaches the GaSb energy gap. The spike seen at 0.8 eV has been explained by Rediker [52] as being caused by a surface trapping mechanism since this is a dc phenomenon occuring when the incident light is chopped at 13 Hz, but disappearing when the chopping rate was increased to 800 Hz. The theory is supported by the observation that the spike was eliminated by surface etching treatments.

These photovoltaic effects may be explained by assuming that carriers are not created to any appreciable extent by incident photons having an energy less than the InAs band gap. Between 0.4 and 0.8 eV electron–hole pairs are generated on the InAs side of the junction. The electrons produced in the p-type InAs should therefore tunnel through the barrier, but many will recombine because of the low tunnelling probability. If the junction is open-circuit it will bias itself in the forward direction to counteract the current produced by the small number of photogenerated carriers which tunnel through the barrier to produce a small photovoltage. At photon energies greater that the forbidden gap of GaSb, electron-hole pairs are created in the n-type material. Holes produced within a diffusion length of the junction are therefore swept into the InAs region and are responsible for the large photovoltaic response above 0.8 eV. This explanation therefore implies that electron-hole pairs produced

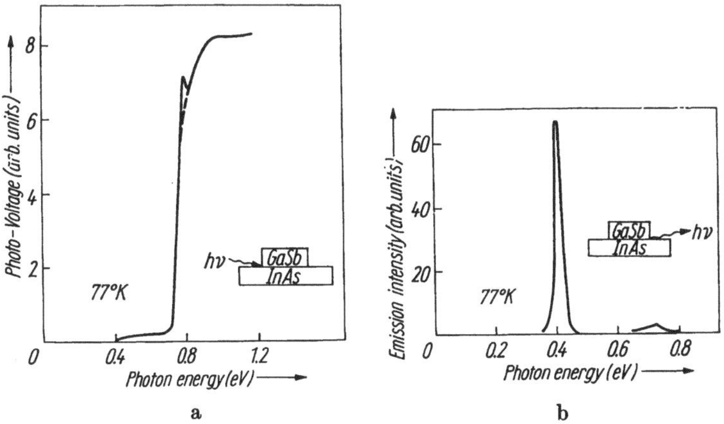

Fig. 23. a) The photo voltaic spectral response of an InSb-GaASb p-n junction at 77 °K with incident radiation directed at the junction.
b) The spectral response from the edge of the forward biased p-n InAs-GaSb junction at 77 °K (the same junction as Fig. 23a).
N_a InAs ≈ 6×10^{17} cm^{-3}, N_d GaSb ≈ 1×10^{17} cm^{-3}. (After Rediker et al. [53])

on the InAs side of the junction cause an electron current which is reduced by the notch, and that a pair production on the GaSb side produces hole current which is not impeded.

Rediker et al. [52] have also shown that, when an InAs–GaSb p-n heterojunction is forward biased, infrared emission from the edge of the junction takes place. This is shown in Fig. 23b. The large emission intensity at 0.4 eV (the forbidden energy gap of InAs) indicates that injection of electrons from the n-type GaSb into the p-type InAs occurs when radiative recombination takes place. A small amount of emission is also obtained at an energy near that of the GaSb band gap. Rediker [52] has explained these results by postulating a barrier in the conduction band at the interface. Under forward bias conditions electrons tunnel through the barrier from the GaSb to the InAs where they recombine to produce radiation corresponding to the InAs band gap. Since the energy of the radiation is substantially less than that of the forbidden energy gap of GaSb, the absorption coefficient in this material is low and allows the GaSb to act as a transparent substrate.

It is interesting to note that the wide gap emission observed by Rediker et al. [52] is probably that first proposed by Kroemer [1]. At the present time this type of emission has not been observed in other types of alloyed heterojunctions.

5. Conclusions

In this review an attempt has been made to present a summary of the information available on the physics and metallurgy of alloyed semiconductor heterojunctions and to discuss some of the mechanisms postulated for the interpretation of their properties.

At present the technologies employed in the preparation of heterojunctions are relatively simple in concept, but the metallurgical mechanisms involved can be very complex especially where ternary or quaternary systems are encountered. Of the three major alloying processes, interface alloying is probably the easiest both to accomplish and to understand, but its use is limited by the small range of compatible materials and the lack of control over the composition of the recrystallised layer. The solution growth process has probably more scope for development since the composition of the regrown layers can be controlled to some extent, but a big disadvantage is the complicated metallurgy involved and the risk of diffusion from the alloy constituents. The vapour transport-interface alloy has only limited advantages and is thus probably only of use for specific applications.

The electrical and opto-electrical characteristics of abrupt heterojunctions are strongly affected by the notch created by the discontinuity in the conduction band edge. Because a barrier exists for electrons flowing between two different semiconductors, only a small fraction of those excited in, or within a diffusion length of, the transition region actually traverse the junction to give a photoresponse. Instead electrons are trapped in the notch so that photocurrent can result from electrons in the notch entering the wider band gap material by thermal stimulation or by modulation of the normal diode current by trapped electrons causing a modification of the energy band profile.

The presence of a barrier and notch in the conduction band edge has been shown to greatly affect quantum efficiency. Unless the discontinuities can be eliminated by the use of graded junctions or the use of materials where the wide band gap semiconductor has a larger electron affinity than the narrow gap material, the applications of heterojunctions will be severely limited.

At present the rather soft reverse characteristics of alloyed heterojunctions are a limitation to their use as fast switching diodes. However, the development of techniques and the improvements in design may in the future remove these limitations.

The alloyed heterojunction can in some cases provide a sensitive photon detecting device, and in the future this may well be the most useful application for these devices.

Acknowledgement

The author would like to thank Mr. H. C. Wright for many interesting and stimulating discussions.

References

[1] H. Kroemer, Bull. Amer. Phys. Soc. **2**, 143 (1956).
[2] H. Kroemer, RCA Rev. **18**, 1332 (1957).
[3] H. Kroemer, Proc. IRE **45**, 535 (1957).
[4] J. Tauc, Rev. mod. Phys. **21**, 307 (1957).
[5] A. Armstrong, Proc. IRE **46**, 1307 (1958).
[6] T. K. Lakshmanan, Proc. IRE **48**, 1646 (1960).
[7] M. Wolf, Proc. IRE **48**, 1246 (1960).
[8] K. Lehovec, Proc. IRE **40**, 1408 (1952).
[9] P. R. Emtage, J. appl. Phys. **33**, 1950 (1962).
[10] R. J. Keyes and T. M. Quist, Proc. IRE **50**, 1822 (1962).
[11] M. I. Nathan, W. P. Dumke, G. Burns, F. H. Dill, and G. Lasher, J. appl. Phys. Letters **1**, 62 (1962).
[12] R. H. Rediker, T. M. Quist, and B. Lax, Proc. IEEE **51**, 218 (1963).
[13] R. F. Rutz, Proc. IEEE **51**, 470 (1963).
[14] F. F. Fang and H. N. Yu, Proc. IEEE **51**, 860 (1963).
[15] A. G. Foyt, Proc. IEEE **51**, 852 (1963).
[16] J. R. A. Beale and P. C. Newman, private communication.
[17] R. Ruth, J. C. Marinace, and W. Dunlap, J. appl. Phys. **31**, 995 (1960).
[18] R. L. Anderson, IBM J. Res. Developm. **4**, 283 (1960).
[19] R. L. Anderson, Solid State Electronics **5**, 341 (1962).
[20] R. L. Anderson, Trans. IRE ED **9**, 509 (1962).
[21] J. C. Marinace, IBM J. Res. Developm. **4**, 280 (1960).
[22] S. S. Perlman, D. L. Feucht, and R. M. Williams, Trans. IRE ED **9**, 509 (1962).
[23] W. S. Oldham, and A. G. Milnes, Trans. IRE ED **9**, 509 (1962).
[24] W. S. Oldham and A. G. Milnes, Solid State Electronics **6**, 121 (1963).
[25] S. S. Perlman, Trans. IRE ED **10**, 335 (1963).
[26] W. S. Oldham and A. G. Milnes, Trans. IRE ED **10**, 335 (1963).
[27] W. S. Oldham et al., J. Electrochem. Soc. **110**, 536 (1963).
[28] M. Weinstein, R. O. Bell, and A. A. Menna, J. Electrochem. Soc. **111**, 674 (1964).
[29] F. A. Pizzarello, J. Electrochem. Soc. **109**, 226 (1962).
[30] San-Mei Ku, J. Electrochem. Soc. **110**, 991 (1963).
[31] H. C. Theuerer, J. Electrochem. Soc. **108**, 649 (1961).
[32] H. T. Minden, J. Electrochem. Soc. **112**, 300 (1965).
[33] R. H. Rediker, S. Stopek, and J. H. R. Ward, Solid State Electronics **7**, 621 (1964).

[34] N. K. Kiselava, Soviet Phys.—Cryst. **9**, 365 (1964).
[35] J. Shirafiji and N. Nakayama, Japan. J. appl. Phys. **3**, 801 (1964).
[36] J. Shewchun and L. Y. Wei, J. Electrochem. Soc. **111**, 1145 (1964).
[37] E. D. Hinkley, R. H. Rediker, and M. C. Lavine, Solid State Res. Group, Lincoln Laboratories, MIT **2**, No 23 (1964).
[38] J. R. Dale and M. J. Josh, Solid State Electronics **8**, 1 (1965).
[39] J. Brownson, J. appl. Phys. **35**, 1356 (1964).
[40] W. G. Oldham and A. G. Milnes, Solid State Electronics **7**, 153 (1964).
[41] W. T. Read Jr, Phil. Mag. **45**, 775 (1954).
[42] R. L. Anderson, IBM J. Res. Developm. **4**, 280 (1960).
[43] M. J. Hampshire and G. T. Wright, Brit. J. appl. Phys. **15**, 1331 (1964).
[44] F. F. Fang and W. E. Howard, J. appl. Phys. **35**, 612 (1964).
[45] E. Spenke, Electronic Semiconductors, McGraw-Hill Publ. Comp., 1958.
[46] H. K. Henisch, Rectifying Semiconductor Contacts, Oxford University Press 1957.
[47] A. Lopez and R. L. Anderson, Solid State Electronics **7**, 695 (1964).
[48] B. Agusta and R. L. Anderson, J. appl. Phys. **36**, 206 (1965).
[49] T. S. Moss and R. D. F. Hawkins, Infrared Phys. **1**, 111 (1961).
[50] A. van der Ziel, Solid State Physical Electronics, Prentice-Hall, Englewood Cliffs 1957 (p. 284).
[51] J. Grossman, J. Electrochem. Soc. **110**, 1065 (1963).
[52] R. H. Rediker, S. Stopek, and E. D. Hinkley, Trans. ASM **233**, 463 (1965).
[53] R. S. Mroczkowski, M. C. Lavine, and H. C. Gatos, Trans. ASM **233**, 456 (1965).
[54] K. Lonsdale, Phil. Trans. **A 240**, 219 (1947).
[55] R. C. Hanneman, R. E. Ogilvie, and A. Modrziejewski, J. appl. Phys. **33**, 1429 (1962).
[56] R. W. James, The Optical Principles of the Diffraction of X-rays, Chap. VIII, Bell, London 1958.
[57] Y. Takeda, T. Hirai, and M. Hirao, J. Electrochem. Soc. **112**, 363 (1965).
[58] C. Kolm, S. A. Kulin, and B. L. Averbach, Phys. Rev. **108**, 965 (1957).
[59] J. Brownson, Trans. AIME **233**, 450 (1965).
[60] A. Levitas, C. C. Wang, and B. H. Alexander, Phys. Rev. **95**, 846 (1955).
[61] E. R. Johnson and S. M. Christian, Phys. Rev. **95**, 560 (1954).
[62] F. Herman, Phys. Rev. **95**, 847 (1955).
[63] H. M. James and A. S. Ginsberg, J. phys. Chem. **57**, 90 (1953).
[64] R. Braunstein, Phys. Rev. **130**, 869 (1963).
[65] R. Braunstein, Phys. Rev. **130**, 879 (1963).
[66] I. E. Gorshkov and N. A. Goryunova, Zh. neorgan. Khim. **3**, 668 (1958).
[67] M. Hansen, Constitution of Binary Alloys, McGraw-Hill, Publ. Comp. 1958.
[68] E. D. Hinkley, R. H. Rediker, and D. K. Jadus, Appl. Phys. Letters **6**, 144 (1965).
[69] C. Hilsum and A. C. Rose-Innes, Semiconducting III—V Compounds, Pergamon Press 1961.
[70] P. M. Hamilton, Semiconductor Products and Solid State Technics **3**, 15 (1964).
[71] Yu. M. Shaskov, Metallurgy of Semiconductors, Pitman and Sons, Ltd., London 1960.
[72] S. I. Novikova, Soviet Phys. — Solid State Phys. **3**, 129 (1961).
[73] G. A. Wolff, L. Tolman, N. J. Field, and J. C. Clark, Proc. Internat. Colloquium on Semiconductors and Phosphors, Garmisch 1956, (p. 463).
[74] Tables de Constantes et Données Numeriques, 12 -Semi-conducteurs, Pergamon Press, Paris 1961.
[75] J. R. Dale and M. J. Josh, Solid State Electronics **7**, 177 (1964).
[76] H. J. van Hook and E. S. Lenker, Trans. AIME **227**, 220 (1963).
[77] J. C. Woolley and B. A. Smith, Proc. Phys. Soc. **70B**, 153 (1957).
[78] J. C. Woolley and B. A. Smith, Proc. Phys. Soc. **72**, 214 (1958).
[79] H. C. Wright and T. L. Tansley, private communication.

[80] T. L. TANSLEY and P. C. NEWMAN, to be published.
[81] M. S. ABRAHAMS, R. BRAUNSTEIN, and F. D. ROSI, J. Phys. Chem. Solids **10**, 204 (1959).
[82] O. G. FOLBERTH, Z. Naturf. **10a**, 502 (1955).
[83] W. G. SPITZER and J. M. WHELAN, Phys. Rev. **114**, 59 (1959).
[84] L. SPITZER JR. and R. HARM, Phys. Rev. **89**, 977 (1953).
[85] W. G. SPITZER and H. Y. FAN, Phys. Rev. **106**, 882 (1957).
[86] J. R. DALE, to be published.
[87] D. L. KEMDALL, Appl. Phys. Letters **4**, 67 (1964).
[88] H. DIEDRICH, private communication and German Pat. No. 1178948 (1960).
[89] W. G. PFANN, Trans. AIME **203**, 961 (1955).
[90] W. G. PFANN, K. E. BENSON, and J. H. WERNICK, J. Electronics **2**, 597 (1957).
[91] J. H. WERNICK, J. chem. Phys. **25**, 47 (1957).
[92] A. J. MLAVSKI and M. WEINSTEIN, J. appl. Phys. **34**, 2885 (1963).
[93] J. BROWNSON, Semicond. Symposium, Electrochem. Soc., New York 1962.
[94] S. S. PERLMAN and D. L. FEUCHT, Solid State Electronics, **7**, 911 (1964).
[95] P. J. PRICE, Proc. Conf., Phys. of Semiconductors, Inst. of Phys. and Phys. Soc., Exeter 1962 (p. 99).
[96] L. J. VAN RUYVEN, Thesis, TH Eindhoven 1964.
[97] A. G. CHYNOWETH and K. G. MCKAY, Phys. Rev. **106**, 418 (1957).
[98] K. G. MCKAY, Phys. Rev. **94**, 877 (1954).
[99] N. M. VORONKOVA and D. N. NASLEDOV, Soviet Phys. — Solid State Phys. **6**, 1736 (1965).
[100] R. BRAUNSTEIN, et al., Appl. Phys. Letters **3**, 31 (1963).
[101] T. B. RAMACHANDRAM and W. J. MORONEY, Proc. IEEE **52**, 1358 (1964).
[102] P. C. NEWMAN, Electronics Letters **1**, 265 (1965).

(Received January 19, 1966)

Original Papers

phys. stat. sol. **16**, 389 (1966)

Institute of Physical Chemistry, Czechoslovak Academy of Sciences, Prague

Some Remarks about Electronic Correlation in Crystal Surfaces: The Ideal (100) Surface of a Semi-Infinite Diamond-Like Crystal[1])

By

M. Tomášek

The simple qualitative method for treating electronic correlation in crystal surfaces, suggested recently by the author and by Davison and Amos, is discussed in detail. As an example, a study is made of the effect of interelectronic repulsion on the behaviour of electrons localized near an ideal (100) surface of a diamond-like crystal. The spin-degeneracy of the electron surface states is found to be split and possible physical consequences of this splitting are mentioned. Surface states lying in "forbidden" bands differ from volume states in their sensitivity to the magnitude of the interelectronic repulsion included. Surface states with energies lying within the volume state energy bands seem to have the same sensitivity to the interelectronic repulsion as the volume states.

Eine einfache qualitative Methode für die Untersuchung der elektronischen Korrelation in Kristalloberflächen, die kürzlich von dem Autor dieser Arbeit sowie von Davison und Amos vorgeschlagen wurde, wird ausführlich diskutiert. Als Beispiel wird der Einfluß der interelektronischen Abstoßung auf die Eigenschaften der Elektronen, die an der idealen (100)-Oberfläche eines Kristalls vom Diamant-Typ lokalisiert sind, studiert. Es zeigt sich, daß eine Aufspaltung von Energieniveaus der Oberflächenzustände von Elektronen mit Spin ↑ und ↓ möglich ist. Oberflächenzustände, welche in „verbotenen" Energiebändern liegen, unterschieden sich von Volumenzuständen durch ihre Empfindlichkeit für interelektronische Abstoßung. Oberflächenzustände, die in den Energiebändern der Volumenzustände liegen, haben die gleiche Empfindlichkeit wie die Volumenzustände.

1. Introduction

However strange it may look like, the methods of one-electron theory of electronic structure seem to work better in solid state physics, especially of non-transition metals than with molecules. One might think this impression arose because, at present, more precise knowledge of the electronic structure is needed in the latter case to interpret experimental data than with solids. But this is probably not the case. An example is the success of the one-electron theory of Fermi surfaces of non-transition metals. It seems to us that wherever in solids, the "molecular" character is more pronounced than the "solid state" one, we are less justified to use the simple one-electron theory and have to rely upon more sophisticated methods, which take explicitly into account the interelectronic repulsion, i.e. we are forced to consider the so-called "correlation problem".

In the case of solid surfaces, we are faced just with such a situation. The crystal surface can be considered as something between a molecule and a real crystal bulk and those electronic properties, which depend on the surface behaviour or are mainly determined by it, may fail to be described well by one-electron methods. This refers before all to the electronic surface states.

---

[1]) Presented at the Semiconductor Surface Physics Meeting, Institute of Semiconductors, Ukr. Akad. Nauk, Kiev, 1st to 3rd June 1965.

These rather loose considerations resemble some of the results of a recent investigation [1], in which surface and chemisorption electronic states were treated on the (111) plane of a diamond-like crystal from the point of view of correlation. It has been found that electronic states localized near the surface are more sensitive to the inclusion of interelectronic repulsion than volume states, i.e. the electronic states in the bulk. This means that correlation effects should play a more important role in the surface than in the bulk of the crystal. Although this effect can be explained formally by the screening of the interelectronic repulsion on the surface, which differs from that in the bulk, nevertheless it is interesting to see that it might be just this difference in screening which separates the two cases of "molecules" and "solids" we are speaking here so freely about.

A very simple and qualitative approach to the correlation problem, recently suggested and utilized by the present author [1] and also, in a slightly less general form, by Davison and Amos [2], is commented below, being first briefly described and is used afterwards to analyse the role of correlation on the ideal (100) plane of a diamond-like crystal.

2. General Comments on the Method

The mathematics of the proposed method [1] is based on the technique of one-electron Green's function in the localized functions representation [3], i.e. all calculations are in fact carried out in the LCAO approximation. It is not without interest to note that, though correlation is a many-electron effect, we use the one-electron picture and the one-electron language. This is an outstanding feature of the Löwdin alternant molecular orbital (AMO) method [4], upon the basic idea of which the present approach is founded. However, this also imposes limitations: We can study only crystals of alternant character, i.e. crystals, in which all lattice points can be divided into two completely equivalent and interpenetrating sublattices $\mathcal{A}$ and $\mathcal{B}$.

In the Green's function technique, the knowledge of wave functions of an ideal infinite crystal as well as of corresponding energies is quite indispensable. It is by no means easy to obtain correlated wave functions [5] even in the case of ideal crystals.

To lower the correlation energy, electrons with antiparallel spins try to avoid each other and assume disjoint parts of space, e.g. electrons with up ($\uparrow$) spin are more probable to be found on atoms A of sublattice $\mathcal{A}$, electrons with down ($\downarrow$) spin on atoms B of sublattice $\mathcal{B}$ of an alternant crystal. The correlated wave function of an electron of either spin will exhibit therefore an alternant character [4].

Atoms A seem to attract $\uparrow$ spin electrons and repel $\downarrow$ spin electrons and mutatis mutandis with atoms B. Therefore, atoms A influence an arbitrary electron by another potential than atoms B. Note that an $\uparrow$ spin electron of an infinite crystal sees just the same form of the crystal potential as a $\downarrow$ spin electron; it is only necessary to exchange atoms A and B in there places. Hence, volume state energy bands will be the same for both $\uparrow$ and $\downarrow$ spin electrons.

As to the potential, the above situation resembles a binary analogue of the alternant crystal under consideration: We calculate therefore the infinite crystal wave functions and energy spectrum of the binary analogue of our original system, and this we do in such a way that they depend explicitly on the difference between the potential of atoms A and B. They are used as the correlated wave functions and the corresponding energies of the alternant crystal in our simplified approach. To be concrete, the binary analogue to the diamond-like lattice is the sfalerite type lattice. The correlated wave functions of an $\uparrow$ spin electron are the same as those of a $\downarrow$ spin electron but lie in real space with a coordinate system now fitted to the $\mathcal{A}$ sublattice. By changing the difference between the potentials of atoms of either sublattice, we can control the extent to which electrons with antiparallel spins avoid each other and also the degree of mixing of electrons with antiparallel spins on the same atom.

Let us provide these ideas with a formal mathematical framework (compare [4]). Suppose we have a set ($l = 1, \ldots, t$) of localized orbitals φ_l on each atom of an alternant crystal, which belong to valence electrons of an incomplete outer shell, and each of them is orthogonal to any other localized orbital on all atoms of the crystal. These orbitals may be Löwdin orthogonalized orbitals or Wannier functions. The one-electron LCAO approximation gives for the crystal orbital of an arbitrary electron the following ansatz:

$$\Psi_{j,\eta}(\boldsymbol{r}) = \sum_{l=1}^{t} \left[c_{j,l}^{\mathcal{A}}(\boldsymbol{\eta})\, \Phi_{l,\eta}^{\mathcal{A}}(\boldsymbol{r}) + c_{j,l}^{\mathcal{B}}(\boldsymbol{\eta})\, \Phi_{l,\eta}^{\mathcal{B}}(\boldsymbol{r})\right], \tag{1}$$

where

$$\Phi_{l,\eta}^{\mathcal{C}}(\boldsymbol{r}) = N^{-3/2} \sum_{n} \mathrm{e}^{i\boldsymbol{\eta}\cdot\boldsymbol{r}_n}\, \varphi_l\,(\boldsymbol{r} - \boldsymbol{r}_n - \boldsymbol{x}_{\mathcal{C}})$$

is a Bloch-like function of l-type orbitals on sublattice $\mathcal{C}$ ($\mathcal{C} = \mathcal{A}, \mathcal{B}$), $c_{j,l}^{\mathcal{C}}(\boldsymbol{\eta})$ are unknown coefficients underlying determination, $\boldsymbol{r}_n$ is the radius vector of the center of one of the elementary cells of the crystal containing two atoms A and B placed in positions $\boldsymbol{x}_{\mathcal{A}}$ or $\boldsymbol{x}_{\mathcal{B}}$ with respect to the center of the cell, $\boldsymbol{\eta}$ is the wave vector of the first Brillouin zone, N the number of elementary cells of an edge of the large cube with respect to which Born-Kármán conditions are defined and j is the index of the energy band ($j = 1, \ldots, 2\,t$).

After the corresponding secular problem for the infinite cyclic alternant crystal has been solved, we find that the energy bands can be divided into two groups, one having wave functions $\psi_{j,\eta}^{(+)}(\boldsymbol{r})$ of the form (1), the other of the form

$$\Psi_{j,\eta}^{(-)}(\boldsymbol{r}) = \sum_{l=1}^{t} \left[c_{j,l}^{\mathcal{A}}(\boldsymbol{\eta})\, \Phi_{l,\eta}^{\mathcal{A}}(\boldsymbol{r}) - c_{j,l}^{\mathcal{B}}(\boldsymbol{\eta})\, \Phi_{l,\eta}^{\mathcal{B}}(\boldsymbol{r})\right] \tag{2}$$

j being now the index of the "pair" of bands, $j = 1, \ldots, t$. The alternant character of one-electron wave functions is obtained, following Löwdin [4], by taking linear combinations of functions (1) and (2):

$$\Psi_{j,\eta}^{\uparrow} = a_{j,\eta}\, \Psi_{j,\eta}^{(+)} + b_{j,\eta}\, \Psi_{j,\eta}^{(-)} = g_{j,\eta}\, \Phi_{j,\eta}^{\mathcal{A}} + s_{j,\eta}\, \Phi_{j,\eta}^{\mathcal{B}}, \tag{3}$$

$$\Psi_{j,\eta}^{\downarrow} = a_{j,\eta}\, \Psi_{j,\eta}^{(+)} - b_{j,\eta}\, \Psi_{j,\eta}^{(-)} = s_{j,\eta}\, \Phi_{j,\eta}^{\mathcal{A}} + g_{j,\eta}\, \Phi_{j,\eta}^{\mathcal{B}}, \tag{4}$$

where

$$\Phi_{j,\eta}^{\mathcal{C}} = \sum_{l=1}^{t} c_{j,l}^{\mathcal{C}}(\boldsymbol{\eta})\, \Phi_{l,\eta}^{\mathcal{C}}, \qquad g_{j,\eta} = a_{j,\eta} + b_{j,\eta}, \qquad s_{j,\eta} = a_{j,\eta} - b_{j,\eta}. \tag{5}$$

An $\uparrow$ spin electron is "attracted" by atoms of sublattice $\mathcal{A}$ and "repelled" by atoms of sublattice $\mathcal{B}$. Therefore, in the ground state, the amplitudes of its wave functions will be great ($g_{j,\eta}$) on sublattice $\mathcal{A}$ and small ($s_{j,\eta}$) on sublattice $\mathcal{B}$. For a $\downarrow$ spin electron, the situation will be reversed. This is clearly seen from equations (3) and (4). Note, that the linear variant (3) or (4) is just the LCAO crystal orbital ansatz of the binary analogue of our alternant crystal. Our scheme is therefore justified. These considerations certainly have to be completed by showing that our "different orbitals for different spins" determinantal wave function is at least an approximate eigenstate of S^2; this question as well as that of the relation of our simple approach to the full many-electron theory are discussed in a forthcoming paper.

It appears, that the procedure can be applied to any crystal of alternant character, irrespective, of how many valence electrons its atoms may have, but the following condition seems to be essential, namly, that only the lower half of the set of volume state energy levels availlable should be occupied in the ground state [6]. The situation is particularly clear, when each atom has only one valence electron in the incomplete outer shell. For the case of more electrons than one, however, a situation arises, where two or more valence electrons with parallel spins may be found on the same atom. Here, it is Hund's rules which guarantee that our scheme works well. The guarantee is complete for large interatomic distances, i.e. almost isolated atoms. For finite interatomic distances of real crystals it will be only approximate. Another guarantee is that electrons with parallel spins are sufficiently well correlated already in the restricted Hartree-Fock approximation, i.e. in the one-electron theory.

Surface states, with correlation included, are treated as follows: We delimit the binary analogue crystal by the respective delimiting plane and for homoatomic surfaces we suppose that it consists of B atoms (without loss of generality and in agreement with [1, 7, 8]). We solve the problem both for the case of electronegative B atoms and electropositive B atoms using standard procedures described in [7] or [8]. The solution for the electronegative case will describe the behaviour of $\downarrow$ spin electrons on the investigated surface of the alternant crystal with correlation included, the solution for the electropositive case will describe the behaviour of $\uparrow$ spin electrons. Details can be found in Section 3. Among other applications of the procedure we mention the work on the problem of surface states of σ- and π-electrons in planar graphite which is now in progress, or the problem of chemisorption localized states on antiferromagnetic transition metals.

However, two comments are necessary with respect to this way of calculating surface states. First of all, it would be better to treat surface effects of a semi-infinite crystal directly as a many-body problem, i.e. from first principles. This is a rather difficult task and is being considered now by the author. Here, we proceed by fitting correlated wave functions of volume state electrons to one-electron functions of either vacuum zero potential field (pure surface) or to one-electron functions of adsorbed atoms and molecules (chemisorption). How much this is justified is the question of further investigations. Secondly, we use orthogonal localized orbitals, and it is well known that for the degenerate case such as p- or d-type functions or equivalent sp^3 hybrides, the corresponding orthogonal orbitals are not sufficiently strongly localized. Therefore, our boundary condition (see e.g. (15)) may differ a little bit from the natural one.

Although so far [1, 2] the approach has been used, similarly as in Section 3 of this paper, as a simple qualitative method for treating correlation, it is not excluded that, in principle, it could be exploited as a rough semi-quantitative tool to estimate the importance of correlation in surface effects of concrete crystals. To reach this goal, it is necessary to know the ratio of the coefficients $g_{j,\eta}$ and $s_{j,\eta}$ for concrete crystals. Such an information can be supplied by a variational calculation using the classical AMO method; there exists only one calculation of this type for crystals [9] at present but new results are believed to appear very soon [10]. Another possibility is to follow Coulson's suggestion [11 to 14] and using uncorrelated one-electron wave functions and energies to try to calculate this ratio by perturbation method. The effectiveness of the latter recipe for crystals is now investigated.

3. Example

Now, we are ready to analyse the case of surface states of the ideal (100) plane of a diamond-like crystal with correlation included. The mathematics of this case is obviously parallel to that of sfalerite type lattice and the latter has been already carried out in [7]. Hence, we shall only formulate the problem and shall take the solution from [7].

Let us suppose that our set of orthogonal localized orbitals is based on four sp^3 hybrides, i.e. we have four orthogonal localized orbitals of sp^3 character on each A or B atom of the crystal. Suppose that the only non-zero matrix elements of the one-electron Hamiltonian $H^\downarrow$ describing the motion of $\downarrow$ spin electrons are the following ones: The Coulomb integrals α_A or α_B of an orbital placed on the A or B atom, respectively, the resonance integrals γ_A or γ_B of two orbitals placed on an A or B atom, respectively, and the resonance integral γ of two orbitals lying along a bond connecting two neighbouring atoms A and B. The values of parameters with indices A (α_A, γ_A) generally differ from the values of those parameters with indices B (α_B, γ_B). To follow Section 2, we assume $|\alpha_B| > |\alpha_A|$, $|\gamma_B| > |\gamma_A|$. For obvious reasons, the corresponding matrix elements of the Hamiltonian $H^\uparrow$ which describes the behaviour of $\uparrow$ spin electrons will be obtained from those of $H^\downarrow$ by interchanging the indices A and B, e.g. the quantity α_B now describes the Coulomb integral of an orbital placed on atom A, the quantity α_A corresponds to a Coulomb integral of an orbital located on atom B, etc. The ground state one-electron wave functions of $\uparrow$ and $\downarrow$ spin electrons will then have the properties required in Section 2, i.e. they will be correlated.

Let us consider explicitly only the case of volume and surface states of $\downarrow$ spin electrons, which is formally identical with the case of the sfalerite lattice of [7]. Results for $\uparrow$ spin electrons can be obtained from those of $\downarrow$ spin electrons by interchanging the indices A and B. In agreement with [7], let us take the wave function of a $\downarrow$ spin electron in the infinite cyclic crystal in the form (compare equation (3) of [7])

$$\Psi_{j,\eta}^{\downarrow}(\boldsymbol{r}) = \sum_{p=1}^{8} c_{j,p}^{\downarrow}(\boldsymbol{\eta})\, \Phi_{p,\eta}(\boldsymbol{r}) \,, \tag{6}$$

where

$$\Phi_{p,\eta}(\boldsymbol{r}) = N^{-3/2} \sum_{n} e^{i\boldsymbol{\eta}\cdot\boldsymbol{r}_n} \varphi_p(\boldsymbol{r} - \boldsymbol{r}_n) \,.$$

The unit vector associated with component η_3 of the wave vector $\boldsymbol{\eta} = (\eta_1, \eta_2, \eta_3)$ is chosen to be perpendicular to the crystal surface, i.e. two of the primitive translations of the direct crystal space, $\boldsymbol{b}_1$ and $\boldsymbol{b}_2$, lie in the (100) plane (see [7, 15]). In comparison with (1), we have replaced the index l by p, where $p = 1, \ldots, 4$ corresponds to orbitals on atom A, $p = 5, \ldots, 8$ to orbitals on atom B. Using the properties of our matrix elements, (6) can be seen to be equivalent to (4).

Volume state energy bands of $\downarrow$ spin electrons based on the wave function (6) are determined by the secular equation

$$\Delta^{\downarrow} = 0 \,, \tag{7}$$

where

$$\Delta^{\downarrow} = \begin{vmatrix} M_A & Z \\ \tilde{\tilde{Z}} & M_B \end{vmatrix} \tag{8}$$

and

$$\left.\begin{aligned} (M_C)_{jk} &= (\alpha_C - W - \gamma_C)\,\delta_{jk} + \gamma_C\,, \\ Z_{jk} &= \gamma\left[e^{i\eta_{4-j}}(1-\delta_{j,2}) + e^{i(\eta_{4-j}+\eta_{4-j+1})}\delta_{j,2}\right]\delta_{4-j+1;k}\,, \\ C &= A, B; \quad j, k = 1, \ldots, 4; \quad \xi_0 = 0 \end{aligned}\right\} \tag{9}$$

and where W is the energy. By using new variables

$$\left.\begin{aligned} &x_1 = (W - \alpha_A + \gamma_A)\,\gamma^{-1}\,, && x_2 = (W - \alpha_B + \gamma_B)\,\gamma^{-1}\,, \\ &\gamma_1 = \gamma_A\,\gamma^{-1}\,, && \gamma_2 = \gamma_B\,\gamma^{-1}\,, \\ &\varphi_1' = (x_1 x_2 - 4\gamma_2 x_1 - 1)\,\gamma_2^{-1}\,, && \varphi_2' = (x_1 x_2 - 4\gamma_1 x_2 - 1)\,\gamma_1^{-1}\,, \\ &1 + e^{i\eta_k} = \nu_k\,e^{i\alpha_k}\,, \quad \nu_k = \nu_k^*\,, && k = 1, 2\,, \end{aligned}\right\} \tag{10}$$

equation (8) can be written as

$$\Delta^{\downarrow} = \gamma_1\gamma_2\gamma^8 (x_1 x_2 - 1)^2 \left[\varphi_1'\varphi_2' - \nu_1^2 - \nu_2^2 - 2\nu_1\nu_2\cos(\eta_3 - \alpha_1 + \alpha_2)\right]. \tag{11}$$

The solution of (7) is the same as in [7, 8] and it was analysed in detail in [8], pp. 337–8. We use here the same notations: "Allowed" energy bands of volume states are denoted as I, II, III, IV, "forbidden" energy bands are D_1, C_1,, B, C_2, D_2, in order of increasing energy (see Fig. 2, 3). The bands I, II are valence bands, III and IV are conduction bands. Note that (11) is invariant under the interchange of the indices A and B. Volume state energy bands of ↑ spin electrons are therefore identical with those of ↓ spin electrons, as stated in Section 2.

It is highly interesting to find that nothing similar holds for surface states. The proof is as follows. Let us write the ↓ spin electron wave function in the semi-infinite crystal, delimited by the (100) plane, in the form [7, 15]

$$\Psi^{\downarrow}_{j,\eta_1,\eta_2}(\boldsymbol{r}) = \sum_{p=1}^{8}\sum_{n_3=1}^{\infty} d^{\downarrow}_{j;pn_3}(\eta_1,\eta_2)\,\Phi'_{p;\eta_1,\eta_2}(\boldsymbol{r} - n_3\boldsymbol{b}_3)\,, \tag{12}$$

where

$$\Phi'_{p;\eta_1,\eta_2}(\boldsymbol{r} - n_3\boldsymbol{b}_3) = N^{-1}\sum_{n_1,n_2} e^{i(n_1\eta_1+n_2\eta_2)}\,\varphi_p(\boldsymbol{r} - n_1\boldsymbol{b}_1 - n_2\boldsymbol{b}_2 - n_3\boldsymbol{b}_3)\,.$$

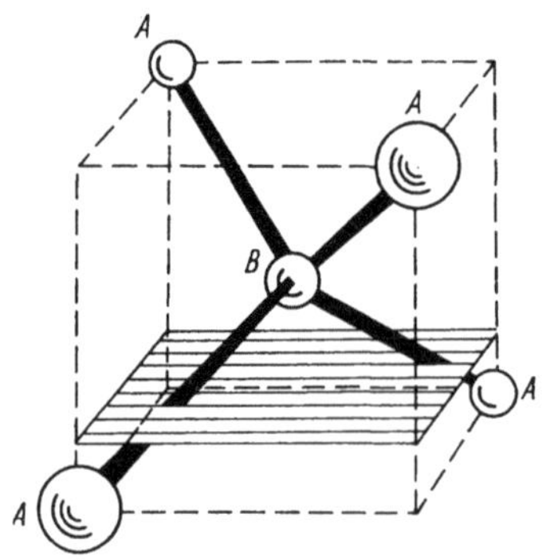

Fig. 1. An ideal (100) surface plane of a diamond-like crystal. The shaded plane represents the crystal surface. Surface orbitals projecting towards vacuum may be seen

n_1, n_2, n_3 are integers defining crystal lattice points with position vectors $\boldsymbol{r}_n = n_1 \boldsymbol{b}_1 + n_2 \boldsymbol{b}_2 + n_3 \boldsymbol{b}_3$. The (100) surface plane is characterized by the relation $n_3 = 1$. It cuts two bonds, connecting a surface B atom with its two neighbouring A atoms (see Fig. 1). The two B atom surface orbitals projecting towards vacuum will be denoted by indices $p = 7,8$.

According to the general theory [3], the coefficients $d^{\downarrow}_{j;pn_3}$ and the surface state energy of $\downarrow$ spin electrons are determined by the following system of equations [7, 15]:

$$d^{\downarrow}_{j;pn_3} = -\gamma\, L^{\downarrow}_{pn_3;10}\, d^{\downarrow}_{j;81} - \gamma\, L^{\downarrow}_{pn_3;20}\, e^{i\eta_2}\, d^{\downarrow}_{j;71}\,, \tag{13}$$

where

$$L^{\downarrow}_{pn_3;p'n'_3} = \frac{(-1)^{p'+p+1}}{2\pi} \int\limits_0^{2\pi} \frac{\Delta^{\downarrow}_{p'p}}{\Delta^{\downarrow}}\, e^{i(n_3-n'_3)\eta_3}\, d\eta_3\,. \tag{14}$$

$\Delta^{\downarrow}_{p'p}$ is the determinant $\Delta^{\downarrow}$ with the p'-th row and the p-th column omitted. We respect the boundary conditions on the crystal surface by taking

$$d_{j;10} = 0\,, \qquad d_{j;20} = 0\,. \tag{15}$$

Putting $n_3 = 0$, $p = 1,2$ in (13), we therefore have two equations:

$$\left.\begin{aligned} -\gamma\, L^{\downarrow}_{10;10}\, d^{\downarrow}_{j;81} - \gamma\, L^{\downarrow}_{10;20}\, e^{i\eta_2}\, d^{\downarrow}_{j;71} &= 0\,,\\ -\gamma\, L^{\downarrow}_{20;10}\, d^{\downarrow}_{j;81} - \gamma\, L^{\downarrow}_{20;20}\, e^{i\eta_2}\, d^{\downarrow}_{j;71} &= 0 \end{aligned}\right\} \tag{16}$$

with the following solubility condition:

$$L^{\downarrow}_{10;10}\, L^{\downarrow}_{20;20} - L^{\downarrow}_{10;20}\, L^{\downarrow}_{20;10} = 0\,, \tag{17}$$

which determines the energies of surface states. Equation (17) has been treated in [7]. By lengthy transformations, it has been rearranged and finally split in two equations (see [7] equations (20), (21); compare also equations (18), (19) of [15]):

$$x_2 = 0\,, \tag{18}$$

$$(\nu_2^2 - 4)\, \nu_1\, \nu_2^{-1}\, z_1 = \varphi'_1\, \varphi'_2 + 4 - \nu_1^2 + 2\,(x_2\, \varphi'_1 + x_2^{-1}\, \varphi'_2)\,, \tag{19}$$

where

$$z_1 = \frac{A}{2\,\nu_1\,\nu_2}\left[1 - \left(1 - \frac{4\,\nu_1^2\,\nu_2^2}{A^2}\right)^{1/2}\right], \tag{20}$$

$$A = \varphi'_1\, \varphi'_2 - \nu_1^2 - \nu_2^2\,.$$

Now, if we go over to the case of $\uparrow$ spin electrons, we must interchange the indices A and B. This means that the indices 1 and 2 of the quantities x_1, x_2, φ'_1, and φ'_2 must be interchanged. By doing this in (18) and (19), we obtain new equations, differing from those for $\downarrow$ spin electrons. Hence, the energies of surface states of $\uparrow$ and $\downarrow$ spin electrons are generally different. We see that the inclusion of interelectronic repulsion leads to the splitting of surface state energy levels, the wave functions of which are localized near the crystal surface. For uncorrelated wave functions of the usual one-electron theory, $\uparrow$ and $\downarrow$ spin electrons possessed a common system of surface states, each level of which could be simultaneously occupied by both an $\uparrow$ and a $\downarrow$ spin electron [15]. General physical reasons for the occurrence of this splitting have been discussed elsewhere [1].

Let us show on equations (18) and (19) how this splitting depends upon the amount of correlation included. (19) will be solved graphically for special values of the respective parameters (see Fig. 2 and 3). We introduce new quantities:

$$\left.\begin{aligned} x &= \frac{1}{2}(x_1 + x_2)\,, \\ \delta &= \frac{1}{2}(x_1 - x_2) = \frac{1}{2\gamma}(\alpha_B - \alpha_A) + \frac{1}{2}(\gamma_1 - \gamma_2)\,, \end{aligned}\right\} \tag{21}$$

where x is a new variable and δ is a parameter, controlling the extent to which $\uparrow$ and $\downarrow$ spin electron wave functions are correlated. Note that for real situations not very large values of $|\delta|$ are typical. As to the behaviour of our scheme for large $|\delta|$ ($|\delta| \to \infty$), see [1]. For the same physical situation the $\uparrow$ spin electron case is characterized by the value $-|\delta|$ of the quantity δ; the value $\delta = |\delta|$ belongs to $\downarrow$ spin electrons.

Using (21), (18) can be written as

$$x = \delta\,. \tag{22}$$

For that root of (19), which may be found in the interval B, i.e. in the gap, we see the dependence of surface state energy levels on the quantity δ from Fig. 2 and 3, which have been drawn for special values of ν_1 and ν_2. Surface state

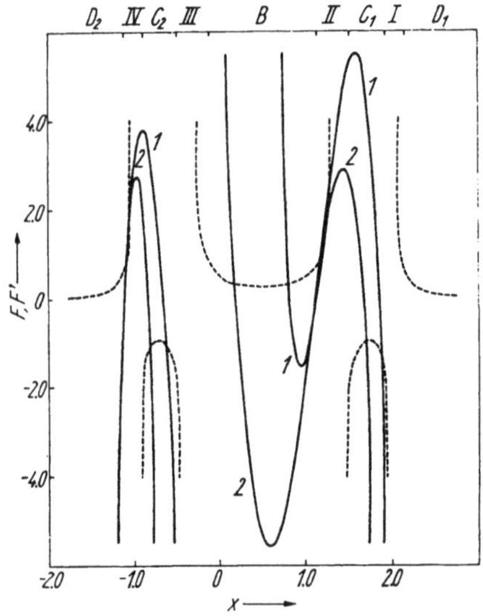

Fig. 2

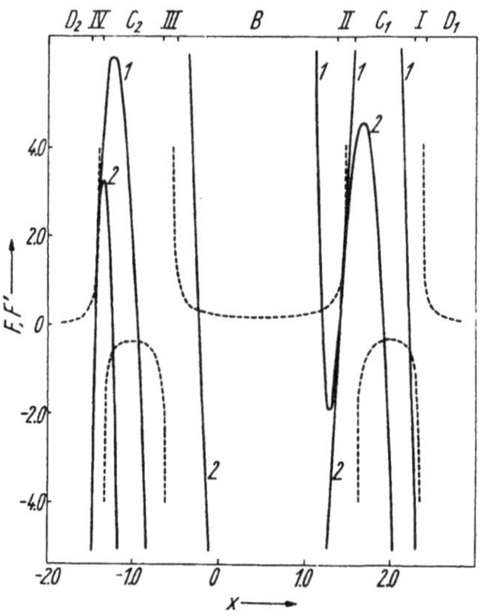

Fig. 3

Fig. 2. The dashed curve represents the function $F = 2\,\nu_1 \nu_2 z_1$ for $|\delta| = 0.5$. The solid curves 1 and 2 represent the function $F' = 2\,\nu_2^2\,(\nu_2^2 - 4)^{-1}\,[\varphi_1' \varphi_2' + 4 - \nu_1^2 + 2\,(x_2\,\varphi_1' + x_{-\frac{1}{2}}\,\varphi_2')]$ for $\delta = 0.5$ (the case of $\downarrow$ spin electrons) and $\delta = -0.5$ (the case of $\uparrow$ spin electrons), respectively. For these cases the values $\gamma_1 = \gamma_2 = 0.25$, $\nu_1 = 2$, $\nu_2 = 1$ have been chosen. The cooridnates of the intersections of the dashed curve with some of the solid curves give the surface state energy in x units. The real energy can be obtained by using (21) and (10). On the top of the figure the limits of volume state energy bands may be found

Fig. 3. The dashed curve represents the function F for $|\delta| = 1$. The solid curves 1 and 2 correspond to the function F' for $\delta = 1$ ($\downarrow$ spin electrons) and $\delta = -1$ ($\uparrow$ spin electrons), respectively. Again, the values $\gamma_1 = \gamma_2 = 0.25$, $\nu_1 = 2$, $\nu_2 = 1$ have been chosen

energies are represented by the coordinates x of the intersections of dashed curves with solid curves. The physical interpretation of surface states (18), as well as of those of (19) lying in the interval B of Fig. 2 and 3 is the same as in the usual one-electron theory without correlation [15].

In Fig. 2 and 3 other roots of equation (19), both for ↑ and ↓ spin electrons, may be found. These are surface states placed in "forbidden" intervals C_1 and C_2 and furthermore those located completely in the bands II and IV of volume states. The latter kind of surface states, which is connected with changes in the character of the bonds between subsurface atoms of the crystal, resembles very much the volume states. It is interesting to see that the ↑ and ↓ spin electron levels of these states practically coincide in energy and that they are sensitive with respect to the interelectronic repulsion in a similar way as volume states.

Let us consider what happens in the gap when the system is in the ground state. If the correlation is small, the surface state bands of both ↑ and ↓ spin electrons of equation (19) will probably be well separated from those of (18), and as they will lie at lower energies, they will be completely occupied. Surface state bands of (18) will be empty. For larger correlations, however, the ↓ spin electron band of (18) may be found to lie lower in energy than the ↑ spin electron band of (19) and, therefore, only ↓ spin electron bands will be completely occupied, whereas the ↑ spin electron ones will be empty. Therefore, in this case, the ideal surface should exhibit something like a "ferromagnetic" character. This "ferromagnetism" should gradually disappear by going to non-zero temperature, i.e. due to excitations. Similar results have been found for the (111) surface of the same crystal type [1]. It would be interesting to see whether these effects could be detected experimentally by either the ferromagnetic resonance or neutron diffraction methods.

References

[1] M. Tomášek, Surface Science, to be published.
[2] S. G. Davison and A. T. Amos, J. chem. Phys. **43**, 2223 (1965).
[3] J. Koutecký, Phys. Rev. **108**, 13 (1957).
J. Kontecký and M. Tomášek, Phys. Rev. **120**, 1212 (1960).
[4] P. O. Löwdin, Symp. Molecular Phys., Nikko, 1953 (p. 13); Phys. Rev. **97**, 1509 (1955); Adv. Phys. **5**, 1 (1956); J. appl. Phys., Suppl. **33**, 251 (1962) and other papers.
[5] P. O. Löwdin, Adv. chem. Phys. **2**, 209 (1959).
[6] J. C. Slater and G. F. Koster, Phys. Rev. **94**, 1498 (1954).
[7] M. Tomášek, Czech. J. Phys., to be published.
[8] J. Koutecký and M. Tomášek, Surface Science **4**, 333 (1965).
[9] G. Dermit, Phys. Rev. **127**, 1110 (1962).
[10] J. L. Calais, private communication.
[11] C. A. Coulson, Tetrahedron **19**, Suppl. II, 49 (1963).
[12] J. de Heer, Revs. Mod. Phys. **35**, 631 (1963).
[13] J. D. Swalen and J. de Heer, J. chem. Phys. **40**, 378 (1964).
[14] E. Kapuy, Theoret. chim. Acta **3**, 379 (1965).
[15] J. Koutecký, Czech. J. Phys. **B12**, 184 (1962).

(Received December 10, 1965)

phys. stat. sol. **16**, 399 (1966)

Institute of Physical Chemistry, Polish Academy of Sciences, Warsaw

Quantum-Statistical Calculations on the Work Function of Metals

By

M. Dubejko and S. Olszewski[1])

Work functions of metals with atomic numbers below 40 are calculated using the statistical atomic theory. Good agreement with experiment is obtained for the average value of the work function for each element. For a metallic solid solution with a small heat of mixing the expected work function is rather close to that of the component having the lowest work function and rather insensitive to changes in atomic concentrations.

Es wird die Austrittsarbeit von Metallen mit Atomzahlen unterhalb 40 mit der statistischen Atomtheorie berechnet. Gute Übereinstimmung mit dem Experiment wird bei Metallen für den Mittelwert der Austrittsarbeit erhalten. Für eine metallische Legierung mit einer kleinen Lösungswärme wird eine Austrittsarbeit erwartet, die ziemlich genau dem Wert der Komponente mit der niedrigsten Austrittsarbeit entspricht und die ziemlich unempfindlich gegen Änderungen der Konzentrationen ist.

1. The Model

Let us consider the electron energy of a metal in two ways: first, when the metal is neutral, and secondly, when one electron is removed from the metal block. The work function W is the difference in electron energy between the two cases.

The calculations of W have been performed in a set of papers by means of both wave-mechanical and quantum-statistical methods (see references [1] and [2] for the respective reviews). A characteristic point is in both theories that they concern numerically only metals on the left of the periodic table of elements. This is the consequence of the general assumption made in either theory that a metal cell is composed of a pointlike positive core surrounded by only a few quasi-free electrons.

Our purpose is to calculate W in statistical approximation irrespective of the position of the metal atomic number Z in the periodic table. The model of the metal is the statistical one of the solid valid for arbitrary Z [3, 4]. In this model the crystal is composed of cells called "crystal atoms", the electron distribution around a given nucleus in the cell being determined by the statistical equation with appropriate boundary conditions.[2]) The electrons of the crystal atom are all treated similarly, and thus no core other than the atomic nucleus exists in the cell.

With this model, the energy calculations seem to fit metals best, because it is based on the free electron assumptions. The comparison between theory and

---

[1]) Postal address: Doc. Dr. S. Olszewski, Instytut Chemii Fizycznej PAN, Warszawa, Kasprzaka 44/52.

[2]) Similar ideas of "independent" neutral atoms in the metal, or alloy, developed by Ziman [5], may be considered an important support to our theory.

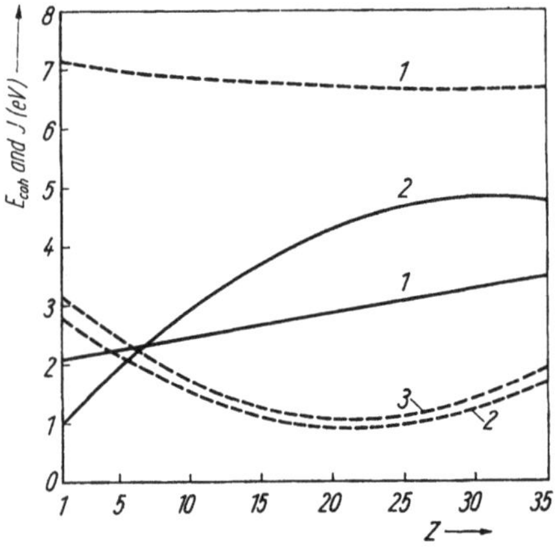

Fig. 1. Parabolas of equation (1).
Solid lines:
1 — experimental cohesive energy of metals,
2 — theoretical ionization potential with exchange correlations.
Broken lines:
1 — experimental ionization potential of free metal atoms,
2 — theoretical cohesive energy with exchange correlations,
3 — theoretical ionization potential with exchange and Coulomb correlations

experiment for the cohesive energies of metals is shown in Fig. 1. This figure may be compared with Fig. 2 of [4], where the data from the statistical model are matched with those from experiment for all kinds of monoatomic solids, and not only metals.[3]) Because the free electron assumptions underlie the model, the statistical calculations are expected to be relatively accurate for the transitional elements, where the number of the metal (free) electrons is particularly high.

The removal of one electron from the solid (metal) can be considered as equal to the removal of an electron from a metal cell, because (i) the other cells of the metal are essentially neutral, and thus have virtually no potential capable of influencing the electron during removal, and (ii) polarization of the cells neighbouring on that becoming positive upon removal of the electron is not very marked. (It is characteristic that the electron density at the boundary of the crystal cell — crystal atom — remains the same irrespective of whether the cell is or is not ionized [3, 4]). We finally assume the cells at the metal boundary to be not much different from those deeper inside and the contribution of the dipole term on the metal surface to be small in comparison with the volume-dependent part of W, so that it can be neglected.

Thus W is approximately equal to the difference in energy between the crystal atom and the corresponding crystal ion with the charge $+1\,e$. The method of calculating the electron energy and distribution in the crystal atoms and ions is given in [3] for the case where only the exchange correlations are considered, and in [4] for that where both exchange and Coulomb correlations are taken into account.

It should be emphasized that our calculation of W is purely theoretical, i.e. no parameters beyond the fundamental atomic constants were included into the framework. This makes our treatment rather relevant to the Gombás theory of W (see pp. 310 and 330 of [2a]) for alkali metals than to that for these metals which is based on the cellular approximation of Wigner and Seitz, since in the last scheme the core potential established from the empirical data is used [6]. The treatment of W with non-empirical cellular potential (e.g. like that by ortho-

[3]) The ranges of the variable Z in Fig. 2 and Fig. 3 in [4] are erroneously given as between 0 and 30; they should be between 1 and 30.

gonalized plane waves) gives quite unrealistic results in comparison with experiment [7, 8].[4])

In addition, the empirical value for the ionization potential of the free atom is generally applied in determining W by means of the cellular method (see [6], p. 399 and [1], pp. 230 and 231). On the other hand, the Gombás work is not free from an inconsistency. The point is that constant density of valence electrons in the cell is assumed as well as finite volume of positive cores. This volume of the core amounts in case of Na $\approx 25\%$ of the cell volume [2 a].

2. Results

The roughness of the model used in calculations is apparent, so only a general insight into the dependence of W on Z of the metal can be obtained. We calculate $W(Z)$ only up to $Z \approx 35$, because the accuracy of the numerical calculations decreases as Z increases. This accuracy is limited mainly by the finite number of digits which can be accepted by the computer for $\varphi'(0)$, or $\chi'(0)$, the initial slopes of the solutions of the statistical equations (cf. [3] and [4]).

All the physical quantities investigated in the present paper: W, the cohesive energy E_{coh}, and the *free* metal atom ionization potential I are expressed in terms of Z as

$$Q(Z) = \alpha + \alpha' Z + \alpha'' Z^2 , \tag{1}$$

where Q is a given physical quantity. The coefficients α, α', and α'' are obtainable by the method of least squares [10].

The theoretical $Q(Z)$ were calculated from the data obtained with Z equal to 2, 10, 18, 28, and 36. The experimental relations in equation (1) have been determined from the measured data for the following values of Z:

$$3, 4, 11 \text{ to } 13, 19, 20, 22 \text{ to } 30, 37, 38, 40, 42, 45 \text{ to } 48, 50, 55 \text{ to } 60, 62, \\ 73 \text{ to } 83, \text{ and } 92. \tag{2}$$

(Calculations of $I^{\text{exp}}(Z)$ were made without the numbers 73, 75, 77 in (2).) The experimental data for E_{coh} are taken from [11], and those for W^{exp} and I^{exp} from [12].[5])

The resulting relations in (1) are plotted in Fig. 1 and 2, whereas the average values of W and E_{coh} per one element (one metal) are given in Table 1.

It is characteristic that $W(Z)$ calculated with exchange correlations becomes rather constant upwards of $Z \approx 20$. (A similar behaviour of $W(Z)$ calculated with exchange and Coulomb correlations is suggested for large Z.) This constancy derives support from the relatively small variations of the experimental W over rather wide ranges of Z.

---

[4]) Good results for cohesive energies of alkali metals obtained recently by means of the orthogonalized plane waves method with a non-empirical potential [9] imply that the method can be successful also for W, since the calculations of cohesive energy and W are related (see [6], Chap. 11). However, reasonable results for the cohesive energy are not necessarily accompanied by similar agreement for W [7, 8]. One should notice still that the lattice parameters in [9] were not determined on the theoretical basis, and no calculations of W in [9] are reported.

[5]) For the experimental W of a given Z the average value of the corresponding photoelectric data in [12] was taken into account. Where no photoelectric data were available, the thermionic W was used.

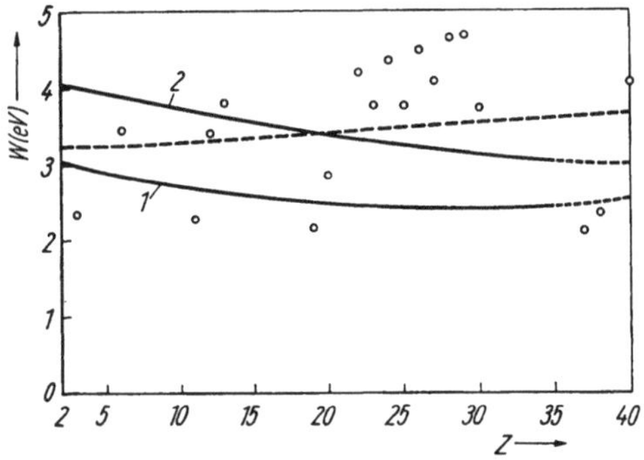

Fig. 2. Work function parabolas of equation (1).
Solid lines:
1 — theoretical with exchange correlations,
2 — theoretical with exchange and Coulomb correlations.
Broken line: Experimental.
Circles: Experimental values of W (circle $z = 7$ must be shifted to $z = 4$)

Table 1

Average values of work function and cohesive energy per one metal element with Z smaller than about 40

Work function (eV)				Cohesive energy (eV per atom)			
Theoretical		Experimental		Theoretical		Experimental	
Exch.	Exch. + Coul.			Exch.	Exch. + Coul.		
2.8	3.5	3.4	3.5*	3.7	4.6_5	2.8	3.0*

The values in columns 1 to 3, and 5 to 7 of the table are obtained by integration of the corresponding parabolas in Fig. 2 and Fig. 1. The integration limits were $Z = 2$ and $Z = 36$ for the data in columns 1 to 3, and $Z = 1$ and $Z = 36$ for columns 5 to 7. The asterisked numbers are the arithmetical means of appropriate experimental quantities taken for Z of (2) as from 1 to 40.

For small Z, the apparent increase in the theoretical E_{coh} with Z is accompanied by a corresponding decrease in I. The complementarity of $E_{coh}(Z)$ and $I(Z)$ has already been noted by Wigner and Seitz [11], who worked with experimental data. The increase of the theoretical I with the larger Z, and the corresponding slight decrease in the theoretical E_{coh}, result from the approximation in the representing $I(Z)$ and $E_{coh}(Z)$ by equation (1) rather than from the theory itself.

Considering the absolute values of the calculated I and W, agreement with experiment is seen to be for I much poorer than for W (Fig. 2). This implies the conclusion — already formulated by Sommerfeld and Bethe [13] — that the statistical approximation is more suitable to treat solids than free atoms. The reason is that areas of very small electron densities, poorly tractable by purely statistical methods, are important in free atomic objects, not in solids.

A characteristic point of the theory is that only in the case of $Z = 1$ it appears impossible to solve the statistical equations of [3] and [4] for the positive crystal ion, though a solution clearly exists for the *free* positive ion.

3. Work Function for Metallic Solid Solutions

Along the above line of thought we may gain some insight into the work function W^a of metallic solid solutions. Our basic point is that the metal cells entering into the solution are not much different from those composing the corresponding pure metals. This is due to the fact established by theory that the electron density at the boundary of the crystal atom (or crystal ion) is the same for all metals (all Z), and thus no drastic jumps in electron density appear in the alloy when it is composed of cells with different kinds of pure metals. At the same time the minimum of electron energy in the alloy is attained, because each component cell is at its own energy minimum, equal to the minimum for the cell in the pure metal, and the total alloy energy can be considered as the sum of the energies of these cells.[6])

Essentially, the last statement is true only when the energies of removal of electrons from different cells are equal. This means — according to Section 1 — the case when work functions of the metal components of the alloy are approximately the same. If the work functions of the components differ, then the electrons from the cells of smaller W tend to pass into those of the higher work function, and a net positive charge tends to appear in the former cells. However — basing on the results of the band theory —, this charge is immediately screened by the metal electrons [15]. The screening is particularly marked for metals having large densities of states near the Fermi level (transition metals), thus, often practically no field of the net charge exists outside the area of the cell where the additional charge is produced [16]. In result, as long as the mutual perturbations are not too strong, we have merely some kind of rearrangement of charges inside the cells of different W when they enter into an alloy, rather than marked changes in the magnitude of the charge in cells. The rearrangement involves only a small part of the total charge in the cell, and its influence on the electron energy in the cell is expected to be small.

In consequence, bearing in mind that the work function of a solid metal is determined mainly by the electron distribution in an individual cell (Section 1 of this paper), we expect for a metallic solution: (i) no drastic changes in W^a to accompany rather large changes in the atomic concentrations of components, since these changes do not influence the cells much; and (ii) the work function of the alloy to be close to that of the component which has the lowest W.

Recent results of the measurements of W^a for copper-nickel [17] and nickel-iron [18] alloys seem to confirm these views. However, when the degree to which the cells differ from those in pure metals is important (the changes taking place especially when the phase of the metallic system is changed), the rules (i) and (ii) will no more be obeyed. These rules can be expected to hold above all for metal solid solutions with small heats of mixing (for heats of mixing of different metal solutions see e.g. Mott [19]).

Acknowledgement

The use of the digital computers Ural II of the Polish Academy of Sciences and GIER of the University of Warsaw is gratefully acknowledged.

---

[6]) The elastic deformations of cells and the respective changes in energy produced during the alloying are regarded as negligible [14].

References

[1] C. HERRING and M. H. NICHOLS, Rev. mod. Phys. **21**, 185 (1949).
[2] P. GOMBÁS, Encyclopedia of Physics, Ed. S. FLÜGGE, Vol. 36, Springer-Verlag, Berlin 1956.
[2a] P. GOMBÁS, Die statistische Theorie des Atoms und ihre Anwendungen, Springer-Verlag, Wien 1949.
[3] S. OLSZEWSKI, phys. stat. sol **3**, 2221 (1963).
[4] S. OLSZEWSKI and M. DUBEJKO, phys. stat. sol. **9**, 181 (1965).
[5] J. M. ZIMAN, Adv. Phys. **13**, 89 (1964).
[6] F. SEITZ, Modern Theory of Solids, McGraw-Hill Book Comp., New York 1940 (Chap. 9 to 11).
[7] C. HERRING and A. G. HILL, Phys. Rev. **58**, 132 (1940).
[8] B. DONOVAN, Phil. Mag. (7) **43**, 868 (1952).
[9] J. CALLAWAY, Phys. Rev. **119**, 1012 (1960).
J. CALLAWAY and D. F. MORGAN, Phys. Rev. **112**, 334 (1958).
J. CALLAWAY and E. L. HAASE, Phys. Rev. **108**, 217 (1957).
[10] H. MARGENAU and G. M. MURPHY, The Mathematics of Physics and Chemistry, D. Van Nostrand Comp., New York 1959 (Chap. 13.37).
[11] E. P. WIGNER and F. SEITZ, Solid State Phys. **1**, 97 (1955).
[12] Handbook of Chemistry and Physics, Vol. 2, Ed. C. D. HODGMAN, Cleveland, Ohio 1955/56.
[13] A. SOMMERFELD and H. BETHE, Handbuch der Physik, Vol. 24/2, Ed. H. GEIGER and K. SCHEEL, Springer-Verlag, Berlin 1933 (p. 427).
[14] S. OLSZEWSKI, to be published.
[15] G. F. KOSTER and J. C. SLATER, Phys. Rev. **96**, 1208 (1954).
[16] J. FRIEDEL, Nuovo Cimento Suppl. **7**, 287 (1958).
[17] W. M. H. SACHTLER, G. J. H. DORGELO, and R. JONGEPIER, J. Catalysis **4**, 100 (1965).
[18] R. DUŚ, private communication.
[19] N. F. MOTT in Metallic Solid Solutions, Ed. A. GUINIER and J. FRIEDEL, W. A. Benjamin, New York 1963.
[20] B. Ch. DIUBUA, O. K. KULTASHEV, and L. V. GORSHKOVA, Fiz. tverd. Tela **8**, 1105 (1966).
[21] B. Ch. DIUBUA, O. K. KULTASHEV, and I. A. TSYGANOVA, Radiotekhnika i elektronika **9**, 2061 (1964).
[22] B. Ch. DIUBUA, A. I. PEKAREV, B. N. POPOV, and M. A. TYLKINA, Radiotekhnika i elektronika **7**, 1566 (1962).

(Received March 12, 1966)

Note added in proof:

Work functions for solid solutions of Mo and Ta with tungsten have recently been investigated experimentally [20]. For wide ranges of atomic concentrations (from zero to about 80 percent of atomic concentration of the component having the lower work function) W^a of the metallic solid solution is almost constant and equal to the the work function of the pure component with the lower W. The minimum of W^a which is below the work function of the component with the lower W is suggested to be explainable by the surface effect of the dipoles of the two metals in the solution.

The W^a of the other two-metal systems that have been investigated experimentally [21, 22] hardly lend themselves for treatment by the present theory in view of the multi-phase systems formed during the alloying.

phys. stat. sol. **16**, 405 (1966)

Institute of Radio Engineering and Electronics, Czechoslovak Academy of Sciences, Prague

On the Theory of the Superexchange Interaction in Ionic Crystals: $KNiF_3$

By

Z. Šroubek

Studies are made of superexchange interaction in ionic crystals using the Heitler-London theory and with particular reference to $KNiF_3$. The generalized Heitler-London quantitative theory of covalency allows a detailed study to be made of the transfer of an electron from cation to cation and the transfer of two electrons from an anion to the cations. No further assumptions are made and the calculations are performed numerically. The antiferromagnetic superexchange constant given by this calculations is in good agreement with the experimental superexchange constant.

Die Supraaustauschwechselwirkung wird in Ionenkristallen mit der Heitler-Londonschen Theorie untersucht, wobei $KNiF_3$ besonders berücksichtigt wird. Die verallgemeinerte Heitler-Londonsche quantitative Theorie der kovalenten Bindung gestattet eine detaillierte Untersuchung des Elektronenüberganges von Kation zu Kation und des Übergangs von zwei Elektronen von einem Anion zum Kation. Ohne weitere Annahmen werden numerische Berechnungen durchgeführt. Die so berechnete antiferromagnetische Supraaustauschkonstante zeigt gute Übereinstimmung mit der experimentell bestimmten.

1. Introduction

The problem of superexchange has been the subject of many articles. The most complete attempts were made by Yamashita and Kondo [1] and Keffer and Oguchi [2] in the old context of the Heitler-London method. It was nearly impossible for them to come to any quantitative conclusions, except as to order of magnitude, because of the large number of uncertain parameters which entered into the calculation. Anderson [3] suggested a new approach to the theory of superexchange based on the Hartree-Fock solution of the magnetic electron wavefunction (antibonding orbital) in the diamagnetic lattice. He has shown that only two effects appear to be of any quantitative significance in problems involving d-electrons. The "kinetic exchange" always antiferromagnetic, represents the second order effect of virtual electron transfer between magnetic ions and the ferromagnetic "potential" exchange. The aim of this theory is to point out the quality of the effect.

In our paper we would like to solve the problem of superexchange within the framework of the generalized Heitler-London method with configuration interaction. This method takes into account the rearrangement among the virtually transferred electron and d-electrons and can be used for relatively successful quantitative calculation of the covalency in ionic crystals [4]. We believe that the other transfer effects, important in the superexchange theory, can be also calculated with the same accuracy.

To overcome the difficulties with the ab-initio calculation from atomic wavefunctions, we define some semiempirical parameters and thus obtain simple

expressions for antiferromagnetic energy stabilization. Important and for numerical calculation favourable seem to be the two-electron transfers:
a) one electron is transferred from the anion to the cation and the other from the next cation to the vacated anion orbital,
b) one electron is transferred from the anion to the cation on the left, and at the same time another electron is transferred from the anion to the cation on the right.

Such excitations have been previously proposed by Yamashita and Kondo [1], Pratt [5], Goodenough and Loeb [6], and Nesbet [7]. The first mechanism resembles the "kinetic exchange" of the Anderson theory of superexchange.

The theoretical superexchange constant is in good agreement with experiment, thus the other superexchange effects are probably not so important and are only briefly discussed.

2. The Definition of the "Ground" and "Excited" Configurations

We are dealing with the most simple model of the linear cation—anion—cation (e.g. Ni^{2+}—F^-—Ni^{2+}) group in the cubic perovskite crystal $KNiF_3$. The important role in superexchange is performed by four electrons only: one magnetic electron with orbital wavefunction ψ_a on the cation A (Ni^{2+}), two electrons with antiparallel spins in the orbital wavefunction ψ_d on the diamagnetic anion D (F^-) and one magnetic electron with the wave-function ψ_b on the other cation B (Ni^{2+}). The wavefunctions of the other electrons are denoted by one symbol ψ_i.

The two possible many-electrons ground state wavefunctions (Slater determinants) are then (+ and − are spin eigenfunctions of S_z)

$$M = \{\psi_a^+ \psi_d^+ \psi_d^- \psi_b^- \psi_i\}, \quad N = \{\psi_a^- \psi_d^+ \psi_d^- \psi_b^+ \psi_i\}. \tag{1}$$

The wavefunctions ψ_a and ψ_d, ψ_b and ψ_d overlap slightly with the overlap integral $S = \langle\psi_a|\psi_d\rangle = \langle\psi_b|\psi_d\rangle$. Then the wavefunctions (1) are also not orthogonal and have the overlap integral $\langle M|N\rangle = -S^4$.

We can now define four simple excited configurations which are characterized by electron transfer from the anion to the cation. The respective wavefunctions are of the following form:

$$\left.\begin{aligned} P &= \{\psi_a'^+ \psi_d'^+ \psi_a'^- \psi_b^- \psi_{ip}'\}, \\ R &= \{\psi_a^+ \psi_b'^+ \psi_d'^- \psi_b'^- \psi_{ir}'\}, \\ Y &= \{\psi_a'^- \psi_a'^+ \psi_d'^- \psi_b^+ \psi_{iy}'\}, \\ T &= \{\psi_a^- \psi_d'^+ \psi_b'^- \psi_b'^+ \psi_{it}'\}. \end{aligned}\right\} \tag{2}$$

The ψ_i' are now the wavefunctions appropriate to the new ionicities. Further exists this relation: $\psi_{ip}' = \psi_{iy}'$, $\psi_{ir}' = \psi_{it}'$.

The Hamiltonian $\mathcal{H}$, which includs the electron-electron and electron-nuclear interactions, has appreciable matrix elements (of equal value) between M and P, R and between N and Y, T. The perturbed ground state wavefunctions M' and N' are

$$M' = M + \gamma (R + P), \quad N' = N + \gamma (Y + T). \tag{3}$$

The expression for γ is well known [4], [8], [9] ,[10]

$$\gamma = -\frac{\langle M|\mathcal{H}|P\rangle - S\langle M|\mathcal{H}|M\rangle}{\langle P|\mathcal{H}|P\rangle - \langle M|\mathcal{H}|M\rangle}. \tag{4}$$

Besides the configurations (2) there exist three configurations of higher excited states. The most important seem to be the configurations with two electrons on the D-ion and two electrons on one cation.

The two appropriate wavefunctions are of the following form:

$$Q = \{\psi_a'^{+} \psi_d^{+} \psi_a'^{-} \psi_d^{-} \psi_{iq}'\}, \quad W = \{\psi_d^{+} \psi_b'^{+} \psi_d^{-} \psi_b'^{-} \psi_{iw}'\}. \tag{5}$$

The last possible configuration is:

$$Z = \{\psi_a'^{+} \psi_a'^{-} \psi_b'^{+} \psi_b'^{-} \psi_{iz}'\}. \tag{6}$$

3. The Energy Separation between Parallel and Antiparallel Spin Configurations

The wavefunctions M' and N' are, of course, not the eigenfunctions of total spin. If we denote the eigenfunction of $S^2 = 0$ by Φ_A and the eigenfunction of $S^2 = 1\,(1 + 1) = 2$ by Φ_P, the two appropriate linear combinations of M' and N' are given by

$$\Phi_A = 2^{-1/2} A\,(M' - N'), \quad \Phi_P = 2^{-1/2} B\,(M' + N'). \tag{7}$$

A and B are the normalization constants equal to $(1 - S_{M'N'})^{-1/2}$ and $(1 + S_{M'N'})^{-1/2}$ respectively and $S_{M'N'} = \langle M'|N'\rangle$. The problem resembles now closely that one of triplet-singlet separation in hydrogene molecules. To find the energies of parallel and antiparallel spin configurations, we have to solve the secular equation

$$\begin{vmatrix} \langle M'|\mathcal{H}|M'\rangle - E & \langle M'|\mathcal{H}|N'\rangle - S_{M'N'}\,E \\ \langle M'|\mathcal{H}|N'\rangle - S_{M'N'}\,E & \langle N'|\mathcal{H}|N'\rangle - E \end{vmatrix} = 0. \tag{8}$$

The energy separation δE between parallel and antiparallel spin configurations is then to first order of $S_{M'N'}$

$$\delta E = 2\,(\langle M'|\mathcal{H}|N'\rangle - S_{M'N'}\langle N'|\mathcal{H}|N'\rangle). \tag{9}$$

The expression (9) contains in generalized form all the important terms discussed by Keffer and Oguchi [2] (Anderson, Anderson-Hasegawa exchange terms etc.). Unfortunately, the exact calculation is difficult because δE represents a small value achieved by the difference between large numbers, and some integrals are not precisely determined. Preliminary calculation for $KNiF_3$ shows that even the sign of δE is not certain.

On the other hand the wavefunctions (5) and (6) are the eigenfunctions of $S^2 = 0$. Due to this fact the Hamiltonian $\mathcal{H}$ can mix the wavefunctions (5) and (6) only with Φ_A, and only the energy of the antiferromagnetic state is decreased. The main purpose of this paper is to calculate this antiferromagnetic stabilization with highest possible precision using the result of the modern covalency theory [4], [9], [10].

The perturbed wavefunction Φ_A' of antiparallel spin configuration is

$$\Phi_A' = \Phi_A + \alpha\,Q + \beta\,W + \xi\,Z. \tag{10}$$

The energy shift caused by the mixing of Q and W configurations is denoted by ΔE_1 and can be calculated by varying α and β to minimize the total energy of the system. We obtain

$$\Delta E_1 = \frac{(\langle\Phi_A|\mathcal{H}|Q\rangle - S_{QA}\langle\Phi_A|\mathcal{H}|\Phi_A\rangle)^2}{U_1} + \frac{(\langle\Phi_A|\mathcal{H}|W\rangle - S_{AW}\langle\Phi_A|\mathcal{H}|\Phi_A\rangle)^2}{U_1}. \tag{11}$$

S_{QA} and S_{WA} are the overlap integrals between Q and Φ_A, and W and Φ_A, respectively. U_1 is the energy difference between the Q or W and Φ_A state and represents the total energy required to remove an electron for example from the B cation and put it in the half-filled ψ_a orbital of the A cation. The expression (11) can be considerably simplified if we assume the existence of the following relations between the matrix elements of $\mathcal{H}$:

$$\left.\begin{aligned} \langle M|\mathcal{H}|Q\rangle &= \langle N|\mathcal{H}|Q\rangle = \langle M|\mathcal{H}|W\rangle = \langle N|\mathcal{H}|W\rangle = \\ &= \langle Q|\mathcal{H}|R\rangle = \langle Q|\mathcal{H}|T\rangle = \langle W|\mathcal{H}|P\rangle = \langle W|\mathcal{H}|Y\rangle = 0\,, \\ \langle Q|\mathcal{H}|P\rangle &= -\langle Q|\mathcal{H}|Y\rangle = \langle W|\mathcal{H}|R\rangle = -\langle W|\mathcal{H}|T\rangle\,. \end{aligned}\right\} \quad (12)$$

Substituting (3), (7), and (12) into (11) we obtain a simple expression

$$\Delta E_1 = 4\,\gamma^2\,\frac{(\langle Q\,|\mathcal{H}|\,P\rangle - S\,\langle M\,|\mathcal{H}|\,M\rangle)^2}{U_1}\,. \quad (13)$$

It is favourable to rewrite the expression (13) in the following way:

$$\Delta E_1 = \frac{4\,\gamma^2}{U_1}\,[(\langle Q|\mathcal{H}|P\rangle - S\langle P|\mathcal{H}|P\rangle) + S(\langle P|\mathcal{H}|P\rangle - \langle M|\mathcal{H}|M\rangle)]^2. \quad (14)$$

Now $\langle Q|\mathcal{H}|P\rangle - S\langle P|\mathcal{H}|P\rangle$ is the transfer integral between the P and Q configurations and can be compared with the transfer integral $\langle P|\mathcal{H}|M\rangle - S\langle M|\mathcal{H}|M\rangle$ which is essential in the theory of covalency.

The energy shift caused by the mixing of Z is denoted by ΔE_2. By varying ξ to minimize the total energy of the system we obtain

$$\Delta E_2 = \frac{(\langle \Phi_A\,|\mathcal{H}|\,Z\rangle - S_{ZA}\,\langle \Phi_A\,|\mathcal{H}|\,\Phi_A\rangle)^2}{U_2}\,. \quad (15)$$

S_{ZA} is the overlap integral between Z and Φ_A. U_2 is the energy difference between Z and Φ_A.

As in the previous case, following simplifying relations exist

$$\begin{aligned} &\langle M|\mathcal{H}|Z\rangle = \langle N|\mathcal{H}|Z\rangle = 0\,, \\ &\langle P|\mathcal{H}|Z\rangle = \langle R|\mathcal{H}|Z\rangle = -\,\langle Y|\mathcal{H}|Z\rangle = -\,\langle T|\mathcal{H}|Z\rangle\,. \end{aligned} \quad (16)$$

In analogy to (14) we get

$$\Delta E_2 = \frac{8\,\gamma^2}{U_2}\,[(\langle Z|\mathcal{H}|P\rangle - S\,\langle P|\mathcal{H}|P\rangle) + S\,(\langle P|\mathcal{H}|P\rangle - \langle M|\mathcal{H}|M\rangle)]^2. \quad (17)$$

4. Numerical Evaluation of ΔE_1 for $KNiF_3$

For simplicity we neglect the contribution of F^- 2 s electrons. Consequently ψ_d is the $|z\rangle$ function of F^- and ψ_a and ψ_b are the $|3\,z^2 - r^2\rangle$ wavefunctions of Ni^{2+}. All the important integrals have been given by Sugano and Shulman [11]. Using their results, Freeman and Watson [9], Šimánek and Šroubek [10] and Sugano and Tanabe [4] have calculated the transfer integral $\langle P|\mathcal{H}|M\rangle - S\,\langle M|\mathcal{H}|M\rangle$.

They got (in atomic units)

$$\langle P|\mathcal{H}|M\rangle - S\,\langle M|\mathcal{H}|M\rangle = -\,0.032\,. \quad (18)$$

It is important to note that the calculation has been performed with the ground state wavefunctions ψ_a, ψ_b, ψ_d, ψ_i and that the real transfer integral is greater.

The integral (18) can be expressed explicitly in the form

$$-\langle P|\mathcal{H}|M\rangle + S\langle M|\mathcal{H}|M\rangle = S\langle\psi_d|h|\psi_d\rangle - \langle\psi_a|h|\psi_d\rangle + S\langle ad||ad\rangle + S\langle bd||bd\rangle + S\langle dd||dd\rangle - \langle aa||ad\rangle + \langle dd||ad\rangle - \langle ba||bd\rangle . \quad (19)$$

In this equation the following abbreviation has been used:

$$\langle ab||cd\rangle = \int\int d\tau_1\, d\tau_2\, r_{12}^{-1}\, \psi_a(1)\, \psi_b(2)\, \psi_c(1)\, \psi_d(2) . \quad (20)$$

The effective Hamiltonian h involves the kinetic energy operator, core potential, and the interaction between electrons, except of the four electrons in question.

In analogy to (19) we can write:

$$-\langle Q|\mathcal{H}|P\rangle + S\langle P|\mathcal{H}|P\rangle = S\langle\psi_b|h|\psi_b\rangle - \langle\psi_b|h'|\psi_d\rangle + S\langle db||db\rangle + 2S\langle ab||ab\rangle - 2\langle ad||ab\rangle - \langle dd|db\rangle . \quad (21)$$

We have neglected the change of the fluorine wavefunctions in h and h' and the change of wavefunctions in the integrals (20). Such simplification does not modify the results substantially.

We denote by h' the new effective Hamiltonian appropriate to new ionicities. It contains typical Coulomb interaction operators of the type

$$\int d\tau_2\, \psi_i(2)\, \psi'_{iq}(2)\, r_{12}^{-1}$$

in contrast to h which contains the operators

$$\int d\tau_2\, \psi_i(2)\, \psi_i(2)\, r_{12}^{-1} .$$

ψ'_{iq} differs from ψ_i by including of more concentrated nickel wavefunctions appropriate to Ni^{3+} on the B site.

We use the following approximation:

$$\langle\psi_b|h|\psi_d\rangle \doteq \frac{S}{2}\left(\langle\psi_b|h'|\psi_b\rangle + \langle\psi_d|h'|\psi_d\rangle\right) \quad (22)$$

and

$$\langle\psi_b|h'|\psi_b\rangle \doteq \frac{1}{2}\left(\langle\psi_k|h''|\psi_b\rangle + \langle\psi_b|h|\psi_b\rangle\right) , \quad (23)$$

where h'' contains operators

$$\int d\tau_2\, \psi'_{iq}(2)\, \psi'_{iq}(2)\, r_{12}^{-1} .$$

From (22) and (23) we get

$$\langle\psi_b|h'|\psi_d\rangle - \langle\psi_b|h|\psi_d\rangle = \frac{1}{4} S\left(\langle\psi_b|h''|\psi_b\rangle - \langle\psi_b|h|\psi_b\rangle\right) . \quad (24)$$

$\langle\psi_b|h|\psi_b\rangle$ and $\langle\psi_b|h''|\psi_b\rangle$ are the energies of an electron on Ni^{2+} when the nickel electrons are in appropriate Ni^{2+} wavefunctions in the first case and in Ni^{3+} wavefunctions in the second case. The numerical value of $\langle\psi_b|h''|\psi_b\rangle$ is not known. But when we use the Ni^{1+} wavefunctions in the effective Hamiltonian, (denoted by $\bar{\bar{h}}$) the difference is approximately [4]

$$\langle\psi_b|\bar{\bar{h}}|\psi_b\rangle - \langle\psi_b|h|\psi_b\rangle = -0.4 .$$

By linear extrapolation we get

$$\langle\psi_b|h''|\psi_b\rangle - \langle\psi_b|h|\psi_b\rangle = +0.4 . \quad (25)$$

Substituting (25) into (24) we obtain

$$\langle\psi_b|h'|\psi_d\rangle - \langle\psi_b|h|\psi_d\rangle = 0.006\,. \tag{26}$$

The approximation is, of course, very rough. The only aim is to show that the change of wavefunctions does not change ΔE_1 considerably as (26) is small in comparison with the other terms in the transfer integral (21).

We can now compare the transfer integrals (19) and (21). To do it, we must first notify the following apparent relations between the two-centre integrals and the transfer integrals

$$\langle dd||bd\rangle = \langle dd||ad\rangle\,,$$

$$\langle ad||ab\rangle = \langle ba||bd\rangle\,,$$

$$\langle\psi_a|h|\psi_d\rangle = \langle\psi_b|h|\psi_d\rangle\,.$$

The terms in (19) and (21), which are multiplied by the overlap integral S are the energies (in ionic approximation) of the fluorine p-electron and nickel d-electron, respectively. The energy of the d-electron was calculated by Sugano and Shulman [11] and is equal to 0.120. The energy of p-electron is equal to 0.110, as calculated by Šimánek and Šroubek [10].

Using these results we find

$$\begin{aligned}(-\langle Q|\mathcal{H}|P\rangle + S\langle P|\mathcal{H}|P\rangle) - (-\langle P|\mathcal{H}|M\rangle + S\langle M|\mathcal{H}|M\rangle) = \\ = (\langle\psi_b|h|\psi_d\rangle - \langle\psi_b|h'|\psi_d\rangle) + \langle aa||ad\rangle - \\ - \langle ad||ab\rangle + S\,(0.120 - 0.110)\,.\end{aligned} \tag{27}$$

Substituting the numerical values from (18) and (26) and the numerical values of the interaction integrals from [11] into (27) we get

$$-\langle Q|\mathcal{H}|P\rangle + S\langle P|\mathcal{H}|P\rangle = +\,0.056\,. \tag{28}$$

The numerical value of the energy difference $\langle P|\mathcal{H}|P\rangle - \langle M|\mathcal{H}|M\rangle$ can be calculated using the well-known experimental ionisation potentials of Ni^{2+} and F^-. The interaction integrals are given by Shulman and Sugano [11]. Subtracting the small polarization energy (~ 0.05) we obtain

$$\langle P|\mathcal{H}|P\rangle - \langle M|\mathcal{H}|M\rangle = 0.3\,. \tag{29}$$

The energy U_1 can be estimated as the sum of (29) and the one — electron energy difference between F^- and Ni^{2+} in the new crystal field. Alternatively one could get U_1 directly as the energy difference between Ni^{3+} — Ni^{1+} and Ni^{2+} — Ni^{2+} for free ions, subtracting the nearest—neighbour repulsion and polarization corrections.

The result is

$$U_1 = 11\text{ eV}\,. \tag{30}$$

The covalency parameter $\gamma = 0.140$ is known from experiment. Substituting (28), (29), (30) and γ into (14) we obtain

$$\Delta E_1 = 2.65\times 10^{-4}\text{ at.u.} \tag{31}$$

in comparison with the experimental value

$$\Delta E = 4.32\times 10^{-4}\text{ at.u.}\,.$$

5. Numerical Evalution of for ΔE_2 $KNiF_3$

To find the numerical value of (17) we must first estimate the transfer integral $\langle Z|\mathcal{H}|P\rangle - S\langle P|\mathcal{H}|P\rangle$.

We can proceed in a similar way as in the previous case. The explicit form of $\langle Z|\mathcal{H}|P\rangle - S\langle P|\mathcal{H}|P\rangle$ is

$$\begin{aligned} -\langle Z|\mathcal{H}|P\rangle + S\langle P|\mathcal{H}|P\rangle = {} & S\langle\psi_d|h|\psi_d\rangle - \langle\psi_d|\bar{h}|\psi_b\rangle + \\ & + 2S\langle ad||ad\rangle + S\langle db||db\rangle - \\ & - 2\langle ad||ab\rangle - \langle bb||db\rangle . \end{aligned} \quad (32)$$

By the symbol $\bar{h}$ we denote the effective Hamiltonian which contains typical Coulomb interaction operators of the type

$$\int d\tau_2\, \psi'_{iz}(2)\, \psi'_{ip}(2)\, r_{12}^{-1}$$

ψ'_{iz} contains the more concentrated wavefunctions of F^{1+} and the spreaded wavefunctions of Ni^{1+}. For simplicity we suppose the cancellation of both wavefunction deformation effects and thus

$$\langle\psi_d|\bar{h}|\psi_b\rangle \doteq \langle\psi_d|h|\psi_b\rangle .$$

Comparing (32) and (19) and (18) we obtain

$$\begin{aligned} -\langle Z|\mathcal{H}|P\rangle + S\langle P|\mathcal{H}|P\rangle = {} & 0.032 - S[E(F^-) - E(F^\circ)] + \\ & + S\langle db||db\rangle + \langle dd||ad\rangle - \langle ad||ab\rangle . \end{aligned} \quad (33)$$

The difference of the fluorine energies $E(F^-) - E(F^\circ)$ is known from experiment (0.490 at.u.) and the interaction integrals are given by Sugano and Shulman [11]. The numerical result is

$$-\langle Z|\mathcal{H}|P\rangle + S\langle P|\mathcal{H}|P\rangle = 0.046 . \quad (34)$$

The energy U_2 is the sum of (29) and of the energy required to remove the electron from the ψ_d orbital of F° and put it in the ψ_b orbital of Ni^{2+}(B). The energy of electron transfer from F° to Ni^{2+} is approximately equal to the sum of (29) plus the difference between the ionisation potentials of F^- and F° and the interaction of the transferred electron with the electron on the ion A. The calculation is straightforward with the result

$$U_2 = 25\text{ eV} . \quad (35)$$

Substituting these values in (17) we get

$$\Delta E_2 = 1.5 \times 10^{-4}\text{ at.u.} . \quad (36)$$

The resulting energy difference ΔE between ferromagnetic and antiferromagnetic states of the $Ni^{2+} - Ni^{2+}$ ion pair in $KNiF_3$ is equal to the total sum of ΔE_1 and ΔE_2

$$\Delta E = \Delta E_1 + \Delta E_2 = 3.9 \times 10^{-4}\text{ at.u.}$$

which is nearly equal to the experimental value

$$\Delta E = 4.3 \times 10^{-4}\text{ at.u.} .$$

6. Conclusion

The calculated stabilization of the antiferromagnetic state of $KNiF_3$ by double-electron transfers is in good agreement with the experiment. We have neglected the role of other superexchange effects described by the expression (9), because their importance is not evident and their numerical evalution more difficult.

The accuracy of the given calculation is of the same order as the similar calculations in the modern theory of covalency. We may suppose that the error of the U_1 (11 eV) and U_2 (25 eV) values does not exceed much $\pm$ 20%. More serious is the neglecting of the deformation of the fluorine wavefunctions in the calculation of the transfer integrals. We take it into account only in the calculation of (32) but in a very simple and inaccurate way. Nevertheless the approximation seems to be satisfactory as the so calculated value of γ is in agreement with experiment. A more accurate calculation is at present impossible as the interaction integrals with deformed wavefunctions are not yet available.

The author thanks especially Dr. E. Šimánek and Dr. Nai Li Huang for valuable correspondence and sugestions, Dr. K. Ždánský for many interesting discussions and Dr. S. Krupička for his kind interest in this work.

References

[1] J. YAMASHITA and J. KONDO, Phys. Rev. **109**, 730 (1958).
[2] F. KEFFER and T. OGUCHI, Phys. Rev. **115**, 1428 (1959).
[3] P. W. ANDERSON, Solid State Phys. **14**, 99 (1963).
[4] S. SUGANO and Y. TANABE, J. Phys. Soc. Japan., **20**, 1155 (1965).
[5] G. W. PRATT, JR., Phys. Rev. **97**, 926 (1955).
[6] J. B. GOODENOUGH and A. L. LOEB, Phys. Rev. **98**, 391 (1955).
[7] R. K. NESBET, Phys. Rev. **119**, 658 (1960).
[8] F. KEFFER, T. OGUCHI, W. O'SULLIVAN, and J. YAMASHITA, Phys. Rev. **115**, 1553 (1959).
[9] R. E. WATSON and A. J. FREEMAN, Phys. Rev. **134**, A 1526 (1964).
[10] E. ŠIMÁNEK and Z. ŠROUBEK, phys. stat. sol. **4**, 251 (1964).
[11] S. SUGANO and R. G. SHULMAN, Phys. Rev. **130**, 517 (1963).

(Received March 31, 1966)

phys. stat. sol. **16**, 413 (1966)

Physikalisch-Technisches Institut der Deutschen Akademie der Wissenschaften zu Berlin, Abteilung Kristallelektronik

Zur Zeitabhängigkeit raumladungsbegrenzter Injektionsströme in Halbleitern

Von

H. LEMKE

Es wird die Zeitabhängigkeit von Raumladungsströmen unter Berücksichtigung eines Haftterms untersucht. Für die Füllung und Entleerung der Haftstellen werden Näherungslösungen angegeben, aus denen die Einfangszeit τ_c und Emissionszeit τ_e bestimmt werden können. Bei hinreichend kleiner Spannung ändert sich der Injektionsstrom exponentiell mit der Zeitkonstanten τ_c. Nach Abschalten der Injektionsspannung werden die freien Ladungsträger von dem durch haftende Raumladung verursachten Feld aus dem Kristall herausgeschwemmt, wodurch die thermische Ladungsträgerkonzentration p_0 auf den Minimalwert $p = (\tau_R/\tau_e)\, p_0$ (τ_R dielektrische Relaxationszeit) absinken kann. Bei starker Füllung des Haftterms gegenüber dem thermischen Gleichgewicht ändert sich dabei die haftende Raumladungskonzentration exponentiell mit der Zeitkonstanten τ_e. Bei schwacher zusätzlicher Haftstellenfüllung ist die zeitliche Änderung der haftenden Raumladung konstant und hängt von der im thermischen Gleichgewicht vorhandenen Besetzung ab.

An investigation is made of the time dependence of space-charge-limited currents taking trapping into account. Approximations are made concerning transitions involving the traps which give the trapping time τ_c and the emission time τ_e. At sufficient low voltages the injection current changes exponentially with a time constant of τ_c. If the injection voltage is turned off, the free carriers are swept out of the crystal by the field of the trapped space charge whereby the carrier concentration can fall to a minimum value $p = (\tau_R/\tau_e)\, p_0$ (p_0 thermal equilibrium concentration, τ_R dielectric relaxation time). With a much higher concentration of filled traps than at thermal equilibirum, the trapped space-charge changes exponentially with a time constant of τ_e. If the concentration of filled traps is only marginally greater than at thermal equilibrium the trapped space-charge changes linearly with time at a rate which depends on the equilibrium population.

1. Einleitung

Experimentelle Untersuchungen der Majoritätsträgerinjektion in Isolatoren und Halbleitern können zur Bestimmung zahlreicher Störstellen- und Kristallparameter herangezogen werden [1, 2]. Über die Theorie der stationären Injektionsströme und über an einer Reihe von anorganischen und organischen Kristallen erhaltene experimentelle Ergebnisse wurde in einem zusammenfassenden Artikel von Lampert [3] berichtet.

Von Rose [1] wurde erstmalig darauf hingewiesen, daß aus der Zeitabhängigkeit des Injektionsstroms der Wirkungsquerschnitt des Haftterms für Majoritätsladungsträger bestimmt werden kann. Gegenüber Lebensdauermessungen haben derartige Untersuchungen den Vorteil, daß andere für die Kinetik maßgebende Störstellenparameter wie Aktivierungsenergie und Konzentration der Haftzentren gleichzeitig aus der stationären Strom-Spannungs-Charakteristik ermittelt werden können. Da bei der Majoritätsträgerinjektion nur eine Umbesetzung der Störstellen über das Majoritätsträgerband erfolgt, entfallen weiterhin einige für die Auswertung von Lebensdauermessungen notwendige Voraussetzungen [4, 5] über die Größe der Störung des thermischen Gleichgewichts und über die Lage des Ferminiveaus.

Für die Zeitabhängigkeit raumladungsbegrenzter Injektionsströme existiert zur Zeit noch keine vollständige Theorie. Das dem Stromtransport zugrunde liegende System nichtlinearer partieller Differentialgleichungen wurde bisher ausführlich von Many und Rakavy [6] für die Injektion untersucht. Eine einfache Behandlung des Problems ist möglich, wenn der Injektionsvorgang, der durch die Driftzeit beschrieben wird, wesentlich schneller als die Haftstellenumladung abläuft und wenn die injizierte Ladungsträgerkonzentration hinreichend groß gegenüber der thermischen Ladungsträgerkonzentration ist, so daß Relaxationseffekte vernachlässigbar werden. In diesem Fall können Drift- und Einfangzeit der Haftstellen aus der Zeitabhängigkeit des Injektionsstroms bestimmt werden. Diese Voraussagen konnten bei Messungen an organischen Kristallen bestätigt werden [7, 8, 9].

Wegen des komplizierten Gleichungssystems wurde in den folgenden Untersuchungen davon abgesehen, allgemeine analytische Lösungen zu finden. Ziel der Untersuchungen ist dagegen die Ableitung von Näherungsbeziehungen, aus denen die Einfangzeit τ_c und die Emissionszeit τ_e bestimmt werden können. Die hierzu erforderlichen Voraussetzungen sollten physikalisch verständlich und experimentell leicht realisierbar sein.

Ausgangspunkt der Überlegungen bildeten experimentelle Befunde zum zeitlichen Abbau haftender Raumladung in mit tiefen Donatoren dotiertem Silizium [10]. In dieser Arbeit konnten aus der Messung der Halbwertszeiten für die Haftstellenentleerung in Abhängigkeit von der Temperatur die Aktivierungsenergien bestimmt werden. Eine Berechnung der Wirkungsquerschnitte unter der Annahme, daß die Halbwertszeiten mit den Emissionszeiten größenordnungsmäßig übereinstimmen, führte jedoch zu falschen Resultaten. Wie im folgenden gezeigt werden soll, ist diese Diskrepanz auf die Vernachlässigung der im thermischen Gleichgewicht vorhandenen Haftstellenbesetzung zurückzuführen.

2. Beschreibung des verwendeten Modells

Den Berechnungen wurde das in Fig. 1 dargestellte Bändermodell zugrunde gelegt. Es wird ein homogen dotierter p-leitender Kristall mit N_A flach liegenden Akzeptoren ($E_{AV} \approx 0{,}04$ eV) und N_D tiefliegenden Donatoren ($E_{DV} \approx$ $\approx 0{,}4$ eV) betrachtet.[1]) Die Ausdehnung der Anreicherungskontakte, die z. B. durch Eindiffusion einer hohen Akzeptorkonzentration hergestellt werden können, soll klein gegenüber der Kristalldicke L sein. Die Feldverteilung in diesen Zonen wird im folgenden vernachlässigt. Es wird ein Temperaturbereich vorausgesetzt, in dem die Akzeptoren im Kristallvolumen vollständig ionisiert sind, so daß an der Umladung nur der Donatorterm beteiligt ist. Für die wirksame Haftstellenkonzentration $H - H_0^+$ erhält man somit bei $N_D > N_A$ und hinreichend tiefer Temperatur den Wert $N_D - N_A$, wobei $H = N_D$ und $H_0^+ \approx N_A$.

Das Verhalten der stationären Strom-Spannungs-Charakteristik und auch die Zeitabhängigkeit der Injektionsströme hängt wesentlich von der Größe der

[1]) Eine derartige Dotierung ist bei Silizium für eine Untersuchung der Majoritätsträgerinjektion unter Wirkung eines Haftterms geeignet [10].

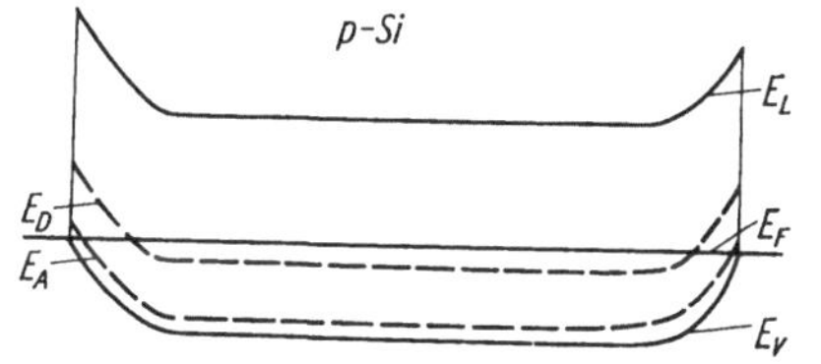

Fig. 1. Energieschema des Modellkristalls

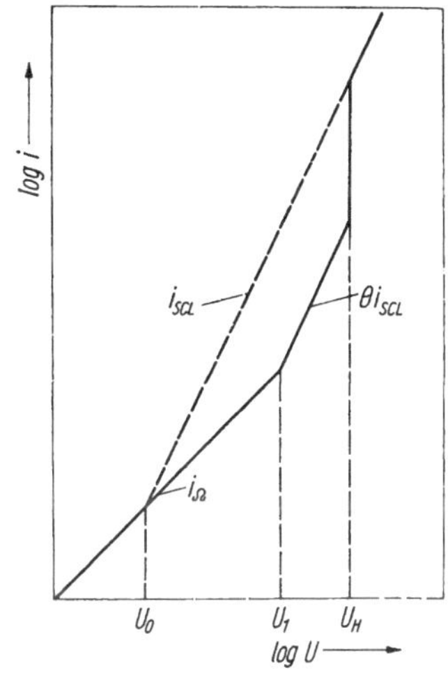

Fig. 2. Schematische Strom-Spannungs-Charakteristik für einen Halbleiterkristall mit einem Haftterm nach Lampert [2]

mittleren injizierten Löcherkonzentration $\overline{\Delta p}$ im Vergleich zur thermischen Ladungsträgerkonzentration p_0 ab. Im einfachsten Fall für einen haftstellenfreien Isolator gilt [11]

$$\Delta p(x) = \frac{3}{4} \frac{\varepsilon \varepsilon_0 U}{e L^2} \left(\frac{L}{x}\right)^{1/2} \quad \text{und} \quad \overline{\Delta p} = 2\, \Delta p(L) \tag{1}$$

($\varepsilon \varepsilon_0$ Dielektrizitätskonstante, e Elementarladung, U Spannung). Der zu dieser injizierten Löcherkonzentration zugeordnete Strom [12] ist durch

$$i_{\mathrm{SCL}} = \frac{9}{8} \varepsilon \varepsilon_0 \mu \frac{U^2}{L^3} \tag{2}$$

gegeben. Unter der Wirkung eines Hafterms wird die Verteilung (1) bei schwacher Füllung der Haftstellen nicht geändert und lediglich um den Faktor $\Theta = K/(H - H_0^+)$ bei tieferen Temperaturen abgesenkt (K Massenwirkungskonstante des Hafterms).

Der qualitative Verlauf der Strom-Spannungs-Charakteristiken [2] ist in Fig. 2 für einen Halbleiter dargestellt. Ein stationärer Raumladungsstrom von der Größe $\Theta\, i_{\mathrm{SCL}}$ kann bei $\Theta\, \overline{\Delta p} \gtrsim p_0$ beobachtet werden ($\Theta \ll 1$). Ohne Umladung der Haftstellen wird der Mott-Gurney-Strom i_{SCL} für $\overline{\Delta p} \gtrsim p_0$ größer als der ohmsche Strom. Die Raumladungsverteilung (1) wird für $\overline{\Delta p} < p_0$ infolge Relaxation und bei $\overline{\Delta p} \approx H - H_0^+$, wenn die Haftstellen bis zum Kathodenkontakt homogen aufgefüllt werden, wesentlich geändert. Für die folgenden Ableitungen wurde daher $p_0 \leqq \overline{\Delta p} \ll H - H_0^+$ vorausgesetzt. Diese Bedingung kann experimentell leicht durch Wahl einer geeigneten Dotierung, Temperatur und Injektionsspannung realisiert werden. Die „Grenz"konzentrationen p_0, p_0/Θ und $H - H_0^+$ entsprechen in der Fig. 2 den charakteristischen Spannungen U_0, U_1 und U_{H}.[2])

[2]) Würde der Halbleiter nur einen Akzeptorterm enthalten, so wäre die Konzentration p_0 (im Störleitungsbereich) identisch mit der Haftstellenkonzentration, und die Spannungswerte U_1 und U_{H} würden zusammenfallen.

Zur Berechnung der Zeitabhängigkeit des Raumladungsstroms müssen Kontinuitätsgleichung, Poisson-Gleichung und Bilanzgleichung simultan gelöst werden:

$$j(t) = e\,\mu\,p\,E + \varepsilon\,\varepsilon_0 \frac{\partial E}{\partial t}\,, \tag{3}$$

$$\frac{\partial}{\partial t}(p + H^+) = -\,\mu\,\frac{\partial}{\partial x}(p\,E)\,, \tag{4}$$

$$\frac{\partial E}{\partial x} = \frac{e}{\varepsilon\,\varepsilon_0}(H^+ - H_0^+ + p - p_0)\,, \tag{5}$$

$$\frac{\partial H^+}{\partial t} = -\,r_e\,H^+ + r_c\,p\,(H - H^+)\;; \qquad r_e = r_c\,K \tag{6}$$

(j Gesamtstrom, E Feldstärke, p Löcherkonzentration, H^+ Konzentration der ionisierten Donatoren, r_e Emissionswahrscheinlichkeit, r_c Einfangkoeffizient). Der vom Ort unabhängige Gesamtstrom setzt sich aus dem Teilchenstrom und dem Verschiebungsstrom zusammen und kann im Außenkreis gemessen werden. In den Beziehungen (3) und (4) wurde der Diffusionsstrom vernachlässigt, da hierdurch die Behandlung des Problems wesentlich erleichtert wird. Bei Berücksichtigung der Diffusion ist selbst die Berechnung der stationären Ströme für einen haftstellenfreien Isolator wesentlich komplizierter. Die exakte Lösung wird jedoch für Spannungen $U \gtrapprox 10\,kT/e$ nur noch unwesentlich durch den Diffusionsstrom beeinflußt [13].

3. Injektion der Raumladung und Haftstellenfüllung

Wird an den Kristall zur Zeit $t = 0$ ein hinreichend steiler Spannungsimpuls angelegt, so sind zur Lösung des Gleichungssystems folgende Zusatzbedingungen zu stellen [6]:

$$E(0, t) = 0\,, \tag{7a}$$

$$E\,(x > 0, 0) = \frac{U}{L}\,, \tag{7b}$$

$$\int_0^L E(x, t)\,\mathrm{d}x = U\,, \tag{7c}$$

$$p(x, 0) = p_0\,, \tag{8a}$$

$$H^+(x, 0) = H_0^+\,. \tag{8b}$$

Die Bedingungen (8a, 8b) sind die Anfangswerte der Konzentrationen in den Haftstellen und im Valenzband. Die Bedingung (7a) enthält die Forderung, daß die virtuelle Anode bei $x = 0$ liegt, während die Bedingungen (7b, 7c) besagen, daß zur Zeit $t = 0$ das Kristallvolumen raumladungsfrei und die Spannung an den Kristallelektroden zeitlich konstant ist.

Wenn die Driftzeit t_D klein gegen die Einfangzeit $\tau_c = [r_c\,(H - H_0^+)]^{-1}$ ist, können die Injektion der Raumladung und die Haftstellenumladung unabhängig voneinander behandelt werden [6].

Für $H^+ = H_0^+$ erhält man aus den Beziehungen (3) und (5) für den Gesamtstrom

$$j(t) = \frac{\mu\,\varepsilon\,\varepsilon_0}{2}\frac{\partial}{\partial x}E^2 + \varepsilon\,\varepsilon_0\frac{\partial E}{\partial t} + e\,\mu\,p_0\,E\,,$$

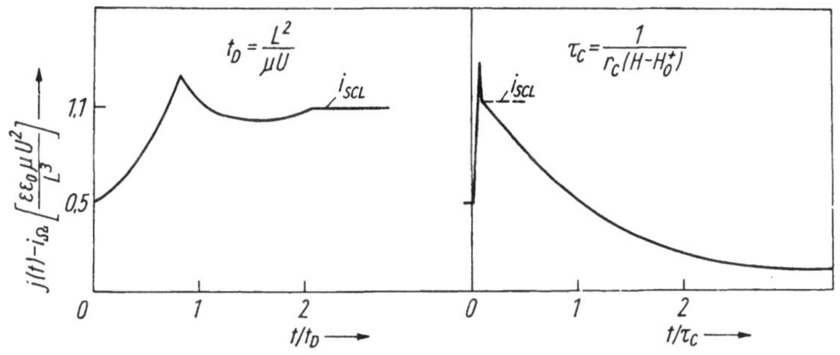

Fig. 3. Zeitlicher Verlauf des Raumladungsstroms während der Injektion und Haftstellenumladung

woraus durch Integration über x unter Berücksichtigung von (7a) und (7c) für den Strom am Kathodenkontakt

$$j(t) = \frac{\mu\,\varepsilon\,\varepsilon_0}{2\,L} E^2 + e\,\mu\,p_0\,\frac{U}{L} \quad \text{für} \quad x = L \tag{9}$$

folgt. Durch Eliminieren von j aus der Beziehung (3) erhält man mit $\alpha = e\,\mu/(\varepsilon\,\varepsilon_0)$ die Differentialgleichung

$$\frac{\partial E}{\partial t} = \frac{\mu}{2\,L} E^2 + \alpha\,p_0\left(\frac{U}{L} - E\right) - \alpha\,\Delta p\,E \quad \text{für} \quad x = L\,. \tag{10}$$

Solange die driftende positive Raumladung den Kathodenkontakt noch nicht erreicht hat ($\Delta p = 0$), erhält man aus der Lösung der Gleichung (10) für die Feldstärke und gemäß der Beziehung (9) für den Raumladungsstrom monoton ansteigende Funktionen [6]. Für $\Delta p > 0$ wurde die Differentialgleichung von Many und Rakavy [6] numerisch gelöst. Wie die qualitative Darstellung in Fig. 3 zeigt, nimmt der Raumladungsstrom bei $t_1 \approx 0{,}8\,t_D$ einen Maximalwert und bei $t \gtrapprox 2\,t_1$ seinen stationären Wert an.

Da die injizierten Löcher für $0 < t < t_1$ den Gegenkontakt noch nicht erreicht haben, fließt der Raumladungsstrom am Kathodenkontakt als reiner Verschiebungsstrom (1. Term in der Beziehung (9)), der zu einer negativen Aufladung der Gegenelektrode führt (siehe Fig. 4). Bei einem Halbleiter ist nach der Beziehung (10) wegen $E \gtreqq U/L$ die Feldstärkeänderung und damit die Aufladung schwächer als bei einem Isolator. Physikalisch wird dieses Verhalten durch den Relaxationsstrom verursacht, der ständig einen Teil der negativen Flächenladung neutralisiert, so daß auch der stationäre Strom nicht den vollen nach der Beziehung (2) zu erwartenden Wert annehmen kann. Nach numerischen Ergebnissen von Many und Rakavy ist jedoch der Einfluß der Relaxation vernachlässigbar,

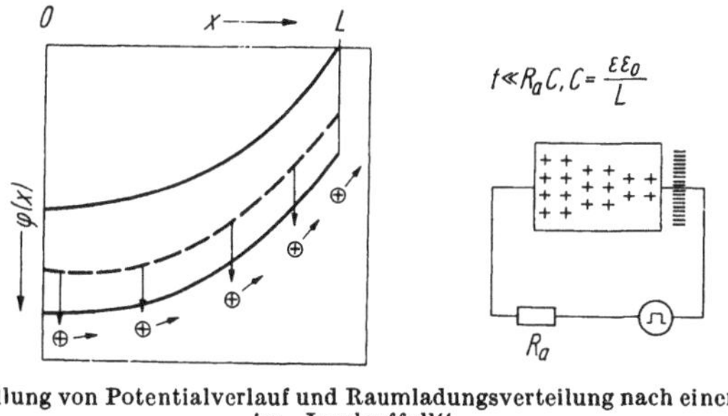

Fig. 4. Qualitative Darstellung von Potentialverlauf und Raumladungsverteilung nach einer Raumladungsinjektion im „Leerlauffall"

wenn $\overline{\Delta p} \gtrapprox p_0$ oder gleichbedeutend damit $t_D \gtrapprox \tau_R$ gilt. Für Silizium bei $T = 300\ °K$ erhält man unter Verwendung der Zahlenwerte $\mu = 420\ cm^2/Vs$, $\alpha = 6 \times 10^{-5}\ cm^3\ s^{-1}$, $p_0 \lessapprox 10^{11}\ cm^{-3}$, $U \gtrapprox 10$ V und $L \approx 1 \times 10^{-2}$ cm die Zeitkonstanten $t_D \lessapprox 3 \times 10^{-8}$ s und $\tau_R \gtrapprox 2 \times 10^{-7}$ s, während die Einfangzeiten für $r_c \approx 10^{-8}\ cm^3 s^{-1}$ und $H - H_0^+ \approx 10^{13}\ cm^{-3}$ bei $\tau_c \approx 10^{-6}$ s liegen.[3])

Zur Berechnung der Zeitabhängigkeit des Raumladungsstroms während der Haftstellenumladung kann man bei $t_D \ll \tau_c$ als Anfangswerte für die injizierte Löcherkonzentration die Beziehung (1) oder für den Strom[4]) die Beziehung (2) verwenden. Gilt an jeder Stelle im Kristall

$$\Delta H^+(x, t) = \Delta p(x, 0) - \Delta p(x, t)\,, \qquad (11)$$

so ist die Feldstärkeverteilung zeitunabhängig und die Lösung der Bilanzgleichung (6) (siehe Anhang 1) kann in der Form $\Delta p(x, t) = \Delta p(x, 0) \cdot f(t)$ geschrieben werden. Durch Eliminieren von $\Delta p(x, 0)$ mit der Poisson-Gleichung (5) und Einsetzen in die Stromgleichung (3) kann der Strom entsprechend der Beziehung (9) am Kathodenkontakt berechnet werden. Für $H_0^+ \ll H$ erhält man bei schwacher Füllung der Haftstellen (siehe Anhang 1)

$$j(t) = i_{SCL}(0) \frac{1}{1 + \Theta} \left[\exp\left(-\frac{1 + \Theta}{\tau_c} t \right) + \Theta \right] + i_\Omega \quad \text{für } i_{SCL}(0) \gtrapprox i_\Omega \qquad (12)$$

($i_{SCL}(0)$ von Haftstellen unbeeinflußter stationärer Raumladungsstrom, i_Ω ohmscher Strom).

Hierbei wurde angenommen, daß der erste Term in Gleichung (9), der bei vollständiger Aufladung des Kristalls einen reinen Teilchenstrom darstellt, mit dem Raumladungsstrom entsprechend der Beziehung (2) übereinstimmt. Den qualitativen Verlauf des Stroms zeigt Fig. 3. Bei $\Theta \ll 1$ erhält man aus der Beziehung (12) die Einfangzeit τ_c, aus der bei Kenntnis der Haftstellenkonzentration der Wirkungsquerschnitt berechnet werden kann. Eine von der Haftstellenkonzentration unabhängige Bestimmung des Wirkungsquerschnitts ist durch Messung der Emissionswahrscheinlichkeit bei zwei unterschiedlichen Temperaturen möglich. Im folgenden Abschnitt soll daher untersucht werden, unter welchen Bedingungen eine Ermittlung von r_e aus der Haftstellenentleerung nach Abschalten der Injektionsspannung möglich ist.

4. Haftstellenentleerung

Bei hinreichend langer Wirkung der Injektionsspannung stellen sich im Kristall stationäre Verteilungen freier und haftender Löcher ein, die ebenso wie in den Beziehungen (8a) und (8b) als Anfangswerte zu verwenden sind. Mit diesen Konzentrationen sind auch die Anfangswerte der Feldstärke gemäß der Beziehung (7b) vorgegeben.[5]) Die analytischen Ausdrücke sind jedoch schon bei Berücksichtigung eines Haftstellenterms kompliziert [2].

Für die im folgenden durchgeführten Näherungsrechnungen wird vorausgesetzt, daß die Ortsabhängigkeit der Raumladung durch die Beziehung (1) be-

---

[3]) Die Zahlenwerte wurden den experimentellen Verhältnissen bei p-Si angepaßt: $N_D \approx 10^{13}\ cm^{-3}$, $N_A \approx H_0^+ \approx 10^{11}$ bis $10^{13}\ cm^{-3}$, Wirkungsquerschnitt des neutralen Donatorterms $\approx 10^{-15}\ cm^2$ [10].

[4]) Der Strom muß zur Zeit t mit $2\, t_D \lessapprox t \ll \tau_c$ gemessen werden.

[5]) Die rechte Seite von (7b) erhält man durch Integration der Poisson-Gleichung unter Verwendung der stationären Werte für ΔH^+ und Δp.

schrieben werden kann. Bei der Vorgabe der Randbedingung für die Potentiale muß die Umladung der Kristallelektroden über den Außenkreis berücksichtigt werden.

Nach Abschalten der Injektionsspannung zur Zeit $t = 0$ liegt im Kristall eine Potentialverteilung vor, wie sie in Fig. 4 qualitativ dargestellt worden ist. Die negative Flächenladung am Kathodenkontakt kann bei $t > 0$ über den Außenwiderstand R_a teilweise zum Anodenkontakt zurückfließen, wobei nach dem Maschensatz

$$U(t) + j(t)\, R_a = 0 \tag{13}$$

gilt. Integriert man (3) über $0 \leqq x \leqq L$, so erhält man unter Berücksichtigung der geometrischen Kapazität $C = \varepsilon\, \varepsilon_0/L$ (Elektrodenfläche $= 1\ \mathrm{cm}^2$) und der Beziehung (13) die Differentialgleichung

$$\frac{\mathrm{d}U}{\mathrm{d}t} = -\frac{1}{R_a\, C}\, U - \frac{\alpha}{L}\int\limits_0^L p\, E\, \mathrm{d}x\,. \tag{14}$$

Bei kleiner RC-Konstante des Außenkreises und kleiner Löcherkonzentration überwiegt der erste Term, und man erhält als Lösung die bekannte Beziehung für die Kondensatorentladung

$$U(t) = U(0)\exp\left(-\frac{t}{R_a\, C}\right). \tag{15}$$

Bei großer RC-Konstante wird ein Teil der Spannung durch den im Kristall fließenden Löcherstrom abgebaut. Da eine exakte Berechnung dieses Terms schwierig ist, soll zur Abschätzung angenommen werden, daß im Kristall eine homogene Verteilung freier Löcher $\overline{p}$ und haftender Raumladung $\overline{\Delta H^+}$ vorliege, wobei $\overline{\Delta H^+}$ zunächst zeitlich konstant sei. Durch Integration der Kontinuitätsgleichung (4) erhält man unter Berücksichtigung von

$$\frac{\mathrm{d}\overline{\Delta H^+}}{\mathrm{d}t} = 0 \quad \text{und} \quad \frac{\mathrm{d}\overline{p}}{\mathrm{d}x} = 0$$

die Konzentration $\overline{p}\,(t)$ und durch Einsetzen in (14) bei $R_a\, C = \infty$ die Beziehung

$$U(t) = U(0)\,\frac{\overline{\Delta H^+} - \overline{p}(0)}{\overline{\Delta H^+} - \overline{p}(0)\,\exp\,\{-\,\alpha\,[\overline{\Delta H^+} - \overline{p}(0)]\,t\}}\,. \tag{16}$$

Durch die innere Entladung kann daher, solange die Haftstellen nicht entleert werden, die Spannung $U(0)$ maximal um den Faktor $[\overline{\Delta H^+} - \overline{p}(0)]/\overline{\Delta H^+}$ abgesenkt werden. Hierbei werden die im Kristall vorhandenen freien Ladungsträger in der Zeit $t_2 \approx (\alpha\overline{\Delta H^+})^{-1}$ zum Kathodenkontakt gesaugt.[6]) Beschreibt man die Entleerungszeit der Haftstellen durch die Halbwertszeit $t_{1/2}$, so ergeben sich zwei Möglichkeiten für die Festlegung der Potentiale. Im Leerlauffall $t_{1/2} \ll R_a\, C$ wird die Spannung zwischen den Kristallelektroden erst nach vollständiger Entleerung der Haftstellen abgebaut, wobei die Bedingung (7a) erhalten bleibt. Die Zeitabhängigkeit der integralen haftenden Raumladung kann in diesem Fall aus einer Spannungsmessung bestimmt werden.

---

[6]) Diese Zeitkonstante ist für den Prozeß der Löcherherausschwemmung bei $\overline{p} \ll \overline{\Delta H^+}$ charakteristisch.

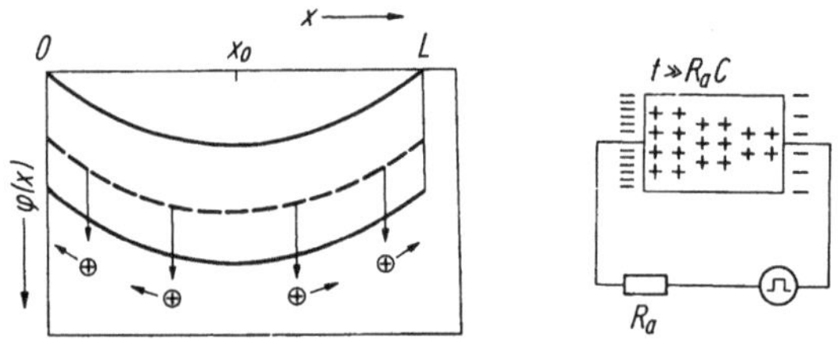

Fig. 5. Qualitative Darstellung von Potentialverlauf und Raumladungsverteilung im „Kurzschlußfall"

Im Kurzschlußfall $t_{1/2} \gg R_a C$ erfolgt die Haftstellenentleerung unter kurzgeschlossenen Elektroden, wobei die Lage des Potentialminimums $x_0(t)$ innerhalb der Zeit $R_a C$ zur Kristallmitte hin verschoben wird. Da dieser Fall im allgemeinen bei einer Strommessung vorliegt und auch experimentell leichter realisiert werden kann, soll er im folgenden näher untersucht werden. Zur Zeit $t = 0$ wächst $E(x)$ auf Grund der Beziehung (5) monoton mit x, wobei $E(x_0 = 0) = 0$ gilt. Sobald der Anodenkontakt für $t > 0$ negativ aufgeladen wird, gilt $E(0) < 0$. Da während dieser Zeit die Raumladung und damit die Monotonie der Feldstärke nicht wesentlich geändert wird, verschiebt sich die Nullstelle x_0 in das Kristallvolumen. Fur die Zeitabhängigkeit erhält man aus

$$E(x, t) = \frac{e}{\varepsilon \varepsilon_0} \int_{x_0(t)}^{x} (\Delta H^+ + \Delta p) \, dx \tag{17}$$

durch Differentiation nach der Zeit unter Verwendung von (4) und (3)

$$\varepsilon \varepsilon_0 \frac{\partial E}{\partial t} + e \mu p E = -e (\Delta H^+ + \Delta p) \Big|_{x = x_0} \frac{dx_0}{dt} = j(t) \,. \tag{18}$$

Da für $t \gg R_a C$ nach Beziehung (15) $U(t)$ und damit $j(t)$ (Beziehung (13)) gegen Null geht, wird auch nach Ablauf dieser Zeit eine Gleichgewichtslage des Potentialminimums erreicht, wobei für den inneren Strom folgende Beziehung gilt:

$$\int_0^{x_0} p E \, dx = - \int_{x_0}^{L} p E \, dx \,. \tag{18a}$$

Die Entladung über den Außenkreis führt daher zu einer Potentialverteilung, wie sie in Fig. 5 qualitativ dargestellt wurde. Die genaue Lage von x_0 hängt von der Raumladungsverteilung zur Zeit $t = 0$ ab. Bei einer homogenen Auffüllung der Haftstellen gilt $x_0 = L/2$, während man für eine Ortsabhängigkeit wie in der Beziehung (1) $x_0 = 4/9\, L$ erhält.[7])

Für die Zeitabhängigkeit der Feldstärke erhält man im Kurzschlußfall $t \gg R_a C$ aus der Gleichung (18)

$$E(x, t) = E(x, t_0) \exp \left[- \alpha \int_{t_0}^{t} p(x, t') \, dt' \right] \quad \text{für} \quad t_0 \gg R_a C \,. \tag{19}$$

Hieraus erkennt man, daß im allgemeinen während der Haftstellenentleerung $p < p_0$ gilt. Wäre z. B. $p = p_0$, so müßte die Änderung der Feldstärke durch

[7]) Die Lage des Potentialminimums erhält man unter Vorgabe der Raumladungsverteilung durch Integration der Poisson-Gleichung mit der Bedingung $\varphi(L) - \varphi(0) = 0$ (φ Potential).

die Relaxationszeit bestimmt sein. Da jedoch nach der Beziehung (17) die Feldstärke von der haftenden Raumladung abhängt, muß man auf Grund der Bilanzgleichung (6) eine Änderung der haftenden Raumladung und damit der Feldstärke mit der Zeitkonstanten τ_e erwarten.[8]) Aus der Gleichung (19) folgt dann, daß an jeder Stelle im Kristall für die Löcherkonzentration $p = r_e/\alpha = p_0\,\tau_R/\tau_e$ gelten muß.

Bisher wurde angenommen, daß während der Umladung der Kristallelektroden und der Herausschwemmung der Löcher durch das Raumladungsfeld die zeitliche Änderung von ΔH^+ vernachlässigbar ist. Beschreibt man die zeitliche Änderung der haftenden Raumladung durch die Zeitkonstante τ_e, so erhält man bei einem Einfangkoeffizienten $r_c \approx 10^{-8}$ cm^3 s^{-1} und einer Haftstellentiefe $E \approx 0{,}4$ eV für $\tau_e \approx 10^{-5}$ s ($T = 300$ °K). Dagegen gilt für die Zeitkonstanten $R_a\, C$ und $(\alpha\, \Delta H^+)^{-1}$, wenn man eine geometrische Kapazität von $C \approx 100$ pF ($L \approx 10^{-2}$ cm, $\varepsilon \approx 12$), einen Arbeitswiderstand $R_a = 100\ \Omega$ und eine injizierte Trägerdichte $\Delta H^+ = 10^{12}$ cm^{-3} voraussetzt,

$$R_a\, C \approx 10^{-8}\,\text{s} \quad \text{und} \quad (\alpha\, \Delta H^+)^{-1} \approx 10^{-8}\,\text{s}\,.$$

Man kann daher annehmen, daß unter Voraussetzung hinreichend großer Emissionszeiten oder geeignet tiefer Temperaturen eine wesentliche Umladung der Haftstellen erst dann einsetzt, wenn sich im Kristall eine Potentialverteilung entsprechend der Fig. 5 eingestellt hat. Sobald die freien Löcher zum größten Teil aus dem Kristallvolumen herausgeschwemmt worden sind, verhält sich die verbleibende Konzentration quasistationär zu $\Delta H^+(t)$, da jedes aus den Haftstellen emittierte Loch mit einer gegenüber der Retrapping-Wahrscheinlichkeit großen „Relaxationswahrscheinlichkeit" zu den Kristallelektroden abgesaugt wird.

Mit der Näherung $\frac{\partial p}{\partial t} \ll \frac{\partial \Delta H^+}{\partial t}$ erhält man daher für die Kontinuitätsgleichung (4)

$$\frac{\partial \Delta H^+}{\partial t} = -\,\mu\, \frac{\partial}{\partial x}\,(p\,E)\,, \tag{20}$$

die an der Stelle $x = x_0$ wegen $E(x_0) = 0$ durch Berücksichtigung von (5) in der Form

$$\frac{\partial \Delta H^+}{\partial t} = -\,\alpha\, p\,(\Delta H^+ + p - p_0) \quad \text{für} \quad x = x_0 \tag{21}$$

geschrieben werden kann.

Durch Eliminieren von p in den Gleichungen (6) und (21) kann die Bilanzgleichung in eine gewöhnliche Differentialgleichung für $\Delta H^+(t)$ übergeführt werden (siehe Anhang 2).

Die wesentlichen Eigenschaften der Lösung können leicht erkannt werden. Solange die haftende Raumladung groß ist, wird der größte Teil der emittierten Löcher durch das Raumladungsfeld zu den Elektroden abgesaugt, wodurch der 2. Term in der Bilanzgleichung (6) vernachlässigbar ist. Als einfache Lösung folgt unter dieser Voraussetzung in Übereinstimmung mit der in Anhang 2 angegebenen „exakten" Lösung

$$\frac{\Delta H^+(t) + H_0^+}{\Delta H^+(0) + H_0^+} = \exp\,(-\,r_e\, t) \tag{22}$$

[8]) Für $p = p_0$ geht die Bilanzgleichung in $\frac{d\Delta H^+}{dt} \approx -\,r_e \Delta H^+$ über.

oder

$$\Delta H^+(t) = \Delta H^+(0) \exp(-r_e t) \quad \text{für} \quad \Delta H^+ \gg H_0^+ \tag{22a}$$

und

$$\Delta H^+(t) = \Delta H^+(0) - r_e H_0^+ t \quad \text{für} \quad \Delta H^+ \ll H_0^+ . \tag{22b}$$

Die Lösungen (22a) und (22b) gehören zu bestimmten Spannungsbereichen der stationären Kennlinie (siehe Fig. 2). Für $U_0 < U < U_1$ gilt

$$\Delta p \approx \Theta \, \Delta H^+ \ll p_0 = \Theta \, H_0^+ \quad \text{und damit} \quad \Delta H^+ \ll H_0^+ ,$$

während man für $U_1 < U < U_H$

$$\Delta p \approx \Theta \, \Delta H^+ \gg p_0 \quad \text{und} \quad \Delta H^+ \gg H_c^+$$

erhält.

Aus den Beziehungen (22a) und (22b) kann die Zeitabhängigkeit der Löcherkonzentration berechnet werden. Für $\Delta H^+ \gg |p - p_0|$ folgt aus der Gleichung (21)

$$p(t) = -\frac{1}{\alpha} \frac{d}{dt} \ln \Delta H^+ \tag{23}$$

oder

$$p = \frac{r_e}{\alpha} \quad \text{für} \quad \Delta H^+ \gg H_0^+ , \tag{23a}$$

$$p = \frac{r_e}{\alpha} \frac{H_0^+}{\Delta H^+(0) - r_e H_0^+ t} \quad \text{für} \quad \Delta H^+ \ll H_0^+ . \tag{23b}$$

Die nach Abschalten des Injektionsimpulses bei großer Raumladungskonzentration vorhandene Löcherkonzentration r_e/α kann bei Silizium unterhalb der Eigenleitungskonzentration liegen. Für das schon angegebene Modell erhält man z. B. $r_e/\alpha \approx 10^9 \text{ cm}^{-3}$, während die Eigenleitungskonzentration bei 10^{10} cm^{-3} und p_0 bei 10^{11} cm^{-3} liegt ($T = 300\,°\text{K}$). Durch eine haftstellenbeeinflußte Majoritätsträgerinjektion kann daher der Stromtransport in einem Halbleiterkristall im Impulsbetrieb weitgehend gesperrt werden.

Wenn die haftende Raumladung auf die Größe der thermischen Trägerkonzentration abgesunken ist, wird das Raumladungsfeld und damit der „Relaxationsstrom" zu den Kontakten klein, so daß die Löcherkonzentration im Kristall wieder ansteigen kann. Wenn der Kristall in einen quasineutralen Zustand übergeht, gilt $p \approx p_0 - \Delta H^+$, woraus als Näherungslösung der Bilanzgleichung (6) eine Exponentialfunktion mit der Zeitkonstanten τ_c folgt (siehe

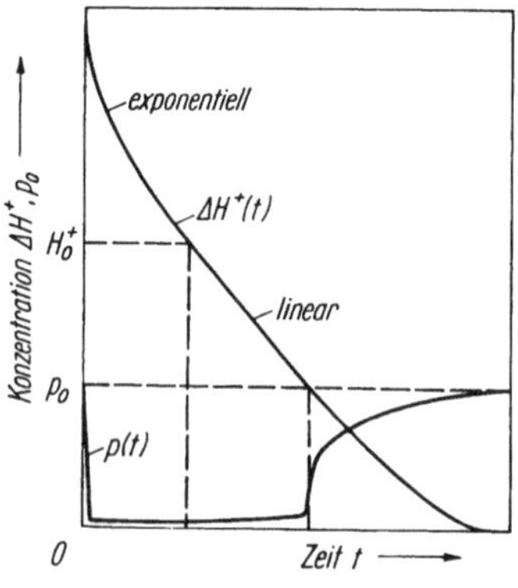

Fig. 6. Qualitative Darstellung der Zeitabhängigkeit der haftenden Raumladung und freien Löcherkonzentration

Fig. 7. Qualitative Darstellung der Abhängigkeit des Injektionsstroms von der Haftstellenentleerung

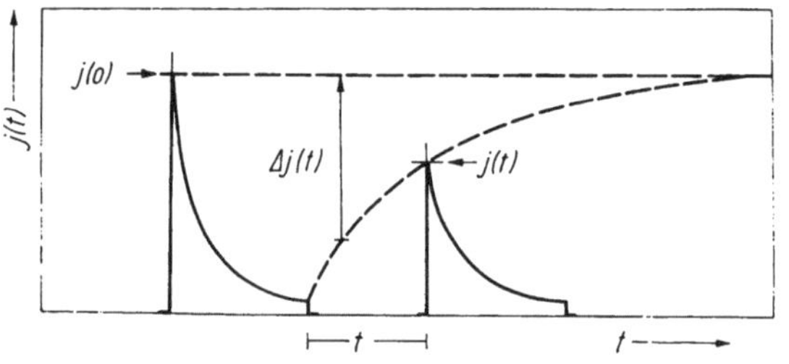

Anhang 2, Beziehung (A6a)). Ein Wiederanstieg kann schon früher einsetzen, wenn $\Delta H^+ \lessapprox r_c (H - H_0^+)/\alpha > p_0$ gilt. In diesem Fall ist die Retrapping-Wahrscheinlichkeit so groß, daß nur noch ein Teil der aus den Haftstellen emittierten Löcher zu den Kontakten driften kann. Die haftende Raumladung verhält sich dann wie eine freie Raumladung, die mit der Relaxationswahrscheinlichkeit $\alpha\, p_0$ abklingt (siehe Anhang 2, Beziehung (A6b)). Der zeitliche Verlauf von haftender Raumladung und Löcherkonzentration ist in Fig. 6 qualitativ dargestellt.

Zur Bestimmung des Wirkungsquerschnitts aus der Emissionswahrscheinlichkeit r_e ist die Beziehung (22a) geeignet. Solange $\Delta H^+ \gg H_0^+$ gilt, ist die relative zeitliche Änderung der Raumladung unabhängig von der vorher angelegten Injektionsspannung, so daß der Wirkungsquerschnitt und gleichzeitig die Haftstellentiefe aus der Messung der Zeitkonstanten in Abhängigkeit von der Temperatur bestimmt werden können.

Wenn $\Delta H^+ \ll H_0^+$ gilt, hängt die Halbwertszeit entsprechend der Beziehung (22b) von der injizierten Trägerkonzentration und damit von der Spannung ab,

$$t_{1/2} = \frac{\Delta H^+(0)}{2\, H_0^+}\, \tau_e\,. \tag{24}$$

Aus der Messung der Halbwertszeiten bei konstanter Injektionsspannung in Abhängigkeit von der Temperatur kann daher nur die Dissoziationsenergie des Haftterms ermittelt werden. Zur Bestimmung des Wirkungsquerschnitts muß zusätzlich der Quotient $\Delta H^+(0)/H_0^+$ bekannt sein.

Die Zeitabhängigkeit der haftenden Raumladung kann im Kurzschlußfall aus einer gegenüber der Injektion um die Zeit t verzögerten 2. Strommessung ermittelt werden (siehe Fig. 7). Wenn der Kristall noch haftende Raumladungen enthält, ist der Strom $j(t)$ an der Vorderflanke des Meßimpulses kleiner als der Strom $j(0)$ an der Vorderflanke des Injektionsimpulses. (Exakter wäre die Bezeichnung $j\,(-\,T)$, wobei T Dauer des Injektionsimpulses.) Da die thermischen Ladungsträger in den Kristall hineindriften müssen, erreicht auch der ohmsche Strom erst nach Ablauf etwa einer Driftzeit seinen vollen Wert. Für die injizierte freie Raumladung $\Delta p(x, t)$ gilt die Beziehung

$$\Delta H^+(x, t) + \Delta p(x, t) = \Delta H^+(x, 0) + \Delta p(x, 0) = \frac{\varepsilon\, \varepsilon_0}{e} \frac{\partial E}{\partial x}\,. \tag{25}$$

Bei Vernachlässigung von $\Delta p(x, 0)$ ($\Theta \ll 1$) erhält man durch Eliminieren von $\Delta H^+(x, t)$ mit der Beziehung (22)

$$\Delta p(x, t) = \left(\frac{\varepsilon\, \varepsilon_0}{e} \frac{\partial E}{\partial x} + H_0^+\right) [1 - \exp\,(-\, r_e\, t)]\,. \tag{26}$$

Durch Einsetzen in (3) und Integration über $0 \leqq x \leqq L$ folgt für den Strom an der Vorderflanke des Meßimpulses

$$j(t) = \left[\frac{\mu\, \varepsilon\, \varepsilon_0}{2\, L}\, E^2(L) + e\, \mu\, H_0^+ \frac{U}{L}\right] [1 - \exp\,(-\, r_e t)] + i_\Omega\,.$$

Für $\Delta j = j(0) - j(t)$ erhält man unter Berücksichtigung der Beziehung (9)

$$\Delta j = i_{\mathrm{SCL}} \exp\,(-\, r_e\, t) - e\, \mu\, H_0^+ \frac{U}{L} [1 - \exp\,(-\, r_e\, t)]\,. \tag{27}$$

Den qualitativen Verlauf für $\Delta j(t)$ zeigt Fig. 7. Wegen der für die Beziehung (22) notwendigen Voraussetzung $\Delta H^+(x, t) \gg p_0$ ist der Gültigkeitsbereich der Gleichung (27) auf $\Delta j \gg i_\Omega$ eingeschränkt.

Da die Zeitgesetze in (27) und (22) übereinstimmen, kann die Zeitabhängigkeit der haftenden Raumladung direkt aus der Messung von Δj bestimmt werden.

Die Ableitung einer Beziehung für $\Delta j(t)$ bei kleiner haftender Raumladung ($\Delta H^+ \lesssim p_0$) ist wesentlich komplizierter, da bei der Berechnung der Zeitabhängigkeit der haftenden Raumladung die Löcherkonzentration berücksichtigt werden muß. Wegen der inhomogenen Verteilung der freien Löcher im Kristallvolumen kann auch die lokale Lösung der Bilanzgleichung (Anhang 2) nicht mehr für den ganzen Bereich der Ortskoordinate $0 \leqq x \leqq L$ verwendet werden. Weiterhin muß man annehmen, daß bei kleiner Raumladungsfeldstärke ein wesentlicher Teil des Stromtransports durch Diffusion erfolgt, so daß die Vernachlässigung des Diffusionsstroms während der Haftstellenentleerung nicht mehr erlaubt ist. Bei experimentellen Untersuchungen an Silizium [14] konnte bisher die Zeitabhängigkeit (27) gut bestätigt werden.

Herrn Dr. G. O. Müller und Herrn Dr. E. Schnürer möchte ich für wertvolle Anregungen und zahlreiche Diskussionen danken.

Anhang 1

Lösung der Bilanzgleichung für die Haftstellenfüllung

Durch Eliminieren von ΔH^+ in den Beziehungen (6) und (11) erhält man eine gewöhnliche Differentialgleichung zur Bestimmung von Δp,

$$- y' = \gamma\, y^2 + [1 + \Theta\,(1 + \gamma_0) - \gamma]\, y - \Theta\,(1 + \gamma_0)\,, \tag{A1}$$

wobei

$$\gamma_0 = \frac{H_0^+}{H - H_0^+}\,, \quad \gamma = \frac{\Delta p(x, 0)}{H - H_0^+}\,, \quad y = \frac{\Delta p(x, t)}{\Delta p(x, 0)}\,,$$

$$\Theta = \frac{r_e}{r_c\,(H - H_0^+)} \quad \text{und} \quad t \to r_c\,(H - H_0^+)\,t\,.$$

(Für die relative Zeit wird das gleiche Symbol verwendet.) Als Lösung erhält man hieraus

$$y(t) = \frac{1}{2\gamma}\,\frac{[1 + \Theta(1 + \gamma_0) - \gamma + R]\,A\exp(-R t) - [1 + \Theta(1 + \gamma_0) - \gamma - R]}{1 - A\exp(-R t)}\,, \tag{A2}$$

wobei

$$R = \{[1 + \Theta\,(1 + \gamma_0) - \gamma]^2 + 4\,\Theta\,\gamma\,(1 + \gamma_0)\}^{1/2}\,,$$

$$A = \frac{1 + \Theta\,(1 + \gamma_0) + \gamma - R}{1 + \Theta\,(1 + \gamma_0) + \gamma + R}\,.$$

Entsprechend Fig. 2 können die Parameter $\Theta\,\gamma_0$, γ_0 und γ mit den Spannungen U_0, U_1 und U in Verbindung gebracht werden,

$$\Theta\,\gamma_0 \approx \frac{U_0}{U_H}\,, \quad \gamma_0 \approx \frac{U_1}{U_H}\,, \quad \gamma \approx \frac{U}{U_H}\,.$$

Bei schwacher Haftstellenfüllung gilt $\gamma \ll 1$, und man erhält durch Entwicklung des Wurzelausdrucks $R(\Theta, \gamma)$

$$y(t) = \frac{1}{1 + \Theta(1 + \gamma_0)}\,\{\exp[-(1 + \Theta)\,t] + \Theta\,(1 + \gamma_0)\}\,. \tag{A2a}$$

Diese Beziehung entspricht für $\gamma_0 \ll 1$ dem von Many und Rakavy [6] für einen Isolator abgeleiteten Ausdruck.

Eine von Gregory und Jordan [15] angegebene Lösung erhält man für $\Theta \ll 1$ oder für den Fall, daß die Reemission aus den Haftstellen in das Majoritätsträgerband klein gegen die Einfangrate ist,

$$y(t) = \frac{[1 - \gamma - \Theta(1 + \gamma_0)(1 - \gamma)^{-1}] \exp[-(1 - \gamma)t] + \Theta(1 + \gamma_0)(1 - \gamma)^{-1}}{1 - \gamma \exp[-(1 - \gamma)t]}. \quad \text{(A2b)}$$

Die Gültigkeit dieser Beziehung ist jedoch für größere Werte von γ in Frage gestellt. Bei $\gamma = 0{,}5$ sind z. B. die Haftstellen etwa bis zur Kristallmitte vollkommen gefüllt, so daß schon eine merkbare Verschiebung der Raumladungsverteilung zum Kathodenkontakt hin auftreten muß. Diese Verschiebung bedeutet aber, daß während der Umladung der Haftstellen der Strom ortsabhängig wird, wodurch eine lokale Lösung unter der in (11) getroffenen Annahme zu falschen Resultaten führen kann.

Anhang 2

Lösung der Bilanzgleichung für die Haftstellenentleerung am Ort des Potentialminimums

Aus der Bilanzgleichung (6) und der Kontinuitätsgleichung (21) erhält man durch Eliminieren von p eine gewöhnliche Differentialgleichung zur Bestimmung von $\Delta H^+(x_0, t)$,

$$\frac{d\eta}{dt} = -\frac{1}{2\Theta}[(1 + 2\Theta)\eta + \Theta + \varepsilon] + \\ + \{[\eta + (1 + 2\Theta)\varepsilon - \Theta]^2 + 4\Theta\varepsilon(1 + \Theta)(1 - \varepsilon)\}^{1/2}, \quad \text{(A3)}$$

wobei

$$\eta = \frac{H^+(x_0, t)}{H_0^+}, \quad \Theta = \frac{p_0}{H_0^+}, \quad \varepsilon = \frac{r_c(H - H_0^+)/\alpha}{H_0^+}$$

und t die relative Zeit $r_e t$ bedeutet. In der Bilanzgleichung wurde hierbei $r_c(H - H^+) = r_c(H - H_0^+)$ gesetzt. Diese Voraussetzung ist gerechtfertigt, da der Retrapping-Term erst bei kleiner Konzentration ΔH^+ mit dem Emissionsterm in der Beziehung (6) vergleichbar wird. Dieses Verhalten wird durch die bei hoher Raumladungskonzentration stark herabgesetzte freie Löcherkonzentration verursacht.[9])

Eine explizite Lösung $t = f(\eta)$ kann durch Partialbruchzerlegung und Integration gefunden werden,

$$t = -\frac{1 + \Theta - \varepsilon}{1 + \Theta} \ln \frac{(1 + 2\Theta)\eta - R(\eta) + \varepsilon + \Theta}{(1 + 2\Theta)\eta_0 - R(\eta_0) + \varepsilon + \Theta} - \\ - \frac{1}{2(1 + \Theta)} \ln \frac{\eta + R(\eta) + \varepsilon(1 + \Theta) - \Theta(1 - \varepsilon)}{\eta_0 + R(\eta_0) + \varepsilon(1 + \Theta) - \Theta(1 - \varepsilon)} + \\ + \frac{1 - 2\varepsilon}{2(1 + \Theta)} \ln \frac{[\varepsilon(1 + 2\Theta) - \Theta]\eta - (\varepsilon + \Theta)[R(\eta) - (\varepsilon + \Theta)]}{[\varepsilon(1 + 2\Theta) - \Theta]\eta_0 - (\varepsilon + \Theta)[R(\eta_0) - (\varepsilon + \Theta)]}, \quad \text{(A4)}$$

wobei

$$R(\eta) = \{[\eta + (1 + 2\Theta)\varepsilon - \Theta]^2 + 4\Theta\varepsilon(1 + \Theta)(1 - \varepsilon)\}^{1/2}$$

---

[9]) Die Löcherkonzentration beginnt wieder zu steigen, wenn $\Delta H^+ \lesssim p_0$. Wegen $p_0 = \Theta H_0^+$ und $\Theta \ll 1$ gilt dann auch $\Delta H^+ \ll H_0^+$.

und

$$\eta_0 = \frac{\Delta H^+(x_0, 0)}{H_0^+}.$$

Die Parameter Θ und ε kennzeichnen die Konzentrationen p_0 und $r_c\,(H - H_0^+)/\alpha$. Die letzte Größe beschreibt dabei diejenige haftende Raumladungskonzentration, bei der die „Relaxationswahrscheinlichkeit" $\alpha\,\Delta H^+$ durch das innere Feld den Wert der Einfangwahrscheinlichkeit $r_c\,(H - H_0^+)$ annimmt. Bei nicht zu hoher Haftstellenkonzentration liegt diese Konzentration unter der Inversionsdichte von Silizium ($T = 300$ °K). Für $H - H_0^+ \approx 10^{13}$ cm^{-3} und $r_c \approx 10^{-8}$ cm^3 s^{-1} erhält man $r_c\,(H - H_0^+)/\alpha \approx 10^9$ cm^{-3}, so daß neben $\Theta \ll 1$ auch im allgemeinen $\varepsilon \ll 1$ gilt.

Wenn $\eta \gg \Theta$ und $\eta \gg \varepsilon$, erhält man durch Entwicklung

$$\ln \frac{\eta + 1}{\eta_0 + 1} = -t. \tag{A5}$$

Diese Beziehung kann auch direkt aus der Bilanzgleichung (6) bei Vernachlässigung der Einfangrate abgeleitet werden. Für $\eta \ll \frac{\varepsilon + \Theta}{2}$ folgt die Näherungsbeziehung

$$\ln \frac{\eta}{\eta_0} = -\frac{1 + \Theta}{\varepsilon + \Theta}\, t, \tag{A6}$$

die für $\Theta \ll 1$ in

$$\ln \frac{\eta}{\eta_0} = -\frac{1}{\Theta}\, t \quad \text{für} \quad \varepsilon \ll \Theta, \tag{A6a}$$

$$\ln \frac{\eta}{\eta_0} = -\frac{1}{\varepsilon}\, t \quad \text{für} \quad \varepsilon \gg \Theta \tag{A6b}$$

übergeht.

Literatur

[1] A. Rose, Phys. Rev. **97**, 1538 (1955).
[2] M. A. Lampert, Phys. Rev. **103**, 1648 (1956).
[3] M. A. Lampert, Rep. Progr. Phys. **27**, 329 (1964).
[4] W. Shockley und W. T. Read, Phys. Rev. **87**, 825 (1952).
[5] G. Bemski, Proc. IRE **46**, 990 (1958).
[6] A. Many und G. Rakavy, Phys. Rev. **126**, 1980 (1962).
[7] W. Helfrich und P. Mark, Z. Phys. **166**, 370 (1962).
[8] A. Many, S. Z. Weisz und M. Simhony, Phys. Rev. **126**, 1989 (1962).
[9] M. Silver, M. Swicord, R. C. Jarnigan, A. Many, S. Z. Weisz und M. Simhony, J. Phys. Chem. Solids **23**, 419 (1962).
[10] H. Lemke, phys. stat. sol. **3**, 2083 (1963).
[11] F. Stöckmann, Halbleiterprobleme (Vieweg, Braunschweig) **6**, 279 (1961).
[12] N. F. Mott und R. W. Gurney, Electronic Processes in Ionic Crystals, Clarendon Press, Oxford 1940.
[13] J. Lindmayer, J. Reynolds und C. Wrigley, J. appl. Phys. **34**, 809 (1963).
[14] H. Lemke, phys. stat. sol. **16**, 427 (1966).
[15] B. L. Gregory und A. G. Jordan, Phys. Rev. **134**, 1378 (1964).

(Received April 7, 1966)

phys. stat. sol. **16**, 427 (1966)

Physikalisch-Technisches Institut der Deutschen Akademie der Wissenschaften zu Berlin, Abteilung Kristallelektronik

Zeitabhängigkeit raumladungsbegrenzter Injektionsströme in Fe-dotiertem p-Silizium[1])

Von

H. LEMKE

Es wird über Impulsmessungen raumladungsbegrenzter Ströme bei p^+-p-p^+-Kristallen mit eindiffundierten Anreicherungskontakten berichtet. Aus der Zeitabhängigkeit des Injektionsstroms kann die Einfangwahrscheinlichkeit $r_c\,(H - H_0^+)$ bestimmt werden. Bei Vorliegen einer gegenüber der thermischen Löcherkonzentration großen haftenden Raumladungskonzentration erfolgt die Haftstellenentleerung bei stark abgesenkter Löcherkonzentration, wodurch der Retrapping-Einfluß vernachlässigbar ist. Für die zeitliche Änderung der haftenden Raumladungskonzentration ist das Verhältnis von injizierter getrappter Trägerdichte $\Delta H^+(0)$ zu der im thermischen Gleichgewicht vorhandenen Zentrenbesetzung H_0^+ maßgebend. Bei hoher zusätzlicher Besetzungskonzentration erfolgt die Entleerung nach einer Exponentialfunktion, wodurch eine direkte Bestimmung der Emissionswahrscheinlichkeit r_e ermöglicht wird. Bei kleiner injizierter Trägerdichte ist die zeitliche Änderung der haftenden Raumladung konstant und proportional zu $r_e\,H_0^+$. Aus den Emissionswahrscheinlichkeiten erhält man für den Wirkungsquerschnitt (Löcher) des Eisenzentrums $g \cdot \Theta_p = 0{,}7 \times 10^{-15}\ \mathrm{cm}^2$ (g statistischer Gewichtsfaktor).

Pulse measurements of space-charge-limited currents on p^+-p-p^+-type crystals with enriched contacts are reported. From the time dependence of injection current the capture probability $r_c\,(H - H_0^+)$ can be determined. If the trapped space-charge is large compared with the thermal hole concentration the trap emptying process strongly reduces the hole concentration whereby the influence of re-trapping can be neglected. The time dependence of the trapped space-charge concentration is determined by the ratio of trapped injected carrier density $\Delta H^+(0)$ to the population of the centres in thermal equilibrium H_0^+. In the case of a high additional population emptying obeys an exponential law whereby a direct determination of the emission probability r_e is possible. If the injected carrier concentration is low the rate of change of the trapped space charge is constant and proportional to $r_e H_0^+$. From the emission probabilities the cross section (holes) of an Fe-centre is found to be $g\,\theta = 0.7 \times 10^{-15}\ \mathrm{cm}^2$ (g statistical weighting factor).

1. Einleitung

Als Modellsubstanz zur Untersuchung von Majoritätsträgerinjektionen unter der Wirkung eines Haftterms ist reines Silizium mit spezifischen Widerständen über 1 kΩ cm bei einer zusätzlichen Dotierung mit tiefen Donatoren [1] geeignet. Wird die Donatorenkonzentration N_D größer als die Konzentration N_A der flach liegenden Akzeptoren gewählt, so ist die bei geeignet tiefen Temperaturen wirksame Haftstellenkonzentration durch $H - H_0^+ \approx N_D - N_A$ gegeben.

---

[1]) Teilweise vorgetragen auf der Arbeitstagung des Fachausschusses „Halbleiter" der Deutschen Physikalischen Gesellschaft in Freudenstadt, 1965.

Die in einer früheren Veröffentlichung [1] mitgeteilten Ergebnisse orientierender Untersuchungen sollen in der vorliegenden Arbeit für Fe-dotiertes Silizium erweitert und präzisiert werden. Dabei soll vor allem die Zeitabhängigkeit der Injektionsströme näher untersucht werden, da die bisher mitgeteilten Ergebnisse zu einer fehlerhaften Bestimmung der Wirkungsquerschnitte geführt haben.[2]) Ursache hierfür war der nicht berücksichtigte Besetzungsgrad der Haftzentren im thermischen Gleichgewicht. Bei Anwesenheit eines Raumladungsfeldes ist die zeitliche Änderung der injizierten haftenden Löcherkonzentration durch den gesamten Emissionsstrom $r_e (H_0^+ + \Delta H^+)$ bestimmt, so daß die Emissionswahrscheinlichkeit r_e und damit der Wirkungsquerschnitt im allgemeinen nur bei Kenntnis von H_0^+ ermittelt werden kann [2].

Die im folgenden beschriebene relativ einfache Methode zur Messung von Wirkungsquerschnitten für Majoritätsträger kann auch für andere tiefliegende Donatorzentren verwendet werden. Dem Auswerteverfahren wurden Näherungsbeziehungen zugrunde gelegt [2], deren Voraussetzungen leicht experimentell realisiert werden können.

Die Zeitabhängigkeit von Raumladungsströmen wurde für den Fall der Injektion ausführlich von Many und Rakavy [3] untersucht. Die von den Autoren abgeleitete Beziehung für die zeitliche Änderung der Haftstellenbesetzung bei Isolatorkristallen ist, wie die im folgenden mitgeteilten Ergebnisse zeigen, auch für Halbleiterkristalle bei starker Injektion ($\Delta H^+ \gtrapprox p_0$) anwendbar.

Bei der Auswertung der experimentellen Ergebnisse zur Haftstellenentleerung wurde davon ausgegangen, daß der Kristall nach Abschalten der Injektionsspannung als geladener Kondensator [4] vorliegt, dessen Ladestrom bei Anlegen eines zweiten Spannungsimpulses von der noch vorhandenen haftenden Raumladung abhängt.

Über die Bestimmung des Wirkungsquerschnitts aus der Einfangwahrscheinlichkeit bei Majoritätsträgerinjektionen in Silizium berichteten Gregory und Jordan [5]. Die Messungen wurden an mit flachen Donatoren kompensiertem Silizium bei $T = 4{,}2\ °K$ durchgeführt, wobei der Hafteffekt durch negativ geladene Borzentren verursacht wurde.

2. Versuchsdurchführung

Die Untersuchungen wurden an Sandwich-Anordnungen mit einem Elektrodenabstand von $L \approx 2 \times 10^{-2}$ cm durchgeführt. Die Eindiffusion von Bor zur Herstellung der Anreicherungskontakte und die Dotierung mit Eisen erfolgte gleichzeitig bei 1200 °C (15 min) unter Argon oder Vakuum. Zur Messung der von Haftstellen unbeeinflußten Raumladungsströme wurden Spannungsimpulse von 1 bis 100 V mit Anstiegszeiten unter 200 ns verwendet. Die stationären Ströme wurden aus den stationären Impulsströmen oder bei niedrigen Spannungen aus Gleichstrommessungen ermittelt. Das Zeitverhalten der Haftstellenentleerung wurde mit einem Doppelimpulsverfahren untersucht[3]), bei dem die Verzögerungszeit des 2. (Abtast-)Impulses gegen den Haftstellenfüllimpuls moduliert wurde.

[2]) In [1] wurde angenommen, daß die für die Haftstellenentleerung gemessenen Halbwertszeiten größenordnungsmäßig mit den Emissionszeiten r_e^{-1} übereinstimmen. Die hierfür notwendige Bedingung $\Delta H^+/H_0^+ \gg 1$ war jedoch bei den früheren Untersuchungen nicht erfüllt.

[3]) Für die Entwicklung der Impulsmethode möchten wir Herrn Dr. G. O. Müller und für den Aufbau der Generatoren Herrn G. Schulz danken.

3. Experimentelle Ergebnisse

Fig. 1 zeigt den typischen Verlauf von Stromimpulsen bei Spannungen, bei denen die Driftzeit kleiner als die Relaxationszeit ist, so daß Raumladungen in das Kristallvolumen injiziert werden können. Der zweite zeitlich verzögerte Impuls findet den Kristall praktisch noch vollständig „geladen" vor, so daß im wesentlichen nur der stationäre Strom fließen kann. Für die Aufnahme der Impulsstromkennlinien mußte daher die Wiederholungsfrequenz so niedrig gewählt werden, daß während der Impulspausen ein vollständiger Abbau des Raumladungsfeldes erfolgen konnte.

Der Strom an der Vorderflanke des ersten Impulses ist eine Überlagerung des ohmschen Stroms und des Raumladungsstroms [3],

$$i(0) = i_{\mathrm{SCL}} + i_{\Omega} = \frac{9}{8} \varepsilon \varepsilon_0 \mu \frac{U^2}{L^3} + e \mu p_0 \frac{U}{L} \quad \text{für} \quad i_{\mathrm{SCL}} \gtrsim i_{\Omega}$$

($\varepsilon \varepsilon_0$ Dielektrizitätskonstante, μ Beweglichkeit, L Elektrodenabstand, p_0 thermische Löcherkonzentration, e Elementarladung, U Spannung).

Nach theoretischen Ergebnissen von Many und Rakavy [3] müßte der Strom nach Ablauf der Zeit $t = 0{,}8\, t_{\mathrm{D}}$ (t_{D} Driftzeit) kurzzeitig um etwa 20% über dem Strom i_{SCL} liegen. Dieses „Überschwingen" des Stroms konnte wegen der zu kleinen Zeitauflösung der von uns verwendeten Meßanordnung und der zu geringen Flankensteilheit der Spannungsimpulse bisher nicht beobachtet werden. Für $U \gtrsim 5$ V, $L = 2 \times 10^{-2}$ cm und $\mu \gtrsim 400$ cm²/Vs liegen die Driftzeiten unter 10^{-7} s.

Aus der zeitlichen Änderung des Stroms bei anliegender Injektionsspannung kann die Einfangwahrscheinlichkeit der Haftstellen r_{c} $(H - H_0^+)$ bestimmt werden [3]. Hierzu ist erforderlich, daß der ohmsche Stromanteil hinreichend klein ist und daß für die Emissionswahrscheinlichkeit $r_{\mathrm{e}} \ll r_{\mathrm{c}} (H - H_0^+)$ gilt. In Fig. 2 ist der zeitliche Verlauf der Stromdifferenz zwischen Injektionsstrom und stationärem Strom für verschiedene Spannungen aufgetragen. Die Zeitkonstante τ_{c} wurde aus den bei kleinen Spannungen vorliegenden Exponentialfunktionen berechnet. Für den in Fig. 2 angeführten Kristall erhält man $\tau_{\mathrm{c}} = 2{,}5\ \mu$s. Da die Konzentration H_0^+ bei hinreichend tiefen Temperaturen nur

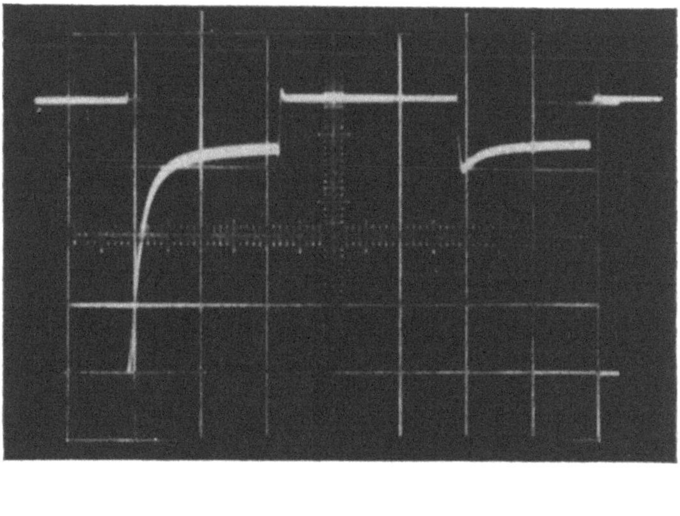

Fig. 1. Injektionsstrom in einem Fe-dotierten p^+–p–p^+-Siliziumkristall (10 µs/cm)

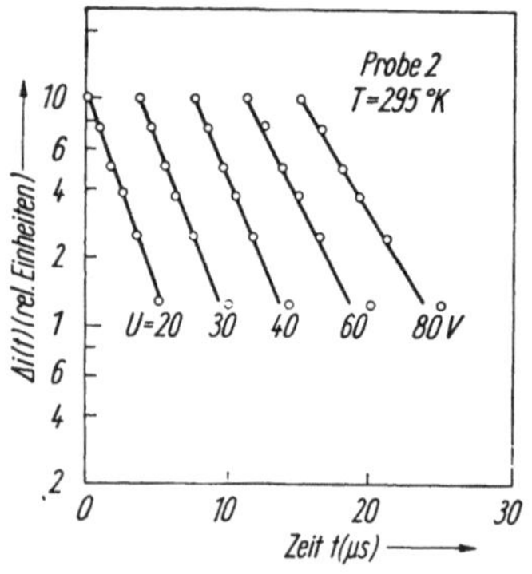

Fig. 2. Zeitabhängigkeit des Injektionsstromes für verschiedene Spannungen

noch unwesentlich von der Temperatur abhängt, ist auch die Einfangzeit weitgehend unabhängig von der Temperatur.[4])

Würde Eisen nur ein Donatorniveau bilden, so wäre bei tieferen Temperaturen $H_0^+ \approx N_A^-$. Diese Bedingung bedeutet, daß der größte Teil der von den Akzeptoren emittierten Löcher von den Donatorzentren eingefangen worden ist. Auf Grund der Komplexbildung zwischen Eisen und Bor in Silizium [6, 7] wird jedoch ein zweites, oberhalb der Bandmitte gelegenes Donatorniveau ($E_{CD} = 0{,}55$ eV) wirksam, so daß der Besetzungsgrad des Donatorniveaus bei $E_{DV} = 0{,}4$ eV im allgemeinen kleiner ist. Da die Paarbildung schon bei Zimmertemperatur [7] mit merklicher Geschwindigkeit stattfindet, kann im allgemeinen nach Ablauf einiger Tage schon eine beträchtliche Umbesetzung der Störstellen beobachtet werden, die durch Messung der Leitfähigkeit in einfacher Weise verfolgt werden kann. Unmittelbar nach der Dotierung ist die Konzentration der Eisen-Bor-Komplexe klein, und die Kompensation der Löcherkonzentration $p_0 = N_A^-$ des Ausgangskristalls wird im wesentlichen durch das Eisenniveau bei 0,4 eV bewirkt. Wenn $N_D \gg N_A$, liegt das Ferminiveau in der Nähe dieses Donatorniveaus. Mit zunehmender Paarbildung wird die Kompensation stärker, und das Ferminiveau wird zur Bandmitte hin verschoben.

Im folgenden werden experimentelle Ergebnisse für zwei Kristallproben mit unterschiedlichen Ferminiveaulagen angeführt. Beide Kristalle stammen aus derselben Dotierungscharge. Das Meßprogramm wurde so durchgeführt, daß während einer Meßreihe keine wesentliche Änderung des Ferminiveaus auftrat. Bei längeren Pausen mußten hierfür die Proben mit kleiner Konzentration an Eisen-Bor-Komplexen gekühlt werden.

Fig. 3 zeigt die Impulsstromkennlinie und stationäre Kennlinien der Probe 1 mit einem Ferminiveau bei 0,49 eV ($T = 300$ °K). Die Impulsstrommeßwerte liegen weitgehend auf der eingezeichneten Geraden mit der Steigung 2, die dem theoretischen, aus der Probengeometrie berechneten Injektionsstrom entspricht.[5]) Die stationären Kennlinien zeigen erst oberhalb 70 V ein von einer ohmschen Charakteristik abweichendes Verhalten. Die Temperaturabhängigkeit des ohmschen Stroms ist durch das Donatorniveau bei 0,4 eV bestimmt.

Fig. 4 zeigt dagegen entsprechende Kennlinien einer bei Zimmertemperatur eigenleitenden Probe. Die stationären Kennlinien enthalten bei Spannungen von

[4]) Für $H_0^+(T) \ll H$ ist die Haftstellenkonzentration auch bei höheren Temperaturen von T unabhängig.

[5]) Für die Driftbeweglichkeit wurden dabei Meßergebnisse von Zerbst und Heywang [8] verwendet.

0,1 bis 1 V einen sublinearen Ast, der durch die Extraktion der Minoritätsträger verursacht wird. Bei tieferen Temperaturen macht sich die Elektronenextraktion weniger bemerkbar, da das Ferminiveau in Richtung zum Valenzband hin verschoben wird. Im Bereich von 1 bis 10 V sind die Kennlinien wieder annähernd linear, wobei jedoch die zugehörige Trägerkonzentration niedriger liegt. Oberhalb von 10 V gehen die Kennlinien in einen quadratischen Ast über. Die durch Rechtecke gekennzeichneten Meßpunkte sind stationäre Impulsstromwerte. Die Gleichstrommeßwerte für beide Polungen wurden durch Kreise bzw. Punkte markiert.

Bei beiden Kristallproben stimmen die Impulsströme weitgehend mit den theoretischen Injektionsströmen überein. Die Kristalle unterscheiden sich im Injektionsgebiet im wesentlichen durch ihre stationären Kennlinien. Für die mittlere injizierte Trägerdichte folgt aus den Kennlinien

$$\text{Probe 1:}\quad \overline{\Delta p} = \Theta\,\overline{\Delta H^+} \ll p_0 = \Theta\,H_0^+ \to \overline{\Delta H^+} \ll H_0^+\,,$$

$$\text{Probe 2:}\quad \overline{\Delta p} = \Theta\,\overline{\Delta H^+} \gg p_0 = \Theta\,H_0^+ \to \overline{\Delta H^+} \gg H_0^+$$

($\overline{\Delta p}$ mittlere Konzentration freier injizierter Löcher, $\overline{\Delta H^+}$ mittlere Konzentration haftender Löcher, $\Theta = \dfrac{r_e}{r_c\,(H - H_0^+)}$).[6] Je nach dem Größenverhältnis von

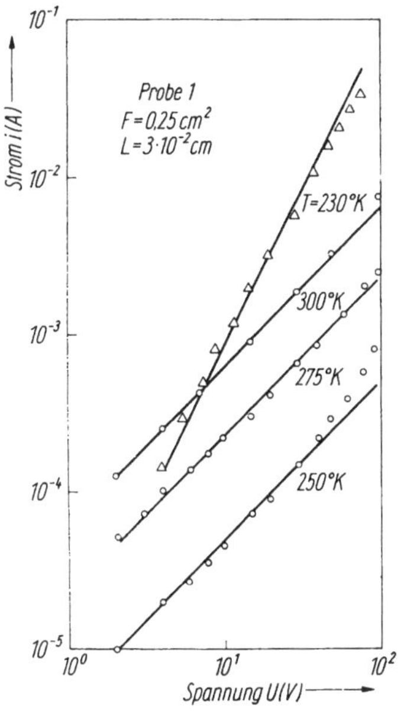

Fig. 3. Impulsstromkennlinie und stationäre Kennlinien für einen Kristall mit starker Besetzung des Haftstellenterms im thermischen Gleichgewicht. △———△ Impulsstrom, ○———○ stationärer Strom (Gleichstrommessung)

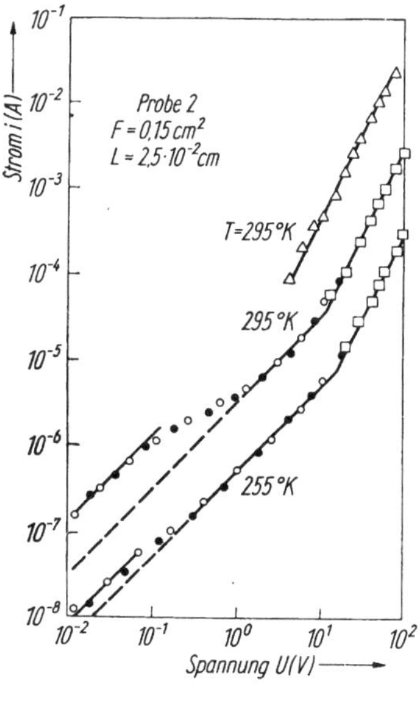

Fig. 4. Impulsstromkennlinie und stationäre Kennlinien für einen Kristall mit schwacher Besetzung des Haftstellenterms im thermischen Gleichgewicht. △———△ Impulsstrom, ○———○ stationärer Strom (Gleichstrommessung), ●———● umgekehrte Polung, □———□ stationärer Impulsstrom

[6]) Die Beziehungen gelten für $\Theta \ll 1$. Für die injizierten Ladungsträger besteht bei Stationarität dieselbe Gleichgewichtsbedingung wie im thermischen Gleichgewicht.

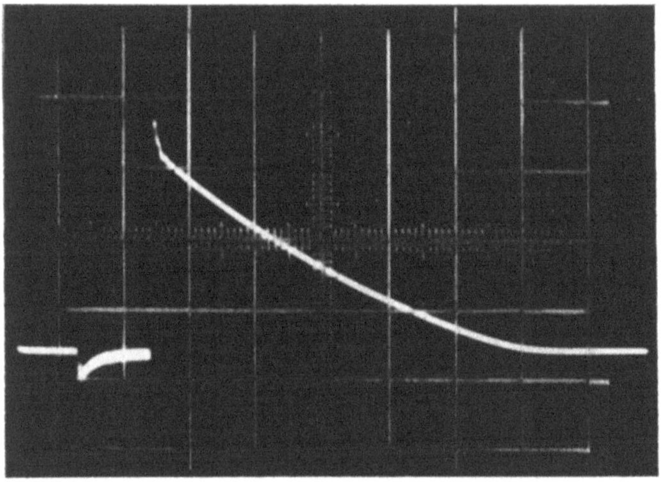

Fig. 5. Aufladespannung am Kristall nach einer vorangehenden Injektion. Als Arbeitswiderstand wurde im Außenkreis eine Schaltdiode verwendet (10 μs/cm)

$\overline{\Delta H^+}/H_0^+$ hat man bei beiden Proben einen unterschiedlichen zeitlichen Abbau der haftenden Raumladung zu erwarten [2].

Die Zeitabhängigkeit der haftenden Raumladung kann durch Messung der Spannung an den Kontakten nach der Injektion oder durch Strommessung mit einem Doppelimpuls untersucht werden. Bei der ersten Methode muß die Zeitkonstante des Außenkreises hinreichend groß gegenüber der Entleerungszeit der Haftstellen sein. Unter dieser Voraussetzung kann die während der Injektion auf den Kathodenkontakt gebrachte negative Ladung nicht über den Außenkreis abfließen, so daß die Spannung von der zeitlichen Änderung der haftenden Löcherkonzentration abhängt [2]. Bei hinreichend starkem Raumladungsfeld driften die aus den Zentren emittierten Löcher ohne wesentliches Retrapping zum Kathodenkontakt und bewirken einen Abbau der Spannung. Fig. 5 zeigt die zeitliche Änderung der Spannung, die an einer als Arbeitswiderstand verwendeten Schaltdiode gemessen wurde. Während der Injektion (negativer Impuls) liegt an der Diode die Flußspannung von 1 bis 2 V, während an der Rück-

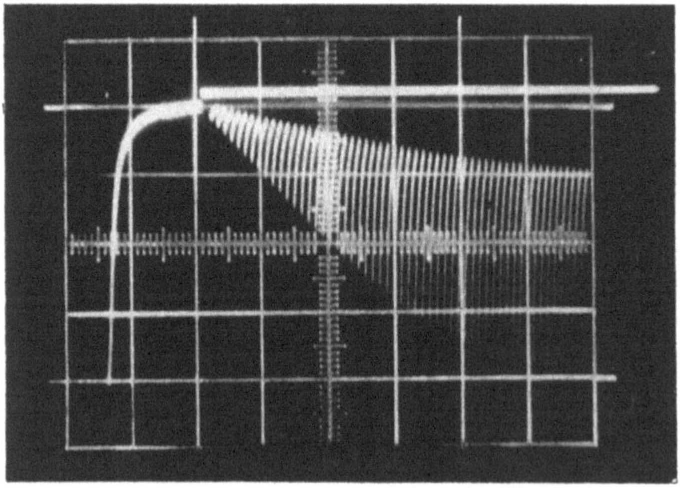

Fig. 6. Injektionsstrom in Abhängigkeit von der Haftstellenentleerung bei schwacher Umbesetzung der Zentren. $\Delta H^+ \ll H_0^+$ (200 μs/cm)

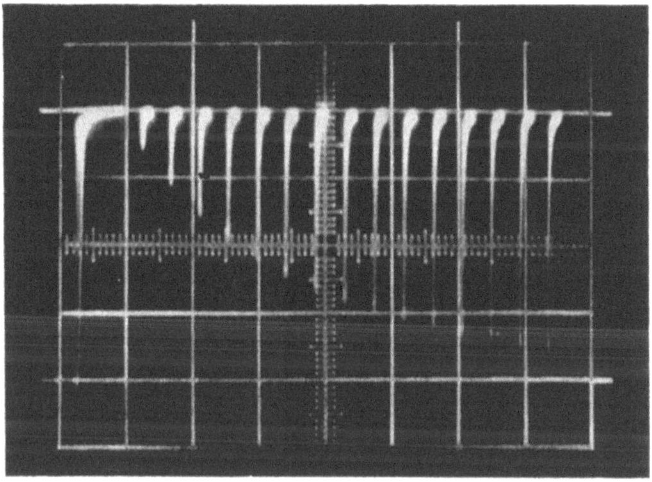

Fig. 7. Injektionsstrom in Abhängigkeit von der Haftstellenentleerung bei starker Umbesetzung der Zentren. $\Delta H^+ \gg H_0^+$ (200 µs/cm)

flanke des Injektionsimpulses etwa die volle Injektionsspannung gemessen wird. Enthält die Probe einen wesentlichen Teil freier Ladungsträger, so wird ein Teil der Spannung relativ schnell abgebaut, während der übrige Teil mit der zeitlichen Änderung der haftenden Löcherkonzentration abklingt. Zur Messung großer Entleerungszeiten $\tau_e \gtrsim 1$ ms muß jedoch der Sperrwiderstand der Diode bei Kristallkapazitäten von 10 pF über $10^8\ \Omega$ liegen.

Meßtechnisch einfacher ist dagegen, die Haftstellenentleerung durch eine Strommessung zu untersuchen. Bei einem Außenwiderstand von 100 Ω verschwindet innerhalb von etwa 10^{-9} s die Potentialdifferenz an den Kontakten, und die aus den Haftstellen emittierten Ladungsträger werden je nach ihrer Lage im Kristallvolumen zum Anoden- bzw. Kathodenkontakt abgesaugt [2].

Fig. 6 und 7 zeigen die mit einem Doppelimpuls gemessenen zeitlichen Stromänderungen für zwei Proben mit unterschiedlichem Zeitverhalten. Bei der Aufnahme der Oszillogramme wurde die Verzögerungszeit mit einer Sägezahnspannung moduliert, so daß die untere Hüllkurve, die durch den verzögerten Impuls beschrieben wird, direkt den interessierenden Zeitvorgang wiedergibt.[7]) Bei den Auswertungen wurde angenommen, daß die Stromdifferenz $\Delta i(t) = i_0 - i(t)$ proportional zur haftenden Löcherkonzentration ist [2] (i_0 Strom an der Vorderflanke des Injektionsimpulses, $i(t)$ Strom an der Vorderflanke des verzögerten (Abtast-)Impulses, der Zeitnullpunkt liegt an der Rückflanke des Injektionsimpulses).

In [2] wurde gezeigt, daß bei Vorhandensein eines Raumladungsfeldes die Konzentration der freien Ladungsträger unter ihren thermischen Gleichgewichtswert abgesenkt wird. Dieser Prozeß setzt nach Abschalten des Injektionsimpulses ein und verläuft im allgemeinen schnell gegenüber der Haftstellenentleerung. Für die maßgebende Zeitkonstante erhält man z. B. bei $T = 300$ °K unter Verwendung der Zahlenwerte $\overline{\Delta H^+} \approx 10^{12}\ \mathrm{cm}^{-3}$, $\mu \approx 400\ \mathrm{cm}^2/\mathrm{Vs}$, $\varepsilon = 11{,}6$ den Wert $\left(\frac{e\,\mu}{\varepsilon\,\varepsilon_0}\,\overline{\Delta H^+}\right)^{-1} \approx 10^{-8}$ s, der unterhalb der von uns verwendeten Zeitauflösung liegt.

[7]) Die obere Hüllkurve wird durch die Länge des bei der Aufnahme verwendeten Helltastimpulses verursacht.

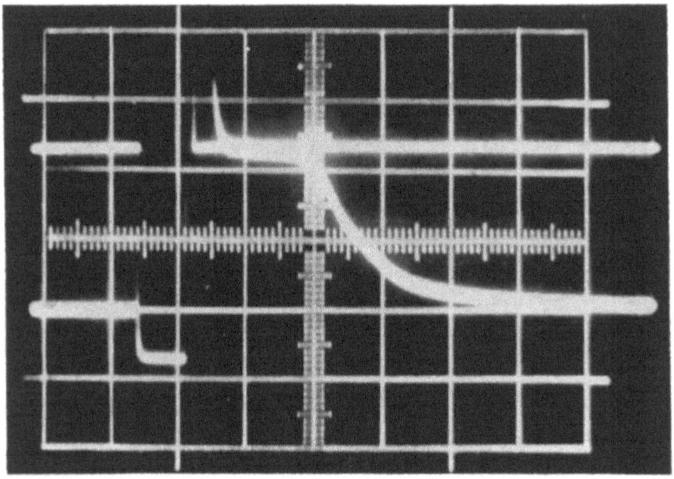

Fig. 8. Löcherextraktion nach einer vorangehenden Injektion. Die Nullinien für Strom und Spannung liegen in der oberen Schirmbildhälfte übereinander (20 μs/cm)

Diese Löcherextraktion kann durch eine zusätzlich an die Probe gelegte Gleichspannung nachgewiesen werden.[8]) Bei dem in Fig. 8 gezeigten Oszillogramm wurde gleichzeitig der Spannungsimpuls und der durch die Probe fließende Strom aufgenommen. Die Null-Linien für Strom und Spannung liegen in der oberen Hälfte des Schirmbildes übereinander. Vor dem Einsatz des Injektionsimpulses fließt durch den Kristall ein zur thermischen Ladungsträgerkonzentration proportionaler Gleichstrom, der nach Abschalten der Injektionsspannung für eine Zeit t_S weitgehend gesperrt wird (die Lage des Stromimpulses ist durch den auf dem Oszillogramm sichtbaren Spannungsimpuls gekennzeichnet). Der Strom umgekehrter Polarität hinter der Rückflanke des Injektionsimpulses wird durch die Umladung im Außenkreis verursacht, bei der die während der Injektion auf den Kathodenkontakt aufgebrachte negative Ladung teilweise zum Anodenkontakt zurückfließt. Die haftende Raumladung treibt die freien Löcher aus dem Kristall heraus, wodurch der Gleichstrom bis zu dem Zeitpunkt unterbrochen wird, bei dem die mittlere Raumladung so weit abgebaut worden ist, daß sich im Kristall ein quasineutraler Zustand einstellen kann. Der Wiederanstieg der Löcherkonzentration setzt für $t \gtrsim t_S$ relativ steil ein, wobei t_S durch die Beziehung $\overline{\Delta H^+}(t_S) \approx p_0$ bestimmt ist. Solange das Raumladungsfeld im Kristall hinreichend groß ist, ändert sich die Löcherkonzentration nur langsam im Vergleich zur haftenden Raumladung. Diese Quasistationarität der freien Ladungsträgerkonzentration hat zur Folge [2], daß die Zeitabhängigkeit der haftenden Raumladung durch eine einfache Beziehung näherungsweise beschrieben werden kann:

$$\ln \frac{\Delta H^+(t) + H_0^+}{\Delta H^+(0) + H_0^+} = - r_e \, t \,, \tag{1}$$

woraus die beiden Grenzfälle

$$\Delta H^+ = \Delta H^+(0)\, e^{-r_e t} \quad \text{für} \quad \Delta H^+(t) \gg H_0^+ \,, \tag{1a}$$

$$\Delta H^+ = \Delta H^+(0) - r_e H_0^+ t \quad \text{für} \quad \Delta H^+(t) \ll H_0^+ \tag{1b}$$

folgen.

[8]) Die Spannung wird so gewählt, daß der entsprechende stationäre Strom im ohmschen Ast der Kennlinie liegt.

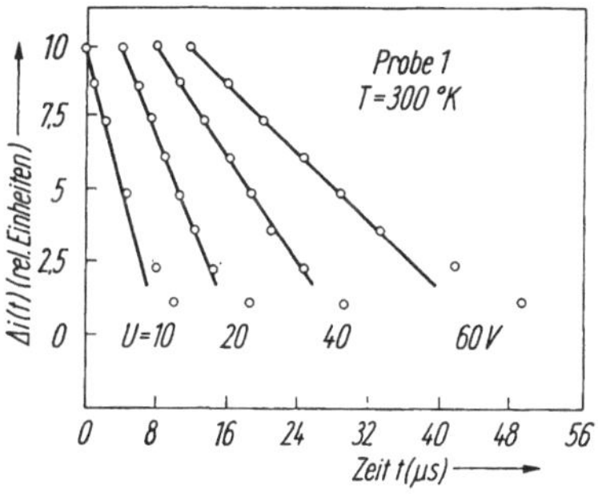

Fig. 9. Lineare Darstellung der Stromänderung nach einer vorangehenden Injektion für die Probe 1 bei verschiedenen Spannungen

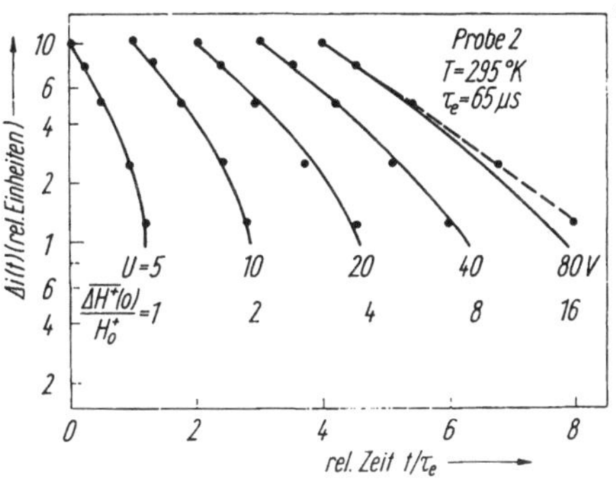

Fig. 10. Logarithmische Darstellung der Stromänderung nach einer vorangehenden Injektion für die Probe 2 bei verschiedenen Spannungen. Die ausgezogenen Kurven wurden aus dem Besetzungsverhältnis $\overline{\Delta H^+}(0)/H_0^+$ berechnet (Beziehung (1))

Beide Zeitgesetze können aus den Oszillogrammen in den Fig. 6 und 7 abgelesen werden. Bei der Probe 1 (Fig. 6) ändert sich der Strom über 70% der Gesamtamplitude weitgehend linear mit der Zeit, während bei der Probe 2 (Fig. 7) eine exponentielle Änderung vorliegt.

Kennzeichnet man die Entleerungszeit der Haftstellen durch die Halbwertszeit $t_{1/2}$, so erhält man für $\Delta H^+ \ll H_0^+$ zunächst eine lineare Zunahme mit der injizierten Trägerdichte gemäß

$$t_{1/2} = \frac{\overline{\Delta H^+}(0)}{2\, H_0^+}\, r_e^{-1}\,, \tag{2}$$

die für $\Delta H^+ \gg H_0^+$ gegen den von der Injektionsspannung unabhängigen Grenzwert $\ln 2/r_e$ übergeht.

Die zeitliche Änderung des Stroms ist für die Probe 1 in Fig. 9 linear über der Zeit und für die Probe 2 in Fig. 10 logarithmisch über der relativen Zeit $r_e\, t$ dargestellt. Während die Meßwerte für die Probe 2 bei hohen Spannungen durch eine Exponentialfunktion mit der Zeitkonstanten r_e^{-1} (gestrichelte Gerade) beschrieben werden können, sind die entsprechenden Stromänderungen der Probe 1 weitgehend linear, wobei die Tangenten von der Spannung abhängen. In beiden Darstellungen liegt der Zeitnullpunkt bei dem ersten Meßpunkt jeder Kurve.

Da bei der Probe 1 die haftende Raumladung zeitlinear abnimmt, muß auch die Sperrzeit t_S linear von der injizierten Trägerdichte bzw. Spannung abhängen. Aus $\overline{\Delta H^+}(t_S) \approx p_0$ folgt

$$t_S \approx \frac{\overline{\Delta H^+}(0) - p_0}{r_e\, H_0^+} = 2\, t_{1/2} - \tau_c\,. \tag{3}$$

In Fig. 11 sind die Zeiten t_S und $t_{1/2}$ über der Spannung bzw. über der mittleren inji-

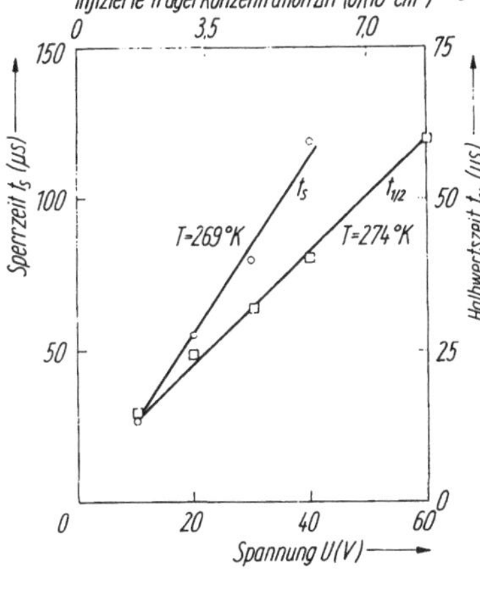

Fig. 11. Halbwertszeit und Sperrzeit der Probe 1 in Abhängigkeit von der mittleren injizierten Raumladungskonzentration

zierten Löcherkonzentration aufgetragen. Hierbei wurde angenommen, daß zwischen der Spannung und der injizierten Löcherkonzentration der gleiche Zusammenhang besteht wie für den haftstellenfreien Isolator [9].

4. Bestimmung des Wirkungsquerschnitts

Eine direkte Bestimmung des Wirkungsquerschnitts Θ_p ohne Kenntnis der Konzentration H_0^+ kann zunächst nur für die Probe 2 durchgeführt werden. Aus der Messung von τ_e bei zwei unterschiedlichen Temperaturen können aus der Beziehung

$$r_e = r_c K_{DV}, \tag{4}$$

wobei $K_{DV} = g N_V \exp\left(-\frac{E_{DV}}{k T}\right)$ und $r_c = v_{th} \Theta_p$, der Einfangkoeffizient r_c und die Dissoziationsenergie E_{DV} bestimmt werden (K_{DV} Massenwirkungskonstante, g statistisches Gewicht des Donatorniveaus, N_V effektive Zustandsdichte des Valenzbandes, E_{DV} Dissoziationsenergie, v_{th} mittlere thermische Geschwindigkeit).

Aus der Auftragung der Meßwerte τ_e über $1/T$ in Fig. 12 (durch Kreise markiert) erhält man für die Aktivierungsenergie des Donatorniveaus $E_{DV} = 0{,}4$ eV in Übereinstimmung mit den Ergebnissen von Collins und Carlson [6]. Die Aktivierungsenergie kann auch aus dem Temperaturgang der Leitfähigkeit der Proben 1 und 2 bestimmt werden. Da die Probe 2 bei Zimmertemperatur eigenleitend ist, mußte hierzu die Leitfähigkeit bei tieferen Temperaturen gemessen werden, bei denen der Einfluß der Minoritätsträger vernachlässigbar ist.

Unter Verwendung des Zahlenwertes $m_{eff} = 0{,}59\, m_0$ [10] für die effektive Löchermasse erhält man für die Massenwirkungskonstante bei Zimmertemperatur $K_{DV}/g = 2{,}3 \times 10^{12}$ cm^{-3}, woraus für den Einfangkoeffizienten und den Wirkungsquerschnitt die Zahlenwerte

$$g\, r_c = 1{,}1 \times 10^{-8}\ \text{cm}^3\,\text{s}^{-1}$$

und

$$g\, \Theta_p = 0{,}7 \times 10^{-15}\ \text{cm}^2$$

folgen.

Bei der Probe 1 können die Emissionszeiten und damit der Wirkungsquerschnitt entsprechend der Beziehung (2) nur bei Kenntnis von $\Delta H^+(0)/H_0^+$ aus den Halbwertszeiten berechnet werden.[9])

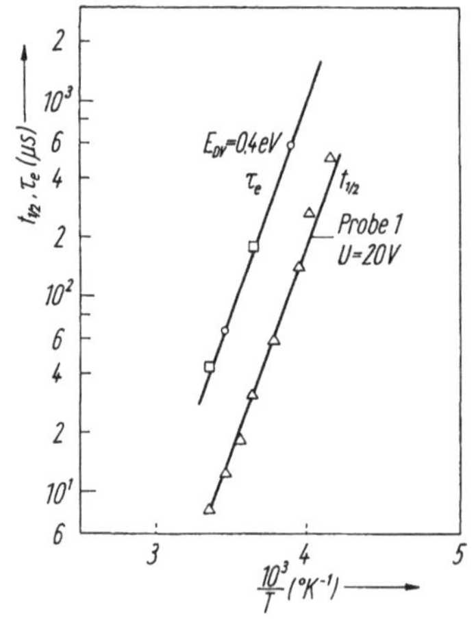

Fig. 12. Entleerungszeiten τ_e für die Proben 1 und 2 in Abhängigkeit von der Temperatur. ○——○ gemessene τ_e-Werte für die Probe 2, □——□ berechnete τ_e-Werte für die Probe 1, △——△ Halbwertszeiten der Probe 1 bei $U = 20$ V

[9]) Prinzipiell wäre auch bei der Probe 1 die Messung der Entleerungszeiten τ_e durch Erhöhung der injizierten Ladungsträgerkonzentration möglich, so daß entsprechend der Beziehung (1a) die Halbwertszeit von H_0^+ unabhängig wird. Bei höheren Feldstärken oberhalb von etwa 5×10^3 Vcm^{-1} wurde jedoch vielfach der Einsatz einer Doppelinjektion beobachtet, die wahrscheinlich durch Inhomogenitäten im Kathodenkontakt verursacht wird. Dieser Effekt führt zu einer Erhöhung der stationären Stromwerte und zu einer Verkleinerung der Halbwertszeiten.

Die Übereinstimmung der für die Probe 1 erhaltenen Ergebnisse mit denen der Probe 2 kann jedoch auf folgendem Wege überprüft werden: Aus den Einfangzeiten τ_c (siehe Fig. 2) kann mit dem oben angegebenen Einfangkoeffizienten für beide Kristalle die Haftstellenkonzentration $H - H_0^+$ berechnet werden. Wegen $\tau_c^{-1} = r_c\,(H - H_0^+)$ erhält man

$$\text{Probe 1:}\quad \tau_c = 2{,}3 \times 10^{-6}\,\text{s},\quad \frac{1}{g}\,(H - H_0^+) = 3{,}9 \times 10^{13}\,\text{cm}^{-3},$$

$$\text{Probe 2:}\quad \tau_c = 2{,}5 \times 10^{-6}\,\text{s},\quad \frac{1}{g}\,(H - H_0^+) = 3{,}6 \times 10^{13}\,\text{cm}^{-3}.$$

Da bei beiden Proben das Ferminiveau hinreichend weit vom Haftstellenniveau entfernt liegt, gilt $H - H_0^+ \approx H$. Unter Verwendung der Fermiverteilung kann der Besetzungsgrad H_0^+ im thermischen Gleichgewicht für beide Kristalle berechnet werden:

$$H_0^+ = H\,(1 - f) \approx (H - H_0^+)\,g^{-1} \exp\left(\frac{E_{DV} - E_F}{k\,T}\right) = \begin{cases} 1{,}2 \times 10^{12}\,\text{cm}^{-3} \\ 1{,}1 \times 10^{11}\,\text{cm}^{-3} \end{cases}$$

(Probe 1: $E_F = 0{,}49$ eV, Probe 2: $E_F = 0{,}55$ eV; $T = 300$ °K).

Da für beide Kristalle $p_0 \ll H_0^+$ gilt, kann sich H_0^+ unterhalb von Zimmertemperatur nicht mehr wesentlich ändern.

Zur Berechnung der Emissionszeiten aus den Halbwertszeiten muß weiterhin die Abhängigkeit der injizierten Löcherkonzentration von der Spannung bekannt sein. Für die mittlere injizierte Ladungsträgerkonzentration gilt bei einem haftstellenfreien Isolator [9]

$$\overline{\Delta p} = 1{,}5\,\frac{\varepsilon\,\varepsilon_0}{e\,L^2}\,U\,. \tag{5}$$

Mit den Zahlenwerten für $\overline{\Delta H^+}(0) \approx \overline{\Delta p}$ $(\Theta \ll 1)$ und H_0^+ wurde die Zeitabhängigkeit der Haftstellenentleerung entsprechend der Näherungsbeziehung (1) berechnet und mit den Meßwerten verglichen (eingezeichnete Kurven in Fig. 10).

Zur Berechnung der Emissionszeiten für die Probe 1 kann aus der Darstellung der Halbwertszeiten über der mittleren injizierten Ladungsträgerkonzentration (siehe Fig. 11) der Emissionsstrom $r_e\,H_0^+$ entsprechend der Beziehung (2) bestimmt werden. Aus dem Emissionsstrom erhält man mit dem oben angegebenen Zahlenwert für H_0^+ die Emissionszeiten, die in der Fig. 12 durch Quadrate gekennzeichnet wurden. Während die Halbwertszeiten bei einer Injektionsspannung von 20 V wesentlich unter den Emissionszeiten liegen, stimmen die für zwei Temperaturpunkte berechneten τ_e-Werte mit den Meßergebnissen (Probe 2) überein.

Da die Halbwertszeiten entsprechend der Beziehung (2) proportional zu τ_e sind, kann aus ihrer Abhängigkeit von der Temperatur die Aktivierungsenergie des Haftterms bestimmt werden, sofern die injizierte Trägerkonzentration während einer Meßreihe konstant gehalten wird. Wenn der Besetzungsgrad des Haftterms groß gegenüber der injizierten Ladungsträgerkonzentration ist, muß zur Bestimmung des Wirkungsquerschnitts aus den Halbwertszeiten der Faktor $2\,H_0^+/\overline{\Delta H^+}(0)$ berücksichtigt werden. Hiermit kann eine bei früheren Untersuchungen [1] aufgetretene Diskrepanz erklärt werden. Die in der Arbeit angegebenen Wirkungsquerschnitte sind zu groß, da die Zahlenwerte ohne Berücksichtigung dieses Faktors direkt aus den Halbwertszeiten berechnet wurden.

Die Bestimmung von H_0^+ ist im allgemeinen einfach, wenn die Dotierungssubstanz nur einen Donatorterm bildet. In diesem Fall ist der Besetzungsgrad der Donatoren bei tieferen Temperaturen im wesentlichen durch die Akzeptorenkonzentration des Ausgangskristalls bestimmt. Zur Überprüfung dieses Sachverhalts werden zur Zeit entsprechende Untersuchungen an Au-dotiertem Silizium durchgeführt.

Herrn Dr. G. O. Müller und Herrn Dr. E. Schnürer möchte ich für zahlreiche Diskussionen und Anregungen danken.

Literatur

[1] H. LEMKE, phys. stat. sol. **3**, 2083 (1963).
[2] H. LEMKE, phys. stat. sol. **16**, 413 (1966).
[3] A. MANY und G. RAKAVY, Phys. Rev. **126**, 1980 (1962).
[4] M. A. LAMPERT, Rep. Progr. Phys. **27**, 329 (1964).
[5] B. L. GREGORY und A. G. JORDAN, Phys. Rev. **134**, 1378 (1964).
[6] C. B. COLLINS und R. O. CARLSON, Phys. Rev. **108**, 1409 (1957).
[7] W. H. SHEPHERD und J. A. TURNER, J. Phys. Chem. Solids **23**, 1967 (1962).
[8] M. ZERBST und W. HEYWANG, Z. Naturf. **11a**, 608 (1956).
[9] F. STÖCKMANN, Halbleiterprobleme (Vieweg, Braunschweig) **6**, 279 (1961).
[10] J. BLAKEMORE, Semiconductor Statistics, Pergamon Press 1962.

(Received April 7, 1966)

phys. stat. sol. 16, 439 (1966)

Istituto Sperimentale dei Metalli Leggeri, Novara (a) and SORIN, Saluggia (b)

Determination of Diffusion Coefficients in Metals by a Resistometric Method — Application to the Diffusion of Zn in Al

By

S. CERESARA (a), T. FEDERIGHI (a), and F. PIERAGOSTINI (b)

A method is described for determining the diffusion coefficient in dilute solid solutions by measurements of the electrical resistivity. The diffusion equation is solved numerically for the selected experimental conditions and this provides curves suitable for direct comparison with experimental data. Results obtained for the diffusion parameters of Zn in pure Al indicate the accuracy of the method.

Es wird eine Methode zur Bestimmung des Diffusionskoeffizienten durch Messungen des elektrischen Widerstands in verdünnten festen Lösungen beschrieben. Die Diffusionsgleichung wird numerisch für ausgewählte experimentelle Bedingungen gelöst. Die so erhaltenen Kurven sind für einen direkten Vergleich mit experimentellen Daten geeignet. Die Ergebnisse für die Diffusionsparameter von Zn in reinem Al zeigen die Genauigkeit der Methode.

1. Introduction

Among the several experimental techniques employed for studying diffusion in metals (see, e.g., [1] and the references quoted there), the method involving electrical resistivity measurements is one of the oldest. As early as 1911, Bruni and Meneghini [2] were able to correlate electrical resistance changes observed upon annealing of two-layer wires (obtained by electrodeposition) with the formation of metallic solid solutions by diffusion. Later on, the resistometric method was employed by a number of experimenters [3 to 6] to investigate the rate of homogeneization of various metal systems.

In recent years, however, due to the development of direct methods of measurement, first of all radioactive tracer technique, the resistometric method has been confined only to particular diffusion problems, such as diffusion in semiconductors [7] and diffusion of point defects in quenched, cold-worked, or irradiated metals (see e.g. [8]).

The purpose of this paper is to show that the resistometric method, without requiring the sectioning of the specimen, can be a prominent tool in a number of diffusion problems, alternative to other more conventional (and often more complicated) techniques, and is especially useful when radioactive isotopes of the elements of interest are not available.

To show this, a simple geometry has been chosen, namely radial diffusion in a cylinder (i.e. deposited film left to diffuse into a wire), and the related diffusion equation with proper boundary conditions has been solved numerically with the aid of an electronic computer. Then theoretical curves, involving these solutions and suitable for direct comparison with experimental data, have been calculated. Finally, as an application of the method, the diffusion parameters of Zn in Al have been determined.

2. Solution of the Diffusion Equation

The diffusion equation for a long circular cylinder, in which diffusion is radial everywhere, is [9]

$$\frac{\partial c}{\partial t} = \frac{1}{r}\frac{\partial}{\partial r}\left(r\, D\frac{\partial c}{\partial r}\right), \tag{1}$$

where c is the concentration of the diffusing element, which is a function of the radius r and the time t, and D is the diffusion coefficient.

The initial and boundary conditions for the actual problem, where a film of the diffusing substance is deposited on the surface of the cylinder of radius a (the wire sample) at time $t = 0$ and left to diffuse inside the cylinder, are

$$c = \infty\,, \qquad r = a\,, \qquad t = 0\,, \tag{2}$$

$$c = 0\,, \qquad 0 \leqq r < a\,, \qquad t = 0\,, \tag{3}$$

$$c \to c_{\text{f}} \text{ (final concentration)}\,, \qquad t \to \infty\,, \tag{4}$$

$$c \text{ is finite}\,, \qquad 0 \leqq r \leqq a\,, \qquad 0 < t < \infty\,, \tag{5}$$

$$\frac{\partial c}{\partial r} = 0\,, \qquad r = a\,, \qquad t > 0\,. \tag{6}$$

Condition (6) states that no flux of atoms escapes from the surface.

If D is constant, as can be reasonably assumed in the case of diffusion in dilute alloys, the solution of (1) satisfying the conditions (2) to (6) is

$$c(x, \tau) = c_{\text{f}}\left[1 + \sum_{n=1}^{\infty} \frac{J_0(\alpha_n x)}{J_0(\alpha_n)} \exp\left(-\alpha_n^2 \tau\right)\right], \tag{7}$$

where J_0 ordinary Bessel function of the first kind and zero order, α_n positive zeros of the Bessel function of order one, $J_1(z)$, and

$$x = \frac{r}{a}, \tag{8}$$

$$\tau = \frac{D\,t}{a^2}. \tag{9}$$

One can rewrite (7):

$$y(x, \tau) = \frac{c\,(x, \tau)}{c_{\text{f}}} = 1 + \sum_{n=1}^{\infty} \frac{J_0(\alpha_n x)}{J_0(\alpha_n)} \exp\left(-\alpha_n^2 \tau\right),$$

$$J_1(\alpha_n) = 0\,. \tag{10}$$

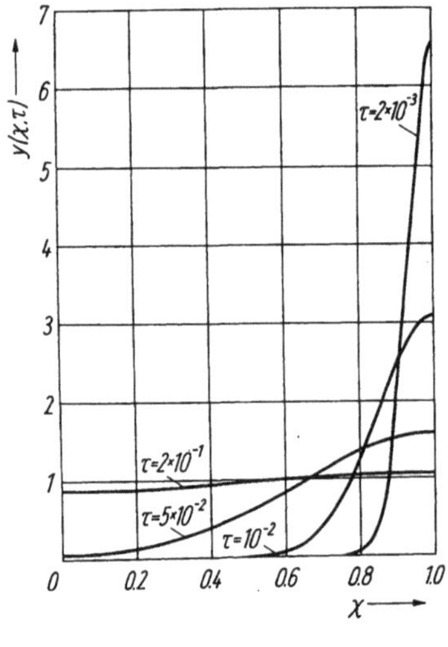

Fig. 1. Concentration distributions for various τ

Equation (10) allows the calculation of the theoretical penetration curve $y(x, \tau)$ for different values of the parameter τ, whatever be the radius of the specimen and the amount of deposited substance.

Unfortunately, the solution (7) is scarcely useful from the numerical standpoint, since it contains a series which converges very slowly for small values of τ, say $\tau < 10^{-2}$. Nevertheless in this case approximate solutions of the problem formulated in (1) to (6), which are suitable for numerical evaluation by means of an electronic computer, can be obtained as described in the Appendix. A plot of some penetration curves is given in Fig. 1.

3. Calculation of the Function $\varphi(\tau)$

To check the penetration curve $y(x, \tau)$ with resistivity data, it is necessary to examine the variation of the resistivity of the wire sample with the time of diffusion.

For this purpose, let us regard the wire as consisting of a series of coaxial cylinders having the cross section $2\pi r\,dr$, and the resistivity

$$\varrho(r) = \varrho_0 + 100\,\alpha\,c(r)\,, \tag{11}$$

where ϱ_0 resistivity of the pure basic metal, α increase in resistivity due to 1% of diffusing substance. The approximation (11) is obviously valid only for dilute solutions: here again we find the same restriction already required for the constancy of D.

The total resistance of the wire is (case of parallel conductors)

$$\frac{1}{R} = 2\pi \int_0^a \frac{r\,dr}{\varrho_0 + 100\,\alpha\,c(r)} = \frac{2\pi a^2}{\varrho_0} I(\tau)\,, \tag{12}$$

where

$$I(\tau) = \int_0^1 \frac{x\,dx}{1 + \gamma\,y(x, \tau)}\,, \tag{13}$$

$$\gamma = \frac{100\,\alpha\,c_f}{\varrho_0}\,. \tag{14}$$

To compare the theory with experimental data it is useful to introduce the normalized function $\varphi(t)$, defined as follows:

$$\varphi(t) = \frac{R - R_i}{R_f - R_i}\,, \tag{15}$$

where R_i and R_f are the resistance values of the specimen before and at the end of the diffusion process, respectively.

The corresponding theoretical function $\varphi(\tau)$ can be easily calculated from (12), keeping in mind that

$$R_i = \frac{\varrho_0}{\pi a^2}\,, \tag{16}$$

$$R_f = \frac{\varrho_0 + 100\,\alpha\,c_f}{\pi a^2}\,, \tag{17}$$

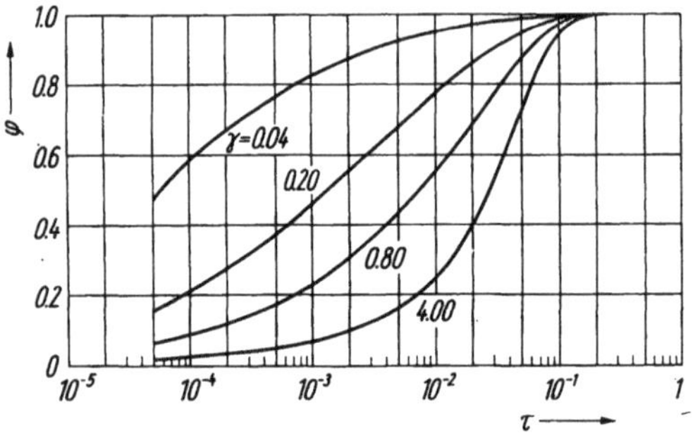

Fig. 2. Theoretical curves $\varphi(\tau)$ for some representative values of γ

so that

$$\varphi(\tau) = \frac{1}{\gamma}\left[\frac{1}{2\,I(\tau)} - 1\right]. \tag{18}$$

Values of $\varphi(\tau)$ as given by (18) are tabulated in Table 1 for several values of γ, chosen in a range of practical interest. The view of the theoretical curves $\varphi(\tau)$ is given in Fig. 2 for some representative values of γ.

From the practical standpoint, for a given experimental curve $\varphi(t)$ (equation (15)) at temperature T, one has an experimental value of γ, determined from the ratio $(R_f - R_i)/R_i$; the corresponding theoretical curve $\varphi(\tau)$ can be easily drawn by interpolation of the data in Table 1. Then, a comparison of theoretical and experimental curves allows to deduce, from (9), the proper value of $\tau/t = D/a^2$, at the temperature T.

4. Diffusion of Zn in Al

As an example for the resistometric method, the diffusion coefficient of Zn in Al has been determined at several temperatures.

4.1 Experimental part

Wire samples of 99.995% purity Al, 1 mm in diameter and 500 mm in length, were wound like a coil and welded to four terminals (two at each end) to carry out electrical resistance measurements in the usual potentiometric arrangement.

Samples, annealed for 8 hours at 600 °C and furnace-cooled, were coated with zinc by chemical plating followed by electrolytic deposition; the deposited layer thickness was about 1 μm.

The annealing treatments were performed in a furnace whose temperature was controlled with an accuracy of ±1 °C. To avoid the presence of quench-induced defects, the specimens were cooled in air from the diffusion temperature.

The resistance measurements after each treatment were carried out at liquid nitrogen temperature, employing a Honeywell Rubicon potentiometer; a dummy sample was used to correct for the variation of the temperature of the nitrogen bath.

4.2 Results

After the determination of the inital value R_i, five diffusion specimens were isothermally annealed at temperatures chosen in a wide range, as shown in Table 2. After sufficiently long annealing time, each sample was aged at 600 °C

Table 1

Values of the function $\varphi(\tau)$ for several values of γ

γ	$\tau=5\times10^{-1}$	$\tau=2\times10^{-1}$	$\tau=10^{-1}$	$\tau=5\times10^{-2}$	$\tau=2\times10^{-2}$	$\tau=10^{-2}$	$\tau=5\times10^{-3}$	$\tau=2\times10^{-3}$	$\tau=10^{-3}$	$\tau=5\times10^{-4}$	$\tau=2\times10^{-4}$	$\tau=10^{-4}$	$\tau=5\times10^{-5}$
0.04	0.9999	0.9999	0.9982	0.9907	0.9740	0.9522	0.9250	0.8745	0.8247	0.7636	0.6697	0.5837	0.4748
0.06	0.9999	0.9999	0.9971	0.9864	0.9610	0.9308	0.8919	0.8245	0.7608	0.6872	0.5805	0.4938	0.3811
0.09	0.9999	0.9998	0.9955	0.9801	0.9427	0.9008	0.8481	0.7612	0.6841	0.6003	0.4871	0.4055	0.3001
0.13	0.9999	0.9997	0.9938	0.9721	0.9203	0.8647	0.7975	0.6927	0.6055	0.5166	0.4040	0.3308	0.2390
0.18	0.9999	0.9996	0.9917	0.9626	0.8944	0.8239	0.7436	0.6249	0.5322	0.4428	0.3353	0.2708	0.1941
0.20	0.9999	0.9995	0.9909	0.9589	0.8846	0.8099	0.7245	0.6018	0.5082	0.4194	0.3145	0.2528	0.1814
0.25	0.9999	0.9994	0.9890	0.9502	0.8617	0.7761	0.6817	0.5522	0.4578	0.3714	0.2731	0.2172	0.1568
0.30	0.9999	0.9993	0.9872	0.9419	0.8406	0.7459	0.6446	0.5111	0.4175	0.3343	0.2421	0.1907	0.1390
0.40	0.9999	0.9992	0.9840	0.9268	0.8029	0.6937	0.5833	0.4469	0.3568	0.2800	0.1986	0.1537	0.1145
0.60	0.9999	0.9989	0.9785	0.9009	0.7410	0.6131	0.4945	0.3610	0.2795	0.2139	0.1481	0.1112	0.0861
0.80	0.9999	0.9987	0.9741	0.8792	0.6917	0.5530	0.4325	0.3053	0.2318	0.1746	0.1192	0.0873	0.0697
1.00	0.9999	0.9985	0.9705	0.8607	0.6511	0.5059	0.3862	0.2659	0.1990	0.1483	0.1003	0.0720	0.0588
1.20	0.9999	0.9984	0.9674	0.8447	0.6169	0.4678	0.3501	0.2364	0.1750	0.1293	0.0869	0.0613	0.0510
1.40	0.9999	0.9983	0.9648	0.8306	0.5875	0.4362	0.3210	0.2133	0.1565	0.1149	0.0768	0.0535	0.0450
1.60	0.9999	0.9982	0.9625	0.8181	0.5619	0.4094	0.2970	0.1947	0.1418	0.1035	0.0689	0.0474	0.0403
1.80	0.9999	0.9981	0.9605	0.8068	0.5393	0.3864	0.2768	0.1794	0.1299	0.0944	0.0625	0.0426	0.0366
2.00	0.9999	0.9980	0.9588	0.7967	0.5192	0.3663	0.2595	0.1665	0.1199	0.0868	0.0573	0.0387	0.0334
2.20	0.9999	0.9980	0.9572	0.7875	0.5011	0.3486	0.2445	0.1555	0.1115	0.0804	0.0529	0.0355	0.0308
2.40	0.9999	0.9979	0.9559	0.7790	0.4847	0.3328	0.2314	0.1460	0.1042	0.0749	0.0491	0.0328	0.0286
2.60	0.9999	0.9979	0.9546	0.7713	0.4697	0.3187	0.2198	0.1377	0.0979	0.0702	0.0459	0.0305	0.0266
2.80	0.9999	0.9978	0.9535	0.7641	0.4560	0.3060	0.2094	0.1303	0.0924	0.0661	0.0431	0.0285	0.0249
3.00	0.9999	0.9978	0.9525	0.7575	0.4434	0.2944	0.2001	0.1238	0.0875	0.0625	0.0406	0.0267	0.0235
3.20	0.9999	0.9977	0.9515	0.7513	0.4317	0.2839	0.1917	0.1179	0.0832	0.0592	0.0384	0.0252	0.0221
3.40	0.9999	0.9977	0.9507	0.7455	0.4209	0.2742	0.1840	0.1126	0.0792	0.0563	0.0365	0.0238	0.0210
3.60	0.9999	0.9977	0.9499	0.7401	0.4108	0.2652	0.1770	0.1079	0.0757	0.0537	0.0347	0.0226	0.0199
3.80	0.9999	0.9977	0.9492	0.7351	0.4013	0.2570	0.1706	0.1035	0.0725	0.0513	0.0331	0.0215	0.0189
4.00	0.9999	0.9976	0.9485	0.7303	0.3925	0.2493	0.1647	0.0995	0.0695	0.0492	0.0317	0.0205	0.0181

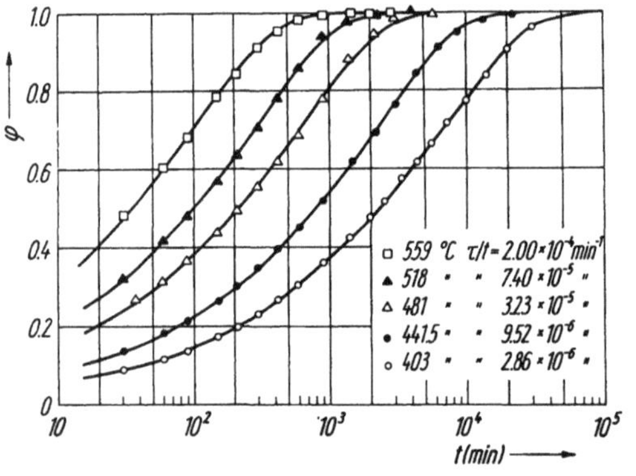

Fig. 3. Diffusion of Zn in Al (99.995%). Experimental points $\varphi(t)$ at the indicated temperatures and theoretical curves $\varphi(\tau)$

Table 2
Summary of experimental data

Sample	Annealing temperature (°C)	γ	τ/t (min^{-1})	$D \times 10^{10}$ (cm^2/s)
1	559	0.7681	2.00×10^{-4}	83.33
2	518	0.8087	7.40×10^{-5}	30.83
3	481	0.7513	3.23×10^{-5}	13.45
4	441.5	0.8266	9.52×10^{-6}	39.66
5	403	0.7358	2.86×10^{-6}	11.91

until its resistance remained constant; this final value was taken as R_f. The experimental values of the function $\varphi(t)$ (formula (15)) are represented by the points in Fig. 3. In the same figure, the theoretical curves $\varphi(\tau)$ calculated for the proper value of γ according to (18) are represented by the full lines: the agreement of the experimental results with the theory is excellent. The values of the ratio τ/t for which the experimental and theoretical φ coincide, are given in Table 2. From these values and from (9) the diffusion coefficients D which are also given in Table 2, have been calculated. Fig. 4 shows that the coefficients D satisfy an Arrhenius-type law

$$D = D_0 \exp(-Q/k\,T)\,, \qquad (19)$$

where D_0 is (0.9 ± 0.3) cm^2/s, and Q, the activation energy, is (1.32 ± 0.02) eV. Both values are in close agreement with the values $D_0 = (1.1 \pm 0.4)$ cm^2/s, $Q = (1.34 \pm 0.03)$ eV, observed for the diffusion of Zn in Al by the standard radioisotope technique [10].

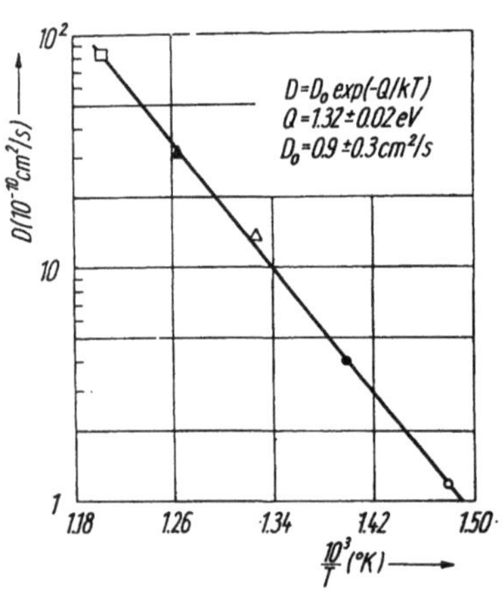

Fig. 4. Diffusion of Zn in Al (99.995%). Determination of the diffusion energy

5. Conclusions

The results obtained in the above section show that the resistometric method enables a very accurate determination of the diffusion coefficient of the elements in pure metals.

In general, the method can be well applied whenever resistivity changes can be surely connected with the diffusion of the element to be studied. This seems to be also the case with the diffusion of an element in homogeneous and stable solid solutions, i.e. in the absence of clustering phenomena.

The sensitivity of the method can be discussed using (16) and (17):

$$\frac{R_f - R_i}{R_i} = \frac{\varrho_f - \varrho_0}{\varrho_0} = \frac{\Delta\varrho}{\varrho_0}. \tag{20}$$

Now, for evaluating, with a good precision, isothermal annealing curves such as in Fig. 3, it is sufficient that $\Delta\varrho/\varrho_0$ is greater than a certain limit, which, as experience indicates, is of the order of 2% for measurements at liquid nitrogen temperature.

As an example, in the case of Al, one has ϱ_0 (– 195 °C) = 0.23 μΩ cm; if one takes for α a mean value of 0.5 μΩ cm, one obtains for the final concentration of the alloy, c_f, a value of about 8×10^{-5}. In other words, the method appears to be "a priori" very sensitive.

Acknowledgements

The authors are grateful to Prof. C. Panseri, Director of I.S.M.L., and to the Direction of SORIN for the permission to publish these results.

The mathematical analysis summarized in the Appendix has been carried out by A. Ascari of SORIN; the numerical computations have been performed under the supervision of T. Zucca, also of SORIN, with the aid of the computer ELEA 6001 of the SORIN Saluggia Centre.

Appendix

As remarked in Section 2, the analytical expression (10)

$$y(x, \tau) = 1 + \sum_{n=1}^{\infty} \frac{J_0(\alpha_n x)}{J_0(\alpha_n)} \exp(-\alpha_n^2 \tau), \quad J_1(\alpha_n) = 0 \tag{A1}$$

does not lend itself to the numerical computation of the function $y(x, \tau)$ for small or intermediate values of τ, say $\tau < 10^{-2}$.

Indeed, for $\tau = 0$, $x = 1$ the series in (A1) diverges to $+\infty$, as required by the initial boundary condition (2). Yet the series diverges, in an oscillatory way, also for $\tau = 0$, $x = 0$, since the sign of $J_0(\alpha_n)$ is $(-1)^n$ and $|J_0(\alpha_n)|^{-1}$ steadily increases with n. Although the series $\sum [J_0(\alpha_n)]^{-1}$ turns out to be "summable" to -1 by a proper summation process, thus agreeing with condition (3), the above circumstances suffice to anticipate that the convergence of (A1) will be very poor for small τ.

This difficulty may be overcome with the aid of a method suggested by Carslaw and Jaeger [11].

The first step in the solution of the problem formulated in (1) to (6) is, in every case, a Laplace transformation of $y(x, \tau)$ with respect to τ:

$$\bar{y}(x, s) = \int_0^\infty e^{-s\tau} y(x, \tau) \, d\tau .$$

The resulting boundary-value problem for $\bar{y}(x, s)$ is most quickly solved by the standard eigenvalue technique: the solution is obtained as a series whose inverse Laplace transform is exactly (A1).

However, the same problem may be solved also by means of the Green's function technique. The result is

$$\bar{y}(x, s) = \frac{I_0(s^{1/2} x)}{2 s^{1/2} I_1(s^{1/2})}, \tag{A2}$$

where $I_0(z)$, $I_1(z)$ are the modified Bessel functions of the first kind, of orders zero and one. The Laplace inversion yields again (A1); but if, before performing it, one substitutes for I_0, I_1 in (A2) their asymptotic expansions for large argument (i.e. for large s), inversion will give an approximate expression for $y(x, \tau)$ valid for small τ.

In applying the outlined procedure, it is convenient to treat $x = 0$ and $x > 0$ as separate cases.

For $x = 0$ the inversion of the asymptotic expansion of (A2) (where the numerator is now unity) leads to the so-called Whittaker function $W_{k,m}(z)$, or more properly to its asymptotic expansion. After various reductions one obtains

$$y(0, \tau) \approx \frac{1}{2\tau} (1 + \tau) \exp\left(-\frac{1}{4\tau}\right) + O(\tau^2), \tag{A3}$$

an expression which, despite its simplicity, is fairly adequate for numerical computation in the range $0 < \tau < 10^{-2}$.

For $0 < x \leq 1$ the inverse transform of the asymptotic expansion of (A2) can be written

$$y(x, \tau) \approx \frac{1}{2} x - \frac{1}{2}\left[(\pi\tau)^{-1/2} \exp\left\{-\frac{(1-x)^2}{4\tau}\right\} + \right.$$
$$\left. + \sum_{n=0}^{\infty} a_n(x) (4\tau)^{n/2} i^n \operatorname{erfc} \frac{1-x}{\sqrt{\tau}}\right], \tag{A4}$$

where the functions $a_n(x)$ are polynomials in x^{-1} of degree $n + 1$ which may be recursively constructed starting from the asymptotic expansions of the Bessel functions, the first ones being

$$a_0(x) = \frac{1}{8}(3 + x^{-1}), \quad a_1(x) = \frac{3}{128}(11 + 2x^{-1} + 3x^{-2}),$$

$$a_2(x) = \frac{3}{1024}(83 + 11x^{-1} + 9x^{-2} + 25x^{-3}), \text{ etc.,}$$

and $i^n \operatorname{erfc} z$ denotes the repeated integral of the complementary error function, defined by

$$i^n \operatorname{erfc} z = \int_z^\infty i^{n-1} \operatorname{erfc} t \, dt ,$$

$$i^0 \operatorname{erfc} z = \operatorname{erfc} z = \frac{2}{\sqrt{\pi}} \int_z^\infty e^{-t^2} dt .$$

The expressions for the first $a_n(x)$ show that (A4) should not be used for too small x, say $x < 0.1$; on the contrary it is well suited also for $x = 1$, where it becomes a series in $\sqrt{\tau}$ the first terms of which are

$$y(1, \tau) \approx \frac{1}{2} (\pi \tau)^{-1/2} + \frac{1}{4} + \frac{3}{8} \pi^{-1/2} \tau^{1/2} + \frac{3}{16} \tau + \frac{21}{64} \pi^{-1/2} \tau^{3/2} + \cdots . \qquad (A5)$$

It is apparent that the condition $y(1, 0) = \infty$ is fulfilled by this expression, as well as $y(0, 0) = 0$ is satisfied by (A3).

References

[1] D. Lazarus, Solid State Phys. **10**, 71 (1960).
[2] G. Bruni and D. Meneghini, Atti Reale Accad. naz. Lincei **20**, 671 (1911).
[3] I. Runge, Z. anorg. Chem. **115**, 293 (1921).
[4] A. E. van Arkel, Metallwirtschaft, Metallwiss., Metalltech. **7**, 656 (1928).
[5] S. Tanaka and C. Matano, Mem. Coll. Sci., Kyoto Imp. Univ. **13**, **343** (1930).
[6] W. Seith and O. Kubaschewski, Z. Elektrochem. **41**, 551 (1935).
[7] W. C. Dunlap Jr., Phys. Rev. **94**, 1531 (1954).
[8] T. Broom, Adv. Phys. **3**, 26 (1954).
[9] J. Crank, The Mathematics of Diffusion, Clarendon Press, Oxford 1956.
[10] J. E. Hilliard, B. L. Averbach, and M. Cohen, Acta metall. **7**, 86 (1959).
[11] H. S. Carslaw and J. C. Jaeger, Operational Methods in Applied Mathematics, Oxford University Press 1948.

(Received April 13, 1966)

phys. stat. sol. 16, 449 (1966)

Brookhaven National Laboratory, Upton, New York (a)
and Columbia University, New York (b)

A Field Ion Microscope Study of Neutron Irradiated Tungsten[1])

By

M. ATTARDO (a) and J. M. GALLIGAN (b)

A study of the damage introduced into tungsten by fast neutrons using field ion microscopy, is described. This consisted of examining the type of damage as reflected in the vacancy content of the material and the removal of the vacancy content after various annealing treatments. Evidence is presented which shows that vacancies are removed from tungsten in stage IV. The observations also indicate the presence of depleted zones in the irradiated tungsten.

Das Strahlungsdamage durch schnelle Neutronen in Wolfram wird mit dem Feld-Ionenmikroskop untersucht. Dabei wird gefunden, daß sich der Typ des Damage im Lückengehalt des Materials widerspiegelt und daß sich die Lücken durch verschiedene Temperungsbehandlungen beseitigen lassen. Es wird der Hinweis gegeben, daß die Lücken aus Wolfram im Zustand IV entfernt werden. Die Untersuchungen zeigen außerdem die Anwesenheit von Verarmungsgebieten im bestrahlten Wolfram.

1. Introduction

The recovery of physical properties, upon annealing, in irradiated materials is of importance in understanding the nature of defects in crystals as well as radiation damage itself [1]. In this regard various studies of properties of metals by quenching, irradiation and cold work have revealed a wide range of stages of removal of the defects introduced by the various treatments, with very little comprehensive understanding of the specific defects involved [2]. This is especially true of the recovery stages associated with cold work and irradiation. It is also worth mentioning that the recovery stages in body-centered and face-centered cubic metals are quite similar in this respect.

Some previous field ion microscope observations by Müller on tungsten irradiated with ions while directly observing the damage have revealed the presence of vacancies and interstitials [3]. Somewhat similar results have been obtained by Brandon, Southon, and Wald [4], who used particle irradiation. Another investigation of interest to the present study is that reported by Bowkett, Hren, and Ralph [5] who have irradiated tungsten with neutrons to an integrated flux of 5×10^{17} nvt and mostly observed vacancy and vacancy clusters. The present investigation was specifically undertaken to study the removal of isolated vacancies by direct observation with a field ion microscope and these results will be reported on here.

In this regard extensive measurements of the annealing of damage have been made by Schultz [6] and Moteff and Smith [7]. This work bears directly on the conclusions drawn from the above mentioned work.

[1]) This work was performed under the auspices of the United States Atomic Energy Commission.

2. Experimental Procedure

Tungsten wire of commercial grade (General Electric Grade 218 with a quoted purity of 99.98% W) was irradiated in the Brookhaven National Laboratory graphite reactor to an integrated flux of 10^{18} nvt, counting only fast neutrons ($E > 1$ MeV). Prior to irradiation the material was annealed for three hours at 1500 °C in vacuum to remove any interfering defects and part of this wire was subsequently examined after the annealing treatment. This material served as a blank in all subsequent experiments and also served to establish the vacancy concentration of well annealed material. Ten tips of this material were examined in a field ion microscope, with the tip in good thermal contact with a liquid nitrogen reservoir and He gas at a pressure of approximately 10^{-3} Torr as the imaging gas. A typical micrograph from this series of experiments is shown in

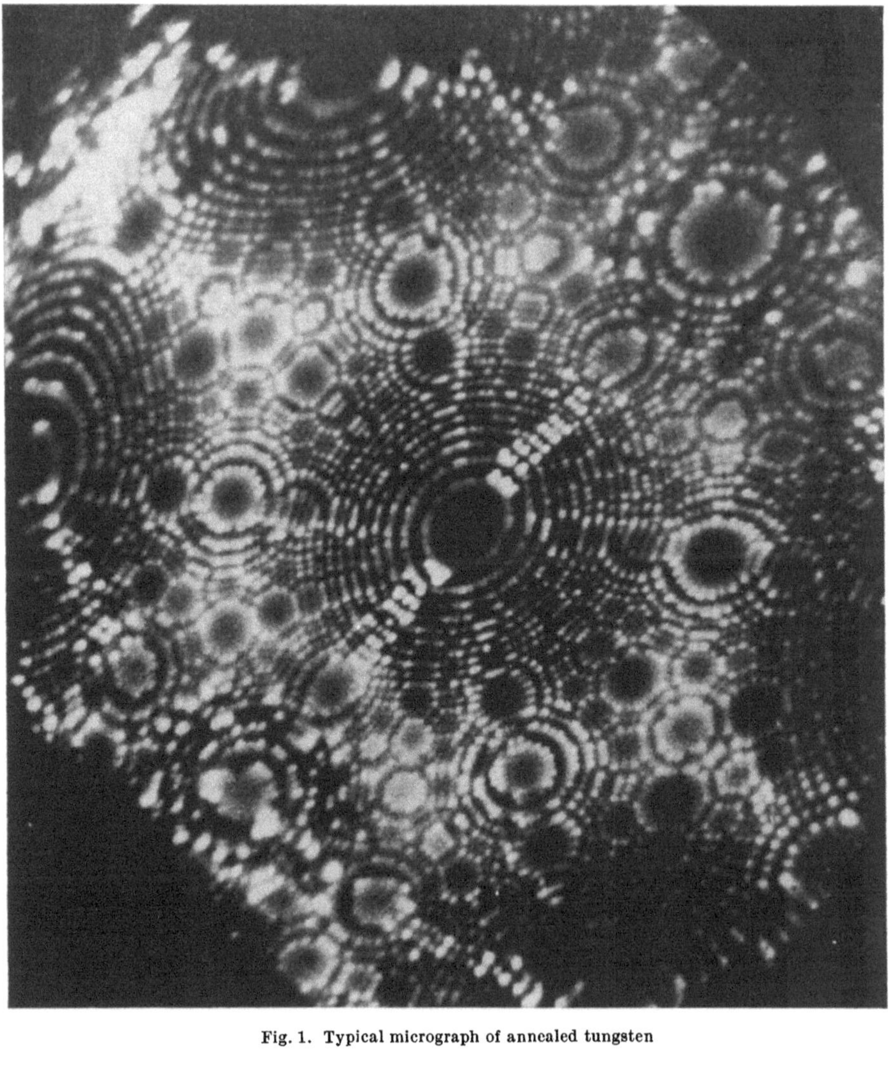

Fig. 1. Typical micrograph of annealed tungsten

Fig. 1. Repeated examination of this material given the above mentioned treatment has shown that the fraction of vacant lattice sites is considerably less than 10^{-5}. This examination consists of field evaporation of a single layer and examining the specific arrangement of atoms on planes such as (233), (122), and (433). It should also be noted that observations were undertaken after the tip had been field evaporated until a stable end form had been obtained by field evaporation. (This procedure has been fully discussed by Müller [8], and will not be considered here.)

From this material, characterized in this way, two sets of specimens were irradiated, one set while the specimen was maintained at 70 °C and the other set at a temperature of 400 °C.

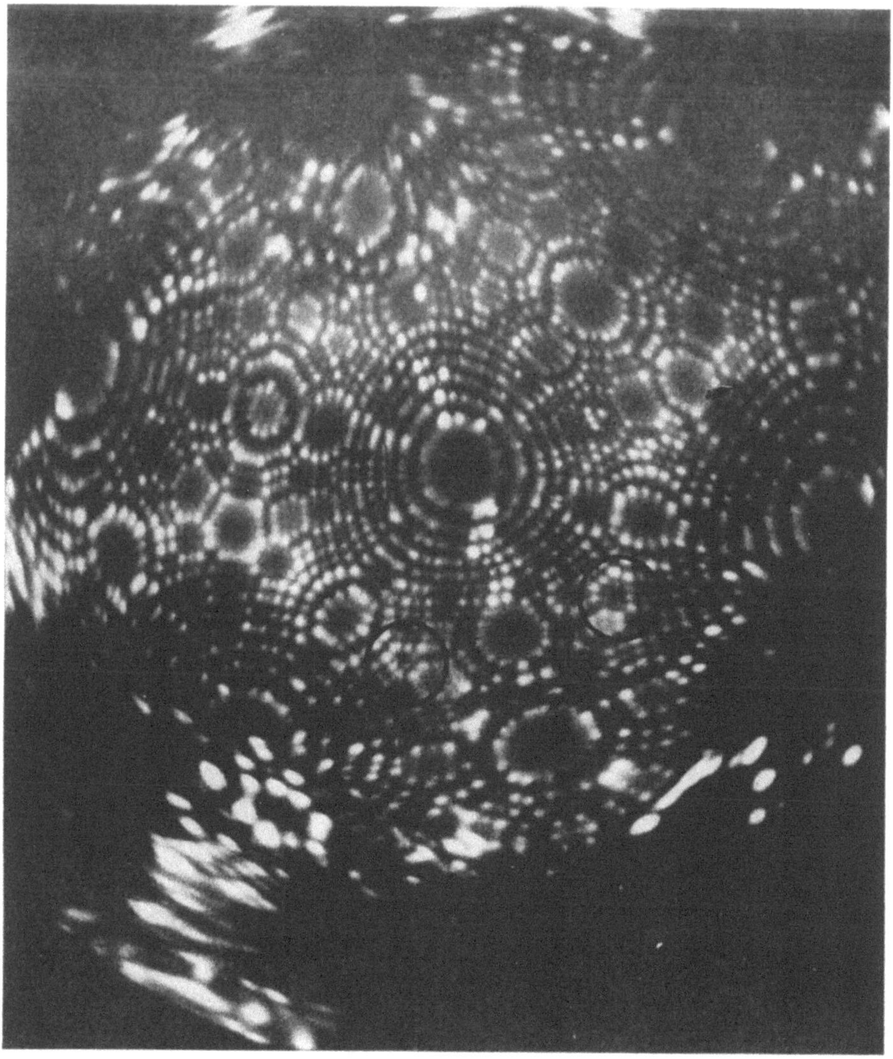

Fig. 2a. Micrographs of single volume vacancies in irradiated tungsten

3. Results and Discussion

Some field ion micrographs of the tungsten irradiated at 70 °C are shown in Fig. 2a and b, 3, 4, and 5 and are typical of observations made on this material. In these photographs it is quite apparent that vacancy clusters and single vacancies are observed in agreement with previous observations. Fig. 2a and b show typical observations of single vacancies; in Fig. 3, a trivacancy within a closed net plane; Fig. 4 and 5 show larger clusters, some being roughly 10 to 15 Å in diameter.

It is interesting to consider first the upper limit of the size of such vacancy clusters. According to the Kinchin-Pease analysis [9], the number of Frenkel

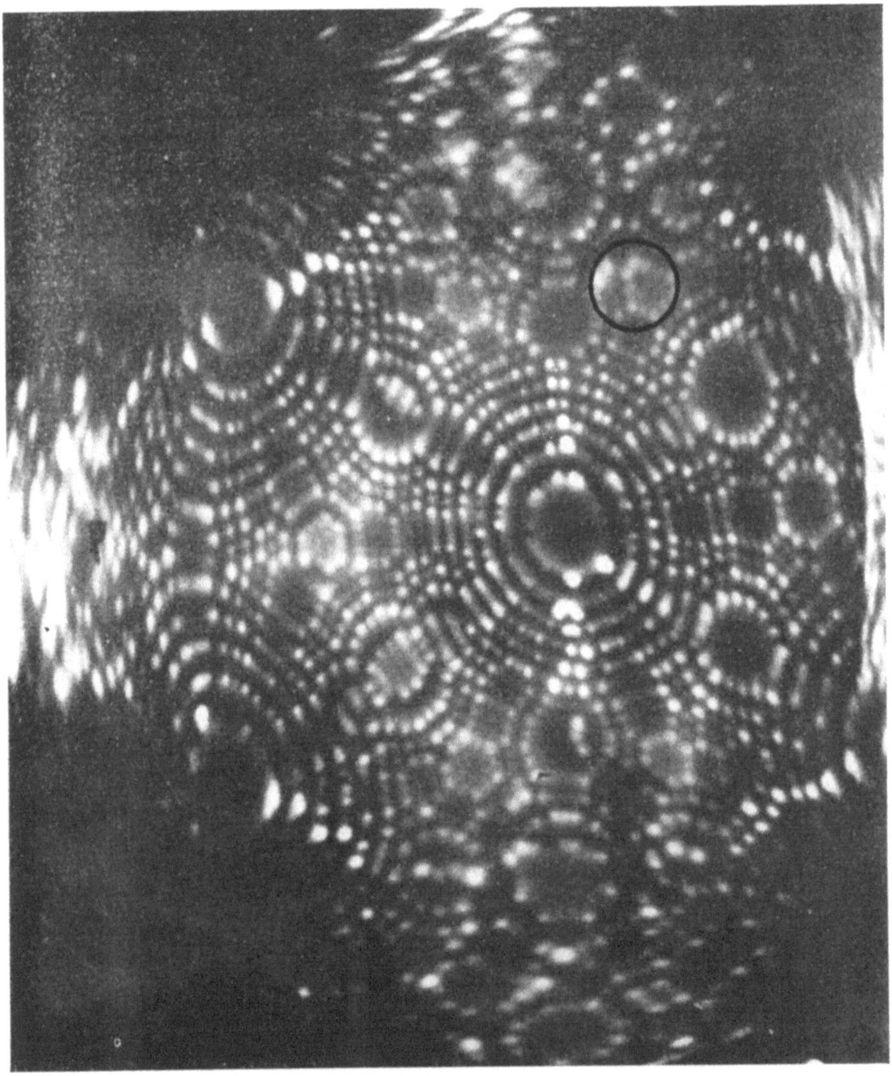

Fig. 2b. Micrographs of single volume vacancies in irradiated tungsten

pairs obtained from collision of a 2 MeV neutron with a tungsten atom is given by

$$V_D = \frac{E}{2\,E_d}\,.$$

E energy transferred to the primary displaced atom by a 2 MeV neutron,
E_d energy of displacement of tungsten atom in a tungsten matrix (50 eV).
If we use the values appropriate for this experiment, we get approximately the same number of vacancies as observed in some of the larger clusters. Since these clusters have been formed from the local deposition of a large amount of energy which locally raises the temperature of many of the atoms involved, one expects some recovery to occur even at the lowest specimen temperatures. Such

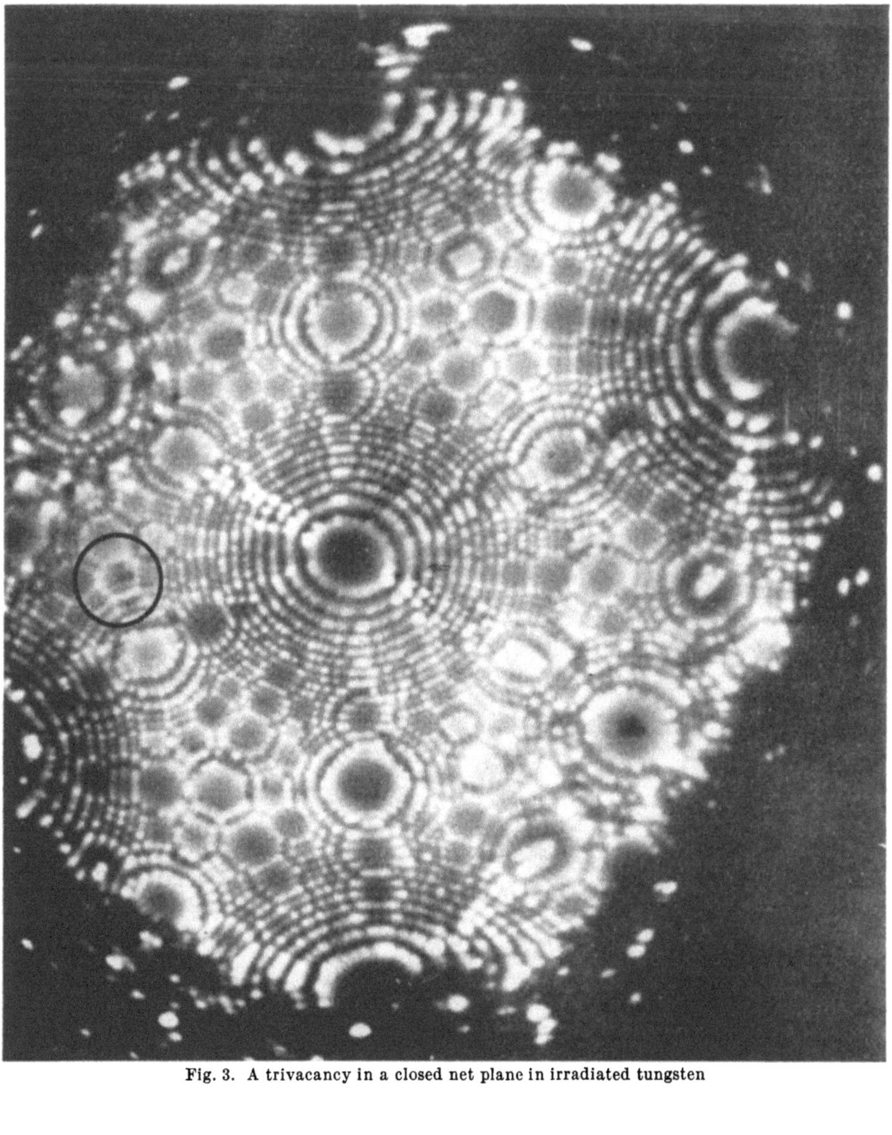

Fig. 3. A trivacancy in a closed net plane in irradiated tungsten

recovery processes have been considered by various authors, but in any event no matter which of these recovery processes is dominant, some recovery is to be expected [10]. If one admits the logical necessity of some recovery process after the damage event, then the size of the clusters should be accordingly less than the value given by the Kinchin-Pease analysis. Leibfried [10], for example, has argued that in order to match the experimentally observed damaged rate, theoretical estimates of the damage must be reduced by about five to bring experiment into agreement with theory. Since our observations indicate that the size of the larger clusters can be commensurate with the value predicted by the Kinchin-Pease analyses, three distinct possibilities could account for the observation:

A. Some defect is mobile below 70 °C which *increases* the size of vacancy clusters following a damage event.

B. *No* recovery has occurred — immediately after the damage is introduced by the primary displacement atom.

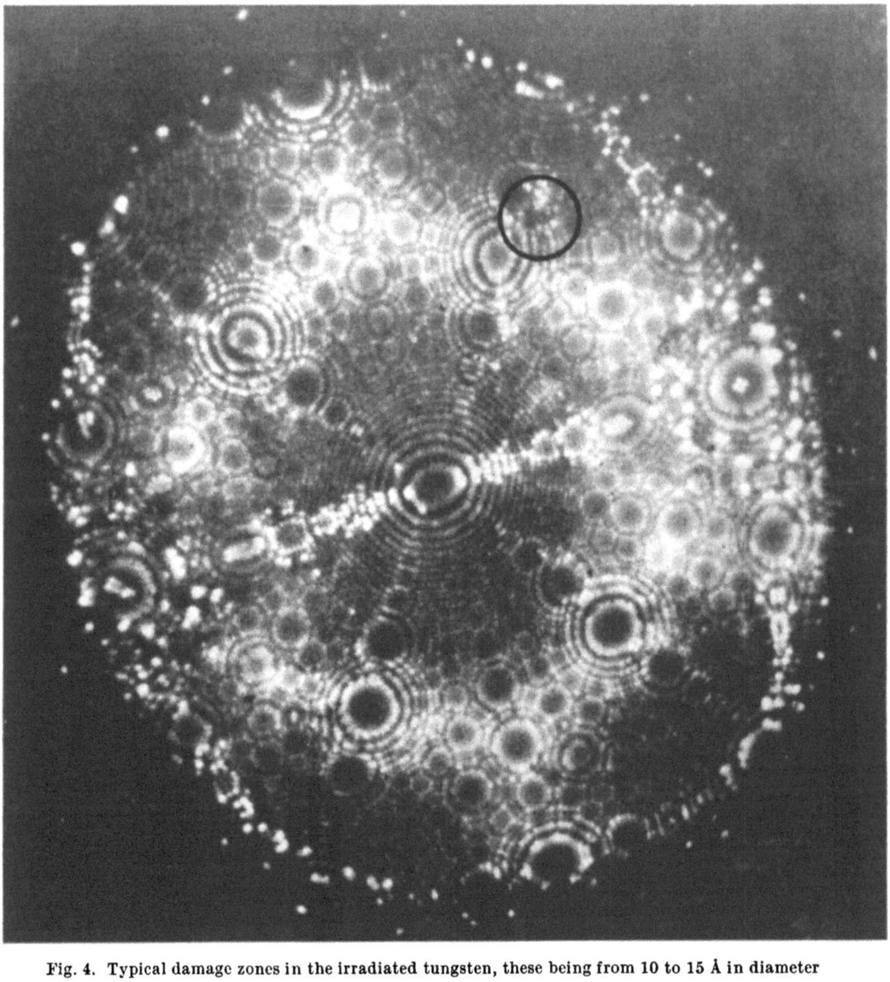

Fig. 4. Typical damage zones in the irradiated tungsten, these being from 10 to 15 Å in diameter

C. The observed vacancy cluster size represents a depleted zone [11] from which loosely bound atoms are immediately pulled away when observations are made with the field ion microscope.

Consider first the relevant arguments with respect to possibility A. Since two recovery stages occur below the temperature of irradiation (70 °C) [12], the defect which increases the vacancy cluster size might be expected to have an activation energy of less than 1 eV. Considering also the known energy of formation of single vacancies in W [13] as well as the energy of self diffusion [14, 15], single vacancies can be reasonably ruled out as a possible mobile defect in the required temperature range.

Another possibility is a multiple vacancy of some sort which is mobile below 70 °C and can be adsorbed by a small vacancy cluster, thereby increasing the

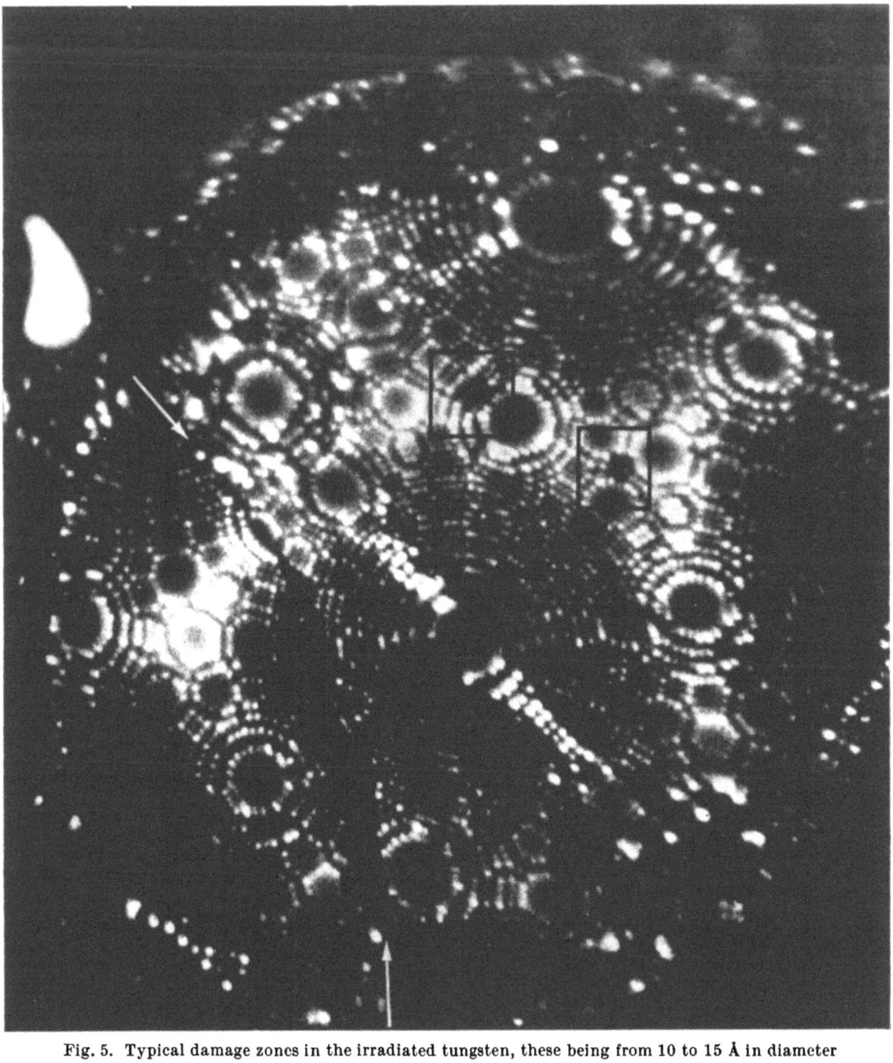

Fig. 5. Typical damage zones in the irradiated tungsten, these being from 10 to 15 Å in diameter

vacancy cluster size. Such a multiple vacancy could be a divacancy, or perhaps a trivacancy, but closeknit movement of larger clusters would involve too many correlations to move at such a low temperature. If the defect is a divacancy, this would accordingly require a relatively low energy of motion for a divacancy in a material with a high binding energy which is at variance with experience but of course not disproved. In line with this argument, we think it unreasonable to require divacancies or trivacancies to move at such a temperature.

Possibility B, as outlined above, would require that no recovery of any sort occurs during and immediately after the damage event. If this is so, a large body of indirect experiments would require other interpretations than they have been given, i.e. the resistivity change associated with the radiation damage at the lowest temperatures of bombardment [10]. In fact, recovery processes were introduced not only as a logical necessity but to bring the whole theoretical picture of radiation damage into accord with experiment, especially the low temperature data.

From these arguments we are left with alternative C, for which there is, again, considerable indirect evidence, in particular the mechanical property changes associated with irradiation [16].

Another interesting aspect of this whole problem which we have investigated has been the recovery stage in tungsten corresponding to the removal of single vacancies. Stage III recovery in tungsten occurs at approximately 370 °C [6, 7] and stage IV is the recovery occurring above 450 °C. The tungsten irradiated at 400 °C therefore corresponds to an irradiation above stage III recovery. The speciments irradiated at 400 °C again contained single vacancies indicating that stage III recovery is not the stage corresponding to single vacancy migration. This conclusion has been corroborated by annealing the material irradiated at 70 °C at successively higher temperatures, the wires being annealed in an inert gas atmosphere of argon for one and one-half hours at each temperature. After each anneal some specimens were formed from the treated material and the fraction of vacant lattice sites was again determined by field evaporation. The results of this experiment are shown in Fig. 6. From the high temperature irradiation experiment and from the annealing studies we can say that stage III does not correspond to the removal of single volume vacancies. We can also say that the removal of single volume vacancies predominantly occurs in stage IV recovery of irradiated tungsten.

These conclusions on the removal of single vacancies have previously been suggested through experiments by Schultz [6] and Moteff and Smith [7] and is directly proved by field ion microscopy.

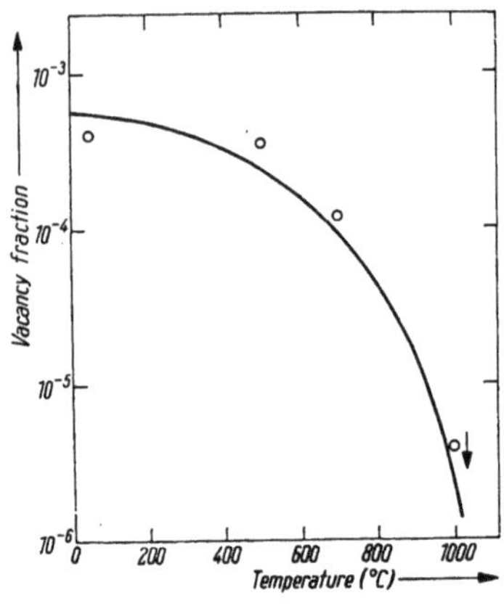

Fig. 6. Vacancy concentration as a function of annealing temperature of irradiated tungsten (10^{18} nvt)

4. Summary and Conclusions

From this work the following conclusions can be made without any auxiliary arguments:

1. Single volume vacancies are observed after irradiation of tungsten with fast neutrons (integrated fast flux of 10^{18} nvt, $E > 1$ MeV) at concentration levels in excess of 10^{-2} at%.
2. Stage III recovery in tungsten does not correspond to removal of vacancies since they are stable above stage III.
3. Vacancies are substantially removed above 700 °C, which corresponds to stage IV recovery.

A conclusion which requires some auxiliary arguments is the following:

4. Depleted zones exist in fast neutron irradiated tungsten.

Acknowledgement

We wish to acknowledge the help of D. Gurinsky and J. Sadofsky throughout this investigation.

References

[1] H. G. van Bueren, Imperfections in Crystals, North-Holland Publ. Comp. Amsterdam 1961.
[2] J. S. Koehlen and F. Seitz, Advances in Sol:d State Physics 2, 307 (1956).
[3] M. K. Sinha and E. W. Müller, J. appl. Phys. 4, 35 (1964); 4, 1256 (1964).
[4] D. Brandon, M. Southon, and M. Wald in Properties of Reactor Materials and the Effects of Radiation Damage, Ed. D. J. Littler, Butterworth, London 1962.
[5] H. Bowkett. J. Hren, and B. Ralph, Third European Regional Conf. Electron Microscopy, Prague, Vol. A, 191 (1964).
[6] H. Schultz, Acta metall. 12, 649 (1964).
[7] F. Moteff and J. P. Smith, ASTM Symposium on Flow and Fracture Behavior of Metals and Alloys in Nuclear Environments, American Society for Testing and Materials, Philadelphia 1965 (p. 171).
[8] E. W. Müller, Field Ionization and Field Ion Microscopy in Adv. Electronics and Electron Phys. 13, 135 (1960).
[9] G. H. Kinchin and R. S. Pease, Rep. Progr. Phys. 18, 1 (1955).
[10] G. Leibfried, The Interaction of Radiation with Solids, North-Holland Publ. Comp. Amsterdam 1964.
[11] A. Seeger, Proc. 2nd Internat. Conf. Peaceful Uses of Atomic Energy.
[12] M. W. Thompson, Phil. Mag. 5, 278 (1960).
[13] H. Schultz, Acta metall. 12, 761 (1964).
[14] F. A. M. van Liempt, Recueil Trav. chim. Pays-Bas, 64, 239 (1945).
[15] R. L. Andelin, J. D. Knight, and M. Kohn, Trans. AIME 233, 19 (1965).
[16] A. Seeger and U. Essman, Proc. Internat. School of Physics „Enrico Fermi", Course XVIII: Radiation Damage in Solids, Academic Press, New York 1962.

(Received April 25, 1966)

phys. stat. sol. 16, 459 (1966)

State University of Chernovtsy

On the Theory of Anisotropic Thermoelectric Power in Semiconductors

By

A. G. SAMOILOVICH, M. V. NITSOVICH, and V. M. NITSOVICH

Possible mechanisms for the appearance of an anisotropic thermoelectric power in semiconductors are examined using simple models. It is shown that an anisotropic thermoelectric power may arise both as a result of combined phonon and impurity scattering, and in the presence of two types of carriers if at least one type of carrier has an anisotropic effective mass. The effect of different parameters on the magnitude of the anisotropy in the thermoelectric power is examined and it is shown that for reasonable values of these parameters the anisotropy may be significant.

В данной работе на простых моделях изучается механизм возникновения анизотропии термоэдс в полупроводниках. Показано, что анизотропия термоэдс может возникнуть как в результате совместного рассеяния на фононах и примесных ионах, так и при наличии двух типов проводимости, если, по крайней мере, один из типов носителей тока обладает анизотропной эффективной массой. Исследовано влияние различных параметров на величину анизотропии термоэдс, показано, что при разумных значениях параметров анизотропия термоэдс может достигать значительной величины.

Thermoelectric phenomena in anisotropic media are relatively insufficiently studied. It has been recently shown, however, that anisotropic compounds have some advantages for constructing thermoelectric generators and refrigerators [1]. Therefore the study of the mechanism for the apperance of a thermoelectric power becomes of high importance.

As it is shown in [2] in the presence of any scattering mechanism the thermoelectric power tensor is reduced to a scalar. The thermoelectric power will be isotropic independently of the strength of crystal anisotropy. This is true both for single-valley and many-valley models. This shows that the appearance of an anisotropic thermoelectric power requires some special conditions.

It will be demonstrated that an anisotropic thermoelectric power may arise in the presence of two or several scattering mechanisms, as well as in the presence of two kinds of current carriers.

1. Anisotropic Thermoelectric Power in the Presence of Different Scattering Mechanisms

Let us examine the influence of a combined scattering by phonons and impurities on the anisotropic thermoelectric power.

We examine a simple model of an anisotropic semiconductor with four energy valleys, which are situated on a plane (Fig. 1). The surfaces of constant energy are assumed to be rotational ellipsoids. Non-degeneracy of the electron gas is supposed throughout this paper.

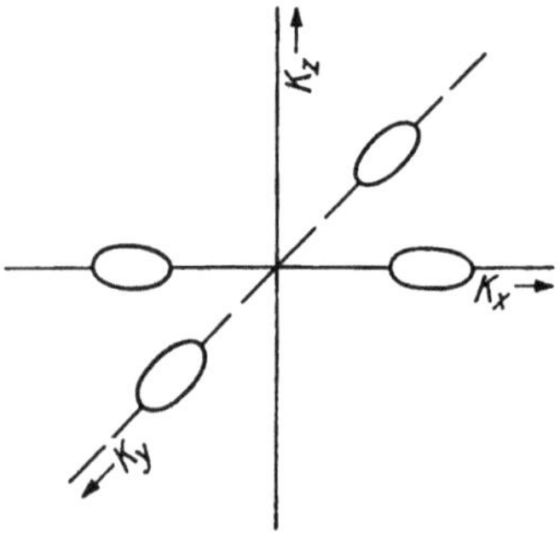

Fig. 1

The thermoelectric power tensor for one energy valley has the form

$$\alpha_{ii} = \frac{k}{e}\left[\frac{\int \tau_{ii}\frac{\partial n}{\partial \varepsilon} v_i \varepsilon \, d\tau_p}{k\,T\int \tau_{ii}\frac{\partial n}{\partial \varepsilon} v_i^2 \, d\tau_p} - \mu^*\right]. \quad (1)$$

Hence, we obtain from (1) after summing up over all valleys

$$\alpha_{||} = \frac{k}{e}\left[\frac{\int \tau_\perp e^{-\frac{\varepsilon}{kT}} \varepsilon^{5/2} \, d\varepsilon}{k\,T\int \tau_\perp e^{-\frac{\varepsilon}{kT}} \varepsilon^{3/2} \, d\varepsilon} - \mu^*\right], \quad (2)$$

$$\alpha_\perp = \frac{k}{e}\left[\frac{\frac{2}{3}\frac{m_{||}}{m_\perp}\int \tau_\perp e^{-\frac{\varepsilon}{kT}} \varepsilon^{5/2} \, d\varepsilon + \int \tau_{||} e^{-\frac{\varepsilon}{kT}} \varepsilon^{5/2} \, d\varepsilon}{k\,T\left(\frac{2}{3}\frac{m_{||}}{m_\perp}\int \tau_\perp e^{-\frac{\varepsilon}{kT}} \varepsilon^{3/2} \, d\varepsilon + \int \tau_{||} e^{-\frac{\varepsilon}{kT}} \varepsilon^{3/2} \, d\varepsilon\right)} - \mu^*\right]. \quad (3)$$

As the scattering anisotropy due to impurities predominates over that of phonons, we shall neglect the latter. Then the relaxation time tensor can be found from the formula

$$\frac{1}{\tau_{||,\perp}} = \frac{1}{\tau^{f}} + \frac{1}{\tau^{i}_{||,\perp}}, \quad (4)$$

where [3, 4]

$$\tau^{f} = \frac{\pi\, C_{11}^{1}\, \hbar^4}{k\,T\, C_1^2 \sqrt{2\, m_\perp^2\, m_{||}\, \varepsilon}}, \quad (5)$$

$$\tau^{i}_{||,\perp} = \frac{\tau_0}{F_{||,\perp}}, \quad (6)$$

$$\left.\begin{aligned}
\tau_0 &= \frac{8\,\beta^3\, m_\perp\, \varkappa^2\, \varepsilon^{3/2}}{3\,\pi\, N\, e^4 \sqrt{2\, m_{||}}}\,;\\
F_{||} &= 2\left[f(\beta,\gamma^2) - \frac{\beta}{1+\beta^2}\ln\frac{1+\gamma^2}{\gamma^2}\right],\\
F_\perp &= (\beta^2-1)\left[f(\beta,\gamma^2) + \frac{\beta}{\beta^2-1}\ln\frac{1+\gamma^2}{\gamma^2} - \frac{(\pi-2\,x)\,\beta}{(\beta^2-1)\,\varrho}\right];\\
\beta^2 &= \frac{m_{||}-m_\perp}{m_\perp}, \quad \gamma^2 = \frac{\hbar^2}{8\, m_{||}\, R_0^2\, \varepsilon}, \quad a = \operatorname{arctg}\beta,\\
\varrho &= \frac{\sqrt{1+\gamma^2(1+\beta^2)}}{\beta}, \quad x = \operatorname{arctg}\varrho\,,\\
f(\beta,\gamma^2) &= \frac{\pi-2\,x}{2}\ln\frac{\beta\,\varrho+1}{\beta\,\varrho-1} - \left(2\,a+\frac{\pi}{2}\right)\ln 2 +\\
&\quad + L\,(x+a) - L\,(x-a) + L\left(\frac{\pi}{2}-2\,a\right) + 2\,L(a)\,.
\end{aligned}\right\} \quad (7)$$

ε is the electron energy, R_0 the screening radius, N the ion concentration, $\varkappa$ the dielectric constant, and $L(x)$ the Lobachevskii function. Substituting this in (2) and (3) and introducing

$$C_{\|,\perp} = \int_0^\infty \frac{e^{-x} x^4 \, dx}{x^2 + a^2_{\|,\perp}}, \quad D_{\|,\perp} = \int_0^\infty \frac{e^{-x} x^3 \, dx}{x^2 + a^2_{\|,\perp}}, \tag{8}$$

where

$$a^2_{\|,\perp} = \frac{\pi^2 C^1_{11} \hbar^4 N e^4}{2 m^2_\perp C^2_1 \varkappa^2 (k T)^3} F_{\|,\perp} \tag{9}$$

we get

$$\alpha_\| = -\frac{k}{e} \frac{C_\perp}{D_\perp}, \quad \alpha_\perp = -\frac{k}{e} \frac{\frac{2}{3}\frac{m_\|}{m_\perp} C_\perp + C_\|}{\frac{2}{3}\frac{m_\|}{m_\perp} D_\perp + D_\|}. \tag{10}$$

Hence, we get the following expression for the anisotropic thermoelectric power:

$$\Delta\alpha = \alpha_\| - \alpha_\perp = -\frac{k}{e} \frac{C_\perp D_\| - C_\| D_\perp}{D_\perp \left(\frac{2}{3}\frac{m_\|}{m_\perp} D_\perp + D_\|\right)}. \tag{11}$$

This theoretical result for the anisotropic thermoelectric power is plotted in Fig. 2, 3, and 4 as a function of temperature for different parameter values. The following conclusions can be drawn from the analysis auf these curves.

The presence of maxima on all curves has a simple meaning. As anisotropy of the thermoelectric power is absent for non-combined scattering, the maximum should lie in that temperature range where both scattering mechanisms act most effectively.

Fig. 2. 1: $N = 10^{17}$ cm^{-3}, 2: $N = 10^{16}$ cm^{-3}, 3: $N = 10^{15}$ cm^{-3}; $\beta^2 = 9$, $m_\perp = 10^{-1} m_0$, $m_\| = m_0$

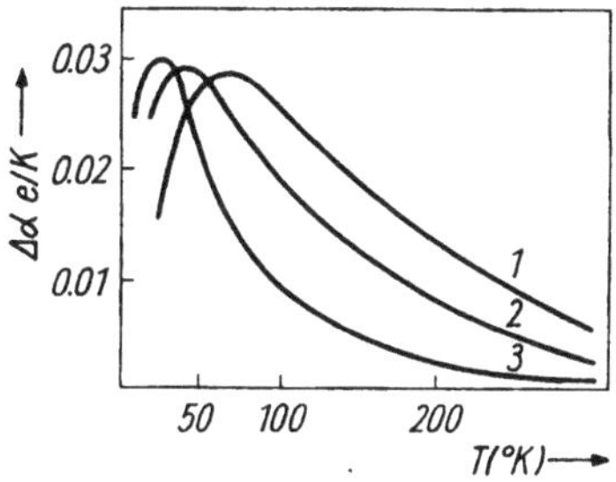

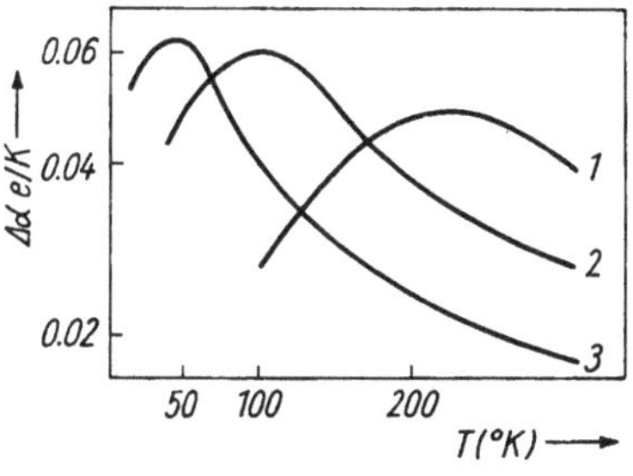

Fig. 3. 1: $N = 10^{17}$ cm^{-3}, 2: $N = 10^{16}$ cm^{-3}, 3: $N = 10^{15}$ cm^{-3}; $\beta^2 = 99$, $m_\perp = 10^{-1} m_0$, $m_\| = 10 \, m_0$

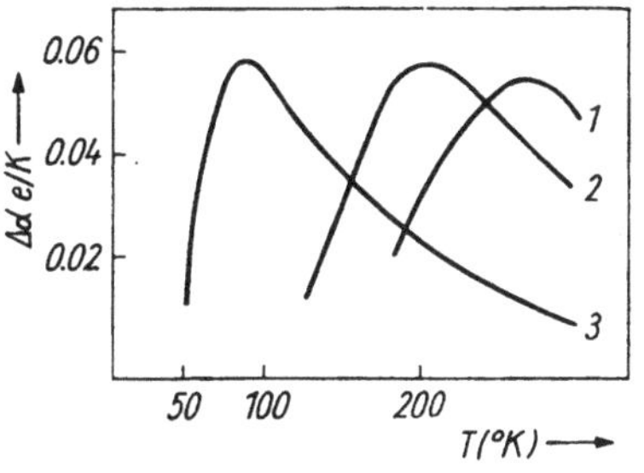

Fig. 4. 1: $N = 10^{17}$ cm^{-3}, 2: $N = 10^{16}$ cm^{-3}, 3: $N = 10^{15}$ cm^{-3}; $\beta^2 = 99$, $m_\perp = 10^{-2} m_0$, $m_\| = m_0$

A shift of the maxima to the left with decreasing impurity concentration is due to the fact that phonon scattering begins to predominate earlier with decreasing impurity concentration.

Shift of the maxima to the left with an increase of mass anisotropy accounts for the fact that the efficiency of impurity scattering decreases. This is clear from the expression for τ_0 (formula (7)), as τ_0 is proportional to β^3. Thus, increase in mass anisotropy has the same effect as decrease of impurity concentration.

Comparing Fig. 3 and 4 we see that the reduction of $m_\perp$ leads to a decrease of the anisotropic thermoelectric power. This is related to the fact that $m_\perp$ greatly increases with the decrease of $a^2_{\|,\perp}$. As a result the difference in the numerator of formula (11) is reduced.

The shift of the maxima to the region of higher temperatures is due to the fact that τ_0 is proportional to $m_\perp$ (see formula (7)), so that the increase of τ_0, following the increase of β^3, is compensated by the decrease of $m_\perp$; as a result the impurity scattering becomes somewhat more effective.

The value of the anisotropic thermoelectric power in the presence of a combined scattering mechanism is relatively small. With the mentioned parameter values it lies within the limits (5 to 25) μV/°K.

2. Anisotropic Thermoelectric Power in the Presence of Two Types of Carriers

Let us examine the most simple case of a uniaxial crystal, in which two types of carriers are present: One type (holes) possesses an isotropic mass, contributes a fraction σ to the electroconductivity, and has the thermoelectric power α_n; the other type of carriers (electrons) has an anisotropic mass, thermoelectric power α_e and contributes $\sigma_\|$, $\sigma_\perp$ to the electroconductivity. Then

$$\alpha_\| = \frac{\alpha_n \sigma + \alpha_e \sigma_\|}{\sigma + \sigma_\|}, \quad \alpha_\perp = \frac{\alpha_n \sigma + \alpha_e \sigma_\perp}{\sigma + \sigma_\perp}. \tag{12}$$

Hence

$$\Delta\alpha = \alpha_\| - \alpha_\perp = \frac{(\alpha_n - \alpha_e)\,\sigma\,(\sigma_\| - \sigma_\perp)}{(\sigma + \sigma_\|)\,(\sigma + \sigma_\perp)}. \tag{13}$$

Considering acoustic phonon scattering we have

$$\alpha_n = \frac{k}{e}(2 - \mu_n^*), \quad \alpha_e = -\frac{k}{e}(2 - \mu_e^*), \tag{14}$$

where μ_n^* and μ_e^* are reduced Fermi levels (chemical potentials) of holes and electrons, respectively.

Thus

$$\alpha_n - \alpha_e = \frac{k}{e}(4 - \mu_n^* - \mu_e^*). \tag{15}$$

On the other hand it is known that

$$\mu_n^* + \mu_e^* = -\Delta E^*, \tag{16}$$

where ΔE^* is the reduced energy gap. Therefore

$$\alpha_n - \alpha_e = \frac{k}{e}(4 + \Delta E^*). \tag{17}$$

Consequenty formula (13) may be rewritten

$$\Delta\alpha = \frac{k}{e}(4 + \Delta E^*)\frac{\sigma(\sigma_{\parallel} - \sigma_{\perp})}{(\sigma + \sigma_{\parallel})(\sigma + \sigma_{\perp})}. \tag{18}$$

Introduction of impurities may change the concentration of one of the carriers, leaving the concentration of the other unchanged. To make the task more definite, i.e. to have in the formula the smallest number of parameters, we assume that the condition

$$\frac{\partial \Delta\alpha}{\partial \sigma} = 0 \tag{19}$$

is fulfilled. Hence,

$$\sigma = \sqrt{\sigma_{\parallel}\sigma_{\perp}}, \tag{20}$$

or

$$\frac{n_n}{n_e} = \frac{\sqrt{u_{\parallel}u_{\perp}}}{u_n}, \tag{21}$$

where n is the concentration of carriers, u the mobility. It is possible to achieve the fulfilment of conditions (20) and (21) either by changing the concentration of one of the carriers, or by changing the mobility of one of the carriers, which is sometimes possible [5].

It is evident from formula (21) that if the hole mobility is approximately equal to the electron mobility, the maximum anisotropy of the thermoelectric power can be obtained by working near the region of intrinsic conductivity.

With the conditions (20) and (21) the maximum anisotropic thermoelectric power is given by

$$\Delta\alpha = -\frac{k}{e}(4 + \Delta E^*)\frac{\sqrt{K} - 1}{\sqrt{K} + 1}, \tag{22}$$

where the parameter of anisotropy is

$$K = \frac{\sigma_{\perp}}{\sigma_{\parallel}} = \frac{u_{\perp}}{u_{\parallel}}. \tag{23}$$

It should be emphasized that in the theory of anisotropic scattering usually a different anisotropy parameter is examined which has the same form as (23), but $u_{\perp}$ and $u_{\parallel}$ mean the components of the electron mobility for one energy valley.

The mobility components $u_{\perp}$, $u_{\parallel}$ used here are averaged over all energy valleys. The anisotropy parameter (23) is much smaller than the usual one. To estimate the value of $\Delta\alpha$ one should first estimate K, which can be done in a definite model. The simplest model consists of anisotropic holes in the centre of the Brillouin zone and of four energy valleys for electrons with rotational ellipsoids as constant energy surfaces (Fig. 5).

This model can be treated with the formulae of [6]. Considering the phonon spectrum to be

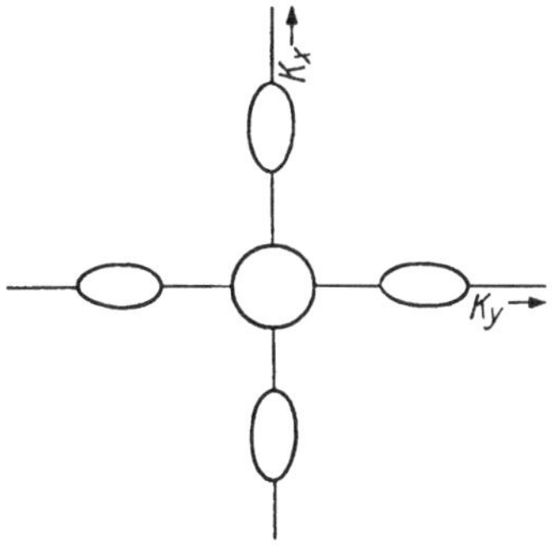

Fig. 5

isotropic, we get for the anisotropy parameter

$$K = \frac{S_0 + S_1}{2\,S_0}\,, \tag{24}$$

where

$$\begin{aligned} S_0 = 1 &+ 2\,\frac{c_2}{c_1}\,\frac{m_{||}}{m_\perp}\,\frac{1}{\beta^2}\left(1 - \frac{3}{\beta^3}\,\mathrm{arctg}\,\beta\right) + \\ &+ \frac{m_{||}}{m_\perp}\,\frac{c_2^2}{c_1^2}\,\frac{1}{\beta^4}\left\{\frac{m_{||}}{m_\perp}\left(1 - \frac{6}{\beta^2} - \frac{3}{2}\,\frac{m_\perp}{m_{||}}\,\frac{1}{\beta^2} + \frac{15}{2\,\beta^3}\,\mathrm{arctg}\,\beta\right) + \right. \\ &\left. + \frac{c_{11}^1}{c_{44}^1}\left[2 + \frac{15}{2\,\beta^2} - \frac{3}{2\,\beta^3}\,(3\,\beta^2 + 5)\,\mathrm{arctg}\,\beta\right]\right\}, \end{aligned} \tag{25}$$

$$\begin{aligned} S_1 = 1 &+ \frac{c_2}{c_1}\,\frac{m_{||}}{m_\perp}\,\frac{1}{\beta^2}\left(1 + \frac{3}{\beta^2} - 3\,\frac{m_{||}}{m_\perp}\,\frac{1}{\beta^3}\,\mathrm{arctg}\,\beta\right) + \\ &+ \frac{c_2^2}{c_1^2}\,\frac{m_{||}}{m_\perp}\left\{\frac{m_{||}}{m_\perp}\left[1 + \frac{15}{4\,\beta^2} - \frac{3}{4\,\beta^3}\,(3\,\beta^2 + 5)\,\mathrm{arctg}\,\beta\right] - \right. \\ &\left. - \frac{c_{11}^1}{4\,c_{44}^1}\left[13 + \frac{15}{\beta^2} - \frac{3}{\beta^3}\,(\beta^4 + 6\,\beta^2 + 5)\,\mathrm{arctg}\,\beta\right]\right\}. \end{aligned} \tag{26}$$

c_{11}^1, c_{44}^1 are elastic constants, c_1, c_2 are deformation potential constants.

Fig. 6 shows the depencence of $\frac{\sqrt{K} - 1}{\sqrt{K} + 1}$ on the ratio $\frac{c_1}{c_2}$ for different parameter values. We can see from the diagram that, although the parameter values considerably influence the shape of the curves, the most probable value of $\frac{\sqrt{K} - 1}{\sqrt{K} + 1}$ is of the order 0.1.

The value of $\Delta\alpha$ essentially depends on ΔE^*. For CdSb we have $\Delta E = 0.5$ eV [7]. In the region of 300 °K, i.e. in the region of intrinsic conductivity and of an anisotropic thermoelectric power, $\Delta E^* = 20$.

Thus, one expects a value of

$$\Delta\alpha \approx 206\ \mu\text{V}/°\text{K}\,.$$

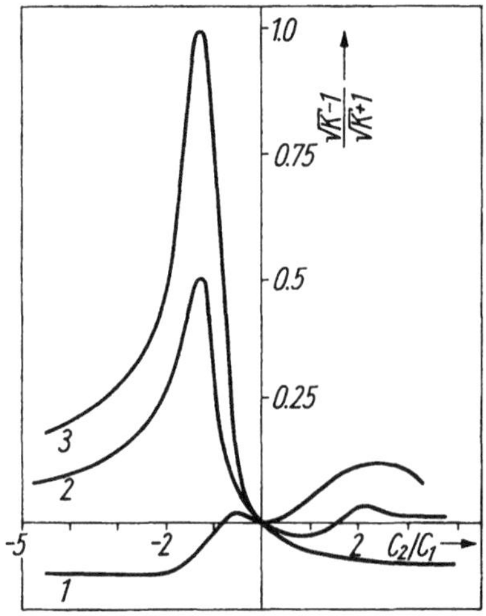

The experimental value is 150 to 200 μV/°K. Consequently, in this case theory very well agrees with experiment.

Unfortunately, scantiness of experimental data does not allow at present a further check of the theory. The above calculations, however, show that in the case of two types of carriers one can expect anisotropic thermoelectric powers ranging from a few fractions to a few units of k/e.

Fig. 6. 1: $\beta^2 = 20$, $\frac{c_{11}^1}{c_{44}^1} = 1$; 2: $\beta^2 = 100$, $\frac{c_{11}^1}{c_{44}^1} = 4$; 3: $\beta^2 = 100$, $\frac{c_{11}^1}{c_{44}^1} = 1$

References

[1] T. C. Harman and T. M. Honig, J. appl. Phys. **33**, 3178 (1963); **33**, 3188 (1963); **34**, 189 (1963).
T. C. Harman, T. M. Honig, and B. M. Tarmy, J. appl. Phys. **34**, 2215 (1963); **34**, 2225 (1963).
[2] A. G. Samoilovich and M. I. Klinger, Fiz. tverd. Tela **2**, 141 (1959).
[3] I. V. Dakhovskii, Fiz. tverd. Tela **5**, 2332 (1963).
[4] D. G. Andrianov, I. V. Dakhovskii, E. M. Omelyanovskii, and V. I. Fistul, Fiz. tverd. Tela **6**, 2825 (1964).
[5] S. V. Airapetyants, V. A. Efimov, T. S. Stavitskaya, L. S. Stilbons, and L. M. Sysova, Zh. teor. Fiz. **27**, 2167 (1957).
[6] A. G. Samoilovich, I. Ya. Korenblit, I. V. Dakhovskii, and V. D. Iskra, Fiz. tverd. Tela **3**, 3285 (1961).
[7] I. M. Pilat and L. I. Anatiichuk, Fiz. tverd. Tela **5**, 3618 (1963); **6**, 1528 (1964).

(Received April 27, 1966)

phys. stat. sol. **16**, 467 (1966)

Philips Research Laboratories, N. V. Philips' Gloeilampenfabrieken, Eindhoven

Magnetic and Structural Characteristics of Some Equiatomic Rare-Earth Germanides

By

K. H. J. Buschow and J. F. Fast

Magnetic and structural data are given for eleven rare-earth germanium intermetallic compounds represented by the formula RGe. The FeB type of structure is observed for R = La and Ce, while the CrB type is found for R = Pr, Nd, Sm, Gd, Tb, Dy, Ho, Er, and Y. The compounds LaGe and YGe are Pauli paramagnets. The compounds PrGe and NdGe are found to become ferromagnetically ordered upon cooling, whereas the compounds CeGe, SmGe, GdGe, TbGe, DyGe, and ErGe order antiferromagnetically. A field-induced transition from the antiferromagnetic state to a ferrimagnetic state is observed for the compound HoGe at 4.2 °K when studied in moderate fields; when pulsed field up to 80.000 Oe is used a second transition point is observed. The susceptibility at high temperatures of most of the compounds follows the Curie-Weiss law with effective moments close to those of the trivalent free rare earth ions. In the case of SmGe, a contribution to the susceptibility from higher multiplets is observed.

Für elf intermetallische Verbindungen zwischen Seltenen Erden und Germanium, die durch die Formel RGe dargestellt werden, werden die magnetischen und strukturellen Daten angegeben. Die FeB-Struktur wird für R = La und Ce, die CrB-Struktur für R = Pr, Nd, Sm, Gd, Tb, Dy, Ho, Er und Y gefunden. Die Verbindungen LaGe und YGe sind Paulische Paramagnetika. Die Verbindungen PrGe und NdGe werden bei Abkühlung ferromagnetisch, während CeGe, SmGe, GdGe, TbGe, DyGe und ErGe sich antiferromagnetisch ordnen. In mittleren Feldern wird bei 4,2 °K ein feldinduzierter Übergang vom antiferromagnetischen zum ferrimagnetischen Zustand für die Verbindung HoGe beobachtet. In Impulsfeldern von 80 kOe wird ein zweiter Übergangspunkt gefunden. Bei hohen Temperaturen folgt die Suszeptibilität der meisten Verbindungen dem Curie-Weiss'schen Gesetz mit effektiven Momenten, die nahe bei denen der dreiwertigen freien Ionen der Seltenen Erden liegen. Für SmGe wird ein Beitrag von höheren Multipletts zur Suszeptibilität gefunden.

1. Introduction

Investigations dealing with the structural and magnetic properties of equiatomic intermetallic compounds in which one component is a rare-earth metal have received considerable attention in recent years.

The present study is concerned with compounds of the formula RGe in which R represents a rare-earth metal or yttrium. Measurements of the electric conductivity showed that these compounds will behave essentially as metallic in character. Due to the small radial extent of the 4f wave functions it may be assumed that the direct exchange will be small and that an indirect exchange via the conduction electrons will predominate. In this case one would expect that the magnetic coupling will be dependent on the number of conduction electrons. It is shown below that the crystal structures of most of the compounds RGe are isotopic with those of RNi and RGa. As the number of valence electrons is different for each series of compounds, a comparison of their magnetic behaviour might be interesting.

2. Experimental

The rare-earth elements used in this investigation had a purity of 99.9%, germanium had a purity of at least 99.99%. The compounds were prepared by melting the stoichiometric proportions of the elements together in an arc-furnace. The samples were vacuum-annealed subsequently in ThO_2 crucibles. After annealing the samples were powdered and examined by X-ray diffraction methods using a Philips X-ray powder diffractometer type PW 1050/30.

CuK_α-radiation was used for all compounds investigated. The magnetic measurements were carried out for temperatures ranging from 4.2 to 600 °K with applied fields up to 30 kOe. Incidentally the magnetic behaviour was studied in a pulsed field up to more than 80 kOe [2].

Depending on the required sensitivity the magnetizations or susceptibilities were measured by a Faraday method with a null coil pendulum magnetometer of the type described before [1]. In cases that a higher sensitivity was needed use was made of a Curie balance. The measurements were performed on powdered single-phase materials.

3. Results and Discussion

As was found from X-ray investigation, the main part of the RGe compounds crystallises with the CrB-type of structure. Only LaGe and CeGe belong to the FeB-type which is, however, closely related [3] to the CrB-type. In both structure types the R atoms have coordination numbers of 7, and the FeB- and CrB-types are often considered as arising as a sort of transition structure between the cubic NaCl-type (coordination number 6) and the cubic CsCl-type (coordination number 8) [4].

The lattice constants for these compounds are given in Table 1. The structures and lattice constants for a few of the compounds listed above have already been reported in the literature. The present results are in close agreement with the data given by Parthé et al. [5] though they disagree slightly with those observed by Baenziger and Moriarty [6].

In the sequence RNi a similar change-over in structure from the FeB-type to the CrB-type is observed [7, 8]. It is worth noting, however, that for the nickel compounds the first rare-earth elements give rise to the CrB-type whereas

Table 1

Compound	Lattice constants (Å)			Unit cell volume (Å^3)	d_x	Type of structure
	a	b	c			
LaGe	8.474	4.118	6.097	212.8	4.121	FeB
CeGe	8.343	4.081	6.031	205.3	4.298	FeB
PrGe	4.479	11.083	4.050	201.0	4.406	CrB
NdGe	4.444	10.99	4.027	196.7	4.575	CrB
SmGe	4.376	10.87	3.995	190.0	4.869	CrB
GdGe	4.318	10.77	3.966	184.4	5.161	CrB
TbGe	4.285	10.70	3.942	180.8	5.316	CrB
DyGe	4.256	10.65	3.921	177.7	5.488	CrB
HoGe	4.235	10.60	3.911	175.7	5.607	CrB
ErGe	4.208	10.58	3.897	173.6	5.728	CrB
YGe	4.259	10.67	3.932	178.6	3.755	CrB

Table 2

Compound	Observed transition temp. (°K)	Asymptotic Curie temp. (°K)	Effective moment (μ_B) observed	Effective moment (μ_B) calculated
LaGe	—	Pauli paramagnetic		
CeGe	10	~ 0	2.50	2.54
PrGe	39	+30	3.69	3.58
NdGe	28	+18	3.74	3.62
SmGe	40	No Curie-Weiss behaviour		
GdGe	62	−13	8.21	7.94
TbGe	48	− 5	9.76	9.72
DyGe	36	~ 0	10.60	10.63
HoGe	18	− 5	10.77	10.60
ErGe	7	−15	10.0	9.60
YGe	—	Pauli paramagnetic		

the FeB-type is observed for the heavier rare-earth elements. This sequence in structure type is opposite to that of the RGe compounds suggesting that the size effect is of minor importance for these structure types. Both types of structure have the same number of atoms per unit cell. From the smooth decrease in unit cell volume of the RGe compounds with increasing atomic number of R it can be derived that the cerium atom in the compound CeGe will occur in the trivalent state. The results of the magnetic measurements are summarized in Table 2 and Fig. 1 to 5. With the exception of SmGe good agreement of the magnetic susceptibility with the Curie-Weiss law is observed for all compounds above, say, 75 °K. The effective (paramagnetic) moments determined from the slope of the curve of the reciprocal susceptibility vs. temperature are listed in Table 2. They agree rather well with the effective moments calculated for the free trivalent rare-earth ions (last column of Table 2), which means that in the compounds RGe the rare-earth atoms have the same configuration in their 4f orbitals as the free tripositive ions. As was already anticipated from crystallographic work these

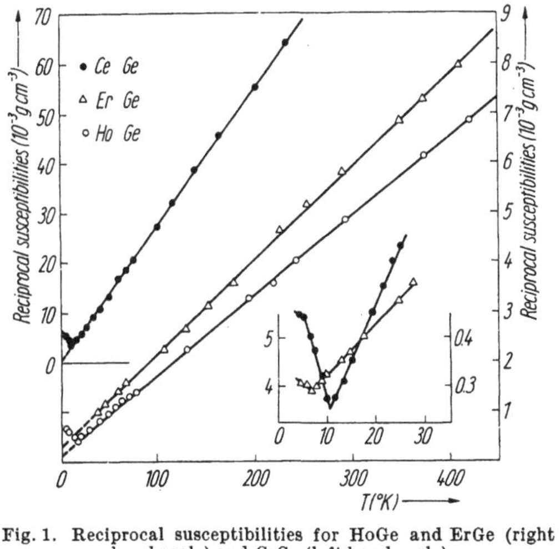

Fig. 1. Reciprocal susceptibilities for HoGe and ErGe (right hand scale) and CeGe (left hand scale)

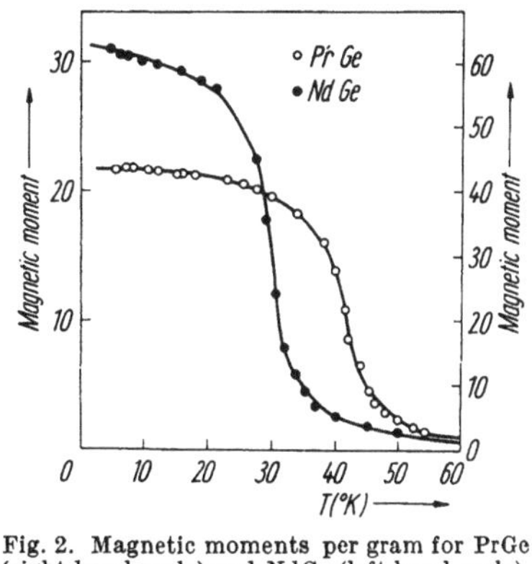

Fig. 2. Magnetic moments per gram for PrGe (right hand scale) and NdGe (left hand scale), in an applied field of 6000 Oe

data confirm the fact that cerium is trivalent in the compound RGe. It is interesting to note that in the sequence of compounds RNi the compound CeNi acts differently in so far as magnetic results have shown it to be tetravalent [7, 8]. Although no effective moment could be derived in the case of SmGe the crystallographic data of Table 1 indicate that also in this compound the rare-earth metal is trivalent.

From Fig. 1 to 5 it is seen that at low temperatures a magnetic transition point is observed for almost all compounds. The compound CeGe (Fig. 1) has a sharp minimum in the χ^{-1} vs. T plot and very likely becomes an antiferromagnet below 10 °K. Susceptibilities have been determined only up to 300 °K in this case as the CeGe was subject to heavy oxidation on further heating. The compounds PrGe and NdGe are ferromagnetic below 39 and 28 °K respectively (Fig. 2). The value for the saturation moments are 1.73 μ_B for PrCe and 1.64 μ_B for NdGe; the latter value was obtained by plotting the magnetization vs. H^{-2} and extrapolating to $H \to \infty$. These moments are considerably smaller than those expected for the free trivalent ions i.e. 3.20 μ_B for PrGe and 3.27 μ_B for NdGe. This behaviour is, however, rather common for intermetallic compounds of the rare-earth elements and may be due in most cases to a partial quenching of the orbital angular momentum on cooling by the crystal field of the surrounding ions.

Results for SmGe are shown in Fig. 3. The magnetic behaviour above 50 °K can easily be understood in terms of van Vleck's treatment of the Sm^{3+} ion [9]. The deviation from the Curie-Weiss law is due in this instance to the rather small interval between the lowest multiplets so that for higher temperatures we may expect a relatively large contribution to the effective moment from higher lying levels. The behaviour of SmGe at low temperatures is quite similar to that of the compounds GdGe, TbGe, and DyGe (Fig. 3): all four compounds give rise to a minimum in the χ^{-1} vs. T plot at 40, 62, 48 and 36 °K respectively, which may point to an antiferromagnetic ordering below these temperatures. The initial rise of the reciprocal susceptibility below the Néel point is followed in each case, however, by a decrease in χ^{-1} at still lower temperatures. This behaviour suggests that below the Néel temperature a new contribution to the susceptibility arises which increases on cooling and, when superimposed on the decreasing antiferromagnetic susceptibility, will give rise to the observed maxima in the χ^{-1} vs. T curve. Within experimental error, no remanence was observed at 4.2 °K.

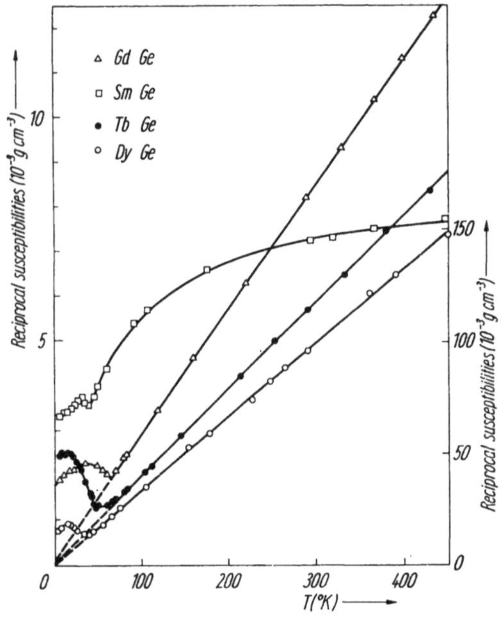

Fig. 3. Reciprocal susceptibilities for SmGe (right hand scale) and GdGe, TbGe, DyGe (left hand scale)

This indicates that the additional susceptibility contribution is of paramagnetic rather than ferromagnetic nature. The enhancement in susceptibility is too strong and occuring at too high temperatures to be caused by traces of paramagnetic oxide impurities that escaped detection by X-ray diffraction.

It is possible that the enhanced susceptibility observed on cooling below the Néel temperature is of a supraparamagnetic or superantiferromagnetic origin. As indicated by Néel such phenomena may be observed in bulk antiferromagnetic materials [10, 11]. The susceptibility curve should in this case be dependent on the conditions of preparation of the sample. Observations on GdGe samples support this view. It is interesting to note that similar maxima in the χ^{-1} vs. T curves have also been observed for the RAg compounds [12, 13].

The results for ErGe and HoGe are shown in Fig. 1, 4, and 5. For these compounds sharp minima are observed in the reciprocal susceptibility plots so that these compounds very likely are antiferromagnetic at low temperatures. The susceptibility of the compound HoGe proved to be field dependent near the magnetic transition point. A study of the magnetization curves for various field strengths showed that the observed peaks are slightly shifted to lower temperatures as the field is increased (Fig. 4). This means that a higher field tends to prevent the antiferromagnetic coupling. Fig. 5 shows the magnetic results obtained for HoGe at 4.2 °K in a pulsed field up to about 80000 Oe. The observed behaviour indicates a field-induced transition from an antiferromagnetic configuration at small fields to a ferromagnetic configuration for very high fields.

It is seen, however, that the final ferromagnetic state is not reached immediately; the occurrence of two critical fields instead of one may be evidence for the presence of a third magnetic configuration existing between the fields H_{c_1} and H_{c_2}. Here only part of the antiferromagnetic ordering has been overcome and hence this intermediate state is best arranged under the broad classification of ferrimagnetism. The magnetic moment reached at the critical field H_{c_2} is

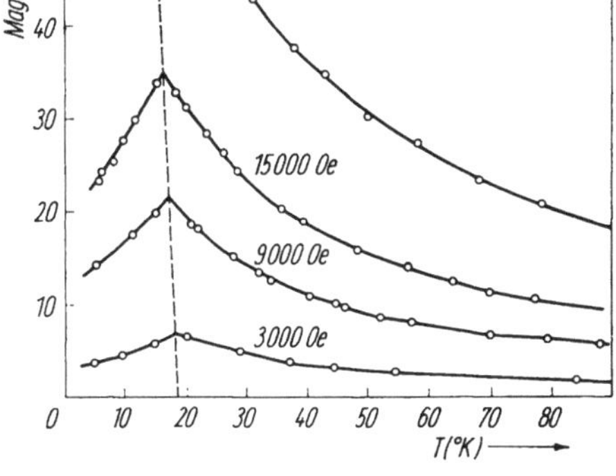

Fig. 4. Magnetic moments per gram vs. temperature for HoGe in various applied fields

Fig. 5. Magnetic moment per gram vs. applied field for HoGe at 4.2 °K

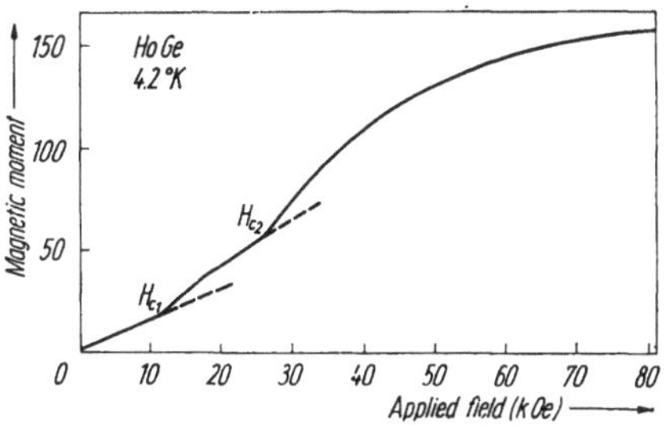

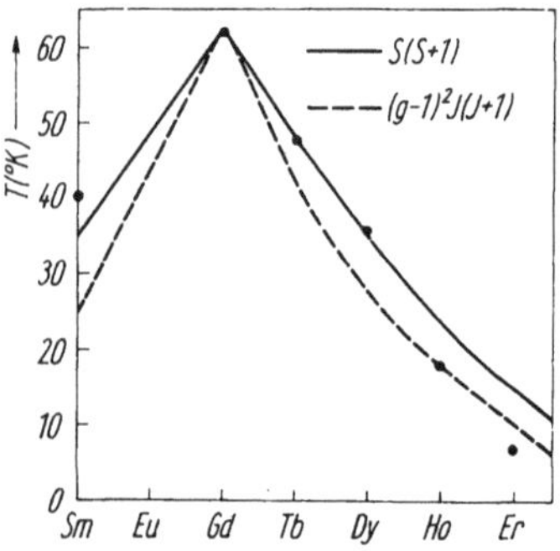

Fig. 6. Normalized values of $(g-1)^2 J(J+1)$ and $S(S+1)$ for some trivalent elements as compared with the transition temperatures of the corresponding RGe compounds

approximately 1/3 of the final ferromagnetic moment. This would suggest that if magnetic configurations less complicated than the screw-type are accepted, the number of sublattices in the intermediate state is a multiple of 3.

For a complete determination of the magnetic structure of the compound HoGe in various fields proper neutron diffraction work will be needed. The final moment reached in the ferromagnetic state beyond 80 kOe corresponds to 6.5 μ_B. This value is rather low compared to $g J = 10\ \mu_B$. The reason for this is probably the quenching of part of the orbital momentum by the crystal field, as was also observed for the compounds PrGe and NdGe.

Measurements on ErGe show that this compound has a specific resistivity of about 170 $\mu\Omega$ cm which is approximately twice the value observed for pure erbium metal. The compound ErGe can thus be considered as metallic and it is reasonable to assume that the other compounds RGe will behave similarly. In that case the main contribution to the magnetic coupling of the rare-earth moments will proceed by an indirect exchange via the conduction electrons.

This implies that the strength of the interaction between the magnetic moments should be proportional to $(g-1)^2 J(J+1)$ or to $S(S+1)$ for the spin only case [14]. In molecular field theory the magnetic transition temperature for an antiferromagnet may in general be expressed as a linear function of the strengths of the magnetic interactions. That means that for compounds giving rise to the same type of magnetic ordering and having about the same lattice parameters the antiferromagnetic transition temperature should also vary as $(g-1)^2 J(J+1)$ or as $S(S+1)$. In Fig. 6 the ordering temperatures of the isotypic compounds which on cooling become antiferromagnetic are plotted against atomic number. The values of $(g-1)^2 J(J+1)$ and $S(S+1)$ are added for comparison. As already mentioned the compounds RGe are isostructural with the compounds RNi and RGa. Contrary to the compound RGe we encounter ferromagnetism in most of the compounds RNi [7, 8] and RGa [15]. A comparison of the lattice parameters of the compounds RNi, RGa, and RGe shows that those observed for RGe lie between those of RNi and RGa. This means that if a Ruderman-Kittel-Yosida type of exchange interaction is accepted to occur in these compounds the main reason for the difference in magnetic behaviour is probably due to the difference in the number of valence electrons.

Acknowledgement

The authors want to thank Dr. Ir. H. Zijlstra for his help concerning the pulsed field experiments.

References

[1] U. Enz, J. F. Fast, S. van Houten, and J. Smit, Philips Res. Rep. **17**, 451 (1962).
[2] L. A. Ellenkamp and H. Zijlstra, Z. angew. Phys. **16**, 400 (1963).
[3] K. Schubert, Kristallstrukturen zweikomponentiger Phasen, Springer-Verlag 1964.
[4] O. Schob and E. Parthé, Acta cryst. **19**, 214 (1965).
[5] E. Parthé, D. Hohnke, W. Jeitschko, and O. Schob, Z. Naturf. **52**, 155 (1965).
[6] N. C. Baenziger and J. S. Moriarty, Acta cryst. **14**, 946 (1961).
[7] R. E. Walline and W. E. Wallace, J. chem. Phys. **41**, 1587 (1964).
[8] S. C. Abrahams, J. Z. Bernstein, R. C. Sherwood, J. H. Wernick, and H. J. Williams, J. Phys. Chem. Solids **25**, 1069 (1964).
[9] J. H. van Vleck, The Theory of Electric and Magnetic Susceptibilities, Oxford University Press 1932.
[10] L. Néel, J. Phys. Soc. Japan **17** S, B-I 676 (1962).
[11] I. S. Jacobs and C. P. Bean, in Magnetism, Vol. III, Ed. G. T. Rado and H. Suhl, Academic Press 1963 (p. 294).
[12] J. Pierre and R. Pauthenet, C. R. Acad. Sci. (France) **260**, 2739 (1965).
[13] R. E. Walline and W. E. Wallace, J. chem. Phys. **42**, 604 (1965).
[14] P. G. de Gennes, C. R. Acad. Sci. (France) **247**, 1836 (1958).
[15] K. H. J. Buschow, unpublished results.

(Received May 2, 1966)

phys. stat. sol. **16**, 475 (1966)

Edgar C. Brain Laboratory for Fundamental Research, United Steel States Corporation, Research Center, Monroeville, Pennsylvania

Ferromagnetic Curie Temperatures of Some Iron–Zinc Solid Solutions

By

H. A. Wriedt and S. Arajs

The preparation of homogeneous iron–zinc alloys by zinc-vapor impregnation of iron is described. Ferromagnetic Curie temperatures of the b.c.c. solid solutions containing 11.2, 14.4, and 20.1 at% zinc were determined to be 981.3, 957.0, and 925.0 °K, respectively. These measurements are in good agreement with the results obtained by Fallot but not with those of other investigators. Zinc lowers the Curie temperature of iron linearly by about 5.8 deg/at% zinc. This lowering is inconsistent with the simple molecular field predictions.

Es wird die Herstellung von homogenen Eisen–Zink-Legierungen durch eine Zinkdampf-Behandlung des Eisens beschrieben. Für 11,2; 14,4 und 20,1 at% Zink wurde die ferromagnetische Curie-Temperatur zu 981,3; 957,0 und 925,0 °K bestimmt. Die Messungen befinden sich mit den Fallotschen Ergebnissen in guter Übereinstimmung, nicht aber mit denen anderer Autoren. Zink erniedrigt die Curie-Temperatur von Eisen linear um etwa 5.8 grd/at% Zink. Diese Erniedrigung stimmt nicht mit Vorhersagen der einfachen Molekularfeldtheorie überein

1. Introduction

The ferromagnetic Curie temperature, denoted by T_F, of a ferromagnetic substance is the temperature at which spontaneous magnetization disappears. This temperature is not always easy to determine experimentally because of the difficulty in determining the spontaneous magnetization. For this reason, and sometimes also because of inhomogeneous alloys, one often finds considerable scatter among the reported Curie temperatures for a particular alloy system.

Recently [1] we used a simple technique for accurate determination of the ferromagnetic Curie points of binary iron alloys containing germanium, silicon, molybdenum, and manganese. In this paper we present the results of a similar study on iron-rich iron–zinc solid solutions and discuss their significance.

2. Experimental Considerations

Three homogeneous, coarse-grained iron–zinc alloys, containing 11.2, 14.4, and 20.1 at% zinc, were made by impregnating purified iron strips with zinc from the vapor phase in evacuated Vycor capsules. The impurities found in the iron strips before the alloying with zinc are reported in Table 1. To produce alloys free from gross porosity, the solid b.c.c. iron was converted directly, without intermediate stages of Γ phase or liquid, to b.c.c. solutions of the required compositions. The procedure which excluded phase changes in the alloys required, as a first step, the separate preparation of some Γ phase alloy with 72 wt% zinc. This was made from purified zinc granules (99.99% Zn) and the purified iron rolled to sheet 0.0005″ thick. The metals were sealed in a tubular Vycor capsule with 100 Torr hydrogen (room temperature); a local constriction of the capsule cross-section kept the iron from contact with the pool of liquid zinc which formed when the capsule was heated to 1040 °K for reacting. At this temperature the zinc pool disappeared between the 10th and the 14th day, when the capsule was water-quenched. In being converted to Γ phase alloy, the 0.0005″ sheets had shrunk, crinkled, and sintered somewhat. This alloy was crushed, divided, and resealed with 0.012″ strips of the purified iron in a set of three evacuated Vycor capsules. The alloy and the pure iron were kept apart by a neck in the capsule as before. The impregnation anneal proceeded first at 1040 °K for 9 days; when the possibility of liquid forming on the alloy surfaces had passed, the temperature was raised to 1160 °K. The capsules were oil-quenched after one month at 1160 °K. Bright surfaces on the quenched alloys indicated that very little oxygen had diffused through the capsule walls.

The zinc contents of the alloys were determined with the electron microprobe, by lattice parameter measurement with X-rays, and by chemical analysis. Using the microprobe to determine the zinc contents at the core and near the surface of each strip at several locations remote from the edges, we established that all three alloys are chemically homogeneous to $\pm$ 0.1 at% zinc.

The ferromagnetic Curie temperatures were determined by use of a technique described elsewhere [2]. Since it was impossible to make solid spheres from the above-described iron-zinc alloys, the specimens for the Curie point determinations were prepared by breaking samples of the alloys into small pieces and sealing these pieces with 15 Torr argon (room temperature) in a silica capsule. The alloy occupied an approximately hemispherical space (ca. 0.5 cm ⌀) at the lower end of the capsule; the remainder of the capsule was filled with a solid silica rod. After completion of the magnetization measurements, which required about 4 h per curve of the type shown in Fig. 1, no significant condensation of zinc was found on the walls of any capsule. Fig. 1 shows a typical curve of magnetization (σ) as a function of absolute temperature (T). The ferromagnetic Curie temperature appears as a break in the σ vs. T plot which is similar to those occurring when solid spherical specimens are used [1].

Table 1

Impurity content of the purified iron before alloying with zinc

Impurity element	Impurity content (ppm)	Impurity element	Impurity content (ppm)
Ag	< 1	Mo	10
Al	5	N	3
As	<10 ND	Ni	35
B	< 5	O	16
Be	< 5 ND	P	7
C	50	Pb	5
Ca	1	S	<10
Cd	< 5 ND	Sb	<10 ND
Co	40	Si	10
Cr	30	Sn	5
Cu	8	Ti	20
Ga	<50 ND	V	2
Mg	< 1	W	<10 ND
Mn	4		

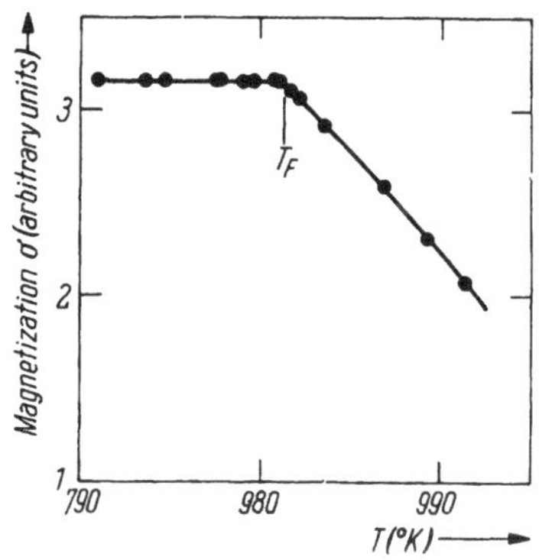

Fig. 1. Determination of the ferromagnetic Curie temperature for the iron +11.2 at% zinc alloy with H_a = 90 Oe

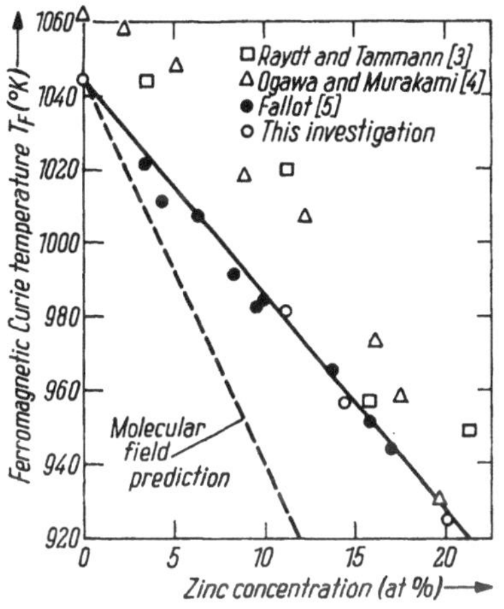

Fig. 2. Ferromagnetic Curie temperatures of iron-zinc solid solutions

3. Results and Discussion

The ferromagnetic Curie temperatures of the iron–zinc solid solutions studied in this investigation are summarized in Table 2. The quantity T_F for pure iron in this table was determined with chips from the same stock of iron that had been used previously in making a sphere for which T_F was found to be 1044.1 °K. The iron chips were enclosed in a silica capsule in the same way as were the iron–zinc alloys chips. Fig. 2 shows our Curie temperature data for iron–zinc alloys (through which the solid line is drawn) with that of previous studies [3 to 5] for which numerical values have been published. Our measurements are in very good agreement with the results obtained by Fallot [5] but not with those of the other investigators. Fallot's data and ours indicate that zinc lowers the ferromagnetic Curie point of iron linearly by about 5.8 °K/at% zinc.

From simple molecular field considerations [1], the ferromagnetic Curie temperature should be proportional to $(1 - c)$, where c is the atomic fraction of the "non-magnetic" solute atoms. This prediction, which, for small values of c, is supported also by more advanced theoretical treatments based on the localized magnetic moment model (briefly discussed in [1]) is shown as a dotted line in Fig. 2. Obviously, the experimental behavior of T_F for iron–zinc solid solutions does not follow this simple dilution behavior. In this respect the iron–zinc system is similar to the b.c.c. solid solutions of iron with silicon, germanium, and molybdenum. The latter alloys also exhibit large deviations from the simple molecular field theory expectations. Theoretical understanding of the variation of T_F with solute concentration in iron alloys is thus still very incomplete.

Table 2

Ferromagnetic Curie temperatures of iron–zinc alloys

Zinc concentration (at%)	T_F (°K)
0	1044.0 ± 0.5
11.2 ± 0.2	981.3 ± 0.5
14.4 ± 0.2	957.0 ± 1.0
20.1 ± 0.3	925.0 ± 1.0

Acknowledgements

The authors are grateful to the following members of this laboratory for their contributions to this study: H. A. Hughes, who assisted in preparing the alloys; W. L. Doyle, J. A. Gula, and L. Zwell, who conducted the wet chemical, microprobe, and X-ray methods, respectively, of the alloy analysis, G. P. Wray and J. W. Conroy, who assisted with the magnetization measurements.

References

[1] S. Arajs, phys. stat. sol. **11**, 121 (1965).
[2] S. Arajs and R. V. Colvin, J. appl. Phys. **35**, 2424 (1964).
[3] U. Raydt and G. Tammann, Z. anorg. Chem. **83**, 257 (1913).
[4] Y. Ogawa and T. Murakami, Tech. Rep. Tohoku Univ. **8**, 53 (1929).
[5] M. Fallot, Ann. Phys. (France) **7**, 420 (1937).

(Received May 2, 1966)

phys. stat. sol. **16**, 479 (1966)

Tube Investments Research Laboratories, Hinxton Hall, Cambridge

Interrupted Vapour Deposition of Epitaxial Gold Films Grown Inside an Electron Microscope

By

M. J. Stowell and T. J. Law

During the growth of epitaxial gold films inside an electron microscope, the gold vapour stream was interrupted in order to study changes in the deposit structure which occur in the absence of incident vapour. It is shown that rapid shape changes of the deposit islands characteristic of 'liquid-like' coalescence continue when deposition is stopped and that further coalescences do occur. The results of these experiments give added confirmation to the thesis of Pashley et al. [1] that the 'liquid-like' shape changes of coalescing deposit islands are brought about by surface self-diffusion motivated by capillarity forces.

Um die Strukturänderungen epitaxial aufgedampfter Goldschichten bei Abwesenheit des einfallenden Dampfes zu untersuchen, wurde während des Wachstums im Elektronenmikroskop der Golddampfstrahl unterbrochen. Es werden schnelle Formänderungen der aufgedampften Inseln, kennzeichnend für ‚liquid-like' Verschmelzungen, sowie weitere Verschmelzungen gefunden, wenn die Aufdampfung unterbrochen wird. Diese experimentellen Ergebnisse ergeben eine zusätzliche Bestätigung der These von Pashley et al. [1], daß die ‚liquid-like' Formänderungen der verschmelzenden Inseln durch Oberflächenselbstdiffusion infolge von Kapillarkräften verursacht werden.

1. Introduction

In a previous paper from this laboratory on the growth of (111) gold films inside an electron microscope [1], evidence was presented to show that the 'liquid-like' coalescence of deposit islands most probably takes place by surface self-diffusion of the deposited material. Some preliminary observations on the effect of interrupting the vapour stream were reported, and these provided support for this conclusion. It was noted that, in the final stages of growth of the discontinuous film, channels which had begun to fill in just before the gold vapour supply was stopped, continue to fill in afterwards. The purpose of this paper is to report further observations of deposit changes which take place in the absence of incident vapour.

2. Experimental

Oriented (111) gold films were grown inside a Siemens Elmiskop I electron microscope, the modified specimen chamber of which has been described in detail by Pashley et al. [1]. The electron image was recorded from the microscope viewing screen on Kodak Tri-X ciné film and the camera was run at 8 frames/s and 16 frames/s. At various stages of the film growth, a shutter was placed between the off-axis vapour source and the substrate to prevent any further direct deposition of gold, while still allowing the electron image to be observed. Most of the experiments were performed with the substrate temperature at 400 °C; in one case, a temperature of 475 °C was employed.

3. Results

Before considering the effect of interrupting the vapour stream, we will briefly mention the structural changes which take place while the deposit is continually growing. During the early stages of growth, deposit islands slowly increase in size until they touch a neighbour. A neck having high surface curvature then forms between the islands and, initially, grows very rapidly. This growth is accompanied by marked shape changes of the composite island. The rate of neck growth decreases rapidly as the neck increases and the final stage of coalescence corresponds to adjustment of the island to a crystallographic equilibrium shape determined by anisotropy of the surface free energy. During the last stages of growth of a discontinuous deposit long, channels ≈ 100 Å across and ≈ 300 to 400 Å high are rapidly filled in once they have been bridged, and this process is also accompanied by pronounced shape changes of the surrounding deposit.

Changes in island topography which take place in the absence of incident vapour will now be considered. Three main results evolve from our studies: 1. Most of the islands exhibit no detectable topographic change over periods up to about 30 s after deposition stops; in certain instances, very small changes can be detected. 2. The shape changes arising from a coalescence event which takes place just before the vapour stream is cut off continue in the absence of incident vapour. 3. These shape changes often cause an island to touch and coalesce with a neighbour. The time-scale of these latter coalescences, which occur in the absence of incident vapour, is comparable with those which take

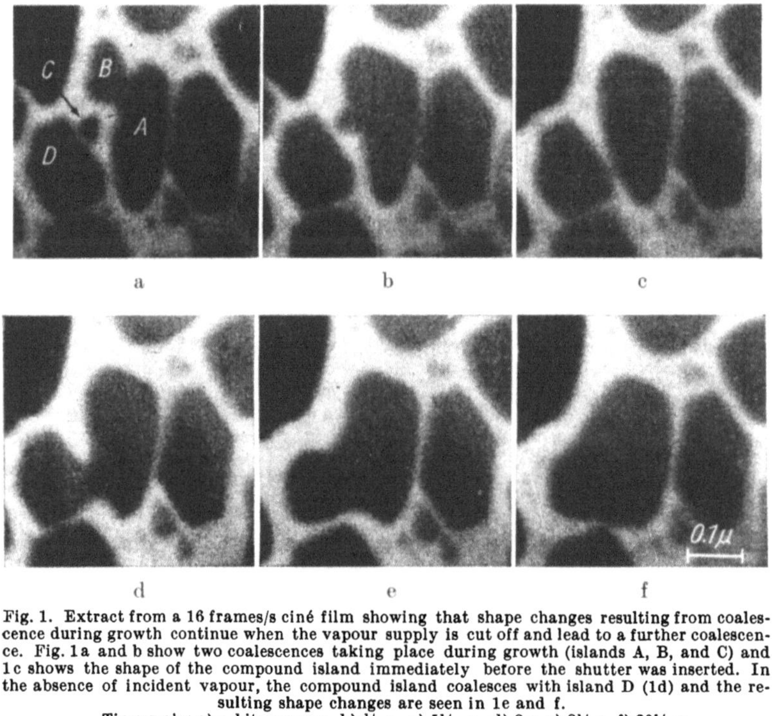

Fig. 1. Extract from a 16 frames/s ciné film showing that shape changes resulting from coalescence during growth continue when the vapour supply is cut off and lead to a further coalescence. Fig. 1a and b show two coalescences taking place during growth (islands A, B, and C) and 1c shows the shape of the compound island immediately before the shutter was inserted. In the absence of incident vapour, the compound island coalesces with island D (1d) and the resulting shape changes are seen in 1e and f.
Time scale: a) arbitrary zero, b) $^1/_2$ s, c) $5^1/_{16}$ s, d) 8 s, e) $8^1/_2$ s, f) $20^1/_4$ s

place between islands of similar sizes when vapour is continually being supplied to the deposit. These results are illustrated by means of the three following sequences taken from ciné films.

Fig. 1a to c show changes which take place while vapour is incident on the specimen and the remainder of the sequence, Fig. 1d to f, illustrate the changes occurring after the shutter has been inserted. In Fig. 1a islands A and B have just coalesced. The composite island changes shape and neck growth between the two original islands causes interaction with the small island C, leading to the coalescence shown in Fig. 1b. Fig. 1c, which is the frame immediately before the shutter was inserted, shows how the shape of the island A has changed. A close frame-by-frame inspection of the ciné film shows that this shape change is continuous. Some three seconds after the shutter is inserted, islands A and D coalesce (Fig. 1d) (the two islands were seen to be separate on the frame preceeding Fig. 1d) and half a second later, the composite island shows a marked change in shape (Fig. 1e). After this initially rapid change, which eliminates regions of very high surface curvature, the rate of change of shape is more gradual. The last frame before the shutter is removed is seen in Fig. 1f ($11^3/_4$ s after Fig. 1e) and it is clear that further changes have taken place. The island is gradually tending towards a crystallographic shape although this state is not achieved completely.

A second example of a coalescence taking place in the absence of incident vapour is shown in Fig. 2. Island B, which is the product of a coalescence which took place before the shutter was inserted, is changing shape by contraction in a vertical direction and expansion in the orthogonal direction (cf. Fig. 2a and b). The coalescence between islands A and B takes place $1^5/_8$ s after the shutter is inserted and the sequence of Fig. 2 illustrates the continuous nature of the topographic changes during the time that no vapour was incident on the specimen.

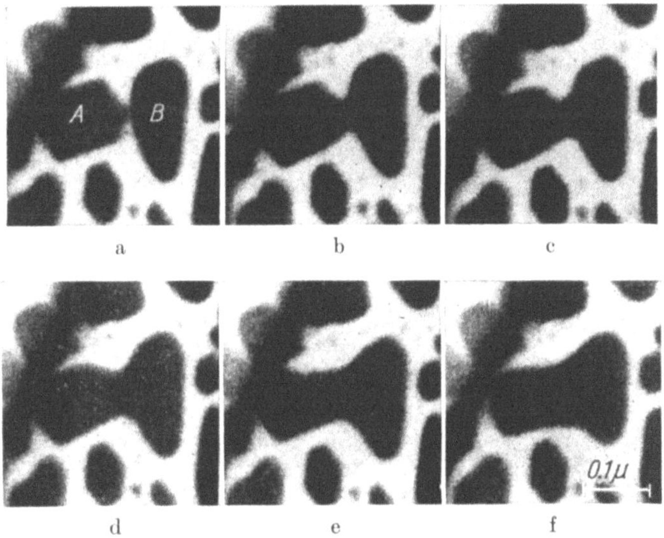

Fig. 2. Ciné film sequence which shows the coalescence of islands A and B in the absence of incident vapour. Notice that island A loses its crystallographic shape on coalescence and that island B shows a pronounced change of length in the vertical direction during coalescence.
Time scale: a) arbitrary zero, b) $^1/_{16}$ s, c) $^1/_8$ s, d) $^1/_2$ s, e) $1^7/_8$ s, f) $5^1/_4$ s

Such events occur at all stages of growth of a discontinuous film. The coalescence of very large islands and the elimination of channels take place in a 'liquid-like' manner when a film is being continuously grown and it would be expected that in the event of interruption of the vapour supply some structural modifications would continue in the manner just illustrated. Fig. 3 shows that this is indeed so. A channel C between two large islands, which had coalesced prior to insertion of the shutter, continues to fill in when no atoms are incident from the vapour. Because of the smaller surface curvatures which appear at this stage of growth, the changes are less dramatic than at an earlier stage (Fig. 3a, b). As the channel is eliminated, small islands which have grown from secondary nuclei (see [1]) coalesce with the main deposit mass, as shown in Fig. 3c, d, and e. These interactions lead to an accelerated and jerky motion of the growing front. However, even without these secondary coalescences, re-entrant regions continue towards a state of lower surface curvature, as may be seen by comparing Fig. 3e and f.

It was mentioned earlier in this section that most islands exhibit no detectable change in shape when the vapour supply is stopped. One of the interesting features of this deposit–substrate combination is that, during continuous growth inside an electron microscope, crystallographically shaped islands become rounded after coalescence and later revert to a crystallographic form [2]. This reversion to a regular shape has never been observed during the time that the beam-limiting shutter has been in position, even for very small islands. We conclude that, in the poor vacuum conditions in which our films are grown, a continuous supply of vapour is necessary for the final shape changes to occur.

Although we have illustrated the more drastic shape changes which take place without incident vapour, we emphasize that many small changes are observed.

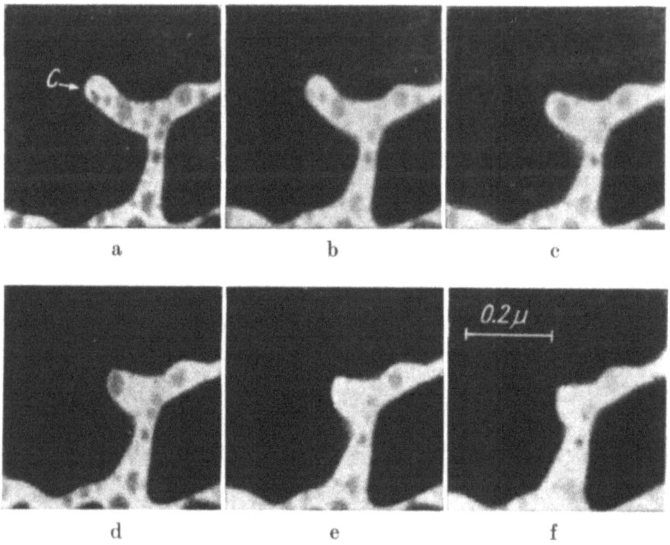

Fig. 3. A channel C which was filling in when vapour was being deposited, continued to be eliminated after the vapour limiting shutter had been inserted. Motion of the highly curved extremity of the channel is clear in 3a and b. Further advancement of this front causes coalescences with small secondary nuclei (3c, d, and e) and continued filling in of the channel is apparent in 3f.
Time scale: a) arbitrary zero, b) 2 s, c) 6 s, d) 8 s, e) 10 s, f) 22 s

The fact that most of the islands do not exhibit detectable changes is probably due in part to the limitations of our image recording technique. By using a very fast emulsion for ciné recording the electron image from the electron microscope screen, much of the detail at island edges is lost and we would not expect to detect very small topographic changes.

4. Discussion

It has been shown in the preceeding section that the shape changes resulting from coalescence of deposit islands continue to take place when the supply of vapour to the growing film is stopped. The changes can lead to further 'liquid-like' coalescences and the rates of neck growth and morphological changes during these coalescences are comparable with those which take place in the presence of incident vapour. These results confirm qualitatively the thesis put forward by Pashley et al. [1] that 'liquid-like' coalescence is brought about by surface self-diffusion of the deposit material, motivated by capillarity forces. It is also evident that the vapour supply has little influence on the rapid topographical changes we observe during continuous growth. This may be deduced from the geometry of coalescing particles. If neck growth was due entirely to the collection of atoms directly from the vapour and from the bare substrate regions, the coalescing islands would grow in a manner suggested by the addition of a near uniform 'skin'. The growth of a neck between the islands would be observed but no marked changes in profile would occur in regions far from the neck. The rate of neck growth would be dependent solely on the deposition rate and would not be temperature dependent. In addition, shrinkage of islands on the substrate would not be observed and, of course, no changes what-so-ever would be observed when the vapour supply was stopped. It should be mentioned, however, that in our present experiments, and in those reported previously [1, 2], deposition rates were less than about 1 Å/s. At much higher rates of deposition, it is highly likely that necks between coalescing islands can never be completely eliminated while the deposit is discontinuous. This is a very important factor in the determination of the domain size of multiply positioned films and of the grain size of polycrystalline films [3]. During the later stages of growth at the slow rates we employ, this effect is clearly illustrated. The very large islands (several thousands of Ångstroms across) are rarely allowed to revert to crystallographic shapes (i.e. to coalesce completely) before uniform growth leads to another coalescence with an island of a comparable size, and, further, it is often observed that re-entrant regions are not completely eliminated before another coalescence takes place.

During the experiments reported here no serious attempt was made to investigate long term changes in deposit structure with no incident vapour (that is for times greater than about 30 s). The reason for this is that carbonaceous and residual gas contamination in the electron microscope is not entirely negligible, even at substrate temperatures approaching 500 °C. Although the rate at which contamination builds up at 400 °C is one or more orders of magnitude less than the deposition rate of gold under our normal experimental conditions, we do not consider that contamination can be neglected in the type of experiment discussed here. Hence, no extensive work has been performed on long anneals of deposited films. It is not surprising that most of the islands showed no detectable shape change while no vapour was incident on the deposit or that changes to crystallographic shapes were not seen. During continuous growth the final rounding off and transition of an island to a crystallographic shape takes place quite slowly and a time lapse of tens to hundreds of seconds measured from coalescence is needed for these changes to be complete. The effect on this process of a continuously increasing contamination layer could be to reduce continuously the surface mobility of deposit atoms and to inhibit the final stages of coalescence altogether after a sufficiently long time. There is evidence of marked changes of the surface self-diffusion coefficient D_s of copper [4] and silver [5] with gaseous environment. Although several gases appear to increase D_s, it is reasonable to assume that solid contaminants would decrease D_s in a pronounced manner. These arguments do not alter our conclusions on the mode of mass transfer during 'liquid-like' coalescence. The conclusions may possibly be modified if the contaminating molecules are forced, by some unknown mechanism, always to reside on the deposit surface. In this case, we would expect D_s to decrease as the film grew, thus reducing the role played by surface self-diffusion in the coalescence phenomena. We have no evidence to contradict this possibility but all our results indicate that the effect is small at the temperatures we have used in our studies. There is obviously a great need for 'in situ' studies of thin film growth to be carried out under ultra high vacuum conditions.

In the earlier account of our thin film growth experiments [1], it was concluded that, in the final stages of growth of a discontinuous deposit, the role of incident vapour in 'liquid-like' behaviour becomes more important with increasing island size. We wish to point out here that this does not infer that surface self-diffusion is unimportant. Indeed, it is the most important mechanism for bringing about the observed shape changes. But, as the size of the coalescing islands increases, the surface curvatures developed at the base of re-entrant regions generally decrease and the driving forces for the observed topographic changes are much lower than for smaller islands. Consequently, shape changes take place more gradually and the effect of contamination build-up is more serious. We consider that one important influence of the incident vapour is to maintain a relatively clean surface during growth so that surface diffusion remains the prominent mode of mass transport.

These latest observations modify two of our previous conclusions, which were based upon much less evidence. It was originally stated [1] that "... no new coalescences can start once the vapour supply ceases ...". Clearly, this is not true, since island shape changes brought about by surface self-diffusion can lead to fresh coalescences (e.g. Fig. 1 and 2). Further, it was also concluded in [1] that, for the coalescence of large islands, the final stage of neck growth "... results mainly from the freshly arriving vapour ...". Our latest results indicate that the surface self-diffusion mechanism is the prominent one right up to the final elimination of a neck.

Finally, we deal briefly with the possibility of obtaining data on surface self-diffusion from the rates at which these small islands coalesce. Because of theoretical difficulties and of the wide range of island geometries encountered, only order-of-magnitude estimates can be made. Preliminary analysis of our data leads to the conclusion that the value of D_s pertinent to the process we observe is probably a factor of ten or more greater than that used in [1] for the calculations of sintering rates of gold spheres at 400 °C. This is not unreasonable, as it appears that impurity effects hamper mass transport by surface self-diffusion even at temperatures near the melting point (Dr. N. A. Gjostein, private communication). Detailed considerations on this point will be published in due course.

Acknowledgements

We are grateful to Dr. D. W. Pashley for his continued interest and for stimulating discussion. Mr. A. G. Doggett assisted with the experimental work. This paper is published by permission of the Chairman of Tube Investments Limited.

References

[1] D. W. Pashley, M. J. Stowell, M. H. Jacobs, and T. J. Law, Phil. Mag. **10**, 127 (1964).
[2] D. W. Pashley and M. J. Stowell, Proc. 5th Internat. Conf. Electron Microscopy, Philadelphia 1962, Ed. S. S. Breese, Academic Press, New York 1962 (paper GG-1).
[3] M. J. Stowell and T. J. Law, phys. stat. sol. **16**, 117 (1966).
[4] F. J. Bradshaw, R. A. Brandon, and C. Wheeler, Acta metall. **12**, 1057 (1964).
[5] G. E. Rhead, Acta metall. **13**, 223 (1965).

(Received May 3, 1966)

phys. stat. sol. 16, 487 (1966)

Institut für Theoretische Physik (II) der Universität Marburg/Lahn

Energy Band Structures of Selenium and Tellurium (Kohn-Rostoker Method)[1]

By

J. TREUSCH and R. SANDROCK

The band structures of hexagonal Se and Te are calculated using a group theoretical treatment based on the Kohn-Rostoker method. With certain assumptions the method yields continuous wave functions with continuous derivatives. The energy gaps agree qualitatively with experiment. The polarization dependence of optical transitions can be explained by the present band models. Hole effective masses are estimated, and it is found that $m_\perp$ is about four times as large as $m_\parallel$ in both substances. The variation of the Se energy gap with pressure is investigated and found to fit experiments.

Die Bandstrukturen von hexagonalem Se und Te werden mit Hilfe der Kohn-Rostoker-Methode berechnet. An einem Beispiel wird die Ausnutzung der Kristallsymmetrie dargelegt. Es wird gezeigt, daß die Methode unter bestimmten Annahmen zu stetigen und stetig differenzierbaren Wellenfunktionen führt. Die rechnerisch bestimmten Bandabstände stimmen qualitativ mit dem Experiment überein. Die Polarisationsabhängigkeit der optischen Übergänge an der Bandkante läßt sich mit dem gegebenen Bändermodell erklären. Die effektiven Löchermassen werden abgeschätzt, $m_\perp$ ist für Se und Te ungefähr viermal so groß wie $m_\parallel$. Für Selen wird die Änderung des Bandabstandes mit dem Druck untersucht. Das Ergebnis entspricht dem Experiment.

Se and Te share a peculiar, highly anisotropic chain structure with a comparatively low degree of symmetry. For this reason, investigations of their band model have so far been confined to tight-binding approaches, considering only nearest neighbour interactions within one chain. It seemed worthwhile, therefore, to try a calculation that takes the exact crystal structure into account. For the sake of its good convergence, we chose the Green's function method by Kohn and Rostoker [1], which had not yet been applied to crystals other than cubic.

1. Method

The mean features of this method shall be described as applied to a Bravais-lattice. The latter be defined by the vectors

$$\boldsymbol{R}_n = n_1 \boldsymbol{\tau}_1 + n_2 \boldsymbol{\tau}_2 + n_3 \boldsymbol{\tau}_3 , \tag{1}$$

$\boldsymbol{\tau}_i$ being the fundamental translation vectors of the lattice. The reciprocal lattice is then given by

$$\boldsymbol{K}_n = n_1 \boldsymbol{\varkappa}_1 + n_2 \boldsymbol{\varkappa}_2 + n_3 \boldsymbol{\varkappa}_3 \tag{2}$$

with

$$\boldsymbol{\varkappa}_i \boldsymbol{\tau}_j = 2 \pi \delta_{ij} . \tag{3}$$

---

[1]) A short note concerning a part of this work has already been published in Solid State Commun. 3, 361 (1965).

The Schrödinger equation (in atomic units)

$$\big(-\triangle + V(\boldsymbol{r})\big)\,\Psi_{\boldsymbol{k}}(\boldsymbol{r}) = E(\boldsymbol{k})\,\Psi_{\boldsymbol{k}}(\boldsymbol{r}) \tag{4}$$

is to be solved with a periodic potential, that means

$$V(\boldsymbol{r} + \boldsymbol{R}_n) = V(\boldsymbol{r})\,. \tag{5}$$

The wave function consequently satisfies the Bloch condition. Equation (4) is interpreted as an inhomogeneous differential equation of second order (with $V(\boldsymbol{r})\,\Psi(\boldsymbol{r})$ as inhomogeneity) and solved using a Green's function,

$$\Psi_{\boldsymbol{k}}(\boldsymbol{r}) = \int G_{\boldsymbol{k}}(\boldsymbol{r}, \boldsymbol{r}')\,V(\boldsymbol{r}')\,\Psi(\boldsymbol{r}')\,\mathrm{d}\boldsymbol{r}'\,. \tag{6}$$

(The solutions of the homogeneous equation — plane waves $\exp(i\sqrt{E}\,r)$ — cannot satisfy the Bloch condition unless $E = (\boldsymbol{K}_n + \boldsymbol{k})^2$ and are neglected therefore.) Equation (6) is replaced by an equivalent variational principle:

$$\delta\Lambda = \delta \int \Psi^*(\boldsymbol{r})\,V(\boldsymbol{r})\,\{\Psi(\boldsymbol{r}) - \int G(\boldsymbol{r}, \boldsymbol{r}')\,V(\boldsymbol{r}')\,\Psi(\boldsymbol{r}')\,\mathrm{d}\boldsymbol{r}'\}\,\mathrm{d}\boldsymbol{r} = 0\,. \tag{7}$$

To make possible the integration a muffin-tin potential is assumed:

$$V(\boldsymbol{r}) = \begin{cases} V(r) & \text{for } r < \varrho \\ V_0 = \text{const} & \text{for } r \geqq \varrho \end{cases} \tag{8}$$

i.e. the atoms are surrounded with spheres inside which the potential is spherically symmetrical. Outside these spheres the potential is constant and can be set equal to zero by shifting the energy scale. For $r < \varrho$, $\Psi(\boldsymbol{r})$ can be expanded in the series

$$\Psi_{\text{in}}(\boldsymbol{r}) = \sum_{l=0,m}^{L} C_{lm}\,R_l(r)\,Y_{lm}(\boldsymbol{r})\,, \tag{9}$$

the $R_l(r)$ being solutions of the radial part of the Schrödinger equation

$$\left(-\frac{1}{r^2}\frac{\mathrm{d}}{\mathrm{d}r}\,r^2\,\frac{\mathrm{d}}{\mathrm{d}r} + \frac{l\,(l+1)}{r^2} + V(r) - E\right) R_l(r) = 0\,. \tag{10}$$

$G(\boldsymbol{r}, \boldsymbol{r}')$ can be expanded as [1, equation (3.13)]

$$G(\boldsymbol{r}, \boldsymbol{r}') = \sum_{lm}\sum_{l'm'} \{A_{lm\,l'm'}\,j_l(\varkappa r)\,j_{l'}(\varkappa r') + \varkappa\,\delta_{ll'}\,\delta_{mm'}\,j_l(\varkappa r)\,n_l(\varkappa r')\}\times$$
$$\times\; Y_{lm}(\boldsymbol{r})\,Y^*_{l'm'}(\boldsymbol{r}') \tag{11}$$

with $\varkappa = \sqrt{E}$.

When (9) and (11) are substituted into (7) and Green's formula is used twice to carry out the spherical integrals one obtains a system of linear equations for the coefficients C_{lm}, soluble only if

$$\det\left|A_{lm\,l'm'} + \varkappa\,\delta_{ll'}\,\delta_{mm'}\,\frac{n_l'(\varkappa r) - n_l(\varkappa r)\,L_l(r)}{j_l'(\varkappa r) - j_l(\varkappa r)\,L_l(r)}\bigg|_{r=\varrho}\right| = 0 \tag{12}$$

with $L_l(r) = \mathrm{d}\ln R_l(r)/\mathrm{d}r$, $l, l' \leqq L$.

Solving equation (12) one finds the required connection between E and $\boldsymbol{k}$, since the $A_{lm\,l'm'}$ are functions of E and $\boldsymbol{k}$. The accuracy of the solutions depends on the choice of L, but as described in former publications [2, 3] the essential advantage of the Kohn-Rostoker method is its rapid convergence.

Another way to solve our problem is based on a method given by Korringa [4, 5]. Starting from (6) one can introduce the condition of continuity of $\Psi(\boldsymbol{r})$ at $r = \varrho$, instead of using a variational principle. Since $V(\boldsymbol{r}) = 0$ for $r \geqq \varrho$ one finds

$$\Psi_{\text{out}}(\boldsymbol{r}) = \int_{r' \leqq \varrho} G(\boldsymbol{r}, \boldsymbol{r}')\, V(r')\, \Psi_{\text{in}}(\boldsymbol{r}')\, d\boldsymbol{r}' \tag{13}$$

that means

$$\Psi_{\text{out}}(\boldsymbol{r}) = \int_{r' = \varrho} \left\{ G(\boldsymbol{r}, \boldsymbol{r}') \frac{d\Psi(\boldsymbol{r}')}{dr'} - \frac{dG(\boldsymbol{r}, \boldsymbol{r}')}{dr'} \Psi(\boldsymbol{r}') \right\} d\boldsymbol{r}' , \tag{14}$$

the last equation following by use of Green's formula. Together with (9) and (11) we find from (14)

$$\Psi_{\text{out}}(\boldsymbol{r}) = \sum_{l m}^{\infty} \left\{ \sum_{l' m'}^{L} \varrho^2\, C_{l' m'}\, A_{l m l' m'}\, (j_{l'}\, R'_{l'} - j'_{l'}\, R_{l'})_{r' = \varrho}\, j_l(\varkappa\, r) + \right.$$
$$\left. + \varkappa\, \delta_{l l'}\, \delta_{m m'}\, (j_l\, R'_l - j'_l\, R_l)_{r' = \varrho}\, n_l(\varkappa\, r) \right\} Y_{l m}(\boldsymbol{r}) \,. \tag{15}$$

The condition of continuity at the surface of the atomic spheres can now be formulated as

$$\lim_{r \to \varrho^-} \Psi_{\text{in}}(\boldsymbol{r}) = \lim_{r \to \varrho^+} \Psi_{\text{out}}(\boldsymbol{r}) \,. \tag{16}$$

Introducing (9) and (15) into (16) one sees that the latter is again equivalent to a system of linear equations for $C_{l m}$, the secular determinant of which turns out to be the same as the one already known following from the variational procedure. Calculating the determinant one makes use of the equation

$$\big(n'_l(\varkappa\, r)\, j_l(\varkappa\, r) - n_l(\varkappa\, r)\, j'_l(\varkappa\, r)\big)_{r = \varrho} = 1/\varkappa\, \varrho^2 \,. \tag{17}$$

It should be pointed out that this argumentation only holds for $L = \infty$, since otherwise the equation system for $C_{l m}$ is insoluble. No statement can be made concerning the differentiability of $\Psi(\boldsymbol{r})$ at $r = \varrho$.

A possibility to improve our knowledge of the wave function is based on the following fact [6]: In equation (10) $V(r)$ is negligible if $l > L$, L chosen such that

$$\frac{L\,(L + 1)}{r^2} \gg V(r) \quad \text{for all values of } r \leqq \varrho. \tag{18}$$

In this case we can use as a trial function

$$\Psi_{\text{in}}(\boldsymbol{r}) = \sum_{l, m}^{L} C_{l m}\, R_l(r)\, Y_{l m}(\boldsymbol{r}) + \sum_{\substack{l = L+1 \\ m}}^{\infty} C_{l m}\, j_l(\varkappa\, r)\, Y_{l m}(\boldsymbol{r}) \,. \tag{19}$$

(19) combined with (14) results in

$$\Psi_{\text{out}}(\boldsymbol{r}) = \sum_{l m}^{L} \left\{ \sum_{l' m'}^{L} \ldots \text{ as in (15)} \right\} +$$
$$+ \sum_{\substack{l = L+1 \\ m}}^{\infty} \left\{ \sum_{l' m'}^{L} \varrho^2\, C_{l' m'}\, A_{l m l' m'}\, (j_{l'}\, R'_{l'} - j'_{l'}\, R'_{l'})_{r = \varrho} \right\} j_l(\varkappa\, r)\, Y_{l m}(\boldsymbol{r}) \,. \tag{20}$$

Substituting (19) and (20) into (16) we find a new system of linear equations, the solution of which is given by the solution of the L-dimensional secular equation (12) and by the condition

$$C_{lm} = \sum_{l'm'}^{L} \varrho^2 \, C_{l'm'} \, A_{lm\,l'm'} \, (j_{l'} \, R'_{l'} - j'_{l'} \, R_{l'})_{r=\varrho} \quad \text{for} \quad l > L \, . \tag{21}$$

That means, to determine the eigenvalues exactly it is sufficient to determine the L-dimensional secular determinant (12) with L now *fixed* by (18). The interspace wave function is then given completely. The coefficients C_{lm} with $l > L$ of the series representing $\Psi_{\text{in}}(\boldsymbol{r})$ are given only by the condition of continuity. The variational procedure (for $l \leqq L$ leading to the same secular equation (12)) does not yield an information for $l > L$.

Besides the continuity now also the differentiability is granted. For $l > L$ the two series (19) and (20) are identical and therefore the derivatives are equal, too.[2]) For $l \leqq L$ the condition

$$\lim_{r \to \varrho^-} \frac{\mathrm{d}\Psi_{\text{in}}}{\mathrm{d}r} = \lim_{r \to \varrho^+} \frac{\mathrm{d}\Psi_{\text{out}}}{\mathrm{d}r} \tag{22}$$

leads once more to the known secular equation (12). (The summation is because of its finiteness now interchangeable with the differentiation.)

The result to be pointed out is the following: If one neglects $V(r)$ for $l > L$ (an assumption very well justified for L about 10 in most cases) the Kohn-Rostoker method yields continuous wave functions with continuous derivatives all over the space. This fact, that makes the method still more attractive, seems not to be noticed until now.

All the arguments described above can readily be extended to "complex" lattices, as suggested by Segall [8]. Here the term "complex" denotes a crystal containing in its unit cell several atoms with position vectors $\boldsymbol{a}_j$. Applying Ewald's summation procedure to hexagonal structures yields the following formula for the coefficients $A_{lm\,l'm'}$:

$$A_{lm\,l'm'}^{jj'} = 4\,\pi\, i^{l-l'} \sum_{LM} i^{-L} \, D_{LM}^{jj'} \, C_{LM\,lm\,l'm'} \tag{23}$$

with

$$C_{LM\,lm\,l'm'} = \int \mathcal{Y}_{LM} \, Y^*_{lm} \, Y_{l'm'} \, \mathrm{d}\Omega \, , \tag{24}$$

---

[2]) The convergence of the infinite sum contained in (20) — and thus the convergence of (19), too — becomes evident, if one substitutes the $A_{lm\,l'm'}$ from [1, eq. (A 2.5)]. After some transformations we get

$$\Psi_{\text{out}}(\boldsymbol{r}) = \frac{4\,\pi\,\varrho^2}{\tau} \sum_{l'm'}^{L} \frac{i^{l'}}{j_{l'}(\varkappa\, s)} \, C_{l'm'} \sum_n \frac{\exp\,[i\,(\boldsymbol{K}_n + \boldsymbol{k})\,\boldsymbol{r}]}{(\boldsymbol{K}_n + \boldsymbol{k})^2 - E} \times$$
$$\times \, j_{l'}\,(|\boldsymbol{K}_n + \boldsymbol{k}|\, s) \;\, Y_{l'm'}\,(\boldsymbol{K}_n + \boldsymbol{k})\,(j_{l'}\, R'_{l'} - j'_{l'}\, R_{l'})_{r=\varrho} \, ,$$

where s is arbitrary, $0 \leqq s < r$. τ is the unit cell volume. The series is summed up in $\boldsymbol{k}$-space according to Ewald [7 eq. (4.10), (7.11)]. The same holds for the differentiated sum.

and

$$D_{LM}^{jj'} = D_{LM}^{jj'}(1) + D_{LM}^{jj'}(2) + D_{00}(3)\,\delta_{L0}\,\delta_{jj'}\,. \tag{25}$$

$$D_{LM}^{jj'}(1) = -\frac{4\pi}{\tau}\frac{i^L}{\varkappa^L}\sum_n \frac{\exp\left[i\,(\boldsymbol{K}_n+\boldsymbol{k})\,(\boldsymbol{a}_j-\boldsymbol{a}_{j'})\right]}{(\boldsymbol{K}_n+\boldsymbol{k})^2-E}\times$$
$$\times\,|\boldsymbol{K}_n+\boldsymbol{k}|^L\,\mathcal{Y}_{LM}^*\,(\boldsymbol{K}_n+\boldsymbol{k})\exp\left\{\frac{-(\boldsymbol{K}_n+\boldsymbol{k})^2+E}{\eta}\right\}, \tag{26}$$

$$D_{LM}^{jj'}(2) = -\frac{1}{\sqrt{\pi}}\frac{2^{L+1}}{\varkappa^L}\sum_{s\neq 0}\exp\,(i\,\boldsymbol{k}\,\boldsymbol{R}_s)\,|\boldsymbol{R}_s-\boldsymbol{a}_j+\boldsymbol{a}_{j'}|^L\,\mathcal{Y}_{LM}\,(\boldsymbol{R}_s-\boldsymbol{a}_j+\boldsymbol{a}_{j'})\times$$
$$\times\int\limits_{\sqrt{\eta}/2}^{\infty}\exp\left[-(\boldsymbol{R}_s-\boldsymbol{a}_j+\boldsymbol{a}_{j'})^2\,\xi^2+\frac{E}{4\,\xi^2}\right]\xi^{2L}\,\mathrm{d}\xi\,, \tag{27}$$

$$D_{00}(3) = -\frac{\sqrt{\eta}}{2\pi}\sum_{v=0}^{\infty}\frac{\left(\frac{E}{\eta}\right)^v}{v!\,(2v-1)}\,. \tag{28}$$

Here $Y_{lm}(\boldsymbol{r})$ denotes a spherical harmonic, $\mathcal{Y}_{lm}(\boldsymbol{r})$ a hexagonal harmonic (cf. the following chapter). The parameter η was chosen so that the so-called "structure constants" $D_{LM}^{jj'}$ could be granted for to be exact to four figures at least, if $D_{LM}^{jj'}(2)$ was neglected at all. More details may be found in the authors' theses (Marburg 1965).

2. Benefits of Symmetry

As developed above, the method holds for any solid. It shall now be applied to hexagonal semiconductors Se and Te. Both of these elements have the same crystal structure, a characteristic feature of which are long spiral chains [9, cf. Fig. 1]. The underlying space group D_3^4 is generated by the following symmetry elements (combined with the primitive translations of the hexagonal lattice): one threefold screw axis $\mathscr{C}_3 = \left\{C_3\middle|\frac{1}{3}\,\mathbf{c}\right\}$, $\mathscr{C}_3^2 = \left\{C_3^2\middle|\frac{2}{3}\,\mathbf{c}\right\}$ and, in a plane perpendicular to it, three twofold axes $\mathscr{C}_2 = \{C_2|0\}$, $\mathscr{C}_2' = \left\{C_2'\middle|\frac{\mathbf{c}}{2}\right\}$, $\mathscr{C}_2'' = \left\{C_2''\middle|-\frac{\mathbf{c}}{2}\right\}$, forming angles of 120°.

Thus, our structure exhibits a comparatively low degree of symmetry, the point group containing six elements only. Nevertheless they can be used to reduce considerably the number of independent trial functions in equation (9) and consequently the dimension of the secular determinant (12). This may be done in two different ways:

1. In equation (9), the wave function is not expanded in ordinary spherical harmonics but in so-called hexagonal harmonics [10]. These are linear combinations

$$\mathcal{Y}_{l\mu} = \sum_m \alpha_{lm}^{(\mu)}\,Y_{lm}$$

which transform according to one of the irreducible representations compiled

Table 1

Character table for Γ

	E	$\mathscr{C}_3$	$\mathscr{C}_3^2$	$\mathscr{C}_2$	$\mathscr{C}_2'$	$\mathscr{C}_2''$
Γ_1	1	1	1	1	1	1
Γ_2	1	1	1	-1	-1	-1
Γ_3	2	-1	-1	0	0	0
	$\begin{pmatrix} 1 & \\ & 1 \end{pmatrix}$	$\begin{pmatrix} -\frac{1}{2} & \frac{\sqrt{3}}{2} \\ -\frac{\sqrt{3}}{2} & -\frac{1}{2} \end{pmatrix}$	$\begin{pmatrix} -\frac{1}{2} & -\frac{\sqrt{3}}{2} \\ \frac{\sqrt{3}}{2} & -\frac{1}{2} \end{pmatrix}$	$\begin{pmatrix} -1 & \\ & 1 \end{pmatrix}$	$\begin{pmatrix} \frac{1}{2} & -\frac{\sqrt{3}}{2} \\ -\frac{\sqrt{3}}{2} & -\frac{1}{2} \end{pmatrix}$	$\begin{pmatrix} \frac{1}{2} & \frac{\sqrt{3}}{2} \\ \frac{\sqrt{3}}{2} & -\frac{1}{2} \end{pmatrix}$

in Table 1. Then the Ansatz for the wave function in an atomic sphere

$$\Psi_{\text{in}}^j(\boldsymbol{r}) = \sum_{l\mu} C_{l\mu}^j \, R_l(r_j) \, \mathscr{Y}_{l\mu}(\boldsymbol{r}_j) \, ; \qquad \boldsymbol{r}_j = (\boldsymbol{r} - \boldsymbol{a}_j) \quad \text{with} \quad r_j \leqq \varrho$$

will include only those hexagonal harmonics which behave correctly under all rotations leaving this atom at its place.

2. The space-group D_3^4 is non-symmorphic and thus contains operations which do not leave invariant any atom in the unit cell. These operations, however, transform different atoms into each other. Hence, they connect the values of the wave function in different atomic spheres and establish relations between coefficients $C_{l\mu}^j$ with different j.

Example: Suppose we want to calculate an eigenvalue belonging to representation Γ_1. In this case, the entire wave function in the unit cell must be transformed into itself by all the symmetry operations listed in Table 1. Only one of these — i.e. $\mathscr{C}_2$, the twofold rotation round the x-axis — leaves one atom unshifted (number one in Fig. 1) and makes possible the following statement: Among the trial functions, which are superposed in (9) to give the wave function $\Psi_{\text{in}}^1(\boldsymbol{r})$ in sphere 1, there may only be hexagonal harmonics invariant under $\mathscr{C}_2$. Owing to this restriction the number of unknown coefficients $C_{l\mu}^j$ is reduced to about one half. (In the same way, a wave function with symmetry type Γ_2 is built up by those functions only, which change sign under $\mathscr{C}_2$). At first sight, no such restriction seems to hold concerning the spheres 2 and 3 which are not left

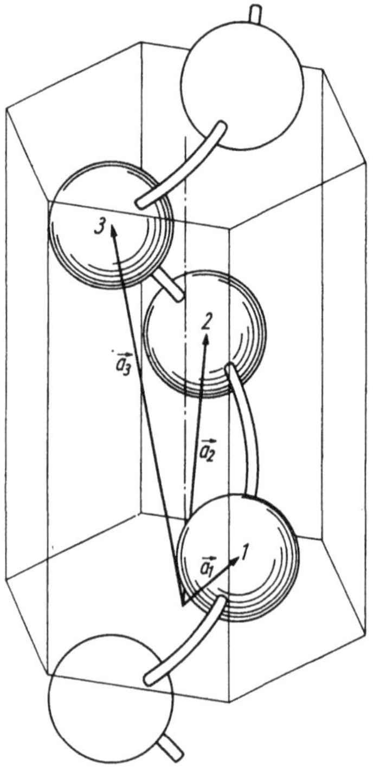

Fig. 1. Unit cell of the selenium structure with atomic position vectors

Fig. 2. Brillouin zone

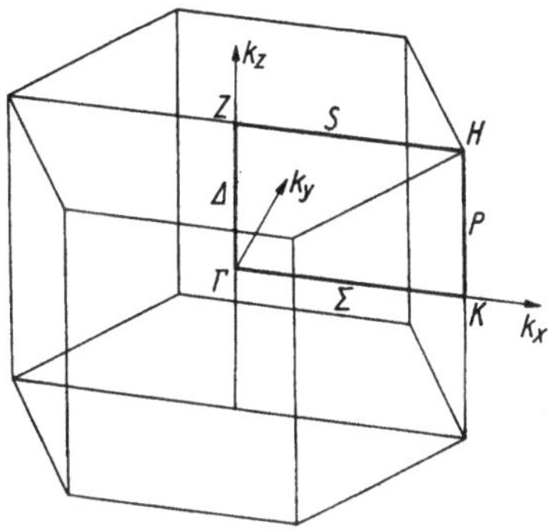

invariant by any operation. Yet we can make use of the threefold screw axis that transforms atom 1 into atom 2. Hence the Γ_1-type wave function in sphere 2 must arise from the function in sphere 1 by applying $\mathscr{C}_3$, and this may already be taken into account in the ansatz (9):

$$\Psi_{\text{in}}^2(\boldsymbol{r}) = \mathscr{C}_3\, \Psi_{\text{in}}^1(\boldsymbol{r}) = \Psi^1\left(\left\{C_3 \left| \frac{\boldsymbol{c}}{3}\right.\right\}^{-1} \boldsymbol{r}\right) = \Psi^1\left(C_3^{-1}\, \boldsymbol{r} - \frac{\boldsymbol{c}}{3}\right) =$$
$$= \sum_{l\mu} C_{l\mu}^1\, R_l\left(\left|C_3^{-1}\, \boldsymbol{r} - \frac{\boldsymbol{c}}{3} - \boldsymbol{a}_1\right|\right) \mathscr{Y}_{l\mu}\left(C_3^{-1} \boldsymbol{r} - \frac{\boldsymbol{c}}{3} - \boldsymbol{a}_1\right) =$$
$$= \sum_{l\mu} C_{l\mu}^1\, R_l(r_2)\, \mathscr{Y}_{l\mu}(C_3^{-1}\, \boldsymbol{r}_2)\,.$$

The function $\Psi_{\text{in}}^3(\boldsymbol{r})$ in sphere 3 is obtained quite analogously by applying $\mathscr{C}_3^2$ to Ψ^1. Thus the number of unknown coefficients $C_{l\mu}^j$ is once more reduced to one third.

In order to calculate an electron state with symmetry type Γ_3, we must use in sphere 1 two different expansions, either transforming according to one row of this two-dimensional representation. Then the corresponding sums in spheres 2 and 3 are again determined by applying the threefold screw axis as above and, additionally, by forming those linear combinations which are prescribed by the matrices of $\mathscr{C}_3$ and $\mathscr{C}_3^2$ (cf. Table 1).

As in Γ, we may utilize the full point-group symmetry in Z, K, and H (Fig. 2). Along the axes Δ and P, the wave vector group includes only the threefold screw axis, making possible to reduce the dimension of the secular determinant to one third. Along the axes Σ and S, the twofold rotation $\mathscr{C}_2$ allows a reduction to one half, compared with the general case.

3. Numerical Calculations

In performing the calculations on an IBM 7090 computer, the following scheme was observed: 1. The radial Schrödinger equation was solved for the interior of the atomic spheres. 2. The structure constants were computed. 3. The secular determinants were set up and evaluated.

1. To find a numerical solution of the radial equation (10), we utilized the Noumerov technique [11]. Inside the atomic spheres we used free atom potentials, worked out self-consistently by Herman and Skillman [12]. By way of trial, we also employed a superposition of neighboring atom potentials, but the

final results were not much affected by such a modification. However, varying the constant potential V_0 outside did not only shift the band system as a whole, but also changed the interband distances. We chose V_0 to be about $V(\varrho) + 1.25$ ryd for both Se and Te, which seems to give an appropriate band model, although the gap could be adjusted even better.

2. As mentioned earlier, our structure constants were tested to be exact to at least four figures. This might look poor, compared with Ham and Segall's six figure precision. Because of the more complicated structure, however, and since no selection rules hold for the non-diagonal structure constants, we had to evaluate 10 up to 25 times as many of them, depending on the point considered in $\boldsymbol{k}$-space. Moreover, one should mention that an error of 10^{-4} in the determinantal elements results in an error of only 10^{-3} in a determinant of dimension 10 for example. Altogether 100000 structure constants were figured out for either substance, taking two thirds of the total computing time. In cubic crystals, it is sufficient to calculate the structure constants once and for all, since different substances belonging to the same space group have identical arrangements of atoms — apart from a scale factor. This is not true with hexagonal Se and Te because of the different c/a ratios.

3. The determinants were evaluated at a sequence of energies with intervals of 0.01 ryd. This was found to be sufficient in general to locate by interpolation the zeros with an uncertainty of 0.03 eV. All of our calculations were performed using an expansion (9) of the wave function including $l \leqq 2$; this led to determinants of dimension 5, 4, 9 for Γ_1, Γ_2, Γ_3, respectively. The same values hold for the symmetry points Z, K, H. Along the axes, the corresponding dimensions are: 9 for Δ and P, 13 and 14 for Σ_1, S_1 and Σ_2, S_2, respectively. To test the convergence, several eigenvalues at Γ were determined once more including $l \leqq 3$ and proved to be shifted by less than 0.08 eV. Energy levels were calculated at the following points in $\boldsymbol{k}$-space: Γ, Z, K, H, furthermore three points on each of the axes Δ and Σ and two points on P and S near to H.

4. Results and Comparison with Experiment

Fig. 3 and 4 show the calculated band structures for Se and Te, respectively. The bands are grouped into three triplets, all originating from atomic p-states (4p for Se, 5p for Te). Such triplets have also been found by means of tight-

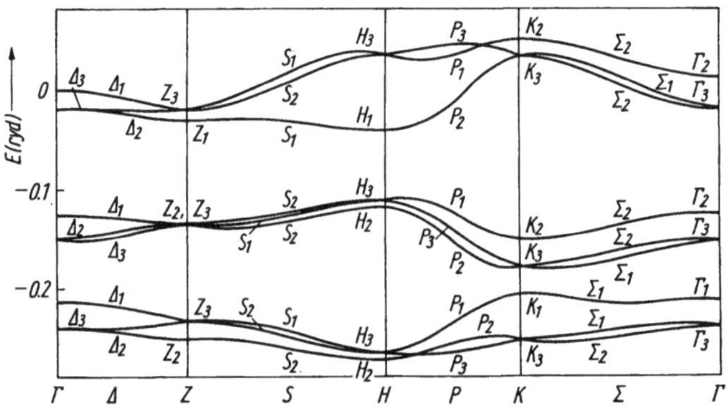

Fig. 3. Band structure of selenium

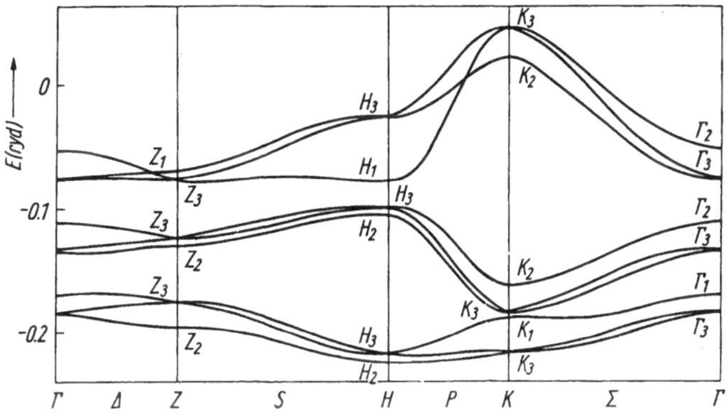

Fig. 4. Band structure of tellurium

binding calculations by Reitz [13] and Olechna and Knox [14], who considered the hexagonal Δ-axis only. We found, however, that for both elements the gap should be situated at, or in the neighborhood of, the Brillouin Zone corner H as proposed by Hulin [15]. The forbidden band width $E_G = 1$ eV for Se is in the correct order of magnitude (experiment: 1.8 eV); for Te, the agreement is even better ($E_G = 0.3$ eV, experiment: 0.35 eV).

The lowest conduction band is nondegenerate with minimum at H_1. On the other hand, the uppermost valence band is doubly degenerate (H_3). Proceeding to the axes, the degeneracy is removed. The valence band edge seems to be located on the P-axis, one sixth of the axis length off the point H (for Se). The selection rules for direct optical transitions are the same, no matter whether the maximum is in H or along P (Table 2). They enable us to explain the

Table 2

Allowed electrical dipole transitions

a) Point H (also valid for Γ, Z, K)

Without spin

	H_1	H_2	H_3
H_1	0	$\parallel$	$\perp$
H_2	$\parallel$	0	$\perp$
H_3	$\perp$	$\perp$	$\parallel, \perp$

With spin

	H_4	H_5	H_6
H_4	0	$\parallel$	$\perp$
H_5	$\parallel$	0	$\perp$
H_6	$\perp$	$\perp$	$\parallel, \perp$

b) Axis P (also valid for Δ)

Without spin

	P_1	P_2	P_3
P_1	$\parallel$	$\perp$	$\perp$
P_2	$\perp$	$\parallel$	$\perp$
P_3	$\perp$	$\perp$	$\parallel$

With spin

	P_4	P_5	P_6
P_4	$\parallel$	$\perp$	$\perp$
P_5	$\perp$	$\parallel$	$\perp$
P_6	$\perp$	$\perp$	$\parallel$

polarization dependence of the absorption edge as observed on Se by Stuke and Keller [16]. The lowest energy transition ($H_3 \to H_1$ or $P_1 \to P_2$) exists only for light polarized perpendicular to the hexagonal axis, the next one ($H_2 \to H_1$) only for parallel polarization. Now Stuke and Zimmerer [17] are repeating the above-mentioned measurements at liquid air temperature and find several weak additional maxima in their reflectivity curves which seem to be explainable, if we take spin–orbit splitting qualitatively into account. Further transitions, observed with Se at elevated energies, cannot take place in H, the second conduction band lying too high at this point. They should probably occur in Z.

As to Te, the spin–orbit splitting of the upper valence band H_3 (giving $H_4 + H_5 + H_6$) is very likely to be responsible for the wellknown infrared 11 μm band. Its selection rule — perpendicular forbidden — can only be accomplished by an $H_4 \to H_5$ transition, the corresponding levels in Γ and Z being degenerate by time reversal. Korovin and Firsov [18] try to explain the 11 μm absorption maximum without considering spin. Then the polarization dependence requires assuming a transition between levels which belong to different band triplets and thus leads to an unadequately high energy.

If we replace our energy bands near H by parabolae, we may roughly estimate the effective hole masses. We obtain for the uppermost valence band in Se: $m_\perp \approx 4\, m_0$, $m_\parallel \approx m_0$ (with somewhat smaller but similar values for the lower valence bands). For Te, the corresponding results are: $m_\perp \approx 1.5\, m_0$, $m_\parallel \approx \approx 0.3\, m_0$. In view of the small number of points calculated in $\boldsymbol{k}$-space, these values should not be taken too seriously. In fact, they are rather upper limits for the effective masses, stronger curvatures being possible as well. The ratio $m_\perp/m_\parallel \approx 4$, however, should be reliable. It agrees well with experiments of Stuke [17] on Se and Grosse [19] on Te, though it contradicts the measurements of Mendum and Dexter [20], who found $m_\perp/m_\parallel < 1$ from cyclotron resonance in Te. Our calculations result in a two- or four-valley model with oblate spheroids of constant energy (two if the valence band edge is in point H, four if it is on the P-axis). The extremely flat conduction band of Te is to be mentioned.

An interesting property of selenium is the uncommonly strong pressure dependence of its gap. As an example we have recalculated the energy levels in point H using lattice parameters which belong to an hydrostatic pressure of 25000 at. The corresponding decrease in the forbidden band width is experimentally found to be 20% [21]; our calculation yielded 25%. According to compressibility measurements of Vereshchagin et al. [22], the above-mentioned pressure causes a change in the lattice constants by $\Delta a/a = -10\%$, $\Delta c/c = = +0.8\%$, i.e. the selenium chains are moved together essentially undeformed. Consequently, the effect of pressure could hardly have been described correctly by any method that considers nearest neighbors within one chain only. On the other hand our results show that the interaction between different chains is taken into account fairly well by the Kohn-Rostoker method.

Acknowledgements

We wish to express our gratitude to Professor O. Madelung for many helpful suggestions and his continuous interest in this work. We are also indebted to Dr. J. Stuke for making available to us his results prior to publication, and to the staff of the "Deutsches Rechenzentrum, Darmstadt" for technical assistance.

References

[1] W. Kohn and N. Rostoker, Phys. Rev. **94**, 1111 (1954).
[2] B. Segall, Phys. Rev. **124**, 1797 (1961).
[3] F. S. Ham, Phys. Rev. **128**, 82 (1962).
[4] J. Korringa, Physica **13**, 392 (1947).
[5] P. M. Morse, Proc. Nat. Acad. Sci. USA **42**, 276 (1956).
[6] H. Bross, Phys. kondens. Materie **3**, 119 (1964).
[7] F. S. Ham and B. Segall, Phys. Rev. **124**, 1786 (1961).
[8] B. Segall, Phys. Rev. **105**, 108 (1957).
[9] R. H. Asendorf, J. chem. Phys. **27**, 11 (1957).
[10] D. G. Bell, Rev. mod. Phys. **26**, 311 (1954).
[11] G. W. Pratt, Phys. Rev. **88**, 1217 (1952).
[12] F. Herman and S. Skillman, Atomic Structure Calculations, Prentice Hall Inc., Englewood Cliffs (N.J.) 1963.
[13] J. R. Reitz, Phys. Rev. **105**, 1233 (1957).
[14] D. J. Olechna and R. S. Knox, Phys. Rev. **140**, A 986 (1965).
[15] M. Hulin, J. Phys. Chem. Solids **27**, 441 (1966).
[16] J. Stuke and H. Keller, phys. stat. sol. **7**, 189 (1964).
[17] J. Stuke, private communication.
[18] L. I. Korovin and Iu. A. Firsov, Soviet Phys.-J. techn. Phys. **3**, 2219 (1958).
[19] P. Grosse, Z. Phys. **193**, 318 (1966).
[20] J. H. Mendum and R. N. Dexter, Bull. Amer. Phys. **9**, 632 (1964).
[21] H. L. Suchan, S. Wiederhorn, and H. G. Drickamer, J. chem. Phys. **31**, 355 (1959).
[22] L. F. Vereshchagin, et al., Dokl. Akad. Nauk SSSR **165**, 297 (1965).

(Received May 5, 1966)

phys. stat. sol. **16**, 499 (1966)

Institut für Magnetische Werkstoffe der Deutschen Akademie der Wissenschaften zu Berlin, Jena

Untersuchungen über Ummagnetisierungsprozesse in Rechteckferriten[1])

Von

E. STEINBEISS

Es wurde der Einfluß einer einachsigen Anisotropieenergie auf die Feldstärkeabhängigkeit der Ummagnetisierungszeit untersucht und die Ergebnisse im Zusammenhang mit einigen Modellvorstellungen über Ummagnetisierungsprozesse diskutiert.

The effect of an uniaxial anisotropy energy on the field dependence of the switching time is investigated. Several models for magnetization reversal are discussed.

1. Einleitung

Ferrite mit rechteckförmiger Hystereseschleife werden in großen Stückzahlen als Schalt- und Speicherelemente für Datenverarbeitungsanlagen benötigt. Im Hinblick auf diesen Anwendungszweck ist besonders das Impulsverhalten von Interesse, worunter man die Abhängigkeit der Ummagnetisierungszeit (Schaltzeit) τ von der Impulsfeldstärke H versteht. Fig. 1 zeigt einen typischen Verlauf der Funktion $1/\tau\,(H)$. Diese Kurve kann man in drei Gebiete unterteilen, für die jeweils die Beziehung

$$\frac{1}{\tau} = \frac{1}{S_{\mathrm{w}i}} (H - H_{0i}) \; ; \qquad i = 1, 2, 3 \tag{1}$$

gilt (S_{w} Switchingkoeffizient, H_0 Startfeldstärke).

Nach Gyorgy [1] sollen in diesen drei Feldstärkebereichen verschiedene Ummagnetisierungsprozesse wirksam sein, im Gebiet I Wandverschiebungen, im Gebiet II inhomogene und im Gebiet III homogene Drehungen der Magnetisierung. Über die bei Impulsexperimenten in Rechteckferriten vorkommenden Ummagnetisierungsprozesse sind in der Literatur verschiedene Ansichten geäußert und zum Teil mit Experimenten belegt worden [1 bis 10].

Der folgende Abschnitt gibt einen kurzen Überblick über die bisher diskutierten Modelle.

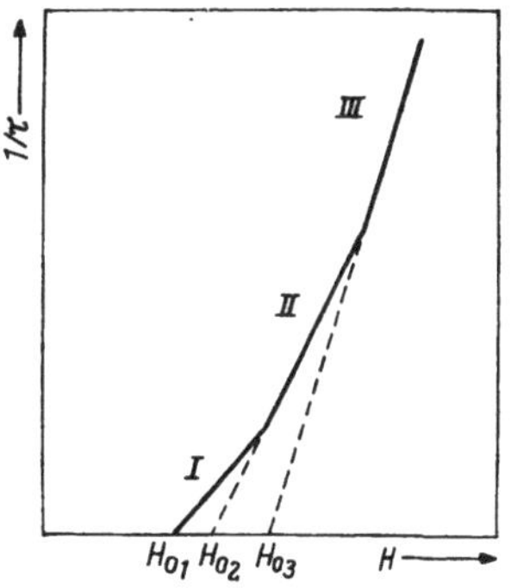

Fig. 1. Typischer Verlauf der Funktion $1/\tau(H)$ für Rechteckferrite

[1]) Diese Arbeit ist ein Auszug aus der Dissertation [19].

2. Ummagnetisierungsmodelle

2.1 Wandverschiebungen

Für Impulsfeldstärken in der Nähe der Koerzitivkraft wird allgemein das Modell der Ummagnetisierung durch Wandverschiebungen angenommen. Der theoretische Zusammenhang zwischen der Schaltzeit τ und der Feldstärke H wurde für dieses Modell zuerst von Kittel [11] angegeben. Er erhielt für den Switchingkoeffizienten S_{w_1} im Gebiet I den Ausdruck

$$S_{\mathrm{w}_1} = \frac{\bar{d}}{R} = \frac{\bar{d}\,\alpha}{2\gamma'}\,\frac{1}{\sqrt{A}} \int\limits_{\Theta_0}^{\Theta_0+\pi} [g(\Theta) - g(\Theta_0)]^{1/2}\,\mathrm{d}\Theta \sim \frac{\bar{d}\,\alpha}{\delta}\,, \tag{2}$$

wobei R die Wandbeweglichkeit, $\bar{d}$ die während der Schaltzeit τ von jeder Wand im Mittel zurückgelegte Strecke, $\gamma' = \gamma/(1+\alpha^2)$ (γ gyromagnetisches Verhältnis), A die Austauschenergiekonstante, δ die Blochwanddicke, α eine Dämpfungskonstante und $g(\Theta)$ die Winkelabhängigkeit der magnetischen Kristallanisotropie-Energiedichte bedeuten.

2.2 Inhomogene Drehungen

Auf Grund der Tatsache, daß S_{w_2} im Feldstärkebereich II von der Art des untersuchten Materials relativ unabhängig ist, wurde von Gyorgy das Modell der inhomogenen Drehungen eingeführt [7]. Er berechnete mit Hilfe der Landau-Lifshits-Gleichung

$$\dot{\boldsymbol{M}}_{\mathrm{s}} = -\gamma'\,(\boldsymbol{M}_{\mathrm{s}} \times \boldsymbol{H}) - \frac{\alpha\,\gamma'}{M_{\mathrm{s}}}\big(\boldsymbol{M}_{\mathrm{s}} \times (\boldsymbol{M}_{\mathrm{s}} \times \boldsymbol{H})\big) \tag{3}$$

den Switchingkoeffizienten für ein spezielles Spiralmodell (Fig. 2). Dabei wird der polykristalline Ferritringkern durch einen unendlich langen Zylinder in z-Richtung ersetzt. Um die Streufeldenergie bei der Ummagnetisierung klein zu halten, soll sich die Magnetisierung auf einer Spirale mit der Periodenlänge l anordnen. In dem Ansatz für die wirksame Feldstärke

$$\boldsymbol{H} = -\left(\frac{8\,\pi^2\,A}{l^2\,M_{\mathrm{s}}^2}\right)[M_x\,\boldsymbol{i} + M_y\,\boldsymbol{j}] - H_{\mathrm{a}}\,\boldsymbol{k} \tag{4}$$

wurde die Anisotropieenergie vernachlässigt.

Die Lösung der Gleichung (3) führt mit diesen Ansätzen auf die Beziehung

$$\tau\,(H_{\mathrm{a}} - H_0) = S_{\mathrm{w}_2} = \frac{1+\alpha^2}{\alpha\,\gamma}\left[3 - \frac{\dfrac{8\,\pi^2\,A}{l^2\,M_{\mathrm{s}}}}{H_{\mathrm{a}}}\cdot \ln \frac{H_{\mathrm{a}} + 0{,}9\left(\dfrac{8\,\pi^2\,A}{l^2\,M_{\mathrm{s}}}\right)}{H_{\mathrm{a}} - 0{,}9\left(\dfrac{8\,\pi^2\,A}{l^2\,M_{\mathrm{s}}}\right)}\right] \tag{5}$$

mit

$$H_0 = \frac{64\,\pi^4\,A^2}{l^4\,M_{\mathrm{s}}^2\,H_{\mathrm{a}}}\,. \tag{6}$$

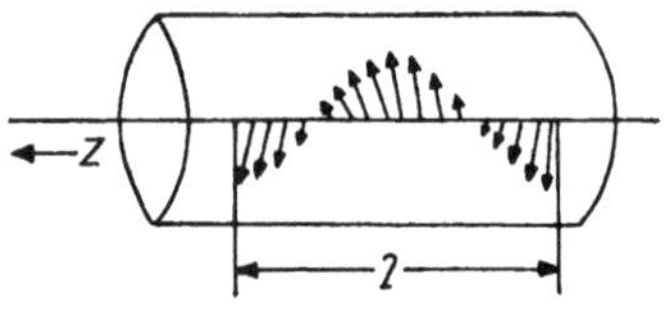

Fig. 2. Spiralmodell nach Gyorgy

Für ein geeignet gewähltes l sind im Feldstärkebereich II inhomogene Drehungen möglich und der Switchingkoeffizient wird für $H_a \gg H_0$

$$S_{w_2} \approx 3 \frac{1 + \alpha^2}{\alpha \gamma}. \tag{7}$$

2.3 Homogene Drehungen

In Analogie zu dem Schaltverhalten dünner magnetischer Schichten bei großen Impulsfeldern wurde von Gyorgy und Hagedorn [8] für den dritten Feldstärkebereich ein Modell der Ummagnetisierung durch homogene Drehungen vorgeschlagen. Die Rechnungen sind für ein spezielles Modell durchgeführt worden, wonach der Ringkern wieder als langer Zylinder aufgefaßt wird, der aus dünnen Schichten bestehen soll, die zu Schläuchen mit abnehmendem Durchmesser aufgewickelt sind (Fig. 3).

Die Gleichung (3) konnte unter Vernachlässigung von Anisotropiefeldern für dieses Modell in eine Form gebracht werden, die direkt mit einem von Kikuchi [12] für dünne Schichten angegebenen Gleichungssystem vergleichbar ist, dessen Lösung übernommen wurde. Kikuchis Ausdruck für die Schaltzeit lautet

$$\tau = \frac{\alpha}{\gamma} \left(\frac{2C}{H_a} \right) + \frac{1}{\alpha \gamma} \left(\frac{I}{M_s} \right). \tag{8}$$

C ist durch die Anfangslage der Magnetisierung festgelegt, und I ist ein von Kikuchi angegebenes Integral, das nur schwach von H_a abhängt. Elfant und Friedländer [4] geben für den Bereich III die Formel

$$S_{w_3} = 3 \frac{\alpha}{\gamma} \tag{9}$$

an.

3. Einfluß einer einachsigen Anisotropieenergie auf das Schaltverhalten von ringförmigen Ferritproben

Es ist in der Literatur mehrfach versucht worden, durch geeignete Experimente die oben diskutierten Ummagnetisierungsmodelle voneinander zu unterscheiden. Während die benutzten Methoden (partielles Schalten, Einfluß transversaler Hilfsfelder auf S_w [1, 4, 8]) vorwiegend qualitativen Charakter tragen und außerdem meist mehrere Deutungsmöglichkeiten zulassen, soll im folgenden eine weitere Methode diskutiert werden, die in erster Näherung einer theoretischen Behandlung zugänglich ist. Man kann in Ferriten mit negativer Magneto-

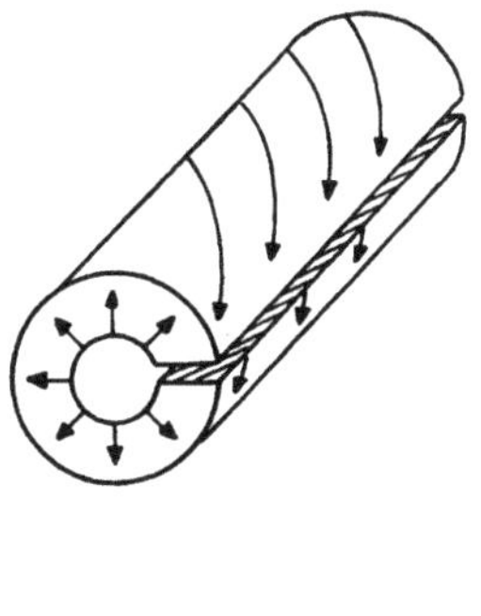

Fig. 3. Spezielles Rotationsmodell nach Gyorgy und Hagedorn

striktionskonstante λ_s durch eine äußere Druckspannung σ eine einachsige Anisotropieenergie $K_u \approx \lambda_s \sigma$ in Ringrichtung erzeugen. Der Einfluß dieser Anisotropieenergie auf die verschiedenen Switchingkoeffizienten S_{w_i} soll im folgenden berechnet werden.

3.1 Wandverschiebungen

Aus Gleichung (2) geht hervor, daß der Switchingkoeffizient für das Gebiet der Wandverschiebungen von der Wanddicke und dem mittleren Wandabstand bzw. von der Anzahl der Ummagnetisierungskeime abhängt. Beide Größen werden durch eine einachsige Anisotropieenergie beeinflußt. Eine Näherungsformel für den Switchingkoeffizienten, die den Einfluß von K_u auf die Wanddicke berücksichtigt, gaben bereits Menyuk und Goodenough [2] an:

$$S_{w_1} \approx \frac{\bar{d}\,\alpha}{A^{1/2}\,\gamma'} \sqrt{K_1 + \lambda_s \sigma}\,. \tag{10}$$

Zum anderen wird die Größe $\bar{d}$ durch eine einachsige Anisotropieenergie vergrößert, weil die Anzahl der Ummagnetiserungskeime von der Streufeldenergie an den Korngrenzen abhängt und diese durch den Ausrichtungseffekt der einachsigen Anisotropie herabgesetzt wird. Beide Effekte führen zu einer Vergrößerung des Switchingkoeffizienten mit zunehmendem K_u.

3.2 Inhomogene Drehungen

In der von Gyorgy [7] abgeleiteten Beziehung (5) für inhomogene Drehungen (Spiralmodell) sind entmagnetisierende und Anisotropiefelder nicht berücksichtigt.

Liegt nun in Richtung der Achse des unendlich langen Zylinders (z-Achse) die Vorzugsrichtung einer einachsigen Anisotropieenergie, dann hat das Anisotropiefeld die Form [13]

$$\boldsymbol{H}_k = -\frac{2\,K_u}{M_s^2}\,[M_x\,\boldsymbol{i} + M_y\,\boldsymbol{j}]\,.$$

Bei der Lösung der Gleichung (3) für den Fall des Spiralmodells ist also für das effektive Feld statt Gleichung (4) folgende Summe anzusetzen:

$$\boldsymbol{H} = -\left(\frac{8\,\pi^2\,A}{l^2\,M_s^2} + \frac{2\,K_u}{M_s^2}\right)[M_x\,\boldsymbol{i} + M_y\,\boldsymbol{j}] - H_a\,\boldsymbol{k}\,. \tag{11}$$

Die Schaltzeitbeziehung für dieses Problem erhält man dann aus Gleichung (5) und (6), indem man den Faktor $\frac{8\,\pi^2\,A}{l^2\,M_s^2}$ durch die Summe $\left(\frac{8\,\pi^2\,A}{l^2\,M_s^2} + \frac{2\,K_u}{M_s^2}\right)$ ersetzt:

$$\tau\,(H_a - H_0) =$$

$$= S_{w_2} = \frac{1+\alpha^2}{\alpha\,\gamma}\left[3 - \frac{M_s\left(\frac{8\,\pi^2\,A}{l^2\,M_s^2} + \frac{2\,K_u}{M_s^2}\right)}{H_a}\cdot\ln\frac{H_a + 0{,}9\,M_s\left(\frac{8\,\pi^2\,A}{l^2\,M_s^2} + \frac{2\,K_u}{M_s^2}\right)}{H_a - 0{,}9\,M_s\left(\frac{8\,\pi^2\,A}{l^2\,M_s^2} + \frac{2\,K_u}{M_s^2}\right)}\right] \tag{12}$$

mit

$$H_0 = \frac{\left(\frac{8\,\pi^2\,A}{l^2\,M_s} + \frac{2\,K_u}{M_s}\right)^2}{H_a}\,. \tag{13}$$

Für genügend große Schaltfelder wird S_{w_2} eine Konstante, die nicht von K_u abhängt (vgl.(7)).

Bei dünnen Permalloyschichten ergaben Messungen der Schaltzeitkurve $1/\tau(H)$, daß in dem diskutierten Feldstärkebereich nur die Startfeldstärke von K_u abhängt und der Switchingkoeffizient entsprechend den obigen Ergebnissen konstant bleibt (vgl. [1]).

3.3 Homogene Drehungen

Für $l \to \infty$ geht die inhomogene Rotationsmode (Spiralmodell) in eine homogene über. In diesem Falle hat das Austauschfeld keine x- und y-Komponenten. Dafür sind die Entmagnetisierungsfaktoren in x- und y-Richtung mit 2π einzusetzen. Damit lautet die Lösung der Gleichung (3) [14]

$$\tau(H_a - H_0) = \frac{1+\alpha^2}{\alpha\gamma}\left[3 - \frac{M_s}{H_a}\left(2\pi + \frac{2K_u}{M_s^2}\right) \cdot \ln \frac{H_a + 0{,}9\, M_s\left(2\pi + \frac{2K_u}{M_s^2}\right)}{H_a - 0.9\, M_s\left(2\pi + \frac{2K_u}{M_s^2}\right)}\right]. \qquad (14)$$

Der Switchingkoeffizient stimmt mit dem für das Spiralmodell überein, jedoch ist

$$H_0 = \frac{M_s^2}{H_a}\left(2\pi + \frac{2K_u}{M_s^2}\right)^2 \qquad (15)$$

wesentlich größer. Für den Fall der MgMn-Rechteckferrite sind z. B. solche Drehprozesse erst für Impulsfelder > 1500 Oe möglich. Das Vorkommen einer solchen Rotationsmode ist daher äußerst unwahrscheinlich.

4. Meßmethode

Wie im Abschnitt 3 gezeigt wurde, unterscheiden sich Wandverschiebungsprozesse von Drehprozessen in der Abhängigkeit der Ummagnetisierungszeit von einachsigen Kristallanisotropien. Wenn die verschieden steilen Bereiche der Schaltzeitkurve $1/\tau(H)$ dadurch zustande kommen, daß in ihnen unterschiedliche Ummagnetisierungsprozesse vorherrschen, so sollten Messungen der Switchingkoeffizienten S_w in Abhängigkeit von K_u eine Entscheidung über die Richtigkeit dieser Annahme zulassen.

Wir haben daher die Funktion $1/\tau(H)$ an ringförmigen Ferritproben ($\varnothing$ = 5 mm) in Abhängigkeit von K_u gemessen. Die Proben hatten die Zusammensetzung $Mn_{1,4}Fe_{1,6}O_4$ und eine Magnetostriktionskonstante λ_s von ca. -2×10^{-5}. Eine einachsige Anisotropieenergie genügender Größe wurde entweder durch Einschmelzen der Ringkerne in Polystyrol oder mit Hilfe einer hydraulischen Presse erzeugt.

Im Feldstärkebereich I wurden die Feldimpulse mit einem in [15] beschriebenen Impulsgenerator erzeugt, der mit Elektronenröhren arbeitet.

Zur Erzeugung genügend steiler Impulse in den Feldstärkebereichen II und III haben wir die Methode der Entladung eines Koaxialkabels mit Hilfe eines Quecksilberrelais (Typ HGX 1003 der Firma Clare) verwendet.

Die in einer Sekundärwindung induzierte Signalspannung wurde auf dem Schirm eines Breitbandoszillographen sichtbar gemacht. Als Schaltzeit wurde die Signalbreite zwischen zwei Punkten gemessen, an denen die Spannung 20% ihres Maximalwertes beträgt. Die Eichung der Zeitbasis erfolgte mit Hilfe von Verzögerungskabeln.

4.1 Meßergebnisse

Die Schaltzeitkurven im Feldstärkebereich I sind für polykristalline Ringkerne der obengenannten Zusammensetzung in Fig. 4 dargestellt.

Die durch Spannungen der Größe $\sigma = 1100\ \mathrm{kp/cm^2}$ erzeugte einachsige Anisotropieenergie hat eine Änderung des Switchingkoeffizienten um einen Faktor 2,7 von 0,66 auf 1,78 µsOe zur Folge.

In Fig. 5 sind die an der gleichen Probe gemessenen Schaltzeitkurven für die Bereiche II und III eingezeichnet.

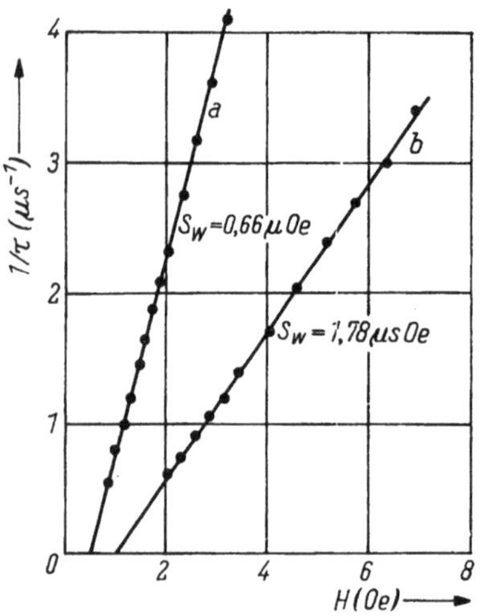

Fig. 4. Schaltzeitkurven von Mn-Ferrit ($Mn_{1,4}Fe_{1,6}O_4$) im Feldstärkebereich I.
Kurve (a): Ohne äußere Druckspannung
Kurve (b): Bei einer Druckspannung von ca. 1100 kp/cm²

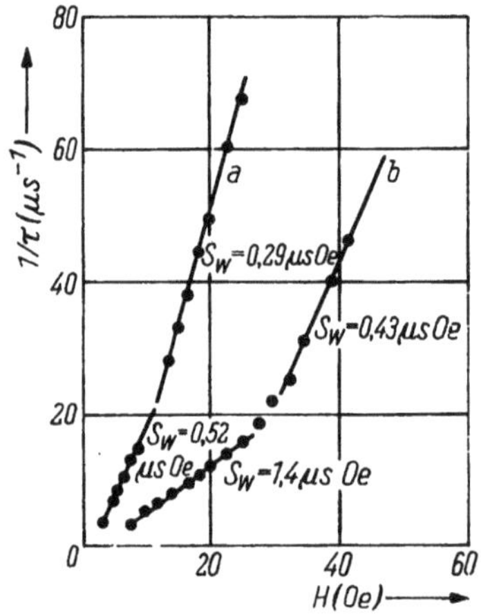

Fig. 5. Schaltzeitkurven von Mn-Ferrit ($Mn_{1,4}Fe_{1,6}O_4$) in den Feldstärkebereichen II und III.
Kurve (a): Ohne äußere Druckspannung
Kurve (b): Bei einer äußeren Druckspannung von ca. 1100 kp/cm²

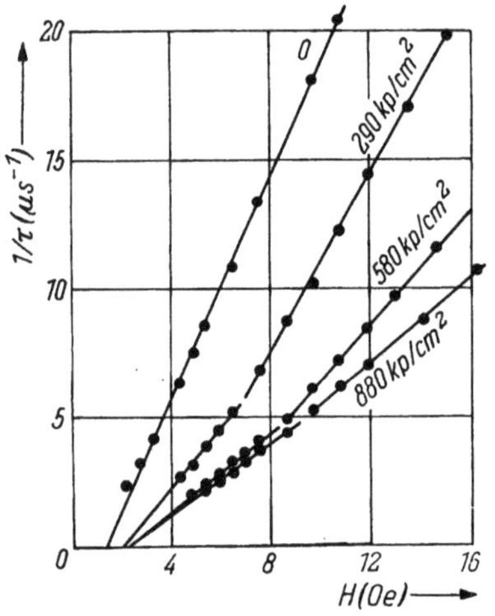

Fig. 6. Schaltzeitkurven von Mn-Ferrit ($Mn_{1,4}Fe_{1,6}O_4$) im Feldstärkebereich II bei verschiedenen Druckspannungen

Im Bereich II wird der Switchingkoeffizient durch die äußere Spannung um den gleichen Faktor vergrößert wie im Bereich I. Im Feldstärkebereich III beobachtet man jedoch nur noch eine Änderung um den Faktor 1,5.

Die durch die Polystyroleinbettung erzeugten Spannungen von ca. 1100 kp/cm^2 wurden durch Vergleich der Schaltzeitkurven mit den entsprechenden Messungen in der hydraulischen Presse bestimmt.

Fig. 6 zeigt die Schaltzeitkurven $1/\tau(H)$ im Feldstärkebereich I und II für einige Spannungen zwischen 0 und 900 kp/cm^2. Eine weitere Steigerung der Spannung bringt keine merkliche Änderung der Switchingkoeffizienten in den Bereichen I und II mit sich, obwohl der mittlere Wandabstand noch nicht die Probendimensionen erreicht hat. Außerdem nehmen die Remanenz und die Rechteckigkeit der Hystereseschleifen merklich ab.

4.2 Diskussion der Meßergebnisse

Der beobachtete Einfluß einer einachsigen Anisotropieenergie auf den Switchingkoeffizienten weist darauf hin, daß auch in den Feldstärkebereichen II und III noch mit Wandverschiebungen zu rechnen ist. Schließt man eine Änderung der Wandbeweglichkeit mit wachsender Feldstärke aus, dann würden die verschiedenen Steigungen der Funktion $1/\tau(H)$ in den drei Feldstärkebereichen durch eine Änderung der effektiven Wandkonfiguration (Verkleinerung von $\bar{d}$ mit wachsender Feldstärke) zu erklären sein. Dieser Deutungsversuch setzt voraus, daß die Wachstums- und Bildungsfeldstärken von Ummagnetisierungskeimen in dem untersuchten Feldstärkebereich liegen, also kleiner als 20 Oe sind.

Die Existenz von Keimen an Kristallitgrenzen und Poren ist experimentell gesichert (siehe z. B. [6]). In der Literatur sind Abschätzungen angegeben, wonach ellipsoidförmige Kristallitgrenzenkeime bei Feldern von ca. 6 bis 10 Oe wachstumsfähig sind [16] und wonach langgestreckte Keime an Poren und Einschlüssen bei Feldstärken von

$$H_w \approx \left[0{,}9\left(\frac{\sigma_w}{r\,M_s^2}\right) + 0{,}05\right] M_s \tag{16}$$

bzw.

$$H_w > M_s\left[0{,}005 + 0{,}13\,\frac{r_c}{r}\right] \tag{17}$$

wachsen können [17] (σ_w spezifische Wandenergie). Dabei gilt die Beziehung (16) für Keime an kleinen Poren mit einem Radius $r \gtrsim r_c$, während (17) für Poren mit $r \gg r_c$ abgeleitet wurde (r_c ist der kritische Porenradius, bei dem im Feld $H_a = 0$ eine Keimbildung energetisch gerade möglich ist).

Die Beziehung (16) führt bei dem untersuchten Manganferrit auf Feldstärken von ca. 10 Oe, die Beziehung (17) auf 2 Oe $< H_w <$ 20 Oe je nach der Größe von r_c/r.

An vielen Poren und Korngrenzen sind im diskutierten Feldstärkebereich auch Keimbildungen möglich, wie das Vorkommen von solchen Keimen beim Feld Null zeigt.

Die Schwierigkeit besteht offenbar darin, daß zwischen dem Anfangs- und Endzustand einer solchen Keimbildung eine hohe Energiebarriere liegt. Daher können sich nur an den Stellen Keime bilden, an denen durch Defekte im Gitter (z. B. Gebiete mit kleinen K_1-Werten) diese Energiebarriere genügend abgeflacht ist [18].

Da die Keimbildung durch Drehprozesse erfolgt, ist der durch Drehungen ummagnetisierte Teil der Probe um so größer, je mehr Keime gebildet werden und je größer das Volumen eines einzelnen Keimes ist. Die relativ starke Abnahme des Switchingkoeffizienten im Bereich III gegenüber $S_{w_2}(S_{w_2}/S_{w_3} \approx 2)$ und die deutlich geringere Abhängigkeit von äußeren Spannungen weisen darauf hin, daß im Feldstärkebereich III derartige Drehprozesse merklich an der Ummagnetisierung beteiligt sind.

Herrn Prof. Dr. M. Steenbeck sowie den Herren Dr. G. Vogler und Dr. L. Michalowsky bin ich für wertvolle Anregungen und Diskussionen sehr zu Dank verpflichtet.

Literatur

[1] E. M. Gyorgy, J. appl. Phys. **31**, 110 S (1960).
[2] N. Menyuk und J. P. Goodenough, J. appl. Phys. **26**, 8 (1965).
[3] J. E. Knowles, Proc. Phys. Soc. **78**, 233 (1961).
[4] R. E. Elfant und F. J. Friedländer, J. Phys. Soc. Japan **17** (SB-I), 640 (1962).
[5] M. K. Haynes, J. appl. Phys. **29**, 472 (1958).
[6] J. E. Knowles, Proc. Phys. Soc. **75**, 885 (1960).
[7] E. M. Gyorgy, J. appl. Phys. **29**, 1709 (1958).
[8] E. M. Gyorgy und F. B. Hagedorn, J. appl. Phys. **30**, 1368 (1959).
[9] L. Shevel, J. appl. Phys. **30**, 47 S (1959).
[10] R. F. Elfant, J. appl. Phys. **34**, 1112 (1963).
[11] Ch. Kittel, Phys. Rev. **80**, 918 (1950).
[12] R. Kikuchi, J. appl. Phys. **27**, 1352 (1950).
[13] P. R. Gilette und K. Oshima, J. appl. Phys. **29**, 529 (1958).
[14] L. Michalowsky, private Mitteilung.
[15] L. Michalowsky und G. Schmidt, Nachrichtentechnik **12**, 323 (1962).
[16] G. Vogler, Z. angew. Phys. **13**, 185 (1961).
[17] Ch. Kittel, Rev. mod. Phys. **21**, 541 (1949).
[18] C. Abraham und A. Aharoni, Phys. Rev. **128**, 2496 (1962).
[19] E. Steinbeiss, Dissertation, Jena 1965.

(Received May 10, 1966)

phys. stat. sol. 16, 507 (1966)

Institut für Reaktorwerkstoffe, Kernforschungsanlage Jülich

On the Mechanism of Sputtering

By

Chr. Lehmann and P. Sigmund

It is shown that spot patterns in single crystal sputtering do not necessarily arise as a consequence of focused collision sequences. Spots can be obtained if it is assumed that only the surface of a solid has a regular structure. The spot width is determined by the energy distribution in the collision cascade and the potential barrier at the surface. The contribution of focusons to sputtering is negligible at low bombarding energies because of the small penetration depth of the ions. Recent experimental results indicate that the average pa thlength offocusons contributing to sputtering is extremely small even at high bombarding energies, so that the proposed sputtering mechanism should be dominant over the whole energy range. This result appears to be consistent with collision cascade theory if forward scattering is taken into account.

Показано, что картина пятен при распылении монокристалла не обязательно является результатом последовательных фокусирующих соударений. Пятна могут возникать в предположении, что только поверхность твёрдого тела имеет регулярную структуру. Ширина пятна определяется распределением энергии в каскаде столкновений и потенциальным барьером на поверхности. При малых энергиях бомбардировки можно пренебречь вкладом фокусонов в механизм распыления из-за малой глубины проникновения ионов. Как показывают новейшие опыты, роль фокусонов в механизме распыления весьма незначительна даже при высоких энергиях бомбардировки, так что предлагаемый механизм распыления следует считать преобладающим во всём диапазоне энергий. Этот результат оказывается в согласии с теорией каскадов столкновений , если учитывать рассеяние вперёд.

1. Introduction

Atom ejection patterns from single crystal sputtering show characteristic spots corresponding to ejection along certain low-indexed crystallographic directions. This effect has been discovered by Wehner [1] who bombarded various b.c.c. and f.c.c. metals with 40 to 300 eV ions. Since then, spots have been found by many investigators with numerous metals and semiconductors at various bombarding energies. It has become customary to consider these spots as a direct evidence for focused collision sequences (focusons) which were postulated by Silsbee [2]. In the "focuson model of sputtering" it is assumed that a majority of atoms sputtered away from the surface receive their energy from primary collisions several lattice distances beneath the surface via collision sequences along close-packed directions. Harrison et al. [3] have recently performed computer simulations of the sputtering process and found that spot patterns do not necessarily arise as a consequence of focusons, and that it might be possible to explain the magnitude of the sputtering yield by those primary collision events taking place quite close to the surface. In the present investigation we come to similar conclusions from a consideration of experimental results. Furthermore, a mechanism of sputtering is proposed which explains spot patterns and which appears to be consistent with collision cascade theory and various experimental facts.

2. Low Energy Sputtering

Let us first discuss the case of *low* bombarding ion energy ($E_0 < 1$ keV). At these energies the penetration depth of heavy ions is generally so small, of the order of one lattice constant, that one would expect sputtering to be a pure surface effect. Focusons travelling towards the surface will have a maximum chain length of one or two lattice units. This is hardly sufficient to explain spots observed at $E_0 \gtrapprox 50$ eV [1]. Furthermore, energy spectra of atoms sputtered perpendicularly to the (100), (110), and (111) surfaces of single crystals by 600 eV Hg^+ ions are almost identical with the spectrum for a polycrystal [4]. From the focuson model one would expect a marked difference in the spectra, especially at ejection energies greater than 10 eV.

We conclude that spots arise as a consequence of regular surface structure. Due to dissipation of the energy of the impinging ion, some surface atoms will receive a certain amount of energy which enables them to overcome the potential barrier at the surface. Quite generally, the energy distribution of these atoms will be strongly peaked at small energies, no matter whether it is determined by the collision density of a random medium or of a lattice. This has the immediate consequence that the sputtered intensity has pronounced maxima in those crystal directions where ejection requires a minium of energy. We mention two most evident cases:

1. Atoms in the layer just *beneath* the surface have to penetrate more or less symmetric rings of surrounding atoms before being sputtered (Fig. 1a). Ejection should be easiest along the axis of a ring. Spots in such directions are commonly ascribed to "assisted focusing" [5]. Because of the relatively high potential barriers (≈ 10 eV or more) these spots occur at several hundred eV bombarding ion energy.

2. Sputtering of an atom in the *uppermost* surface layer should be easiest by means of a collision started by one of its close neighbours beneath the surface (Fig. 1b). If the surface atom receives the total energy from its neighbour by a *central* collision it has best chances to overcome the surface binding energy

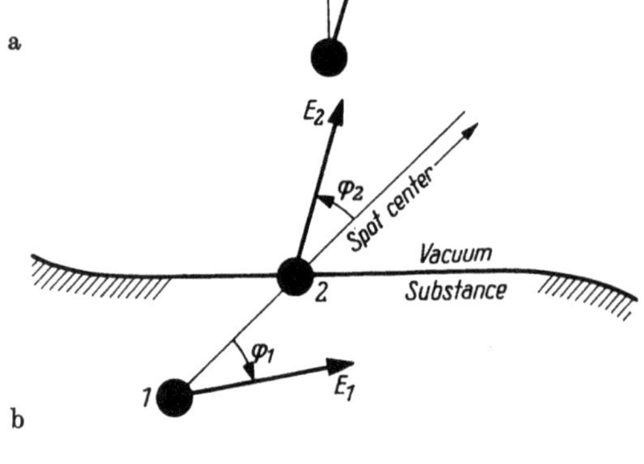

Fig. 1. Mechanisms of spot formation.
a) Mechanism 1, b) Mechanism 2

U (≈ 5 eV). Spots in nearest neighbour directions are commonly ascribed to "pure Silsbee" (isolated chain) focusons.

Note that these two mechanisms are different from single event focusing as proposed by Hasiguti et al. [6] and others. The width of the spots is primarily determined by the energy distribution, and it is not necessary to claim *focusing* ring or two-particle collisions.

3. Example: ⟨110⟩ Spot in Cu

As an example, we give a rough estimate of the spot width in close neighbour sputtering (mechanism 2). For random slowing-down the energy distribution of atoms 1 just beneath the surface (Fig. 1 b) may be approximated by a power law

$$g(E_1) \sim E_1^{-n} \tag{1}$$

with [7] $3/2 \lessapprox n \lessapprox 3$. This is consistent with the measured energy spectra [4] and calculations of collision density in a random infinite medium [8]. For the present purpose it is sufficient to assume the velocity distribution to be isotropic. Scattering between the two atoms in question may be expressed by $\varphi_2 = \Lambda(E_1, \varphi_1)\,\varphi_1$, where $\Lambda(E_1, \varphi_1)$ is the well-known focusing parameter. For Born-Mayer interaction Λ may be assumed to be independent of φ_1 for not too large angles [9]. From this, we obtain the angular distribution $S(\varphi_2)$ of atoms 2 sputtered around the considered direction, assuming a threshold energy $U(\varphi_2) \approx U$ for small angles φ_2:

$$S(\varphi_2)\,\mathrm{d}\Omega_2 = \int g(E_1)\,\mathrm{d}E_1\,\mathrm{d}\Omega_1 \approx \mathrm{d}\Omega_2 \int_W^\infty \mathrm{d}E_1\, g(E_1)/\Lambda^2(E_1)\,. \tag{2}$$

The limits of the first integral are defined by

$$(E_2 =) \qquad E_1 \cos^2[\varphi_1\,(1 + \Lambda)] > U\,, \qquad \varphi_2 < \Lambda\,\varphi_1 < \varphi_2 + \mathrm{d}\varphi_2\,.$$

In the second integral, the lower limit is

$$W = U\left(1 + \varphi_2^2\,[1 + \Lambda^{-1}\,(U)]^2\right).$$

In the second step we have made an expansion up to second order in φ_2. Let us use a simple but plausible form for Λ [9, 10]:

$$\Lambda(E_1) = (E_1/E_f)^\alpha\,, \tag{3}$$

where E_f is the focusing energy and the exponent α is close to 1/2. Inserting (1) and (3) into (2) and taking $n = 2$ we obtain

$$S(\varphi_2) \sim U^{-1-2\alpha}\,(1 - [1 + 2\,\alpha]\,\varphi_2^2\,[1 + (E_f/U)^\alpha]^2)\,. \tag{4}$$

As expected above, we obtain a definite spot already for the case $\alpha = 0$ (no focusing). For $\alpha = 1/2$, the intensity has a half-width given by

$$\varphi_2^{\text{half}} = 1/2\,[1 + (E_f/U)^{1/2}]^{-1}\,.$$

For the Cu model of Gibson et al. [10] one has $E_f \approx 10\,U$, so that one obtains $\varphi_2^{\text{half}} = 7°$. Observed half-widths are greater. Broadening is provided by temperature vibrations and deviations from ideal surface structure. Of course, this estimate only shows that our mechanism gives the right order of magnitude, while detailed calculations would have to take into account the actual (anisotropic) energy distribution, a more accurate focusing parameter, angular variation of the binding energy U, detailed structure of the surface, and influence of neighbouring surface atoms.

4. High Energy Sputtering

It appears natural to assume this sputtering mechanism to be valid also at *high* bombarding energies. This would explain the sharp spots observed in hexagonal metals [6, 11] near the $\langle 20\bar{2}3\rangle$ directions, were atoms are arranged on zig-zag rows so that the Silsbee focusing mechanism should not operate. Note that $\langle 20\bar{2}3\rangle$ is a close neighbour direction. "Streaks" observed in the ejection pattern of f.c.c. metals [12] could be explained by the assumption that ejection requires relatively low energy along certain lattice *planes*.

Our model also explains relative minima of the average energy of sputtered atoms in the spots as observed by Weijsenfeld [13]. This can be shown by a similar calculation as for $S(\varphi_2)$ in section 3.

The question arises whether focusons contribute an essential part to the sputtering yield and so provide an *additional* mechanism leading to spot patterns.

From collision cascade theory, one expects that focusons with small ranges (initial energy $\ll E_f$) are much more frequent than those having long ranges. In calculations of energy spectra [14] and sputtering yields [15] the fraction of long-range focusons arriving at the surface has been overestimated by assuming a production rate of focusons corresponding to a hard sphere collision spectrum [16]. Using more realistic energy spectra, it is possible to show that the majority of focusons contributing to the sputtering yield have their origin only one or two atomic distances beneath the surface [7]. Furthermore, several experimental results indicate that the average focuson range is only a couple of lattice distances:

1. In hexagonal substances, the sputtering yield has the same magnitude for (0001) and $(10\bar{1}0)$ surfaces (Southern [11]). Focusons may only contribute in the latter case.

2. Sputtering yields of many b.c.c. and f.c.c. metals depend only slightly on target temperature between 300 and 1000 °K [17]. The range of focusons, however, is expected to drop by more than a factor of two in the same temperature range [14, 18].

3. There is a striking similarity in the behaviour of sputtering yield and reciprocal sublimation energy as a function of atomic number of the material, both for high and low energy sputtering [19]. It is generally accepted that the sublimation energy is closely related to the surface binding energy U. If long range focusons would dominate, the surface binding energy should be of minor importance [15].

4. The shape of the energy spectrum of gold atoms sputtered into a $\langle 110\rangle$ direction by 40 keV A^+ ions [20] corresponds to what one would expect from random slowing-down [8]. In fact, it has a sharp peak at 6 eV ($\approx U$) and falls off at higher energies as E_2^{-n} with $n \approx 2$. If focusons of moderate range would be important, one would expect a spectrum falling off less rapidly.

5. Conclusions

As the average path length travelled by focusons contributing to sputtering is so extremely small, one arrives at the following conclusions:

1. It is not possible to explain the observed spot patterns *only* by focusons, even at high ion energies.

2. Sputtering yields calculated with and without focusons will have the same order of magnitude.

As a consequence, one does not expect a significant difference between sputtering yields of polycrystalline and amorphous substances, provided that all other quantities, such as atomic number, density, and surface binding energy, are similar.

As spot patterns have been the main evidence for focusons until now, it seems worthwhile to re-examine the problem of their experimental verification. One possibility would be to measure the cut-off energy in the time-of-flight spectrum of sputtered atoms [20] for a great variety of ion energies. Note that a relatively sharp cut-off in the time-of-flight spectrum is not evidence enough for focusons. Also Wehner's spectra [4] and the Maxwell distribution function reveal this feature.

Acknowledgements

It is a great pleasure to thank H. H. Andersen, G. Leibfried, J. Lindhard, M. T. Robinson, J. B. Sanders, W. Schilling, and R. Sizmann for inspiring discussions on the physics of sputtering.

References

[1] G. K. Wehner, J. appl. Phys. **26**, 1056 (1955); Phys. Rev. **102**, 690 (1956).
[2] R. H. Silsbee, J. appl. Phys. **28**, 1246 (1957).
[3] D. E. Harrison, J. P. Johnson, and N. S. Levy, Appl. Phys. Letters **8**, **33** (1966).
[4] R. V. Stuart and G. K. Wehner, J. appl. Phys. **35** 1819 (1964); see also R. A. Ashmyanskii, M. B. Beniaminovich, and V. I. Veksler, Soviet Phys. — Solid State Phys. **7**, 1314 (1965), translated from Fiz. tverd. Tela **7**, 1623 (1965).
[5] R. S. Nelson, and M. W. Thompson, Proc. Roy. Soc. **A259**, 458 (1961).
[6] R. R. Hasiguti, R. Hanada, and S. Yamaguchi, J. Phys. Soc. Japan **18**, Suppl. III, 164 (1963).
[7] P. Sigmund, to be published.
[8] M. T. Robinson, Phil. Mag. **12**, 145 (1965); **12**, 741 (1965).
[9] G. Duesing and G. Leibfried, phys. stat. sol. **9**, 463 (1965).
[10] J. B. Gibson, A. N. Goland, M. Milgram, and G. H. Vineyard, Phys. Rev. **120**, 1229 (1960).
[11] B. Perović, Proc. 5th Internat. Conf. Ion. Phen. Gases, North-Holland Publ. Comp. 1962 (p. 1172). A. L. Southern, ORNL-3676 (1964) (p. 74).
[12] R. S. Nelson and M. W. Thompson, Phys. Letters **2**, 124 (1962).
[13] C. H. Weijsenfeld, 7th Internat. Conf. Ion. Phen. Gases, Belgrade 1965.
[14] R. S. Nelson, M. W. Thompson, and H. Montgomery, Phil. Mag. **7**, 1385 (1962).
[15] M. W. Thompson, Proc. 5th Internat. Conf. Ion. Phen. Gases, North-Holland Publ. Comp. 1962 (p. 85).
Yu. V. Bulgakov, Bull. Acad. Sci. USSR **28**, 1377 (1964), translated from Izv. Akad. Nauk SSSR, Ser. fiz. **28**, 1474 (1963).
[16] G. Leibfried, J. appl. Phys. **30**, 1388 (1959); **31**, 117 (1960).
[17] R. S. Nelson, Phil. Mag. **11**, 291 (1965).
C. E. Carlston, G. D. Magnuson, A. Comeaux, and P. Mahadevan, Phys. Rev. **138A** 759 (1965).
[18] J. B. Sanders and J. M. Fluit, Physica **30**, 129 (1964).
[19] O. Almén and G. Bruce, Nuclear Instrum. and Methods **11**, 257 (1961).
G. K. Wehner, Phys. Rev. **108**, 35 (1957).
[20] M. W. Thompson, Phys. Letters **6**, 24 (1963).

(Received May 11, 1966)

phys. stat. sol. 16, 513 (1966)

Department of Physics, Banaras Hindu University, Varanasi

Lattice Vibrations of Aluminium on the Basis of Krebs's Model

By

M. M. SHUKLA and B. DAYAL

The vibration spectrum of aluminium is studied using Krebs's model. The phonon dispersion curves for the three symmetry directions [100], [110] and [111] are presented and are found to be in very good agreement with neutron scattering measurements. The $(\theta - T)$-curves calculated from Krebs's model agree very well with experiment using a screening parameter deduced from the Bohm-Pines theory.

Das Schwingungsspektrum von Aluminium wird unter Benutzung des Krebsschen Modells untersucht. Die gefundenen Dispersionskurven für die drei symmetrischen Richtungen [100], [110] und [111] befinden sich in sehr guter Übereinstimmung mit Neutronenstreuungsmessungen. Die mit dem Krebsschen Modell berechneten $(\Theta - T)$-Kurven stimmen sehr gut mit den Experimenten überein, wenn ein Abschirmparameter benutzt wird, der aus der Theorie von Bohm und Pines hergeleitet wird.

1. Introduction

Krebs [1] has recently proposed a new elastic force model for the study of the lattice dynamics of metals. While the volume forces have been introduced on the lines of de Launay's work [2], Lax's [3] suggestion has been followed to keep the frequencies periodic in the reciprocal space. Krebs [4] has studied the lattice vibrations of sodium, potassium and lithium on his model. This model has been recently applied for a study of several metals in this laboratory. Shukla and Dayal [5, 6, 7] have studied the case of noble metals. They find that this model reproduces the experimental C_v remarkably well if the screening parameter λ is taken as an empirical constant which varies systematically with the ionic radii between the values given by the theories of Bohm and Pines and Thomas and Fermi. The same degree of success has been possible in explaining neutron scattering results in the dispersion curves of copper, the only substance of the group for which such data are available. Mahesh and Dayal [8, 9] have studied the lattice vibrations of alkali metals [8] and transition metals [9] belonging to the body-centered cubic structure. They have also obtained an excellent agreement with the experimental results.

In view of these successes of the Krebs's model it was thought proper to investigate a slightly more complicated case. Aluminium is a trivalent metal and the position regarding the screening constant is not very clear. Extensive data about the dispersion curves and C_v are available. A number of workers have carried out measurements of the dispersion curves of this metal by diffuse X-ray scattering as well as by cold neutron scattering. Among such investigations the work of Stedman and Nilsson [10] about cold neutron scattering is most recent and perhaps most accurate. This metal, therefore, appears to be quite suitable for testing the applicability of Krebs's model.

In the present paper we have applied Krebs's model to study the lattice dynamics of aluminium. The screening parameter has been taken to be the same as given by the Bohm-Pines theory. The model has, however, been extended to include the effect of the first three neighbours. It has been found that dispersion curves and the specific heats are very satisfactorily reproduced by this theory.

2. Secular Determinant

The secular determinant used for the present computation is given by

$$|M_{ij} - m\,\omega^2 I| = 0 \,, \tag{1}$$

where $i, j = 1, 2, 3$ and I is the unit matrix of order three. m is the mass of the atom and ω the angular frequency. Each element M_{ij} is the sum of two parts, the ionic coupling coefficients $(M_{ij})^i$ and the electronic contributions to the elements of the dynamical matrix, I_{ij}. Since we have included the effect of the third nearest neighbour also, the ionic coupling coefficient is a bit different from that given by Krebs. The new expressions for typical diagonal and non-diagonal elements are given by

$$[M_{11}]^i = 2\,\alpha_1\,\{2 - C_1\,(C_2 + C_3)\} + 4\,\alpha_2\,S_1^2 + \\ + 8\,\alpha_3\left\{1 - \frac{2}{3}\,C_2\,C_3\,(2\,C_1^2 - 1) - \frac{1}{6}\,C_1\,C_3\,(2\,C_2^2 - 1) - \frac{1}{6}\,C_1\,C_2\,(2\,C_3^2 - 1)\right\}, \tag{2}$$

$$[M_{12}]^i = 2\,\alpha_1\,S_1\,S_2 + \frac{4}{3}\,\alpha_3\,S_1\,S_2\,\{(2\,C_3^2 - 1) + 4\,C_3\,(C_1 + C_2)\}\,, \tag{3}$$

where S_i and C_i are $\sin \pi\,k_i\,a$ and $\cos \pi\,k_i\,a$ respectively and k_i $(i = 1, 2, 3)$ are the components of the wave vector $\boldsymbol{k}$, and a is the lattice parameter. α_1, α_2, and α_3 are Hooke's force constants for the first, second and third nearest neighbours respectively. These force constants are related to the elastic constants by the following equations:

$$\left.\begin{aligned} a\,c_{11} &= 2\,\alpha_1 + 4\,\alpha_2 + 12\,\alpha_3 + a\,k_e\,, \\ a\,c_{12} &= a_1 + 6\,\alpha_3 + a\,k_e\,, \\ a\,c_{44} &= \alpha_1 + 6\,\alpha_3\,, \end{aligned}\right\} \tag{4}$$

where k_e the bulk modulus of the electron gas.

The expression for the electronic coupling coefficient has been given by one of us (Shukla [5]) in an earlier paper.

3. Numerical Computation

The calculation of the frequency distribution has been performed exactly in the same way as described earlier by Dayal and Sharan [11] by solving the secular determinant (1).

The force constants α_1, α_2 and α_3 are calculated from three different equations relating them with the elastic constants $(c_{11} - c_{12})$, c_{44}, and an experimentally observed frequency obtained from neutron scattering. In this case we have taken this vibration frequency to be the one of the boundaries of the transverse branch of [110]. The elastic constants at 80 °K have been taken from the work of Kamm and Alers [12]. The data used are given below:

$$\begin{aligned} c_{11} &= 11.373 \times 10^{11}\ \text{dyn/cm}^2, \\ c_{12} &= 6.191 \times 10^{11}\ \text{dyn/cm}^2, \\ c_{44} &= 3.128 \times 10^{11}\ \text{dyn/cm}^2, \\ a &= 4.04\ \text{Å}, \\ \nu &= 8.75 \times 10^{12}\ \text{Hz}. \end{aligned}$$

The calculated values of the force constants are found to be

$$\alpha_1 = 16.0874 \times 10^3 \text{ dyn/cm},$$
$$\alpha_2 = 2.0748 \times 10^3 \text{ dyn/cm},$$
$$\alpha_3 = -0.5751 \times 10^3 \text{ dyn/cm}.$$

The computed dispersion curves for aluminium along the three symmetry directions [100], [110], and [111] are shown in Fig. 1 to 3. The continuous curve represents the calculated frequencies while the crosses, solid circles, and hollow circles represent the experimental results of Stedman and Nilsson [10].

The C_v have been calculated by numerical sampling from histograms. For this purpose the frequencies have been divided into intervals of 0.4×10^{12}. The calculated $(\Theta - T)$ curve is plotted in Fig. 4 as a continuous line and crosses represent the experimental points. The experimental C_v have been taken from the work of Giauque and Meads [13]. The electronic heat capacity has been taken to be 3.27×10^{-4} cal mol^{-1} deg^{-2} from the work of Howling et al. [14]. The calculated and observed values of C_v are reported in Table 1.

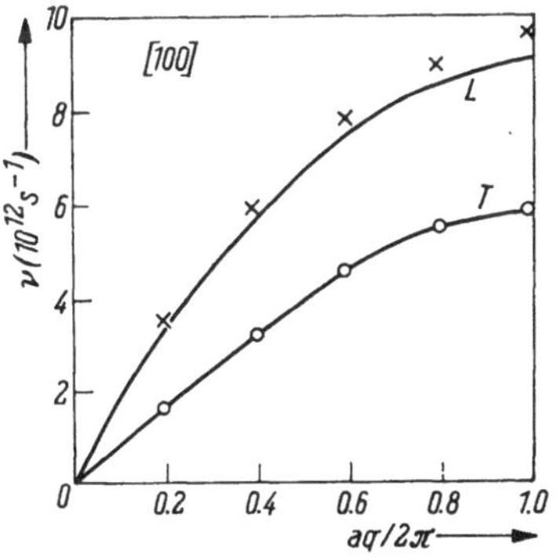

Fig. 1. Dispersion curve for aluminium in [100] direction.
——— Calculated
× × × ○ ○ ○ } Experimental points

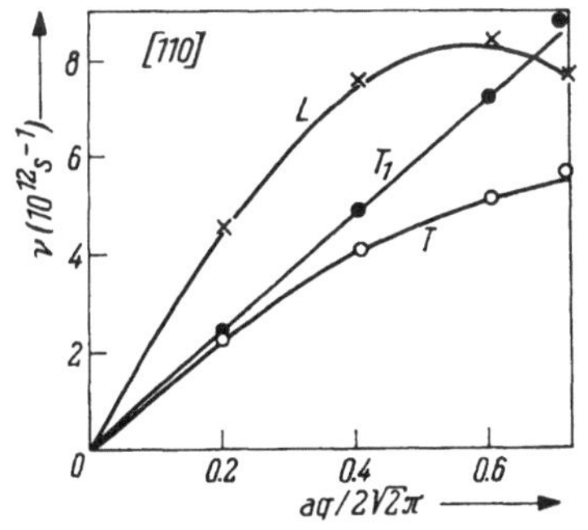

Fig. 2. Dispersion curve for aluminium in [110] direction.
——— Calculated
× × × ○ ○ ○ ● ● ● } Experimental points

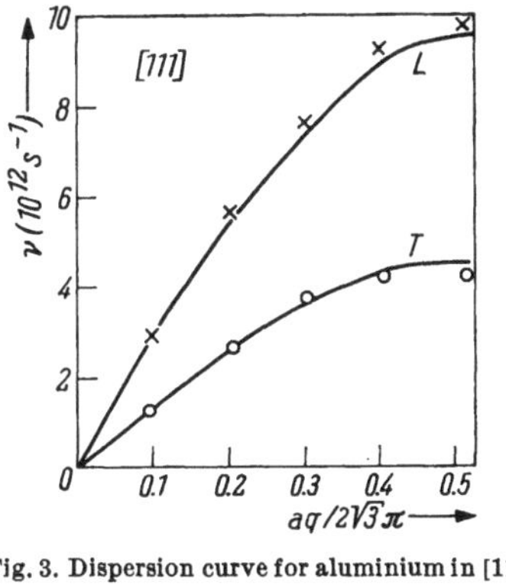

Fig. 3. Dispersion curve for aluminium in [111] direction.
——— Calculated
× × × ○ ○ ○ } Experimental points

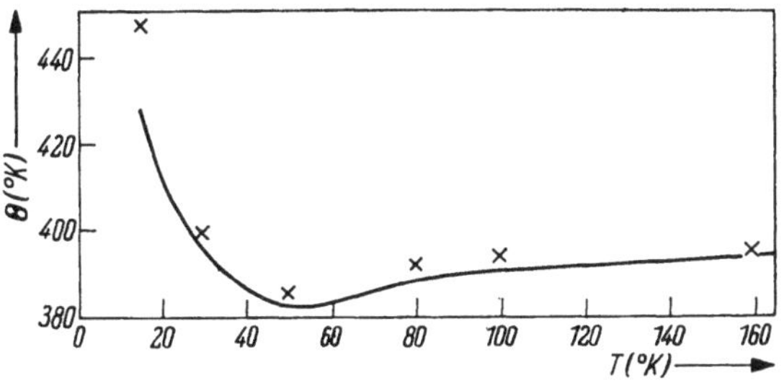

Fig. 4. $(\theta - T)$ curve for aluminium.
——— Calculated
× × × Experimental points

Table 1

Specific heats of aluminium in cal mol^{-1} deg^{-2}

Temperature (°K)	Calculated C_v	Experimental $(C_v - \gamma T)$
15	0.020	0.017
30	0.210	0.193
50	0.926	0.906
80	2.303	2.272
100	3.092	3.062
160	4.491	4.484

4. Discussion

It is evident from the study of Fig. 1 to 3 that the calculated and experimental results are found to be in excellent agreement. At low vectors the calculated frequencies almost reproduce the experimental results but some disagreement occurs at the boundaries of the Brillouin zone.

It is clear from Fig. 4 that Krebs's model is able to reproduce the course of the experimental $(\Theta - T)$ curve also very satisfactorily. The calculated curve is about one per cent below the experimental one. In the study of the specific heats of noble metals by Krebs's model we found that the screening parameter varies between the values given by Bohm and Pines and by Thomas and Fermi. It was observed that the screening parameter of those metals whose ionic radii are small follow the Bohm-Pines expression. The present study supports this conclusion because aluminium has got a small ionic radii and its screening parameter is the same as that given by the Bohm-Pines theory.

Acknowledgement

The authors are grateful to Dr. Reginold Stedman of Atomenergi, Studsvik, Sweden, for providing them with his unpublished tables of the vibrational frequencies of aluminium.

References

[1] K. KREBS, Phys. Letters (Netherlands) **10**, 12 (1964).
[2] J. DE LAUNEY, J. chem. Phys. **21**, 1979 (1953).
[3] M. LAX, Proc. Internat. Conf. on Lattice Dynamics, Copenhagen 1963 (paper A24).
[4] K. KREBS, Phys. Rev. **138**, 143 (1965).
[5] M. M. SHUKLA, phys. stat. sol. **7**, K11 (1964).
[6] M. M. SHUKLA, phys. stat. sol. **8**, 475 (1965).
[7] M. M. SHUKLA and B. DAYAL, J. Phys. Chem. Solids **26**, 1343 (1965).
[8] P. S. MAHESH and B. DAYAL, phys. stat. sol. **9**, 351 (1965).
[9] P. S. MAHESH and B. DAYAL, Phys. Rev. **143**, 443 (1966).
[10] R. STEDMAN and G. NILSSON, Inelastic Scattering of Neutrons, Vol. 1, International Atomic Energy Agency, Vienna 1965 (p. 211).
[11] B. DAYAL and B. SHARAN, Proc. Roy. Soc. **A 259**, 361 (1960).
[12] G. N. KAMM and G. A. ALERS, J. appl. Phys. **35**, 327 (1964).
[13] W. F. GIAUQUE and P. F. MEADS, J. Amer. Chem. Soc. **61**, 1897 (1941).
[14] D. H. HOWLING, V. MENDOZA, and Z. E. ZIMMERMAN, Proc. Roy. Soc. **A 229**, 86 (1955).

(Received May 12, 1966)

phys. stat. sol. **16**, 517 (1966)

Centre National de la Recherche Scientifique, Faculté des Sciences, Paris

Émission photoluminescente visible du gadolinium inclus dans le réseau cristallin du sulfure de zinc

Par

M. Bancie-Grillot et P. Bourtayre

Nous précisons ici certains caractères de l'émission de raies visibles du sulfure de zinc activé par du gadolinium, que nous avons récemment découverte [1]. Elle est constituée de 46 raies fines, s'étendant entre 4664 et 7057 Å; six d'entre elles, formant un système secondaire, disparaissent par adjonction de minimes traces de cobalt et sont atténuées dans les cristaux mixtes de zinc et de cadmium. Cette émission ne peut être attribuée à l'ion trivalent Gd^{3+}, seule forme considérée chimiquement stable de ce lanthanide. Il est en effet connu que le premier état excité de cet ion est, énergétiquement, à près de 4 eV au-dessus de son état fondamental correspondant à la configuration électronique $4f^7$. Dans le réseau cristallin assez fortement covalent du sulfure de zinc, le gadolinium capterait solidement un électron, son état stable non excité étant probablement $4f^7\,5d^1$.

Es wird das kürzlich in [1] veröffentlichte Emissionsspektrum von Gadolinium-dotiertem Zinksulfid im sichtbaren Spektralbereich genauer diskutiert. Das Emissionsspektrum besteht aus 46 schmalen Linien im Bereich zwischen 4664 und 7057 Å. Sechs davon bilden ein sekundäres System, sie verschwinden bei Zusatz einer sehr kleinen Kobalt-Konzentration und werden in ZnS-CdS-Mischkristallen geschwächt. Diese Emission kann nicht dem dreiwertigen Gd^{3+}-Ion zugeschrieben werden, das als einzige stabile Formdieses Lanthanids angesehen wird. Der erste angeregte Zustand dieses Ions liegt bekanntlich bei 4 eV oberhalb des Grundzustandes, der der $4f^7$-Konfiguration entspricht. Im hinreichend stark kovalenten Zinksulfid-Gitter kann das Gadolinium ein Elektron fest einfangen. Sein stabiler nicht-angeregter Zustand ist wahrscheinlich $4f^7\,5d^1$.

1. Introduction

L'étude de l'émission luminescente spécifique des ions lanthanides trivalents inclus comme impuretés dans les cristaux ioniques montre que la position médiane que le gadolinium occupe dans ce groupe d'éléments lui confère des propriétés particulières. La configuration $4f^7$ des ions Gd^{3+} rend chacun de ces électrons plus fortement lié que dans les autres ions des terres rares et par conséquent il faut leur fournir une énergie plus grande pour les amener au premier état excité. C'est pourquoi les raies d'émission fluorescente de tous les autres ions lanthanides se situent dans le domaine visible du spectre, tandis que celles de l'ion Gd^{3+} apparaissent seulement dans l'ultraviolet, vers 3000 Å. Par exemple, dans le chlorure de sodium, le gadolinium fournit deux groupes de raies centrées sur 3130 Å (7 composantes) et sur 3080 Å (3 composantes) [2]. De même, dans le fluorure de calcium [3], les seules raies observées sont également centrées sur 3150 Å. L'appoint d'énergie pour atteindre son état excité le plus bas est donc d'environ 4 eV.

Les cristaux ioniques ci-dessus mentionnés étant transparents à des photons d'énergie supérieure à cette valeur, il n'est pas difficile de provoquer cette fluorescence. Mais si l'on considère au contraire un cristal comme le sulfure de zinc dont la largeur de la bande interdite n'est que 3,7 eV, on ne peut espérer atteindre ce premier état excité de Gd^{3+}, ni directement, ni par transfert d'énergie par l'intermédiaire du réseau. Si, incidemment, cet état était atteint, le cristal serait opaque aux photons émis. Il semble donc bien que le gadolinium, dont seul l'ion trivalent est connu stable, doive être exclus des activateurs possibles du sulfure de zinc.

Il n'y a eu jusqu'ici qu'un nombre extrêmement restreint de publications sur les cristaux de sulfure de zinc activés par les éléments du groupe des lanthanides, probablement en raison des difficultés de leur obtention. Ce composé n'est en effet fusible qu'à pression très élevée et la diffusion de l'impureté dans le réseau cristallin déjà constitué est beaucoup plus malaisée que pour les activateurs classiques comme le cuivre ou l'argent. Une récente étude d'ensemble de V. L. Levchin et ses collaborateurs [4] a montré que, sous excitation cathodique, la plupart des ions lanthanides y donnent une émission fluorescente. Mais ces auteurs n'en ont effectivement pas trouvé pour le gadolinium.

Depuis quelques temps, nous avons également entrepris des recherches sur le comportement des ions lanthanides dans le réseau du sulfure de zinc, en vue d'une exploration détaillée de son champ intracristallin. Nos premiers résultats sur l'étude de leurs propriétés photoluminescentes ont été présentés en 1965 [1]. Nous n'en avons pas exclus à priori le gadolinium. Le caractère fortement covalent des liaisons entre les atomes du cristal-hôte pouvait en effet beaucoup plus profondément perturber que prévu les états excités des ions activateurs. Avec des photons excitateurs d'énergie moindre de 4 eV ($\lambda \approx 3600$ Å du Hg), l'émission habituelle de Gd^{3+} ne pouvait être envisagée. Mais nous avons ainsi découvert [1] une nouvelle émission du gadolinium s'étendant sur tout le domaine visible du spectre, dont nous précisons ci-dessous quelques caractères.

Il ne s'agit pas d'un résultat accidentel. Nous avons préparé un nombre assez grand d'échantillons, dans différentes conditions, et avons retrouvé exactement ce spectre du gadolinium, très différent de ceux de tous les autres lanthanides.

2. Résultats expérimentaux

Dans le travail préliminaire cité ci-dessus [1], nous avions réalisé l'enregistrement de l'émission à l'aide d'un photomultiplicateur RCA 1P21, dont la faible sensibilité au-dessus de 6000 Å nuisait à la précision des résultats. Nous avons donc poursuivi cette étude en perfectionnant l'appareillage. Nous présentons ici ce spectre enregistré à l'aide d'un photomultiplicateur EMI 9558 A, de bonne sensibilité dans le rouge et le proche infrarouge jusqu'à 0,9 µm. Ces courbes se rapprochent davantage de la répartition énergétique spectrale. On y observe 46 raies fines.

La courbe du haut de la Fig. 1 reproduit ce spectre d'émission de ZnS (Gd) entre 4500 et 5500 Å; celle de la Fig. 2 le reprend entre 5000 et 7000 Å, avec des fentes d'entrée et de sortie du monochromateur deux fois plus fines. La comparaison des hauteurs du groupe de raies situées vers 5400 Å montre bien que la première figure se rapporte à des flux lumineux sensiblement quatre fois supérieurs à ceux de la seconde. Plusieurs raies rouges sont manifestes, au delà du groupe centré vers 6250 Å, le plus intense, qui était précédemment [1] notre limite, mais ne représente donc pas la transition provenant du niveau excité le plus bas. Le spectre est pur jusque vers 6900 Å: en effet, la lumière excitatrice, filtrée par un verre de Wood, est bien exempte de radiations visibles; mais au

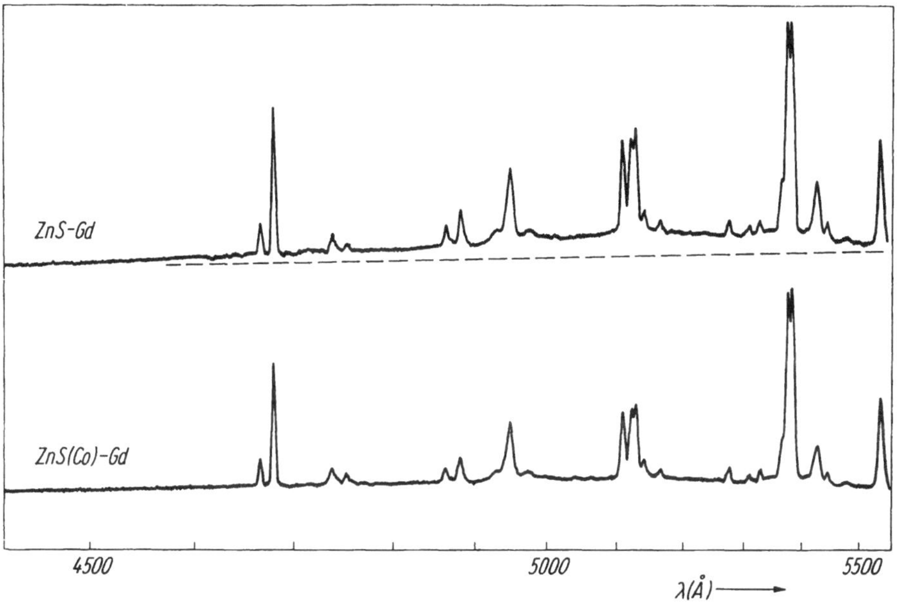
ZnS-Gd
ZnS(Co)-Gd
4500
5000
5500
λ(Å)

Fig. 1

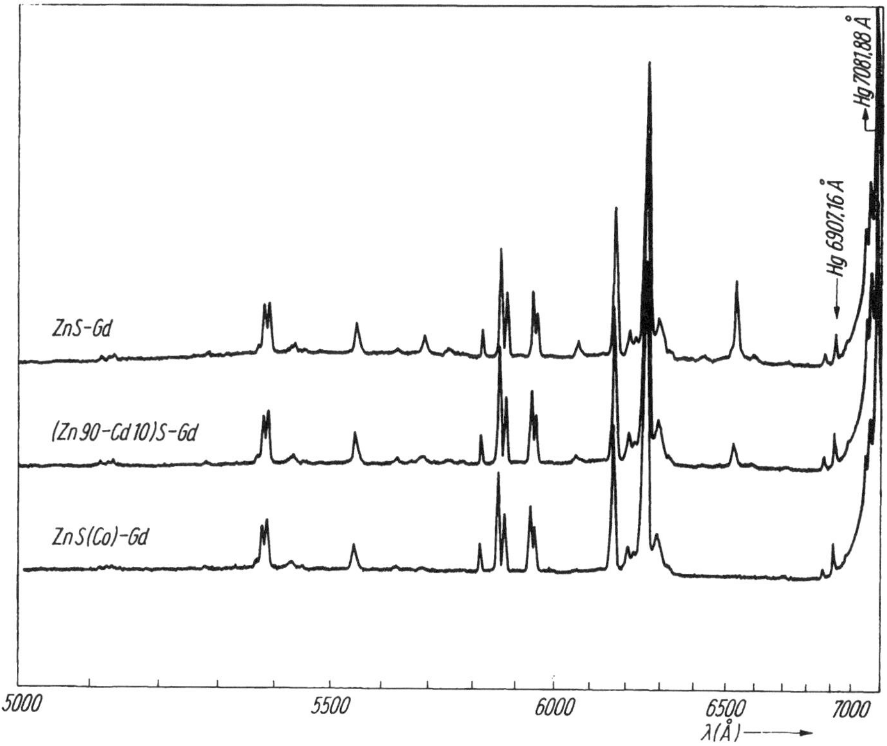
Hg 7081.88 Å
Hg 6907.16 Å
ZnS-Gd
(Zn90-Cd10)S-Gd
ZnS(Co)-Gd
5000
5500
6000
6500
7000
λ(Å)

Fig. 2

delà, la bande et les raies du mercure entre 6900 et 8000 Å n'ont pu être entièrement éliminées et leur réflexion sur les microcristaux explique la montée terminale de la courbe. Néanmoins, en comparant les enregistrements obtenus dans cette région avec des échantillons de sulfure de zinc, soit non activés, soit activés par du gadolinium, nous avons acquis la certitude de l'existence de raies appartenant réellement au système d'émission du gadolinium: ce sont celles non pointées Hg sur la courbe. L'exploration du spectre jusqu'à plus de 9000 Å ne nous a permis de rien révéler au delà de la raie 7057 Å, qui pourrait donc correspondre à la transition de moindre énergie. Celle de plus grande énergie observée est 4664 Å, en deçà de laquelle nous n'avons non plus révélé aucun accident.

Le Tableau 1 énumère les longueurs d'onde des 46 raies d'émission enregistrées. Pour celles d'intensité notable, la précision de la mesure est de 1 Å; pour les autres, faibles ou très faibles, elle est naturellement moins bonne (de l'ordre de 5 Å): on les distingue en écrivant leur valeur entre parenthèses. Leur existence ne fait pas de doute puisqu'on les retrouve dans tous les échantillons de ZnS (Gd). Dans ce tableau, nous avons également porté la valeur en eV de la transition impliquée, ainsi que l'intensité lumineuse relative approximative, en tenant compte de la sensibilité spectrale du photomultiplicateur utilisé et en choisissant une valeur arbitraire de 1000 pour la raie la plus intense $\lambda = 6252$ Å.

On peut remarquer que les pics de la courbe supérieure de la Fig. 1 se détachent d'un fond continu, bien mis en apparence par le tracé de la ligne en pointillés correspondant à l'obscurité. Il s'agit de la bande d'émission du cuivre, à l'état d'infimes traces résiduelles, que l'on ne peut jamais éliminer intégralement [5]. Le rapport des surfaces de la bande et des raies indique qu'entre 4500 et 5500 Å les intensités de ces deux émissions sont du même ordre de grandeur. C'était, de ce point de vue, notre meilleur échantillon. Au cours de la calcination des luminophores il arrive fortuitement très fréquemment que le produit se souille d'une plus grande quantité de cuivre situé dans l'atmosphère ou le creuset en silice ou autre matériau. La concentration de cette impureté luminogène classique est donc aléatoire, et souvent beaucoup plus importante, au point de masquer plus ou moins complètement tout ou partie de l'émission de l'ion lanthanide. Cherchant à dégager le mieux possible les caractéristiques des spectres dûs à ces ions, il n'est bien entendu pas question pour nous d'en rajouter volontairement, comme l'ont proposé certains [6], au titre de soit-disant coactivateur, d'ailleurs parfaitement superflu, mais au contraire de tenter d'éteindre cette émission de bande gènante. Déjà dans notre étude préliminaire [1], nous avions envisagé pour les différents ions lanthanides d'ajouter subsidiairement de minimes traces de cobalt: 1 ou 2×10^{-6}, poison classique de la luminescence du cuivre. Nous avions remarqué, notamment dans le cas du gadolinium, que les pics se retrouvaient dans la même position spectrale et sans modification apparente de leur intensité tandis que le fond lumineux dû au cuivre était pratiquement éliminé. Nous sommes maintenant en mesure de faire quelques observations complémentaires. Les courbes inférieures des Fig. 1 et 2 se rapportent à de tels échantillons avec cobalt. Effectivement, presqu'aucune raie n'est perturbée, ni dans sa position spectrale ni dans son intensité. Il y a cependant restriction pour six d'entre elles: la raie $\lambda = 6526$ Å, d'intensité moyenne pour ZnS–Gd, disparaît complètement par addition de cobalt, en même temps que ses deux satellites $\lambda = 6430$ Å et $\lambda = 6590$ Å, beaucoup plus faibles. Il en est de même des raies peu intenses $\lambda = 6050$ Å et $\lambda = 5743$ Å. Enfin, la raie $\lambda = 5690$ Å est très atténuée.

Tableau 1
Description du spectre de raies de fluorescence du gadolinium dans le réseau du sulfure de zinc

Longueur d'onde (Å)	Énergie de la transition (eV)	Intensité relative approximative
7057	1,7642	210
7039	1,7686	100
6867	1,8129	50
((6700))	1,8581	<20
(6590)*)	1,8891	30
6526*)	1,9077	350
(6430)*)	1,9362	25
(6303)	1,9752	23
6286	1,9805	195
6252	1,9912	**1000**
6216	2,0028	60
6198	2,0086	95
6157	2,0220	480
6050*)	2,0577	45
5953	2,0913	110
5942	2,0952	175
5889	2,1140	165
5867	2,1220	285
5819	2,1395	75
(5743)*)	2,1678	15
5690*)	2,1879	35
(5636)	2,2090	15
5550	2,2432	65
(5483)	2,2706	4
(5453)	2,2830	7
5434	2,2962	22
5382	2,3132	95
5374	2,3167	95
5367	2,3197	22
5326	2,3375	6
5311	2,3441	5
5279	2,3583	8
5171	2,4076	4
5143	2,4206	10
5130	2,4269	36
5124	2,4297	33
5116	2,4335	30
(4980)	2,500	2
4954	2,5130	23
(4937)	2,5217	4
4891	2,5454	12
4873	2,5548	6
4756	2,6177	2
4743	2,6248	5
4682	2,6590	43
4664	2,6693	9

*) Raies qui disparaissent sous l'influence de minimes traces de cobalt et qui s'atténuent dans les sulfures mixtes de zinc et de cadmium.

Nous avons également introduit le gadolinium dans le réseau cristallin de sulfures mixtes de zinc et de cadmium. La courbe intermédiaire de la Fig. 2 donne un exemple des résultats obtenus (composition: 90% ZnS et 10% CdS). L'élargissement de la maille hexagonale ainsi obtenu et les modifications du champ cristallin qui en résultent n'entrainent aucun déplacement sensible des raies d'émission, qu'on retrouve toutes. Cependant, il y a lieu de noter que toutes les transitions qui sont supprimées par addition de cobalt sont également affectées ici: les intensités relatives des raies correspondantes sont notablement diminuées, alors que celles de toutes les autres sont conservées.

3. Discussion

L'étude des sulfures mixtes de zinc et de cadmium, comme celle de l'influence extinctrice du cobalt sur seulement certaines transitions indiquent l'existence, dans le spectre décrit d'un système secondaire de six raies, distinct du système principal et plus constant de 40 raies. Le système variable n'est pas le résultat d'un décalage uniforme vers les grandes longueurs d'onde des raies les plus intenses du système constant: la comparaison des valeurs des énergies des transitions du Tableau 1 le montre. Il y aurait donc deux états différents du gadolinium dans ce réseau cristallin, l'un d'eux résultant probablement de la proximité d'un autre défaut ponctuel, peut-être d'une lacune soufre. Lorsque la concentration en cadmium des sulfures mixtes augmente progressivement, les centres de cette deuxième sorte seraient, soit relativement moins nombreux, soit moins efficaces, du fait de la très grande sensibilité aux minimes effets extérieurs des transferts d'énergie aboutissant à ces centres, démontrée par l'effet des traces de cobalt.

Quoi qu'il en soit, les caractères de cette émission luminescente nouvelle du gadolinium sont, on le voit, tout à fait différents de ceux bien connus de l'ion trivalent Gd^{3+} de configuration $4f^7$ pour le niveau fondamental. Parce qu'ils s'étendent aussi dans le domaine visible, on pourrait seulement rapprocher les spectres ici présentés avec les spectres de flamme des sels de gadolinium et aussi avec ceux de cet élément inclus dans du fluorure de calcium soumis à une forte irradiation γ [8]. Mais au lieu d'observer des raies fines, comme ici, on a au contraire des bandes très larges: dans le dernier cas par exemple, l'émission débute vers 13400 cm^{-1} et est continue jusqu'à plus de 40000 cm^{-1} avec seulement trois maximums très arrondis vers 17000, 25000 et 31000 cm^{-1}. Elle disparaît petit à petit après arrêt de l'irradiation γ, par guérissement du cristal et retour à l'état stable correspondant à la présence de Gd^{3+}. Cependant, la flamme comme le cristal irradié sont des milieux très réducteurs: en présence d'un plasma fortement négatif, il est concevable que l'ion gadolinium refixe au moins un électron. Il peut en être de même dans des cristaux plus covalents qu'ioniques, tels que ceux du type du sulfure de zinc. La probabilité de présence de l'électron de liaison est seulement environ trois fois moindre autour des sites cationiques qu'autour des sites soufre. Dans un tel champ, la charge effective des ions d'impuretés qui se placent en position substitutionnelle est nécessairement bien inférieure à l'électrovalence des ions isolés: ce qui est démontré pour les luminogènes classiques [7] doit être également vrai pour les impuretés lanthanides. Et, spécialement pour le gadolinium, les faits exposés ci-dessus nous amènent à penser que, dans ces conditions l'ion est capable de capter solidement un électron pour atteindre un état stable dans lequel le niveau non excité correspondrait probablement à la configuation $4f^7\ 5d^1$.

Bibliographie

[1] M. Bancie-Grillot et P. Bourtayre, Communication au Colloque de Spectroscopie du Solide, Montpellier, Novembre 1965, J. Phys. Radium (sous presse).
[2] S. S. Skorobogatov, S. A. Sazanova, A. V. Dolgopolova et L. V. Kovaleva, Izv. Akad. Nauk SSSR, Ser. fiz. **29**, 407 (1965).
[3] I. V. Stepanov et P. P. Feofilov, Dokl. Akad. Nauk SSSR **108**, 616 (1956).
P. P. Feofilov, Izv. Akad. Nauk SSSR, Ser. fiz. **26**, 435 (1962).
[4] V. L. Levchin, S. A. Fridman, V. A. Chikhacheva et V. V. Shchaenko, Izv. Akad. Nauk SSSR, Ser. fiz. **29**, 500 (1965).
[5] M. Bancie-Grillot et E. Grillot, C. R. Acad. Sci. (France) **233**, 1062 (1951).
[6] J. D. Kingsley, J. S. Prener et M. Aven, Phys. Rev. Letters **14**, 136 (1965).
[7] E. Grillot, J. Phys. Radium **17**, 624 (1956).

(Received May 12, 1966)

phys. stat. sol. **16**, 525 (1966)

Laboratoire de Physique Générale, Faculté des Sciences, Grenoble

Mesures du temps de retournement ferroélectrique de KH_2PO_4

Par

P. GUYON et J. LAJZEROWICZ

Des mesures du temps de retournement en régime impulsionnel ont été faites et mettent en évidence, à faibles champs, un retournement en deux temps. D'autre part, de nouvelles mesures du champ coercitif sont également présentées.

Measurements of the switching current and switching time are made. At low fields, the switching takes place in two stages. New measurements of the coercive field are presented.

Le dihydrogène phosphate de potassium (KDP) est une substance ferroélectrique connue depuis longtemps [1, 2], dont la transition se fait à 123 °K. Alors que bien des propriétés physiques du KDP sont assez bien connues, aucune étude concernant la dynamique des domaines n'a été publiée à notre connaissance (sauf la mesure du champ coercitif [2, 3]). Nous présentons, d'une part de nouvelles mesures du champ coercitif, d'autre part des résultats concernant le temps de retournement.

1. Techniques expérimentales

1.1 Préparation des échantillons

Les mesures effectuées ont nécessité de nombreux échantillons: Les fortes constantes piézoélectriques rendent très fragiles ces cristaux soumis à des champs électriques rapidement variables. Des lamelles ont été découpées dans un gros monocristal, les faces perpendiculaires à l'axe *c* sont polies avec un aérosol d'alumine, puis métallisées par vaporisation sous vide de différents métaux (aluminium, argent ou or). Les électrodes d'or nous ont donné les meilleurs résultats. Des mesures de capacité nous ont montré que les cristaux sont de bonne qualité diélectrique: Les meilleurs ont une constante diélectrique relative maximum plus grande que 300.000. Pour les mesures de temps de retournement, les dimensions résultent d'un compromis entre deux nécessités: diminuer la capacité afin de minimiser la durée du pic capacitif; augmenter la surface de l'échantillon pour obtenir un courant de retournement important. Finalement, nous avons utilisé des cristaux dont l'épaisseur variait de 0,04 à 0,5 cm, et dont la surface variait de 0,03 à 0,4 cm^2.

1.2 Mesure du champ coercitif

Nous utilisions un double pont de Sawyer et Tower avec un montage dérivé d'un circuit déjà utilisé [4]; la fréquence utilisée est de 50 Hz.

1.3 Mesure des temps de retournement

Nous employons la méthode classique développée par Merz [5]: Des impulsions de tensions carrées, symétriques, sont appliquées sur le cristal, en série, avec une petite résistance aux bornes de laquelle apparaît une tension proportionelle au courant de retournement. Le générateur d'impulsions peut délivrer des signaux à temps de montée très court devant la durée de tous les autres phénomènes (10 à 30×10^{-9} s). La répétition des impulsions est limitée le plus possible. En effet, en premier lieu nous avions utilisé une fréquence de récurence de 1/50 s, ce qui rendait l'observation oscillographique aisée. Mais aucun cristal n'a supporté, à de telle fréquence de retournement, des champs supérieurs à 500 V/cm. Il a donc fallu réduire cette fréquence jusqu'à ne plus faire effectuer au cristal que trois ou quatre retournements successifs espacés d'une seconde, ceci à chaque champ ou à chaque température, suivant la nature du paramètre. D'autre part, l'importance du pic capacitif a nécessité, à forts champs, de générer, avec un circuit *RC* auxiliaire, un pic capacitif pur qui est retranché (dans un amplificateur différentiel) au signal délivré par le cristal. Ce circuit auxilliaire est soumis aux mêmes impulsions que le cristal, et la capacité auxiliaire est telle que le pic correspondant soit égal au pic dû au cristal. L'étude de ces deux signaux à l'oscilloscope différentiel permet d'observer qualitativement le début de la courbe de retournement (Fig. 1).

2. Résultats expérimentaux

2.1 Champ coercitif

La variation, avec la température, généralement admise, du champ coercitif E_c de KDP [2, 3] fait apparaître une valeur environ constante de l'ordre de

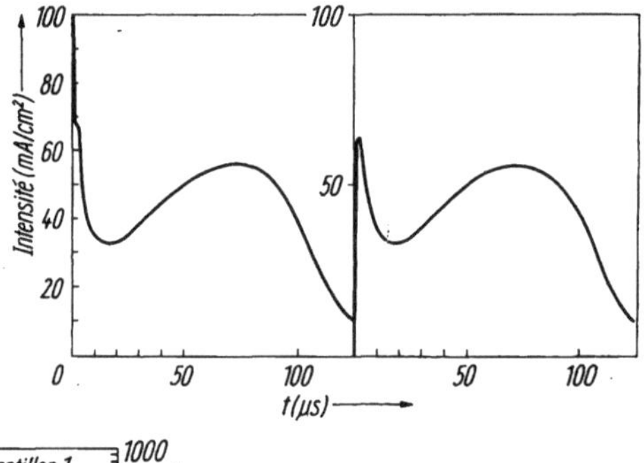

Fig. 1. Allure du courant de retournement à 100 °K (impulsions de 400 V/cm). A gauche: signal issu du cristal; à droite: après soustraction d'une exponentielle $I_0 \exp(-t/RC)$

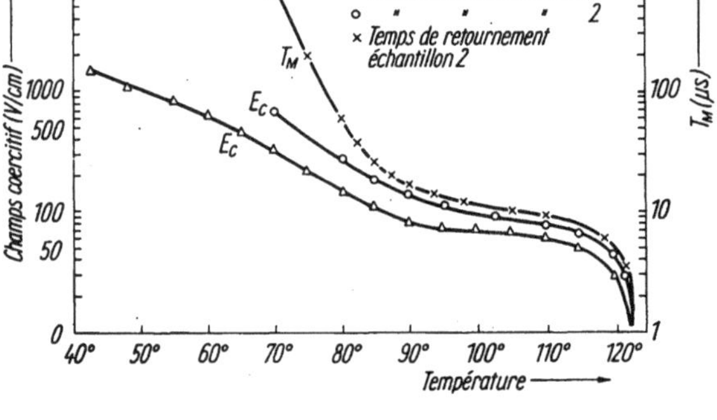

Fig. 2. Champ coercitif de deux échantillons et temps de retournement pour l'un d'eux. E_c est mesure avec un champ de 50 Hz de 2 kV/cm crête et T_M mesuré avec des impulsions de 2 kV/cm

1 à 3 kV/cm entre 105 et 60 °K. Entre 60 et 40 °K, E_c augmenterait considérablement pour atteindre 8 à 12 kV/cm à 40 °K avec un point anguleux à 60 °K. Or, aucune discontinuité dans les variations des autres grandeurs physiques de KDP ne se produit à 60 °K (angle de perte, constante diélectrique, polarisation à saturation, constantes piézoélectriques, etc.). Nous avons donc repris soigneusement la mesure de E_c à 50 Hz entre 120 et 40 °K. (Fig. 2). Nous n'avons pas pu descendre plus bas, car aucun échantillon n'a résisté à des champs de 50 Hz de l'ordre de 2 kV/cm à des températures plus basses que 40 °K. Nous n'avons rencontré aucun point anguleux; d'autre part, la valeur de E_c est considérablement plus basse que les valeurs précédemment proposées (ce qui prouve la pureté de nos cristaux): E_c part de zéro à la température de Curie, croit lentement en restant inférieur à 1000 V/cm, jusqu'à 90 °K, puis augmente plus rapidement; vers 60 °K, E_c vaut entre 600 et 1400 V/cm. Le cristal le meilleur nous a donné un champ de 1400 V/cm à 40 °K, mais n'a pas résisté à des températures plus basses.

2.2 *Temps de retournement*

Nous avons observé le courant de retournement en fonction de la température et du champ appliqué. L'allure du courant (Fig. 3 et 4) montre nettement que le retournement se fait en deux temps: Un premier maximum se produit à quelques 10^{-6} s, puis un deuxième maximum, beaucoup plus tard et plus long, se produit à un temps variable avec la température et la hauteur des impulsions. L'observation du premier maximum est difficile, car il est, en général, masqué par le pic capacitif. Même en retranchant un pic capacitif pur, on ne peut faire qu'une étude qualitative du premier maximum. Toutefois, nous avons pu noter qu'au-dessus de 500 V/cm, il se produit toujours entre 1 ou 2×10^{-6} s lorsque la température ou le champ varient. Nous avons fait une étude plus

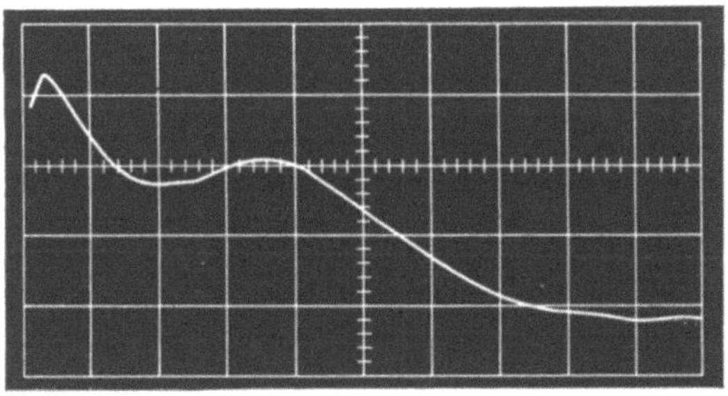

Fig. 3. Forme typique du courant de retournement à 113 °K avec des impulsions faibles de 280 V/cm. 1 cm vertical = = 20 mA/cm², 1 cm horizontal = 18 µs

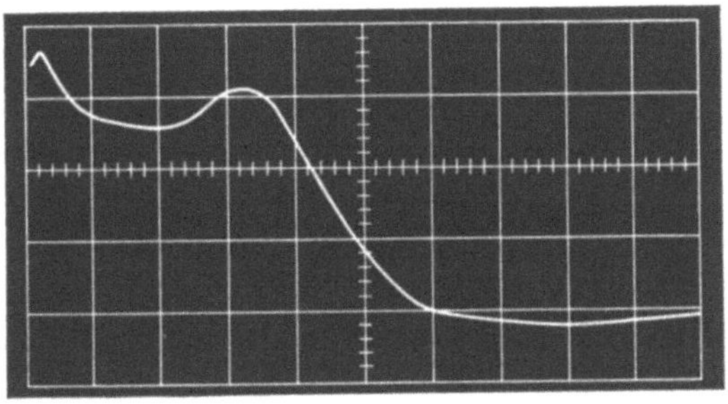

Fig. 4. Retournement à 113 °K. Impulsions de 360 V/cm. 1 cm vertical = 20 mA/cm², 1 cm horizontal = 13 µs

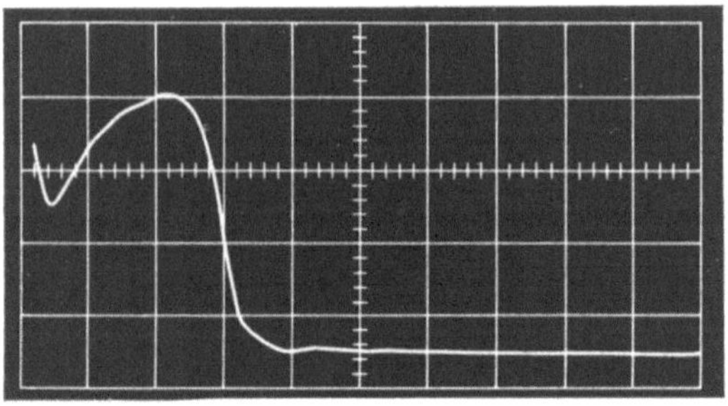

Fig. 5. Retournement à 113 °K. Impulsions de 600 V/cm. 1 cm vertical = = 40 mA/cm², 1 cm horizontal = 10 μs

précise de la position T_M (Fig. 6) et de l'amplitude I_M (Fig. 7) du deuxième maximum avec la température et le champ appliqué. On constate que l'inverse du temps T_M varie de façon analogue à la variation de $1/T_z$ du sulfate de glycocolle. A faible champ, le temps de retournement semble suivre la loi empirique classique du type

$$\frac{1}{T_M} = \frac{1}{t_0} + \frac{1}{t_\infty} e^{-\alpha/E},$$

la présence de $\frac{1}{t_0}$ traduit le fait que $\frac{1}{T_M}$ ne tend pas vers zéro lorsque le champ E tend vers zéro. Il faut remarquer que le courant maximum tend vers zéro avec le champ. Comme l'intégrale du courant de retournement diminue à très faible champ (cristal non complètement polarisé), il n'est pas absurde que le temps T_M ne tende pas vers l'infini lorsque le champ tend vers zéro (le courant I_M, lui, tendant vers zéro, évidemment). Cette loi n'est vérifiée que pour des champs plus petits que 400 V/cm à 110 °K et plus petits que 700 V/cm à 100 °K. Pour les champs petits, le retournement est très lent et la mesure de T_M est imprécise. Nous pouvons, toutefois, proposer une valeur approchée à 102 °K, de t_0, de t_∞

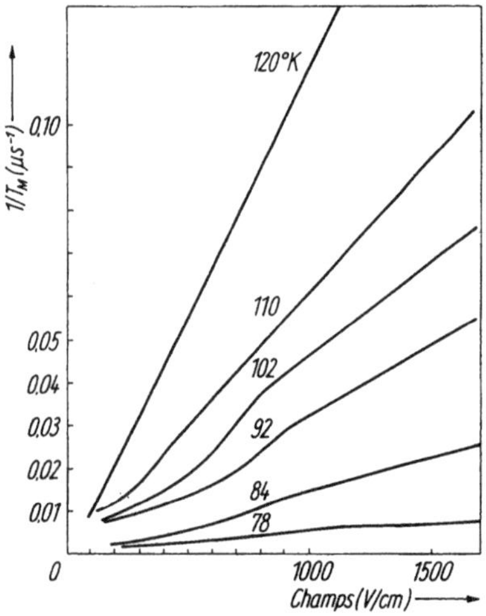

Fig. 6. Inverse du temps de retournement (temps du courant maximum) en fonction du champ des impulsions à differentes températures

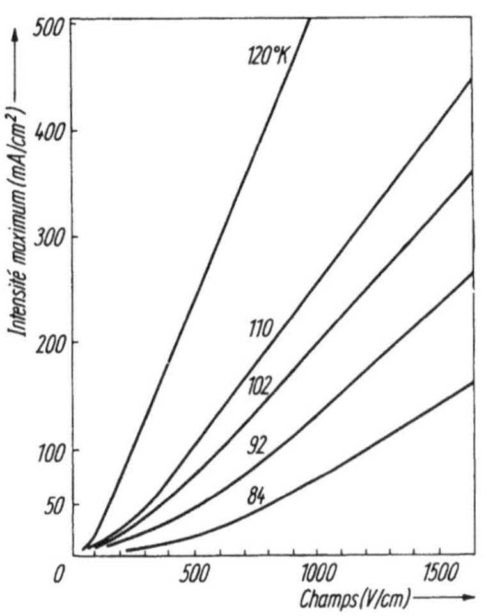

Fig. 7. Intensité maximum du courant de retournement ramenée pour un cristal de 1 cm² en fonction du champ des impulsions à différentes températures

et de α:

$$t_0 = 135\ \mathrm{s}, \quad t_\infty = 150\ \mathrm{s}, \quad \alpha = 200\ \mathrm{V/cm}\ .$$

Pour des champs plus forts, le temps T_M est proportionnel au champ (ainsi que le temps de retour à zéro T_z). Nous n'avons pas pu suivre cette loi pour des champs supérieurs à 2 kV/cm, aucun échantillon n'ayant résisté à des champs supérieurs. Il faut remarquer que 2 kV/cm est déjà considérablement plus grand que le champ coercitif (sauf à température inférieure à 80 °K, mais, alors, les mesures du temps de retournement sont très imprécises, étant donné le lenteur du phénomène). Nous avons tracé l'évolution du temps de retournement T_M en fonction de la température à champ constant: il est intéressant de remarquer que T_M et le champ coercitif restent proportionnels dans une zone de température où $\frac{1}{T_M}$ varie linéairement avec le champ d'impulsions. A températures plus basses, le temps de retournement semble varier comme $T_M = k\,E_c^3$, dans une zone de température où $\frac{1}{T_M}$ suit une loi en $e^{-\alpha/E}$.
(Dans ce qui précède, T_M et E_c étaient mesurés tous les deux avec un champ maximum appliqué de 2 kV/cm.)

3. Conclusion

Il y a une assez grande analogie de comportement entre le KDP et le sulfate de glycocolle (TGS): Ce sont des cristaux ferroélectriques à liaisons hydrogènes et il ne peut y exister que des domaines à 180°. Les champs coercitifs ont des variations analogues, bien que celui de KDP garde une valeur faible sur un intervalle de température beaucoup plus grand. Les variations du temps de retournement sont analogues, bien que le champ à partir duquel $\frac{1}{T_M}$ commence à varier linéairement avec E est becaucoup plus petit pour le KDP (800 V/cm à 100 °K) que pour le TGS (20 kV/cm). L'importante différence centre le TGS et le KDP réside surtout dans la forme des domaines et dans leur évolution au cours d'un retournement [6]. Dans le TGS, on observe des domaines de forme relativement cylindrique (dont l'axe est parallèle à l'axe c) traversant parallèlement à l'axe c l'épaisseur du cristal.

Pour KDP nous avons fait effectuer à des cristaux à électrodes semi-transparentes des retournements progressifs lents, tout en observant optiquement par transparence. Nous avons observé l'existence et la croissance de domaines en pointe, croissant dans un plan perpendiculaire à l'axe c, au début du retournement. La naissance et la croissance de tels domaines semblent responsables du premier maximum dans le courant de retournement en régime impulsionnel, le deuxième maximum étant du à un déplacement latéral des parois.

Remerciements

Nous remercions Monsieur le Professeur Chappelle (Laboratoire de Physique Cristalline de la Faculté des Sciences d'Orsay) qui nous a aimablement fourni le cristal de KDP, ainsi que Madame Fouskova, pour les nombreuses discussions concernant ce problème.

Bibliographie

[1] G. BUSCH et P. SCHERRER, Naturwissenschaften **23**, 737 (1935).
[2] F. JONA et G. SHIRANE, Ferroélectric Crystals, Pergamon Press 1962.
[3] H. M. BARKLA et D. M. FINLAYSON, Phil. Mag. **44**, 109 (1953).
[4] H. DIAMANT, K. DRENCK et R. PEPINSKY, Rev. sci. Instrum. **28**, 30 (1951).
[5] W. J. MERZ, Phys. Rev. **95**, 690 (1954).
[6] A. FOUSKOVA, P. GUYON et J. LAJZEROWICZ, C. R. Acad. Sci. (France) **262**, 907 (1966).

(Received May 12, 1966)

phys. stat. sol. 16, 531 (1966)

University of Essex, Colchester, Essex

Electron Microscopy of Irradiation-Induced Defect Clusters in Magnesium Fluoride[1])

By

D. J. BARBER

The passage of electrons through thin crystals of magnesium fluoride produces point defects. These defects are mobile at temperatures normally attained under the electron beam, and they can give rise both to dislocation loops and to precipitate platelets. The aggregation of these radiation-induced defects can be controlled by using a cooled specimen stage in the electron microscope. This allows the growth and characteristics of the defect clusters to be studied. Dislocation loops lie on $\{011\}$ planes and they sometimes show stacking fault contrast; their Burgers vectors are $^1/_2\langle 01\bar{1}\rangle$ and $\langle 01\bar{1}\rangle$. Prolonged irradiation can produce precipitates on and within the loops. Higher beam intensities produce precipitates directly. The nature of the loops and precipitates is discussed in relation to the production of point defects by ionization mechanisms and irradiation-stimulated migration.

Durch Elektronendurchgang werden in dünnen Magnesiumfluoridkristallen Punktdefekte erzeugt. Die Defekte sind bei den Temperaturen, die normalerweise unter Elektronenstrahlung erreichbar sind, beweglich und bewirken die Bildung von Versetzungsschleifen und Ausscheidungen. Die Ansammlung dieser Strahlungsdefekte kann durch einen gekühlten Objekttisch des Elektronenmikroskops gesteuert werden. Dadurch wird eine Untersuchung des Wachstums und der Charakteristiken der Defekt-cluster möglich. Versetzungsschleifen liegen in $\{011\}$-Ebenen und zeigen manchmal Stapelfehlerkontrast; ihre Burgersvektoren sind $^1/_2\langle 01\bar{1}\rangle$ und $\langle 01\bar{1}\rangle$. Andauernde Bestrahlung kann an und innerhalb der Schleifen Ausscheidungen erzeugen. Höhere Strahlintensitäten erzeugen direkt Ausscheidungen. Die Natur der Schleifen und Ausscheidungen wird in Verbindung mit der Erzeugung von Punktdefekten durch einen Ionisierungsmechanismus und strahlungsstimulierter Wanderung diskutiert.

1. Introduction

Many studies of radiation damage by transmission electron microscopy have been reported in the last few years. Most of the work concerns the production of point defects and defect aggregates by high energy particles such as fast neutrons and accelerated ions of the rare gases. Many nonmetallic compounds are damaged or decomposed by high energy electrons. Studies of this type of damage are important because they should throw light on the mechanisms by which electrons create point defects in crystals, and the ways in which these defects cause chemical changes. The latter may well create a new solid phase and a gaseous phase. Whilst solid phases are often accommodated at structural imperfections such as grain boundaries and dislocations, the liberation or precipitation of gases is more difficult to explain.

The present work stems from the discovery that magnesium fluoride is damaged in the beam of the electron microscope. The results are interesting

[1]) Work carried out while the author was at the National Bureau of Standards, Washington, D.C..

because the effects of the damage differ from those found in the alkali halides and in certain divalent metallic fluorides. Evans' work [1] on fluorite-structured strontium fluoride and calcium fluoride showed that the electron microscope beam produced topotactic changes, apparently liberating fluorine and producing layers of metal atoms within the crystal. Moiré patterns resulted from the diffraction of electrons by the layers of the 'sandwich' crystal. On the other hand, Barber (unpublished work) found that cobalt fluoride, which has the same structure as magnesium fluoride, is not affected by the electron beam. In a previous paper [2], the author mentioned that electron-induced defects in MgF_2 give rise to dislocation loops and precipitates. High levels of irradiation can even produce cavities near the specimen surface. This paper is concerned with the detailed nature of the loops and precipitates, and of the processes which occur in the microscope. After irradiation the samples are coloured, an observation which has led to investigations by Blunt and Cohen [3]. There is some parallel between the present work and that of Forty [4, 5, 6] on lead iodide, which decomposes with the precipitation of lead. Also relevant are the studies of electron-induced damage in potassium chloride by Hibi and Yada [7], and Tanaka et al. [8], in sodium chloride by Yagi and Honjo [9] and in beryllium oxide by Wilks and Clarke [10].

2. Physical Properties of Magnesium Fluoride

MgF_2 crystallizes in the tetragonal rutile structure, with $a_0 = 4.623$ Å and $c_0 = 3.052$ Å so that $c/a = 0.6601$; the structural unit cell is shown in Fig. 1. Van Landuyt et al. [11] have pointed out that the stacking of crystals with the rutile structure can be considered in terms of the stacking of anions, which can be derived from a hexagonal close-packed structure by a slight deformation. The rutile structure is created by filling half of the octahedral interstices with cations.

MgF_2 single crystals are colourless and they are transparent from 7.5 μm in the infrared to 1150 Å in the ultraviolet [12]; they are also birefringent. Barber [2] has investigated the plastic deformation of MgF_2 crystals by means of optical microscopy and chemical etching. Three glide systems were identified. Near room temperature, glide occurred by shears of the type $\{110\}\ \langle 001 \rangle$ and less easily by $\{10\bar{1}\}\ \langle 101 \rangle$; at elevated temperatures (≈ 1000 °C), $\{001\}\ \langle 100 \rangle$ glide was observed. Crystals are not readily attacked by most chemical reagents but they hydrolyse rapidly in air at temperatures above 300 °C.

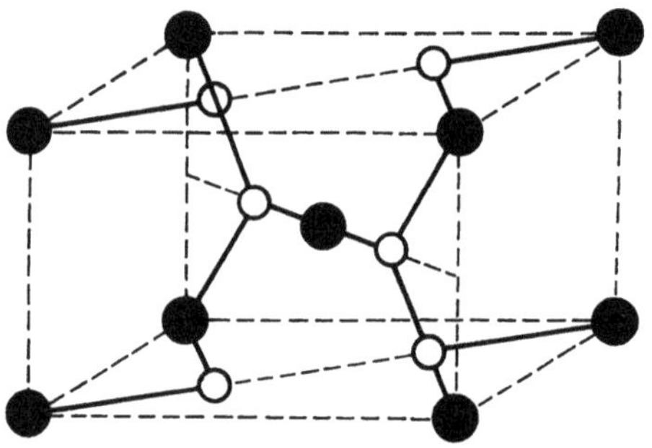

Fig. 1. Unit cell of MgF_2: open circles are fluorine ions, full circles are magnesium ions

3. Experimental Procedure

3.1 Specimen preparation

Samples were formed from sections of single crystal boules grown by the Czochralski technique, as described elsewhere [2]. The starting material was MgF_2 of optical coating grade. The resultant crystals were known to contain MgO as an impurity, and this normally occurred as precipitates decorating grown-in dislocations. Spectrochemical analysis revealed no metallic impurities above 0.001 mol% apart from Al and Cu, which were in the 0.001 to 0.01 mol% range.

Electron microscope samples were made from crystal sections by grinding and shaping followed by chemical dissolution in concentrated sulphuric acid at ≈ 250 °C. Discs of 3 mm diameter and 0.4 mm thick were cut, and a flat-bottomed depression was made in each disc with an ultrasonic grinding tool fed with boron carbide grit. These blanks were suspended in a stirred bath of the hot acid until perforation occurred within the depression. The areas bounding the holes were sufficiently thin to transmit 100 kV electrons but the foils were, nevertheless, robust and could be handled safely by their rims. Foils with $\{100\}$, $\{110\}$, $\{101\}$, and $\{001\}$ orientations were produced. The acid bath was changed frequently to avoid the deposition of precipitate needles onto the sample. Precipitation was particularly troublesome with $\{101\}$ and $\{001\}$ orientations. Thin carbon films were evaporated onto specimens to prevent charging. Metal grids were used to provide added mechanical support, as well as efficient thermal and electrical paths to the specimen cap of the microscope, a JEM 6A.

3.2 Electron microscope technique

The action of the beam was so rapid at normal temperatures that foils became opaque to electrons after a few seconds. To arrest this degradation and enable defect clusters to be studied, a low temperature stage was used, with liquid nitrogen as a coolant. This procedure normally excludes the possibility of tilting the sample, and it was therefore necessary to rely heavily on dark field techniques. To retain optimum resolution, the condenser system (and hence the incident beam) was inclined so that the dark field imaging beam passed down the axis of the objective lens. However, tilting often causes the very lattice reflection of interest to become weak so, to circumvent this problem, the stage was modified to permit tilting by a few degrees about one axis. Examination of contrast due to precipitates was possible at normal temperatures provided low beam intensities were used and observations were restricted to the thinnest parts of the sample (see Section 5.1). For some of this work, therefore, the goniometer stage was used.

4. Review

Contrast from dislocation loops and precipitates

In principle, there are simple methods for determining the habit plane, Burgers vector and nature of dislocation loops. These methods, dependent on the fulfillment of precise diffraction conditions and examination of the resultant contrast, have been reviewed by Amelinckx [13] and others. There are certain practical difficulties when a foil must cooled and cannot be mounted in a normal goniometer stage.

The direction of the Burgers vector of a loop is determined by finding a lattice reflection for which the loop has minimum contrast. Since there is only a slight distortion of the lattice planes parallel to the displacement vector, the loop image formed with electrons diffracted by these planes is weak. This is represented by the condition for zero contrast derived by Hirsch et al. [14], $\boldsymbol{g} \cdot \boldsymbol{b} = 0$, where $\boldsymbol{g}$ and $\boldsymbol{b}$ are respectively the reciprocal lattice vector for the set of planes diffracting, and the Burgers vector of the defect. The character of a loop which appears "open" in projection can be found from the sign of $(\boldsymbol{g} \cdot \boldsymbol{b})\, s$, where s is the interference error (a measure of the deviation from the exact Bragg condition). The image is always within the loop when the product is negative, and vice versa. Ashby and Brown [15] have suggested that the nature of a prismatic loop which is viewed "end-on" can sometimes be found by dark field techniques. This is because the strain contrast images of defects near the upper surface of thick foils are characteristically large, with a black-white asymmetry. The sense of the asymmetry, in relation to the direction of $\boldsymbol{g}$ for the dark field image, should give the nature of the loop. Work by Bell et al. [16] and others has indicated, however, that the application of Ashby and Brown's ideas can sometimes give the wrong result for defects with non-spherical strain fields, such as loops. In other papers Bell and Thomas [17, 18] have identified some of the problems and show how the strain images loops at any depth in the foil can aid in characterization.

Ashby and Brown have also discussed the strain contrast caused by spherical [19] and non-spherical [15] inclusions in thinned foils. In general, when the incident electrons are diffracted strongly by one set of lattice planes, the contrast occurs as two lobes (black or white) about a line of no contrast (l.n.c.). For spherical precipitates the two lobes are symmetrically situated about the l.n.c. and the l.n.c. is normal to the particular $\boldsymbol{g}$ operating, as observed by Philips and Livingston [20] and Ashby and Brown [19]. For platelike precipitates these two conditions only hold when the reciprocal lattice vector $\boldsymbol{g}$ is normal to the trace which the precipitate makes in the foil surface ([17] and this paper). More commonly, the two lobes are skewed about the l.n.c. and there is a characteristic quasi-symmetry line [21, 16] pointing from one to the other. The actual precipitates are only visible when the local orientation of the foil is very close to the Bragg condition ($s \approx 0$).

The dark field strain image of a precipitate is asymmetrical in the case of thick foils and anomalously wide images are formed of defects very close to the top of the foil. These asymmetries in the anomalous images of precipitates and loops can be used to determine their nature (i.e. dilatational or collapse) even though the defect may have a non-radial strain field.

A coherent slablike precipitate will introduce displacements of the matrix normal to the slab, and thus will give rise to fringes similar to stacking-fault fringes. Ashby and Brown further showed that the intensity change due to particles at various depths in the foil may depend on depth. This possible cause of confusion may be removed by considering only defect clusters in regions where $s \approx 0$, as at bend contours.

5. Results

5.1 Observations of electron-induced damage at normal temperatures

The foils contain no grown-in dislocations, these are etched out for the reasons discussed earlier [2]. The best electron-transparent areas are often delicate

Fig. 2. Dot contrast due to defect clusters near two shallow etch pits in a {110} section irradiated at ≈77 °K

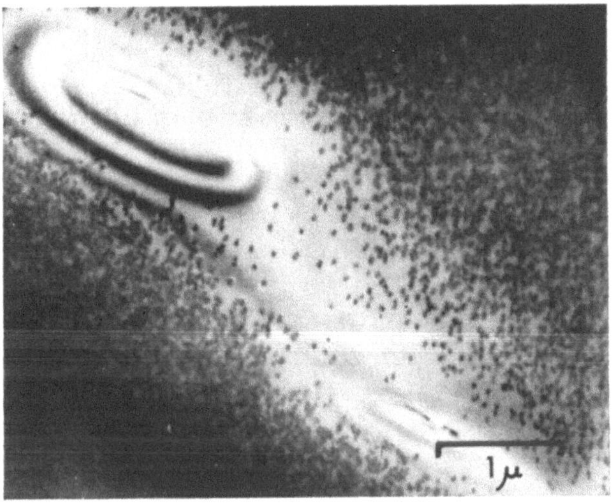

bridges between etch holes. Within two or three seconds of exposing a foil to the beam, myriads of black dots appear. The dots grow rapidly and, after a brief period when some show double-arc contrast, they coalesce, interact, and soon render the samples opaque to electrons. Transparency is maintained only at the very thin edges bounding holes, or at etch pits. Fig. 2 shows an early stage in the development of the damage. Shallow etch pits are revealed by equal thickness fringes. Away from these, the distribution of dots is random (though not necessarily uniform through the section), since there are no pre-existing line defects to act as nuclei.

The changes are hard to record photographically because the whole process is over in a few seconds. Moreover, there is seldom any detectable change in the electron diffraction pattern. Between the transparent periphery of the sample and the interior, however, there exists a transition zone within which precipitate particles can be seen. Tilting shows that the precipitates are platelike, the plates ranging up to an estimated thickness of ≈ 30 Å. The appearance of precipitates in this zone is illustrated by Fig. 3.

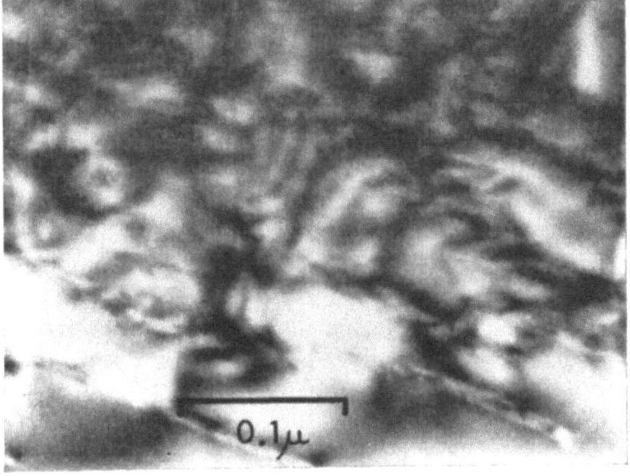

Fig. 3. Coarse precipitates produced in the edge of a {001} sample by room temperature irradiation

After examination in the microscope, the irradiated areas showed a yellow coloration due to the presence of colour centres, an effect which has now been studied by Blunt and Cohen [3] and Hall [22]. High beam intensities (obtained with a large condenser aperture) naturally produce a very rapid decomposition of the crystal. The effects are similar to those reported by Hibi and Yada [7] for electron-irradiated NaCl and KCl. Large lenticular-shaped voids appear and sometimes blisters are even created and burst at the surface, the electrons producing sufficient temperatures to vaporize the sample and produce internal pockets of gas. The voids probably form under less extreme heating by the aggregation of molecular vacancies. None of the damage is caused by ions created in the column of the microscope. This was established in the customary way by inserting a thick carbon film in the condenser aperture.

5.2 Observations of electron irradiation damage at low temperatures

Cooling samples to liquid nitrogen temperatures in the microscope prior to examination retards decomposition. Complete inhibition is dependent on the use of a low beam current and the ability of the irradiated area to dissipate heat introduced by the beam. The latter is largely determined by the morphology of the sample and its degree of contact with the grid and specimen holder, and is not easily controlled.

Point defects are produced by the interaction of electrons with the ions of the crystal. Is the effect of a low temperature to decrease defect production, or does it hinder defect clustering? The following observations have considerable bearing on this question.

An efficiently cooled sample, irradiated with 100 kV/25 mA beam through a 200 μm condenser aperture, shows the slow development of dislocation loops in regions $\approx$ 1000 Å thick. In thinner areas few, if any, appear unless the condenser is focused to give more concentrated heating. The development of defect clusters can thus be controlled. They can be grown to a size such that they can be identified as loops (as distinct from precipitates). Since the loops form by the aggregation of point defects and collapse of the surrounding crystal, they are prismatic, i.e. there is a component of the Burgers vector perpendicular to the loop plane.

Higher flux densities, poorer cooling, and irradiation of thicker parts of the sample all favour the development of precipitates, either directly or at the sites of embryo loops. Prolonged irradiation gives the foil a finely mottled appearance, resulting from surface decomposition and minute unidentified internal defects. Where small loops and precipitates occur together they are easily confused. Moreover, it may be difficult to decide what experiments to conduct, since the nature of the clusters is seldom obvious when they are viewed on the fluorescent screen of the microscope. This problem, and limitations inherent in using the cold stage, explain why some apparently facile experiments have not been performed successfully.

Thick parts of a sample are damaged more easily than thin parts, because a more intense beam is required for penetration, leading to more rapid heating; there is also more absorption. It therefore appears that the higher temperatures generated allow greater defect mobility and, since there are also more defects created, many clusters result.

Clusters do not normally occur in regions of thickness $\lesssim$ 500 Å. Since there is a considerable temperature rise even in very thin areas, the absence of clusters (which should show some contrast even in the kinematic region) must stem from the large surface to volume ratio. Presumably defects are created, but many reach the surface and clusters of visible size do not form.

Samples which have been irradiated carefully at ≈ 77 °K without the creation of damage showed no evidence of loops or precipitates immediately upon inspection in the microscope at room temperature, i.e. either no point defects were produced at ≈ 77 °K or those produced did not form visible aggregates when the sample was warmed without further irradiation.

5.3 Dislocation loops

5.3.1 Loop plane

The limitations of the low-temperature stage prohibited determination of the loop plane by tilting the specimen. A less sophisticated technique was therefore adopted: $\{011\}$, $\{110\}$, and $\{100\}$ foils were examined and the orientations of the projected images of the loops were determined by means of selected area diffraction. In samples irradiated for less than one minute, the images were always either elliptical or bar-like. It was inferred, therefore, that the loops were actually circular and lay in inclined planes. Making this assumption, the directions of the major axes of the images, $\langle hkl \rangle$, are seen to be the traces in the foil plane of the loop planes. The latter could be uniquely determined and verified, therefore, by using the three different foil orientations. The major axes were found to be as follows:

$$\{001\} \text{ section, major axis } \parallel \langle 100 \rangle ,$$
$$\{110\} \text{ section, major axis } \parallel \langle 1\bar{1}1 \rangle ,$$
$$\{100\} \text{ section, major axis } \parallel \langle 011 \rangle \text{ and } \langle 010 \rangle .$$

These results show that the loop plane must be $\{110\}$. Loops in a $\{100\}$ section, with their axes parallel to $\langle 011 \rangle$ and $\langle 010 \rangle$, are shown in Fig. 4. Some of the loops are viewed 'edge-on', while those lying in the two $\{011\}$ planes which are inclined to the beam are 'open'. Fig. 5a and 5b are stereograms to show, respectively, the directions of the loop major axes projected onto $\{100\}$ and $\{110\}$ planes and the corresponding poles of the loop planes. There is a sligh scatter in loop orientations about the ideal loop plane. Measurements derived from Fig. 4 and similar micrographs show that the standard deviation of the loop orientations is $1.2 \pm 0.5°$.

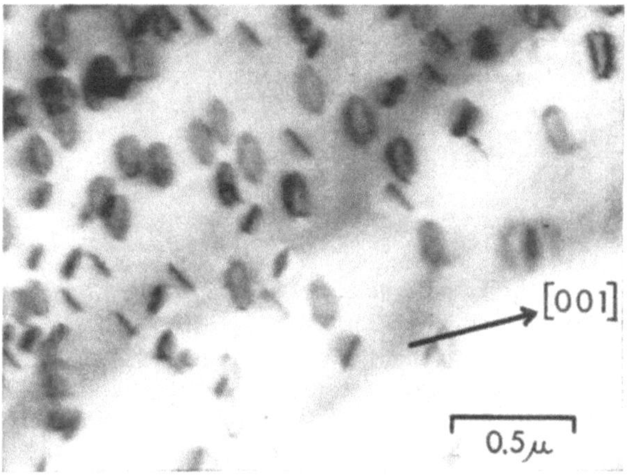

Fig. 4. Dislocation loops lying on {011} planes in a {100} sample irradiated at ≈77 °K

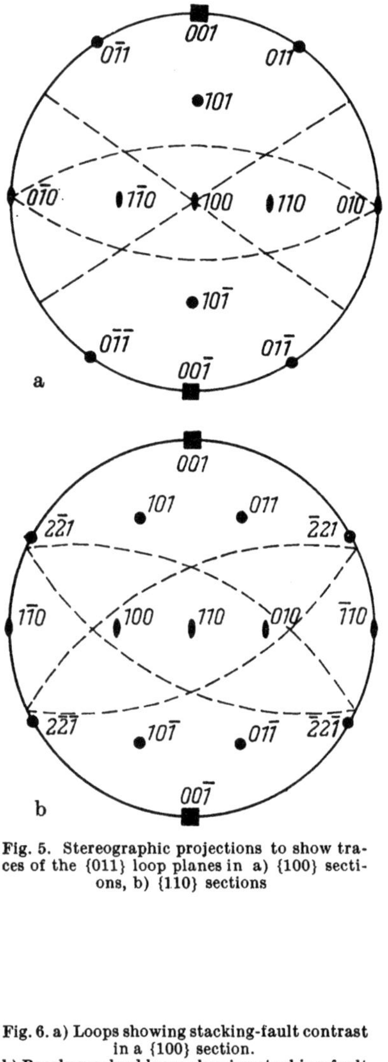

Fig. 5. Stereographic projections to show traces of the {011} loop planes in a) {100} sections, b) {110} sections

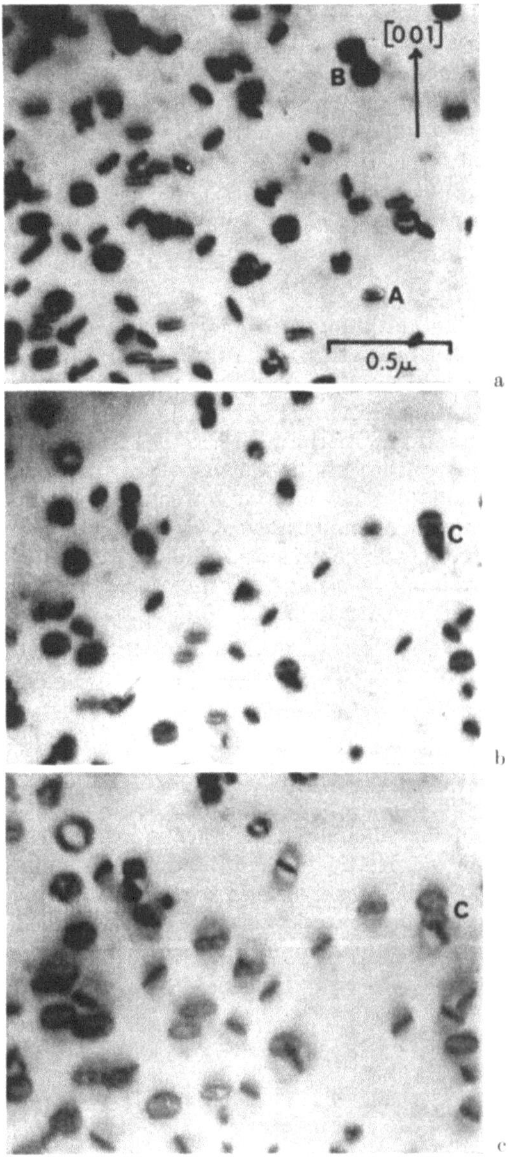

Fig. 6. a) Loops showing stacking-fault contrast in a {100} section.
b) Poorly resolved loops showing stacking-fault contrast in a region adjacent to that in Fig. 6a.
c) Same field as in Fig. 6b, showing transformed loops (without stacking-fault contrast) which contain a singularity

5.3.2 *Burgers vector of loops*

Since the loop plane is {011} it is immediately apparent that the loops cannot be pure prismatic edge dislocations, for there is no lattice translation vector perpendicular to the (011) plane. The [0$\bar{2}$3] vector is nearly normal to the plane; the shortest lattice vectors, [001], [010], and [011], are more steeply inclined. It was simple to establish that the loops did not have ⟨001⟩ and ⟨010⟩ Burgers vectors, since these vectors lie respectively in the '*a*' planes and the '*c*' planes.

All the loops would disappear with the operation of either 020 or 002 lattice reflections, and this was not observed.

Loops lying on the {101} planes at approximately 57° to a {100} foil were sometimes found to exhibit fringes. They could be stacking fault fringes indicating that the loops were partial in character, or they could be fringes caused by thin precipitate platelets within the loops [15]. The question was resolved in favour of stacking faults when fringes were seen to be removed, apparently by the nucleation of a second partial dislocation on the loops. Fig. 6a shows a number of 'open' loops with stacking fault contrast in a {100} foil. One set of loops (for example at A) shows fringes and the bounding dislocation is only weakly imaged; the other set (for example at B) shows fairly uniform contrast within the loop. Fig. 6b shows an area adjacent to that of Fig. 6a under less satisfactory imaging conditions, with the loops still showing fringe contrast. In Fig. 6c fringes are absent from the loops which had distinct fringes in Fig. 6b and the bounding dislocations now have normal contrast. A small feature which is thought to be the second partial can be seen in many of the transforming loops. One set of 'open' loops, less steeply inclined to the electron beam, are largely untransformed. The stacking fault vector most likely to be found in crystals such as TiO_2 and MgF_2 is of the type $^1/_2$ [011]. Reasons for this have been discussed Ashbee et al. [23], Van Landuyt et al. [11], and Barber [2]. The operating reflection for Fig. 6a and 6b was 011, with a weak secondary reflection 002. Let us consider the images of the four representative loops shown in Fig. 7 for a 011 reflection, assigning to the loops Burgers vectors as labelled.

Table 1

$\boldsymbol{b}$	Loop plane	$\boldsymbol{g} \cdot \boldsymbol{b}$	Type of contrast
$\frac{1}{2}$ [011]	(011)	1	normal
$\frac{1}{2}$ $[0\bar{1}1]$	$(0\bar{1}1)$	0	zero
$\frac{}{2}$ [101]	(101)	$\frac{1}{2}$	stacking-fault
$\frac{1}{2}$ $[\bar{1}01]$	$(\bar{1}01)$	$\frac{1}{2}$	stacking-fault

If we recognize the weaker 002 reflection also contributing to Fig. 6a and b, the above scheme is compatible with the observations. Presumably the transformed loops have ⟨011⟩ Burgers vectors.

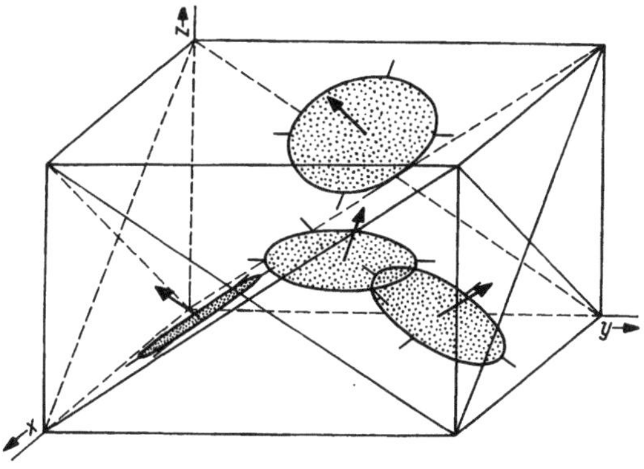

Fig. 7. Diagram to show dislocation loops in the four representative {101} orientations

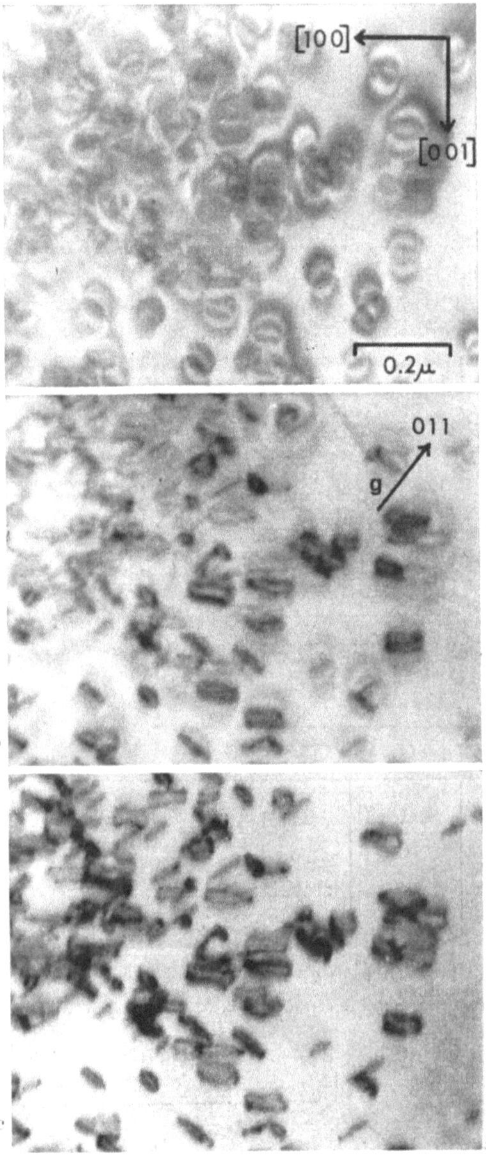

Fig. 8. Showing the diminution of strain contrast and the development of loop contrast at loops. The loops climb as a result of prolonged irradiation

Fig. 8 shows the development of strain and diffraction images in a $\{100\}$ foil, and the start of loop disintegration by climb and precipitation, resulting from prolonged irradiation. The last frame (Fig. 8c), with $\boldsymbol{g} = 011$, also confirms that perfect loops have $\langle 011 \rangle$ Burgers vectors distributed in the manner of the partial vectors in Fig. 7. Many of the 'open' loops have normal contrast and so does one set of the loops viewed 'end-on'. The fourth set shows only weak contrast, indicating that $\boldsymbol{b}$ is $[01\bar{1}]$ or $[01\bar{1}]$. Loss of definition in Fig. 8c results from degradation of the crystal and precipitation on the loops. Confirmation of the deduced Burgers vectors stems from the strain images of loops. Fig. 9 shows strain images in a $\{100\}$ foil formed with an 002 reflection. In an earlier paper [2] it was erroneously stated that these were images of precipitates, not loops. The symmetry of images of loops on the $\{011\}$ and $\{0\bar{1}1\}$ planes shows that the Burgers vectors have no component parallel to $\langle 100 \rangle$. It follows that the Burgers vectors of these loops are $[011]$ and $[01\bar{1}]$, or one half of these values if the loops are partial.

Although it now appears that stacking faults occur in MgF_2, the loop transformations suggest that the stacking fault energy is higher than for isomorphous TiO_2, in which extended faults and partial dislocations are common. Large planar defects have been observed only rarely in MgF_2; it is thought that these are microtwins.

Fig. 10a shows a microtwin and a few clusters. The microtwin makes a $\langle 010 \rangle$ trace in the $\{100\}$ foil and probably lies in a $\{011\}$ plane (as found in TiO_2). Its projected width would then imply a local foil thickness of ≈ 800 Å. The microtwin contains several twinning dislocations which change the fringe

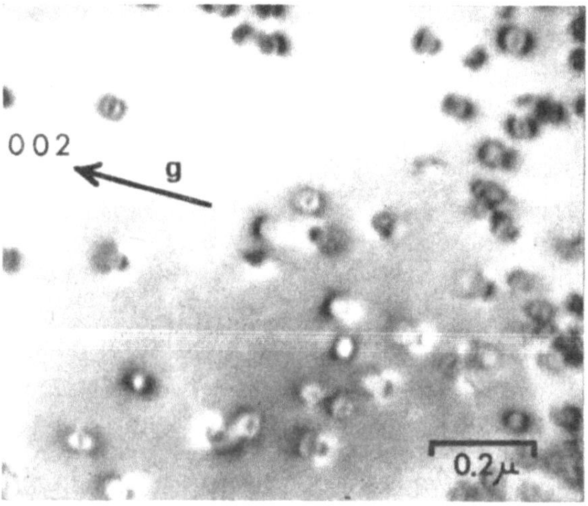

Fig. 9. Strain contrast images of loops in a {110} foil using a 002 reflection

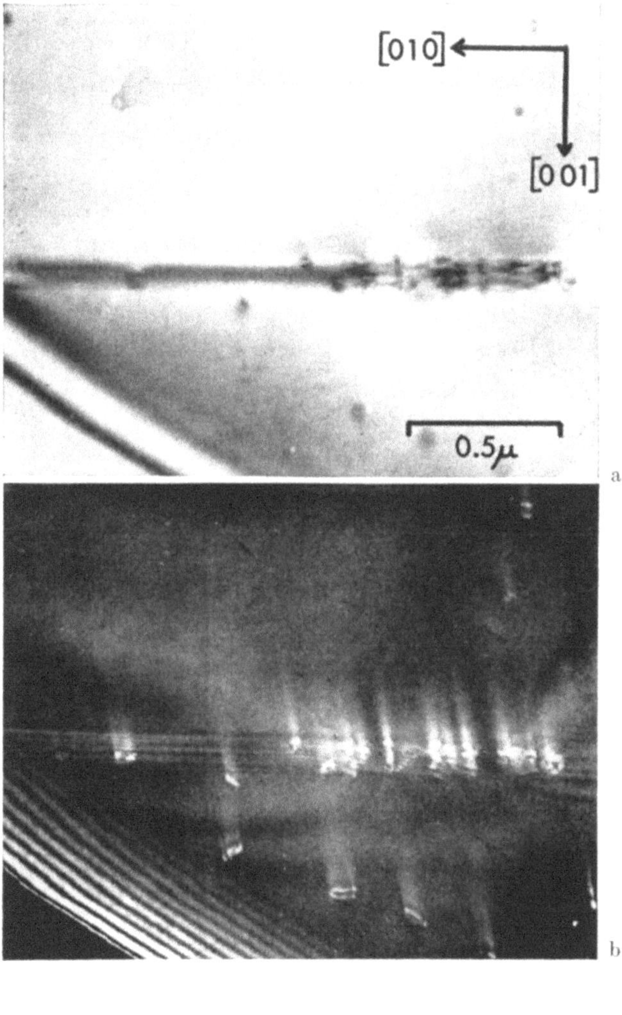

Fig. 10. a) Microtwin and defect clusters in a {100} foil. b) Dark micrograph of a) after further irradiation

pattern, and there are some clusters associated with the dislocations. In Fig. 10b recorded a few seconds later, they show pronounced growth. The nature of the clusters is not clear but their appearance in Fig. 10b, which is a dark field micrograph, indicates that they are probably loops.

5.3.3 *Loop character and loop interactions*

The methods [15, 16, 17, 21] of examining anomalous dark-field strain images could not be used to establish loop character. Bell and Thomas [17] have pointed out that the selection of the diffraction conditions must be made with great care. Correction of foil tilt, after placing a diffracted beam an axis, could not be achieved with the cooled stage. However, very few anomalous images were observed with the cooled foils, which suggests that few clusters are formed within about one extinction distance of the top surfaces. This would be reasonable if most of the defects were vacancies, which could be trapped at the surface. It may also be significant that the mutual annihilation of two loops has not been observed. Constructive interactions have been seen in many samples; these occur only

a

b

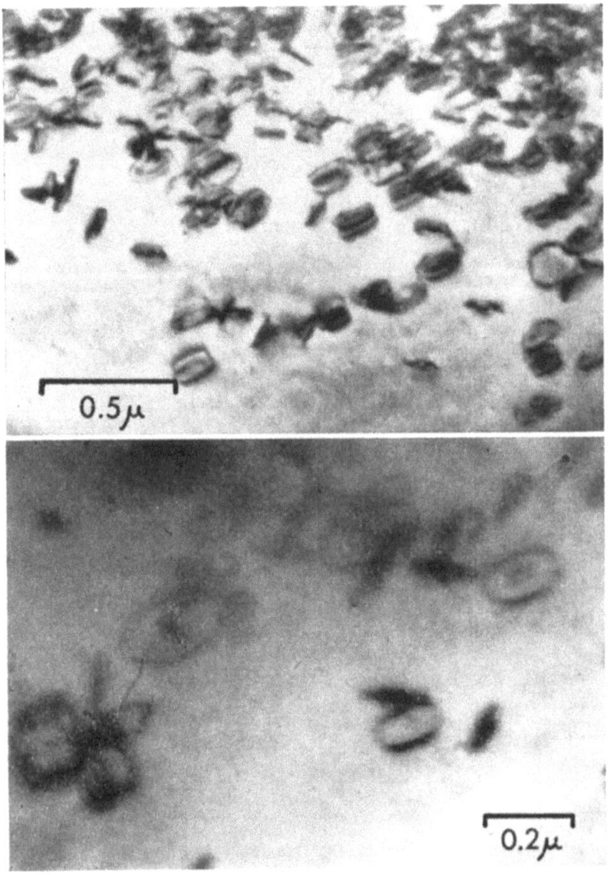

Fig. 11. a) Complex loops including concentric pairs and superposed pairs. b) Loops containing defects

Fig. 12. Hexagonal loops, possibly decorated with precipitates, {100} foil

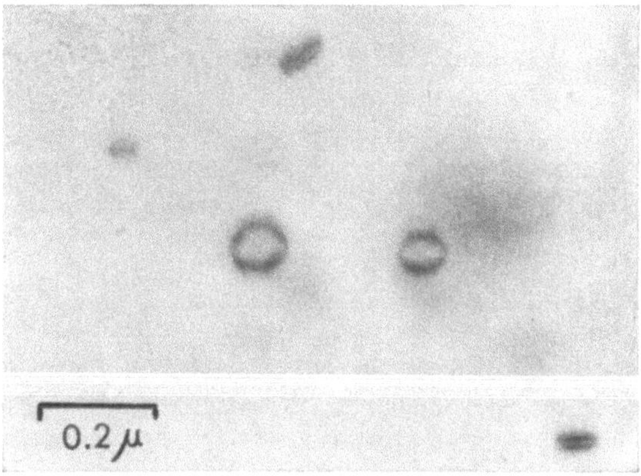

with loops of a similar nature. Frequently two loops combine to form first a figure-eight configuration, and then a more rounded shape. A newly joined pair of loops can be seen at C, Fig. 6c. Loops on neighbouring planes sometimes join in the manner described by Harvey [24] and others, so that the boundary between the loops is a dislocation with a compound Burgers vector, and the figure-eight has a step across its middle. In this case the most likely interaction is $1/2\,[101] + 1/2\,[\bar{1}01] = [001]$, since the '$c$' Burgers vector has the minimum elastic energy. In the later stages of irradiation various complex loop configurations are observed as shown in Fig. 11a. Some of these undoubtedly result from loop-loop interactions but others, which include concentric double loops, are not understood. Some loops are paired or concentric with others, some have become enlarged and circular. Further examples of double loops and loops with unexplained internal features can be seen in Fig. 11b. Prolonged irradiation of a region with a sufficiently low density of loops to preclude interaction, causes the loops to adopt a minimum energy shape which is hexagonal (Fig. 12), and often results in precipitation on and within the loops. This is marked by a thickening and fuzziness of the images and a lessening of any strain contrast; resolvable precipitates are seen on some loops.

5.4 Precipitate clusters

Since both loops and precipitates can cause strain contrast, it is difficult to identify the clusters in cooled foils until they are directly visible. Irradiation at normal temperatures, however, soon creates precipitates and effects due to these can be investigated with the goniometer stage. Large precipitate platelets which are 'end-on' to the beam are clearly visible in foils oriented at the Bragg angle. The precipitates are apparently platelets with their faces parallel to {011} planes. Their nature has been considered using the methods of Ashby and Brown [15, 19], and Bell et al. [16, 17]. Quantitative information about the strains cannot be obtained because the elastic constants for MgF_2 are not known. Fig. 13 shows a dark field micrograph in which several precipitates exhibit anomalous asymmetrical images. Both vacancy- and interstitial-type precipitates are present. They apparently occur in roughly equal numbers. After prolonged irradiation they reach sizes (Fig. 3) at which the coherency contrast diminishes. Moreover, in attaining these sizes the platelets themselves cause the creation of

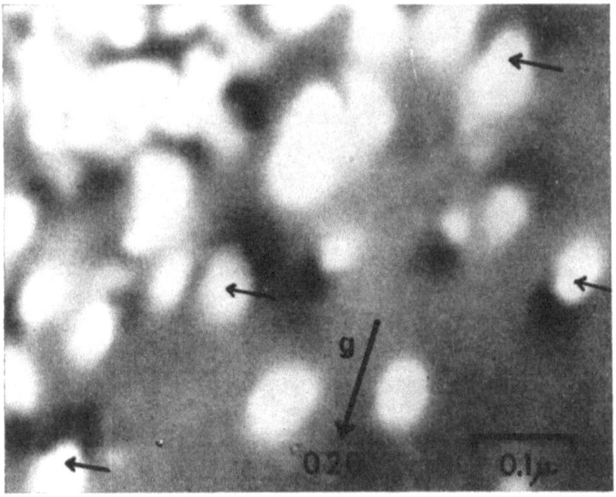

Fig. 13. Dark field micrograph showing anomalously large strain images from some precipitate particles. {100} foil irradiated at room temperature

Fig. 14. Showing the development of precipitates and rows of loops along directions parallel to the precipitate axes in a heavily irradiated {100} foil

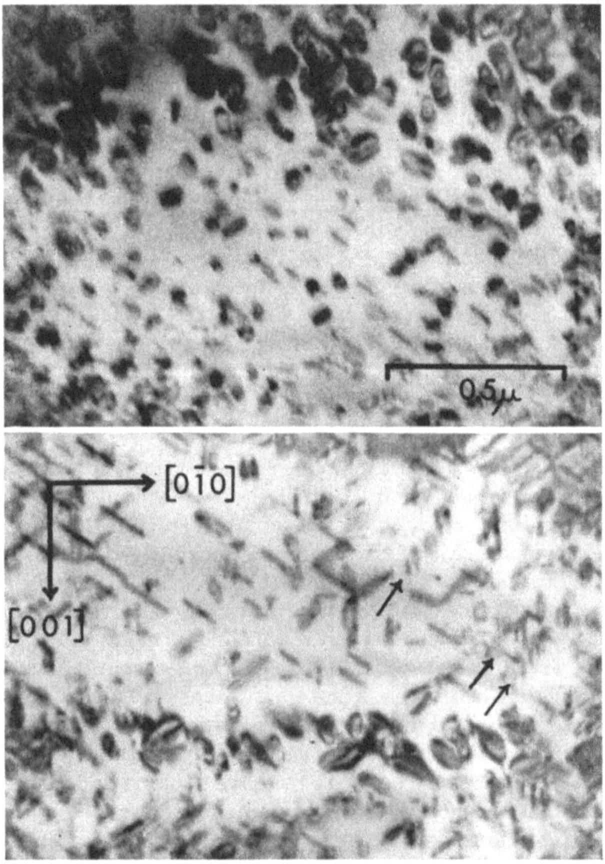

a

b

dislocation loops which occur in groups, like 'smoke rings', in line with the largest precipitates. Possibly they are generated by a prismatic punching mechanism of the type observed by Barber et al. [25] and Jones and Mitchell [26]. Fig. 14 shows the development of a number of bar-like precipitates. In frame b) these are often surrounded by distorted loops. On the axes of some precipitates they are small groups of loops, centred on $\langle 011 \rangle$ directions and apparently lying on $\{010\}$ planes.

6. Discussion

The electrons used for imaging in conventional electron microscopes are not usually sufficiently energetic to produce permanent displacements of fully bonded lattice atoms or ions. Calculation shows that even in 'head-on' collisions, displacements are only feasible for the first few elements in the periodic table. However, there are other ways in which electrons can create point defects and they can also make existing defects mobile. Nelson [27] has pointed out that the stimulated migration of point defects due to electron bombardment can be an important factor in clustering experiments. The activation and aggregation of existing defects alone could not, however, account for all the observed damage.

More than one mechanism can be envisaged for the creation of point defects, but only a few of the mechanisms which have been proposed are mentioned here. There are single ionization and multiple ionization mechanisms (such as the first proposed by Varley [28]), and mechanisms which depend on excited, states. Recent studies [29] have indicated that multiple ionization is unimportant for F-centre production in potassium iodide and possibly other ionic crystals. Pooley [30] has proposed a scheme for the production of anion vacancies and F-centres via excited states which may be applicable to MgF_2. It is clear that electron-induced damage in MgF_2 is intimately related to the creation of anion vacancies because Hall [22] has found that an absorption band, which he attributes to F-centres, cannot be produced by X- or electron irradiation at 77 °K. This correlates with the author's observations on the absence of clusters in foils irradiated at $\approx$ 77 °K. Blunt and Cohen [3] found that colour centres produced in MgF_2 by ionizing radiations are very stable (and therefore immobile) at room temperature. The mobility of point defects at temperatures below 0 °C, inferred from the growth of loops, is therefore surprising unless it is attributed to a mechanism, such as Nelson's [27]. The experiments in which cluster-free irradiated samples were heated without visible clusters appearing show that either few defects are created at 77 °K or that thermal activation is not the factor controlling clustering during irradiation. Hall's work supports the first idea while Blunt and Cohen's observations of stability are evidence for the second one. Since the two ideas are not mutually exclusive, we can envisage a model in which defect production depends on the creation of excited states and is temperature dependent, but clustering is not because it is stimulated by irradiation.

The stacking fault contrast observed within some shows that the original clusters were not collections of molecular vacancies, but either layers of mission fluorine ions or effectively extra layers of magnesium ions in a $\{011\}$ plane. The $\{011\}$ planes can be thought of as the stacking of sandwiches, each of which normally consists of a layer of cations between two layers of anions. It was found that voids were formed when the crystals were subjected to a very intense electron flux, i.e. molecular vacancies are created at higher temperatures than those at which loops form. There are similarities to results of Forty [5, 6] on PbI_2. Dislocation loops and helices were created by low intensity irradiation (uncooled samples); at higher intensities cavities appeared and lead precipitates decorated their walls.

The character of the loops MgF_2 was not established. However, the ease with which halogen has been found to escape from other fluorides [1] and halides favours the idea that the loops are initially formed by lattice collapse around discs of fluorine vacancies. Fluorine diffuses to the surface since there are no other sinks. The stacking fault within the Frank loop is subsequently lost. The details of this process could not be followed.

Intense irradiation (presumably accompanied by higher local temperatures) produces precipitates at loops, and more commonly, in undislocated regions. The precipitates whose characteristics were examined (necessarily those near the top of the foil through which precipitation was probably not uniform), were both vacancy and interstitial. If magnesium precipitates form where fluorine atoms have left the lattice sites (as in CaF_2 and SrF_2) the platelets will exhibite a vacancy-type strain contrast. Some explanation is still needed for the precipitates with interstitial contrast (i.e. the matrix lattice displacements are away from the platelet). It is tentatively suggested that these may be lenticular pockets of fluorine, which would be gaseous for any of the conditions under which precipitates form. Magnesium precipitates could be stabilized by a surface reaction with oxygen in solution or from the microscope column. There is a possibility that some of the platelets are MgO.

Acknowledgements

I wish to thank Dr. R. E. Howard and Dr. J. B. Wachtman, Jr. for useful discussions during the preparation of this paper.

References

[1] T. Evans, Phil. Mag. 8, 1235 (1963).
[2] D. J. Barber, J. appl. Phys. **36**, 3342 (1965).
[3] R. F. Blunt and M. I. Cohen, Nat. Bur. Stand. Rep. No. 8614 (1965).
[4] A. J. Forty, Phil. Mag. **5**, 787 (1960).
[5] A. J. Forty, Phil. Mag. **6**, 895 (1961).
[6] A. J. Forty, private communication.
[7] T. Hibi and K. Yada, J. Phys. Soc. Japan **17**, Suppl. B-11, 212 (1962).
[8] K. Tanaka, M. Mannami, and K. Izumi, J. Phys. Soc. Japan **18**, Suppl. III, 345 (1963).
[9] K. Yagi and G. Honjo, J. Phys. Soc. Japan **19**, 1892 (1964).
[10] R. S. Wilks and F. J. P. Clarke, AERE (Harwell), Rep. R 4435 (1963).
[11] J. Van Landuyt, R. Gevers, and S. Amelinckx, phys. stat. sol. **7**, 307 (1964).
[12] A. Duncanson and R. W. H. Stevenson, Proc. Phys. Soc. (London) **72**, 1001 (1958).
[13] S. Amelinckx, The Direct Observation of Dislocations, Academic Press, New York 1964.
[14] P. B. Hirsch, A. Howie, and M. J. Whelan, Phil. Trans. A **252**, 499 (1960).
[15] M. F. Ashby and L. M. Brown, Phil. Mag. 8, 1649 (1963).
[16] W. L. Bell, D. M. Maher, and G. Thomas, Lattice Defects in Quenched Metals, Academic Press, New York 1965 (p. 739).
[17] W. L. Bell and G. Thomas, Phil. Mag. **13**, 395 (1966).
[18] W. L. Bell and G. Thomas, phys. stat. sol. **12**, 843 (1965).
[19] M. F. Ashby and L. M. Brown, Phil. Mag. 8, 1083 (1963).

[20] V. A. PHILIPS and J. D. LIVINGSTON, Phil. Mag. **7**, 969 (1962).
[21] U. ESSMANN and M. WILKENS, phys. stat. sol. **4**, K53 (1964).
[22] T. P. P. HALL, private communication.
[23] K. H. G. ASHBEE, R. E. SMALLMAN, and G. K. WILLIAMSON, Proc. Roy. Soc. A **276**, 542 (1964).
[24] K. B. HARVEY, Phil. Mag. **8**, 435 (1963).
[25] D. J. BARBER, K. B. HARVEY, and J. W. MITCHELL, Phil. Mag. **2**, 704 (1957).
[26] D. A. JONES and J. W. MITCHELL, Phil. Mag. **3**, 1 (1958).
[27] R. S. NELSON, Phil. Mag. **10**, 723 (1964).
[28] J. H. O. VARLEY, Nature **174**, 886 (1954).
[29] T. P. P. HALL, D. POOLEY, W. A. RUNCIMAN, and P. T. WEDEPOHL, Proc. Phys. Soc. **84**, 719 (1964).
[30] D. POOLEY, Solid State Comm. **3**, 241 (1965); Proc. Phys. Soc., to be published.

(Received May 13, 1966)

phys. stat. sol. **16**, 549 (1966)

Physics Department, Indian Institute of Technology, Kharagpur

On the Electrical and Magnetic Properties of Coloured and Uncoloured Calcium Fluoride Crystals

By

V. V. RATNAM

The electrical and magnetic properties of uncoloured calcium fluoride crystals and of specimens of these crystals coloured by X-rays and electrolysis, are reported. The dc conductivity and dielectric loss measurements of uncoloured and X-ray coloured CaF_2 crystals show that in a freshly prepared crystal 1. the number of free negative ion vacancies is very small and 2. the anti-Frenkel defects which are initially present are in the associated form. These points explain why, for irradiation at room temperature, the α-band (the F-band of this system) is not observed in optical absorption. The nature of the traps is also discussed in the context of the observed changes in the dc conductivity accompanying the thermal bleaching of the colour centres in an X-irradiated crystal. The temperature variation of the magnetic susceptibility indicates that X-irradiation produces both paramagnetic and diamagnetic colour centres. From the results of measurements on electrolytically coloured crystals, the oscillator strength of the F-band is estimated as about 10^{-2}; this also explains why the formation of the α-band is not detected for room temperature X-irradiation.

Es wird über elektrische und magnetische Eigenschaften von nicht verfärbten Kalziumfluoridkristallen und von Proben dieser Kristalle, die durch Röntgenstrahlung und Elektrolyse verfärbt werden, berichtet. Die Gleichstromleitfähigkeit und die Messungen der dielektrischen Verluste zeigen, daß in frisch preparierten Kristallen 1. die Zahl der freien negativen Ionenlücken sehr klein ist und 2. die Anti-Frenkeldefekte, die ursprünglich vorhanden waren, in assozierter Form vorliegen. Diese Punkte erklären, warum bei Bestrahlung bei Zimmertemperatur die α-Bande (die F-Bande dieses Systems) nicht in optischer Absorption beobachtet wird. Die Natur der Haftstellen wird im Zusammenhang mit den beobachteten Änderungen der Gleichstromleitfähigkeit, die beim thermischen Bleichen der Farbzentren in röntgenbestrahlten Kristallen auftreten, diskutiert. Die Temperaturvariation der magnetischen Suszeptibilität zeigt, daß eine Röntgenbestrahlung sowohl paramagnetische wie diamagnetische Farbzentren hervorruft. Aus den Ergebnissen der Messungen an elektrolytisch verfärbten Kristallen wird die Oszillatorenstärke der F-Bande zu 10^{-2} abgeschätzt. Dies erklärt ebenfalls, warum die Bildung der α-Bande durch Röntgenbestrahlung bei Zimmertemperatur nicht nachgewiesen werden kann.

1. Introduction

In spite of the large amount of work [1] on colour centres and related properties of calcium fluoride, a number of problems arising out from the effect of X-rays on this crystal, like the nature of the defects produced in this system by X-irradiation, and the electrical and magnetic properties, were not studied in detail and hence not well understood.

Measurements of the dc conductivity and dielectric loss, before and after X-irradiation have been used to correlate the electrical behaviour of colour centres in the case of alkali halides [2, 3, 4]. Similar measurements are expected to give information about the nature of the charge carriers, their concentration, and their transformation under different experimental conditions. As far back as 1939 Zintl and Udgard [5] showed that YF_3 forms a solid solution with CaF_2, and from X-ray and density measurements they deduced that only anti-Frenkel defects (equal concentrations of negative ion vacancies and interstitials) were present. Ure's measurements [6] of ac conductivity and transference number on pure and YF_3 doped CaF_2 crystals supported these findings.

Similarly, colour centres and lattice imperfections are known to make their characteristic contribution to the magnetic susceptibility, depending on their concentration, electronic state, and valency in which they exist in the system. Information about the effects of radiation on these magnetic centres and about the centres formed by irradiation can be obtained from measurements of the magnetic susceptibility and its temperature variation. This method has been successfully applied to colour centres in alkali halides by Rauch and Heer [7]. This method can, therefore, be profitably applied to other systems also.

The present paper reports results on the electrical properties (dc conductivity and dielectric loss), the magnetic properties of the crystal before and after irradiation, and the magnetic behaviour of electrolytically coloured calcium fluoride crystals. The role of these defects in the formation and behaviour of the colour centres is discussed.

2. Experimental Methods

Single crystals of calcium fluoride obtained from the Harshaw Chemical Co., were used for study. These were of size 1 cm × 1 cm × 1.5 mm for measurements of the electrical properties and 1 cm × 4 mm × 4 mm for magnetic measure-

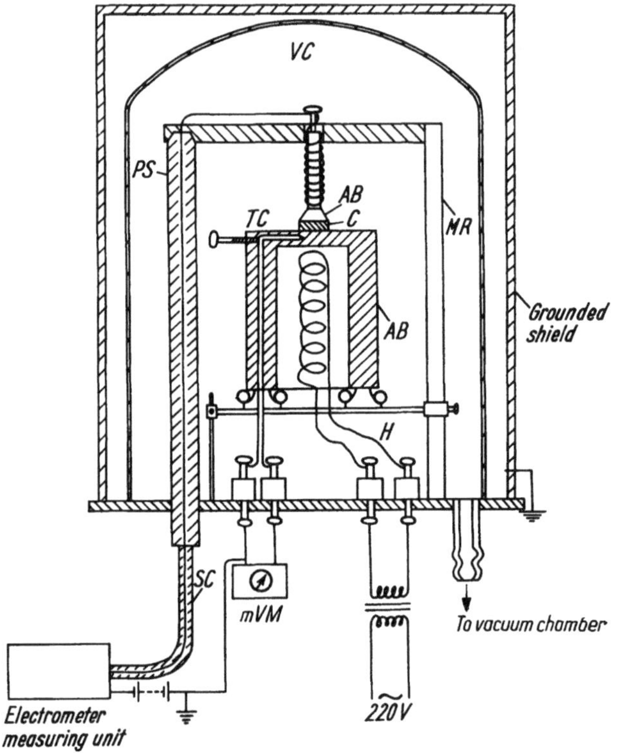

Fig. 1. Schematic diagram of the set-up for the measurement of dc conductivity. TC – thermocouple, mVM – mV meter, AB – aluminium block, C – crystal, H – heater, SC – shielded cable, VC – vacuum chamber, PS – perspex stand, MR – metal rod

ments. The X-irradiation time was 10 h; a TEG 50 Machlette tube was used (operating voltage 35 kV, current 20 mA), with the sample kept at a distance of 3 cm from the target.

The dc conductivity measurements on these crystals were made with the help of an Ekco vibrating reed electrometer which has a limiting sensitivity of 10^{-15} A. The schematic diagram of the experimental set-up is shown in Fig. 1. The measurements were made in dark and in a vacuum of 10^{-4} Torr. Proper precautions were taken to reduce stray electrical pick-up and to provide the necessary insulation for the high tension lead. Good electrical contact of this lead with the top surface of the crystal was ensured by providing spring loading to this contact.

The dielectric loss measurements on this crystal were made on a Marconi circuit magnification meter (type TF 329 G) in the frequency range 50 kHz to 1 MHz. This was specially adopted for measuring small changes in dielectric loss by utilising the well known resonance curve method. The changes in the dielectric loss were obtained in terms of the galvanometer deflections in a manner similar to that adopted by Garlick and Gibson [8], wherein the loss was previously calibrated in terms of an equivalent parallel resistance.

The magnetic susceptibility measurements were carried out on a magnetic micro-balance, the schematic diagram of which is shown in Fig. 2. The balance

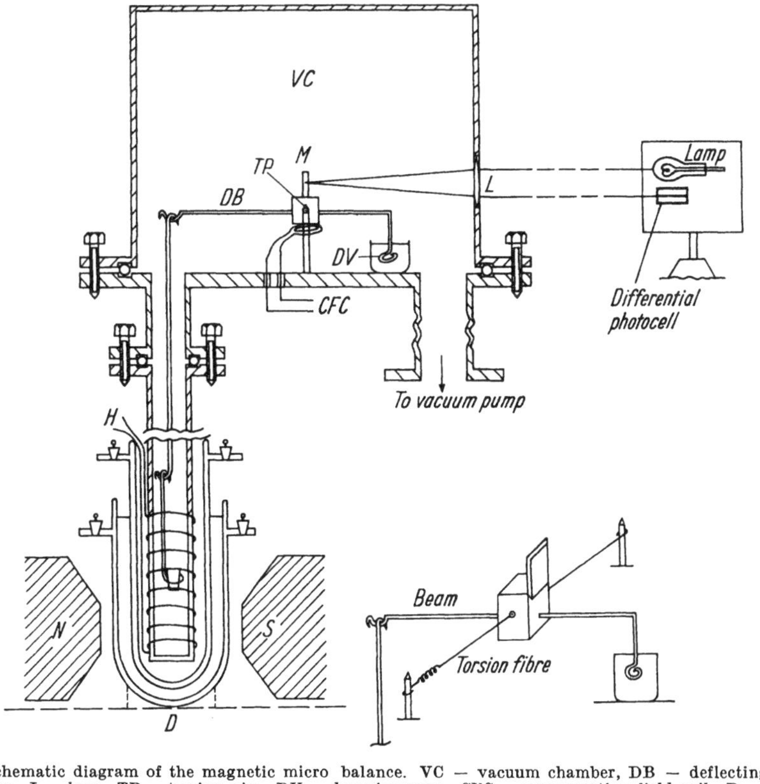

Fig. 2. Schematic diagram of the magnetic micro balance. VC — vacuum chamber, DB — deflecting beam, M — mirror, L — lens, TP — torsion pin, DV — damping vane, CFC — compensating field coil, D — dewar, H — heater

is a modified Faraday type of balance, wherein the vertical pull on the specimen placed in the nonhomogeneous magnetic field of an electromagnet is balanced by the interaction of the magnetic fringe field with a compensating field produced by passing a current through a coil attached to the beam at the junction of the torsion fibre and the deflecting beam. The current in this compensating field coil could be adjusted to bring back the deflected beam and hence the specimen to its original position. The balance was standardised by substances like treble distilled water. Though the absolute values of the measurements were accurate up to 2% only, measurements of the temperature variation of the susceptibility were accurate up to 0.2%; the same holds for the relative values of the room temperature susceptibility of the sample before and after colouration. The temperature variation measurements were made in vacuum in the temperature range 20 to 400 °C.

3. Experimental Results

The dc conductivity measurements indicate the presence of isolated vacancies capable of moving in a dc field, while dielectric loss measurements give information about associated vacancies, vacancy pairs, and complexes.

3.1 dc conductivity

The room temperature conduction current of a fresh crystal was very low ($\approx 10^{-15}$ A) corresponding to a conductivity of $3\times10^{-16}\ \Omega^{-1}\ cm^{-1}$. Its temperature variation in the range 30 to 350 °C (observed under dynamic conditions by noting down the instantaneous values of the current against temperature) is shown in Fig. 3 (curve A). The same figure also shows the curves for the irradiated crystal (curve B) and the bleached crystal (curve C). The following conclusions have been drawn from these studies:

1. Unirradiated crystal. — For this crystal there are three temperature regions for which the log (conductivity) vs. $1/T$ curve is linear; they are: region I from 30 to 175 °C with an activation energy of 0.63 eV , region II from 175 to 300 °C with an activation energy of 0.80 eV, and region III from 300 to 350 °C with an activation energy of 1.98 eV.

2. X-irradiated crystals. — On X-irradiation, a) the room temperature value of the current increases by nearly 100 times; the activation energy for the temperature region 30 to 75 °C is very small (0.28 eV); b) a slight increase in the

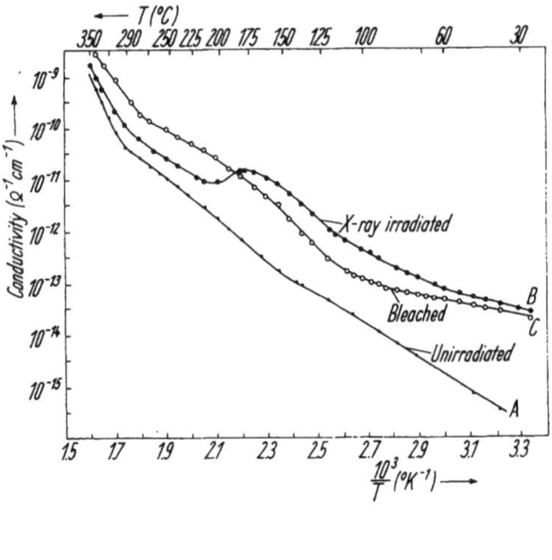

Fig. 3. The temperature dependence of the dc conductivity of an unirradiated, X-irradiated, and bleached CaF_2 crystal

current occurs in the region of the first glow peak followed by a conductivity peak around 180 °C; c) at temperatures above 200 °C, once again, the curve runs parallel to that of the unirradiated crystal.

3. Bleached crystal. — a) The room temperature conductivity of the bleached crystal is not much different from that of the X-irradiated crystal. b) The activation energy in the 30 to 175 °C region is very small (0.22 eV). c) The activation energies for the other two regions are almost the same as those for the unirradiated crystal.

3.2 Dielectric loss

The sensitivity of these measurements is of the order of $10^{-9}\ \Omega^{-1}\ cm^{-1}$ (corresponding to a current of 10^{-8} A). Measurements on the dielectric loss of the unirradiated, irradiated, and the bleached crystals are shown in Fig. 4. The following inferences have been drawn from this study:

1. The dielectric loss of an unirradiated crystal is very small; for the X-irradiated and thermally bleached crystal it is of a measurable value (tg $\delta \approx 10^{-3}$ at 500 kHz).
2. On further X-irradiation this loss decreases.
3. The loss once again increases on subsequent thermal bleaching.

3.3 Magnetic susceptibility measurements

The room temperature diamagnetic susceptibility of a fresh crystal has the value -0.360×10^{-6} cgs emu/g. The temperature variation of the susceptibility of an unirradiated, X-irradiated, and electrolytically coloured crystal is shown in Fig. 5. Curve A, representing the temperature variation of an unirradiated crystal, indicates the presence of paramagnetic impurities. Heat treatment (raising the crystal to about 400 °C and cooling to room temperature) makes the sample more diamagnetic.

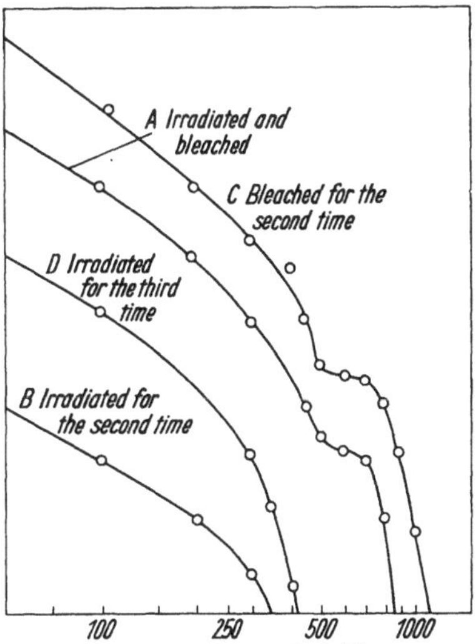

Fig. 4. The dielectric loss of an irradiated and bleached CaF_2 crystal. The effect of cycling through X-irradiation and bleaching is also shown

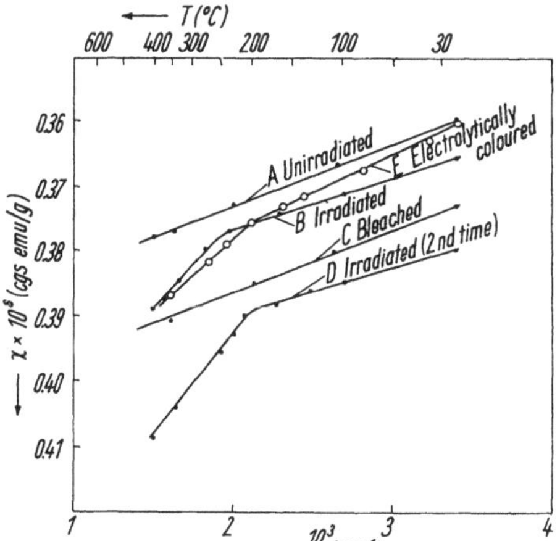

Fig. 5. The temperature dependence of the magnetic susceptibility of an unirradiated, X-irradiated, electrolytically coloured, and bleached CaF_2 crystal. The effect of cycling through X-irradiation and bleaching is also shown

From a study of the above curves, the following observations on the temperature dependence of the susceptibility can be made:

1. The crystal coloured by X-rays becomes more diamagnetic; the χ vs. $1/T$ graph shows two regions: region I between 30 and 200 °C, for which the slope of the curve is less than that for a fresh crystal in the same region, and region II (above 200 °C), in which the slope is much greater than that in region I or that of the uncoloured crystal in region II. This shows that the colour centres which are not bleached up to 200 °C are paramagnetic in nature. The bleached crystal becomes more diamagnetic, but the temperature variation of the susceptibility is similar to that of the uncoloured crystal.

2. In the case of electrolytically coloured crystals the room temperature diamagnetic susceptibility is not much different from that of the uncoloured crystal. The temperature variation of the susceptibility of the coloured crystal shows that the overall behaviour of the colour centres is paramagnetic in both regions. In region II (between 200 and 400 °C), the behaviour of the colour centres in this crystal is similar to that of the X-irradiated crystal.

4. Discussion

4.1 dc conductivity

From conductivity measurements Ure [6] established that the predominant defects in this crystal are anti-Frenkel type defects, involving equal concentrations of negative ion vacancies and interstitials. Since paired vacancies do not contribute to dc conductivity, the current obtained in dc conductivity measurements must be due to the mobility of isolated vacancies or interstitials.

The very low value of the room temperature conductivity in unirradiated crystals indicates that free charge carriers are very few in number. The increase in the slope of the conductivity curve above 175 °C, compared to the slope in the low temperature region, indicates a greater activation energy. From results obtained from optical absorption measurements [9], wherein it was postulated that the associated anti-Frenkel defects break up at about 200 °C, it can be inferred that a greater activation energy is required for this region because that will be the energy required for the dissociation of the anti-Frenkel defect pairs and their mobility as well. The region above 300 °C corresponds to the intrinsic range, where the curves for the unirradiated, irradiated, and bleached crystalsr un parallel.

The large increase in the room temperature conductivity of the irradiated crystal (curve B of Fig. 3) must be due to the availability of a large number of charge carriers. Apart from this, the activation energy in the 30 to 100 °C region is considerably reduced (0.28 eV). This is probably due to the precipitation of vacancies and interstitials along the dislocation lines and (or) pipes, which reduce the resistance of the crystal. Bontinck and Amelinckx [10] have actually observed, by decoration techniques, helixes, and sequences of loops in crystals which have been formed during the growth process itself and have become prominent by annealing in air or by additive colouration. This is perhaps the reason for the low activation energy observed in the bleached crystal in this temperature range.

On heating the irradiated crystals, changes occur in the conductivity accompanying the thermal destruction of the colour centres. The first two glow peaks are accompanied by only a slight increase in conductivity, whereas the third glow peak is accompanied by a large increase in conductivity and a conductivity peak. The small increase in conductivity associated with the first two peaks may be due to the low mobility of the charge carriers, but not due to theirs smaller number, because these glow peaks are fairly intense. From observations of X-ray fluorescence, optical absorption [9], and thermoluminescence [11], which have been explained in terms of recombination of holes with electrons of the Y^{2+} ions (at the Y^{2+} sites), it can be inferred that holes must take part in conductivity. On account of the small conductivity changes accompanying the first two glow peaks, these holes must be localised ones (in the vicinity of the Y^{2+} sites). The third glow peak, accompanied by a conductivity peak, has also been explained [11] in terms of thermally released holes recombining with electrons of the Y^{2+} ions. The conductivity in this region must, therefore, be due to holes, released in the valence band, which travel greater distances before annihilation.

The effect of the fourth glow peak (occurring around 300 °C) is not observed in conductivity since it is most likely that its effect is hidden by the large increase of conductivity in this region due to the thermal generation and increased mobility of a large number of charge carriers (both anions and interstitials).

4.2 Dielectric loss

Compared to the dc conductivity measurements (sensitivity 10^{-15} A), the sensitivity of the ac measurements is limited to currents of the order of 10^{-8} A. Yet dielectric loss measurements have been helpful in giving information about the behaviour of associated or paired vacancies and vacancy complexes. The low value of the dielectric loss of a fresh crystal becomes measurable by X-irradiation and thermal bleaching. Repeating this cycle of operations, the loss increases very much. The simplest form of these dipolar complexes to which the observed dielectric loss is due to, can be an anti-Frenkel defect pair. The decrease in the loss on X-irradiation indicates that the number of such complexes gets reduced by irradiation. Increase in their number on bleaching shows that a large number of them are also formed by X-irradiation. They appear to decrease in number in the irradiated crystal because they have trapped electrons and holes. This information regarding the behaviour of these complexes before and after irradiation has also been useful in explaining the non-detectability of the α-band by X-irradiation at ordinary temperatures [9], since it is now known that the α-band is due to an electron trapped by a free negative ion vacancy.

4.3 Magnetic susceptibility

Magnetic susceptibility measurements show that these crystals contain some paramagnetic (rare earth) impurities. Their concentration is rather too small to be detected in the room temperature susceptibility measurement, but the temperature variation of the susceptibility clearly indicates their presence. The result that the temperature variation of the susceptibility of the bleached crystal is similar to that of the uncoloured crystal indicates that the nature and the number of the paramagnetic impurity centres remain unaffected by the cycle of operations, though the room temperature susceptibility increases after every stage of the cycle of operations, viz., X-raying and bleaching.

Information about the effect of colour centres can then be obtained only from the differences in the slopes of the χ vs. $1/T$ curves of the unirradiated and irradiated crystals. For X-irradiated crystals, these differences show that the overall behaviour of the colour centres bleached by raising the temperature to 200 °C is diamagnetic. The effect of the $4\,d^1$ electrons of the Y^{2+} ions (formed by the conversion of the Y^{3+} ions by irradiation), which is expected to show a paramagnetic behaviour, is not observed in these measurements, perhaps because the effect of these paramagnetic centres is rather small compared to that of the diamagnetic behaviour of the rest of the centres. It may also be due to the fact that the concentration of these Y^{2+} ions, though considerable enough to account for the luminescence and other properties, is too small to be detected in these magnetic measurements.

In electrolytically coloured crystals, however, the room temperature diamagnetic susceptibility is not much different from that of the uncoloured ones, i.e., the change in the lattice diamagnetism is small compared to the changes in the X-irradiated crystal. The temperature variation of the susceptibility of these crystals shows that the colour centres bleached in the temperature region 30 to 200 °C, as well as those bleached in the temperature region 200 to 400 °C are paramagnetic in nature. In fact, in the temperature range 200 to 400 °C the temperature variation of the susceptibility of both X-irradiated and electrolytically coloured crystals is almost similar, indicating that these centres, in crystals coloured by both methods and which are stable up to 200 °C, are similar as far as their paramagnetic behaviour is concerned.

It may be pointed out here that the increase in the net diamagnetic susceptibility with increase in temperature, observed in coloured crystals, is due to a) the temperature variation of the contribution due to the paramagnetic centres present in the crystal and b) the bleaching of the paramagnetic colour centres. In the case of electrolytically coloured crystals, the colour centres are not perceptibly bleached below 200 °C. So the temperature variation of the susceptibility of the crystal in this temperature region can be taken as the contribution of the paramagnetic centres only. Thus a comparison of the slope of the χ vs. $1/T$ curve of the uncoloured with that of the coloured crystal should enable us to estimate the number of paramagnetic centres involved in the temperature variation of the susceptibility in this range. A rough estimate of the oscillator strength for the band can then be made as shown below:

The total diamagnetic susceptibility can be expressed as the sum of the diamagnetism of the lattice and that of the defects and can, therefore, be written as

$$\chi_1 = \chi_0 + \frac{C_1}{T},$$

where χ_0 is the lattice diamagnetism and C_1 the Curie constant associated with the paramagnetic centres. On colouration, the total susceptibility becomes χ_2 which can be expressed as

$$\chi_2 = \chi_0' + \frac{C_1 + C_2}{T},$$

where C_2 corresponds to the Curie constant associated with the number of paramagnetic centres formed by colouration (in excess of the defect centres that are already present).

The difference between the two susceptibilities can be written as

$$\chi_2 - \chi_1 = \frac{C_2}{T},$$

since χ_0' of the electrolytically coloured crystal is not much different from that of the uncoloured crystal. Thus the difference in the slopes of the uncoloured and electrolytically coloured crystal in the region 30 to 200 °C gives C_2 for this region. Using the relation

$$C_2 = N \mu_B^2 \frac{p^2}{3k},$$

N can be calculated if p is known [12]. The data plotted in the graph of Fig. 5 gives

$$C_1 = 9.64 \times 10^{-6} \text{ cgs emu °K/g}$$

and
$$(C_1 + C_2) = 11.28 \times 10^{-6} \text{ cgs emu °K/g},$$

consequently
$$C_2 = 1.64 \times 10^{-6} \text{ cgs emu °K/g},$$

and from
$$N = 3\, k\, C_2/\mu_B^2\, p^2,$$

we get
$$N = 21.3 \times 10^{18}/p^2 \text{ per cm}^3.$$

From the absorption data of the electrolytically coloured crystal [9], the product Nf for the 385 nm band (the F-band of this system) comes out to be 19.6×10^{16} centres cm^{-3}. Combining these two results, f, the oscillator strength, comes out to be 0.03, i.e., of the order 10^{-2}, taking a value of 3 for p^2, since this band is due to unpaired electrons filling up isolated negative ion vacancies. This has to be taken as a rough estimate only, since the other centres, which are bleached along with the 385 nm band, also contribute to changes in the susceptibility, i.e., though it is only the 385 nm band that is known to be due to paramagnetic centres, the effect of other centres, however, cannot be overlooked in this calculation.

This low value of the oscillator strength is perhaps the cause for the non-detectability of this 385 nm band (the α-band) by X-irradiation at room temperature.

Acknowledgement

The author is very much indebted to Professor H. N. Bose, Head of the Department of Physics, for suggesting the problem, for his guidance, and keen interest in the work.

References

[1] References to literature on CaF_2 are given in:
K. Przibram, Irradiation Colours and Luminescence, Pergamon Press 1956.
J. H. Schulman and W. D. Compton, Colour Centres in Solids, Pergamon Press 1962.
P. Görlich, H. Karras, and R. Lehmann, phys. stat. sol. **1**, 389 (1961); **1**, 526 (1961).
[2] D. Dutton and R. Maurer, Phys. Rev. **90**, 126 (1953).
[3] R. G. Breckenridge, J. chem. Phys. **16**, 959 (1948); **18**, 913 (1950).
[4] Y. Haven and A. B. Lidiard, Rep. Bristol Conf. Defects in Crystalline Solids, Phys. Soc., London 1954 (pp. 261 and 283).
[5] E. Zintl and A. Udgard, Z. anorg. allg. Chem. **240**, 150 (1939).
[6] W. Ure, J. chem. Phys. **26**, 1363 (1957).
[7] C. J. Rauch and C. V. Heer, Phys. Rev. **105**, 914 (1957).
[8] G. F. J. Garlick and A. F. Gibson, Proc. Roy. Soc. **188A**, 485 (1947).
[9] V. V. Ratnam, Ph. D. thesis, I.I.T., Kharagpur.
[10] W. Bontinck and S. Amelinckx, Phil. Mag. **2**, 94 (1957).
[11] V. V. Ratnam and H. N. Bose, phys. stat. sol. **15**, 309 (1966).
[12] C. Kittel, Introduction to Solid State Physics, Wiley 1956.

(Received May 16, 1966)

phys. stat. sol. **16**, 559 (1966)

Physics Department, Indian Institute of Technology, Kharagpur

On the Colour Centres and X-Ray Luminescence of Calcium Fluoride Crystals

By

V. V. Ratnam

A study is made of differences in the colouration spectrum of calcium fluoride crystals obtained by X-ray irradiation at different temperatures and by electrolysis. The changes occurring in the X-ray luminescence spectrum due to the use of different irradiation temperatures are also considered. It is shown that the α- and β-bands, usually only observed in additive and electrolytic colouration, can also be produced by X-ray irradiation if the irradiation temperature is between 180 and 250 °C. The formation of the α-band (the F-band of this system) is difficult to detect at lower irradiation temperatures because of 1. the small number of free negative ion vacancies at these temperatures and 2. the low oscillator strength associated with this band. The changes observed in the X-ray luminescence spectrum for different irradiation temperatures are explained and the associated electronic processes are discussed briefly.

Unterschiede im Verfärbungsspektrum von Kalziumfluoridkristallen werden untersucht. Die Verfärbung erfolgt durch Röntgenbestrahlung und durch Elektrolyse, wobei die Unterschiede im Lumineszenzspektrum durch Röntgenbestrahlung bei verschiedenen Temperaturen ebenfalls berücksichtigt werden. Es wird gezeigt, daß die α- und β-Banden, die gewöhnlich nur bei additiver und elektrolytischer Verfärbung beobachtet werden, auch durch Röntgenbestrahlung erzeugt werden, wenn die Bestrahlungstemperatur zwischen 180 und 250 °C liegt. Die Bildung der α-Bande (die F-Bande dieses Systems) ist bei niedrigeren Bestrahlungstemperaturen schwierig nachzuweisen, wegen 1. der geringen Anzahl von freien negativen Ionenlücken bei diesen Temperaturen und 2. der niedrigen mit dieser Bande verbundenen Oszillatorenstärke. Die Unterschiede im Lumineszenzspektrum für verschiedene Bestrahlungstemperaturen werden erklärt und die damit verbundenen elektronischen Prozesse kurz diskutiert.

1. Introduction

Colour centres in calcium fluoride appear to be more complex in nature than those in alkali halides. Different methods of colouration produce dissimilar absorption spectra; X-ray irradiation produces the four band spectrum of Smakula, additive colouration the α- and β-bands of Mollwo, and electrolytic colouration the α- and β-bands and a number of other bands [1, 2, 7].

In spite of the large amount of work [1 to 10] on colour centres and related properties of this crystal, (1) the non-formation of simple colour bands like the F-band of the alkali halides in X-ray colouration and (2) the electronic processes and the emission mechanisms responsible for the X-ray luminescence of this crystal are not well understood.

The present paper reports results of measurements on (1) the optical absorption spectra of crystals coloured by X-rays at different temperatures and by electrolysis and (2) the X-ray fluorescence spectra of the crystals at different temperatures. These results throw considerable light on the above mentioned problems. Some conclusions about the broad nature of the electronic processes taking place inside the crystal are drawn.

2. Experimental Methods

2.1 Crystal material, X-ray colouration and optical absorption

The specimens used were of size 1 cm × 1 cm × 1.5 mm cut from a large single crystal of calcium fluoride obtained from the Harshaw Chemical Co., U.S.A.. During irradiation these were kept at a constant distance of 3 cm from the target of a TEG 50 Machlette tube operated at 35 kV, 20 mA for ten hours, each side of the crystal being exposed to the radiation for half the time. All the measurements on optical absorption were made at room temperature on the Hilger Uvispek spectrophotometer.

2.2 X-ray fluorescence and colouration at elevated temperatures

1. X-ray fluorescence. — The schematic diagram of the experimental set-up for the study is shown in Fig. 1. The crystal is held in the machined cavity of the aluminium block AB and is spring loaded to ensure good thermal contact with the block. The temperature could be kept constant to within 3 °C of any desired temperature as read in terms of a galvanometer deflection, and its constancy checked by recording the temperature photographically on a rotating drum. The light emitted by the crystal was detected by an RCA 1 P 28 photomultiplier and recorded by an Esterline Angus recorder. The spectral nature of this fluorescence was studied by scanning the spectrum after coupling the wavelength drum of a Hilger D246 monochromator to the driving spindle of the recorder. For comparison the spectrum obtained for irradiation at room temperature was also recorded by a quartz spectrograph.

2. X-ray colouration at elevated temperatures. — For colouring the crystal at elevated temperatures, the front face of the crystal (held in the cavity) was covered with a thin aluminium foil to limit effectively the temperature difference between the two faces. After the stoppage of irradiation at each tempera-

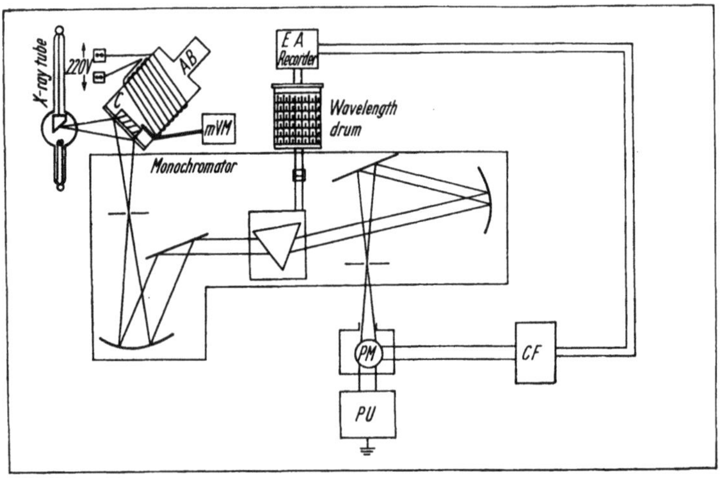

Fig. 1. Schematic diagram of the set-up for the study of the X-ray fluorescence at different temperatures. mVM – mV meter, PM—photomultiplier, PU—power unit, CF—cathode follower, AB—aluminium block

ture the crystal was quenched to room temperature, the optical absorption was measured and before irradiating at the next temperature it was thermally bleached.

2.3 Electrolytic colouration

The crystals were coloured electrolytically in air as well as in vacuum using a pointed platinum cathode and a flat platinum disc as anode. The anode was spring loaded to ensure good electrical contact and also to minimize undue pressure on the crystal. In this case samples of size 1.5 cm × 0.5 cm × 0.5 cm were used. A potential difference of 45 volts was applied and the time required for homogeneous colouration was of the order of 5 minutes at a temperature of 650 °C and about 15 to 20 minutes at 500 °C.

3. Experimental Results

3.1 X-ray colouration

For irradiation at room temperature (30 °C) the four band spectrum of Smakula with peaks at 228, 335, 400, and 580 nm is formed. Its formation is, however, affected by the temperature of irradiation. To study this the crystal was irradiated at different temperatures and the absorption spectra obtained for irradiation at 30, 100, 150, 180, 210, and 250 °C are shown in Fig. 2. For irradiation at and above 300 °C, the colouration is very weak. The same crystal has been used in these measurements. For higher temperatures of irradiation,

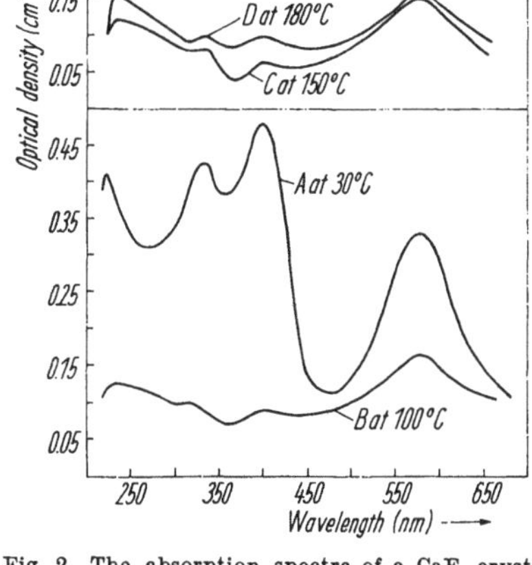

Fig. 2. The absorption spectra of a CaF_2 crystal coloured at different temperatures. After each irradiation the crystal was quenched to room temperature. Measurements were made at room temperature

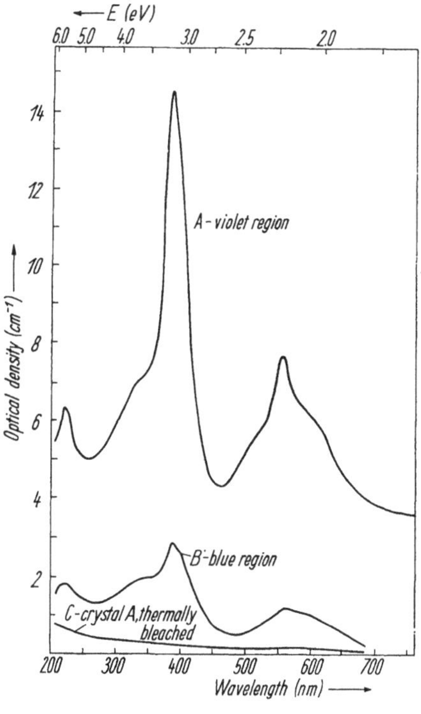

Fig. 3. The absorption spectra of electrolytically coloured CaF_2 crystals. A — violet region; B — blue region, and C — bleached crystal

there is a marked overall reduction in the intensities of the bands and also a change in their relative intensities. Viewing broadly, the positions of the absorption maxima are not appreciably altered by the temperature of irradiation. But for irradiation in the temperature region 180 to 250 °C, the so-called α- and β-bands at 380 and 520 nm respectively make their appearance superimposed on the 400 and 580 nm bands respectively.

3.2 Electrolytic colouration

No appreciable difference was noticed in the absorption spectrum whether the crystal was coloured in vacuum ($\approx 10^{-4}$ Torr) or in air. The colouration, in either case, consisted of two regions- the less dense blue regions on the outer layers and the more dense violet regions in the inner. These absorption spectra with prominent bands at 385 and 555 to 560 nm are shown in Fig. 3. In addition to the above, a number of other bands, including those formed by X-ray irradiation were formed. The band positions and the products $N f$ for these bands, estimated using Smakula's formula, which applied to calcium fluoride becomes

$$N f = 1.14 \times 10^{16} \, K \, W/\text{cm}^3 \, ,$$

are shown in Table 1. In the same table are also shown the corresponding values for an X-ray irradiated crystal for comparison. The centres formed by electrolytic colouration could be bleached by heating the crystals to about 400 °C.

3.3 X-ray fluorescence

The fluorescence spectrum of the crystal under X-irradiation consists of sharp lines as well as broad bands. Fig. 4 shows the microphotometric record of one such spectrum for room temperature irradiation. This was photographed by a quartz spectrograph. The same spectrum scanned by a monochromator with

Table 1

Band positions and concentrations of colour centres in electrolytically coloured and X-irradiated calcium fluoride crystals

Electrolytically coloured crystals. Crystal A (violet coloured)

Band peak (nm)	220	335	385	400	525	555	580	610
Absorption coefficient K (cm^{-1})	14.5	16.2	33.6	28.1	14.0	17.0	14.6	13.8
Half width W (eV)	2.15	2.28	0.51	0.29	0.93	0.54	0.73	0.35
$N f/10^{16} = 1.14\, K\, W$	35.6	42.2	19.6	9.3	14.9	10.5	12.2	5.5

Crystal B (blue coloured)

Band peak (nm)	220	335	385	400	—	560	580	600
Absorption coefficient K (cm^{-1})	4.16	4.40	6.20	5.0	—	2.85	2.62	2.30
Half width W (eV)	2.00	1.91	0.68	0.48	—	0.48	0.52	0.30
$N f/10^{16} = 1.14\, K\, W$	9.5	9.6	4.8	2.7	—	1.6	1.6	0.8

X-irradiated crystal

Band peak (nm)	228	335	—	400	—	—	580	—
Absorption coefficient K (cm^{-1})	2.65	2.95	—	3.34	—	—	1.38	—
Half width W (eV)	1.36	0.92	—	0.74	—	—	0.33	—
$N f/10^{16} = 1.14\, K\, W$	4.1	3.1	—	2.8	—	—	0.5	—

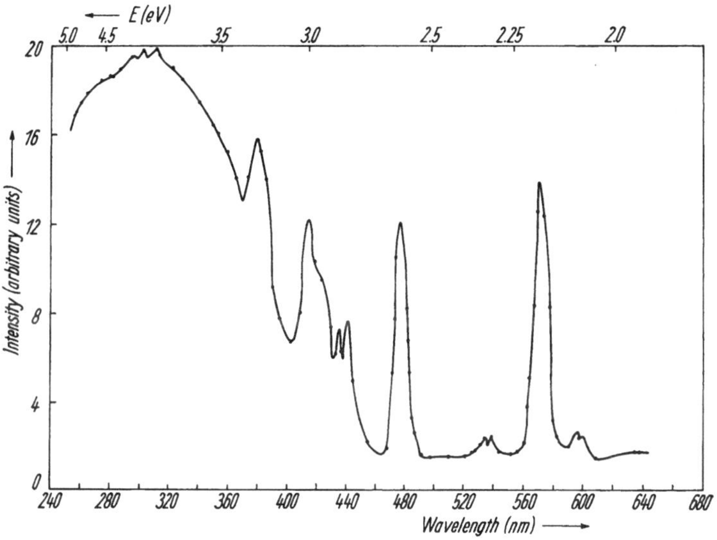

Fig. 4. The microphotometric curve of the X-ray fluorescence spectrum of a CaF_2 crystal as photographed by a quartz spectrograph

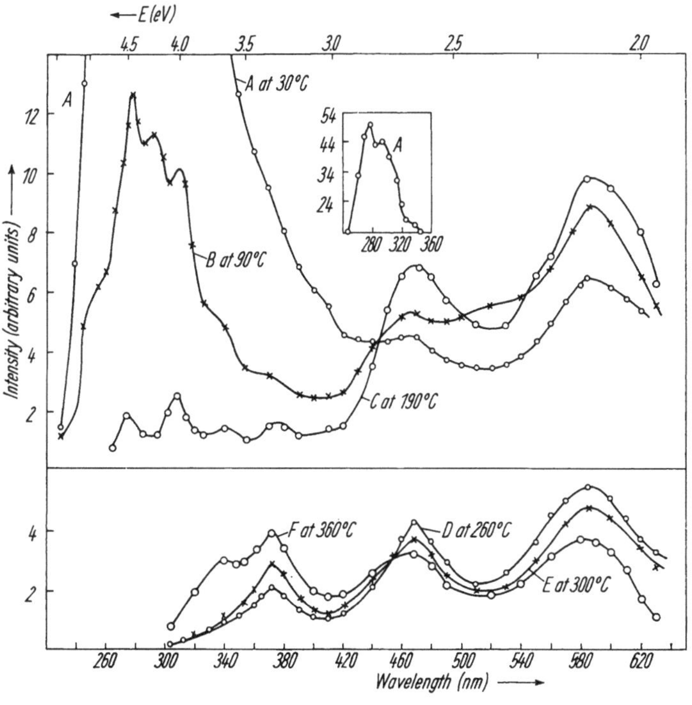

Fig. 5. The X-ray fluorescence spectrum of a CaF_2 crystal at different temperatures as scanned by a monochromator and a photomultiplier. Inset shows the peak intensity of the room temperature spectrum

a comparatively large slit width (to gain adequate intensity) and recorded by a photomultiplier and a recorder is shown in Fig. 5 (curve A). The intensities in the latter case have been corrected for the nonlinear dispersion of the monochromator and the spectral sensitivity of the photomultiplier. Curves B to F of Fig. 5 are the spectra obtained by irradiation at different temperatures.

The essential features of the X-ray fluorescence spectra of this crystals are:

1. The spectra can be broadly divided into four regions — region I from 200 to 300 nm, region II from 300 to 400 nm, region III from 400 to 500 nm, and region IV from 500 to 650 nm.

2. The room temperature luminescence emission is very strong in the 300 to 400 nm; compared to this, emission in other regions is smaller by a factor of 10.

3. Up to 200 °C, with rise in temperature, the emission in the 200 to 300 nm and the 300 to 400 nm regions gradually decreases in intensity, while the emission in the 400 to 500 nm and the 500 to 650 nm regions gradually increase in intensity.

4. Above 200 °C, the emission in the 200 to 300 nm is practically absent; the emission in the 300 to 400 nm region gains in intensity while that in the 400 to 500 nm and 500 to 650 nm regions decreases in intensity with rise in temperature.

5. The luminescence yield retains a measurable value even at 360 °C, the temperature of bleaching of the four band absorption spectrum.

4. Discussion

4.1 Absorption spectra and colour centres

It has been established that commercial crystals of this substance still contain traces of rare earth and other impurities. Working with very pure and activated crystals Görlich [6, 7] and O'Connor [8] have shown that the four band spectrum is due to Yttrium. It has also been shown by O'Connor [8] and Arends [9] that commerical crystals actually contain traces of Yttrium in addition to alkali, alkaline earth and other elements. From the optical and electron spin resonance spectra of the Y^{2+} ions formed by the reduction of the Y^{3+} ions during X-irradiation, the four band spectrum is interpreted as that of the 4 d^1 electron transitions of the Y^{2+} ions.

The spectrum obtained by electrolytic colouration of these commercial crystals is different from that of the X-ray colouration spectrum. Further, the concentration of the colour centres in the former case is many times larger than that produced by X-ray colouration, the concentration of the colour centres being 10^{16} and 10^{17} cm^{-3}, respectively (Table 1). The former consists of the α- and β-bands superimposed on the four band spectrum of X-ray colouration. The formation of the four band spectrum along with the α- and β-bands in the electrolytically coloured crystal incidentally shows that not only the α- and β-bands, but also the bands of the four band spectrum are due to trapped electrons. All these centres can be thermally bleached by heating the crystal to about 400 °C. This result should be taken to emphasize that some chemical decomposition has taken place depositing neutral calcium atoms at dislocations and other defect sites and also on the surface. This is supported by the findings of Bontinck [11, 12] in additively coloured crystals. Naturally a similar effect is expected to occur in X-irradiation and bleaching (especially at elevated temperatures). Similar effects have been observed by Aerts et al. [13] in NaCl. The discussion of the results of the electrical measurements [14] is consistent with this picture, where a) a large increase in the dc conductivity and the dielectric loss of the X-irradiated and bleached crystal and b) a very low activation energy in the range of 30 to 80 °C of the bleached crystal are observed.

The formation of the α- and β-bands has so far been reported in additive and electrolytic colouration only. The present work shows that their formation, which has not been observed so far in X-ray colouration, would be possible by X-irradiation also. Of course, these α- and β-bands could be produced by X-irradiation, not always, but only if, the temperature of irradiation is in the range 180 to 250 °C. The formation of the α-band could not be detected at ordinary temperatures of X-irradiation for the following reasons:

1. Since the 385 nm band (the α-band) in the additively and electrolytically coloured crystal has been identified to be the F-band in this crystal (Arends [9]), its absence in X-ray colouration at ordinary temperatures can be attributed to the very small number, or the absence of free negative ion vacancies. In other words the anti-Frenkel defects normally present in this crystal (Ure [15]) are mostly in the associated form as revealed by the dielectric loss measurements [14], the number of centres associated with free negative ion vacancies is too small to be detected in absorption.

2. This seems to indicate that the oscillator strength for this band which might have been formed by X-irradiation at room temperature is rather low. From the temperature dependence of the magnetic susceptibility before and after colouration, and the optical absorption data of Table I, the oscillator strength associated with this band has been estimated to be of the order of 10^{-2} (results of magnetic measurements [14]).

It is known that anti-Frenkel defects are normally present in this crystal, and a large number of such defects are also formed by X-irradiation at all temperatures. But it seems that they are in the associated form. There should be, therefore, a temperature range in which such paired anti-Frenkel defects become dissociated producing free vacancies and interstitials. X-irradiation in that temperature range is expected to produce the corresponding colour centres. Results of experiments on the colouration of these crystals at elevated temperatures reported earlier show that the α-band can be produced along with the Y^{2+} impurity bands, and the same results could be obtained in a more pronounced manner by electrolysis. Hence at higher temperatures of irradiation, when these associated anti-Frenkel defects break up, the α-centres can be formed in adequate concentration so that the α-band could be detected in absorption in spite of its small oscillator strength.

The origin of the β-band and the other long wavelength bands observed in electrolytic colouration has not been well understood. The relative prominence of these bands and their bleaching characteristics seem to depend on the thermal history of the sample and the method of formation of the colour centres; their bleaching characteristics also vary widely. This may probably be due to some form of colloidal particles of small size. Such particles have been observed by Bontinck in additively coloured crystals [11].

4.2 Fluorescence spectra

The phenomenon of luminescence of this crystal under X-ray irradiation exhibits so many complexities (on account of the presence of a number of impurities) that one cannot reasonably expect to identify the specific emission mechanisms and the electronic processes that operate inside the crystal under the various conditions studied in this investigation. However, a critical analysis of the different observations leads to some general conclusions regarding the broad nature of the emission mechanisms and the electronic processes involved therein.

A large number of electrons and holes are produced inside the crystal during X-irradiation; this number is practically independent of the temperature of the crystal. Emission may take place either by direct excitation of the emitting centres or by the electron hole recombination at suitable lattice sites. Since the crystal does not show measurable luminescence or absorption under UV excitation, it may be concluded that the luminescence under X-ray irradiation is mostly due to the radiative recombination of the electrons and holes at the various emitting centres. The luminescence efficiency of such mechanisms is likely to decrease with rise in temperature.

1. The sharp lines and the luminescence emission in the 200 to 300 nm region are not present in thermoluminescence [16]; this emission which is present in X-ray luminescence decreases in intensity with rise in temperature. These seem to have their origin in the electron hole recombination at the trivalent rare earth impurity ion sites for sharp lines [17] and at the lattice and divalent impurity ion sites for the broad bands [18] in the region of 200 to 300 nm. The large intensities in this region are due to the availability of a large number of recombination sites and centres.

2. The observed increase in the emission intensity in the 400 to 500 and 500 to 650 nm regions, with rise in temperature, under X-ray irradiation, is due to the availability of a large number of free holes for recombination with the electrons of the Y^{2+} ions. At lower temperatures, electrons and holes produced during X-irradiation are quickly trapped in the trapping states before radiative recombination can take place. At higher temperatures (up to 200 °C), such trapping states are thermally unstable and so they are available for radiative recombination. This seems to be the reason for the increase in the intensity of emission in this temperature range for the 400 to 500 and 500 to 650 nm regions.

3. In the temperature range 260 °C and above, the Y^{2+} centres get mostly bleached or become thermally unstable. Thus the intensity of emission due to the recombination of the holes with the Y^{2+} electrons falls in intensity with the rise in temperature. The decrease in the quantum efficiency of radiative recombination may also contribute to the fall in intensity.

4. The emission in the 300 to 400 nm region, however, shows an increase in intensity above 260 °C. At this stage free holes and electrons are present in the crystal; the emission in this region seems to be due to the recombination of such electrons and holes at special lattice defect sites generated by X-rays. The measurements of the dc conductivity and dielectric loss [14] indicate the generation of unassociated anti-Frenkel defects in the crystal by X-rays at these temperatures [14]. It seems natural to assume that the emission in the 300 to 400 nm region is due to the recombination of holes and electron at the lattice vacancies or interstitial ions. If this mechanism for the emission in this region be true, the observed low emission intensities in the region of 30 to 200 °C should mean that such anti-Frenkel defects, i.e., unassociated negative ion vacancies and interstitials are not formed under X-irradiation below 200 °C. The decrease in the intensity of emission is mostly indicative of the predominance of non-radiative recombination with rise in temperature. This is also borne out by the fact the α-band cannot be obtained by X-irradiation unless the temperature of the crystal is around 200 °C.

Acknowledgement

The author hereby wants to express his gratefulness to Prof. H. N. Bose, Head of the Department of Physics for suggesting the problem, for his guidance, and keen interest in the work.

References

[1] E. Mollwo, Nachr. Ges. Wiss. Göttingen **79** (1934).
[2] A. Smakula, Phys. Rev. **77**, 408 (1950); Z. Phys. **138**, 276 (1954).
[3] F. Lüty, Z. Phys. **134**, 596 (1953).
[4] W. Bontinck, Physica **24**, 639 (1958).
[5] W. J. Scouler and A. Smakula, Phys. Rev. **120**, 1154 (1960).
[6] P. Görlich, H. Karras, and R. Lehmann, phys. stat. sol. **1**, 389 (1961); **3**, 98 (1963).
[7] P. Görlich, H. Karras, W. Ludke, H. Mothes, and R. Reimann, phys. stat. sol. **3**, 478 (1963).
[8] J. R. O'Connor and J. H. Chen, Phys. Rev. **130**, 1790 (1963).
[9] J. Arends, phys. stat. sol. **7**, 805 (1964).
[10] V. A. Arkhangelskaya, Ya. E. Kariss, and P. P. Feofilov, Soviet Phys.-Optics and Spectroscopy **18**, 184 (1964).
[11] W. Bontinck and W. Dekeyser, Physica **22**, 595 (1956).
[12] W. Bontinck and S. Amelinckx, Phil. Mag. **2**, 94 (1957).
[13] E. Aerts, S. Amelinckx, and W. Dekeyser, Acta metall. **7**, 29 (1959).
[14] V. V. Ratnam, phys. stat. sol. **16**, 549 (1966).
[15] W. Ure, J. chem. Phys. **26**, 1363 (1957).
[16] V. V. Ratnam and H. N. Bose, phys. stat. sol. **15**, 309 (1966).
[17] H. Haberlandt, B. Karlik, and K. Przibram, Anz. Akad. Wiss. Wien, 7th Dec. 1933 and 11th Jan. 1934; S. B. Akad. Wiss. Wien **143**, 151 (1934); **144**, 77 (1935); **144**, 135 (1935).
[18] K. Przibram, Anz. Akad. Wiss. Wien, 3rd Dec. 1936; Z. Phys. **107**, 709 (1937).

(Received May 16, 1966)

phys. stat. sol. 16, 569 (1966)

Laboratoire d'Électrostatique et de Physique du Métal, Grenoble

Interactions between Domain Walls in Coupled Films[1] [2]

By

F. BIRAGNET, J. DEVENYI, G. CLERC, O. MASSENET, R. MONTMORY, and A. YELON

A study is made of the domain-wall structure of coupled uniaxial ferromagnetic films separated by a thin non-magnetic layer. It is found that the walls in the two films are not always superimposed. After demagnetization in the hard direction they are separated but tend to cross frequently. After demagnetization in the easy direction, they may be separated and parallel or they may be superimposed. An explanation of this behaviour in terms of wall–quasi-wall interactions is considered.

Nous avons étudié les structures de parois dans des ensembles de deux couches ferromagnétiques uniaxiales séparées par un film très mince de matériau non magnétique. Il existe alors un couplage tendant à aligner les aimantations des deux couches extrêmes. On observe que les parois de chaque couche ne sont en général pas superposés. Après désaimantation par un champ appliqué dans la direction de difficile aimantation, elles sont faiblement séparées et se croisent fréquemment. Après désaimantation dans la direction facile, elles peuvent être soit superposées soit séparées et parallèles. Une explication de ce comportement faisant intervenir des interactions entre parois et pseudo-parois est envisagée.

1. Introduction

We present here the results of a study by Lorentz microscopy of the behaviour of the domain walls in two ferromagnetic films, separated by a non-magnetic intermediate layer. It has been found [1] that when the intermediate is sufficiently thin a coupling field tends to align the magnetizations of the two ferromagnetic films parallel.

The first ferromagnetic film has a composition Ni 81% – Fe 19%, the second may be composed of the same material, or of Ni–Fe–Co (45%–10%–45%). If the two films are of different composition, they may have coercive forces which are sufficiently different that one film may be reversed but not the other. We may distinguish them, during observation, by a difference in the contrast of the walls (Ni–Fe–Co has a higher magnetization and therefore a stronger contrast). The intermediate layers may consist of chromium, gold, or SiO, with thicknesses ranging from 20 to 100 Å.

These samples have uniaxial properties, and we can easily determine the axes by observing the behaviour of the ripple when a magnetic field is applied to the sample. This permits us to apply a field along the easy or hard axes. We find that the results obtained by demagnetization with an alternating field are the

[1]) This work has been performed under D.R.M.E. contract No. 64–34–063–00–480–75–01.

[2]) Contribution to the International Colloquium on Magnetic Thin Films, 25th to 28th April in Jena.

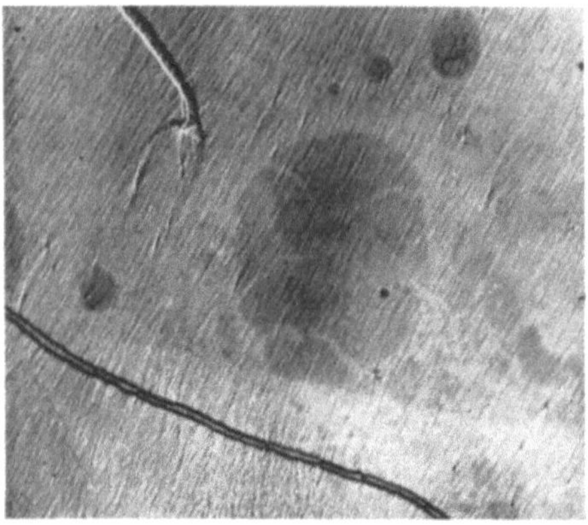

Fig. 1. Domain wall produced by demagnetization along the hard direction

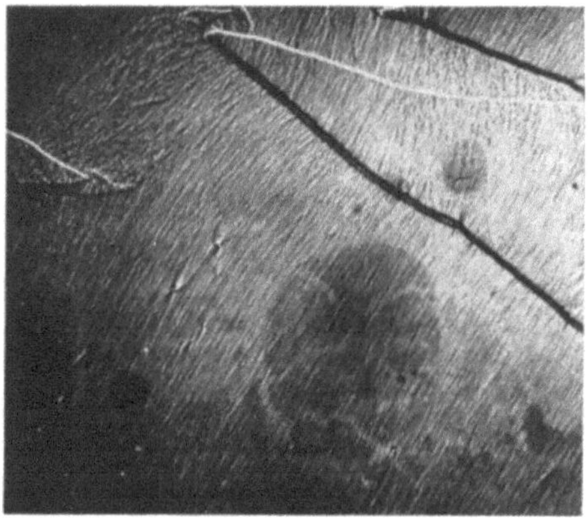

Fig. 2. Domain wall produced by demagnetization along the easy direction

same as those obtained with a continuous field, and so we will not make any distinction between the two methods.

Let us consider a sample consisting of Ni–Fe (250 Å), Cr (50 Å), Ni–Fe–Co (250 Å). The result of a demagnetization of this sample along the hard direction is shown in Fig. 1. We observe a pair of domain walls with black images, clearly separated and crossing regularly along their length. If we perform the same experiment with a field along the easy direction of the film, as shown in Fig. 2, we find that the walls are separated in certain regions and superimposed in others, and that the walls rarely cross. Further, the distance between the walls is not constant and may be substantially larger than in the case of hard-axis demagnetization. In both cases, the distance by which the walls may be separated increases rapidly with the thickness of the intermediate layer.

2. Wall Separation

In a previous publication [2], we have proposed a measurement of the separation between a pair of coupled walls, supposing that the magnetizations of the two films in the region between the walls are antiparallel. It was then shown that the distance between black-wall images and the total width of white-wall images are equal and that these remain constant as the defocalisation varies; this represents the real inside edge to inside edge separation of the walls.

For a series of films of Ni–Fe/Cr/Ni–Fe–Co the separation of the walls as a function of the thickness of the Cr layer is shown in Fig. 3. For a thickness of 20 Å the separation between walls is already of the order of 2000 Å.

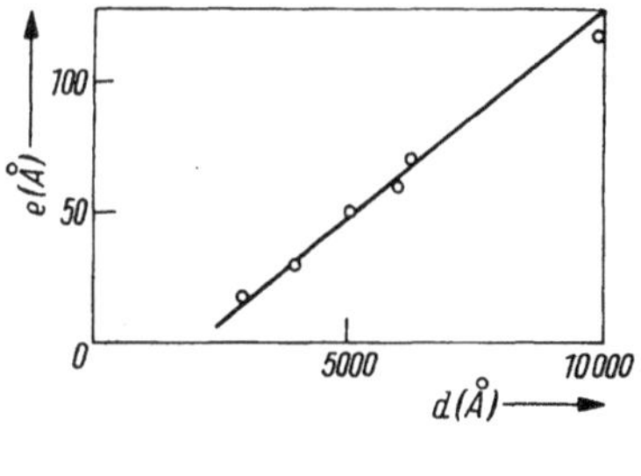

Fig. 3. Variation of the separation of a pair of walls as a function of the intermediate layer thickness (Ni-Fe/Cr/Ni-Fe-Co)

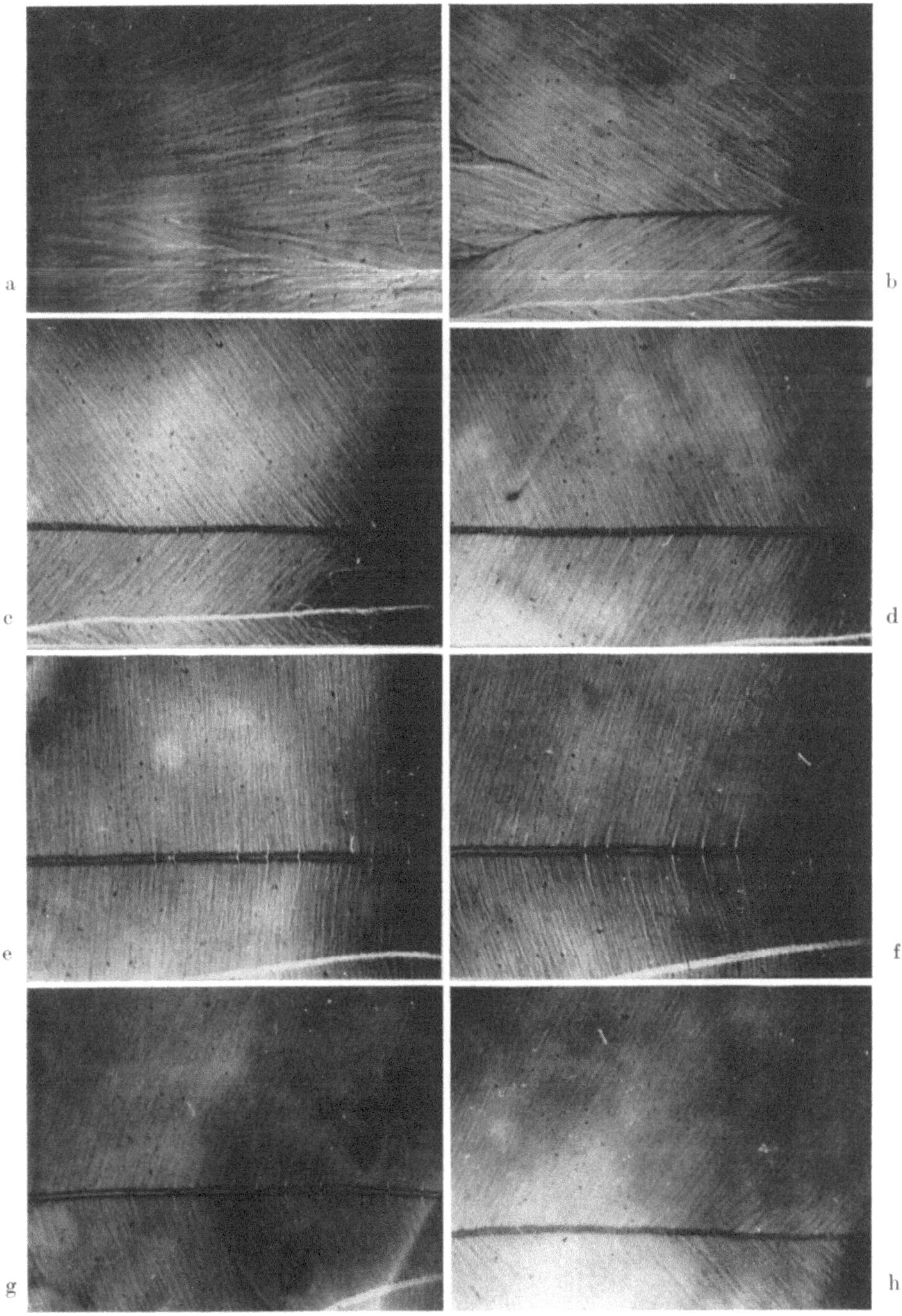

Fig. 4. Behaviour of a pair of walls when a field is applied in the hard direction. a) i_H = 200 mA, b) i_H = = 150 mA, c) i_H = 100 mA, d) i_H = 60 mA, e) i_H = 0, f) i_H = −50 mA, g) i_H = −70 mA, h) i_H = −80 mA

For thicknesses varying between 20 and 100 Å the separation increases linearly with thickness. The substitution of Ni–Fe for Ni–Fe–Co or the change of intermediate material does not modify the nature of the phenomenon. However, for SiO intermediates the separation increases more rapidly. This is not surprising as exchange coupling between the magnetic films decreases more rapidly with thickness for SiO than it does for gold or chromium [3].

In any case, the separation is large even for small intermediate thicknesses. In a subsequent section we shall consider a model to explain this unexpected result.

3. Hard-Axis Demagnetization

The results of demagnetizing a film in the hard direction are illustrated in Fig. 4. A film of Ni–Fe/SiO/Ni–Fe is saturated along the hard axis in Fig. 4a. The field is then decreased to zero in Fig. 4b, c, d, and e, and then reversed in Fig. 4f, g, and h. The domain walls appear and disappear, permitting the observation of the switching of the films. Measurement of the separation of the walls shows that after these appear, the separation, initially small, increases, passes through a maximum of about 1 μm at a negative field and then decreases. The walls produced cross regularly. It is likely that the maximum corresponds to the field at which the net spin direction in the walls reverses [4]. In the following sections we shall attempt to analyse the crossings and to explain the variation of the wall separation.

3.1 Wall crossings

If we observe a pair of black walls, as in Fig. 4e, we observe that at each crossing there is a white line, perpendicular to the walls, which is always on the same side of the crossing. This is illustrated in Fig. 5. For white walls, this line is black. The deflection of the electron beam which produces the line seems to be due to a magnetization at the crosssing point, perpendicular to the normal wall direction. If we assume that the magnetizations of the walls are antiparallel at the crossing, it is extremely difficult to explain the effect.

Further, we find that the crossings are not displaced when one applies a field in the hard direction, in contrast to Bloch lines, which are easily moved. This indicates that the wall magnetizations are the same on both sides of the crossing. Finally, the experiment illustrated in Fig. 6 indicates that a wall has the same magnetization direction along its length. The sample consists of Ni–Fe–Co (250 Å), Cr (34 Å), Ni–Fe (250 Å). The wall in Ni–Fe has weak contrast in Fig. 6a, along the length where the crossings occur, when the field is applied in one sense along the hard axis. The contrast is strong in Fig. 6c, when the field is reversed. In Fig. 6b, when $H = 0$, the situation is intermediate. If the magnetization direction of the wall changed at a crossing, as it does at point A (the contrast is reversed in Fig. 6a and 6c) the contrast would alternate at the crossings.

We have not yet been able to determine the angle at which the walls cross, and therefore, have not been able to index the crossings completely. This angle appears larger for the black walls than for the white walls, but this is uniquely a question of electron optics, as the black-wall images are more widely spaced than the white ones.

Fig. 5. Diagram of the crossing of a pair of Néel walls with parallel magnetization

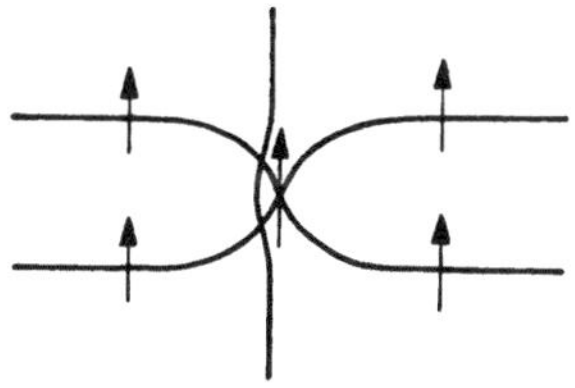

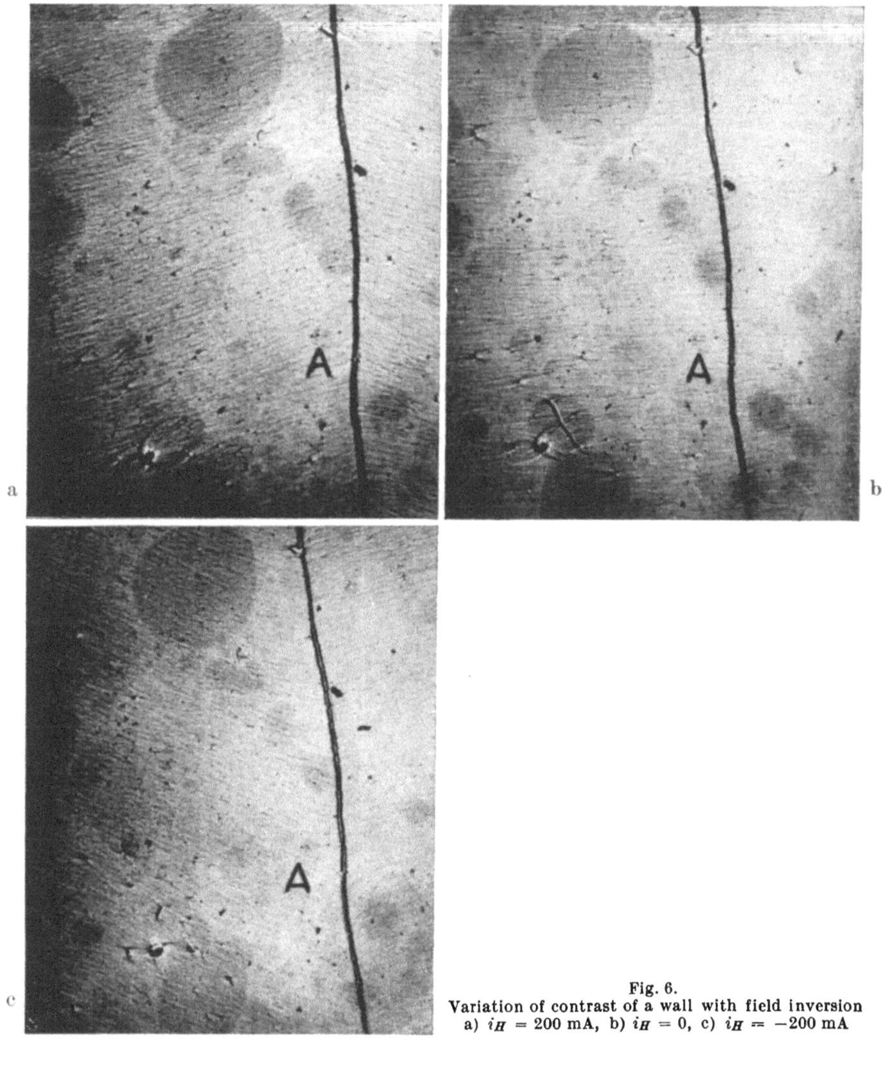

Fig. 6.
Variation of contrast of a wall with field inversion
a) $i_H = 200$ mA, b) $i_H = 0$, c) $i_H = -200$ mA

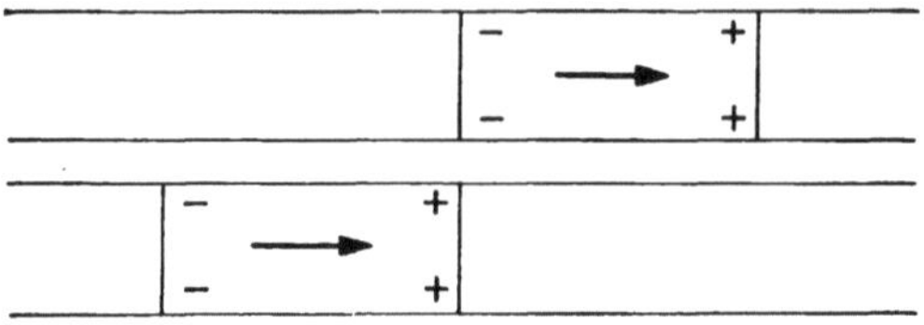

Fig. 7. Theoretical position of two Néel walls with parallel magnetization

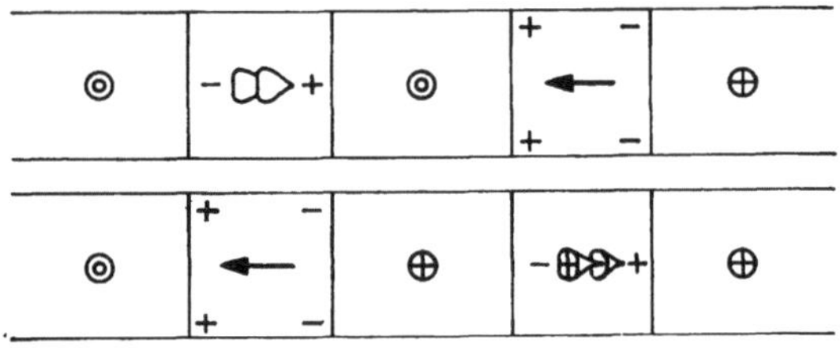

Fig. 8. Model of a pair of Néel walls and of the quasi-Néel walls induced by them

3.2 Model of crossing walls

As a result of the previous considerations, we suppose that the magnetizations of the two walls are parallel and that this direction is constant along their lengths, as shown in Fig. 5. The direction is presumably imposed by the hard-direction field which is applied to produce them. But the magnetostatic interaction between the walls should tend to bring them very close together, as shown in Fig. 7. This is not the case.

One possible explanation of the large separation between walls is that the Néel wall in each film induces a quasi-Néel wall [5] of opposite magnetization in the other film. This is illustrated in Fig. 8. In that case, the wall–quasi-wall interaction within each film will be repulsive, and one may expect a certain equilibrium distance to be established. This interaction will be relatively weak, as the interaction between the wall and quasi-wall will tend to produce flux closure, and leave a low-pole density. The distance will decrease, as we have indicated earlier, when a hard-direction field is applied, as this will provoke a diminution of the pole density and finally the disappearance of the quasi-wall, after which the walls will tend to be very close.

At present, we have no model to explain the crossings; there seem to be no energetic considerations which favour their existence.

4. Demagnetization along the Easy Axis

Fig. 9 shows the effect of applying a field along the easy axis to demagnetize samples of Ni–Fe–Co (250 Å), Cr (50 Å), Ni–Fe (250 Å). We note that in certain regions the walls in the two films are superimposed, indicating that they are antiparallel. In other regions they are separated. The application of a continuos field along the hard axis provokes the superposition of some walls, indicating that they too are antiparallel, while in other regions the walls do not rejoin and are probably parallel, as in the case of hard-axis demagnetization.

For an intermediate layer of thickness 50 Å, the coercive forces of the two films are the same because of the coupling between the films. For slightly greater

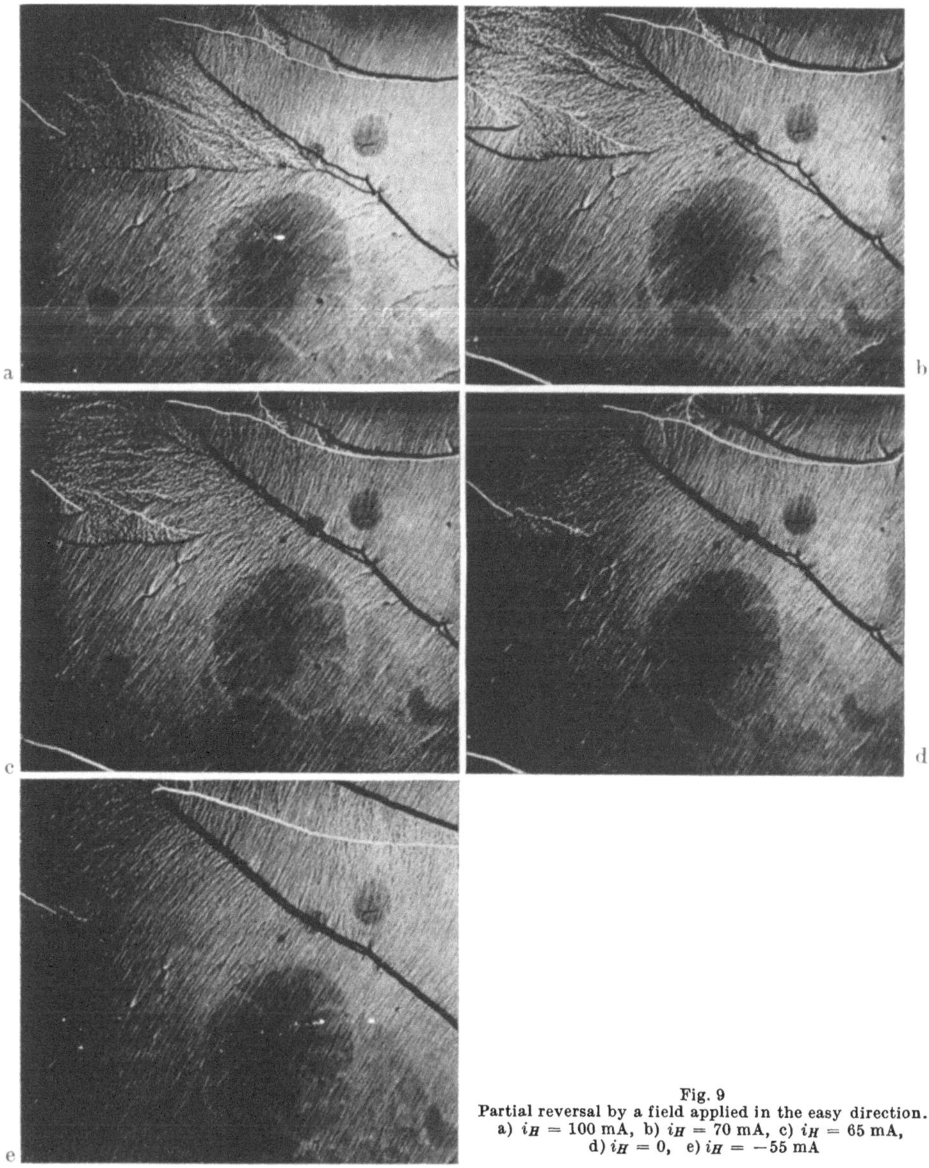

Fig. 9
Partial reversal by a field applied in the easy direction.
a) $i_H = 100$ mA, b) $i_H = 70$ mA, c) $i_H = 65$ mA,
d) $i_H = 0$, e) $i_H = -55$ mA

thicknesses, one can measure two coercive forces. But we can see from Fig. 9 that even at 50 Å some parts of the Ni–Fe film can switch without the Ni–Fe–Co. That is, there is a partial switching of the Ni–Fe film below H_c, although the H_c is the same for both films.

5. Conclusions

We have found that when two films are coupled, their domains are not perfectly superimposed, but their domain walls may be slightly separated. This separation increases with increasing thickness of the intermediate layer. The walls may form in two ways. Their magnetizations may be parallel, and they may cross frequently; this is the situation produced by hard-axis demagnetization. In this case, the separation may be due to wall–quasi-wall interactions. Or, the magnetizations of the two walls may be antiparallel, in which case the walls may be superimposed, or separated, without crossing. This situation is produced by demagnetization in the easy direction. But this last case may also produce walls with parallel magnetization.

References

[1] J. C. BRUYÉRE, O. MASSENET, R. MONTMORY, and L. NÉEL, C. R. Acad. Sci. (France) **258**, 841 (1964); **258**, 1423 (1964).
J. C. BRUYÉRE, G. CLERC, O. MASSENET, D. PACCARD, R. MONTMORY, L. NÉEL, J. VALIN, and A. YELON, IEEE Trans. Magnetics **MAG 1**, 174 (1965).

[2] F. BIRAGNET, J. DEVENYI, G. CLERC, O. MASSENET, L. NÉEL, and A. YELON, to be published in C.R. Acad. Sci. (France.)

[3] J. C. BRUYÉRE, O. MASSENET, R. MONTMORY, and L. NÉEL, IEEE Trans. Magnetics **MAG 1**, 10 (1965).

[4] E. FELDTKELLER, private communication.

[5] J. C. SLONCZEWSKI and S. MIDDELHOEK, Appl. Phys. Letters **6**, 139 (1965).

(Received May 17, 1966)

phys. stat. sol. **16**, 577 (1966)

Institut für Magnetische Werkstoffe der Deutschen Akademie der Wissenschaften zu Berlin, Jena

Zur Theorie des Ferromagnetismus magnetisch anisotroper Schichten mit kubischer Kristallstruktur

II. Berechnung der Curietemperatur für eine Eisenschicht mit (100)-Oberflächenorientierung [1]

Von

W. Brodkorb und W. Haubenreisser

Für eine Eisen-(100)-Schicht werden die ferro- und die paramagnetische Curietemperatur als Funktion der Anzahl der Netzebenen berechnet. Es wird ein Vergleich zwischen den Ergebnissen des Molekularfeld-Modells, der Methode der Greenschen Funktion und experimentellen Daten gegeben.

A calculation is made of the ferromagnetic and the paramagnetic Curie temperature of a Fe-(100)-film as a function of the number of atomic layers. The theoretical results, using a molecular field model, and a Green's function method, are compared with experimental data.

Im Teil I der Arbeit [1] wurden die ferromagnetische und die paramagnetische Curietemperatur in Abhängigkeit von der Schichtdicke für ferromagnetische dünne Schichten mit kubischer Kristallstruktur und (100)-Oberflächenorientierung auf der Grundlage des Heisenberg-Modells berechnet. In dieser Arbeit sollen die in [1] angegebenen Ausdrücke (I, 27) und (I, 32) der ferromagnetischen bzw. paramagnetischen Curietemperatur für eine kubisch raumzentrierte α-Fe-(100)-Schicht diskutiert werden, da die für massives α-Fe nach dem Heisenberg-Modell berechneten magnetischen Eigenschaften gut mit den experimentellen Daten übereinstimmen [2 bis 5].

1. Ferromagnetische Curietemperatur $T_c^f(n)$

Die ferromagnetische Curietemperatur hat nach Formel (I, 27) folgende Gestalt:

$$T_c^f(n) = \frac{2}{3} \frac{S(S+1)}{k_B} \cdot \frac{1}{n} \sum_{\varrho=1}^{n} \left\{ \frac{1}{N} \sum_{\tau=1}^{n} \sum_{h} (T_{\varrho\tau}^{h})^2 \left[\frac{1}{\mathscr{E}_{h\tau}^{(1)}} + \frac{1}{\mathscr{E}_{h\tau}^{(2)}} \right] \right\}^{-1}. \qquad (1)$$

Dabei bedeuten S die Spinquantenzahl pro Atom, n die Anzahl der Netzebenen, k_B die Boltzmann-Konstante und N die Anzahl der Atome in einer Netzebene. h kennzeichnet den Wellenvektor für die Spinanregung innerhalb einer Netzebene. Die Eigenfunktionen $T_{\varrho\tau}^{h}$ ergeben sich aus der Lösung des Eigenwertproblems (I, 16). Dieses Eigenwertproblem läßt sich in analytischer Form nur für $\beta_h = 1$ und einfach kubische Gitter lösen [6, 7]. In allen anderen bekannten Fällen zerfällt das Eigenwertspektrum in sogenannte Raum- und Oberflächenlösungen [8, 9]. Für jeden ebenen Wellenvektor h würde sich eine exakte Lö-

[1]) Teilergebnisse dieser Arbeit sind in [12] enthalten.

sung der Eigenwertgleichung (I, 16) ergeben. In den weiteren Rechnungen werden wir uns bei der Lösung der Eigenwertgleichung auf den Grenzfall $h = 0$ beschränken. In diesem Falle läßt sich β_h bis auf Entwicklungsglieder höherer Ordnung in der Form $\beta_0 = 1 + \frac{3}{4} \frac{n_2^x}{I_2(0)}$ darstellen. β_0 hat für Fe den Wert 1,00031. Aus diesem Grunde können wir den Wert $\beta_0 \approx 1$ verwenden. Diese beiden Näherungen $h = 0$ und $\beta_0 \approx 1$ führen zur Vernachlässigung der Oberflächenlösungen. (Der Beitrag der Oberflächenlösungen zum Eigenwertspektrum läßt sich nur unter großem numerischen Rechenaufwand ermitteln.) Wir erhalten dann folgende Eigenfunktionen und Eigenwerte:

$$\left.\begin{aligned} T^0_{\varrho\tau} &= \left(\frac{2 - \delta_{1\tau}}{n}\right)^{1/2} \cos\left[\frac{\pi}{2n}(\tau - 1)(2\varrho - 1)\right], \\ \bar{x}^0_\tau &= 2\cos\left[\frac{\pi}{n}(\tau - 1)\right], \quad \tau, \varrho = 1, 2, \ldots, n\,. \end{aligned}\right\} \tag{2}$$

Bei Spezialisierung der Ausdrücke $\mathscr{E}^{(1),(2)}_{h\tau}$ (siehe Formel (I, 18a)) in der Beziehung (1) für Fe-(100)-Schichten ($S = 1$) und beim Übergang von der Summation zur Integration über den Wellenvektor h in der ersten Brillouin-Zone ergibt sich

$$\left.\begin{aligned} T^{\mathrm{f}}_{\mathrm{c}}(1) &= \frac{4}{3\,k_{\mathrm{B}}}\, k_{||} \frac{\left(k_{||} + \frac{3}{2} n_1^x\right)}{\left(2\,k_{||} + \frac{3}{2} n_1^x\right)}, \\ T^{\mathrm{f}}_{\mathrm{c}}(n) &= \frac{4}{3\,k_{\mathrm{B}}} \frac{1}{n} \sum_{\varrho=1}^{n} \left\{\sum_{\tau=1}^{n} (T^0_{\varrho\tau})^2\, \Gamma_\tau\right\}^{-1} \quad \text{für } n \geqq 2\,, \end{aligned}\right\} \tag{3}$$

wobei die Abkürzungen

$$\left.\begin{aligned} \Gamma_\tau &= \frac{2}{\pi^2} \frac{1}{I_2(0)} [\mathscr{J}^{(1)}_\tau + \mathscr{J}^{(2)}_\tau]\,, \\ \mathscr{J}^{(i)}_\tau &= \int_0^{\pi/2}\!\!\int \frac{\mathrm{d}y\,\mathrm{d}z}{1 - \gamma^{(i)}_\tau \cos y \cos z}\,, \quad i = 1, 2 \end{aligned}\right\} \tag{4}$$

verwendet wurden. Für die $\gamma^{(i)}_\tau$ erhalten wir folgende Näherungsausdrücke:

$$\left.\begin{aligned} &\gamma^{(1)}_1 = 1 - \frac{3}{4} \frac{n_1^x + 2\,n_2^x}{I_2(0)}\,, \quad \gamma^{(2)}_1 = 1 - \frac{k_{||}}{2\,I_2(0)}\,, \\ &\gamma^{(1)}_\tau = \gamma^{(2)}_\tau \equiv \gamma_\tau = \cos\left[\frac{\pi}{n}(\tau - 1)\right] \quad \text{für } \tau = 2, 3, \ldots, n\,. \end{aligned}\right\} \tag{5}$$

Dabei werden für Fe die Daten benutzt: $I_2(0) = 4\,I$, $I = 3{,}76 \times 10^{-14}$ erg (Austauschintegral für Nächstnachbarwechselwirkung [5, 10], $n_1^x + 2\,n_2^x = 2{,}768 \times \times 10^{-16}$ erg (Beitrag der Dipol-Dipol-Wechselwirkungsenergie nach (I, 12b) für die x-Komponente in Richtung [100] der Schichtebenennormalen). Für die einachsige Volumenanisotropiekonstante $k_{||}$ bei 0° K wurde ein Wert gewählt, der einem Anisotropiefeld von $H_{\mathrm{a}} = 100$ Oe entspricht. Dabei soll die Zuordnung $k_{||} = 2\,\mu_{\mathrm{B}}\,H_{\mathrm{a}}$ gelten, wo μ_{B} das Bohrsche Magneton bedeutet. Die Dipolsummen (I, 10b) für $h = 0$ wurden nach der Methode von Born und Bradburn [11] berechnet: $m^x_{\varrho\varrho}(0) \equiv n_1^x = 10{,}474\,\frac{\mu^2}{a^3}$, $m^x_{\varrho\,\varrho\pm1}(0) = n_2^x = 4{,}172\,\frac{\mu^2}{a^3}$, $m^x_{\varrho\,\varrho\pm2}(0) = = -\,0{,}328\,\frac{\mu^2}{a^3}$, wobei $a = 2{,}86 \times 10^{-8}$ cm (Gitterkonstante für Fe) und $\mu = 2\,\mu_{\mathrm{B}}$

bedeuten. Mit diesen Werten kann man die Größenordnung des Fehlers abschätzen, der bei Berücksichtigung von Dipolwechselwirkungen nur zwischen benachbarten Schichtebenen auftritt: $|n_2^x| \approx 40\%$ von $|n_1^x|$ und $|m_{\varrho\varrho\pm2}^x(0)| \approx 3\%$ von $|n_1^x|$.

Die Durchführung der Integration in Formel (4) liefert für die Γ_τ die folgenden Ausdrücke:

$$\Gamma_1 = \frac{1}{4\pi I}\left[2\ln 2 - \ln\frac{3}{16}\frac{(n_1^x + 2n_2^x)}{I} - \ln\frac{k_{||}}{8I}\right] \quad \text{für} \quad \tau = 1\,,$$

$$\Gamma_\tau = \begin{cases} \dfrac{4}{\pi I}\cdot\dfrac{F(k_\tau^{(1)},\varphi_\tau^{(1)}) - F(k_\tau^{(1)},\varphi_\tau^{(0)})}{\sqrt{1+\gamma_\tau(6+\gamma_\tau)+4\sqrt{\gamma_\tau}(1+\gamma_\tau)}} & \text{für} \quad \gamma_\tau > 0\,, \\ \dfrac{1}{4I} & \text{für} \quad \gamma_\tau = 0\,, \\ \dfrac{2}{\pi I(1-\gamma_\tau)}\left[F\left(k_\tau^{(2)},\dfrac{\pi}{2}\right) - F(k_\tau^{(2)},\varphi_\tau^{(2)})\right] & \text{für} \quad \gamma_\tau < 0 \end{cases} \tag{6}$$

mit $\tau = 2, 3, \ldots, n$, wobei die Abkürzungen

$$\varphi_\tau^{(0)} = \arcsin\left[\frac{1+\gamma_\tau(6+\gamma_\tau)+4\sqrt{\gamma_\tau}(1+\gamma_\tau)}{(1+\gamma_\tau)\,\Omega_\tau^{(0)}}\right]^{1/2},$$

$$\varphi_\tau^{(1)} = \arcsin\left[1 - \frac{(1-\gamma_\tau)^2}{2(1+\gamma_\tau)(1+\sqrt{\gamma_\tau})}\right]^{1/2},$$

$$\varphi_\tau^{(2)} = 2\,\mathrm{arcctg}\left[\frac{\Omega_\tau^{(2)}}{(1-\gamma_\tau)}\right]^{1/2},$$

$$\Omega_\tau^{(0)} = \begin{cases} 3+\gamma_\tau+4\sqrt{\gamma_\tau} \\ 2+\dfrac{4}{\pi}+\gamma_\tau\left(2-\dfrac{4}{\pi}\right)+4\sqrt{\gamma_\tau}\,, \end{cases} \qquad \Omega_\tau^{(2)} = \begin{cases} 2 \\ \dfrac{4}{\pi}+1+\gamma_\tau\left(\dfrac{4}{\pi}-1\right), \end{cases}$$

$$(k_\tau^{(1)})^2 = \frac{8\sqrt{\gamma_\tau}(1+\gamma_\tau)}{1+\gamma_\tau(6+\gamma_\tau)+4\sqrt{\gamma_\tau}(1+\gamma_\tau)}\,, \qquad (k_\tau^{(2)})^2 = -\frac{4\gamma_\tau}{(1-\gamma_\tau)^2}$$

benutzt wurden. $F(k,\varphi)$ ist das elliptische Integral erster Gattung in der Legendreschen Normalform. Bei der Integration ist zu beachten, daß durch die Substitutionen $\mathrm{tg}\, y/2 = t$ und $\mathrm{tg}\, z/2 = s$ die in (4) auftretenden Integrale $\mathcal{J}_\tau^{(i)}$ sich in Integralausdrücke transformieren lassen, deren Integrationsbereich das quadratische Gebiet $-1 \leqq t,\ s \leqq 1$ ist. Bei der weiteren Auswertung der Integrale wurde dieses quadratische Gebiet einmal in der Arbeit [12] durch ein Kreisgebiet mit dem Radius 1 (entspricht den oberen Werten für $\Omega_\tau^{(0)}$ und $\Omega_\tau^{(2)}$) und zum anderen in dieser Arbeit durch einen volumengleichen Kreisbereich (entspricht den unteren Werten für $\Omega_\tau^{(0)}$ und $\Omega_\tau^{(2)}$) ersetzt.

In der unteren Kurve der Fig. 1 (bezeichnet als „sphärisches Modell") haben wir die Abhängigkeit der ferromagnetischen Curietemperatur (3) als Funktion der Anzahl n der Netzebenen dargestellt. (Dabei ist zu bemerken, daß sich die in (6) angeführten Näherungen bei der Berechnung der Größen Γ_τ im Rahmen unserer Rechengenauigkeit nur geringfügig auf den Kurvenverlauf der ferromagnetischen Curietemperatur auswirken.) Man sieht, daß eine meßbare Abhängigkeit von der Schichtdicke $D = (a/2)(n-1)$ erst für $n \lesssim 10$ auftritt. Für den Grenzübergang $n \to \infty$ und bei der Forderung der Translationssymmetrie senkrecht zur Schichtoberfläche $\left(\text{d. h. } (T^{(0)}_{\varrho\tau})^2 \to \frac{1}{n}\right)$ ergibt sich unter Verwendung der Methode von Watson [13]

$$\left.\begin{aligned} T^{\mathrm{f}}_{\mathrm{c}}(\infty) &= \frac{1}{3}\,\frac{S\,(S+1)}{k_{\mathrm{B}}}\,I_2(0)\left[\frac{K^2(k_1)+K^2(k_2)}{\pi^2}\right]^{-1}, \\ k_1^2 &= \frac{1}{2}\left[1-\sqrt{\frac{3}{2}\left(\frac{(n_1^x+2\,n_2^x)}{I_2(0)}\right)}\right], \quad k_2^2 = \frac{1}{2}\left[1-\sqrt{\frac{k_{||}}{I_2(0)}}\right]. \end{aligned}\right\} \tag{7}$$

Dabei bedeutet $K(k)$ das vollständige elliptische Integral erster Gattung.

Mit den oben angegebenen Werten für Fe erhalten wir für die ferromagnetische Curietemperatur des massiven Materials $T^{\mathrm{f}}_{\mathrm{c}}(\infty) = 1056$ °K. Bei $n = 1$ ergibt sich nach Formel (3) der Wert $T^{\mathrm{f}}_{\mathrm{c}}(1) = 0{,}018$ °K. Werden die Anisotropieeffekte in (7) vernachlässigt, so vereinfacht sich diese Formel auf den bekannten Wert $T^{\mathrm{f}}_{\mathrm{c}}(\infty) = 1043$ °K des sphärischen Modells. Dieses Modell liefert für $n = 1$ exakt den Wert $T^{\mathrm{f}}_{\mathrm{c}}(1) = 0$ °K.

2. Paramagnetische Curietemperatur $T^{\mathrm{p}}_{\mathrm{c}}(n)$

Die paramagnetische Curietemperatur hat nach (I, 32) folgende Form:

$$^{(1)}T^{\mathrm{p}}_{\mathrm{c}}(n) = \frac{1}{3}\,\frac{S\,(S+1)}{k_{\mathrm{B}}}\,\frac{1}{2\,n\,N}\sum_{\tau=1}^{n}\sum_{h}\{\mathscr{E}^{(1)}_{h\tau}+\mathscr{E}^{(2)}_{h\tau}\} \tag{8}$$

und ist der nach der Molekularfeldtheorie in erster Näherung berechneten Curietemperatur äquivalent. Bei der Durchführung der h-Summation kann man (8) in der Gestalt

$$^{(1)}T^{\mathrm{p}}_{\mathrm{c}}(1) = \frac{1}{3}\,\frac{S\,(S+1)}{2\,k_{\mathrm{B}}}\left(2\,k_{||}+\frac{3}{2}\,n_1^x\right), \tag{9}$$

$$\begin{aligned} ^{(1)}T^{\mathrm{p}}_{\mathrm{c}}(n) = \frac{1}{3}\,\frac{S\,(S+1)}{2\,k_{\mathrm{B}}}\,\frac{1}{n}\sum_{\tau=1}^{n}\Bigg[&4\,I_2(0)+2\,k_{||}+2\,n_2^x+\frac{3}{2}\,n_1^x+ \\ &+\left(\frac{n_2^x}{2}-2\left(\frac{2}{\pi}\right)^2 I_2(0)\right)\bar{x}^0_\tau\Bigg] \quad \text{für} \quad n \geqq 2 \end{aligned} \tag{10}$$

schreiben. Dabei wurde die Näherung

$$\frac{1}{N}\sum_{h}\left[-2\,I_2(h)+\frac{n_2^x}{2}\right]\bar{x}^h_\tau \approx \bar{x}^0_\tau\left[-2\left(\frac{2}{\pi}\right)^2 I_2(0)+\frac{n_2^x}{2}\right]$$

verwendet, wo

$$I_2(h) = I_2(0)\cdot\cos\left[\frac{h_y\,a}{2}\right]\cdot\cos\left[\frac{h_z\,a}{2}\right]$$

die Fouriertransformierte des Austauschintegrals für die Nächstnachbarwechselwirkung zwischen Atomen einer Netzebene zu denen einer benachbarten Netzebene darstellt. Der durch diese Näherung erhaltene Faktor $\left(\frac{2}{\pi}\right)^2$ kann bei exakter Behandlung des Eigenwertspektrums (I, 16) modifiziert werden. Unter Verwendung der Beziehung $\frac{1}{n}\sum_{\tau=1}^{n}\bar{x}^0_\tau = \frac{2}{n}$ kann man dann den Ausdruck (10) in der Form

$$\frac{{}^{(1)}T_c^p(n)}{{}^{(1)}T_c^p(\infty)} = 1 - \frac{\omega}{n} \quad \text{für} \quad n \geqq 2 \tag{11}$$

schreiben. Dabei bedeuten

$${}^{(1)}T_c^p(\infty) = \frac{1}{3}\,\frac{S\,(S+1)}{2\,k_B}\left[4\,I_2(0) + 2\,k_{||} + 2\,n_2^x + \frac{3}{2}\,n_1^x\right] \tag{12}$$

die paramagnetische Curietemperatur des massiven Materials bei Berücksichtigung der einachsigen Volumenanisotropie und der Formanisotropie und

$$\omega = \left[2\left(\frac{2}{\pi}\right)^2 I_2(0) - \frac{n_2^x}{2}\right]\cdot\frac{1}{3}\,\frac{S\,(S+1)}{k_B\,{}^{(1)}T_c^p(\infty)}\,. \tag{13}$$

Für Fe ergibt sich aus (12) der Wert ${}^{(1)}T_c^p(\infty) = 1454$ °K und aus (13) $\omega = 0{,}404$. In der Molekularfeldtheorie erhält man bei Vernachlässigung der Anisotropieeffekte für ω den Wert 1. Diese Unterschiede sind auf die oben erwähnte h-Summation zurückzuführen, da in der Molekularfeldtheorie für dünne Schichten [16] nur Spinanregungen mit dem Wellenvektor $h = 0$ betrachtet werden. Für $n = 1$ ergibt sich aus (9) ${}^{(1)}T_c^p(1) = 0{,}57$ °K und für den isotropen Fall exakt ${}^{(1)}T_c^p(1) = 0$ °K.

Wenn wir die Gleichung (I, 30) für die paramagnetische Suszeptibilität durch Iteration bis zu Gliedern T^{-3} lösen, wobei wir uns auf den isotropen Fall beschränken und die Näherung $(T^h_{\varrho\tau})^2 \to \frac{1}{n}$ benutzen, erhalten wir die der Molekularfeldtheorie in 2. Näherung äquivalente Curietemperatur

$${}^{(2)}T_c^p(n) = \left[1 - \left(\frac{2}{\pi}\right)^2\frac{1}{n}\right]\frac{{}^{(1)}T_c^p(\infty)}{2}\left[1 + \sqrt{1 - \frac{\left(1 - \frac{8}{n^2}\cdot\left(\frac{2}{\pi}\right)^4\right)\Delta}{\frac{1}{4}\left({}^{(1)}T_c^{(p)}(\infty)\right)^2\left(1 - \left(\frac{2}{\pi}\right)^2\frac{1}{n}\right)^2}}\,\right] \tag{14}$$

mit

$$\Delta = \left[\frac{1}{3}\,\frac{S\,(S+1)}{k_B}\right]^2\cdot\frac{[I_2(0)]^2}{2}\,.$$

Für $n = 1$ folgt ${}^{(2)}T_c^p(1) = 0$ °K und für $n \to \infty$ der Wert ${}^{(2)}T_c^p(\infty) = 1240$ °K Die weitere Durchführung der Iteration liefert eine Reihenentwicklung der paramagnetischen Suszeptibilität $\chi(T)$ nach T^{-1} im Sinne von Kramers und Opechowski [14], wobei sich die Curietemperatur als größte Wurzel aus der Gleichung für die inverse Suszeptibilität $\chi^{-1}(T_c^f) = 0$ ergibt. Die paramagnetische Curietemperatur ${}^{(k)}T_c^p$ folgt als größte Wurzel aus der Gleichung $\chi_k^{-1}({}^{(k)}T_c^p) = 0$, wo der Index k die Ordnung der Iterationsnäherung bezeichnet und angibt, daß die Reihenentwicklung von χ beim Glied $T^{-(k+1)}$ abgebrochen wurde. Der Nachteil der Kramers-Opechowski-Methode zur Bestimmung von T_c^f besteht darin, daß der exakte numerische Wert von $\lim\limits_{k\to\infty} {}^{(k)}T_c^p \to T_c^f$ nicht ermittelt werden kann [15].

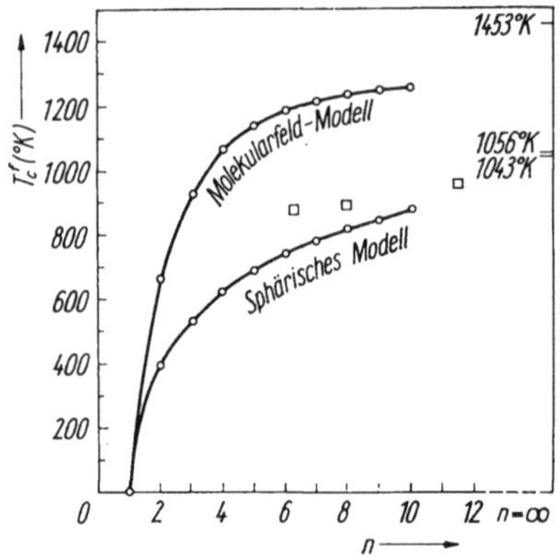

Fig. 1. Abhängigkeit der ferromagnetischen Curietemperatur $T_c^f(n)$ von der Anzahl der Netzebenen n für Fe-(100)-Schichten (□ experimentelle Werte nach Lee, Bolduc und Violet [17], welche mit der Technik des Mößbauer-Effektes erhalten wurden).

3. Diskussion der Ergebnisse

In Fig. 1 werden unsere Ergebnisse über die Schichtdickenabhängigkeit der ferromagnetischen Curietemperatur mit experimentellen Werten [17] und den Resultaten des Molekularfeld-Modells für dünne Schichten verglichen. Die obere Kurve in Fig. 1 stellt die Ergebnisse des Molekularfeld-Modells in erster Näherung für eine magnetisch isotrope Fe-(100)-Schicht dar, wenn man die durch die Säkulardeterminante

$$\left\| s_{ik} - 8 \frac{T_c^{MF}(n)}{{}^{(1)}T_c^p(\infty)} \delta_{ik} \right\| = 0 \,, \quad i, k = 1\,, 2, \ldots, n \tag{15}$$

definierten und von Valenta [16] berechneten größten Eigenwerte $x_n = \frac{T_c^{MF}(n)}{{}^{(1)}T_c^p(\infty)}$ und für ${}^{(1)}T_c^p(\infty)$ den im Grenzfall $n \to \infty$ aus der Säkulardeterminante folgenden Wert ${}^{(1)}T_c^p(\infty) = \frac{2}{3\,k_B} \cdot 8\,I = 1453$ °K benutzt. Dabei kennzeichnet s_{ik} die Anzahl der nächsten Nachbarn eines Atoms der Netzebene i in der Netzebene k, die Zahl 8 die Koordinationszahl des kubisch raumzentrierten Fe und δ_{ik} das Kronecker-Symbol.

Für $n = 1$ folgt aus der Säkulardeterminante (wegen $s_{11} = 0$) der Wert $T_c^{MF}(1) = 0$ °K. Auf den oben erwähnten Sachverhalt bezüglich der Curietemperatur ${}^{(1)}T_c^p(\infty)$ hat bereits Pearson [18] für den Fall des einfach kubischen Gitters hingewiesen. Die mit dem Molekularfeld-Modell berechneten Curietemperaturen für α-Fe haben oberhalb 1173 °K keine physikalische Bedeutung mehr, da die kubisch raumzentrierte α-Phase des reinen Fe in diesem Temperaturbereich in die kubisch flächenzentrierte γ-Phase des Fe übergeht. Aus diesem Grunde sind die mit der Methode der Greenschen Funktionen berechneten ferromagnetischen Curietemperaturen für reines α-Fe physikalisch sinnvoller. Trotzdem ist ein quantitativer Vergleich mit den in der Fig. 1 angegebenen experimentellen Ergebnissen nicht möglich, weil die zur Zeit vorliegenden experimentellen Daten unzureichend sind.

Literatur

[1] W. BRODKORB, phys. stat. sol. **16**, 225 (1966).
[2] G. S. RUSHBROOKE und P. J. WOOD, Mol. Phys. **1**, 257 (1958).
[3] J. GAMMEL, W. MARSHALL und L. MORGAN, Proc. Roy. Soc. **A 275**, 257 (1963).

[4] S. Arajs und R. V. Colvin, J. appl. Phys. **35**, 2424 (1964).
[5] S. Spooner und B. L. Averbach, Phys. Rev. **142**, 291 (1966).
[6] W. Döring, Z. Naturf. **16a**, 1146 (1961).
[7] A. Corciovei, Acta phys. Polon. **24**, 91 (1963).
[8] E. Abbel, Z. Naturf. **18a**, 371 (1963).
[9] R. Jelitto, Z. Naturf. **19a**, 1567 (1964); **19a**, 1580 (1964).
[10] R. I. Joenk, Phys. Rev. **130**, 932 (1963).
[11] M. Born und M. Bradburn, Proc. Cambridge Phil. Soc. **39**, 104 (1943).
[12] W. Brodkorb und W. Haubenreisser, Preprint 1.1 to Internat. Colloquium Magnetic Thin Films, Jena, April 1966.
[13] G. N. Watson, Quaterly J. Math. **10**, 266 (1938).
[14] W. Opechowski, Physica **4**, 181 (1937); **6**, 1112 (1938).
[15] H. A. Brown und J. M. Luttinger, Phys. Rev. **100**, 685 (1955).
[16] L. Valenta, phys. stat. sol. **2**, 112 (1962).
[17] E. L. Lee, P. E. Buldoc und C. E. Violet, Phys. Rev. Letters **13**, 800 (1964).
[18] J. J. Pearson, Phys. Rev. **138**, A 123 (1965).

(Received May 17, 1966)

phys. stat. sol. 16, 585 (1966)

Solid State Physics Department, S.C.K.-C.E.N., Mol

Determination of the Displacement Vector at the Anti-Phase Boundaries in Rutile by Contrast Experiments in the Electron Microscope

By

J. VAN LANDUYT [1])

Contrast experiments are used to determine by transmission electron microscopy and electron diffraction the displacement vector associated with the anti-phase boundaries in rutile. The criteria derived elsewhere [3] for the contrast at π-boundaries are applied, and the conditions for extinction of the contrast are also used to determine the vector. A vector of the type 1/6 ⟨211⟩ is found, which is in disagreement with the 1/2 ⟨101⟩ vector previously proposed [1, 2]. The method used here is general and could be used to find the displacement vector at a stacking fault or an anti-phase boundary in other materials.

Kontrastexperimente werden benutzt, um den Verschiebungsvektor für Gegenphasengrenzen in Rutil durch Elektronenmikroskopie und Elektronenbeugung zu bestimmen. Die Kriterien, die an anderer Stelle [3] für den Kontrast an π-Grenzen hergeleitet wurden, und die Bedingungen für die Löschung des Kontrasts werden für die Bestimmung des Vektors benutzt. Ein Vektor vom Typ 1/6 ⟨211⟩ wird gefunden, der nicht mit dem früher [1, 2] vorgeschlagenen 1/2 ⟨101⟩-Vektor übereinstimmt. Die hier benutzte Methode ist allgemein gültig und könnte zur Bestimmung des Verschiebungsvektors eines Stapelfehlers oder einer Gegenphasengrenze in anderen Materialien benutzt werden.

1. Introduction

A detailed study of anti-phase boundaries in rutile was made in a previous paper [1]. From the observations described there it was concluded that the displacement vector of these boundaries was presumably of the type 1/2 ⟨101⟩ in accord with a proposal initially made by Ashbee and Smallman [2]. Our conclusion was based on the fact that all of the observed fringe patterns exhibited characteristics typical for $\alpha = \pi$ [3]. However, recently Eikum and Smallman [4] presented evidence which ruled out the possibility that the displacement vector should be of the type 1/2 ⟨101⟩. It was therefore worth while to make an attempt to determine the displacement vector unambiguously using different reflections. It turned out that by chance in all our previous observations, diffraction vectors of the type ⟨110⟩ were used, which do give rise to π-fringes. This diffraction vector was automatically favoured because of the high quality fringe pattern it produces and also because of the preferred orientation of the foils which are usually perpendicular to ⟨110⟩.

---

[1]) Also: Solid State Physics Laboratory, Rijksuniversitair Centrum, Antwerpen.

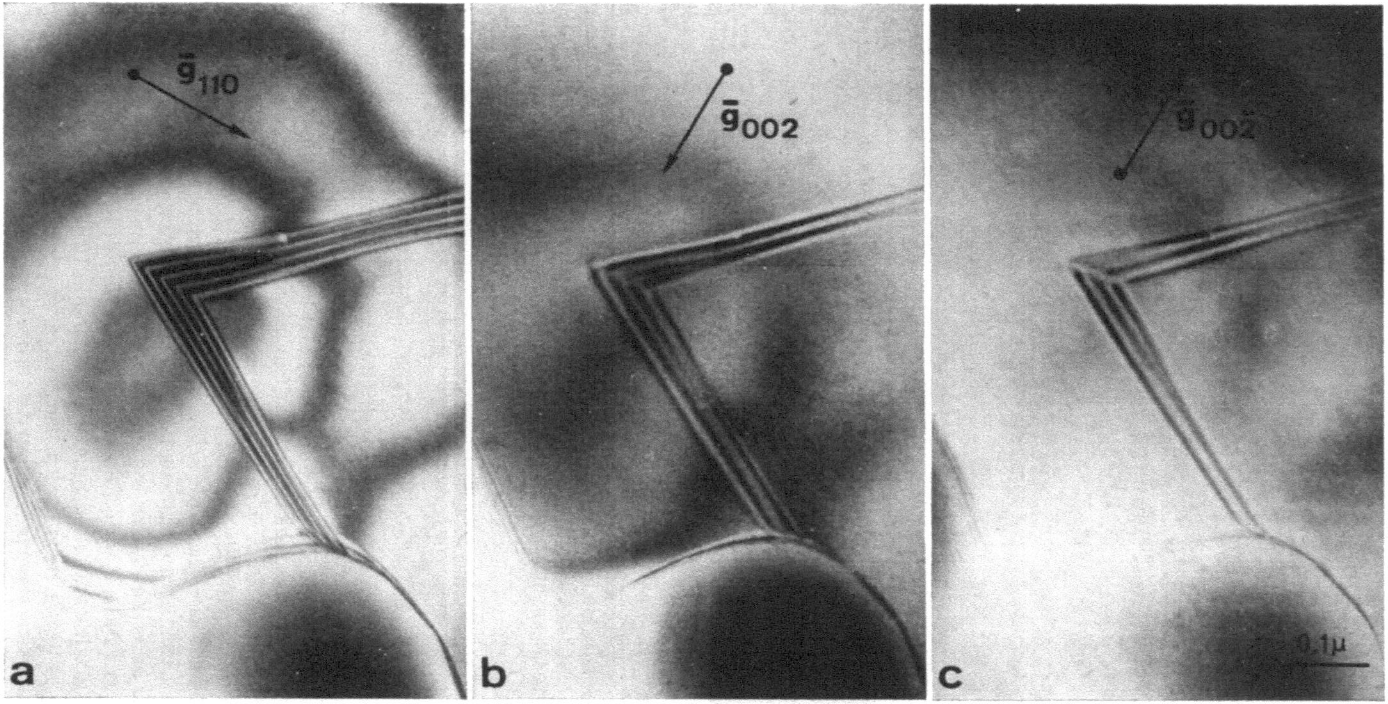

Fig. 1. An anti-phase boundary is shown which presents
a) π-type contrast for $\boldsymbol{g}_{(110)}$ (note the continuity of the fringes at the bend fault and the small spacing of the fringes), b) a contrast different from π-type for $\boldsymbol{g}_{(002)}$, c) the reversal of fringe characteristics for $\boldsymbol{g}_{(00\bar{2})}$

2. New Observations

The new observations are summarized in Table 1 and the photographs illustrating the observations are reproduced as Fig. 1, 2, 3, and 4.

The criteria used to decide whether or not α was equal to π are those derived in [3]; the simplest ones are the continuity of the fringes in a closed loop of boundary, the fringe spacing for $s = 0$ (= 1/2 extinction distance), and the behaviour of the fringes with changing thickness.

Table 1

Diffraction g-vector	Observation	Figure
[101], [210]	$\alpha = \pi$ possible	2
[110]	$\alpha = \pi$ (always)	1, 3, 4
[002], [00$\bar{2}$]	$\alpha \neq \pi$ (presumably $= \pm 120°$)	1, 3
[111], [211]	$\alpha \neq \pi$ (one set of faults out of contrast)	2
[321], [131]	sometimes extinction	4
[004]	uniform contrast or very broad fringes	3

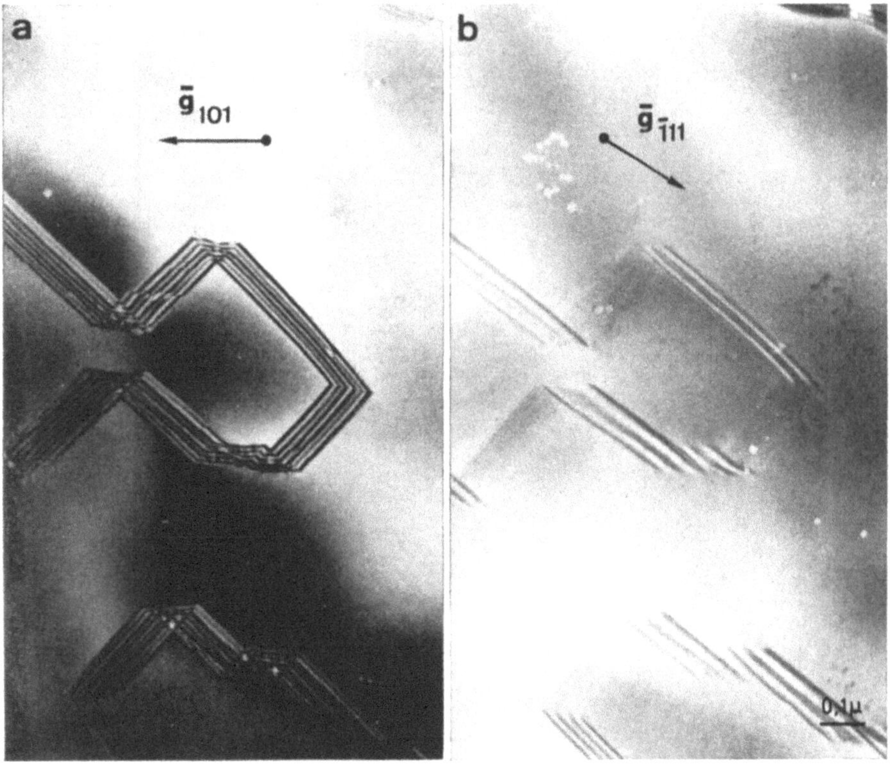

Fig. 2. Anti-phase boundary in rutile showing
a) π-type contrast for $g_{(101)}$, b) extinction of one of the components for a $g_{(111)}$ vector. This observation excludes the 1/6 ⟨212⟩ vector as a possible displacement vector

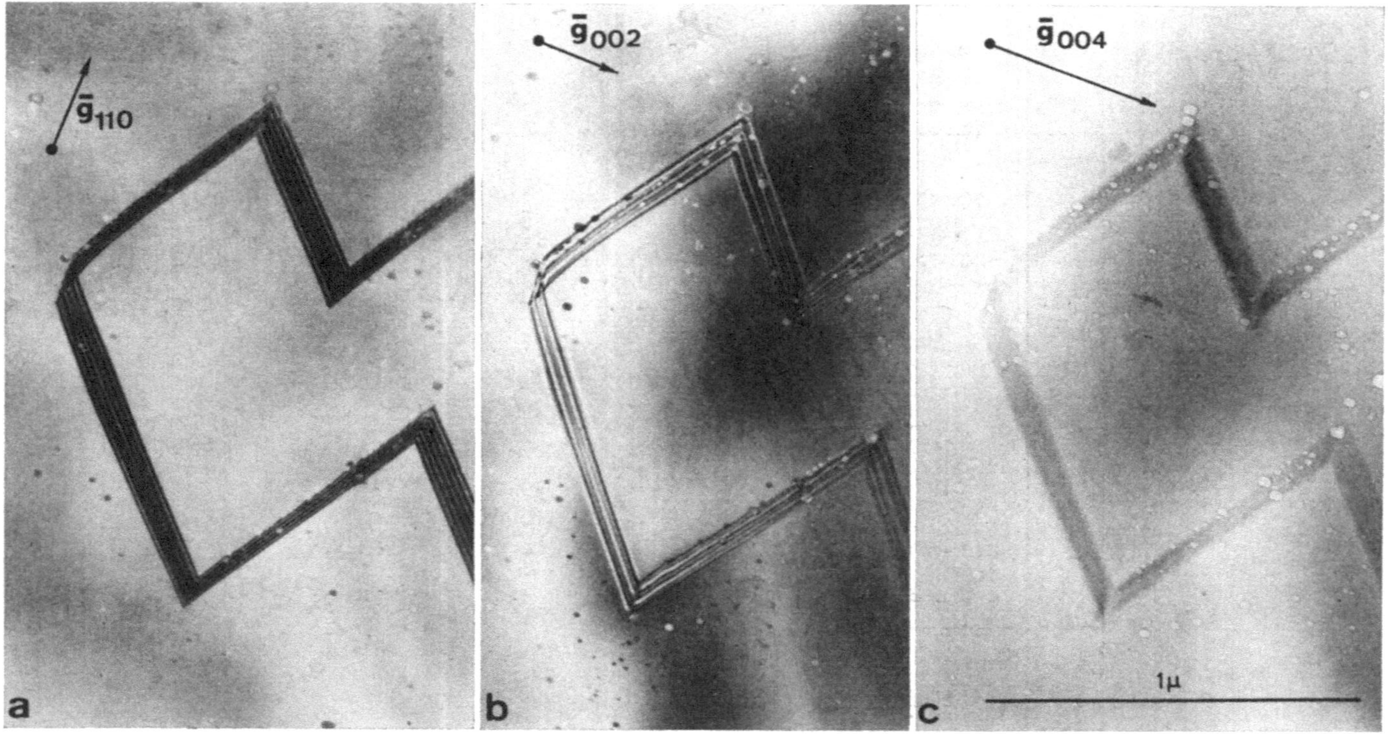

Fig. 3. Anti-phase boundary in rutile

a) $\boldsymbol{g}_{(110)}$, b) $\boldsymbol{g}_{(002)}$, c) $\boldsymbol{g}_{(004)}$. Although both vectors b) and c) should give rise to a contrast of the same type, i.e. $\alpha = \frac{2\pi}{3}$ (see table), the images have a completely different aspect. In the text it is shown that this is due to the difference in extinction distance

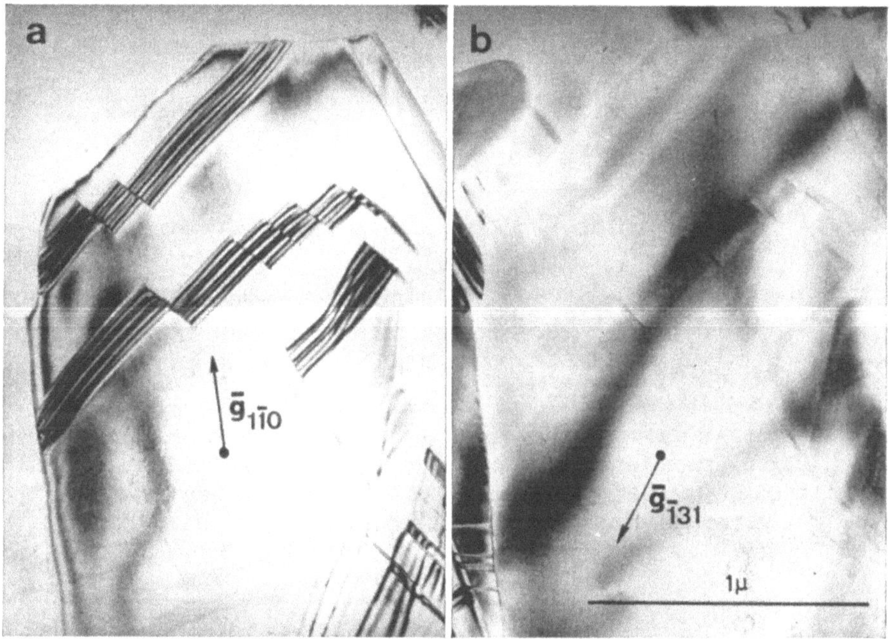

Fig. 4. Two observations are shown illustrating the contrast for a) $g_{(1\bar{1}0)}$, b) $g_{(\bar{1}31)}$ (extinction of one set)

The pronounced difference in aspect of the fringe pattern for [002] and for [004] is due to the large difference in extinction distance: $t_{002} = 662$ Å and $t_{004} = 1975$ Å. Since the average thickness of our foils is around 4000 Å, one would observe about 7 fringes for $\boldsymbol{g}$ = [002] whereas only two broad fringes would be produced for $\boldsymbol{g}$ = [004].

3. Interpretation

We shall write the displacement vector $\boldsymbol{R}$ as $\boldsymbol{R} = 1/n\ [u\ v\ w]$, where u and v are in units a/n and w in units c/n; the diffraction vector is $\boldsymbol{g} = [h\ k\ l]$, where h and k are in units $1/a$ and l in units $1/c$. The phase angle α is then $\alpha = \frac{2\pi}{n}(h\,u + k\,v + l\,w)$, where u, v, w and h, k, and l are integers.

In particular for $\boldsymbol{g} = \langle 002 \rangle$ we have $\alpha = \frac{4\pi w}{n}$. Since for this reflection $\alpha \neq 0, \pi, 2\pi$ we can conclude that w cannot be a multiple of 1/2 or 1/4, i.e. we must exclude $n = 2$ and $n = 4$.

From the fact that always π-contrast is observed for $\langle 110 \rangle$ we conclude that

$$\alpha = \frac{2\pi}{n}(u + v) = \pi \pm 2\,k\,\pi\,.$$

For integers u and v we cannot obtain an odd multiple of π if $n = 3$ or 5, which means that these values for n can be excluded as well. The next simplest proposal is $n = 6$.

From the fact that π-contrast is observed for $\langle 110\rangle$, $\langle 011\rangle$, and $\langle 210\rangle$ follows with a vector $\boldsymbol{R}$ of the type $\boldsymbol{R} = 1/6\,[u\ v\ w]$

$$\frac{2\pi}{6}[\pm u \pm v] = \pm (2 n_1 + 1)\pi ,$$

$$\frac{2\pi}{6}[\pm v \pm w] = \pm (2 n_2 + 1)\pi ,$$

$$\frac{2\pi}{6}[\pm 2u \pm v] = \pm (2 n_3 + 1)\pi ,$$

where n_1, n_2, and n_3 are integers. The simplest solution results for

$$u + v = 3 , \quad v + w = 3 , \quad \text{and} \quad 2u - v = 3 ,$$

i.e., for $u = 2$, $v = 1$, and $w = 1$, or $u = 2$, $v = 1$, and $w = 2$.

In order to distinguish between these two possibilities the characteristics of the fringe pattern due to further diffraction vectors have to be taken into account. A comparison between the characteristics of the pattern to be expected for the two displacement vectors for different active diffraction vectors, and the actually observed pattern is presented in Table 2.

Table 2

$\boldsymbol{g}$	Character for $\boldsymbol{R} = \frac{1}{6}\langle 211\rangle$	Character for $\boldsymbol{R} = \frac{1}{6}\langle 212\rangle$	Observations
[002]	$\alpha = \pm 120°$	$\alpha = \pm 120°$	$\alpha \mp 180°$
[111]	extinction possible	*no* extinction possible	extinction possible
[211]	extinction possible	extinction possible	extinction possible
[101]	$\alpha = 0°, 180°$, or $120°$	$\alpha = 0°, 180°, 120°$	$\alpha = 180°$ or $\mp 180°$
[321]	extinction possible	extinction possible	extinction possible
[131]	extinction possible	*no* extinction possible	extinction possible
[004]	$\alpha = \pm 120°$	$\alpha = \pm 120°$	$\alpha \mp 180°$

From this comparison it is clear that the vector should be $1/6\,\langle 211\rangle$ since two of the observations are in conflict with the alternative assumption.

It should be pointed out that a lattice vector can of course always be added to this vector without changing the conclusion. Although the displacement vector seems to be fairly well established now, we have no detailed picture of the boundary to offer.

References

[1] J. van Landuyt, R. Gevers, and S. Amelinckx, phys. stat. sol. 7, 307 (1964).
[2] K. A. G. Ashbee, R. E. Smallman, and G. K. Williamson, Proc. Roy. Soc. A276, 542 (1963).
[3] J. van Landuyt, R. Gevers, and S. Amelinckx, phys. stat. sol. 7, 519 (1964).
[4] A. Eikum and R. E. Smallman, Internat. Conf. Electron Diffraction and Crystal Defects, Vol. II, Melbourne 1965 (p. D-4).

(Received May 18, 1966)

phys. stat. sol. **16**, 591 (1966)

Department of Physics and Meteorology, Indian Institute of Technology, Kharagpur

Nonradiative Destruction of F-Centres in X-Irradiated KCl Doped with Anionic Impurities

By

M. LUDWIG. MUKHERJEE and H. N. BOSE

A study is made of the thermoluminescent behaviour and room temperature stability of F-centres produced by X-rays in KCl doped with some anionic impurities. A low percentage of KNO_3, KOH, and K_2CO_3 enhances the formation of F-centres but a high percentage inhibits this formation. F-centres in lightly doped samples which are stored in the dark and at room temperature are found to decay with time, in two stages, by some nonradiative processes. The thermoluminescence intensities of these samples are poor. In the case of KNO_3 doping, infrared absorption at 7.1 and 7.7 μm due to $(NO_3)^-$ and $(NO_2)^-$ ions decreases after irradiation by X-rays suggesting decomposition of these complexes. It is suggested that the diffusion of decomposition products with trapped holes is responsible for the decay of F-centres with time at room temperature. The activation energies for the diffusion processes are of the order of 0.98 and 1.05 eV in all the samples. With 1 mol% of KNO_3 doping, the F-centres are almost undetectable. It is estimated that in this case the average distance between one nitrate or nitrite ion and an F-centre is of the order of 12 atomic units.

Die Stabilität von F-Zentren bei Zimmertemperatur und ihr Thermolumineszenzverhalten werden untersucht. Die Zentren werden durch Röntgenbestrahlung von KCl, das mit einigen anionischen Fremdsubstanzen dotiert ist, erzeugt. Ein niedriger Prozentsatz von KNO_3, KOH und K_2CO_3 verstärkt, ein hoher Prozentsatz erniedrigt die Bildung der F-Zentren. In leicht dotierten Proben klingt die Konzentration der F-Zentren bei Zimmertemperatur und im Dunkeln durch einige nichtstrahlende Prozesse in zwei Stufen ab. Die Thermolumineszenz dieser Proben ist schwach. In dotierten Proben (KNO_3) werden durch Röntgenbestrahlung IR-Absorptionsbanden bei 7,1 und 7,7 μm, die durch $(NO_3)^-$- und $(NO_2)^-$-Ionen hervorgerufen werden, abgeschwächt, was auf einen Zerfall der Komplexe hinweist. Es wird angenommen, daß die Diffusion einiger Zerfallsprodukte mit getrapten Löchern für das Abklingen der F-Zentren bei Zimmertemperatur verantwortlich ist. Die Aktivierungsenergien dieser Diffusionsprozesse liegen für alle diese Proben bei 0,98 und 1,05 eV. Mit 1 mol% KNO_3 Dotierung sind fast keine F-Zentren nachweisbar. Es wird vermutet, daß in diesem Falle der mittlere Abstand eines Nitrat- oder Nitrit-Ions vom F-Zentrum von der Größe 12 atomarer Einheiten ist.

1. Introduction

The formation of colour centres and their thermal or optical bleaching characteristics have been investigated by many authors with pure and also impurity activated alkali halides [1]. The presence of intentional, or otherwise, impurities in the crystal has been found to have a profound influence on various properties. Most of the investigations with doped alkali halides have been carried out with cation impurities [2, 3]. Several authors have worked on the photochemical behaviour of the anionic complexes in alkali halide crystals under ultraviolet irradiation [4]. But the effect of anions on the stability of F-centres produced by X-rays has attracted limited attention.

The concentration of F-centres produced by a given dose should depend on the rate of production of F-centers and the rate of destruction by recombination with its positive counterparts (i.e. V-centres) which are also produced during irradiation [5, 6]. The presence of anion impurities can materially affect the rate of electron-hole recombination during the irradiation of the crystal and the equilibrium F-centre concentration for a given X-ray dosage should be perceptibly modified. Hutchinson and Pringsheim have reported the formation of several "new very unstable V-centres" when $KCl:KNO_3$ is X-irradiated at −190 °C [7]. Therefore, experiments on the stability of the F-centres and the corresponding thermoluminescence in alkali halide crystals doped with anion impurities should yield useful informations about the kinetics of the processes involved in the coloration of alkali halide crystals by ionising radiation. The results of investigation on the stability of F-centres and the corresponding thermoluminescence in KCl crystal doped with KNO_3 are reported in this paper.

2. Experimental Procedure

All the samples were grown in our laboratory by the Kyropoulos method from B.D.H. analar grade salts. The samples are $KCl:KNO_3$ (I) (5×10^{-2} mol%), $KCl:KNO_3$(II) (7×10^{-1} mol%), KCl:KOH (5×10^{-1} mol%) and $KCl:K_2CO_3$ (9×10^{-2} mol%) single crystals. Cleaved plates of thickness between 0.5 and 1.0 mm were X-irradiated at room temperature (28 °C) and for comparison of the induced F-band crystal plates of same thickness were always studied. A Philips sealed X-ray tube (Cu-target) was employed for irradiation and a Beckman model DU spectrophotometer for measuring absorptions.

Infrared absorptions of the single crystals of $KCl:KNO_3$(I) before and after X-irradiation were measured with a Baird Infrared Recording spectrophotometer. The samples used were single crystals of dimensions 1 cm × 1 cm × 0.8 cm.

For measuring thermoluminescence of all the samples a RCA IP28 photomultiplier tube was chosen. The temperature of the samples was assumed to be the same as that of the copper sample holder and it is measured by a copper—constantan thermocouple. Both the photomultiplier and the thermocouple response were recorded by the light reflected from two critically damped micro-Moll galvanometers on a rotating drum wrapped with a photographic paper [8]. The thermocouples were previously calibrated by using fixed temperature baths.

3. Experimental Results

The stability of F-centres and the thermoluminescence after X-ray irradiation at room temperature (28 °C) for potassium chloride, pure and doped with different concentrations of potassium nitrate, have been investigated. These crystals show two absorption bands in the infrared region, at 7.1 and 7.7 μm which are presumably due to $(NO_3)^-$ and $(NO_2)^-$ ions. The concentrations of such ions has been estimated with the help of the relation given by Wardzynski [9]

$$N = 3.97\times10^{16}\times K_{max}\, H(\mathrm{cm}^{-3})\,, \tag{1}$$

where K_{max} is the absorption coefficient in cm^{-1}, and H represents the half-width of the absorption band in cm^{-1} (see Table 1). The concentrations of $(NO_3)^-$ and $(NO_2)^-$ ions in the crystal may also be calculated from the amount of KNO_3 added to the melt; the value so obtained is more than twice the amount calculated from the relation given by equation (1). This is in fair agreement with the results reported by Wardzynski [9] for the KCl:KNO_3 system. In the case of KCl:KOH and KCl:K_2CO_3 the concentrations of the impurity ions have been estimated from the amounts added to the melt.

Table 1

Samples	Number of F-centres (10^{16} cm^{-3})	Number of anions calculated from the amount added to the melt (10^{18} cm^{-3})	Number of anions from infrared measurements (10^{18} cm^{-3})
KCl	5.3	—	—
KCl:KNO_3 (I)	8.9	8.4	3.12
KCl:KNO_3 (II)	3.0	120	—
KCl:KOH	12.2	82	—
KCl:K_2CO_3	12.0	16	—

On irradiation with X-rays at 30 kV, 10 mA for 30 min the F-band with a very weak M-band is obtained for pure potassium chloride crystals (Fig. 1). The distance of the sample from the target is nearly 6 cm. The presence of anion impurities in the crystal does not affect the position or half-width of the F-band in any of the cases studied here. The infrared absorption bands at 7.1 and 7.7 μm supposed to be due to $(NO_3)^-$ and $(NO_2)^-$ are considerably reduced in intensity on X-ray irradiation (Fig. 2). When the X-rayed crystals

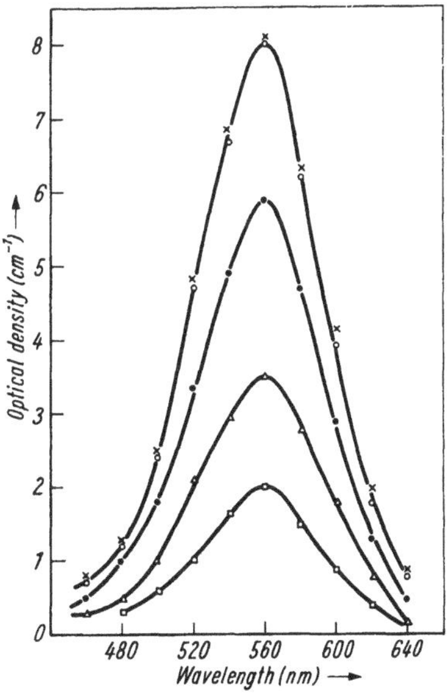

Fig. 1. F-band spectra of pure and impurity doped samples X-rayed at 30 kV, 10 mA for 30 min at room temperature (distance of the samples from the target is 6 cm).
××× KCl : KOH, ○○○ KCl : K_2CO_3, ●●● KCl : KNO_3(I), △△△ KCl "pure", □□□ KCl : KNO_3(II)

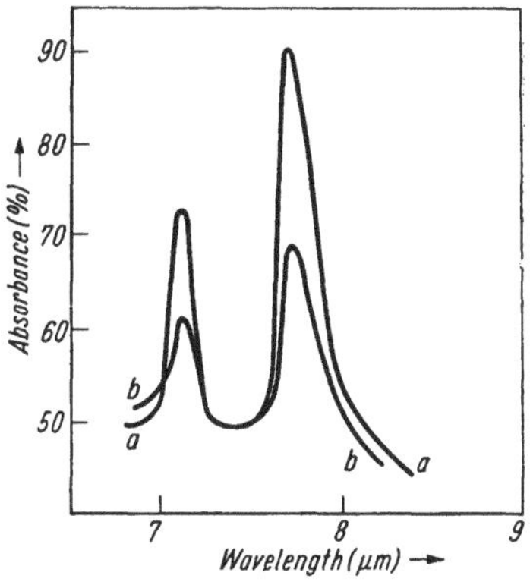

Fig. 2. Infrared absorptions of KCl:KNO_3 (I) (single crystal, thickness 0.8 cm); curve a — before X-raying, curve b — after X-raying (30 kV, 10 mA) for two hours — one hour on each surface. Machlett (target —M_0) X-ray tube has been used.

are irradiated with unfiltered light from an ordinary tungsten filament lamp, the F-band is completely bleached, but the infrared absorption bands do not immediately regain their original intensities. If, however, the bleached crystals are stored at room temperature for about a week, these bands are found to recover most of their initial intensities.

The F-absorption band of various samples under the same dose have been indicated in Fig. 1. The number of F-centres has been calculated from Smakula's formula [5]

$$n = 1.5 \times 10^{16}\,(\mathrm{OD} \cdot \mathrm{cm}^{-1}) \tag{2}$$

where n number of F-centres (cm^{-3}), OD optical density (see Table 1).

The concentrations of F-centres attained with the same dose of X-ray irradiation are found to be higher for $KCl:KNO_3$ crystals than that for pure potassium chloride as long as the concentration of KNO_3 is small. As the KNO_3 concentration in the crystal increases the corresponding F-centre concentration for the same X-ray dosage decreases and when KNO_3 concentrations is greater than 1 mol% the F-band is barely detectable.

Doping the crystals with anion impurities has been found to affect considerably the stability of the F-centres. When the doped crystals are irradiated and stored in the dark at room temperature (28 °C) the F-centre concentration decreases with time. For the pure crystal, there is very little decay of F-centre concentration with time and it seems to attain a stable value within a few minu-

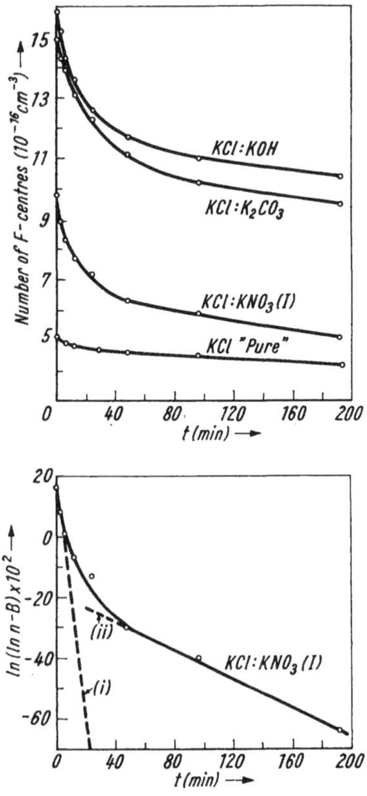

Fig. 3. a) Room temperature (28 °C) bleaching of F-centres with time in KCl:KOH, $KCl:K_2CO_3$, $KCl:KNO_3$(I) and KCl "pure". b) Curve fitting analysis of the room temperature bleaching of $KCl:KNO_3$(I) graph. The slopes (i) and (ii) gives the decay constants α_1 and α_2

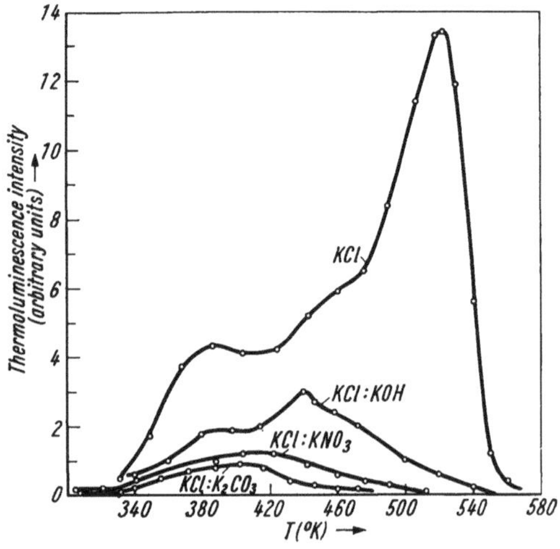

Fig. 4. Thermoluminescence of KCl, KCl:KOH, $KCl:KNO_3$(I), and $KCl:K_2CO_3$ irradiated at room temperature with 30 kV, 10 mA X-rays (KCl is exposed for 15 min, the rest of the samples for 60 min. The rate of heating is 60 °C/s)

tes after the X-ray irradiation at room temperature, whereas for the crystals containing KNO_3, KOH or K_2CO_3 the F-centre concentrations are found to decay over a long period. The results are shown in Fig. 3a. It has been further observed that pure potassium chloride shows a measurable afterglow which can be enhanced by irradiation in the F-band, while in the case of doped crystals the decay of F-centre concentration is not attended by any detectable afterglow. However, both the pure and doped samples luminesce under X-ray irradiation.

Thermoluminescence of KCl, KCl:KNO_3, KCl:KOH and KCl:K_2CO_3 have been measured after X-ray irradiation at room temperature (28 °C). The results are shown in Fig. 4. The glow curve of potassium chloride show three peaks at 386, 460, and 522 °K. Three glow peaks above room temperature have been observed by previous workers [10] and the corresponding trap depths have been estimated as 0.7, 0.83, and 0.95 eV.

The intensity of thermoluminescence of doped samples of potassium chloride is found to be rather poor [4] compared to pure potassium chloride. The doped crystals, in all cases, need about four to five times higher dose of X-ray irradiation than pure potassium chloride before any measurable thermoluminescence can be obtained. This is rather striking in view of the fact that the F-centre concentration in the doped sample is larger than that for pure potassium chloride. It has been further observed that the glow peak at the highest temperature 522 °K is completely suppressed by the presence of KNO_3, KOH, and K_2CO_3 in potassium chloride crystals. The positions of the low temperature peaks do not seem to be much affected by the presence of these anionic impurities.

4. Discussion

The essential results of the present investigation are:

1) Small concentration of anionic impurities enhances, whereas high concentration suppresses the formation of F-centres by X-irradiation at 28 °C.

2) The F-centre concentration of KCl:KNO_3, KCl:KOH, and KCl:K_2CO_3 kept in the dark at 28 °C decays with time; but such decay is not attended with afterglow.

3) The $(NO_3)^-$ and $(NO_2)^-$ centres in KCl:KNO_3 decomposes upon X-ray irradiation.

4) Thermoluminescence yield for the doped crystals is poor compared to that of pure KCl, so much that the high temperature glow peak is completely suppressed by doping.

The enhancement of formation of F-centres in lightly doped NaCl:$NaNO_3$ was reported by Smakula. Similar results were obtained by Hutchinson et al. [7] in case of KCl:KNO_3 X-irradiated at −190 °C.

The isothermal decay of F-centres at room temperature unaccompanied by any afterglow points to some kind of nonradiative process. Since the recombination of thermally released holes or electrons with its counterparts are known to be radiative, it is therefore suggested that the isothermal decay of F-centres is due to the motion of the centres with trapped holes. The diffusion of these centres with trapped holes can take place even at room temperature and when they reach a critical distance [11] from F-centres, they can recombine by some sort of tunnelling process. The presence of unstable V-centres capable of destroying F-centres by tunnelling process in KCl containing $(NO_2)^-$ ions was also suggested by Hutchinson [7] et al.. The decay of the F-centre concentration at room temperature should therefore be controlled by the diffusion mechanism. The rate process is expected to be similar to that in the case of degassing of metals [12] and in such cases the average concentration of diffusing gas follows to a first approximation an exponential law, $m = m_0\,e^{-\alpha t}$, where $\alpha = \frac{\pi^2 D}{h^2}$, h is the distance the centre with trapped hole diffuses before it is annihilated; D is the diffusion constant of the relevant centres.

The decay rate of F-centre concentration may be written as

$$\frac{dn}{dt} = -P \cdot n \cdot m\,,$$

where P is the probability of recombination, n the concentration of F-centres, m the concentration of the centres which can destroy F-centres without producing any detectable glow. Since the F-centre concentration, n, in pure KCl remains practically steady at room temperature, it may be assumed that in doped crystals the F-centre concentration decays only due to non-radiative recombination with the trapped hole centres when these are within interacting distances. Therefore,

$$\frac{dn}{dt} = -P \cdot n \cdot m_0\,e^{-\alpha t},$$

$$\ln n = \ln n_0 + \frac{P}{\alpha}\,m_0\,(e^{-\alpha t} - 1) =$$

$$= A\,e^{-\alpha t} + C\,.$$

When centres of different diffusion coefficients are involved we have

$$\ln n = \sum A_r \exp(-\alpha_r \cdot t) + B\,.$$

They decay curve of F-centre concentration at a fixed temperature is thus expected to be represent a sum of exponential terms plus a constant. The results of the present investigation can be fitted with the relation of the type

$$\ln n = A_1 \exp(-\alpha_1 t) + A_2\,(-\alpha_2 t) + B\,. \qquad (3)$$

The values of α_1 and α_2 are calculated from the slopes (i) and (ii) of the curve of Fig. 3b. An estimate for the diffusion constants can be made assuming the relevant diffusion distances to be of the order of 34 atomic units which is half of the average distance between the anionic impurities. If D_0 is of the order of 1 cm²/s we can find the activation energy from the equation [12] $D = D_0\,e^{-E/kT}$. All these results are given in Table 2. It is interesting to note that the diffusion constants

Table 2

Decay constants, diffusion constants and activation energies associated with two stages of annealing at room temperature ($D_0 \approx 1$ cm²/s)

1s	t stage of annealing	2nd stage of annealing	
α_1	6.6×10^{-4} s^{-1}	α_2	4×10^{-5} s^{-1}
D_1	7.7×10^{-18} cm²/s	D_2	4.7×10^{-19} cm²/s
E_1	0.98 eV	E_2	1.05 eV

estimated in this manner are found to be of the same order of magnitude as is expected for alkali halides at room temperature [13]. The present analysis shows that the destruction of F-centres due to the suggested diffusion process occurs in two stages. Similar two-stage annealing of defects connected with OH-impurities in KCl have been reported by Fröhlich and Grau [14].

Infrared absorption bands at 7.1 μm and 7.7 μm in $KCl:KNO_3$ crystals decrease on X-irradiation, proving thereby that the ions $(NO_3)^-$ and $(NO_2)^-$ are decomposed [7]. Various decomposition products (viz. NO, NO_3^{--}, NO_2^{--}, O_2^+, O_2^-, O^-, N_2^-, N_2^{--}) have been detected by spin resonance experiments [15, 16]. However, the possibilities of diamagnetic ions being produced by X-irradiation cannot be eliminated. It is also to be remembered that the present measurements relating to the isothermal decay of F-centres should give us information about those products which are produced by X-rays but which may be ultimately destroyed by recombination processes. It is therefore rather difficult to identify the diffusing centres responsible for the destruction of the F-centres. Since, however, qualitatively same results are obtained by doping with KNO_3, KOH, and K_2CO_3 it can be assumed that these centres should be associated with the common impurity i.e., oxygen, in these systems.

Thermoluminescence of irradiated alkali halides is known to be due to the radiative recombination of thermally released holes or electrons, the glow peaks representing the various depths. In general, the total intensity of thermoluminescence is closely related to the F-centre concentrations in the alkali halide crystals. The marked decrease in thermoluminescence intensity for potassium chloride due to impurity ions like $(NO_3)^-$, $(CO_3)^{--}$, and $(OH)^-$ in spite of high concentration of F-centres indicates that nonradiative mechanisms are responsible for the destruction of F-centres. The absence of the high temperature glow peak and the marked decrease in the intensity of the other peaks with little effect on their position show that X-ray irradiation can produce in all these doped crystals new unstable centres [7], having higher capture cross-section for holes and such centres are capable of destroying F-centres without any emission. However, a small fraction of the F-centres in impurity doped samples are destroyed by the normal process of radiative recombination of thermally released holes with electrons. This explains why thermoluminescence yield has not been completely quenched in case of impurity doped samples.

It is now understandable why KCl crystal is not easily coloured by X-rays when it contains large amounts of anionic impurities. The approximate concentration of anionic impurities for which almost no F-centre is formed has been estimated as 1 mol%. Assuming a uniform distribution for the impurity anions, the average distance of separation between them would be of the order of 24 atomic units. We are thus led to the conclusion that in this crystal F-centre can be destroyed by a nonradiative recombination with a trapped hole within a distance of the order of 12 atomic units. It is to be pointed out that the trapped holes and electrons are likely to be much closer than this estimated value. X-irradiation produces large concentrations of V-type centres of the type discussed above, within easy interacting distances of the F-centres. It is natural to expect that, in these cases, nonradiative destruction of F-centres by recombination with the V-type centres occurs quickly.

The authors like to thank Dr. J. Sharma for helpful discussions.

Reference

[1] J. H. SCHULMAN and W. D. COMPTON, Color Centers in Solids, Pergamon Press 1963.
[2] E. SONDER and W. A. SIBLEY, Phys. Rev. **140**, A539 (1965).
[3] H. W. ETZEL, Phys. Rev. **87**, 906 (1952).
[4] F. KERKHOFF, Z. Phys. **158**, 595 (1960).
[5] P. V. MITCHELL, D. A. WIEGAND, and R. SMOLUCHOWSKI, Phys. Rev. **121**, 484 (1961).
[6] R. CASLER, P. PRINGSHEIM, and P. YUSTER, J. chem. Phys. **18**, 887, 1564 (1950).
[7] P. PRINGSHEIM, J. chem. Phys. **23**, 369 (1955).
E. HUTCHINSON and P. PRINGSHEIM, J. chem. Phys. **23**, 1113 (1955).
[8] H. N. BOSE, Proc. Phys. Soc. (B) **68**, 249 (1955).
[9] W. WARDZYNSKI, Acta phys. Polon. **17**, 29 (1958).
[10] S. C. JAIN and P. C. MEHENDRU, Phys. Rev. **140**, A 957 (1965).
[11] R. C. FLETCHER and W. I. BROWN, Phys. Rev. **92**, 585 (1953).
[12] P. G. SHEWMON, Diffusion in Solids, McGraw-Hill Book Co. 1963 (p. 18).
[13] D. MAPOTHER, H. CROOKS, and R. MAUER, J. chem. Phys. **18**, 1231 (1950).
[14] F. FRÖHLICH and P. GRAU, phys. stat. sol. **12**, 421 (1965).
[15] C. JACCARD, Phys. Rev. **124**, 60 (1961).
[16] D. SCHOEMAKER and E. BOESMAN, phys. stat. sol. **3**, 1695 (1963).

(Received May 18, 1966)

phys. stat. sol. **16**, 599 (1966)

Laboratoire de Magnétisme et de Physique du Solide, C.N.R.S., Bellevue

Distribution électronique anisotrope dans les semiconducteurs purs au voisinage de l'avalanche

Par

R. Granger

Nous présentons une étude de la distribution électronique dans les semiconducteurs purs soumis à des champs électriques E intenses, permettant de rendre compte du phénomène d'ionisation par choc à travers la largeur de bande interdite ε_i. Dès que la concentration électronique est inférieure à 10^{15} cm^{-3}, les collisions interélectroniques ne suffisent plus pour maintenir la distribution isotrope et la notion de température électronique n'est alors plus valable. D'autres mécanismes de relaxation sont cependant efficaces et nous considérons particulièrement les cas de diffusion par phonons acoustiques ou optiques. Nous montrons que dans le dernier cas, une partie de la distribution (pour des énergies $\varepsilon > \varepsilon_0(E)$), est constituée essentiellement d'électrons dont le moment est dans la direction du champ électrique; ces électrons appelés „balistiques" sont ceux qui, accélérés dans le champ électrique, atteignent une énergie supérieure à ε_i, et causent l'ionisation par choc. Nous séparons la distribution à l'état stationnaire en trois classes: $\varepsilon < \varepsilon_0$, $\varepsilon_0 < \varepsilon < \varepsilon_i$, $\varepsilon > \varepsilon_i$ et calculons les probabilités de transition d'une classe à l'autre. Nous obtenons alors les populations des trois classes en fonction du champ électrique et déduisons enfin la caractéristique courant-tension. Les résultats numériques sont donnés pour plusieurs semiconducteurs, et en particulier le champ de claquage à différentes températures. Si la distribution reste relativement isotrope, la contribution de la partie anisotrope n'en est pas moins efficace, le temps de relaxation des porteurs étant de cinq ordres de grandeur plus faible que leur durée de vie. Nous comparons cette théorie avec les résultats expérimentaux sur InSb.

An analysis is made of the electron distribution function in homogeneous pure semiconductors in high electric fields E, and this is used to interpret experimental results on intrinsic impact ionization across an energy gap ε_i. When the electron concentration is lower than about 10^{15} cm^{-3}, inter-electronic collisions are not dominant and the introduction of an electron temperature is not valid. In these circumstances scattering by acoustic or optical phonons must be considered in detail. In the latter case, a part of the distribution (for energy $\varepsilon > \varepsilon_0(E)$), consists of electrons with momenta along the direction of the electric field; these "balistic" electrons are mainly responsible for the impact ionization, since they are readily accelarated to energies above ε_i. The steady-state electron distribution can be sub-divided in three energy regions: $\varepsilon < \varepsilon_0$, $\varepsilon_0 < \varepsilon < \varepsilon_i$, $\varepsilon > \varepsilon_i$. The transition rates between these regions are calculated. The populations of the three regions are then obtained as functions of the electric field and the current-voltage characteristics are thus deduced. The results are applied to several semiconductors and the breakdown fields are calculated for various temperatures. Although the anisotropy of the distribution is relatively weak, it is by no means negligible, due to a difference of five orders of magnitude between the relaxation time and the lifetime of the carriers. A comparison is made between the theory and experiments on InSb.

1. Introduction

Dans un champ électrique intense, les électrons d'un semiconducteur peuvent acquérir une énergie suffisante pour ioniser, par choc sur des électrons liés, des porteurs de charge supplémentaires, dont la multiplication en „avalanche" conduit à une brusque augmentation de la conductivité au-delà d'un seuil de champ électrique. Ce phénomène revêt un caractère „intrinsèque" ou „extrinsèque" suivant que les porteurs excédentaires sont arrachés à la bande de valence ou à un niveau d'impureté. Le but du présent article est d'en donner une description théorique, en s'attachant plus particulièrement à l'ionistion par choc intrinsèque. Celle-ci n'a été étudiée expérimentalement, dans des échantillons homogènes, que dans l'antimoniure d'indium de haute pureté, où elle est facile à observer à 77 °K ou à plus basse température [1]. Plusieurs théories ont tenté de décrire la distribution électronique en champ électrique intense. La plupart d'entre elles (Wolff [2], Stratton [3], Chuenkov [4] et Bok [5]) admettent que la distribution électronique reste quasi-isotrope, ce qui permet de définir une température électronique T_e. Elles invoquent les collisions interélectroniques comme mécanisme principal d'échange d'énergie restaurant l'isotropie de la distribution malgré la présence du champ électrique. Elles trouvent leur application dans l'intepétation des variations de mobilité avec le champ électrique, et également de l'ionisation par choc extrinsèque. Mais l'ionisation par choc intrinsèque suppose la présence d'électrons de très grande énergie, qui est peu compatible avec une forme de distribution isotrope; et, en effet, elle n'a été observée expérimentalement que dans des échantillons de concentration électronique $n < 10^{15}$ cm^{-3}, où les collisions interélectroniques ne sont plus le mécanisme d'échange d'énergie et de moment prédominant [6]. La notion de température électronique ne s'applique plus, car la distribution électronique devient anisotrope, et ce sont précisément les électrons de grande énergie ayant un moment dirigé dans la direction du champ qui sont responsables de l'ionisation par choc, comme l'ont remarqué Toda et Glicksman [7].

Nous nous sommes proposé d'analyser la distribution électronique en considérant séparément les électrons d'énergie petite et moyenne, qu'on peut considérer comme „isotropes" et les électrons de grande énergie se déplaçant dans la direction du champ, qu'on peut qualifier de „balistiques". C'est parmi ces derniers que se trouvent des électrons d'énergie supérieure au seuil d'ionistaion ε_i (largeur de bande interdite dans le cas du phénomène intrinsèque). La première étape du calcul (Section 2) sera d'évaluer l'énergie ε_0 (et le moment correspondant p_0) au-delà de laquelle commence la population d'électrons „balistiques". Le champ électrique assure un pompage continu d'électrons de la population „isotrope" ($\varepsilon < \varepsilon_0$) à la population „balistique" ($\varepsilon > \varepsilon_0$). Les électrons perdent leur énergie, soit par le processus „utile" d'ionisation par choc (si $\varepsilon > \varepsilon_i$), soit par des processus concurrents. On analysera aux Sections 3 et 4 les probabilités de ces transitions et les flux d'électrons entre les trois populations: $\varepsilon < \varepsilon_0$, $\varepsilon_0 < \varepsilon < \varepsilon_i$, $\varepsilon > \varepsilon_i$. Nous pourrons alors (Section 5) déterminer ces trois populations entre lesquelles se répartissent les électrons en régime stationnaire, et ceci en fonction du champ électrique de façon à faire apparaître le champ critique d'ionisation. L'apparition de l'ionisation par choc au-delà de ce seuil est la conséquence de l'efficacité du pompage par le champ électrique: le temps nécessaire à un électron pour accéder à l'énergie d'ionisation est du même ordre de grandeur que le temps de relaxation (de l'ordre de 10^{-12} s) alors que la durée de vie moyenne des électrons hors d'équilibre est beaucoup plus grande (de l'ordre de 10^{-7} s).

2. Bilan des processus élémentaires d'échange d'énergie

Considérons, à une température de réseau donnée, les électrons d'énergie ε, dont le temps d'un libre parcours est en moyenne $\tau(\varepsilon)$. Comparons l'apport d'énergie par le champ électrique pendant le temps τ à la perte d'énergie subie par l'électron au cours d'une collision. Nous envisageons les deux cas de diffusion par phonons acoustiques ou par phonons optiques.

Le taux d'accroissement de l'énergie dans le champ électrique E est

$$\left(\frac{d\varepsilon}{dt}\right)_E = \frac{q^2 E^2}{m^*} \tau(\varepsilon)$$

(q charge élémentaire, m^* masse effective).

La relaxation par les phonons acoustiques se traduit par une perte d'énergie [6] (par unité de temps)

$$\left(\frac{d\varepsilon}{dt}\right)_{\text{ph.ac.}} = -\,\text{Cte}\,\frac{\varepsilon}{\tau(\varepsilon)}\,.$$

Le temps de relaxation τ est alors proportionnel à $\varepsilon^{-1/2}$, de sorte que

$$\left(\frac{d\varepsilon}{dt}\right)_{\text{ph.ac.}} \sim -\,\varepsilon^{3/2}, \quad \left(\frac{d\varepsilon}{dt}\right)_E \sim \varepsilon^{-1/2}\,.$$

La relaxation par les phonons optiques se traduit par une perte d'énergie sensiblement constante et égale à l'énergie maximale $h\,\nu_0$ des phonons longitudinaux. Ceci s'exprime par un taux de variation de l'énergie

$$\left(\frac{d\varepsilon}{dt}\right)_{\text{ph.opt.}} = -\,\frac{h\,\nu_0}{\tau(\varepsilon)}\,.$$

Le temps de relaxation τ est alors proportionnel à $\varepsilon^{1/2}$, de sorte que

$$\left(\frac{d\varepsilon}{dt}\right)_{\text{ph.opt.}} \sim -\,\varepsilon^{-1/2}, \quad \left(\frac{d\varepsilon}{dt}\right)_E \sim \varepsilon^{1/2}\,.$$

Ces formules sont valables dans le cas d'une bande de conduction „standard" (isotrope, parabolique) qui s'applique aux composés III-V et II-VI que nous envisageons.

La Fig. 1 représente schématiquement, en fonction de ε, la variation de ces taux d'échange d'énergie dans les deux cas de diffusion considérés; les flèches verticales indiquent le sens du bilan d'énergie. Suivant le signe de ce bilan, l'énergie ε de l'électron considéré varie dans un sens qui est indiqué par les

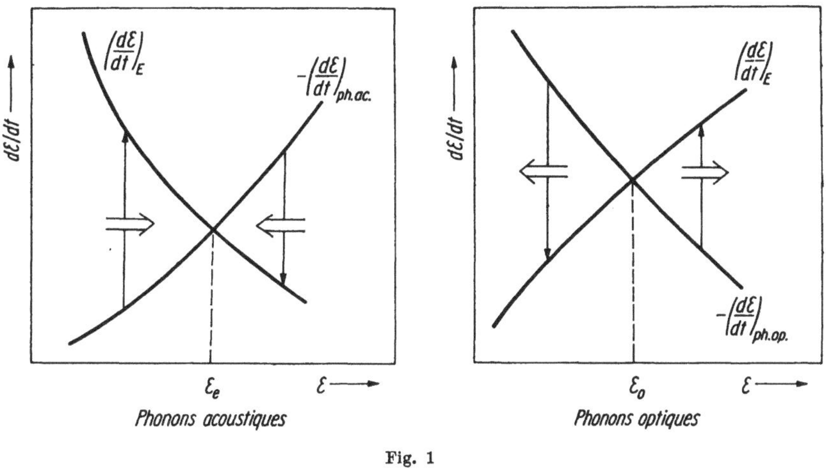

Fig. 1

flèches horizontales. On voit que, dans le cas de diffusion par phonons acoustiques, l'ensemble de la distribution tend à se stabiliser autour d'une énergie d'équilibre ε_e, alors que dans le cas de la diffusion par phonons optiques, les électrons ayant atteint une certaine énergie ε_0 ont tendance à gagner de l'énergie jusqu'à ce qu'un nouveau phénomène, celui d'ionisation par choc par exemple, intervienne pour dissiper cette énergie.

Il apparaît donc que le cas de la diffusion par phonons optiques est le plus favorable pour l'obtention de l'ionisation par choc, car les électrons d'énergie $\varepsilon > \varepsilon_0$ ont un moment dirigé préférentiellement dans la direction du champ électrique et peuvent gagner rapidement l'énergie nécessaire à un choc ionisant. Cette conclusion se rapproche de celle obtenue par Hilsum [8] dans l'analyse du transfert des électrons dans une bande supérieure (effet Gunn).

2.1 Variation de ε_0 avec les paramètres du semiconducteur

La valeur de ε_0 est définie par l'équation

$$\left(\frac{d\varepsilon}{dt}\right)_E = \left|\left(\frac{d\varepsilon}{dt}\right)_{\text{ph.opt.}}\right|$$

qui, en remplaçant τ par $A\ \varepsilon^{1/2}$, s'écrit

$$\frac{h\,\nu_0}{A\ \varepsilon_0^{1/2}} = \frac{q^2\,E^2}{m^*}\,A\ \varepsilon_0^{1/2} \tag{1}$$

avec [9]

$$A = \frac{h}{2^{1/2}\,\pi\,q^2\,\nu_0\,m^{*1/2}}\;\frac{\varkappa_0\,\varkappa_s}{\varkappa_0 - \varkappa_s}\;\frac{\exp\dfrac{h\,\nu_0}{k\,T} - 1}{\exp\dfrac{h\,\nu_0}{k\,T} + 1} \tag{2}$$

($\varkappa_0$ constante diélectrique optique, $\varkappa_s$ constante diélectrique statique).

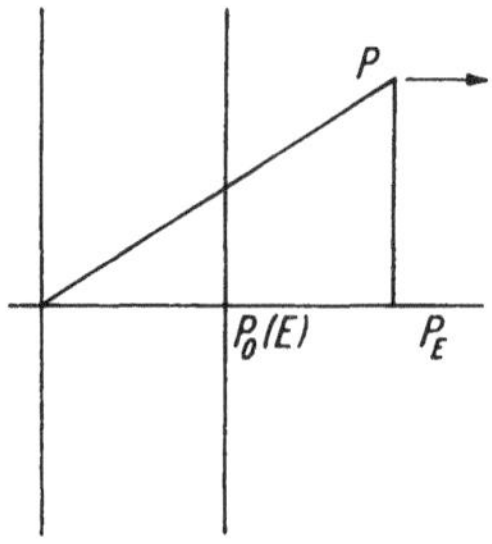

Fig. 2

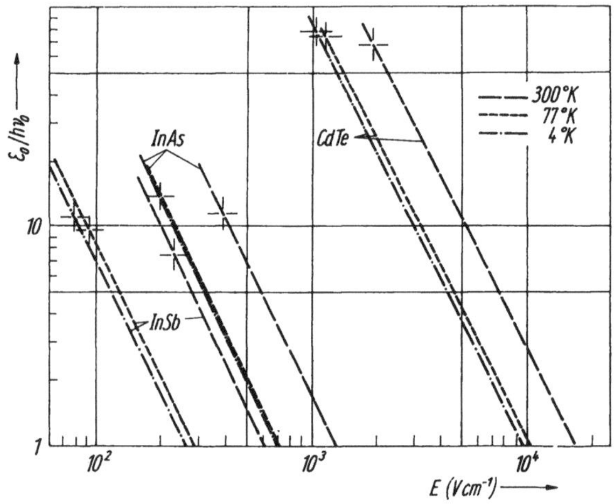

Fig. 3

De (1) nous tirons la valeur de ε_0 et celle du moment p_0 correspondant

$$\varepsilon_0(E) = h\,\nu_0 \frac{m^*}{q^2\,A^2\,E^2}\,, \tag{3}$$

$$p_0(E) = \frac{(2\,h\,\nu_0)^{1/2}\,m^*}{q\,A\,E}\,. \tag{4}$$

Les électrons ayant la projection de leur moment suivant la direction du champ électrique supérieure à $p_0(E)$ (Fig. 2)[1]) ont des échanges d'énergie et de moment avec le réseau suffisamment faibles pour être accélérés dans la direction du champ électrique et constituer la partie anisotrope de la distribution (électrons „balistiques").

Il est important de connaître la variation de ε_0 en fonction du champ électrique, pour une température de réseau donnée. C'est ce que représente la Fig. 3 dans le cas de plusieurs semiconducteurs usuels: InSb, InAs et CdTe. Les valeurs des différents paramètres utilisés sont notées dans le Tableau 1. Celles de A ont été calculées à partir de la mobilité de Hall expérimentale, suivant la formule [10] $\mu_H = 1{,}205\,(q/m^*)\,A\,(k\,T)^{1/2}$, où à partir de l'expression (2). Ces deux méthodes de calcul donnent des résultats très voisins aux températures supérieures ou égales à 77 °K, dans la mesure où la mobilité est principalement limitée par les collisions sur les phonons optiques dans les composés III-V considérés [10] et dans CdTe [11]; pratiquement, nous avons alors choisi la valeur déduite de μ_H. Aux basses températures, A ne peut être estimé qu'à partir de la formule (2), car d'autres mécanismes de relaxation interviennent simultanément (cf. aussi discussion de la Section 6).

La Fig. 3 permet de situer les valeurs de ε_0 par rapport aux deux énergies qui caractérisent les phénomènes physiques principaux mis en jeu: ε_i (égale à la largeur de bande interdite), qui conditionne l'ionisation par choc, et $h\,\nu_0$, qui décrit le mécanisme de relaxation principal.

3. Définition du modèle

Ayant mis en évidence le caractère „balistique" des électrons dont la projection du moment sur l'axe du champ électrique p_E est supérieur à p_0, nous divisons la distribution électronique, dans la bande de conduction, en trois parties (Fig. 4). La partie appelée P_0 comprend tous les électrons pour lesquels $p_E \leq p_0$.

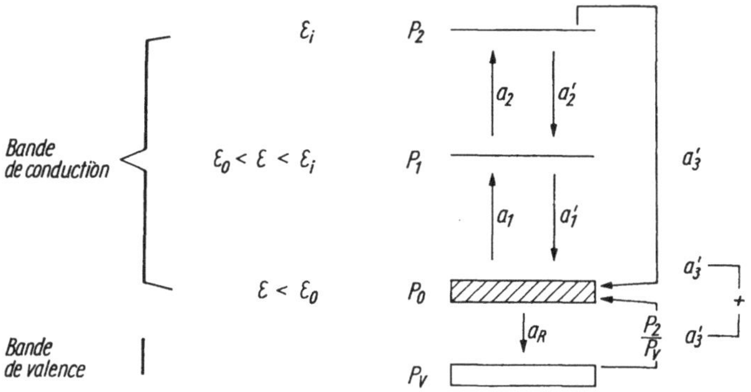

Fig. 4

[1]) Dans les Figures 2, 5 et 6 il faut lire p pour P.

Tableau 1

Composé	InSb			
$\frac{m^*}{m_0}$	0,0143			
$h\nu_0$ (eV)	0,0244			
T (°K)	300	77	20	4
A (s · [eV]$^{-1/2}$)	$4{,}6 \times 10^{-12}$	1×10^{-11}	$1{,}045 \times 10^{-11}$	$1{,}045 \times 10^{-11}$
Composé	InAs			
$\frac{m^*}{m_0}$	0,020			
$h\nu_0$ (eV)	0,031			
T (°K)	300	77	20	4
A (s · [eV]$^{-1/2}$)	$2{,}6 \times 10^{-12}$	$4{,}85 \times 10^{-12}$	$4{,}87 \times 10^{-12}$	$4{,}87 \times 10^{-12}$
Composé	CdTe			
$\frac{m^*}{m_0}$	0,11			
$h\nu_0$ (eV)	0,022			
T (°K)	300	77	20	4
A (s · [eV]$^{-1/2}$)	5×10^{-13}	8×10^{-13}	$8{,}75 \times 10^{-13}$	$8{,}75 \times 10^{-13}$

Considérons que cette population P_0 est isotrope. Les électrons de $p_E \geqq p_0$ sont divisés en deux populations: celle P_1 des électrons d'énergie comprise entre ε_0 et ε_i et celle P_2 des électrons ayant atteint l'énergie d'ionisation ε_i.

Les électrons peuvent passer de la population P_0 à P_1, ou de P_1 à P_2, par gain d'énergie dans le champ électrique. Ils peuvent également être relaxés de P_2 à P_1 ou de P_1 à P_0 par perte d'énergie sur un phonon optique. Nous envisageons également les processus du choc ionisant, par lequel un électron de la population P_2 donne naissance à deux électrons de la population P_0 et à un trou, et enfin la recombinaison d'un électron de la population P_0, hors d'équilibre, avec un trou.

Nous désignerons par a_1, a_2, a_1', a_2', a_3' et a_{R} les probabilités respectives de ces différentes transitions, qui sont égales aux inverses des temps de transit ou de relaxation correspondant (voir Fig. 4).

A l'état stationnaire, le principe du bilan détaillé s'applique à chaque population et conduit aux équations

$$- P_1(E)\, a_2 + P_2(E)\, (a_2' + a_3') = 0\,, \tag{5}$$

$$- P_0(E)\, a_1 + P_1(E)\, (a_2 + a_1') - P_2(E)\, a_2' = 0\,, \tag{6}$$

$$P_0(E)\, a_1 + \delta N(E)\, a_R - P_1(E)\, a_1' - P_2(E)\, 2\, a_3' = 0\,. \tag{7}$$

$\delta N(E)$ est l'accroissement du nombre d'électrons total par rapport à l'équilibre thermodynamique pour lequel $E = 0$ et

$$P_0(0) + P_1(0) + P_2(0) = N \tag{8}$$

(N nombre d'électrons libres dans la bande de conduction). On peut écrire d'autre part

$$P_0(E) + P_1(E) + P_2(E) = N + \delta N(E) \tag{9}$$

ce qui permet d'éliminer $\delta N(E)$ dans (7) sous la forme

$$P_0(E)\, (a_1 + a_R) + P_1(E)\, (a_R - a_1') + P_2(E)\, (a_R - 2\, a_3') = N\, a_R\,. \tag{10}$$

Une quatrième équation faisant intervenir la population P_V de la bande de valence, peut être déduite des trois précédentes car l'ensemble forme un cycle fermé. La combinaison linéaire formée en ajoutant membre à membre (5), (6) et (7) est

$$- P_2(E)\, a_3' + \delta N(E)\, a_R = 0\,.$$

Elle exprime simplement le principe du bilan détaillé, appliqué à la bande de valence.

La résolution du système (5), (6) et (10) nous donne la valeur de chacune des populations des différents niveaux:

$$P_0 = N\, \frac{a_3'(a_2 + a_1') + a_1'\, a_2'}{D}\, a_R\,, \tag{11}$$

$$P_1 = N\, \frac{a_R\, a_1\, (a_2' + a_3')}{D}\,, \tag{12}$$

$$P_2 = N\, \frac{a_1\, a_2\, a_R}{D}\,, \tag{13}$$

où

$$D = a_R\, [a_1\, a_2 + a_2'\, (a_1' + a_1) + a_3'\, (a_1 + a_2 + a_1')] - a_1\, a_2\, a_3'\,.$$

La population totale $P_s = N + \delta N$ est donc

$$P_s = P_1 + P_2 + P_0 = \frac{N}{1 - \dfrac{a_1\, a_2\, a_3'}{a_R\, [a_1\, a_2 + a_2'\, (a_1 + a_1') + a_3'\, (a_2 + a_1 + a_1')]}}\,. \tag{14}$$

Pour E faible $P_s \approx N$, car le second terme du dénominateur de (14) est très faible. Quand E croît, a_R, a_3', a_2' restent constants alors que a_1, a_2, a_1' croissent, de sorte que ce terme croît, ce qui entraîne la croissance de P_s avec E.

4. Evaluation des différentes probabilités

4.1 Probabilités a_1' et a_2'

Le calcul précis des probabilités a_1' et a_2' imposerait de tenir compte de la répartition en énergie de chaque population. Cette complication serait disproportionnée avec la simplicité du modèle à trois populations. Dans la logique de ce dernier, nous utiliserons l'approximation consistant à attribuer aux électrons des populations P_1 et P_2 les énergies moyennes ε_0 et ε_i. Ceci revient à remplacer la probabilité vraie par une limite supérieure qui en est peu différente, dans la mesure où dans chacune de ces populations, les électrons occupent surtout les niveaux d'énergie les plus bas.

Nous écrirons alors pour a_1'

$$a_1' = \frac{1}{\tau(\varepsilon_0)} = \frac{1}{A\,\varepsilon_0^{1/2}}\,,$$

et en utilisant la valeur de ε_0 donnée par (3)

$$a_1' = \frac{q\,E}{(h\,\nu_0\,m^*)^{1/2}}\,. \tag{15}$$

De même pour a_2'

$$a_2' = \frac{1}{\tau(\varepsilon_i)} = \frac{1}{A\,\varepsilon_i^{1/2}}\,. \tag{16}$$

4.2 Probabilité a_2

Nous avons supposé que les électrons ayant $p_E \geqq p_0$ se déplaçaient dans la direction du champ électrique. La probabilité de passage de P_1 à P_2 est l'inverse du temps de transit Δt, nécessaire pour faire croître l'énergie de ε_0 à ε_i sous l'effet du champ électrique (ce qui implique la même approximation qu'à l'alinéa précédent).

La vitesse du porteur en fonction du temps est

$$v = \frac{q\,E}{m^*}\,t + \frac{P_0}{m^*} \quad (t = 0,\ \text{quand } p_E = p_0)\,,$$

d'où

$$\Delta t = \left(\sqrt{\varepsilon_i} - \sqrt{\varepsilon_0}\right)\frac{(2\,m^*)^{1/2}}{q\,E}\,, \qquad a_2 = \frac{1}{\Delta t} = \frac{q\,E}{(2\,m^*)^{1/2}}\,\frac{1}{\varepsilon_i^{1/2} - \varepsilon_0^{1/2}}\,. \tag{17}$$

4.3 Probabilité a_1

Celle-ci traduit le passage des électrons de la partie isotrope de la distribution à la partie balistique. Le problème est de calculer pendant un temps défini, que nous prendrons égal à $\tau(\varepsilon_0)$, le nombre d'électrons dont la projection p_E du moment atteint la valeur p_0. Nous avons utilisé une méthode simple consistant à faire ce calcul rigoureusement dans le cas des champs faibles et à extrapoler le résultat pour les valeurs élevées du champ.

Dans le cas d'une distribution isotrope, les électrons qui auront au temps $\tau(\varepsilon_0)$, $p_E \geqq p_0$ sont ceux pour lesquels, au temps 0, $p_0 - q\,E\,\tau \leqq p_E \leqq p_0$ (cf. bande B de la Fig. 5).

La probabilité a_1 s'écrit alors

$$a_1 = \frac{1}{\tau(\varepsilon_0)} \times \frac{\text{nombre d'électrons tels que } p_0 - q\,E\,\tau \leqq p_E \leqq p_0}{\text{nombre total d'électrons tels que } p_E \leqq p_0}\,.$$

Le nombre d'électrons de $p_E \leqq p_0$ est pratiquement égal, dans les champs faibles, au nombre total d'électrons N.

Nous calculerons donc a_1 par la formule

$$a_1 = \frac{1}{\tau(\varepsilon_0)} \frac{\int\limits_{\text{bande B}} f\, \mathrm{d}^3p}{\int\limits_{\text{tout la distribution}} f\, \mathrm{d}^3p} = \frac{I}{\tau(\varepsilon_0)\, J}\,.$$

La fonction

$$f(p) = C \exp\left[-\frac{(p - m^* \mu E)^2}{2\, m^*\, k\, T}\right]$$

donne une bonne approximation de la distribution dans la bande de conduction, la valeur de $k\, T$ étant faible devant la largeur de bande interdite.

Dans le cas des champs électriques faibles, on a $q\, E\, \tau(\varepsilon_0) \ll p_0$, ce qui permet d'écrire

$$I = q\, E\, \tau(\varepsilon_0)\, K\,,$$

ou K est l'intégrale double de $f(p)$ sur le plan perpendiculaire à E en p_0.

Grâce à la symétrie autour de l'axe E dans l'expace des p, K s'écrit, en prenant comme variable d'intégration l'angle φ (Fig. 6)

$$k = \int\limits_0^{\pi/2} \exp\left[-\frac{(p_0 - m^* \mu E)^2 + p_0 \operatorname{tg}^2 \varphi}{2\, m^*\, k\, T}\right] 2\, \pi\, p_0^2 \operatorname{tg} \varphi\, (1 + \operatorname{tg}^2 \varphi)\, \mathrm{d}\varphi\,.$$

Cette intégrale est aisément calculable:

$$K = 2\, m^*\, k\, T \exp\left[-\frac{(p_0 - m^* \mu E)^2}{2\, m^*\, k\, T}\right].$$

De la même façon,

$$J = \int \exp\left[-\frac{(p - m^* \mu E)^2}{2\, m^*\, k\, T}\, \mathrm{d}^3p = \int \exp\left(-\frac{p_2}{2\, m^*\, k\, T}\right) \mathrm{d}^3p\right.$$

s'exprime, par le calcul usuel, sous la forme

$$J = (2\, \pi\, m^*\, k\, T)^{3/2}\,.$$

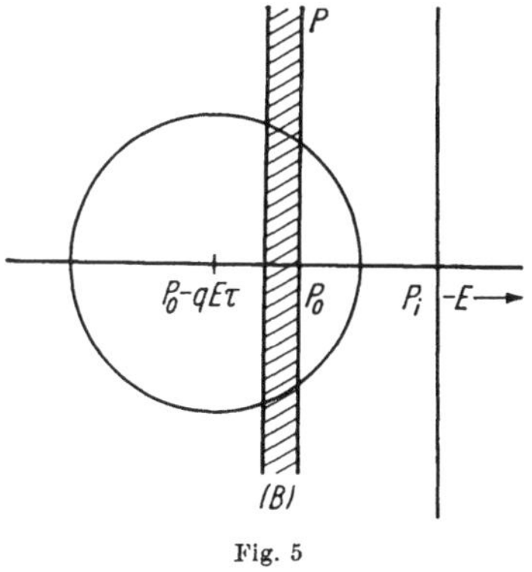

Fig. 5

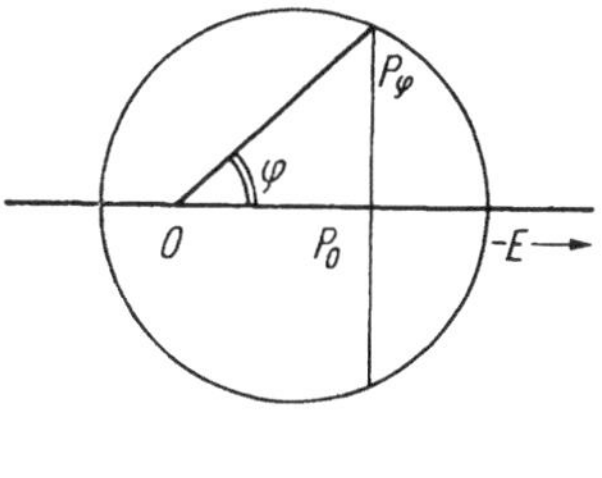

Fig. 6

En conséquence, la valeur de a_1 est

$$a_1 = \frac{q\,E}{(2\,\pi\,m^*\,k\,T)^{1/2}} \exp\left[-\frac{(p_0 - m^*\,\mu\,E)^2}{2\,m^*\,k\,T}\right].$$

Soit, en exprimant la valeur de p_0 d'après (4)

$$a_1 = \frac{q\,E}{(2\,\pi\,m^*\,k\,T)^{1/2}} \exp\left\{-\left[\left(\frac{h\,\nu_0\,m^*}{k\,T}\right)^{1/2} \frac{1}{q\,A\,E} - \frac{m^{*1/2}\,\mu\,E}{(2\,k\,T)^{1/2}}\right]\right\}. \tag{18}$$

4.4 Probabilité a_3'

Les électrons ayant atteint l'énergie correspondant à la largeur de bande interdite sont capables d'induire une ionisation. La section efficace pour l'ionisation d'un électron de la bande de valence décroit rapidement (comme ε^{-2}) en fonction de l'énergie de l'électron incident. Mais on peut penser [12] qu'au voisinage de ε_i, cette valeur est comparable à celle correspondant au phénomène de relaxation sur les phonons optiques. Ceci nous conduirait à admettre pour la probabilité d'ionisation une valeur de l'ordre de $10^{12}\,s^{-1}$. Nous avons cependant utilisé a_3' comme un paramétre, qui recevra différentes valeurs arbitraires: 10^8, 10^{10} et $10^{12}\,s^{-1}$, de façon à estimer l'influence de la variation de la probabilité d'ionisation par choc sur le comportement de l'avalanche.

4.5 Probabilité a_R

Elle est l'inverse de la durée de vie des porteurs hors de l'équilibre thermodynamique. Nous en connaissons une estimation dans le cas de faibles injections, pour InSb et GaAs [13, 14]. La variation de la durée de vie en fonction de l'injection est cependant mal connue. Nous avons choisi pour a_R des valeurs indépendantes du taux d'injection par ionisation, égales à 10^6, 10^8 ou $10^{10}\,s^{-1}$, afin d'étudier la variation du seuil d'ionisation avec la durée de vie.

4.6 Variations des probabilités avec E

Nous avons tracé sur la Fig. 7 les courbes représentant a_1, a_2, a_1', a_2' en fonction du champ électrique pour InSb à 77 °K. La valeur calculée de a_1 aux champs élevés ($E > E_M$) a été tracée en pointillé et n'est pas exacte, car les hypothèses initiales du calcul sont en défaut. Mais en fait, l'ionisation par choc se produit, comme nous le verrons, pour des champs plus petits que E_M.

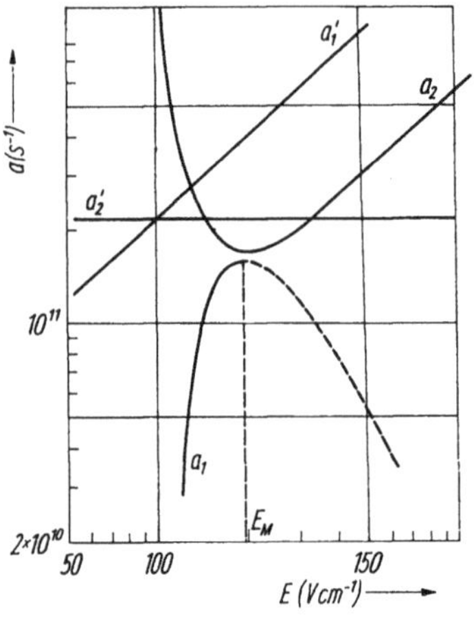

Fig. 7

5. Calcul des populations : Courbes courant-tension

Les formules (11), (12), (13) et (14), dans lesquelles les probabilités ont été remplacées par leur valeur calculée à la section 4, fournissent les populations P_0, P_1, P_2 et leur somme P_s. Les calculs numériques ont été effectués pour les semiconducteurs et aux températures citées dans le Tableau 2.

Tableau 2

Composé semiconducteur	InSb	InAs	CdTe
Températures (°K)		300	300
	77	77	77
	20	20	20
	4	4	

Pour chaque température les valeurs données à a_R sont: 10^6, 10^8, 10^{10} s^{-1} et celles données à a_3': 10^8, 10^{10}, 10^{12} s^{-1}.

Il n'est pas tenu compte dans ce modèle d'ionisation intrinsèque, du gel des électrons sur certains niveaux d'impuretés quand la température décroît.

En effet, l'ionisation par choc des électrons à partir de ces niveaux [15] intervient à des champs beaucoup plus faibles que l'ionisation intrinsèque. Nous donnons en exemple (au Tableau 3) les valeurs des différentes populations en fonction du champ électrique. On voit apparaître très nettement un seuil d'ionisation pour un champ de l'ordre de 110 V/cm. Les populations P_1 et P_2 d'électrons balistiques restent peu nombreuses, même au seuil de l'ionisation. Nous pouvons donc comparer directement les courbes expérimentales représentant la densité de courant en fonction du champ électrique, avec les courbes $P_s(E)$ la contribution des électrons ,,balistiques" au phénomène de conduction étant d'au moins deux ordres de grandeur plus petite que celle des électrons ,,isotropes".

Tableau 3

InSb: 77 °K, $a_3' = 10^{12}$ s^{-1}, $a_R = 10^8$ s^{-1}				
E (V/cm)	$\frac{P_0}{N}$	$\frac{P_1}{N}$	$\frac{P_2}{N}$	$\frac{P_s}{N}$
97	1,0004	$1{,}17\times10^{-7}$	$4{,}17\times10^{-8}$	1,0004
101	1,0028	$9{,}74\times10^{-7}$	$2{,}80\times10^{-7}$	1,0028
105	1,0151	$6{,}12\times10^{-3}$	$1{,}51\times10^{-6}$	1,0152
109	1,0704	$3{,}17\times10^{-5}$	$7{,}04\times10^{-6}$	1.0704
113	1,3217	$1{,}57\times10^{-4}$	$3{,}21\times10^{-5}$	1,3219
117	4,4229	$1{,}79\times10^{-3}$	$3{,}42\times10^{-4}$	4,425

Nous avons tracé les courbes $P_s(E)$ pour InSb à 77 °K, 20 °K, 4 °K (Fig. 8, 9 et 10), InAs à 77 °K (Fig. 11) et CdTe à 300 °K (Fig. 12) pour les différentes valeurs des probabilités d'ionisation a_3 et de recombinaison a_R.

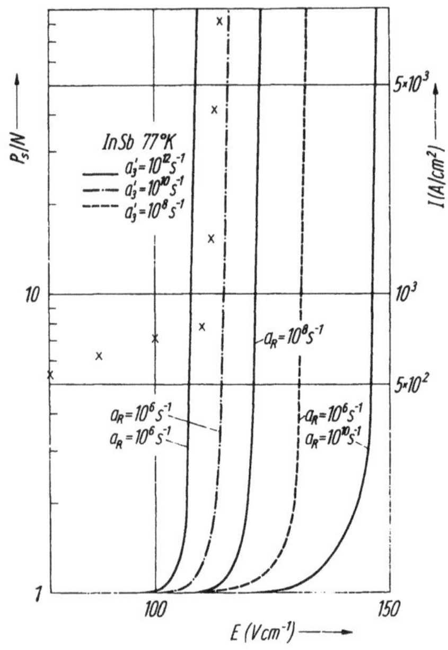

Fig. 8

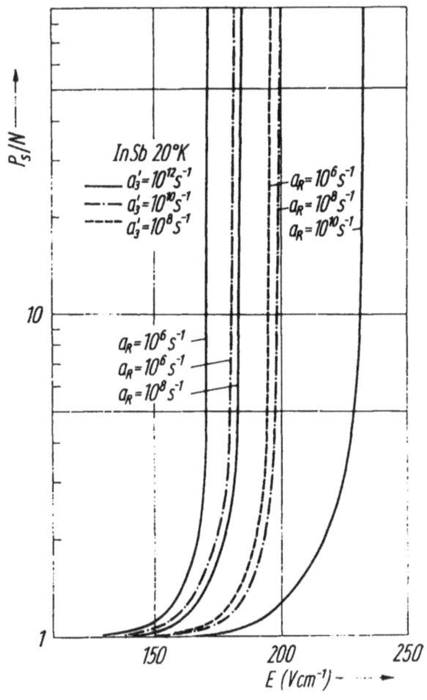

Fig. 9

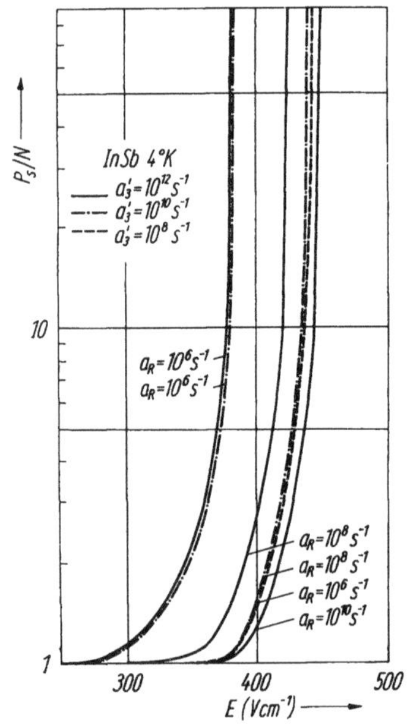

Fig. 10

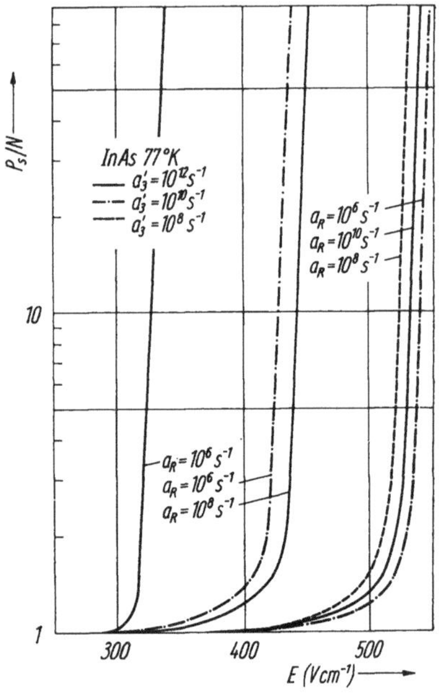

Fig. 11

6. Discussion et comparaison avec l'expérience

Comparons les résultats de la Fig. 8 avec les résultats expérimentaux connus pour InSb à 77 °K. La valeur minimum du seuil d'ionisation déduite des expériences de Glicksman et Steele [1] ou de Kanai [1] se situe au voisinage de 150 V/cm. Bok [5] a cependant trouvé des valeurs nettement plus faibles. Les croix marquées sur la Fig. 8 donnent la courbe d'ionisation — (densité de courant, champ électrique) que nous avons relevée sur un échantillon de type n de concentration 5×10^{13} cm^{-3}. Les valeurs du seuil d'ionisation trouvées par le calcul sont compatibles avec les valeurs mesurées sur des matériaux purs. Nous voyons que la valeur de 10^{12} s^{-1} pour la probabilite d'ionisation a_3' est la plus convenable, ce qui confirme l'hypothèse de Baraff [12] donnant une valeur de section efficace d'ionisation du même ordre de grandeur que celle de la diffusion par les phonons.

Les courbes montrent la dépendance importante du seuil d'ionisation avec la probabilité de recombinaison, qui explique la dispersion des résultats pour des échantillons de même pureté mais dont le degré de compensation et la concentration des pièges peuvent être différents, ce qui se traduit par des probabilités de recombinaisons variant de 10^6 à 10^9 s^{-1}.

Dans le cas d'échantillons à plus forte concentration électronique, le modèle proposé pourrait encore être utilisé en tenant compte de la perte d'énergie dans les collisions interélectroniques. Celle-ci peut se mettre sous la forme [16] $d\varepsilon/dt \sim N\,\varepsilon^{-1/2}$, qui est une fonction décroissante de l'énergie du même type que celle applicable aux collisions avec les phonons optiques.

Le seuil d'ionisation à 4 et 20 °K est supérieur à celui correspondant à 77 °K. Ce résultat, constaté expérimentalement par Bok [5], est dû au fait que la mobilité, décrivant les interactions du gaz d'électrons aux champs faibles, intervient dans l'expression de a_1 (cf. formule (18)). D'autres mécanismes de relaxation (chocs sur les impuretés ionisées par exemple) limitent alors la mobilité, cependant ces processus presque élastiques correspondent à des pertes d'énergie relativement faibles pour les électrons de grande énergie qui nous intéressent. Il n'est donc pas déraisonnable de choisir pour μ, dans la formule (18), la valeur expérimentale, et simultanément pour A, la valeur théorique issue de (2) en ne tenant compte que de la relaxation par phonons optiques. C'est ce que nous avons fait pour $T = 4$ ou 20 °K.

Nous remarquons que la valeur de ε_i, prise égale à celle de la largeur de bande dans nos calculs, intervient peu sur la forme des courbes et même sur la valeur du seuil d'ionisation. En effet, un électron „balistique" (d'energie $>\varepsilon_0$) gagne rapidement de l'énergie dans le champ électrique. La courbe représentant $a_2(E)$

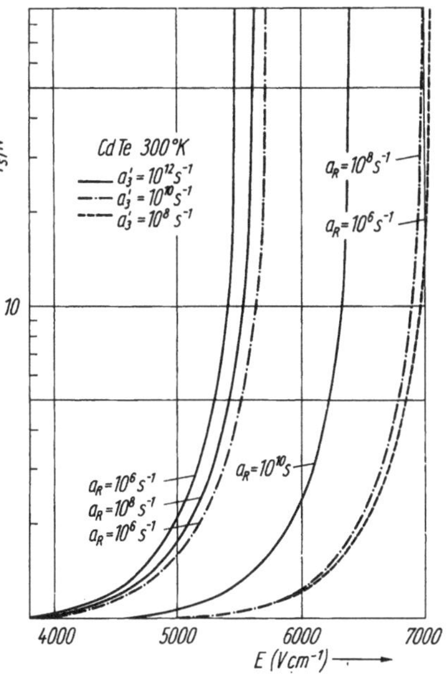

Fig. 12

est peu déformée si on change ε_i; si l'on double ε_i, son minimum a une valeur deux fois plus faible, ce qui donne une variation relative de l'ordre de 5% sur la valeur de P_s au voisinage de l'avalanche.

Dans le cas de composés semiconducteurs comportant une bande d'électrons lourds dont le minimum est à ε_t au-dessus du minimum de la bande d'électrons légers ($\varepsilon_t < \varepsilon_i$), le remplissage de cette bande d'électrons lourds se réalisera avant le phénomène d'ionisation intrinsèque: les électrons ayant alors une masse effective plus grande sont plus difficilement accélérés et l'ionisation par choc débutera pour des champs électriques bien supérieurs.

Il serait intéressant de modifier le modèle proposé pour évaluer la population des électrons lourds. Pour ce calcul, a_3' doit être pris égal à zéro. a_2' traduit le phénomène de relaxation entre les deux bandes électroniques, qui est mal connu; mais on peut penser que a_2' et a_1' sont du même ordre de grandeur. ε_i sera remplacé par ε_t et P_2 désignera alors la population des électrons lourds, dont le seuil de croissance rapide correspondra au champ électrique E_s au-delà duquel apparaît l'effet Gunn [17].

Remerciements

Nous remercions M. Rodot pour ses encouragements et ses fructueuses suggestions, pour la lecture et la correction de ce texte.

Bibliographie

[1] R. BRAY, Bull. Amer. Phys. Soc. **30**, 35 (1955).
M. GLICKSMAN et M. C. STEELE, Phys. Rev. **110**, 1204 (1958); **118**, 474 (1960).
Y. KANAÎ, J. Phys. Soc. Japan **13**, 1065 (1958); **14**, 1302 (1959).
[2] P. A. WOLFF, Phys. Rev. **95**, 1415 (1954).
[3] R. STRATTON, Proc. Roy. Soc. **242**, 355 (1957).
[4] V. A. CHUENKOV, Fiz. tverd. Tela **2**, 799 (1960).
[5] J. BOK, Ann. Radioélectr. **60**, 120 (1960).
[6] H. FRÖHLICH et B. V. PARANJAPE, Proc. Phys. Soc. **B69**, 21 (1956).
[7] H. TODA et M. GLICKSMAN. Phys. Rev. **140**, 1317 (1965).
[8] C. HILSUM, Proc. IRE **50**, 185 (1962).
[9] H. B. CALLEN, Phys. Rev. **76**, 1394 (1949).
[10] C. HILSUM, Proc. Phys. Soc. **76**, 414 (1960); Comm. Conf. Internat. Phys. Semicond., Paris 1964 (p. 1127).
[11] B. SEGALL, M. R. LORENZ et R. E. HALSTEAD, Phys. Rev. **129**, 2471 (1963).
[12] G. A. BARAFF, Phys. Rev. **128**, 6 (1962).
[13] R. N. ZITTER, A. J. STRAUSS et A. E. ATTARD, Phys. Rev. **115**, 266 (1959).
[14] T. KINSEL et I. KUDMAN, Solid State Electronics **8**, 797 (1965).
[15] A. ZYLBERSTEIN, J. Electronics and Control **10**, 429 (1961).
[16] D. PINES et D. BOHM, Phys. Rev. **85**, 338 (1952).
[17] J. GUNN, Solid State Comm. **1**, 88 (1963).

(Received May 18, 1966)

phys. stat. sol. 16, 613 (1966)

Mullard Research Laboratories, Redhill, Surrey

Comparison of Magnetic Measurements and Theoretical Predictions for Nickel–Iron Films over a Wide Composition Range[1])

By

J. A. CUNDALL [2]) and A. P. KING

A comparison is made of the theoretical predictions of several recent models of polycrystalline magnetic films with the experimental results obtained for nickel-iron films in a wide composition range. The models considered are the micro-magnetic and the random anisotropy models. Magnetic measurements were made of the remanence distribution and the easy direction domain wall coercive force. The results suggest that no single theory adequately describes the properties of the films over the whole composition range although all have a limited success for different parts of the range.

On compare les prédictions théoriques de plusieurs modèles récents des couches magnétiques minces polycristallins avec les résultats expérimentaux obtenus sur des couches minces de nickel-fer, couvrant une large gamme de compositions variées. On considére le modèle micromagnétique et le modèle d'anisotropie à directions désordonnées. Des mesures magnétiques son faites sur la distribution de rémanence, et sur le champ coercitif de la direction de facile aimantation, Les résultats suggèrent que les propriétés des minces couches ne peuvent pas être expliquées, sur toute la gamme des compositions par une théorie unique, bien que chaque théorie reussit assez bien à expliquer les résultats pour certaines parties de la gamme.

1. Introduction

A number of models of magnetic thin films have been proposed in the last few years. A successful model must be capable of describing the behaviour of several magnetic quantities: for example the angular dispersion, the field dependence of the rotational hysteresis loss, and the easy direction coercive force. The three models considered in this paper are the random uniaxial anisotropy model [1], the random complex biaxial model [2], and the magnetization ripple model [3, 4].

This paper discusses the comparison of the predictions of these models with experimental measurements over a wide composition range for two series of nickel-iron films 550 Å and 1000 Å thick. The behaviour of the field dependence of the rotational hysteresis loss has been described elsewhere [5].

---

[1]) Contribution to the International Colloquium on Magnetic Thin Films, 25th to 28th April 1966 in Jena.

[2]) Now at Pilkington Brothers Research Laboratories, Lathom, Ormskirk, Lancashire, England.

2. The Angular Dispersion

2.1 The random biaxial model

The random complex biaxial model [2] assumes that a magnetic film may be considered as an array of regions in each of which a local biaxial anisotropy K_b (orientated at random in the film plane) is superimposed on the overall uniaxial anisotropy K_u of the specimen. The energy E of such a region is given by

$$E = K_u \sin^2 \Phi + \frac{K_b}{4} \sin^2 2 (\alpha - \Phi) - M_s H \cos (\theta - \Phi), \tag{1}$$

where Φ is the angle between the magnetization M_s and the overall easy axis, θ is the angle between the applied field H and the easy axis and α is the angle between the local biaxial easy axis and the overall easy axis.

For equilibrium of the magnetization vector

$$\frac{1}{K_u} \frac{\partial E}{\partial \Phi} = 0 = \sin 2 \Phi - \frac{K'}{2} \sin 4 (\alpha - \Phi) - 2 h \sin (\theta - \Phi) , \tag{2}$$

where $K' = K_b/K_u$ and $h = H/H_K = \dfrac{H M_s}{2 K_u}$.

This equation was solved numerically for the angle Φ for a series of values of α in the range $-45° \leqq \alpha \leqq 45°$ (from considerations of symmetry this is a sufficiently wide range) for zero applied magnetic field. The smallest magnitude solution of Φ was assumed when several real solutions occurred. These values of Φ were processed to obtain the remanence distribution as a function of angle from the hard direction.

The angular dispersion α_{90} was then obtained from this distribution for various values of the ratio of the local biaxial anisotropy to the uniaxial anisotropy $K' = K_b/K_u$. This approach is similar to that previously employed [6] in the case of the random uniaxial anisotropy model. The angular dispersion was defined from experimental considerations as being the angle from the hard direction such that after saturation and removal of the saturating field 90% of the magnetization relapses into one direction along the easy axis and 10% into the other; i.e. the easy axis remanent magnetization is 0.80 times the saturation value.

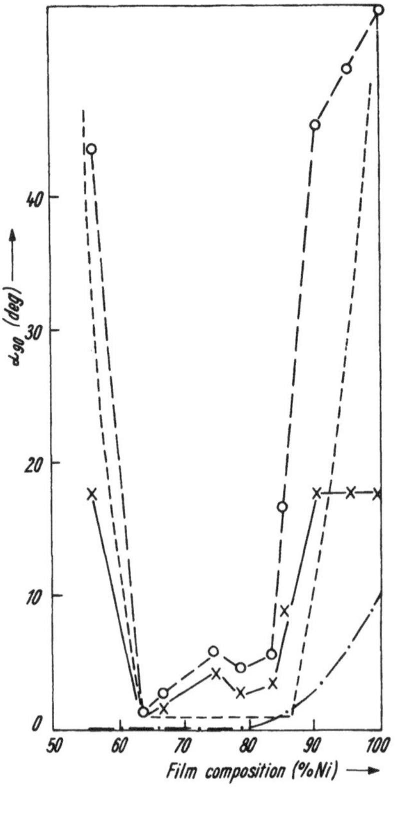

Fig. 1. Theoretical and experimental dependence of the angular dispersion on film composition.
○ Random uniaxial anisotropy model.
× Random complex biaxial model.
— · — Magnetization ripple theory: $t = 1000$ Å, $D = 400$ Å.
— — — Experiment: $t = 1000$ Å, $D = 350$ Å

To facilitate comparison with the experimental observations an 'experimental' value of the reduced anisotropy K' was obtained from the value of the high field cut-off $h'_r = (1 + K')$ in the field dependence of the rotational hysteresis loss, at which the energy loss approached a constant value. The corresponding value for the random uniaxial anisotropy model is $h_r = 1 + K$, where K is the ratio of the local random uniaxial anisotropy to the overall uniaxial anisotropy.

The theoretical values of angular dispersion deduced from the experimental values of K or K' are shown in Fig. 1. The most noticeable feature of this curve is the upper limit of $\alpha_{90} \approx 18°$ for the random complex biaxial anisotropy model. This is a direct result of the fact there is always a resultant easy direction within $\pm 22.5°$ of the overall easy direction irrespective of the magnitude of K'. (For the random uniaxial anisotropy model the resultant easy axis lies within 90° of the overall uniaxial easy axis and, in this case, the limit to α_{90} is 53.4°.)

2.2 Magnetization ripple model

For this model [3, 4] the angular dispersion α_{90} was calculated from an equation derived by Hoffmann [7] for multilayer films:

$$\alpha_{90} = \frac{3}{2}\sqrt{2} \times \frac{1.645}{1.96} \times \frac{D^3 K^2}{8\pi\sqrt{2}\,A\,K_u\,t} \times \left\{1 + \frac{1}{A\,K_u}\left(\frac{M_s^2\,D^2\,K^2\,t^{\frac{1}{2}}}{16\sqrt{2}\,\pi}\right)^{4/5}\right\}^{-1}, \quad (3)$$

where D is the diameter of the grains in the film, K is the local random anisotropy, A is the magnetic exchange energy constant, and t is the film thickness. The local anisotropy K is identified here with the magnitude of the magnetocrystalline anisotropy constant. The factor 1.645/1.96 arises because the present definition of α_{90} is different from that in [6], and the factor $3/2\sqrt{2}$ arises from the effect of ripple hysteresis on α_{90} [8].

Values of α_{90} were calculated from this expression using experimental values of K_u, the film thickness t, and the grain size D, and bulk values of K and M_s. The resulting variation of α_{90} over the composition range 55% to 100% nickel is shown in Fig. 1, although it must be remembered that this expression is strictly applicable only when the magnetization ripple amplitude is small.

Recent modifications by Harte [9] to the magnetization ripple theory include the effect of large angular deviations of the local magnetization directions and have produced an expression for the root mean square amplitude of the ripple δ. The major departure from the previous treatment is the allowance for the field due to charges on the low-angle walls of the ripple structure.

It is apparent from this treatment that the predicted values of α_{90} are of similar magnitude to those predicted by the linear theory for small dispersion films but do not show as fast an increase as the extremes of film composition are approached.

2.3 Experiment

The angular dispersion was measured by a torque magnetometer method, as described previously [5] using an automatic torque magnetometer [10], for two series of films 550 Å and 1000 Å thick deposited on glass substrates at 250 °C in a magnetic field of 30 Oe. A wide minimum value of angular dispersion of $\alpha_{90} \approx 0.8°$ occurs over the composition range 64% to 86% nickel for both series

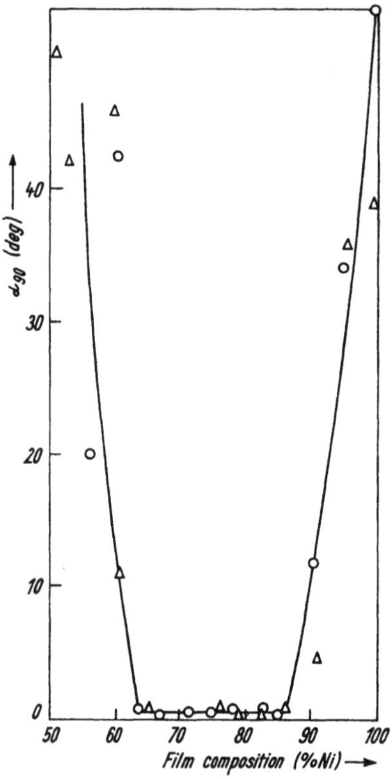

Fig. 2. Experimental dependence of the angular dispersion on film composition.
○ $t = 1000$ Å, $D = 350$ Å.
△ $t = 550$ Å, $D = 175$ Å

of films (Fig. 2). The uniaxial anisotropy evaluated from high field rotating torque curves showed a similar variation with the film composition to that reported by Robinson [11].

3. The Easy Direction Domain Wall Coercive Force

The easy direction domain wall coercive force H_c was calculated as a function of film thickness and film composition for a number of different models and different types of domain wall.

Calculations were made on the surface roughness model [12, 13] for both Néel and Bloch 180° domain walls assuming the surface roughness to be independent of film composition. The relevant expression for the coercive force of Néel walls is

$$H_c = \frac{S}{2\,M_s\,t}\left\{\frac{A\,\pi^2}{q} + \frac{K_u\,q}{2} + \frac{\pi\,M_s^2\,q\,t\,(2\,q+t)}{(q+t)^2}\right\}, \tag{4}$$

where the wall width q is given by

$$-\frac{A\,\pi^2}{q^2} + \frac{K_u}{2} + \frac{\pi\,t^2\,M_s^2}{(q+t)^2} = 0$$

and S is the surface roughness. The expression for Bloch walls is given in the literature [13].

The accuracy of both these calculations was given by the accuracy of solution of the domain wall width, viz. ± 2 Å. The results of these calculations for Bloch walls are shown in Fig. 3, curve A. The coercive force for Néel walls is independent of composition for film thickness greater than about 300 Å. The expressions for the domain wall energy derived by Middelhoek [14] and Prutton [15] for two models of the cross-tie domain wall show that the coercive force is independent of the film composition and that the dependence on film thickness is given by $H_c = H_0\,t^{-1}$, where H_0 is a constant.

Investigations of the coercive force on the magnetization ripple theory [16, 17] have shown the coercive force to be a sensitive function of grain size, film composition, and film thickness. Curve B, Fig. 3, was derived by assuming the local anisotropies are due to magnetocrystalline anisotropy. Curve C was obtained after including a stress contribution to the local anisotropy chosen arbitrarily

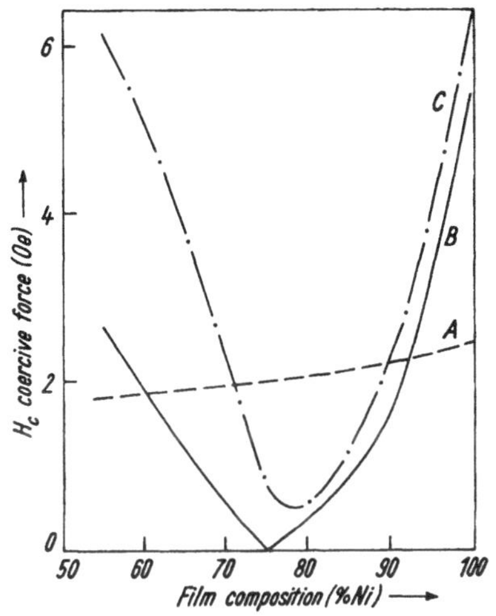

Fig. 3. Theoretical dependence of the coercive force on composition for various models.
Curve A — — — Surface roughness model for Bloch walls: t = 1000 Å (H_c in arbitrary units).
Curve B ——— Joenk: t = 500 Å, D = 350 Å.
Curve C — · — Joenk with stress contribution to local anisotropy

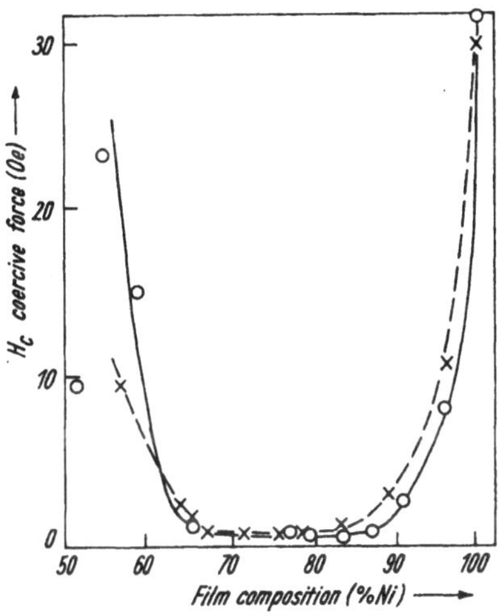

Fig. 4. Experimental dependence of the coercive force on film composition for films of 550 Å and 1000 Å thickness.
○ t = 550 Å, D = 175 Å.
× t = 1000 Å, D = 350 Å

such that the stress contribution at 75% nickel is equal to the magnetocrystalline contribution at 80.3% nickel. The magnitude of the local stress was assumed to be independent of film composition. The easy direction coercive force was measured for the two series of films (550 Å and 1000 Å thick) as a function of composition utilising the Kerr magneto-optic effect. The results are shown in Fig. 4.

4. Discussion

The most important result of this work is the failure of the theoretical predictions to agree quantitatively with the experimental observations of the dependence of the angular dispersion on the film composition.

Both the random uniaxial anisotropy model and the random complex biaxial model predict magnitudes which are larger than the observed angular dispersion over the composition range in which small angular dispersions are found. Although these models both predict an increase in angular dispersion at the extremes of the composition range in question this result is not entirely satisfactory as a semi-phenomenological approach was used in which the validity of the models was assumed in order to allow the deduction of the magnitude of the local random anisotropy from rotational hysteresis measurements. The theoretical upper limit to the magnitude of the angular dispersion (viz. about 18°) in the random complex biaxial anisotropy model was often observed to be exceeded experimentally at the extremes of the composition range. The random complex biaxial anisotropy model shows best agreement with experiment for small angular dispersion specimens and the random uniaxial anisotropy model for high angular dispersion specimens.

The two magnetization ripple approaches [8, 9] predict magnitudes of the angular dispersion that are consistently lower than the experimental values over the whole composition range. The widest discrepancies occur at the extremes of the composition range where the linear treatment is invalid and the large angle theory should be more applicable. Although the experimental increase in angular dispersion at high nickel contents is qualitatively predicted by both theories there is no corresponding increase of similar magnitude at low nickel contents (i.e. less than about 64% nickel).

The position of the theoretical minimum in the compositional dependence of the magnitude of the angular dispersion at 75% nickel is a direct consequence of identifying the local anisotropy with the magnetocrystalline anisotropy. The introduction of a stress contribution would move the position of the minimum to higher nickel contents but is not sufficient to explain the very broad experimental minimum. Accurate calculations are difficult because of the uncertainties of both the magnitude of the stress present and its variation with film composition.

The magnetization ripple theory due to Hoffmann predicts a strong dependence of angular dispersion on the grain size. The experimental grain size was found to be independent of the film composition, but the grain size for the 550 Å thick films (i.e. 175 Å) was considerably smaller than that for 1000 Å thick series of films (350 Å).

The interpretation of the "fall-back" measurement of α_{90} in the random anisotropy models is based upon the assumption that local regions may relapse independently. This is known to be invalid for small dispersion specimens [18]. The representation provided by the magnetization ripple model, viz. relapse of the magnetization into antiparallel orientations in parallel strips, is closer to observations.

The variation of the measured coercive force for the two series of films is remarkably similar showing that the field necessary to move the Bloch walls in the 1000 Å thick films is similar to that required to move the Néel or cross-tie walls in the thinner series of films. The predictions of the surface roughness theories, for both Bloch and Néel walls, is completely at variance with the experimental results. However, the calculations on the ripple theory due to Rother show fair agreement in the shape of the curve which is improved to some extent by a stress contribution to the local anisotropy. The minimum in the calculated curve does not, however, extend over as wide a composition range as do the experimental results.

5. Conclusion

It is apparent that the most detailed approach, the magnetization ripple model, at present fails to predict the correct magnitude of the angular dispersion as a function of film composition. This is linked with the inevitable mathematical approximations necessary to permit numerical solution of the magnetization ripple equations. Although the random anisotropy models are fairly successful in describing the compositional dependence of the angular dispersion the implicit assumption of non-coupled regions is incorrect.

The existing models of domain wall coercive force discussed above are not adequate to explain the observed variation with film composition.

References

[1] R. L. Coren and H. J. Juretschke, Tech. Report. No. 9, Phys. Dept. Polytechn. Inst. Brooklyn 1960.

[2] E. J. Torok, H. N. Oredson, and A. L. Olson, J. appl. Phys. **35**, 3469 (1964).

[3] H. Rother, Z. Phys. **168**, 42 (1962).
[4] H. Hoffmann, J. appl. Phys. **35**, 1790 (1964).
[5] J. A. Cundall and A. P. King, Proc. Internat. Conf. Magnetism, Nottingham 1964, (p. 847).
[6] J. A. Cundall, Mullard Res. Rep. No. 526 (1964).
[7] H. Hoffmann, phys. stat. sol. **5**, 187 (1964).
[8] H. Hoffmann, Z. angew. Phys. **18**, 499 (1965).
[9] K. J. Harte, J. appl. Phys. 37, 1295 (1966).
[10] A. P. King, G. Robinson, J. A. Cundall, and M. J. Hight, J. sci. Instrum. **41**, 766 (1964)
[11] G. Robinson, J. Phys. Soc. Japan **17**, Suppl. B-I 558 (1962).
[12] L. Néel, J. Phys. Radium **17**, 250 (1956).
[13] R. E. Behringer and R. S. Smith, J. Franklin Inst. **272**, 14 (1961).
[14] S. Middelhoek, J. appl. Phys. **34**, 1054 (1963).
[15] M. Prutton, Phil. Mag. **5**, 625 (1960).
[16] H. Rother, Z. Phys. **168**, 283 (1962).
[17] R. J. Joenk, IBM Res. Rep. RC-1281 (1964).
[18] C. D. Olson and A. V. Pohm, J. appl. Phys. **29**, 274 (1958).

(Received May 23, 1966)

phys. stat. sol. 16, 621 (1966)

Chelsea College of Science and Technology, London

The Theory of the Stability of the Solid-Liquid Interface under Constitutional Supercooling (I)

By

R. T. DELVES

This paper discusses the stability of the solid-liquid interface when a crystal is grown from a melt which contains a segregating component. The transient motion of the interface in response to a thermodynamic fluctuation in concentration is calculated using the time dependent diffusion equations. An expansion is made in powers of the amplitude of the fluctuation, the first order terms are found exactly, and some important higher order terms are included. The interface is either stable and returns to its original flat shape or else distorts exponentially with time into the cellular shape. The stability condition is exactly the original constitutional supercooling criterion of Tiller et al.. Physical and mathematical arguments show that the interface is either stable or unstable and that no steady state slightly perturbed interface can exist. This paper is restricted to the special case of infinite thermal diffusivity, and does not include effects on an atomic scale.

Die Arbeit diskutiert die Stabilität der Grenzfläche zwischen der festen und flüssigen Phase, wenn ein Kristall aus einer Schmelze, die eine ausscheidende Komponente enthält, gezüchtet wird. Die nichtstationäre Bewegung der Grenzfläche, durch die thermodynamischen Fluktuationen der Konzentration hervorgerufen, wird mit der zeitabhängigen Diffusionsgleichung berechnet. Eine Entwicklung nach Potenzen der Amplitude der Fluktuationen wird durchgeführt; die Terme erster Ordnung werden exakt gefunden, und einige wichtige Terme höherer Ordnung sind darin enthalten. Die Grenzfläche ist entweder stabil und nimmt wieder ihre ursprüngliche flache Form an oder sie verzerrt sich exponentiell mit der Zeit in die Zellenform. Die Stabilitätsbedingung ist exakt das ursprüngliche Unterkühlungskriterium von Tiller et al.. Physikalische und mathematische Argumente zeigen, daß die Grenzfläche entweder stabil oder instabil ist und daß keine stationäre, schwach gestörte Grenzfläche existieren kann. Die Arbeit beschränkt sich auf den Fall unbegrenzten thermischen Diffusionsvermögens und schließt keine Effekte atomarer Abmessungen ein.

1. Introduction

In 1949 Rutter and Chalmers [1] showed for the first time that if a crystal were grown from a melt and some impurity or a component were rejected by the growing crystal, a layer of supercooled liquid might be formed next to the interface. If the layer forms it usually leads to the breakdown of the originally flat interface into a new shape which varies periodically across the interface. The solid which freezes out has the cellular substructure or is otherwise inhomogeneous. The effect was called 'constitutional supercooling', and has been found to be of great importance to the growth of high quality single crystals from the melt. If inhomogeneous or unsatisfactory crystals are produced it is often due (whether the investigator is aware of it or not) to constitutional supercooling. Two recent reviews of crystal growth from the melt are by Elbaum [2] and Hurle [3].

The original quantitative criterion for instability developed by Tiller, Rutter, Jackson, and Chalmers [4] was that the degree of supercooling in the liquid should increase as the distance from the interface increases. This criterion has proved to be very useful and quantitatively correct in many cases, but is not a complete answer to the problem of stability, as it does not include the redistribution of solute and temperature which must accompany some growing disturbance of the interface, nor the effect of the solid-liquid surface energy. The problem of the shape of the interface is complicated, as it involves two coupled diffusion equations, and two boundary conditions on the interface whose shape is unknown and must be determined as part of the calculation.

Mullins and Sekerka [5] have analysed the stability of the interface in more detail. They considered an interface which at time $t = 0$ has a small sinusoidal perturbation, whose amplitude we shall call B. The solute concentration C and temperature T in the liquid were chosen to satisfy the time independent diffusion equations to first order in B. The concentration of solute in the liquid at the interface had the equilibrium value for the particular temperature and curvature of the interface to first order in B.

Mullins and Sekerka calculated the value of $\dot{B}$ which was required to satisfy the conservation of solute at the interface. They found that $\dot{B}/B$ was independent of time and depended on the physical parameters defining a crystal growing experiment. If $\dot{B}/B$ were greater than zero the amplitude of the perturbation would increase exponentially with time and the interface is unstable. If $\dot{B}/B$ were less than zero the interface would be stable.

There are two difficulties in the calculation of Mullins and Sekerka. First, they set $\partial C/\partial t \equiv 0$ and $\partial T/\partial t \equiv 0$ to simplify the mathematics.

They argued that though this will not be a valid approximation when the interface grows or decays rapidly, the approximation will be justified in the particular case when $\dot{B} \equiv 0$, and the concentration and temperature do not change with time. The object of the calculation was to give just this limit between stability and instability, and the limit should be given correctly by the calculation.

One objection to this calculation is that if $\dot{B} = 0$ the shape of the interface is constant for both $t > 0$ and $t < 0$ and never can have arisen from any other physical state. If $\dot{B}/B < 0$ the interface can have come only from a state with a very large perturbation, and such a state is thermodynamically very unlikely. The interface can come from a likely physical state only when $\dot{B}/B > 0$.

The second objection is the expansion in powers of B used by Mullins and Sekerka when $\dot{B} = 0$. We may imagine in principle continuing the calculation to second and eventually infinite order in B, (though the higher terms are very complicated) and finding the C, T and shape of the interface, all of which are time independent, to order B^n. For each order there are three variables, the fourier coefficients of C and T and of the shape of the interface, and two boundary conditions, and the condition that the shape of the interface does not change with time. This would imply, if the series in B converged for any finite range of B, that we could add to the initial planar interface an arbitary amount of the perturbed interface shape, as B is arbitary, and still get a new stable nonplanar interface. Yet physically one would expect there to be only one solution (or less likely a finite number of solutions) for the shape of the cellular interface besides the planar case.

The answer to this paradox is found by calculating the denominator of a term similar to equation (20) in the paper by Mullins and Sekerka, but of order B^n. This is most easily done for the special case when the solid-liquid surface energy is zero and the thermal diffusivity is infinite. It is found that there is some value of n for which the denominator is zero. The numerator is an excessively complicated function of the experimental conditions, and we can only make this vanish too if we relax the original assumption that the planar interface to which the B terms are added is in the correct position to satisfy the boundary conditions. The B term expansion must be summed and added to the planar interface which is now in a new position. The boundary conditions must be satisfied for total interface shape, total concentration, and total temperature. This condition fixes B, which is no longer an arbitrary parameter.

We can put the problem another way. There are two independent solutions to the equation for the shape of the interface, one planar (an exact solution though only a metastable one) and the cellular shape. If we try to construct the latter as an expansion in powers of some parameter, we cannot take the exact planar state as a starting point.

In this paper we adopt a different approach. We assume that at $t = 0$ the interface is flat and there is a thermodynamic fluctuation in the concentration of solute in the melt. We then calculate the transient behaviour of the interface, (using of course the time dependent diffusion equations) to order B, the amplitude of the fluctuation. The interface initially follows the disturbance, and as $t \to \infty$ the interface either distorts exponentially with time or goes back to its original flat shape. We show that the interface is either stable or unstable and the stationary state assumed by Mullins and Sekerka does not exist. It seems more likely that a fluctuation in concentration will occur, than a fluctuation in the shape of the interface as in this case latent heat in required to melt some parts and freeze others.

We shall examine some of the higher terms in B which seem most likely to alter the stability conditions in section 6. In this paper we limit the problem to the special case of infinite thermal diffusivity and zero interface surface energy. In the second paper these restrictions are removed.

The calculations of this paper give exactly the original criterion of Tiller et al. [4]. This might be expected intuitively, and it could be argued that the more elaborate calculations of this paper are unnecessary. However the problem is of some interest in its own right, and is analogous to the stability of a non-linear mechanical system with an infinite number of degrees of freedom. Secondly there are new techniques for crystal growing such as the thin alloy zone moved by a temperature gradient. In these methods one cannot apply the original criterion of constitutional supercooling with such confidence, and an explicit calculation of the stability of the interface is necessary.

2. Equations for the Restricted Problem

We consider in this paper a special case in which the solid-liquid surface energy is zero, the latent heat is zero, and the thermal diffusivity is infinite. The temperature profile in the liquid is independent of the shape of the interface, and we need consider only the diffusion of solute. For convenience we shall work with distance and time in dimensionless form. The units are D/V and D/V^2

for distance and time respectively, where D is the diffusion constant for solute, and V is the velocity of the temperature profile in the liquid. In the steady state V is also the velocity of the solid-liquid interface. Distances are measured from a reference axis which moves with the temperature profile. The segregation coefficient k is independent of concentration. We consider two directions only, y in the plane of the interface and z the distance normal to the interface. The diffusion equation for solute in the liquid is

$$\frac{\partial^2 C}{\partial z^2} + \frac{\partial^2 C}{\partial y^2} + \frac{\partial C}{\partial z} = \frac{\partial C}{\partial t} \tag{2.1}$$

and at $t = 0$ the concentration is

$$C = (C_L - C_\infty) \exp(-z) + C_\infty + B \cos(2\pi y/\lambda) . \tag{2.2}$$

C_∞ is the concentration of solute at $z = \infty$; $C_L = C_\infty/k$ is the equilibrium concentration of solute at the planar solid-liquid interface, B is the initial amplitude of the fluctuation, whose wavelength is λ. There are no concentration gradients in the x direction. The amplitude of the fluctuation is assumed independent of z. It could be argued that the fluctuation will depend on the local degree of constitutional supercooling at $t = 0$ and so the fluctuation should depend on z. However we expect that the form of the fluctuation at the freezing interface will determine the stability of the interface, and that the form of the fluctuation in the body of the liquid will be unimportant.

p is the operator $\partial/\partial t$. The operational equation for C is

$$\frac{\partial^2 C}{\partial y^2} + \frac{\partial^2 C}{\partial z^2} + \frac{\partial C}{\partial z} - p\, C = -p\, [(C_L - C_\infty) \exp(-z) + C_\infty + B \cos(2\pi y/\lambda)] . \tag{2.3}$$

The solution which satisfies the boundary condition $C \to C_\infty$ as $z \to \infty$ for all time is

$$C = (C_L - C_\infty) \exp(-z) + C_\infty + \frac{p\, B \cos(2\pi y/\lambda)}{p + d} + \\ + \sum_{n,m} {}^{n}A_m\, B^m \cos(2\pi y\, n/\lambda) \exp\left(-Z/2 - Z\sqrt{p + a_c}\right), \tag{2.4}$$

where

$$a_c = \frac{1}{4} + \frac{4\pi^2 n^2}{\lambda^2} \tag{2.5}$$

and

$$d = \frac{4\pi^2}{\lambda^2} .$$

The summation is over all integer values of n but to order B there are only terms with $n = 1$. All operator equations operate on the Heaviside unit function $H(t)$. The interface is described by $z = Z(y, t)$, and when $t = 0$, $Z = 0$ and the interface is planar.

One boundary condition requires the concentration of solute in the liquid at the interface to be

$$C = -\gamma z + C_L \text{ on } z = Z . \tag{2.6}$$

γ is an abbreviation for

$$\frac{dC_L}{dT} \frac{dT}{dz} .$$

The first term in γ is the slope of the liquidus in the phase diagram (without sign) and the second is the temperature gradient in the liquid near the interface. The conservation of solute on the interface gives the boundary condition.

$$-\frac{\partial C}{\partial z} = (1-k)\, C\,(1+p\,Z) \text{ on } z = Z\,. \tag{2.7}$$

$$\begin{aligned}(C_{\mathrm{L}} - C_{\infty}) \exp(-Z) - \sum_n {}^{n}A_m\, B^m \cos(2\pi\, y\, n/\lambda)\left(-\frac{1}{2} - \sqrt{p+a_c}\right) \times \\ \times \exp\left[-\frac{Z}{2} - Z\sqrt{p+a_c}\right] = (1+p\,Z)\,(1-k)\Big[(C_{\mathrm{L}} - C_{\infty})\exp(-Z) + \\ + C_{\infty} + \frac{p\,B\,(\cos 2\pi\, y/\lambda)}{p+d} + \sum_n {}^{n}A_m\, B^m \cos(2\pi\, y\, n/\lambda) \times \\ \times \exp\left(-\frac{Z}{2} - Z\sqrt{p+a_c}\right)\Big].\end{aligned} \tag{2.8}$$

In (2.8) the p of $p\,Z$ operates only on Z.[1]) If we expand Z in a fourier series, which implies an expansion in powers of B.

$$Z = B\,{}^{1}Z_1 \cos(2\pi\, y/\lambda) + B^2 \cos(4\pi\, y/\lambda)\,{}^{2}Z_2 + B^2 \cos(0\, y)\,{}^{0}Z_2\,. \tag{2.9}$$

To first order in B only terms in $\cos(2\pi\, y/\lambda)$ enter[2]) and to this order, omitting the superscript,

$$\begin{aligned}-(C_{\mathrm{L}} - C_{\infty})\, Z_1 - A_1\left(-\frac{1}{2} - \sqrt{p+a_c}\right) = \\ = [1-k]\left[-(C_{\mathrm{L}} - C_{\infty})\, Z_1 + \frac{p}{p+d} + A_1 + C_{\mathrm{L}}\, p\, Z_1\right],\end{aligned} \tag{2.10}$$

$$-\gamma\, Z_1 = -(C_{\mathrm{L}} - C_{\infty})\, Z_1 + \frac{p}{p+d} + A_1\,, \tag{2.11}$$

$$\begin{aligned}\left[\frac{C_{\mathrm{L}} - C_{\infty}}{2} - \gamma\left(\frac{1}{2} - k\right) + (\gamma - C_{\mathrm{L}} + C_{\infty})\sqrt{p+a_c} + (C_{\mathrm{L}} - C_{\infty})\, p\right] Z_1 = \\ = \frac{p\left(-\frac{1}{2} - \sqrt{p+a_c}\right)}{p+d}\,.\end{aligned} \tag{2.12}$$

We introduce a new dimensionless parameter s,

$$\gamma\, s = (C_{\mathrm{L}} - C_{\infty})\,, \tag{2.13}$$

---

[1]) Strictly speaking, the operational method can only be used after the equations have been linearized by expanding in powers of B.

[2]) One should obtain the time dependent solution for A_1 and Z_1 before letting Z in the exponential terms tend to zero. Though this gives the same result for Z_1 the expression (2.4) for C is not well defined unless we take the limit Z tends to zero in the exponential terms after calculated the explicit form of C.

s measures the degree of constitutional supercooling, and $s > 1$ corresponds to the criterion of Tiller et al. [4].

$$Z_1 = \frac{p\left(-\frac{1}{2} - \sqrt{p + a_c}\right)}{\gamma\, s\, G(p)\,(p + d)}\,, \tag{2.14}$$

$$G(p) = p + a_c + \frac{(1-s)}{s}\sqrt{p + a_c} + \frac{1}{s}\left(\frac{s}{2} - \frac{1}{2} - a_c s + k\right) = \tag{2.15}$$

$$= (\sqrt{p + a_c} - \omega_+)(\sqrt{p + a_c} - \omega_-)\,,$$

where

$$\omega = \frac{s - 1 \pm \sqrt{1 + s^2(4\,a_c - 1) - 4\,k\,s}}{2\,s}\,. \tag{2.16}$$

3. Correction of the Operators for Negative z

The operator $\exp(-z\sqrt{p + a_c})$ is only defined for positive values of z. But where the interface distorts some parts of the liquid will be behind the line $z = 0$, and we need to know the concentration of solute in these regions. If the interface is stable we may use the following device. Suppose the calculation is carried out as before to order B and that the maximum value of $|Z_1|$ as a function of time is M_1, assumed to be of the order unity. We shift the reference plane for measuring z to $-M_1$. Equation (2.5) become

$$C = (C_L - C_\infty)\exp(-z + M_1) + C_\infty + \frac{p\,B\cos(2\,\pi\,y/\lambda)}{p + d} +$$

$$+ A_n\cos(2\,\pi\,y\,n/\lambda)\,B^n\exp[-z/2 - z\sqrt{p + a_c}]\,. \tag{3.1}$$

The terms of order B, but independent of y, give ${}^1Z_0$, the position of the interface at $t = 0$,

$$C_L - \gamma(Z_0 - M_1 B) = (C_L - C_\infty)\exp(-Z_0 + M_1 B) + C_\infty\,.$$

Hence $Z_0 = M_1 B$. The terms in $B\cos(2\,\pi\,y/\lambda)$ give

$$-(C_L - C_\infty)Z_1 + \frac{p}{p + d} + A_1\exp\left(-\frac{Z}{2} - Z\sqrt{p + a_c}\right) = -\gamma Z_1\,, \tag{3.2}$$

$$-(C_L - C_\infty)Z_1 - A_1\left(-\frac{1}{2} - \sqrt{p + a_c}\right)\exp\left(-\frac{Z}{2} - Z\sqrt{p + a_c}\right) =$$

$$= (1 - k)\,C_L\,p\,Z_1 - \gamma(1 - k)Z_1\,. \tag{3.3}$$

As to the order B we can set the exponentials equal to one there is no change in the equation for Z_1. M_1 will appear however in the expressions for Z_2 and higher terms.

4. Solution for Z_1 and Stability Conditions

Z_1 may be calculated by turning (2.14) into partial fractions

$$Z_1 = \frac{p}{\gamma\, s\,(p + d)(\omega_+ - \omega_-)}\left[\frac{\left(-\frac{1}{2} - \omega_+\right)}{\sqrt{p + a_c} - \omega_+} - \frac{\left(-\frac{1}{2} - \omega_-\right)}{\sqrt{p + a_c} - \omega_-}\right] \tag{4.1}$$

and as

$$\frac{1}{\sqrt{p + a_c} - \omega} = \frac{\omega}{p + a_c - \omega^2} + \frac{1}{\sqrt{p + a_c}} + \frac{\omega^2}{\sqrt{p + a_c}\,(p + a_c - \omega^2)},$$

$$Z_1 = \frac{1}{\gamma\, s\,(\omega_+ - \omega_-)} \left\{ \left(-\frac{1}{2} - \omega_+ \right) K_1(\omega_+) - \left(-\frac{1}{2} - \omega_- \right) K_1(\omega_-) - (\omega_+ - \omega_-)\, K_2 \right\}. \tag{4.2}$$

$$K_1(\omega) = \frac{\omega\,[\exp(-a_c\, t + \omega^2\, t) - \exp(-dt)]}{(d + a_c + \omega^2)} + \frac{\omega^2 \exp(\omega^2\, t - a_c\, t)}{\sqrt{\pi}\,(d - a_c + \omega^2)} \times$$

$$\times \int\limits_0^t \frac{\exp(-\omega^2\, u)\, du}{\sqrt{u}} - \frac{\omega^2 \exp(-dt)}{\sqrt{\pi}\,(d - a_c + \omega^2)} \int\limits_0^t \exp \frac{(du - a_c u)}{\sqrt{u}}\, du\,, \tag{4.3}$$

$$K_2 = \frac{\exp(-dt)}{\sqrt{\pi}} \int\limits_0^t \frac{\exp(du - a_c\, u)\, du}{\sqrt{u}}\,.$$

It may be verified by a lengthy but straightforward calculation that the solution for Z_1 satisfies the equations (2.10) and (2.11), and this provides the justification for the expansion of the exponentials and the operator manipulations.

We see from 4.3 that if $\mathrm{Re}(\omega^2) > 0$ the conditions for $Z_1 \to 0$ as $t \to \infty$ are either

$$\mathrm{Re}\,(a_c - \omega^2) > 0 \quad \text{or} \quad \mathrm{Re}(\omega) \leqq 0 \tag{4.4}$$

applied to each root. The same result may be obtained by expressing the Laplace transform of Z_1 as a contour integral. The condition for the roots to be real is

$$1 - 4\,k\,s + s^2\,(4\,a_c - 1) \geqq 0\,, \tag{4.5}$$

while if ω is complex the condition for $\mathrm{Re}(\omega^2) > 0$ is

$$1 - 4\,k\,s + s^2\,(4\,a_c - 1) + (1 - s)^2 > 0\,. \tag{4.6}$$

For very long but not infinite wavelengths the conditions (4.5) and (4.6) mark out a narrow strip in the $k - s$ plane, shown Fig. 1. In this region the second condition of (4.4) requires $s < 1$, and the first condition of (4.4) requires $s - 1 + 2\,k\,s > 0$, which is always true for $s > 1$.

When ω is complex and $\mathrm{Re}(\omega^2) < 0$ we see from the asymptotic expansions for K_1 that Z_1 is always stable. As the wavelength

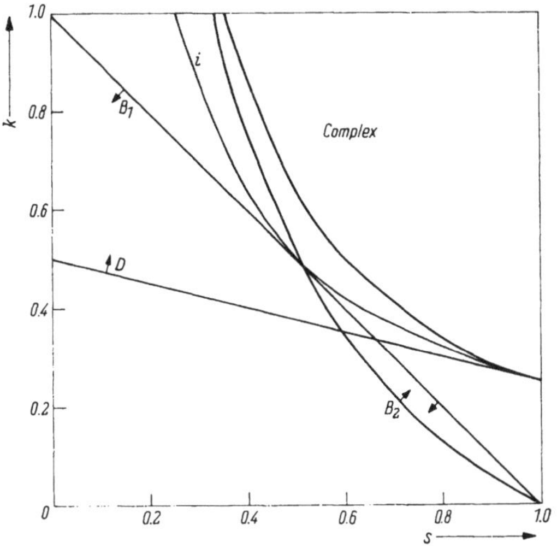

Fig. 1. k against s (equation (2.13)) for very long wavelengths, $a_c = 0.25$. The lines B_1, B_2, and D are the stability conditions of Section 4. The arrows point to the regions of stability. The line i divides real and complex roots

decreases, the region of complex ω moves to the top of the $k - s$ graph, and vanishes when $4\,a_c > 5$, k being less than 1.

We shall show that for a fluctuation of given wavelength terms in Z of order B^n involve the same denominator as in (2.14), but with λ replaced by λ/n. As a necessary, but possibly not sufficient condition for stability, (4.4) must be satisfied for all a_c as n ranges from 1 to ∞. In addition we expect that during a crystal growing experiment, there will be a sufficient probability for a fluctuation with any finite wavelength to occur. We need (4.4) to be true for all values of $4\,a_c - 1$, which ranges from zero to plus infinity.

If ω is real the conditions for stability are

A: $a_c > \omega_+^2$; $s > 1$.
This requires $(s - 1 + 2\,k) > (s - 1)\,2\,a_c^{1/2}$.

B: $a_c > \omega_+^2$; $s < 1$.
This requires $(s - 1 + 2\,k) > (1 - s)\,2\,a_c^{1/2}$, or else $s - 1 + 2\,k\,s > 0$.

C: $\omega_+ < 0$; $s > 1$.
This is never possible.

D: $\omega_+ < 0$; $s < 1$.
This requires $s - 2 + 4\,k > (4\,a_c - 1)\,s$.

E: $a_c > \omega_-^2$; $s > 1$.
This repuires $s - 1 + 2\,k\,s > 0$ which is always true.

F: $a_c > \omega_-^2$; $s < 1$.
This requires both $s - 1 + 2\,k\,s > 0$ and $4\,k\,(s - 1 + k) > (s - 1)^2 \times \times (4\,a_c - 1)$.

G: $\omega_- < 0$; $s > 1$.
This requires $4\,k - 2 + s < (4\,a_c - 1)\,s$.

H: $\omega_- < 0$; $s < 1$.
This is always true.

The conditions for ω_+ are shown in Fig. 1 and 2 for $4\,a_c - 1 = 0$ and $4\,a_c - 1 = 2$ respectively. If $s > 1$ it is not possible to satisfy the conditions for ω_+ for all values of a_c. If $s < 1$ it is always possible to satisfy one or other of the conditions for both roots. The condition for stability is $s < 1$, which is exactly the original condition of Tiller et al. [4]. The interface does not oscillate with time when the roots are complex (which would be true in ordinary mechanical stability problems). To illustrate this Fig. 3 shows $-\gamma\,s\,Z_1$ as a function of time for $4\,a_c = 3$ and three different values of k and s, two of which involve complex roots.

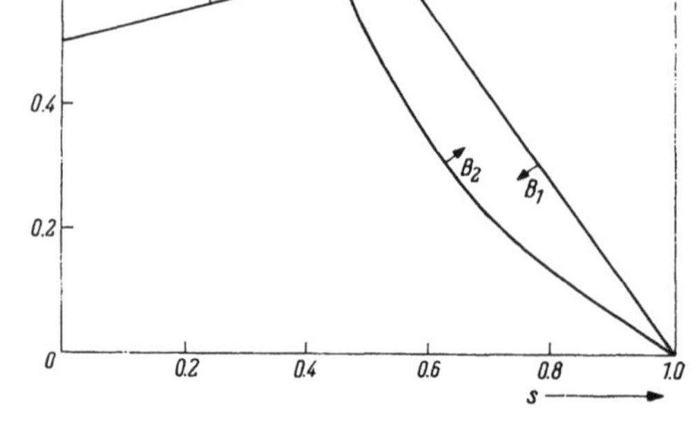

Fig. 2. k against s as in Fig. 1, but for $a_c = 0.75$

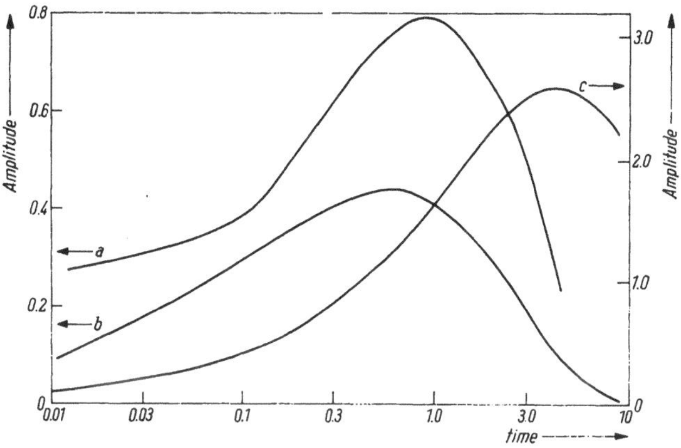

Fig. 3. The amplitude of the sinusoidal perturbation of the interface, $-\gamma s Z_1$, as a function of time, $a_c = 0.75$.
Curve a: Scale on L.H.S., $k = 0.8, s = 0.925$, ω complex.
Curve b: Scale on L.H.S., $k = 0.8$, $s = 0.5$, ω complex.
Curve c: Scale on R.H.S., $k = 0.25$, $s = 1.5$, ω real

If we set the solid-liquid surface energy equal to zero and the thermal diffusivity equal to infinity in equation (20) in the paper by Mullins and Sekerka [5], we obtain their stability condition for the restricted case of the present paper.

$$s - 1 + 2k - 2(s-1)\sqrt{a_c} > 0 . \tag{4.7}$$

If $s > 1$,

$$4k(s-1) + 4k^2 > (s-1)^2 (4a_c - 1) \tag{4.8}$$

which is equivalent to our condition B. If $s < 1$ (4.7) can always be satisfied.

5. The Solution for $s \ll 1$

As s becomes small we get further from the possibility of constitutional supercooling. If $s \to 0$ the negative root ω_- tends to $-1/s$. The terms in $K(\omega_-)$ of (4.3) which involve $\exp(-\omega_-^2 t)$ correspond to the rapid motion of the interface in response to an increase or decrease in solute concentration. The positive root is

$$\omega_+ = \frac{1-2k}{2} + \text{terms order } s .$$

This corresponds to the removal by diffusion of solute built up near the interface during the rapid initial motion. As $k < 1$, $a_c > \omega_+^2$ for all finite wavelengths, but nor for $\lambda = \infty$. The terms in $\exp(-dt)$ in Z_1 correspond to the removal of solute concentration gradients by lateral diffusion in the y direction. When s is not small the three kinds of term become mixed up, but in a rough sense the identification with the three different physical effects remain valid.

6. The Influence of Higher Order Terms

To ensure complete stability the expansion of Z in powers of B should converge for all values of time, and the coefficient of each term in B^n should tend to zero as $t \to \infty$. The higher order terms are very complicated and we shall consider only those terms which seem most likely to upset stability, and deduce necessary but possibly not sufficient conditions for complete stability. It is of course essential in any nonlinear mechanical or electrical problem to consider the influence of higher order terms on stability.

If we suppose that $Z_1, Z_2, \ldots, Z_{n-1}$ have been found as explicit functions of time, we can calculate Z_n as

$$\cos(2\pi y m)\, G(p, m)\, {}^m Z_n = \text{time dependent terms from } Z_1, \ldots, Z_{n-1}\,. \quad (6.1)$$

$G(p, m)$ is the operator of (2.15), but with λ in a_c replaced by λ/m. The terms on the R.H.S. of (6.1) will be very complicated, but we may take a typical term from $Z_1^n/n!$ in the expansion of $\exp(-Z)$. This will contain exponentially decreasing terms, $\exp(-f\,t)$, where f is some pòsitive constant. The condition for stability does not depend on f, but is the condition of (4.4), with the altered value of a_c. We have already used this condition in section 4.

If $\lambda = \infty$ it would appear that the solution for Z_1 in (4.2) tends to a finite limit as $t \to \infty$. $\lambda = \infty$ corresponds to a sudden increase everywhere in the concentration of solute, and the interface must readjust itself to a new position. It is interesting to see how instability can result in this special case. All the higher terms contain the same $G(p)$ operator.

$$Z_2 \sim \frac{1}{G(p)} Z_1\, Z_1 + \text{other terms}\,. \quad (6.2)$$

We may choose one constant term, $\exp(-dt) \equiv 1$ in Z_1

$$Z_2 \sim \frac{1}{G(p)\, G(p)} \exp(-dt) + \cdots \text{etc}\,. \quad (6.3)$$

When there are repeated roots in the denominator the solution contains exponentially decaying terms multiplied by powers of time

$$\frac{1}{[\sqrt{p + a_c} - \omega]^n} \exp(-dt) \sim \frac{(2\,t\,\omega)^{n-1} \exp(-dt)}{(n-1)!\,[\sqrt{p + a_c} - \omega]} + \text{lower powers of } t\,. \quad (6.4)$$

As a result, when the terms $B^n Z_n$ are summed there is the possibility of obtaining time independent terms and exponentially damped terms such as $\exp(-a_c\, t + \omega^2\, t)$ multiplied by an exponentially increasing function of time. In this special case the stability conditions are not valid. On the other hand if λ is finite we can still get repeated roots where we calculate terms such as ${}^1Z_1\, B \times \times \cos\theta$, ${}^1Z_3\, B^3 \cos\theta$, ${}^1Z_5\, B^5 \cos\theta$ etc. However, provided that all terms in Z_1 decay exponentially, it can be shown that there are always at least as many exponentially decaying terms as powers of time in the driving terms

$$G(p)\, {}^1Z_{2n+1} \sim [t \exp(-dt)]^n + \text{other terms}$$

and stability is not upset.

7. Solution for the Boundary between Stable and Unstable States

In this section we examine in more detail the stability of the interface where the stability conditions are just satisfied. We see from the explicit form for Z_1 that $\omega = 0$ as well as $\omega < 0$ gives stability. On the other hand if $a_c - \omega^2 = 0$ or is pure imaginary, Z_1 tends to a constant or sine term as $t \to \infty$. In this case we must consider the influence of higher order terms. We take $a_c - \omega_+^2 = i\, v$. A term in $B^3 \cos(2\pi\, y/\lambda)$, ${}^1Z_3$ is calculated from

$$G(p)\, {}^1Z_3 = \frac{3}{4}\left(\frac{1}{2} - \sqrt{p + a_c}\right) \frac{(Z_1)^3}{3!}\,. \quad (7.1)$$

Z_1 has to be calculated as an explicit function of time before the operator is applied, and the 3/4 comes from $\cos^3\theta = 3/4 \cos\theta +$ other terms. The terms in Z_1 which produce time independent terms in $(Z_1)^2$ are

$$\gamma\, s\, Z_1 = \eta_+ \exp(i\, v\, t) + \eta_- \exp(-\, i\, v\, t)\,, \tag{7.2}$$

where

$$\eta_+ = \frac{\left(-\frac{1}{2} - \omega_+\right)(\omega_+ + |\omega_+|)}{(\omega_+ - \omega_-)\,(d - a_c + \omega_+^2)}$$

and η_- is the same expression but with ω_+ and ω_- interchanged. If Re (ω_+) is > 0, $\left(\mathrm{Re}(\omega_+) < 0\right.$ gives stability$\left.\right)$, and if $\omega_+ = \omega_r + i\,\omega_i$ we find

$$\eta_+\,\eta_- = \frac{2\,(\omega_r^2 + \omega_i^2)}{\omega_i^2\left[\left(\omega_r - \frac{1}{2}\right)^2 + \omega_i^2\right]} \tag{7.3}$$

which is always positive. We can now write the important terms in $(Z_1)^3$.

$$(Z_1)^3 = \frac{6\,\eta_+\,\eta_-\,[2\,Z_1 - \eta_+ \exp(i\,v\,t) - \eta_- \exp(-\,i\,v\,t)]}{2\,(\gamma\, s)^2} + \text{other terms}, \tag{7.4}$$

$$^1Z_3 = \frac{3\,\eta_+\,\eta_-}{8\,(\gamma\, s)^2}\left[\frac{\frac{1}{2} - \sqrt{p + a_c}}{G(p)}\right][2\,Z_1 - \eta_+ \exp(i\,v\,t) - \eta_- \exp(-\,i\,v\,t)]\,. \tag{7.5}$$

In the same way we calculate the most important terms in $^1Z_5$, $^1Z_7$ etc. and find

$$^1Z_{2n+1} = \left[\frac{3\,\eta_+\,\eta_-\left(\frac{1}{2} - \sqrt{p + a_c}\right)}{4\,(\gamma\, s)^2\,(G\,p)}\right]^n \left[Z_1 - \frac{\eta_+}{2}\exp(i\,v\,t) - \frac{\eta_-}{2}\exp(-\,i\,v\,t)\right]. \tag{7.6}$$

If we sum the geometrical series to get the important part of Z

$$Z = \frac{G(p)}{G(p) + \frac{3\,B^2\,\eta_+\,\eta_-}{4\,(\gamma\, s)^2}\left(\frac{1}{2} - \sqrt{p + a_c}\right)}\left[Z_1 - \frac{\eta_+}{2}\exp(i\,v\,t) - \frac{\eta_-}{2}\exp(-\,i\,v\,t)\right]. \tag{7.7}$$

So we have replaced the old operator $G(p)$ by a new one in the denominator of (7.7). The effect is to change $a_c - \omega_+^2$ by adding to it

$$6\,s\,\eta_+\,\eta_-\left(\frac{B}{2\,\gamma\, s}\right)^2\left(2 - s - \frac{3\,s\,B^2\,\eta_+\,\eta_-}{4\,\gamma^2\,s^2}\right). \tag{7.8}$$

As s must be less than 2 when the original $a_c - \omega^2$ is pure imaginary Re $(a_c - \omega^2) > 0$ and this point is stable. It might appear that (7.8) predicts instability for large B. This is not correct as we have omitted in the calculation an infinite set of terms which would give a contribution of order B^4 to 7.8.

In the particular case of this paper we know that $\mathrm{Re}(\omega) < 0$ when $a_c - \omega^2$ is pure imaginary, so strictly speaking the analysis of this section is unnecessary.

In the same way if $s > 1$ and $a_c - \omega^2 = 0$ for some λ we can show that the interface is stable. In the restricted problem of this paper however, there will always be a shorter wavelength perturbation to which the interface will be unstable.

8. Higher Order Terms from the M shift

A further infinite set of higher order terms comes from the M shift of section 3. It can be shown that provided M can be found these terms do not affect the stability conditions. We have a self-consistent equation to solve for M and the value of M becomes larger as B increases, and as we get closer to the values of s, k and λ which produce instability. It is possible when we are close to the instability conditions and when B is too large, that no finite value of M can be found, and the interface is then unstable.

This effect has a physical interpretation. If the interface distorts and forms the grooves found with cellular growth, the solute which is trapped in the grooves cannot diffuse sideways and finds it difficult to diffuse down the grooves into the bulk of the liquid. If the grooves start to form they make the interface more unstable, and the transition from stability to instability is a kind of co-operative effect. It is not possible to make any numerical estimate of the importance of this effect within the restrictions of this paper as we have ignored the solid-liquid surface energy, which determines the stability of the interface to very short wavelength disturbances (Mullins and Sekerka).

9. Conclusions

We have calculated the stability of a freezing interface for the special case in which the thermal diffusivity is infinite. Though no new condition for stability has been derived which differs from the original one of Tiller et al., the mathematical equipment is set out which will enable more complicated crystal growing problems to be tackled. In a second paper we shall remove the restriction that the thermal diffusivity is infinite.

The problem is treated as one in classical mechanics. We have ignored atomic considerations and the possibility that a finite degree of supersaturation is necessary to nucleate solid. Those familiar with plasma dispersion theory or Green's functions in many body problems may detect at times the warmth of a familiar hearth. There are similarities between the Schwarzchild-Krusal instability in a gaseous plasma and constitutional supercooling. One might also draw analogies between Landau damping in the form of Landau's original treatment and the M shift of section 3. The equation for solute diffusion lacks the equivalent of induction or effects from a magnetic field and this makes them qualitatively different and less interesting than the equations which produce plasma dispersion relations.

References

[1] J. W. RUTTER and B. CHALMERS, Canad. J. Phys. **31**, 15 (1953).
[2] C. ELBAUM, Progress in Materials Science, Vol. 8, Pergamon Press 1959.
[3] D. T. J. HURLE, Progress in Materials Sciences, Vol. **10**, Pergamon Press 1961.
[4] W. A. TILLER, J. W. RUTTER, K. A. JACKSON, and B. CHALMERS, Acta metall. **1**, 428 (1953).
[5] W. W. MULLINS and R. F. SEKERKA, J. appl. Phys. **35**, 444 (1964).

(Received May 24, 1966)

phys. stat. sol. **16**, 633 (1966)

Institute of Solid State Physics, Academy of Sciences of the USSR, Moscow

Mössbauer Effect in Some Polyatomic Crystals

By

Ya. A. Iosilevskii

Expressions are obtained for the probability of the Mössbauer effect and the thermal shift in the energy of a γ-photon for binary crystalline compounds AB_{r-1}. These compounds include crystals of the type NaCl or CsCl, crystals of completely ordered alloys ($AuCu_3$, $AlFe_3$ etc.), and monatomic crystals with isolated impurity atoms. It is assumed that the force constants depend only on the differences of the position vectors of the atoms. The treatment includes the two cases in which nucleus A or nucleus B is resonant. The case in which atom A (mass M_0) and atom B (mass M) differ considerably in mass is analysed in greater detail. For $M_0/M \ll 1$ or $M_0/M \gg 1$ the fraction e^{-Z_0} of recoilless emissions (or absorptions) of γ-rays by atom A and the thermal shift δE_0 are independent of M_0/M for given interatomic forces. Values of these quantities are hence determined by the appropriate limiting expressions (M_0 = const, $M \to \infty$ or $M \to 0$). Thus the finite Mössbauer effect for an atoms surrounded by light atoms is independent of the mass of the latter. In addition there exists a limit to the gain in the probability of the Mössbauer effect and in the thermal shift for an atom which can be obtained by surrounding it with increasingly heavier atoms. It is also shown that the probability of the effect and the thermal shift for the atoms A, for the given interactions, are essentially independent of r. The quasiclassical expansions of Z_j and δE_j, for the j-th atom in a unit cell, are obtained for more complicated compounds of the type $A^{(1)}_{r_1} A^{(2)}_{r_2} \ldots A^{(n)}_{r_n}$ $(r_1 + r_2 + \cdots r_n = r)$ by the method of moments.

В предположении, что силовые константы зависят лишь от разности радиус-векторов атомов в решётке, но не от их сорта, получены единые формулы, описывающие вероятность эффекта Мессбауэра и „температурный“ сдвиг линии в бинарных кристаллических соединениях типа AB_{r-1}, к числу которых относятся такие различные, на первый взгляд системы, как кристаллы со структурой типа NaCl и CsCl, кристаллы типа полностью упорядоченных сплавов ($AuCu_3$, $AlFe_3$ и т.д.) и решётки с изолированными примесными атомами. Рассматриваются оба случая, когда активным атомом является атом A или атом B. Детально анализируется случай большого различия масс M_0 и M атомов A и B. Показано, что численные значения вероятности испускания (поглощения) без отдачи e^{-Z_0} и „температурного“ сдвига линии δE_0 для атома A при заданных силах в случаях $M_0/M \ll 1$ и $M_0/M \gg 1$ не зависят ототношения M_0/M и определяются некоторыми простыми предельными выражениями, соответствующими $M_0/M \to 0$ или $M_0/M \to \infty$ (M_0 = const, $M \to \infty$ или $M \to 0$). Это приводит к сохранению конечного эффекта Мессбауэра для атома, находящегося в окружении из сколь угодно лёгких атомов другого сорта, а с другой стороны, кладёт предел выигрышу в эффекте и сдвиге лин ииза счёт окружения из более тяжёлых атомов. Показано также, что вероятность эффекта и сдвиг линии для атомов A при заданных силах, в сущности, не зависят от их относительной концентрации (т.е. от r). Методом моментов получены квазиклассические разложения для Z_j и δE_j, справедливые для любого атома j в элементарной ячейке для более сложных соединений типа $A^{(1)}_{r_1} A^{(2)}_{r_2} \ldots A^{(n)}_{r_n}$ $(r_1 + r_2 + {} + \cdots + r_n = r)$.

1. Introduction

The paper of the present author [1] has been concerned with the lattice dynamics in binary crystalline compounds such as AB_{r-1} with r atoms per unit cell, built by periodically filling the sites of one and the same Bravais lattice by atoms A and B. Under the simplifying, but often well justified, assumption that the effective force constants depend only on the differences between the position vectors of the atoms rather than on the type of atoms, it was possible to solve the above problem in terms of the dynamic problem for a certain monatomic "reference" crystal and to treat, in a unified manner, the vibrations in systems which, at first glance, seem to be very different, such as diatomic crystals of the NaCl or CsCl type [2], completely ordered alloy crystals ($AuCu_3$, $AlFe_3$, etc.), and monatomic lattices with isolated impurity atoms [3 to 5]. This in turn allows to treat the Mössbauer effect in such complicated crystal structures in a unified way. Such a treatment is the object of the present paper.

2. The Dynamic Problem

Consider the main results of the solution of the dynamic problem [1]. We shall deal with two space lattices

$$x_{\boldsymbol{n}}^i = \sum_{p=1}^{3} a_p^i\, n_p \quad \text{and} \quad \xi_{\boldsymbol{\nu}}^i = \sum_{p=1}^{3} \alpha_p^i\, \nu_p$$

and the two reciprocal lattices

$$y_{\boldsymbol{m}}^i = \sum_{p=1}^{3} b_p^i\, m_p \quad \text{and} \quad \eta_{\boldsymbol{\mu}}^i = \sum_{p=1}^{3} \beta_p^i\, \mu_p$$

corresponding to them (the notation is the same as in [1]: $\boldsymbol{a}_p$, $\boldsymbol{\alpha}_p$, $\boldsymbol{b}_p$, $\boldsymbol{\beta}_p$ are the basic vectors, n_p, ν_p, m_p, μ_p are integers, the superscripts denote Cartesian coordinates, $p = 1, 2, 3$), which we shall denote below as X, Ξ Y, and H. The connection between these lattices is established by the connection between their basic vectors

$$\hat{\alpha} = \hat{P}\,\hat{a}\,, \quad \hat{b} = \tilde{\hat{a}}^{-1} = \tilde{\hat{P}}\,\tilde{\hat{\alpha}}^{-1} = \tilde{\hat{P}}\,\hat{\beta}\,, \quad \hat{\beta} = \tilde{\hat{\alpha}}^{-1} = \tilde{\hat{P}}^{-1}\,\hat{b}\,. \tag{2.1}$$

For convenience, the matrixes $a_{pi} \equiv a_p^i$, etc., are introduced here; the wavy line denotes the transponation; $P_{pp'}$ are the given integer coefficients.

Let the sites $\boldsymbol{\xi}_{\boldsymbol{\nu}}$, forming part of the sites $\boldsymbol{x}_{\boldsymbol{n}}$, be occupied by atoms of kind A of mass M_0, and the rest of the sites $\boldsymbol{x}_{\boldsymbol{n}}$ be occupied by atoms of a different kind B of mass M. According to the first relation (2.1) the unit cell volumes $v_\xi = \det \hat{\alpha}$ and $v_x = \det \hat{a}$ of the lattices Ξ and X are connected by the relation $v_\xi = v_x \det \hat{P}$ so that the integer

$$r = \det \hat{P} \tag{2.2}$$

is the number of atoms per unit cell for the structure Ξ.

If the force constants depend only on the differences of the position vectors of the lattice points, the dynamic problem for the structure under consideration is reduced to the system of equations

$$M_{\boldsymbol{n}}\, \omega^2(\boldsymbol{f}, \sigma)\, \psi_{\boldsymbol{f}\sigma}^i(\boldsymbol{x}_{\boldsymbol{n}}) = \sum_{\boldsymbol{n}'} A^{ii'}(\boldsymbol{x}_{\boldsymbol{n}} - \boldsymbol{x}_{\boldsymbol{n}'})\, \psi_{\boldsymbol{f}\sigma}^{i'}(\boldsymbol{x}_{\boldsymbol{n}'})\,, \tag{2.3}$$

$$M_{\boldsymbol{n}} = M\left(1 - \lambda \sum_{\boldsymbol{\nu}} \delta_{\boldsymbol{x}_{\boldsymbol{n}} - \boldsymbol{x}_{\boldsymbol{n}_0}, \boldsymbol{\xi}_{\boldsymbol{\nu}}}\right); \quad \lambda = 1 - \frac{M_0}{M} \quad (-\infty < \lambda < 1) \tag{2.4}$$

and has the solutions determined by the following relations:

$$\psi^i_{f\sigma}(\boldsymbol{x_n}) = \sqrt{\frac{r}{N}}\; v^i_{f\sigma}(\boldsymbol{x_n})\, e^{i\boldsymbol{f}(\boldsymbol{x_n} - \boldsymbol{x_{n_0}})}\,, \tag{2.5}$$

$$v^i_{f\sigma}(\boldsymbol{x_n}) = \frac{1}{\sqrt{M}}\left[\frac{d\ln\omega^2(\boldsymbol{f},\sigma)}{d\lambda}\right]^{1/2} j^k_{f\sigma}\,\frac{\lambda\,\omega^2(\boldsymbol{f},\sigma)}{r} \times$$
$$\times \sum_{\mu}{}^{(0)} \sum_{s=1}^{3} \frac{e^i(\boldsymbol{f}+2\pi\,\boldsymbol{\eta}_\mu, s)\, e^k(\boldsymbol{f}+2\pi\,\boldsymbol{\eta}_\mu, s)}{\omega^2(\boldsymbol{f},\sigma) - \omega_0^2(\boldsymbol{f}+2\pi\,\boldsymbol{\eta}_\mu, s)}\, e^{2\pi i \boldsymbol{\eta}_\mu(\boldsymbol{x_n} - \boldsymbol{x_{n_0}})}\,, \tag{2.6}$$

$$j^i = D^{ik}(\omega^2, \boldsymbol{f})\, j^k\,, \tag{2.7}$$

$$D^{ik}(\omega^2, \boldsymbol{f}) \equiv \frac{\lambda\,\omega^2}{r} \sum_{\mu}{}^{(0)} \sum_{s=1}^{3} \frac{e^i(\boldsymbol{f}+2\pi\,\boldsymbol{\eta}_\mu, s)\, e^k(\boldsymbol{f}+2\pi\,\boldsymbol{\eta}_\mu, s)}{\omega^2 - \omega_0^2(\boldsymbol{f}+2\pi\,\boldsymbol{\eta}_\mu, s)}\,, \tag{2.8}$$

$$\Delta(\omega^2, \boldsymbol{f}) \equiv \det\{\delta^{ik} - D^{ik}(\omega^2, \boldsymbol{f})\} = 0\,, \tag{2.9}$$

$$\frac{d\ln\omega^2(\boldsymbol{f},\sigma)}{d\lambda} = \left\{\frac{\lambda^2\,\omega^4(\boldsymbol{f},\sigma)}{r} \sum_{\mu}{}^{(0)} \sum_{s=1}^{3} \frac{(\boldsymbol{j}_{f\sigma}\,\boldsymbol{e}(\boldsymbol{f}+2\pi\,\boldsymbol{\eta}_\mu, s))^2}{[\omega^2(\boldsymbol{f},\sigma) - \omega_0^2(\boldsymbol{f}+2\pi\,\boldsymbol{\eta}_\mu, s)]^2} - \lambda\right\}^{-1}, \tag{2.10}$$

$$v^i_{f\sigma}(\boldsymbol{x_{n_0}}) = \frac{1}{\sqrt{M}}\left[\frac{d\ln\omega^2(\boldsymbol{f},\sigma)}{d\lambda}\right]^{1/2} j^i_{f\sigma} \neq 0\,, \tag{2.11}$$

$$v^i_{f\sigma}(\boldsymbol{x_n} + \boldsymbol{\xi_\nu}) = v^i_{f\sigma}(\boldsymbol{x_n})\,. \tag{2.12}$$

Here $\boldsymbol{f}$ is a wave vector assuming N/r values in the region $w_\xi = \left(\frac{2\pi}{v_\xi}\right)^3$ (e.g. corresponding to a unit cell or the first Brillouin zone of the lattice H); $\omega^2(\boldsymbol{f},\sigma)$ are the exact solutions of (2.9) (its order is $3r$), $\sigma = 1, 2, \ldots, 3r$ labels various solutions (branches); N is the total number of the X-lattice points (the number of unit cells in the Ξ-lattice is N/r); $\boldsymbol{x_{n_0}}$ is the position vector of the atom A in a unit cell of the structure Ξ containing the origin; $\boldsymbol{j}_{f\sigma}$ is a real unit vector; $\sum_\mu^{(0)}$ denotes the summation over r H-lattice points belonging to the region $w_x = \left(\frac{2\pi}{v_x}\right)^3$, corresponding to a unit cell or the first Brillouin zone of the lattice Y; $\omega_0(\boldsymbol{f}, s)$ and $\boldsymbol{e}(\boldsymbol{f}, s)$ are eigenfrequencies and eigenvectors of the dynamic problem for the "reference" monatomic lattice, i.e. of the problem (2.3) for $M_{\boldsymbol{n}} = M$ $(\lambda = 0)$:

$$\left.\begin{aligned} &M\,\omega_0^2(\boldsymbol{f}, s)\, e^i(\boldsymbol{f}, s) = G^{ii'}(\boldsymbol{f})\, e^{i'}(\boldsymbol{f}, s)\,; \qquad s = 1, 2, 3\,,\\ &e^i(\boldsymbol{f}, s)\, e^i(\boldsymbol{f}, s') = \delta_{ss'}\,, \sum_{s=1}^{3} e^i(\boldsymbol{f}, s)\, e^{i'}(\boldsymbol{f}, s) = \delta^{ii'}\,, \end{aligned}\right\} \tag{2.13}$$

$$G^{ii'}(\boldsymbol{f}) = \sum_{\boldsymbol{n}} A^{ii'}(\boldsymbol{x_n})\, e^{-i\boldsymbol{f}\boldsymbol{x_n}} \tag{2.14}$$

(in (2.13) and (2.14) $\boldsymbol{f}$ varies in the region w_x).

The functions (2.5) and (2.6) satisfy the requirements for orthonormality and completeness

$$\sum_{\boldsymbol{n}} M_{\boldsymbol{n}}\, \psi^{i*}_{f\sigma}(\boldsymbol{x_n})\, \psi^i_{f'\sigma'}(\boldsymbol{x_n}) = \delta_{ff'}\,\delta_{\sigma\sigma'}\,; \qquad \sum_{j=0}^{r-1} M_{\boldsymbol{n}_j}\, v^{i*}_{f\sigma}(\boldsymbol{x}_{\boldsymbol{n}_j})\, v^i_{f\sigma'}(\boldsymbol{x}_{\boldsymbol{n}_j}) = \delta_{\sigma\sigma'}\,, \tag{2.15}$$

$$\sum_{\boldsymbol{f}} \sum_{\sigma=1}^{3r} \psi^{i*}_{f\sigma}(\boldsymbol{x_n})\, \psi^{i'}_{f\sigma}(\boldsymbol{x_{n'}}) = \frac{1}{M_{\boldsymbol{n}}}\,\delta_{\boldsymbol{nn'}}\,\delta^{ii'}\,; \qquad \sum_{\sigma=1}^{3r} v^{i*}_{f\sigma}(\boldsymbol{x}_{\boldsymbol{n}_j})\, v^{i'}_{f\sigma}(\boldsymbol{x}_{\boldsymbol{n}_{j'}}) = \frac{1}{M_{\boldsymbol{n}_j}}\,\delta_{jj'}\,\delta^{ii'}\,, \tag{2.16}$$

where j labels the atoms in a unit cell. Here and below the summation $\sum_f$ or the integration $\int d^3f(\ldots)$ extends over the region w_ξ.

Assuming $j = j' = 0$ in the second equation (2.16), we have the important relations

$$(1-\lambda)\sum_{\sigma=1}^{3r}\frac{d\ln\omega^2(\boldsymbol{f},\sigma)}{d\lambda}(j^i_{\boldsymbol{f}\sigma})^2 = 1\,,\quad \frac{1}{3}(1-\lambda)\sum_{\sigma=1}^{3r}\frac{d\ln\omega^2(\boldsymbol{f},\sigma)}{d\lambda} = 1 \quad (2.17)$$

which hold for any value of $\boldsymbol{f}$.[1])

The derivation of all the above relations is contained in [1]. We may also verify their validity by substituting (2.5) and (2.6) into (2.3) and (2.15). The relation (2.10) is obtained by differentiating both sides of (2.7) with respect to λ, taking into account that $j^i\frac{dj^i}{d\lambda} = \frac{1}{2}\frac{d\boldsymbol{j}^2}{d\lambda} = 0$ because $|\boldsymbol{j}| = 1$ (for similar calculations for an impurity atom see [3 to 5]).

Further we shall also need the following relations referring to the "reference" lattice [1]

$$\frac{1}{r}\sum_\mu^{(0)}\sum_{s=1}^{3}\omega_0^{2m}(\boldsymbol{f}+2\pi\boldsymbol{\eta}_\mu,s)\,e^i(\boldsymbol{f}+2\pi\boldsymbol{\eta}_\mu,s)\,e^j(\boldsymbol{f}+2\pi\boldsymbol{\eta}_\mu,s) =$$
$$= \langle\omega_0^{2m}\rangle^{ij} + \frac{1}{M^m}\sum_{\nu\neq 0}\sum_{\boldsymbol{n}_1\boldsymbol{n}_2\ldots\boldsymbol{n}_m} A^{ii_1}(\boldsymbol{x}_{\boldsymbol{n}_1})\,A^{i_1i_2}(\boldsymbol{x}_{\boldsymbol{n}_2})\ldots A^{i_{m-1}j}(\boldsymbol{x}_{\boldsymbol{n}_m})\times$$
$$\times\, e^{i\boldsymbol{f}\boldsymbol{\xi}_\nu}\,\delta_{\sum_{t=1}^{m}\boldsymbol{x}_{\boldsymbol{n}_t},\,\boldsymbol{\xi}_\nu}\,, \quad (2.18)$$

where

$$\langle\omega_0^{2m}\langle^{ij} \equiv \frac{1}{N}\sum_{\boldsymbol{f}}^{(0)}\sum_{s=1}^{3}\omega_0^{2m}(\boldsymbol{f},s)\,e^i(\boldsymbol{f},s)\,e^j(\boldsymbol{f},s) =$$
$$= \frac{1}{M^m}\sum_{\boldsymbol{n}_1\boldsymbol{n}_2\ldots\boldsymbol{n}_{m-1}} A^{ii_1}(\boldsymbol{x}_{\boldsymbol{n}_1})\,A^{i_1i_2}(\boldsymbol{x}_{\boldsymbol{n}_2})\ldots A^{i_{m-2}i_{m-1}}(\boldsymbol{x}_{\boldsymbol{n}_{m-1}})\,A^{i_{m-1}j}\left(-\sum_{t=1}^{m-1}\boldsymbol{x}_{\boldsymbol{n}_t}\right). \quad (2.19)$$

$\langle\ldots\rangle$ denotes the average over the vibrational spectrum of the "reference" lattice; the summation $\sum_{\boldsymbol{f}}^{(0)}$ or the integration $\int^{(0)} d^3f(\ldots)$ extend over the region w_x. It is rather simple to deduce this relation if, on the left side of (2.18), we take into account that, according to (2.13),

$$\sum_{s=1}^{3}\omega_0^{2m}(\boldsymbol{f},s)\,e^i(\boldsymbol{f},s)\,e^j(\boldsymbol{f},s) = \frac{1}{M^m}[\hat{G}^m(\boldsymbol{f})]^{ij} \quad (2.20)$$

---

[1]) When the wave vector $\boldsymbol{f}$ vanishes or its direction is distinguished by some symmetry element, some of the frequencies $\omega_0(\boldsymbol{f}+2\pi\boldsymbol{\eta}_\mu,s)$ for a given $\boldsymbol{f}$ may coincide with each other, so the order of (2.9) will be less than $3r$. In this case the deficient solutions of (2.2) are

$$v^i(x_{\boldsymbol{n}_0}) = 0\,,\quad \omega^2 = \omega_0^2(\boldsymbol{f}+2\pi\boldsymbol{\eta}_\mu,s)\,.$$

The displacement amplitudes of the remaining atoms can be found from (2.6) by the corresponding limiting process.

and then use (2.14) and the equality

$$\frac{1}{r}\sum_{\mu}{}^{(0)} e^{2\pi i \eta_\mu x_n} = \sum_{\nu} \delta_{x_n, \xi_\nu} \,. \tag{2.21}$$

Let ϱ be a maximum radius of the effective interaction between atoms so that $A^{ii'}(x_n) = 0$ for $|x_n| > \varrho$ and $A^{ii'}(x_n) \neq 0$ for $|x_n| \leqq \varrho$. Then if any sum $\sum_{t=1}^{m} x_{n_t}$, where $|x_{n_t}| \leqq \varrho$, satisfies the condition $\left|\sum_{t=1}^{m} x_{n_t}\right| < |\alpha_p|$ $(p = 1, 2, 3)$, the last term in (2.18) vanishes and hence

$$\frac{1}{r}\sum_{\mu}{}^{(0)} \sum_{s=1}^{3} \omega_0^{2m}(f + 2\pi\,\eta_\mu, s)\, e^i\,(f + 2\pi\,\eta_\mu, s)\, e^j\,(f + 2\pi\,\eta_\mu, s) = \langle\omega_0^{2m}\rangle^{ij} \,. \tag{2.22}$$

Thus, if we take into account interactions (both central and non-central) between neighbouring atoms only, and do not consider linear or planar aggregates of atoms A (cf. [6]), equation (2.22) holds for $m = 1$ in any lattice with any $r \geqq 2$.

Averaging (2.18) over f we have the identity

$$\frac{1}{N}\sum_{f}\sum_{\mu}{}^{(0)} \sum_{s=1}^{3} \omega_0^{2m}\,(f + 2\pi\,\eta_\mu, s)\, e^i\,(f + 2\pi\,\eta_\mu, s)\, e^j\,(f + 2\pi\,\eta_\mu, s) \equiv \langle\omega_0^{2m}\rangle^{ij} \,. \tag{2.19'}$$

Now we shall discuss some examples of the crystal structure under consideration.

1. NaCl or CsCl type structures (see also [1, 2]). — In both cases we have

$$\alpha_p = -a_p + \sum_{s=1}^{3} a_s\,;\quad \sum_{s=1}^{3} a_s = \frac{1}{2}\sum_{s=1}^{3} \alpha_s\,,\quad P_{pp'} = \begin{pmatrix} 0 & 1 & 1 \\ 1 & 0 & 1 \\ 1 & 1 & 0 \end{pmatrix},\quad r = \det \hat{P} = 2\,.$$

α_p, $p = 1, 2, 3$, are the basic vectors of the face-centered cubic lattice in the case of NaCl and the edges of the fundamental cube in the case of CsCl. All the sums $\sum_{\mu}^{(0)}$ contain the two terms corresponding to $\mu = (0, 0, 0)$ and $\mu = (1, 0, 0)$ (or $(0, 1, 0)$ or $(0, 0, 1)$). In this case $\exp\{2\pi i\,\beta_p\,(x_n - x_{n_0})\} = -1$ when n corresponds to atoms B ($n = n_1$).

If we take into account only nearest-neighbour interactions we have [2]

$$\left.\begin{aligned} e^i(f, s) &= e^i\,(f + 2\pi\,\beta_p, s)\,, \\ \omega_0^2(f, s) + \omega_0^2\,(f + 2\pi\,\beta_p, s) &= 2\,\langle\omega_0^2\rangle\,, \end{aligned}\right\} \tag{2.23}$$

$$\langle\omega_0^2\rangle \equiv \frac{1}{3N}\sum_{f}{}^{(0)} \sum_{s=1}^{3} \omega_0^2(f, s)\,. \tag{2.24}$$

(In the case of an NaCl type crystal in the crystallographic coordinate system it is found that $e^i(f, s) = \delta^{is}$; the second relation (2.23) is a special case of (2.22) with $m = 1$ for the model under consideration). In this case we find that $j_{fs} = j_{fs+3} = e(f, s)$ and in the coordinate system with its axes along the vectors $e(f, s)$ $(s = 1, 2, 3)$ (2.9) splits into three independent quadratic equations, so that

$$\omega^2 = \omega^2(\boldsymbol{f}, s + 3\zeta) = \frac{1}{1-\lambda} \times$$

$$\times \left[\left(1 - \frac{1}{2}\lambda\right)\langle\omega_0^2\rangle - (-1)^\zeta \sqrt{\left(1 - \frac{1}{2}\lambda\right)^2 \langle\omega_0^2\rangle^2 - (1-\lambda)\,\omega_0^2\,(2\langle\omega_0^2\rangle - \omega_0^2)}\,\right], \tag{2.25}$$

$$\left.\begin{aligned} v^i_{\boldsymbol{f}, s+3\zeta}(\boldsymbol{x}_{\boldsymbol{n}_0}) &= \frac{1}{\sqrt{M}} e^i(\boldsymbol{f}, s)\left[\frac{\mathrm{d}\ln\omega^2(\boldsymbol{f}, s+3\zeta)}{\mathrm{d}\lambda}\right]^{1/2} = \\ &= e^i(\boldsymbol{f}, s)\sqrt{\frac{1}{2M_0}\frac{\omega^2 - \langle\omega_0^2\rangle}{\omega^2 - \frac{1-\lambda/2}{1-\lambda}\langle\omega_0^2\rangle}}, \\ v^i_{\boldsymbol{f}, s+3\zeta}(\boldsymbol{x}_{\boldsymbol{n}_1}) &= (-1)^\zeta \operatorname{sgn}(\langle\omega_0^2\rangle - \omega_0^2)\, e^i(\boldsymbol{f}, s)\sqrt{\frac{1}{2M}\frac{\omega^2 - \frac{1}{1-\lambda}\langle\omega_0^2\rangle}{\omega^2 - \frac{1-\lambda/2}{1-\lambda}\langle\omega_0^2\rangle}} \end{aligned}\right\} \tag{2.26}$$

($\sigma = s + 3\zeta$; $\zeta = 0$ corresponds to the acoustical branches, and $\zeta = 1$ corresponds to the optical ones; $\omega_0^2 \equiv \omega_0^2(\boldsymbol{f}, s)$. We have used (2.25) and the initial quadratic equation for the calculation of the amplitudes (2.26)).

2. Structures of the type of the completely ordered alloy $AuCu_3$ (AB_3). — In this case X is a face-centered cubic lattice. The corners of the fundamental cube are occupied by atoms A and its face centres are occupied by atoms B, so that

$$\left.\begin{aligned} \boldsymbol{\alpha}_p = -2\boldsymbol{a}_p + \sum_{s=1}^{3}\boldsymbol{a}_s; \quad \sum_{s=1}^{3}\boldsymbol{a}_s = \sum_{s=1}^{3}\boldsymbol{\alpha}_s, \quad P_{pp'} = \begin{pmatrix} -1 & 1 & 1 \\ 1 & -1 & 1 \\ 1 & 1 & -1 \end{pmatrix}, \\ r = \det\hat{P} = 4 \end{aligned}\right\} \tag{2.27}$$

(the $\boldsymbol{\alpha}_p$ coincide with the edges of the fundamental cube). All the sums $\sum\limits_{\boldsymbol{\mu}}^{(0)}$ contain the four terms corresponding to $\boldsymbol{\mu} = (0, 0, 0), (1, 0, 0), (0, 1, 0), (0, 0, 1)$.

3. Structures of the type of the completely ordered alloy $AlFe_3$ (AB_3). — For such structures X is a body-centered cubic lattice. The corners of the unit cubes are occupied by atoms B, and their centres are occupied by atoms A and B alternately (the lattice Ξ is face-centered). In this case

$$\boldsymbol{\alpha}_p = \boldsymbol{a}_p + \sum_{s=1}^{3}\boldsymbol{a}_s; \quad \sum_{s=1}^{3}\boldsymbol{a}_s = \frac{1}{4}\sum_{s=1}^{3}\boldsymbol{\alpha}_s, \quad P_{pp'} = \begin{pmatrix} 2 & 1 & 1 \\ 1 & 2 & 1 \\ 1 & 1 & 2 \end{pmatrix}, \quad r = \det\hat{P} = 4\,. \tag{2.28}$$

All the sums $\sum\limits_{\boldsymbol{\mu}}^{(0)}$ contain the four terms corresponding to $\boldsymbol{\mu} = (0, 0, 0)$, $(1, 0, 0)$, $(-1, 0, 0)$, $(2, 0, 0)$. Instead of the last three values $(0, 1, 0)$, $(0, -1, 0)$, $(0, -2, 0)$ or $(0, 0, 1)$, $(0, 0, -1)$, $(0, 0, 2)$ can be taken.

4. Impurity atoms. — The limit $r \to \infty$ corresponds to a lattice with isolated impurity atoms ($1/r$ is the concentration of the impurities). At the same time the region w_ξ of the permissible values $\boldsymbol{f}$ also tends to zero. Thus we have to put $\boldsymbol{f} = 0$ and $r \to \infty$ in the relations (2.5) to (2.11). In particular, equations (2.7) and (2.8) take the form

$$j^i = D^{ik}(\omega^2, 0)\, j^k \tag{2.29}$$

and

$$D^{ik}(\omega^2, 0) \equiv \frac{\lambda\,\omega^2}{r} \sum_{\mu}^{(0)} \sum_{s=1}^{3} \frac{e^i(2\,\pi\,\eta_\mu, s)\, e^k(2\,\pi\,\eta_\mu, s)}{\omega^2 - \omega_0^2\,(2\,\pi\,\eta_\mu, s)}\,, \tag{2.30}$$

and this coincides exactly with familiar results [3, 4, 5].

3. Probability of the Mössbauer Effect and Thermal Shift

It is well known (see, for example [7, 8]) that the normalized probability W_j of the recoilless emission (resonance absorption) of a γ-ray photon by the j-th nucleus in a unit cell and the second-order Doppler shift in the energy of the γ-ray, δE_j, (also called thermal shift) are given by the expressions

$$W_j = \exp\{-Z_j\}\,, \tag{3.1}$$

$$Z_j = M_j\, R_j \frac{v_\xi}{(2\,\pi)^3} \sum_{\sigma=1}^{3r} \int \mathrm{d}^3 f \frac{|\boldsymbol{\varkappa}\, \boldsymbol{v}_{f\sigma}(\boldsymbol{x}_{n_j})|^2}{\hbar\,\omega(\boldsymbol{f}, \sigma)} \operatorname{ctgh} \frac{\hbar\,\omega(\boldsymbol{f}, \sigma)}{2\,T}\,, \tag{3.2}$$

$$\delta E_j = \frac{E_\gamma}{4\,c^2} \frac{v_\xi}{(2\,\pi)^3} \sum_{\sigma=1}^{3r} \int \mathrm{d}^3 f\, |\boldsymbol{v}_{f\sigma}(\boldsymbol{x}_{n_j})|^2\, \hbar\,\omega(\boldsymbol{f}, \sigma) \operatorname{ctgh} \frac{\hbar\,\omega(\boldsymbol{f}, \sigma)}{2\,T}\,, \tag{3.3}$$

where $R_j = \frac{E_\gamma^2}{2\,M_j c^2}$ is the recoil energy of the free nucleus ($M_j \equiv M_{n_j}$), E_γ and $\boldsymbol{\varkappa}$ are the energy of the γ-ray photon and the unit vector parallel to its momentum, respectively. It will be recalled that the integration $\int \mathrm{d}^3 f(\ldots)$ extends over the region w_ξ. We will discuss these expressions making use of the results obtained in Section 2 and reference [1].

3.1 The emitter (absorber) is the atom A

Let the emitting (absorbing) atom be the atom A ($j = 0$). In this case the formulae (3.2) and (3.3), according to (2.11), take the most simple form

$$Z_0 = (1 - \lambda)\, R_0 \frac{v_\xi}{(2\,\pi)^3} \sum_{\sigma=1}^{3r} \int \mathrm{d}^3 f \frac{(\boldsymbol{\varkappa}\boldsymbol{j}_{f\sigma})^2}{\hbar\,\omega(\boldsymbol{f}, \sigma)} \frac{\mathrm{d} \ln \omega^2(\boldsymbol{f}, \sigma)}{\mathrm{d}\lambda} \operatorname{ctgh} \frac{\hbar\,\omega(\boldsymbol{f}, \sigma)}{2\,T}\,, \tag{3.4}$$

$$\delta E_0 = \frac{E_\gamma}{4\,M\,c^2} \frac{v_\xi}{(2\,\pi)^3} \sum_{\sigma=1}^{3r} \int \mathrm{d}^3 f \frac{\mathrm{d} \ln \omega^2(\boldsymbol{f}, \sigma)}{\mathrm{d}\lambda}\, \hbar\,\omega(\boldsymbol{f}, \sigma) \operatorname{ctgh} \frac{\hbar\,\omega(\boldsymbol{f}, \sigma)}{2\,T}\,. \tag{3.4a}$$

For cubic crystals the expression (3.4) simplifies, yielding

$$Z_0 = (1 - \lambda)\, R_0 \frac{v_\xi}{3\,(2\,\pi)^3} \sum_{\sigma=1}^{3r} \int \mathrm{d}^3 f \frac{1}{\hbar\,\omega(\boldsymbol{f}, \sigma)} \frac{\mathrm{d} \ln \omega^2(\boldsymbol{f}, \sigma)}{\mathrm{d}\lambda} \operatorname{ctgh} \frac{\hbar\,\omega(\boldsymbol{f}, \sigma)}{2\,T}\,. \tag{3.4'}$$

The last expression is independent of $\boldsymbol{\varkappa}$.

The expressions (3.4), (3.4′), and (3.4a) are similar to those obtained for the impurity nucleus [3, 4] and become identical with them in the limiting case $r \to \infty$, $\boldsymbol{f} \to 0$ (see Section 2). On the other hand, for a nearest-neighbour, central, and non-central force model of an NaCl type crystal (as well as of a CsCl type crystal) it is seen from (2.25) and (2.26) that equation (3.4′) for Z_0 coincides exactly with a formula considered in [9].

It is remarkable, however, that the most typical features of the Mössbauer effect can be discussed, by using (3.4) and (3.4a), in a most general form, and they prove to be virtually the same for all the structures under consideration.

3.1.1 $M_0 \ll M$

Then, as is shown in [1], there are three high-frequency optical branches

$$\omega^2(\boldsymbol{f}, 3r+1-s) \approx \frac{\overline{\omega}_0^2(\boldsymbol{f}, s)}{1-\lambda}\left[1 + O(1-\lambda)\right], \qquad (3.5)$$
$$j^i_{\boldsymbol{f}, 3r+1-s} = \overline{e}^i(\boldsymbol{f}, s)\,; \quad s = 1, 2, 3\,,$$

where $\overline{\omega}_0^2(\boldsymbol{f}, s)$ and $\overline{e}^i(\boldsymbol{f}, s)$ are eigenvalues and eigenvectors of the matrix $\frac{1}{r\,M} \sum_\mu^{(0)} G^{ik}(\boldsymbol{f} + 2\pi\,\boldsymbol{\eta}_\mu)$:

$$\left.\begin{aligned} \frac{1}{r} \sum_\mu^{(0)} G^{ik}(\boldsymbol{f} + 2\pi\,\boldsymbol{\eta}_\mu)\,\overline{e}^k(\boldsymbol{f}, s) &= M\,\overline{\omega}_0^2(\boldsymbol{f}, s)\,\overline{e}^i(\boldsymbol{f}, s)\,,\\ \overline{e}^i(\boldsymbol{f}, s)\,\overline{e}^i(\boldsymbol{f}, s') &= \delta_{ss'}\,,\\ \sum_{s=1}^{3} \overline{e}^i(\boldsymbol{f}, s)\,\overline{e}^{i'}(\boldsymbol{f}, s) &= \delta^{ii'}\,; \quad s = 1, 2, 3\,. \end{aligned}\right\} \qquad (3.6)$$

As in this case

$$(1-\lambda)\,\frac{\mathrm{d}\ln\omega^2(\boldsymbol{f}, 3r+1-s)}{\mathrm{d}\lambda} = 1 + O(1-\lambda)\,, \qquad (3.7)$$

the atom A vibrates mainly in these modes; its amplitudes in the rest of the modes are small as $(1-\lambda)$ (see (2.17)).

Making use of (3.5) and (3.7), we obtain the following approximate expressions for Z_0 and δE_0:

$$Z_0 \approx \sqrt{1-\lambda}\,R_0\,\frac{v_\xi}{(2\pi)^3}\sum_{s=1}^{3}\int \mathrm{d}^3 f\,\frac{(\boldsymbol{\varkappa}\,\overline{\boldsymbol{e}}(\boldsymbol{f}, s))^2}{\hbar\,\overline{\omega}_0(\boldsymbol{f}, s)}\,\mathrm{ctgh}\,\frac{\hbar\,\overline{\omega}_0(\boldsymbol{f}, s)}{2\,T\sqrt{1-\lambda}}\,, \qquad (3.8)$$

$$\delta E_0 \approx \frac{1}{\sqrt{1-\lambda}}\,\frac{E_\gamma}{4\,M_0\,c^2}\,\frac{v_\xi}{(2\pi)^3}\sum_{s=1}^{3}\int \mathrm{d}^3 f\,\hbar\,\overline{\omega}_0(\boldsymbol{f}, s)\,\mathrm{ctgh}\,\frac{\hbar\,\overline{\omega}_0(\boldsymbol{f}, s)}{2\,T\sqrt{1-\lambda}}\,. \qquad (3.9)$$

It is characteristic that the integrals on the right sides of (3.8) and (3.9) are determined only by the dynamics of the "reference" lattice.

Introducing the spectral functions normalized to unity

$$\left.\begin{aligned} \overline{g}_0^{(\alpha)}(\omega^2) &= \frac{v_\xi}{(2\pi)^3}\sum_{s=1}^{3}\int\limits_{\overline{\omega}_0^2(\boldsymbol{f}, s) = \omega^2} \frac{(\overline{e}^\alpha(\boldsymbol{f}, s))^2\,\mathrm{d}\Omega_f}{|\nabla\,\overline{\omega}_0^2(\boldsymbol{f}, s)|}\,,\\ \overline{g}_0(\omega^2) &= \frac{1}{3}\sum_{\alpha=1}^{3}\overline{g}_0^{(\alpha)}(\omega^2) = \frac{v_\xi}{3\,(2\pi)^3}\sum_{s=1}^{3}\int\limits_{\overline{\omega}_0^2(\boldsymbol{f}, s) = \omega^2} \frac{\mathrm{d}\Omega_f}{|\nabla\,\overline{\omega}(\boldsymbol{f}, s)|}\,,\\ \int_0^{\overline{\omega}_{0\,m}^{(\alpha)\,2}} \overline{g}_0^{(\alpha)}(\omega^2)\,\mathrm{d}\omega^2 &= \int_0^{\overline{\omega}_{0\,m}^{2}} \overline{g}_0(\omega^2)\,\mathrm{d}\omega^2 = 1\,, \end{aligned}\right\} \qquad (3.10)$$

the expressions (3.8) and (3.9) may be written in the form

$$Z_0 \approx \sqrt{1-\lambda}\,R_0\sum_{\alpha=1}^{3}(\boldsymbol{\varkappa}\,\boldsymbol{j}_\alpha)^2\int_0^{\overline{\omega}_{0\,m}^{(\alpha)\,2}} \frac{\mathrm{d}\,\omega^2}{\hbar\,\omega}\,\overline{g}_0^{(\alpha)}(\omega^2)\,\mathrm{ctgh}\,\frac{\hbar\,\omega}{2\,T\sqrt{1-\lambda}}\,, \qquad (3.8')$$

$$\delta E_0 \approx \frac{1}{\sqrt{1-\lambda}}\,\frac{3\,E_\gamma}{4\,M_0\,c^2}\int_0^{\overline{\omega}_{0\,m}^{2}} \mathrm{d}\omega^2\,\overline{g}_0(\omega^2)\,\hbar\,\omega\,\mathrm{ctgh}\,\frac{\hbar\,\omega}{2\,T\sqrt{1-\lambda}}\,. \qquad (3.9')$$

The integration in (3.10) extends over the part of the surface $\overline{\omega}_0^2(\boldsymbol{f}, s) = \omega^2$ located in the region w_ξ (e.g., in the first Brillouin zone of the H-lattice); $\overline{\omega}_{0m}^{(\alpha)}$ and $\overline{\omega}_{0m} = \text{Max}\{\overline{\omega}_{0m}^{(\alpha)}\}$ are the limiting frequencies of the spectral functions $\overline{g}_0^{(\alpha)}(\omega^2)$ and $\overline{g}_0(\omega^2)$ $\left(\overline{g}_0^{(\alpha)}(\omega^2) \equiv 0 \text{ with } \omega > \overline{\omega}_{0m}^{(\alpha)};\ \overline{g}_0(\omega^2) \equiv 0 \text{ with } \omega > \overline{\omega}_{0m}\right)$. For simplicity, crystals with a symmetry not lower than rhombic are considered here; $\alpha = 1, 2, 3$ labels the orthonormal basic vectors $\boldsymbol{j}_\alpha$ of the crystallographic coordinate system.

If (2.22) holds for $m = 1$ we may write (see (2.20))

$$\frac{1}{r} \sum_{\mu}^{(0)} G^{ik} (\boldsymbol{f} + 2\pi \boldsymbol{\eta}_\mu) = M \langle\omega_0^2\rangle^{ik} . \tag{3.11}$$

Thus

$$\left.\begin{aligned} \overline{\omega}_0^2(\boldsymbol{f}, \alpha) &= \langle\omega_0^2\rangle^{(\alpha)} \equiv \frac{1}{N} \sum_{\boldsymbol{f}}^{(0)} \sum_{s=1}^{3} \left(e^\alpha(\boldsymbol{f}, s)\right)^2 \omega_0^2(\boldsymbol{f}, s) , \\ \overline{e}^{\,i}(\boldsymbol{f}, \alpha) &= j_\alpha^i = \delta^{i\alpha} , \end{aligned}\right\} \tag{3.12}$$

as $\langle\omega_0^2\rangle^{ik}$ has diagonal form in the crystallographic coordinate system.

Substituting (3.12) into (3.8) and (3.9), we obtain

$$Z_0 \approx \sqrt{1-\lambda}\, R_0 \sum_{\alpha=1}^{3} \frac{(\boldsymbol{\varkappa}\, \boldsymbol{j}_\alpha)^2}{\hbar \langle\omega_0^2\rangle^{(\alpha)1/2}} \operatorname{ctgh} \frac{\hbar \langle\omega_0^2\rangle^{(\alpha)1/2}}{2T\sqrt{1-\lambda}} , \tag{3.13}$$

$$\delta E_0 \approx \frac{1}{\sqrt{1-\lambda}} \frac{E_\gamma}{4 M_0 c^2} \sum_{\alpha=1}^{3} \hbar \langle\omega_0^2\rangle^{(\alpha)1/2} \operatorname{ctgh} \frac{\hbar \langle\omega_0^2\rangle^{(\alpha)1/2}}{2T\sqrt{1-\lambda}} \tag{3.14}$$

which agrees completely with the corresponding expressions for a light impurity-atom [3 to 5].

As noted in Section 2, (3.11) is valid already for $r = 2$ (for lattices of any symmetry), if interactions between the coordination spheres, starting with the second one, can be neglected. The greater r, the less rigorous are the restrictions imposed on the value of the interaction radius ϱ.

At the same time, the formulae (3.13) and (3.14) may be considered as approximations of (3.8) and (3.9) still for another reason. Namely, according to (3.6), (2.20) (for $m = 1$), and the identity (2.19′), we have

$$\begin{aligned} \langle\overline{\omega}_0^2\rangle^{(\alpha)} &\equiv \frac{r}{N} \sum_{\boldsymbol{f}} \sum_{s=1}^{3} \overline{\omega}_0^2(\boldsymbol{f}, s) \left(\overline{e}^\alpha(\boldsymbol{f}, s)\right)^2 = \frac{1}{MN} \sum_{\boldsymbol{f}} \sum_{\mu}^{(0)} G^{\alpha\alpha} (\boldsymbol{f} + 2\pi \boldsymbol{\eta}_\mu) = \\ &= \frac{1}{N} \sum_{\boldsymbol{f}} \sum_{\mu}^{(0)} \sum_{s=1}^{3} \omega_0^2 (\boldsymbol{f} + 2\pi \boldsymbol{\eta}_\mu, s) \left(e^\alpha (\boldsymbol{f} + 2\pi \boldsymbol{\eta}_\mu, s)\right)^2 = \langle\omega_0^2\rangle^{(\alpha)} \end{aligned} \tag{3.15}$$

(no summation over α), so that the conversion from (3.8) and (3.9) to (3.13) and (3.14) means a substitution of the average value of the integrand (for each α) by its value at an intermediate point $\boldsymbol{f} = \boldsymbol{f}_0$, which is determined from the equation $\overline{\omega}_0^2(\boldsymbol{f}_0, \alpha) = \langle\omega_0^2\rangle^{(\alpha)}$ (see (3.15)).

It should be emphasized that (3.13) and (3.14) depend neither on details of the crystal structure nor on r. The probability for the Mössbauer effect and the second-order Doppler shift in the energy of the γ-ray are determined only by properties of the "reference" lattice and by the ratio M_0/M.

If the relation (2.22) is valid not only for $m = 1$ but also for several values of $m > 1$, the solutions for the high-frequency optical vibrations may explicity be written, with a greater accuracy,

$$\omega^2(\boldsymbol{f}, 3r + 1 - \alpha) = \omega_{\text{loc}}^{(\alpha)2} + (1 - \lambda)^{\beta(r)}\, \Omega^{(\alpha)}(\boldsymbol{f}), \tag{3.16}$$

where

$$\omega_{\text{loc}}^{(\alpha)2} = \frac{\langle\omega_0^2\rangle^{(\alpha)}}{1 - \lambda} + \frac{\langle\omega_0^4\rangle^{(\alpha)} - \langle\omega_0^2\rangle^{(\alpha)2}}{\langle\omega_0^2\rangle^{(\alpha)2}} + \cdots \tag{3.17}$$

is the frequency of the local mode for the isolated impurity atom (i.e., in the limit $r \to \infty$; cf. [1, 3 to 5]); $\Omega^{(\alpha)}(\boldsymbol{f})$ is a certain function of zero order in $(1 - \lambda)$. The function $\beta(r)$ increases with r $(\beta(r) \to \infty$ when $r \to \infty)$, its specific form is depending on the lattice structure and the radius of the interactions between the atoms. (For instance, if (2.22) holds for $m = 2$, and for $m = 3$ we have to use (2.18), we find $\beta(r) = 1)$.

Making a calculation of the total derivative $\dfrac{\mathrm{d}\ln\omega^2(\boldsymbol{f}, \sigma)}{\mathrm{d}\lambda}$ with the aid of (3.16) and (3.17), one obtains expressions for Z_0 and δE_0 which are more accurate than (3.13) and (3.14).

The expressions of the type (3.13) and (3.14) are analysed in considerable detail in [3 to 5], where it has been noted that for a given vibrational spectrum $g_0^{(\alpha)}(\omega^2)$ such expressions (as well as the more general expressions (3.8) and (3.9)) imply (due to the factor $(1 - \lambda)^{1/2} = \left(\dfrac{M_0}{M}\right)^{1/2}$) a greater probability of the Mössbauer effect, a greater shift at $T = 0$, and a weaker temperature dependence of these values up to

$$T \approx \frac{\hbar\,\langle\omega_0^2\rangle^{(\alpha)\,1/2}}{\sqrt{1 - \lambda}},$$

where the transition to the classical limit occurs.

However, the most interesting result seems to be the following: If we take, as the "reference" lattice, the lattice X_{A} composed entirely of atoms A *with the same interaction between them* the eigenfrequencies $\omega_{0\,0}^2(\boldsymbol{f}, s)$ of such a lattice are connected, according to (2.13), with $\omega_0^2(\boldsymbol{f}, s)$ by the relation

$$\frac{\omega_{0\,0}^2(\boldsymbol{f}, s)}{\omega_0^2(\boldsymbol{f}, s)} = \frac{M}{M_0} = \frac{1}{1 - \lambda}, \tag{3.18}$$

and the expressions (3.13) and (3.14) may be written as

$$Z_0 \approx R_0 \sum_{\alpha=1}^{3} \frac{(\varkappa\, \boldsymbol{j}_\alpha)^2}{\hbar\,\langle\omega_{0\,0}^2\rangle^{(\alpha)1/2}} \operatorname{ctgh} \frac{\hbar\,\langle\omega_{0\,0}^2\rangle^{(\alpha)1/2}}{2\,T}, \tag{3.13'}$$

$$\delta E_0 \approx \frac{E_\gamma}{4\,M_0\,c^2} \sum_{\alpha=1}^{3} \hbar\langle\omega_{0\,0}^2\rangle^{(\alpha)\,1/2} \operatorname{ctgh} \frac{\hbar\,\langle\omega_{0\,0}^2\rangle^{(\alpha)\,1/2}}{2\,T}. \tag{3.14'}$$

Comparing these expressions, say, for $T = 0$

$$Z_0 \approx R_0 \sum_{\alpha=1}^{3} \frac{(\varkappa\, \boldsymbol{j}_\alpha)^2}{\hbar\,\langle\omega_{0\,0}^2\rangle^{(\alpha)\,1/2}}, \quad \delta E_0 \approx \frac{E_\gamma}{4\,M_0\,c^2} \sum_{\alpha=1}^{3} \hbar\langle\omega_{0\,0}^2\rangle^{(\alpha)1/2} \tag{3.19}$$

with the adequate expressions for the "reference" lattice X_A [2])

$$Z_0^{(A)} = R_0 \frac{v_x}{(2\pi)^3} \sum_{s=1}^{3} \int^{(0)} d^3f \frac{(\varkappa\, e(f\, s))^2}{\hbar\, \omega_{00}(f, s)} = R_0 \sum_{\alpha=1}^{3} \frac{(\varkappa\, j_\alpha)^2}{\hbar} \left\langle \frac{1}{\omega_{00}} \right\rangle^{(\alpha)}, \qquad (3.20)$$

$$\delta E_0^{(A)} = \frac{E_\gamma}{4 M_0 c^2} \frac{v_x}{(2\pi)^3} \sum_{s=1}^{3} \int^{(0)} d^3f\, \hbar\, \omega_{00}(f, s) = \frac{E_\gamma}{4 M_0 c^2} \sum_{\alpha=1}^{3} \hbar \langle \omega_{00} \rangle^{(\alpha)} \qquad (3.21)$$

(it will be recalled that the integration $\int^{(0)} d^3f(\ldots)$ extends over the region w_x) we find that the increase in the probability of the Mössbauer effect and the increase in the shift of the recoilless line occurring when the atoms A are surrounded by sufficiently heavy atoms B:

$$Z_0 < Z_0^{(A)}, \qquad \delta E_0 > \delta E_0^{(A)}, \qquad (3.22)$$

are determined only by the inequalities[3])

$$\frac{1}{\langle \omega_{0\,0}^2 \rangle^{(\alpha)\,1/2}} < \left\langle \frac{1}{\omega_{00}} \right\rangle^{(\alpha)}, \qquad \langle \omega_{0\,0}^2 \rangle^{(\alpha)\,1/2} > \langle \omega_{00} \rangle^{(\alpha)} \qquad (3.23)$$

and are independent of the ratio M_0/M.

---

[2]) It follows from symmetry considerations that in the crystallographic coordinate system

$$\frac{v_x}{(2\pi)^3} \sum_{s=1}^{3} \int^{(0)} d^3f \frac{e^\alpha(f, s)\, e^\beta(f, s)}{\omega_{00}(f, s)} = \delta_{\alpha\beta} \frac{v_x}{(2\pi)^3} \sum_{s=1}^{3} \int^{(0)} d^3f \frac{(e^\alpha(f, s))^2}{\hbar\, \omega_{00}(f, s)} \equiv$$

$$\equiv \delta_{\alpha\beta} \langle \omega_{0\,0}^{-1} \rangle^{(\alpha)}$$

and we have (3.20). The result (3.21) is obvious since

$$\sum_{\alpha=1}^{3} \langle \omega_{00} \rangle^{(\alpha)} = 3 \langle \omega_{00} \rangle \equiv \frac{v_x}{(2\pi)^3} \sum_{s=1}^{3} \int^{(0)} d^3f\, \omega_{00}(f, s)\,.$$

[3]) The inequalities

$$\langle \omega^2 \rangle > \langle \omega \rangle^2, \left\langle \frac{1}{\omega^2} \right\rangle > \left\langle \frac{1}{\omega} \right\rangle^2, \quad \langle \omega^2 \rangle \left\langle \frac{1}{\omega} \right\rangle^2 > 1\,, \quad \langle \omega \rangle^2 \left\langle \frac{1}{\omega^2} \right\rangle > 1 \quad (3.23')$$

for the moments of any positive function $g(\omega^2)$, normalized to unity, result from the obvious relations

$$\int_0^{\omega_m^2} (\omega - \langle \omega \rangle)^2 g(\omega^2)\, d\omega^2 \equiv \langle \omega^2 \rangle - \langle \omega \rangle^2 > 0\,,$$

$$\int_0^{\omega_m^2} \left(\frac{1}{\omega} - \left\langle \frac{1}{\omega} \right\rangle \right)^2 g(\omega^2)\, d\omega^2 \equiv \left\langle \frac{1}{\omega^2} \right\rangle - \left\langle \frac{1}{\omega} \right\rangle^2 > 0,$$

$$\int_0^{\omega_m^2} \left(\omega^{1/2} \left\langle \frac{1}{\omega} \right\rangle^{1/2} - \langle \omega \rangle^{1/2} \frac{1}{\omega^{1/2}} \right)^2 g(\omega^2)\, d\omega^2 \equiv 2 \langle \omega \rangle^{1/2} \left\langle \frac{1}{\omega} \right\rangle^{1/2} \left(\langle \omega \rangle^{1/2} \left\langle \frac{1}{\omega} \right\rangle^{1/2} - 1 \right) > 0\,.$$

The latter leads to

$$\langle \omega \rangle \left\langle \frac{1}{\omega} \right\rangle > 1\,,$$

whereby the third and the fourth inequality (3.23′) can be proved immediately with the aid of the first and the second inequality (3.23′).

3.1.2 $M_0 \gg M$

Then the vibrations of an atom A mainly occur in the three acoustic branches [1]

$$\left.\begin{aligned} \omega^2(\boldsymbol{f}, s) &= \frac{\overline{\overline{\omega}}{}_0^2(\boldsymbol{f}, s)}{|\lambda|}\left(1 + O\left(\frac{1}{|\lambda|}\right)\right), \quad j^i_{\boldsymbol{f}s} = \overline{\overline{e}}{}^i(\boldsymbol{f}, s)\ ; \quad s = 1, 2, 3\ , \\ (1 - \lambda)\,&\frac{\mathrm{d} \ln \omega^2(\boldsymbol{f}, s)}{\mathrm{d}\lambda} = 1 + O\left(\frac{1}{|\lambda|}\right), \end{aligned}\right\} \tag{3.24}$$

where $\dfrac{1}{\omega(\boldsymbol{f}, s)}$ and $\overline{\overline{e}}{}^i(\boldsymbol{f}, s)$ are, respectively, the eigenvalues and eigenvectors of the matrix $\dfrac{M}{r} \sum\limits_{\mu}^{(0)} [\hat{G}^{-1}(\boldsymbol{f} + 2\pi\,\boldsymbol{\eta}_\mu)]^{ik}$:

$$\left.\begin{aligned} \frac{1}{r}\sum_{\mu}{}^{(0)} [\hat{G}^{-1}(\boldsymbol{f} + 2\pi\,\boldsymbol{\eta}_\mu)]^{ik}\,\overline{\overline{e}}{}^k(\boldsymbol{f}, s) &= \frac{1}{M\,\overline{\overline{\omega}}{}_0^2(\boldsymbol{f}, s)}\,\overline{\overline{e}}{}^i(\boldsymbol{f}, s)\ , \\ \overline{\overline{e}}{}^i(\boldsymbol{f}, s)\,\overline{\overline{e}}{}^i(\boldsymbol{f}, s') = \delta_{ss'}\ , \quad &\sum_{s=1}^{3} \overline{\overline{e}}{}^i(\boldsymbol{f}, s)\,\overline{\overline{e}}{}^{i'}(\boldsymbol{f}, s) = \delta^{ii'}\ , \end{aligned}\right\} \tag{3.25}$$

$$[\hat{G}^{-1}(\boldsymbol{f})]^{ik} = \frac{1}{M}\sum_{s=1}^{3} \frac{e^i(\boldsymbol{f}, s)\,e^k(\boldsymbol{f}, s)}{\omega_0^2(\boldsymbol{f}, s)}\,. \tag{3.26}$$

This result follows from (2.7) and (2.8) for small ω^2. The relation (3.26) immediately results from (2.13) when multiplied by $\omega_0^{-2}(\boldsymbol{f}, s)\,e^l(\boldsymbol{f}, s)\,[\hat{G}^{-1}(\boldsymbol{f})]^{ij}$ and summed over s and i. According to (2.11) and (2.17), the amplitude of the atom A in the rest of the modes is as small as $1/|\lambda|$.

The expressions (3.4) and (3.4a) with allowance for (3.24) take the form

$$\begin{aligned} Z_0 &\approx |\lambda|^{1/2}\,R_0\,\frac{v_\xi}{(2\pi)^3}\sum_{s=1}^{3}\int \mathrm{d}^3 f\,\frac{(\varkappa\,\overline{\overline{e}}(\boldsymbol{f}, s))^2}{\hbar\,\overline{\overline{\omega}}_0(\boldsymbol{f}, s)}\,\mathrm{ctgh}\,\frac{\hbar\,\overline{\overline{\omega}}_0(\boldsymbol{f}, s)}{2\,T\,|\lambda|^{1/2}} = \\ &= |\lambda|^{1/2}\,R_0 \sum_{\alpha=1}^{3} (\varkappa\,\boldsymbol{j}_\alpha)^2 \int\limits_0^{\overline{\overline{\omega}}{}_{0\,m}^{(\alpha)\,2}} \frac{\mathrm{d}\omega^2}{\hbar\,\omega}\,\overline{\overline{g}}{}^{(\alpha)}(\omega^2)\,\mathrm{ctgh}\,\frac{\hbar\,\omega}{2\,T\,|\lambda|^{1/2}}\ , \end{aligned} \tag{3.27}$$

$$\begin{aligned} \delta E_0 &\approx |\lambda|^{-1/2}\,\frac{E_\gamma}{4\,M_0\,c^2}\,\frac{v_\xi}{(2\pi)^3}\sum_{s=1}^{3}\int \mathrm{d}^3 f\,\hbar\,\overline{\overline{\omega}}_0(\boldsymbol{f}, s)\,\mathrm{ctgh}\,\frac{\hbar\,\overline{\overline{\omega}}_0(\boldsymbol{f}, s)}{2\,T\,|\lambda|^{1/2}} = \\ &= 3\,|\lambda|^{-1/2}\,\frac{E_\gamma}{4\,M_0\,c^2}\int\limits_0^{\overline{\overline{\omega}}{}_{0\,m}^{(\alpha)\,2}} \mathrm{d}\omega^2\,\overline{\overline{g}}_0(\omega^2)\,\hbar\,\omega\,\mathrm{ctgh}\,\frac{\hbar\,\omega}{2\,T\,|\lambda|^{1/2}}\ , \end{aligned} \tag{3.28}$$

where by analogy with (3.10)

$$\left.\begin{aligned} \overline{\overline{g}}{}_0^{(\alpha)}(\omega^2) &= \frac{v_\xi}{(2\pi)^3}\sum_{s=1}^{3}\ \int\limits_{\overline{\overline{\omega}}{}_0^2(\boldsymbol{f}, s) = \omega^2} \frac{(\overline{\overline{e}}{}^\alpha(\boldsymbol{f}, s))^2\,\mathrm{d}\Omega_{\boldsymbol{f}}}{|\nabla\,\overline{\overline{\omega}}{}_0^2(\boldsymbol{f}, s)|}\ , \\ \overline{\overline{g}}_0(\omega^2) &= \frac{1}{3}\sum_{\alpha=1}^{3} g_0^{(\alpha)}(\omega^2) = \frac{v_\xi}{3\,(2\pi)^3}\sum_{s=1}^{3}\ \int\limits_{\overline{\overline{\omega}}{}_0^2(\boldsymbol{f}, s) = \omega^2} \frac{\mathrm{d}\Omega_{\boldsymbol{f}}}{|\nabla\,\overline{\overline{\omega}}{}_0^2(\boldsymbol{f}, s)|}\ , \\ \int\limits_0^{\overline{\overline{\omega}}{}_{0\,m}^{(\alpha)\,2}} \overline{\overline{g}}{}_0^{(\alpha)}(\omega^2)\,\mathrm{d}\omega^2 &= \int\limits_0^{\overline{\overline{\omega}}{}_{0\,m}^{2}} \overline{\overline{g}}_0(\omega^2)\,\mathrm{d}\omega^2 = 1 \end{aligned}\right\} \tag{3.29}$$

are spectral functions determined by the dynamics of the "reference" lattice.

The spectral function $\bar{\bar{g}}_0^{(\alpha)}(\omega^2)$ has to reach a maximum close to the mean frequency $\langle\omega^{-2}\rangle^{(\alpha)-1}$ of the acoustic vibrations in the direction of the α-th crystallographic axis (see (3.24), (3.25), and (3.26)):

$$\langle\bar{\bar{\omega}}_0^{-2}\rangle^{(\alpha)} \equiv \frac{r}{N} \sum_{f \neq 0} \sum_{s=1}^{3} \frac{(\bar{\bar{e}}^{\alpha}(f, s))^2}{\bar{\bar{\omega}}_0^2(f, s)} = \frac{M}{N} \sum_{f \neq 0} \sum_{\mu}{}^{(0)} [\hat{G}^{-1}(f + 2\pi \eta_\mu)]^{\alpha\alpha} =$$

$$= \frac{1}{N} \sum_{f \neq 0}{}^{(0)} \sum_{s=1}^{3} \frac{(e^{\alpha}(f, s))^2}{\omega_0^2(f, s)} \equiv \langle\omega_0^{-2}\rangle^{(\alpha)}. \tag{3.30}$$

The transition to the case of impurity atoms ($r \to \infty$, $f \to 0$) in (3.24) and (3.25) leads to the expression for the quasi-local frequency [3, 4] (see (3.24) to (3.26)):

$$\lim_{\substack{r\to\infty \\ f\to 0}} \frac{1}{r} \sum_{\mu}{}^{(0)} [\hat{G}^{-1}(f + 2\pi \eta_\mu)]^{\alpha\beta} = \lim_{r\to\infty} \frac{1}{r} \sum_{\mu \neq 0}{}^{(0)} [\hat{G}^{-1}(2\pi \eta_\mu)]^{\alpha\beta} =$$

$$= \lim_{r\to\infty} \frac{1}{r} \sum_{\mu \neq 0}{}^{(0)} \sum_{s=1}^{3} \frac{e^{\alpha}(2\pi\eta_\mu, s)\, e^{\beta}(2\pi\eta_\mu, s)}{\omega_0^2(2\pi\eta_\mu, s)} = \langle\omega_0^{-2}\rangle^{(\alpha)} \delta^{\alpha\beta}, \tag{3.31}$$

what completely agrees with the behavior of $\bar{\bar{g}}_0^{(\alpha)}(\omega^2)$.

Using the generalised mean-value theorem, we shall substitute the integrands in (3.27) and (3.28), with the exception of $\bar{\bar{g}}_0^{(\alpha)}(\omega^2)$, by their values at the points $\omega^2 = \langle\omega_0^{-2}\rangle^{(\alpha)-1}$ (3.30) for each α, and shall take into account the normalization condition (3.29) for $\bar{\bar{g}}_0^{(\alpha)}(\omega^2)$. This leads to the expressions

$$Z_0 \approx R_0 \sum_{\alpha=1}^{3} \frac{(\varkappa j_\alpha)^2}{\hbar \omega_q^{(\alpha)}} \operatorname{ctgh} \frac{\hbar \omega_q^{(\alpha)}}{2T}, \tag{3.32}$$

$$\delta E_0 \approx \frac{E_\gamma}{4 M_0 c^2} \sum_{\alpha=1}^{3} \hbar \omega_q^{(\alpha)} \operatorname{ctgh} \frac{\hbar \omega_q^{(\alpha)}}{2T}, \tag{3.33}$$

where

$$\omega_q^{(\alpha)2} = \frac{1}{|\lambda| \langle\omega_0^{-2}\rangle^{(\alpha)}} \approx \frac{1}{\langle\omega_{00}^{-2}\rangle^{(\alpha)}} \tag{3.34}$$

is the quasi-local frequency.

As is seen, the general characteristics of the expressions (3.27) and (3.28) are independent of the detailed crystal structure and in the approximation (3.32), (3.33) the numerical values of Z_0 and δE_0 are independent of r. The following properties of the above expressions are most characteristic: a small (as compared with the lattice consisting of atoms A only and having the same spectrum as X_B) probability of the Mössbauer effect and a small shift of the recoilless line at $T = 0$ (due to the factor $|\lambda|^{1/2}$), a rather strong temperature dependence of these values as well as the transition to the classical limit at temperatures as low as

$$T \gtrsim \hbar \operatorname{Max} \{\omega_q^{(\alpha)}\}. \tag{3.35}$$

On the other hand, it is seen by comparing (3.32) and (3.33) for $T = 0$ with (3.20) and (3.21) that *for the same forces* the decrease in the magnitude of the effect and of the shift

$$Z_0 > Z_0^{(A)}, \quad \delta E_0 < \delta E_0^{(A)} \tag{3.36}$$

due to a surrounding consisting of light atoms is determined only by the inequality (see footnote 3)

$$\left\langle \frac{1}{\omega_{00}^2} \right\rangle^{(\alpha)1/2} > \left\langle \frac{1}{\omega_{00}} \right\rangle^{(\alpha)} \quad \text{and} \quad \left\langle \frac{1}{\omega_{00}^2} \right\rangle^{(\alpha)-1/2} < \langle \omega_{00} \rangle^{(\alpha)} \tag{3.37}$$

and is independent of the ratio M_0/M.

For cubic crystals in all the above expressions we have to take

$$g_0^{(\alpha)}(\omega^2) = g_0(\omega^2)\,, \quad \bar{g}_0^{(\alpha)}(\omega^2) = \bar{g}_0(\omega^2)\,, \quad \bar{\bar{g}}_0^{(\alpha)}(\omega^2) = \bar{\bar{g}}_0(\omega^2)\,.$$

Owing to the first equality all the partial moments have to be replaced by the full ones

$$\langle \omega_0^s \rangle^{(\alpha)} = \langle \omega_0^s \rangle \equiv \int_0^{\omega_{0m}^2} \omega^s \, g_0(\omega^2) \, d\omega^2 \,; \qquad s = -2, -1, 0, 1, 2, \ldots .$$

The numerical values for certain moments in the Debye model $\left(g_0(\omega^2) = \frac{3}{2\,\omega_{0m}^3} \sqrt{\omega^2}\right)$ are

$$\left\langle \frac{\omega_{0m}^2}{\omega^2} \right\rangle = 3\,, \quad \left\langle \frac{\omega_{0m}}{\omega} \right\rangle = \frac{3}{2}\,, \quad \left\langle \frac{\omega}{\omega_{0m}} \right\rangle = \frac{3}{4}\,, \quad \left\langle \frac{\omega^2}{\omega_{0m}^2} \right\rangle = \frac{3}{5}\,.$$

In connection with the result obtained we should like to focus attention on numerical calculations of the probability of the Mössbauer effect in diatomic crystals such as NaCl [9] and for an impurity atom in the Bravais cubic crystal [4]. (In both cases a nearest-neighbour, central, and non-central force model has been used.)

3.2 The emitter (absorber) is the atom B

When the emitting (or absorbing) atom is the atom B, the resulting intensity of the recoilless γ-emission (resonance absorption) is determined by the value

$$W = \frac{1}{r-1} \sum_{j=1}^{r-1} W_j\,. \tag{3.38}$$

This expression may be slightly simplified for a cubic crystal, in which all the atoms B occupy equivalent lattice positions which may be obtained one from another by symmetry transformations, or in the case when $Z_j \ll 1$ (the latter also refers to polycrystals). Then

$$Z = R \frac{v_\xi}{3\,(r-1)\,(2\,\pi)^3} \sum_{\sigma=1}^{3r} \int d^3f \, \frac{M \sum_{t=1}^{r-1} |\boldsymbol{v}_{f\sigma}(\boldsymbol{x}_{n_j})|^2}{\hbar\,\omega(\boldsymbol{f},\sigma)} \operatorname{ctgh} \frac{\hbar\,\omega(\boldsymbol{f},\sigma)}{2\,T} =$$

$$= R \frac{v_\xi}{3\,(r-1)\,(2\,\pi)^3} \sum_{\sigma=1}^{3r} \int d^3f \, \frac{1}{\hbar\,\omega(\boldsymbol{f},\sigma)} \left[1 - \frac{M_0}{M} \frac{d \ln \omega^2(\boldsymbol{f},\sigma)}{d\lambda}\right] \operatorname{ctgh} \frac{\hbar\,\omega(\boldsymbol{f},\sigma)}{2\,T}\,. \tag{3.39}$$

Here we have used the orthonormalization relations (2.15) for the vectors $\boldsymbol{v}_{f\sigma}(\boldsymbol{x}_{n_j})$ and (2.11).

Similarly, if the atoms B occupy equivalent positions in a lattice of any symmetry, thus giving rise to the same value of the thermal shift, the expression

(3.3) may be written in the form

$$\delta E_j = \delta E = \frac{1}{r-1} \sum_{t=1}^{r-1} \delta E_t = \frac{E_\gamma}{4\, M\, c^2} \times$$

$$\times \frac{v_\xi}{(r-1)\,(2\,\pi)^3} \sum_{\sigma=1}^{3\,r} \int \mathrm{d}^3 f \left[1 - \frac{M_0}{M} \frac{\mathrm{d} \ln \omega^2(\boldsymbol{f}, \sigma)}{\mathrm{d}\lambda}\right] \hbar\, \omega(\boldsymbol{f}, \sigma) \operatorname{ctgh} \frac{\hbar\, \omega(\boldsymbol{f}, \sigma)}{2\, T}. \quad (3.40)$$

To calculate (3.39) and (3.40) we have to know the whole vibrational spectrum of the crystal and not only the vibrational spectrum of the atoms A. However, the results obtained may be interpreted qualitatively in a rather simple manner.

When $M_0 \ll M$, the expression in the brackets in (3.39) and (3.40) is small for the high-frequency optical branches ($\sigma = 3\,r - 2, 3\,r - 1, 3\,r$), according to (2.17) and (3.7). Thus the low-frequency branches $\sigma = 1, 2, \ldots, 3\,r - 3$ contribute mainly to Z and δE. This leads to a smaller probability of the effect, a smaller shift at $T = 0$, and a stronger temperature variation of Z and δE as compared with the values for the "reference" lattice X_B.

In contrast to this in the case $M_0 \gg M$, the expression in the brackets in (3.39) and (3.40) is very small for the low-frequency acoustic branches $\sigma = 1, 2, 3$ (cf. (2.17) and (3.24)). Thus the optical branches ($\sigma = 4, 5, \ldots, 3\,r$) contribute mainly to Z and δE. This results in a greater probability of the effect, a greater shift at $T = 0$, and a slower temperature variation of Z and δE as compared with the values for the "reference" lattice X_B.

4. Quasi-Classical Limit

One more result may be deduced very simply in the most general form for arbitrary relative values of M_0 and M without referring to (2.4), viz. the quasi-classical limit of (3.2) and (3.3). Indeed, expanding $\operatorname{ctgh} \frac{\hbar\,\omega}{2\,T}$ in a series for a sufficiently high temperature:

$$\operatorname{ctgh} x = \frac{1}{x} + \frac{x}{3} - \frac{x^3}{45} + \cdots; \qquad x \equiv \frac{\hbar\,\omega}{2\,T}, \quad 0 < x < \pi, \quad (4.1)$$

we are led to the problem of calculating the moments

$$\mu^{i\,k}_{(j)\,m} = \frac{r}{N} \sum_{\boldsymbol{f}} \sum_{\sigma=1}^{3\,r} M_j\, v^{i\,*}_{\boldsymbol{f}\sigma}(\boldsymbol{x}_{\boldsymbol{n}_j})\, v^{k}_{\boldsymbol{f}\sigma}(\boldsymbol{x}_{\boldsymbol{n}_j})\, \omega^{2\,m}(\boldsymbol{f}, \sigma)\,; \quad m = -1, 0, 1, 2, \ldots, \; M_j \equiv M_{\boldsymbol{n}_j} \quad (4.2)$$

of the partial spectral function, describing the vibrations of the j-th atom.[4])

[4]) Such a function may be defined as (cf. (3.10) and (3.29))

$$g^{i\,k}_{(j)}(\omega^2) = \frac{v_\xi\, M_j}{(2\,\pi)^3} \sum_{\sigma=1}^{3\,r} \int_{\omega^2(\boldsymbol{f},\sigma) = \omega^2} \frac{v^{i\,*}_{\boldsymbol{f}\sigma}(\boldsymbol{x}_{\boldsymbol{n}_j})\, v^{k}_{\boldsymbol{f}\sigma}(\boldsymbol{x}_{\boldsymbol{n}_j})\, \mathrm{d}\Omega_f}{|\nabla\, \omega^2(\boldsymbol{f}, \sigma)|},$$

$$\int_0^{\omega_m^2} g^{i\,k}_{(j)}(\omega^2)\, \mathrm{d}\omega^2 = 1\,, \qquad g^{i\,k\,*}_{(j)}(\omega^2) = g^{i\,k}_{(j)}(\omega^2)\,.$$

The normalization relation follows from (2.16); the fact that $g^{i\,k}_{(j)}(\omega^2)$ is real follows from the relation $v^{i\,*}_{\boldsymbol{f}\sigma}(\boldsymbol{x}_{\boldsymbol{n}_j}) = v^{i}_{-\boldsymbol{f}\sigma}(\boldsymbol{x}_{\boldsymbol{n}_j})$.

For $m = 0$, according to (2.16), we have

$$\mu^{i\,k}_{(j)\,0} = \delta^{i\,k}\,.$$

To obtain the general expression for $\mu^{i\,k}_{(j)\,m}$ with $m \geqq 1$ we substitute $\boldsymbol{n}$ by $\boldsymbol{n}_m$ and i by i_m in equation (2.3), and then multiply it by

$$\frac{1}{M_{\boldsymbol{n}_2} M_{\boldsymbol{n}_3} \ldots M_{\boldsymbol{n}m}} \psi^{i_1'*}_{\boldsymbol{f}\sigma}(\boldsymbol{x}_{\boldsymbol{n}_j}) A^{i_1 i_2}(\boldsymbol{x}_{\boldsymbol{n}_j} - \boldsymbol{x}_{\boldsymbol{n}_2}) \times$$
$$\times A^{i_2 i_3}(\boldsymbol{x}_{\boldsymbol{n}_2} - \boldsymbol{x}_{\boldsymbol{n}_3}) \ldots A^{i_{m-1} i_m}(\boldsymbol{x}_{\boldsymbol{n}_{m-1}} - \boldsymbol{x}_{\boldsymbol{n}_m})\,.$$

Taking into account (2.15) and (2.16) we have (after summations over $\boldsymbol{n}_2$ $\boldsymbol{n}_3, \ldots, \boldsymbol{n}_m, i_2, i_3, \ldots, i_m, \boldsymbol{f}$ and σ)

$$\mu^{i_1' i_1}_{(j)\,m} = \frac{1}{M_j} \sum_{\boldsymbol{n}_2 \boldsymbol{n}_3 \ldots \boldsymbol{n}m} \frac{1}{M_{\boldsymbol{n}_2}, M_{\boldsymbol{n}_3} \ldots M_{\boldsymbol{n}m}} A^{i_1 i_2}(\boldsymbol{x}_{\boldsymbol{n}_j} - \boldsymbol{x}_{\boldsymbol{n}_2}) \times$$
$$\times A^{i_2 i_3}(\boldsymbol{x}_{\boldsymbol{n}_2} - \boldsymbol{x}_{\boldsymbol{n}_3}) \ldots A^{i_m i_1'}(\boldsymbol{x}_{\boldsymbol{n}_m} - \boldsymbol{x}_{\boldsymbol{n}_j})\,. \tag{4.3}$$

In particular, in the crystallographic coordinate system $\mu^{\alpha\,\beta}_{(j)\,1}$ takes the form

$$\mu^{\alpha\,\beta}_{(j)\,1} = \frac{1}{M_j} A^{\alpha\,\beta}(0) = \frac{M}{M_j} \langle\omega_0^2\rangle^{(\alpha)} \delta^{\alpha\,\beta}\,. \tag{4.4}$$

Multiplying (2.3) by $\omega^{-2}(\boldsymbol{f}, \sigma)\, \psi^{i_1'*}_{\boldsymbol{f}\sigma}(\boldsymbol{x}_{\boldsymbol{n}_j})\, [\hat{A}^{-1}(\boldsymbol{x}_{\boldsymbol{n}_j} - \boldsymbol{x}_{\boldsymbol{n}})]^{i_1\,i}$ and summing over $\boldsymbol{f}$ (over the region w_ξ), σ, $\boldsymbol{n}$, and i, we get

$$\mu^{i_1' i_1}_{(j)-1} = \frac{r}{N} \sum_{\boldsymbol{f}\neq 0} \sum_{\sigma=1}^{3\,r} \frac{M_j\, v^{i_1'*}_{\boldsymbol{f}\sigma}(\boldsymbol{x}_{\boldsymbol{n}_j})\, v^{i_1}_{\boldsymbol{f}\sigma}(\boldsymbol{x}_{\boldsymbol{n}_j})}{\omega^2(\boldsymbol{f}, \sigma)} = M_j\, [\hat{A}^{-}(0)^1]^{i_1 i_1'} =$$
$$= \frac{M_j}{N\,M} \sum_{\boldsymbol{f}\neq 0}^{(0)} \sum_{s=1}^{3} \frac{e^{i_1'}(\boldsymbol{f}, s)\, e^{i_1}(\boldsymbol{f}, s)}{\omega_0^2(\boldsymbol{f}, s)} \tag{4.5}$$

or

$$\mu^{\alpha\,\beta}_{(j)-1} = M_j [A^{-1}(0)]^{\alpha\,\beta} = \frac{M_j}{M} \langle\omega_0^{-2}\rangle^{(\alpha)} \delta^{\alpha\,\beta} \tag{4.6}$$

in the crystallographic coordinate system.

So we obtain finally

$$Z_j = 2\,R\,T\,\hbar^{-2} \sum_{\alpha=1}^{3} (\varkappa\,\boldsymbol{j}_\alpha)^2 \langle\omega_0^{-2}\rangle^{(\alpha)} + \frac{R_j}{6\,T} -$$
$$- \frac{R_j\,\hbar^2\,M}{360\,T^3\,M_j} \sum_{\alpha=1}^{3} (\varkappa\,\boldsymbol{j}_\alpha)^2 \langle\omega_0^2\rangle^{(\alpha)} + \cdots, \tag{4.7}$$

$$\delta E_j = \frac{3\,E_\gamma\,T}{2\,M_j\,c^2} \left(1 + \frac{\hbar^2}{12\,T^2} \frac{M}{M_j} \langle\omega_0^2\rangle \ldots\right) \tag{4.8}$$

($\langle\omega_0^2\rangle$ is determined by (2.24)). The form of these expressions depends neither on r nor on the specific structure of the crystal. (Expressions such as (4.7) and (4.8) have been obtained previously in [10, 11, 4], especially for the case of an impurity atom.) The dependence on the details of structure appears only in the next terms of the series, which have not been written explicitly.

Relation (2.4) was not used in deriving (4.7) and (4.8), and therefore the last expressions are valid for any γ-active atom in any crystal structure of the type $A^{(1)}_{r_1} A^{(2)}_{r_2} \ldots A^{(n)}_{r_n}$ $(r_1 + r_2 + \cdots + r_n = r)$, provided the initial assumption on the force constants is justified.

References

[1] YA. A. IOSILEVSKII, Fiz. tverd. Tela **8**, 2025 (1966).
[2] YA. A. IOSILEVSKII, Fiz. tverd. Tela **7**, 885 (1965).
[3] YU. KAGAN and YA. A. IOSILEVSKII, Zh. eksper. teor. Fiz. **42**, 259 (1962).
[4] YU. KAGAN and YA. A. IOSILEVSKII, Zh. eksper. teor. Fiz. **44**, 284 (1963).
[5] YA. A. IOSILEVSKII and YU. KAGAN, Zh. eksper. teor. Fiz. **46**, 2165 (1964).
[6] YA. A. IOSILEVSKII, Zh. eksper. teor. Fiz. **51**, 201 (1966).
[7] YU. KAGAN, Zh. eksper. teor. Fiz. **40**, 312 (1961).
[8] YU. KAGAN, Zh. eksper. teor. Fiz. **47**, 366 (1964).
[9] YU. KAGAN and V. A. MASLOV, Zh. eksper. teor. Fiz. **41**, 1296 (1961).
[10] A. A. MARADUDIN and P. A. FLINN, Phys. Rev. **126**, 2059 (1962).
[11] A. A. MARADUDIN, P. A. FLINN, and S. RUBY,, Phys. Rev. **126**, 9 (1962).

(Received May 23, 1966)

phys. stat. sol. 16, 651 (1966)

Irkutsker Pädagogisches Institut, Irkutsk

Der Einfluß einachsiger elastischer Spannungen auf die Ummagnetisierungsprozesse dünner Permalloyschichten[1])

Von

V. A. BURAVIKHIN, V. G. KAZAKOV, V. S. KHRISTOSENKO und P. I. KRUKOVER

Es wurde der Einfluß äußerer einachsiger Spannungen auf die kritischen Felder untersucht, die die Prozesse der quasistatischen Ummagnetisierung von Permalloyschichten charakterisieren: die Koerzitivkraft H_c, das Anisotropiefeld H_k, das Keimbildungsfeld H_z, das Startfeld eines Keimes H_s, das minimale Feld der Wandverschiebung H_0 und das Feld der Zerstörung der Bereichsstruktur H_p. Es wurde festgestellt, daß die Veränderungen der Felder H_z und H_s unter der Wirkung einer äußeren Belastung in starkem Maße vom Randprofil der Schichten abhängen. Die Felder H_0 und H_p nehmen bei Deformation für Schichten dünner als $\approx$ 1000 Å zu; für Schichten dicker als $\approx$ 1000 Å ändern sie sich praktisch nicht. Die Koerzitivkraft läßt sich (in Abhängigkeit vom Randprofil der Schicht und von der Größe der Deformation) als mittleresWandverschiebungsfeld wie auch als Keimbildungsfeld bestimmen.

Исследовалось влияние внешних одноосных напряжений на критические поля, характеризующие процессы квазистатического перемагничивания пермаллоевых пленок: поле коэрцитивной силы H_c, поле анизотропии H_k, поле зародышеобразования H_z, поле старта зародыша H_s, минимальное поле смещения границ H_0 и поле разрушения доменной сртуктуры H_p. Установлено, что на изменение полей H_z и H_s под действием внешней нагрузки в большей мере оказывает влияние форма краев пленок. Поля H_0, H_p для пленок тоньше $\approx$ 1000 Å растут при деформации, а для пленок толще $\approx$ 1000 Å — практически не изменяются. Коэрцитивная сила может определяться (в зависимости от формы края и величины деформации) как средним полем смещения границ, так и полем зеродышеобразования.

1. Einleitung

Gegenwärtig gibt es eine Reihe von Arbeiten über die Untersuchung des Einflusses elastischer Spannungen auf die Prozesse der quasistatischen Ummagnetisierung, die Koerzitivkraft und das Anisotropiefeld dünner ferromagnetischer Schichten [1 bis 7].

Es wurde experimentell gefunden, daß man bei Dehnung der Schichten mit positiver Magnetostriktion längs der Achse der leichten Magnetisierung eine Änderung der Koerzitivkraft beobachtet. Eine befriedigende Erklärung dieser Erscheinung gibt es aber noch nicht. Einige Autoren, z. B. Feldtkeller [8], betrachten die Koerzitivkraft als Wandverschiebungsfeld. Einer Wandverschiebung geht aber immer die Bildung von Ummagnetisierungskeimen voraus, und, wie in den Arbeiten [9 und 10] gezeigt ist, kann die Koerzitivkraft in einigen Fällen durch das Keimbildungsfeld bestimmt sein.

---

[1]) Beitrag zum Internationalen Kolloquium über magnetische dünne Schichten, 25. bis 28. April 1966 in Jena.

Theoretisch wurde der Keimbildungsprozeß in Schichten in den Arbeiten [11 und 12] betrachtet. Ilicheva, Kanavina und Shishkov [13] erhielten eine lineare Abhängigkeit zwischen dem Keimbildungsfeld und der Größe des Anisotropiefeldes, die experimentell von Ilicheva für Schichten von weniger als 1000 Å Dicke bei Änderung der Anisotropiefeldstärke durch schräges Aufdampfen bestätigt wurde.

Wir unternahmen den Versuch, den Ummagnetisierungsprozeß von Schichten genauer zu untersuchen, indem wir die Anisotropiefeldstärke durch einachsige Spannungen änderten. Es wurden folgende Größen gemessen: das effektive Anisotropiefeld H_k, die Koerzitivkraft H_c, das Keimbildungsfeld H_z, das Startfeld eines Keimes H_s, das minimale Feld H_0 für die Wandverschiebung aus dem entmagnetisierten Zustand der Schicht und das maximale Wandverschiebungsfeld H_p, das wir Feld zur Zerstörung der Bereichsstruktur nennen. Unter Berücksichtigung der Tatsache, daß Randeffekte die Größe des Keimbildungsfeldes und der Koerzitivkraft beeinflussen können [6, 14], wurden kreisförmige Schichten mit unterschiedlichen Randprofilen untersucht.

2. Experimentelle Methode

Permalloyschichten mit einer Dicke von 150 bis 3000 Å wurden durch thermisches Verdampfen der Legierung 25% Fe-75% Ni auf Mikroskopdeckgläschen der Abmessungen $32 \times 40 \times 0{,}17$ mm³, die auf 200 °C erhitzt wurden, in einem Vakuum von $\approx 10^{-5}$ Torr hergestellt. Während der Schichtherstellung wurde ein konstantes Magnetfeld von 50 Oe angelegt. Die Feldrichtung fiel mit der Schichtebene zusammen.

Eine Maske gewährleistete die gleichzeitige Herstellung von zwei Schichten mit 3 mm Durchmesser, die 3 mm voneinander entfernt auf einer Unterlage waren und unterschiedliche Randprofile hatten. Die Schicht mit steilem Rand hatte eine Abfallzone der Schichtdicke von etwa 0,015 mm und die Schicht mit schrägem Rand eine solche von 0,15 mm.

Bei dem Herstellungsprozeß der Schichten wurde ein normales Auftreffen der Atome auf die Unterlage gewährleistet. Der Abstand von der Dampfquelle zur Unterlage betrug 150 mm. Die Schichtdicken wurden mit einer Interferenzmethode gemessen.

Die Spannungen in den Schichten wurden durch Verbiegung der Unterlagen in einer Spezialvorrichtung mit Hilfe von zwei Paaren von parallelen Schneiden hervorgerufen. Die Konstruktion des Biegemechanismus gab die Möglichkeit, homogene Spannungen in der Schicht parallel zur Spannungsrichtung zu erhalten, und erlaubte es, den Biegungspfeil der Unterlagen bis auf 5×10^{-3} mm zu bestimmen. Die Verlängerung der Schicht wurde nach der Formel

$$\frac{\Delta l}{l} = \frac{4 D h}{L^2}$$

bestimmt, wobei D die Dicke der Glasunterlage, L der Abstand zwischen den zwei inneren Schneiden und h der Biegungspfeil sind.

Das ummagnetisierende Feld wurde mit Hilfe von Helmholtz-Spulen erzeugt. Die Untersuchung des Keimbildungsfeldes H_z, des Startfeldes für einen Keim H_s, und der Wandverschiebungsfelder H_0 und H_p wurde sowohl mit Hilfe des magnetooptischen Kerr-Effektes als auch mit der Methode der Pulvermuster durchgeführt. Die Größe des effektiven Anisotropiefeldes H_k und die Koerzitivkraft H_c

wurden aus den Hystereseschleifen bestimmt, die magnetooptisch in der Richtung der Achsen der leichten und schweren Magnetisierung der Schichten aufgenommen wurden. In allen Fällen wurden die Schichten längs der Achse der leichten Magnetisierung gedehnt.

3. Experimentelle Ergebnisse

Beim Experiment wurde entdeckt, daß der Einfluß der elastischen Spannungen auf die magnetischen Größen der Schichten in bedeutendem Maße von der Dicke und dem Randprofil abhängt.

In Fig. 1 ist die Abhängigkeit von H_k, H_c, H_z, H_s, H_0 und H_p von der Größe der relativen Deformation ($\Delta l/l$) zweier Schichten von 160 Å, die unterschiedliche Randprofile hatten, angegeben. Aus der Zeichnung ersieht man, daß mit wachsender Dehnung der Schichten alle magnetischen Größen zunehmen. Zu beachten ist die Tatsache, daß das Verhalten der untersuchten Größen fast nicht von der Randform abhängt; das ist für relativ dünne Schichten (<500 bis 600 Å) typisch.

Die Kurve, die die Abhängigkeit H_c von $\Delta l/l$ darstellt, liegt zwischen dem Minimum H_0 und dem Maximum H_p des Wandverschiebungsfeldes; das erlaubt es, die Koerzitivkraft der Schichten mit der angegebenen Dicke als Mittelwert der Wandverschiebungsfelder zu betrachten.

Wie die Beobachtungen der Bereichsstrukturänderungen zeigten, hängt der Charakter der Ummagnetisierungsprozesse von der Größe der angelegten Spannung ab. Bei kleinen Belastungen beginnt der Ummagnetisierungsprozeß der Schicht mit dem Auftreten von Ummagnetisierungskeimen in Feldern wesentlich unterhalb der Wandverschiebungsfelder. Mit zunehmendem Feld wachsen die Keime durch kleine Sprünge, und bei dem Feld $H = H_0$ zerfällt die Schicht in eine Reihe von Bereichen. Bei großen Belastungen ist der Keimbildungsprozeß erschwert, und die Schicht wird durch einige relativ große Sprünge ummagnetisiert. Die weitere Zunahme der Belastung führt dazu, daß H_z größer als H_0 wird, und die Schicht wird durch einen Sprung ummagnetisiert.

Mit zunehmender Dicke beginnt die Randform der Schichten einen spürbaren Einfluß auf die Spannungsabhängigkeit einer Reihe von magnetischen Größen auszuüben. Schon bei einer kleinen Belastung wurden Schichten von 900 bis 1400 Å Dicke mit schrägem Rand in einem magnetischen Wechselfeld praktisch nicht entmagnetisiert, deshalb fehlt in der Fig. 2 die Wandverschiebungskurve.

In Fig. 2 ist die Abhängigkeit der magnetischen Größen von $\Delta l/l$ für zwei Schichten von 1120 Å Dicke angegeben. Der Figur entnimmt man, daß H_c

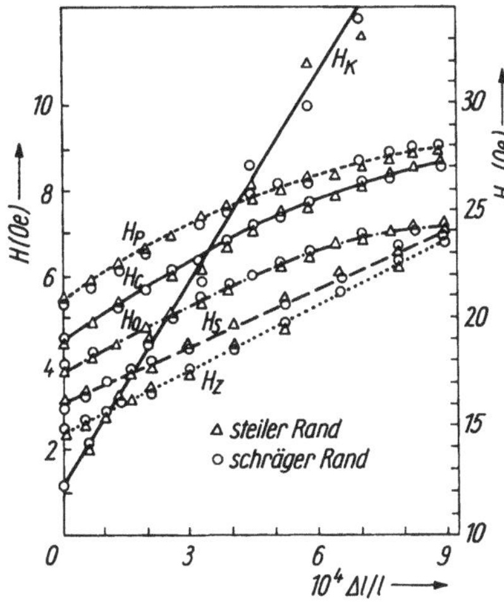

Fig. 1. Abhängigkeit der magnetischen Größen von der relativen Deformation von Fe–Ni-Schichten von 160 Å Dicke

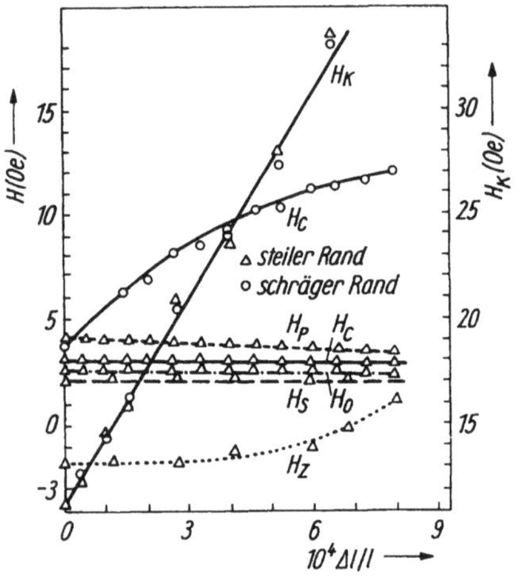

Fig. 2. Abhängigkeit der magnetischen Größen von der relativen Deformation von Fe–Ni-Schichten von 1120 Å Dicke

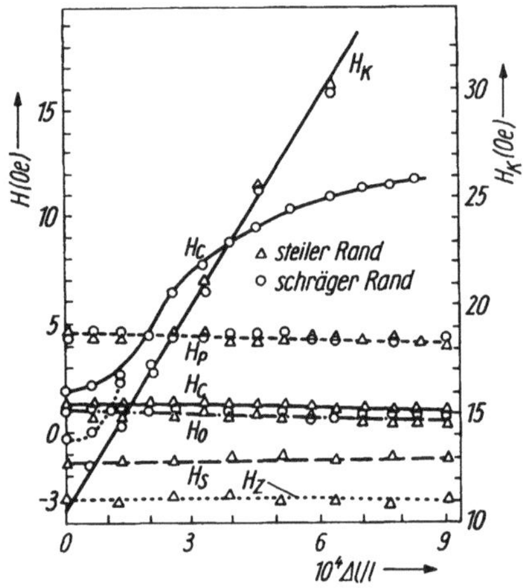

Fig. 3. Abhängigkeit der magnetischen Größen von der relativen Deformation von Fe–Ni-Schichten von 1960 Å Dicke

bei einer Deformation der Schicht durch die Form des Randes bestimmt wird. Wenn für eine Schicht mit steilem Rand H_c durch die Wandverschiebungsfeldstärke bestimmt ist und fast nicht von der Größe der Deformation abhängt, so hängt H_c für eine Schicht mit schrägem Rand nicht von den Wandverschiebungsfeldern ab und wächst mit der Dehnung der Schicht wesentlich an. In der gegebenen Schicht ist der Prozeß der Keimbildung ohne Anlegen einer Deformation sogar erschwert.

Für eine Schicht der gegebenen Dicke mit steilen Rändern bilden sich Keime in einem negativen Feld. Mit zunehmender Belastung geht der Keimbildungsprozeß allmählich in das Gebiet positiver Felder über, und das Feld H_z nähert sich dem Feld H_0. Es ist möglich, daß bei großen Deformationen, wenn H_z größer als H_0 wird, die Ummagnetisierung durch einen Sprung ebenso wie in Schichten mit schrägen Rändern verwirklicht wird; das ruft seinerseits ein starkes Anwachsen der Koerzitivkraft hervor.

In Fig. 3 sind die Abhängigkeiten der untersuchten magnetischen Kenngrößen für Schichten von 1960 Å Dicke ebenfalls von der relativen Dehnung der Schichten dargestellt. In dem gegebenen Fall hängen für Schichten mit steilem Rand alle magnetischen Größen außer H_k von der Größe der Dehnung in den Grenzen der angegebenen Belastungen fast nicht ab.

Ein anderes Bild beobachtet man für Schichten mit schrägen Rändern. Mit zunehmender Belastung beobachtet man ein starkes Anwachsen der Koerzitivkraft; im Unterschied zu den Schichten geringerer Dicke wächst aber die Kurve der H_c-Änderung nicht sofort steil nach oben an, sondern hat eine charakteristische Krümmung am Anfang. Das wird, wie die Beobachtungen der Änderung der Bereichsstruktur zeigen, durch die Existenz von Keimen bei kleinen Deformationen und durch das schnelle Anwachsen des Keimbildungsfeldes mit wachsender Deformation erklärt.

Eine weitere Zunahme der Schichtdicke führt zu einer quantitativen Änderung des Verhaltens einiger untersuchter Größen bei einer Deformation der Schichten. Im Unterschied zu den sehr dünnen Schichten haben die Felder

Fig. 4. Abhängigkeit der magnetischen Größen von der relativen Deformation von Fe–Ni-Schichten von 3000 Å Dicke

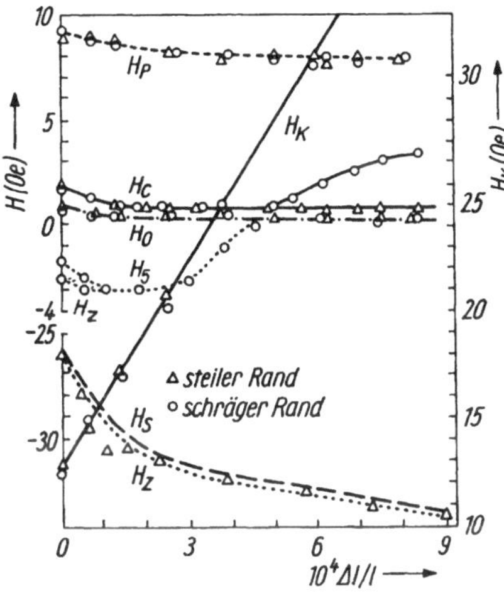

H_c, H_0, H_p, H_z, H_s (für eine Schicht von 3000 Å Dicke mit steilem Rand, Fig. 4) die Tendenz, mit zunehmender Deformation kleiner zu werden. Die Felder H_s und H_c für eine Schicht mit schrägem Rand nehmen zunächst ab und wachsen dann an, die Felder H_0 und H_p fallen aber mit den entsprechenden Feldern für die Schichten mit steilem Rand zusammen.

Das Verhalten des effektiven Anisotropiefeldes bei Dehnung der Schichten hängt von der Schichtdicke wenig und von dem Randprofil gar nicht ab. In allen Fällen wird eine lineare Abhängigkeit von H_k von der Dilatation beobachtet, wobei die Größe des magnetoelastischen Parameters $\Delta H_k/(\Delta l/l)$ in dem ganzen Dickenintervall annähernd konstant bleibt.

4. Schlußfolgerungen

1. Das Verhalten der Wandverschiebungsfelder H_0 und H_p bei elastischer Deformation der Schichten hängt von ihrer Dicke ab. Für Schichten dünner als 600 bis 700 Å wachsen die Wandverschiebungsfelder mit zunehmender Deformation parallel zur Achse der leichten Magnetisierung. Die Wandverschiebungsfelder für Schichten dicker als 900 bis 1000 Å hängen von der Größe der Deformation nicht wesentlich ab, und für Schichten mit einer Dicke von der Größenordnung 3000 Å beobachtet man einen Rückgang dieser Felder mit zunehmender Deformation.

2. Auf das Verhalten von H_c bei einer Dehnung der Schichten parallel zur Achse der leichten Magnetisierung haben die Dicke und die Randform der Schichten einen großen Einfluß. In allen Fällen kann man, wenn $H_z < H_0$ ist, die Koerzitivkraft als mittleres Wandverschiebungsfeld betrachten, und wenn $H_z > H_0$ ist, dann wird die Koerzitivkraft durch das Verhalten des Keimbildungsfeldes bei der Deformation der Schicht bestimmt.

3. Die Größe des Keimbildungsfeldes nimmt für Schichten mit steilen Rändern bei kleiner Dicke ($<$1000 Å) bei Deformation der Schichten zu und für Schichten größerer Dicke ($>$2000 Å) ab. Die Keimbildung in Schichten mit schrägen Rändern wird bei hinreichender Deformation gehemmt.

4. Der Charakter des Verhaltens des Anisotropiefeldes hängt bei Deformation der Schichten von ihrer Dicke und den Randprofilen praktisch nicht ab.

Literatur

[1] S. Middelhoek, Z. angew. Phys. **13**, 151 (1961).
[2] E. N. Mitchell, G. J. Lykken und G. D. Babcock, J. appl. Phys. **34**, 715 (1963).
[3] T. S. Crowther, J. appl. Phys. **34**, 580 (1963).

[4] H. L. PINCH und A. A. PINTO, J. appl. Phys. **35**, 828 (1964).
[5] V. A. BURAVIKHIN und V. G. KAZAKOV, Fiz. Metallov i Metallovedenie **19**, 45 (1965).
[6] V. A. BURAVIKHIN, V. G. KAZAKOV und V. I. POPOV, Izv. Akad. Nauk SSSR, Ser. fiz. **29**, 659 (1965).
[7] V. A. BURAVIKHIN und V. G. KAZAKOV, Izv. Akad. Nauk SSSR, Ser. fiz. **29**, 668 (1965).
[8] E. FELDTKELLER, Z. angew. Phys. **15**, 206 (1963).
[9] S. METHFESSEL, S. MIDDELHOEK und H. THOMAS, J. appl. Phys. **32**, 294 S (1961).
[10] E. N. ILICHEVA, Dissertation, Phys. Fakultät der Moskauer Staatl. Universität 1965.
[11] C. SHTRIKMAN und D. TREVES, J. appl. Phys. **32**, 245 S (1961).
[12] W. ANDRÄ, Z. angew. Phys. **13**, 141 (1961).
[13] E. N. ILICHEVA, N. G. KANAVINA und A. G. SHISHKOV, Fiz. Metallov i Metallovedenie **21**, 22 (1966).
[14] M. BECKERMAN und K. H. BEHRNDT, IBM J. Res. Developm. **4**, 198 (1960).

(Received May 25, 1966)

phys. stat. sol. 16, 657 (1966)

Irkutsker Pädagogisches Institut, Irkutsk

Über die Struktur und die Polarität von Bereichsgrenzen in ferromagnetischen Schichten[1])

Von

V. A. Buravikhin und V. I. Popov

Es wird experimentell gefunden, daß sich unter der Wirkung einachsiger elastischer Spannungen die Orientierung der Bereichsgrenzen ändert. Dabei wird die Grenze verschieden polar, wenn der Winkel 2φ den Wert $\approx 180°$ erreicht. Die Grenzen wechselnder Polarität kommen in einem großen Dickenintervall vor. Die Struktur dieser Grenzen ist in Schichten verschiedener Dicke verschieden. Der Übergang durch eine Hauptgrenze ist nur bei dünnen Schichten (< 300 Å) ein reiner Néel-Übergang. In Schichten größerer Dicke tritt beim Übergang durch die Hauptgrenze außer einem Néel-Anteil auch ein Bloch-Anteil auf, wobei die Größe des letzteren mit der Schichtdicke zunimmt.

Из приведенных экспериментов следует, что под действием одноосных напряжений изменяется градусность границ. При том, как только угол 2φ достигает $\approx 180°$ граница становится разнополярной. Границы с переменной полярностью встречаются в очень широкой области толщин. Структура таких границ в пленках разных толщин различна. Переход через основную границу является чисто неелевским только в пленках малых толщин (< 300 Å). В пленках же больших толщин переход через основную границу, нагряду с неелевской компанентой, содержит и блоховскую, величина которой увеличивается по мере увеличения толщины пленок.

Dem Studium der Struktur und der Polarität von Bereichsgrenzen dünner ferromagnetischer Schichten sind die Arbeiten [1 bis 9] gewidmet, in denen die Struktur und die Polarität entweder ohne Einwirkung äußerer Faktoren auf die Schichten oder unter dem Einfluß eines äußeren Feldes untersucht wurden. In der Arbeit [10] sind die experimentellen Untersuchungen über den Einfluß elastischer Spannungen auf die Polarität der Wände mit Hilfe der Pulvertechnik beschrieben.

In der vorliegenden Arbeit werden die Resultate über den Einfluß elastischer Spannungen auf die Struktur der Wände ferromagnetischer Schichten mitgeteilt. Es werden auch die Struktur und die Polarität der Bereichsgrenzen in Schichten unterschiedlicher Dicke erörtert.

Die untersuchten Schichten wurden aus der Legierung 75%Ni-25%Fe, die eine positive Magnetostriktion besitzt, durch thermisches Aufdampfen in einem Vakuum von 10^{-5} Torr hergestellt. Als Unterlagen wurden frische Spaltflächen von NaCl-Einkristallen oder Deckgläschen mit einer vorher aufgedampften dünnen NaCl-Zwischenschicht benutzt. Während der Herstellung der Schichten wurde ein konstantes Magnetfeld angelegt, dessen Richtung mit der Schichtebene zusammenfiel. Die Struktur der Bereichsgrenzen wurde mit der Schatten-

---

[1]) Beitrag zum Internationalen Kolloquium über magnetische dünne Schichten, 25. bis 28. April 1966 in Jena.

methode mit Hilfe der Elektronenmikroskope EM-7 und UEMW-100 untersucht, für die die Spezialvorrichtungen angefertigt wurden, die es erlauben, eine Schicht zu deformieren und während der Untersuchung ein vertikales Magnetfeld anzulegen. Die quantitative Größe der Deformation wurde nicht bestimmt.

In relativ dünnen Schichten (150 bis 400 Å) wurden nach ihrer Entmagnetisierung durch ein Wechselfeld parallel zur Achse der schweren Magnetisierung oft Néel-Wände beobachtet, wobei der Drehwinkel des Magnetisierungsvektors durch die Wände wesentlich von 180° verschieden sein kann. In solchen Fällen ist die Richtung des Magnetisierungsvektors im mittleren Teil der Wand über ihre ganze Länge gleich, d. h., es werden Néel-Wände mit konstanter Polarität beobachtet.

Unter der Einwirkung einer allmählich größer werdenden Spannung, die mit der Wandrichtung zusammenfällt, streben die Magnetisierungsvektoren in den Bereichen danach, die Richtung der Spannung einzunehmen (Fig. 1 b, c). Sobald der Drehwinkel des Magnetisierungsvektors (2φ) durch die Wand 180° wird,

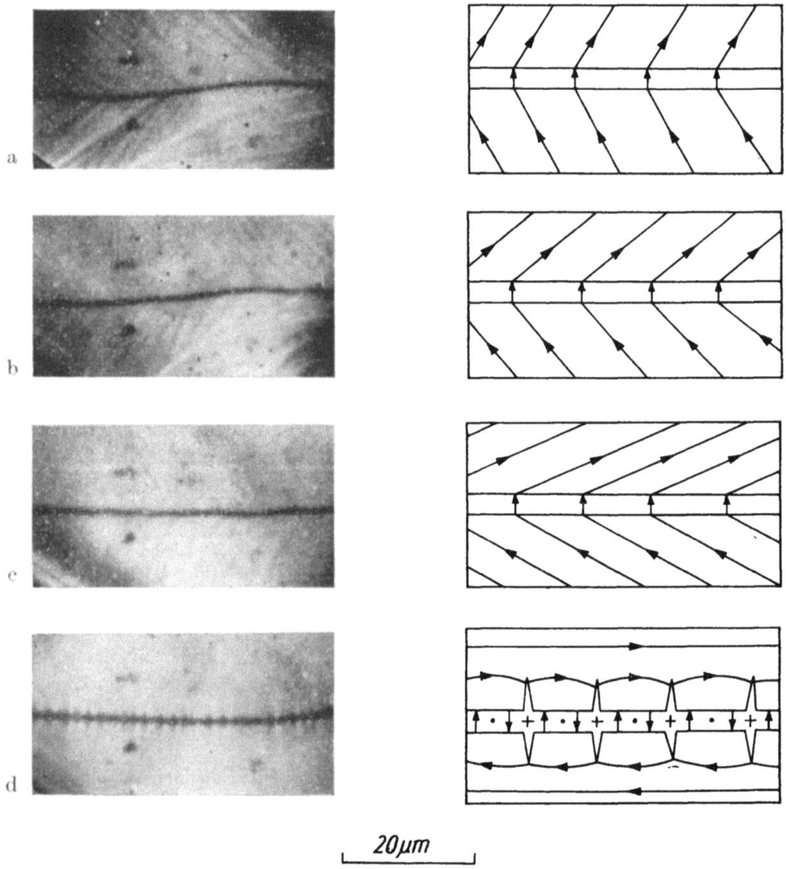

Fig. 1. Änderung der Struktur der Bereichsgrenze einer Fe–Ni-Schicht von 170 Å Dicke unter der Wirkung von Zugspannungen und schematische Darstellung der Magnetisierungsvektoren. Unterlagentemperatur 300 °C. Richtung der Zugspannung und Vorzugsachse horizontal. Die Spannung wächst in dieser Zeichnung wie in Fig. 2 von a) bis d)

zeigt sich eine Feinstruktur der Wand (Fig. 1d), d. h., der Magnetisierungsvektor in einzelnen Wandbereichen ändert seine Richtung in die antiparallele. Im gegebenen Fall besteht die Wand aus Néel-Stücken mit entgegengesetzter Magnetisierungsrichtung, die durch Bloch-Linien getrennt sind. In der Nähe von Kreuz-Bloch-Linien ändert sich die Magnetisierungsrichtung stark; dadurch wird das Auftreten von senkrechten Wandstücken hervorgerufen.

Die Entstehung von Néel-Wänden mit wechselnder Polarität unter der Wirkung angelegter Spannungen kann man qualitativ auf folgende Art erklären: Das Streufeld der Wand (H_0) wächst bei einer Zunahme des Winkels 2φ [11], und sobald H_0 größer als das effektive Anisotropiefeld H_k wird, dreht sich die

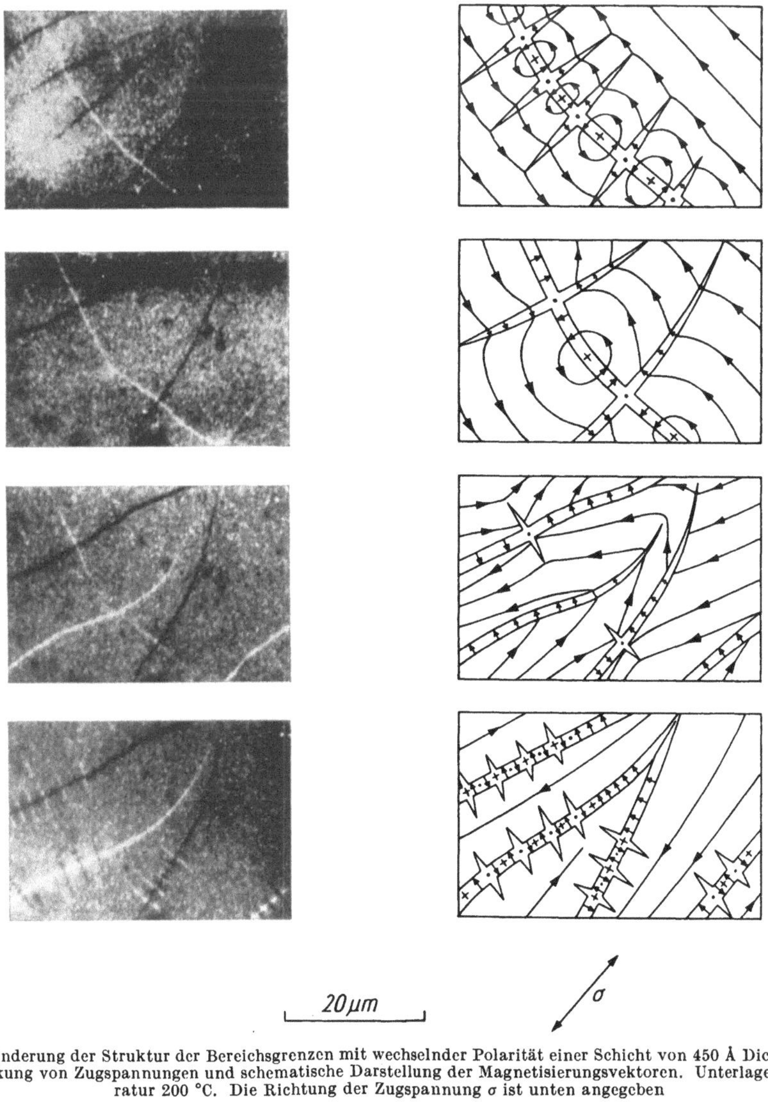

a b c d

Fig. 2. Änderung der Struktur der Bereichsgrenzen mit wechselnder Polarität einer Schicht von 450 Å Dicke unter der Wirkung von Zugspannungen und schematische Darstellung der Magnetisierungsvektoren. Unterlagentemperatur 200 °C. Die Richtung der Zugspannung σ ist unten angegeben

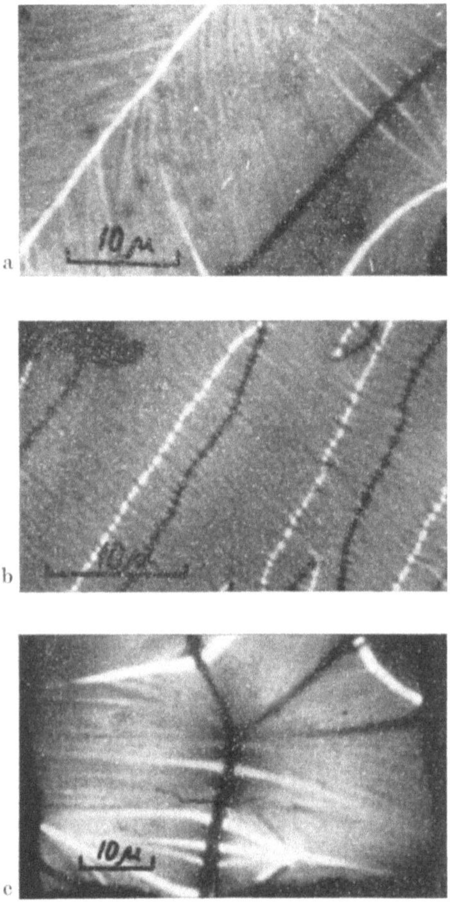

Fig. 3. Bereichsstruktur von Fe-Ni-Schichten unterschiedlicher Dicke (D), die auf einer bis auf 200 °C erhitzten Unterlage aufgedampft wurden. a) D = 580 Å, b) D = 160 Å, c) D = 950 Å

Magnetisierung in einzelnen Wandteilen in die Richtung von H_0. Im gegebenen Fall muß man beachten, daß in H_k auch das Anisotropiefeld eingeht, das durch die angelegten Spannungen bedingt ist.

Oft beobachtet man nach der Entmagnetisierung von Eisen-Nickel-Schichten Wände mit wechselnder Polarität vor Anlegen der Spannung (Fig. 2a). Wenn die Spannung senkrecht zur Hauptwand angelegt wird, beobachtet man eine Zunahme der Länge der Querwände und auch eine sprunghafte Vergrößerung des Abstandes zwischen ihnen (Fig. 2b). Mit zunehmender Spannung nimmt der Winkel 2φ, um den der Magnetisierungsvektor bei dem Durchgang durch die Hauptwand gedreht wird, ab, und die Wand verschwindet allmählich. In Fig. 2c ist zu sehen, daß die Hauptwand fast ganz zerstört ist. Der Winkel 2φ für die Wände, die aus den Querwänden entstehen, nimmt mit zunehmender Spannung zu. Die Änderung der Magnetisierungsrichtung zwischen diesen Wänden wird stärker, und bei einer bestimmten Spannung entstehen zusätzliche Wände (in Fig. 2c die hellen Wände). Eine weitere Zunahme der Spannung ruft, wie im vorigen Fall, die Bildung von Wänden mit wechselnder Polarität hervor (Fig. 2d).

Die experimentellen Daten zeigen, daß man die Bereichsgrenzen wechselnder Polarität in einem großen Dickenintervall antreffen kann und daß wahrscheinlich der entscheidende Faktor, der den Charakter einer Wand beeinflußt, nicht die Schichtdicke, sondern der Winkel ist, um den die Magnetisierungsrichtung beim Durchgang durch die Wand geändert wird. Als Beispiel sind in Fig. 3 Photographien der Bereichsstruktur von Schichten verschiedener Dicke gezeigt. Fig. 3a entnimmt man, daß die Wände auf einer Schicht gleicher Dicke entweder unipolar oder mit wechselnder Polarität in Abhängigkeit vom Drehwinkel 2φ sein können.

Es wurde ein Magnetfeld senkrecht zur Schichtebene angelegt, um die Normalkomponente in den Wänden mit wechselnder Polarität festzustellen. Wenn die zur Schichtebene senkrechten Komponenten der Magnetisierungsvektoren auf beiden Seiten der Bloch-Linien fehlen oder sehr klein sind, kann man die Verschiebung der Bloch-Linien nicht beobachten. Im entgegengesetzten Fall

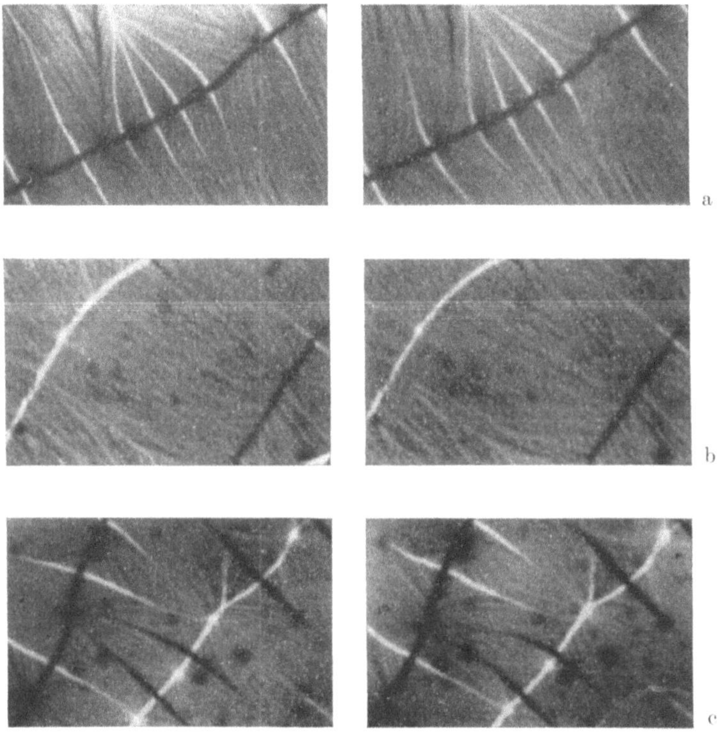

Fig. 4. Einfluß eines Magnetfeldes, das senkrecht zur Ebene von Fe–Ni-Schichten unterschiedlicher Dicke angelegt wurde, auf die Struktur ihrer Wände. Unterlagentemperatur 200 °C. a) und c) $D = 580$ Å, b) $D = 190$ Å. Linke Teilbilder ohne Magnetfeld. Rechte Teilbilder: a) 10,5 Oe, b) 29 Oe, c) 7,2 Oe

müssen die Wandteile, in denen die Normalkomponenten der Magnetisierung günstig bezüglich des angelegten Feldes orientiert sind, zunehmen. In Schichten mit einer Dicke größer als 350 Å beobachtet man unter der Wirkung eines vertikalen Feldes eine Verschiebung der Bloch-Linien, wobei bei dickeren Schichten ein Feld von kleinerer Feldstärke notwendig ist, um eine spürbare Verschiebung zu erhalten (Fig. 4a, c).

Es ist zu bemerken, daß die Richtung der Verschiebung der Bloch-Linien in nebeneinander liegenden Wänden entgegengesetzt und in ein und derselben Wand über ihre ganze Länge gleich ist. Dieser Effekt wird auch beobachtet, wenn das Magnetfeld senkrecht zu einer Wand, aber parallel zur Schichtebene angelegt wird. Bei der Durchführung der Experimente war die horizontale Komponente des Magnetfeldes offenbar unwesentlich, weil eine Drehung der Probe in ihrer Ebene um 180° relativ zum Magnetfeld die Richtung der Verschiebung der Bloch-Linien bezüglich der Schicht nicht beeinflußte. Um den Einfluß der Durchbiegung der Schicht auf dem Trägernetz nachzuprüfen, wurde die Richtung der Verschiebung der Bloch-Linien in allen Quadraten des Arbeitsteils des Trägers geprüft. Es zeigte sich, daß die Richtung der Verschiebung der Bloch-Linien längs einer Wand entlang der ganzen Länge immer gleich war, mit Ausnahme der Stellen am Rand der Öffnung, wo eine merkliche Schrägstellung der Schicht spürbar war. In Schichten mit einer Dicke kleiner als 350 Å wurde keine Verschiebung der Bloch-Linien beim Anlegen unseres maximalen Feldes (36 Oe) normal zur Schichtoberfläche bemerkt.

Literatur

[1] E. E. HUBER, D. O. SMITH und J. B. GOODENOUGH, J. appl. Phys. **29**, 294 (1958).
[2] S. METHFESSEL, S. MIDDELHOEK und H. THOMAS, J. appl. Phys. **31**, 302 S (1960).
[3] H. RUBINSTEIN und R. SPAIN, J. appl. Phys. **31**, 306 S (1960).
[4] L. V. KIRENSKII und V. A. BURAVIKHIN, Dokl. Akad. Nauk SSSR **136**, 575 (1961).
[5] L. V. KIRENSKII und V. A. BURAVIKHIN, Izv. Akad. Nauk SSSR, Ser. fiz., **25**, 569 (1961).
[6] E. FELDTKELLER, Z. angew. Phys. **13**, 157 (1961).
[7] E. FUCHS, Z. angew. Phys. **14**, 203 (1962).
[8] W. LIESK, Z. angew. Phys. **14**, 200 (1962).
[9] V. A. BURAVIKHIN und A. G. SHISHKOV, Uch. Zap. Irkutskogo Ges. Ped. Inst. **21**, 19 (1964).
[10] V. A. BURAVIKHIN und V. G. KAZAKOV, Kristallografiya **9**, 558 (1964).
[11] S. MIDDELHOEK, J. appl. Phys. **34**, 1054 (1963).

(Received May 25, 1966)

phys. stat. sol. **16**, 663 (1966)

Physikdepartment der Technischen Hochschule München

Electrical Properties of Anthracene Single Crystals at Very Low Temperatures

By

P. THOMA and G. VAUBEL

Herrn Professor Dr. N. RIEHL zum 65. Geburtstag gewidmet

The dark current characteristics of anthracene single crystals are measured for a wide range of field strengths up to breakdown at temperatures of 4, 77, and 293 °K. Above the trap-filled limit the logarithmic current-voltage characteristic has slopes of the order of 5 to 7. It is shown that this part of the characteristic can be described by a simple expression. The photovoltages under the action of uv excitation are measured and represented analytically. Using a glow-curve method it is shown that anthracene crystals have quasicontinuous trap distributions for electrons and holes. Because of their lower concentration the distribution of electron traps is observed as a sequence of discrete trap levels.

Dunkelstromcharakteristiken von Anthrazeneinkristallen werden zwischen einer sehr tiefen Bezugsfeldstärke und der Durchschlagsfeldstärke bei Temperaturen von 4, 77 und 293 °K gemessen. Oberhalb des Grenzpunktes des "trap filled limit" besitzt die Strom-Spannungscharakteristik im doppelt logarithmischen Maßstab Steigungen zwischen 5 und 7 sowie höhere Steigungen. Dieser Teil der Charakteristik kann durch einen einfachen Ausdruck beschrieben werden. Die Photospannungen bei UV Anregung werden gemessen und für ihre Beschreibung Ausdrücke abgeleitet. Mit Hilfe von Glowkurvenmethoden wird gezeigt, daß Anthrazenkristalle quasikontinuierliche Haftstellenverteilungen für Elektronen und Löcher besitzen. Wegen ihrer geringeren Konzentration kann die Elektronenhaftstellenverteilung als eine Folge von diskreten Hafttermen beobachtet werden.

1. Introduction

In anthracene single crystals there exist point defects, i.e. narrow regions containing small lattice deformations, which constitute discrete traps for excited states, charge carriers or complexes of both, as shown in a previous paper [1]. Helfrich and Lipsett have discussed the properties of such excited states [2]. In two other papers [3, 4] we have discussed evidence of these traps from conductivity measurements. The present paper is devoted to a more detailed description of these experiments as well as to other studies of the electrical properties of anthracene single crystals at low temperatures.

2. The Crystals

The same four types of crystals were used in the investigations reported here that have been described in the preceding paper.

a) Monocrystalline platelets with surface area 0.5 to 2 cm², 10 to 80 μm thick, grown from the vapour phase.

b) Monocrystalline platelets, also grown from the vapor phase, with nearly the same dimensions. The raw substance, however, was aceotropically distilled and zone refined 150 times. These crystals had a slightly stronger fluorescence than type A crystals.

c) Large single crystals grown from the melt which were kindly supplied by Dr. W. G. Schneider of NRC, Ottawa.

d) Large single crystals (Harshaw—scintillation grade crystals) 1″ ∅, 0.5 to 2 mm thick.

3. Experimental Details

The low temperature equipment has already been described [1]. Different types of electrodes were used to apply electric fields to the crystals:

a) Glass plates on which transparent electrodes had been deposited by evaporation of Ag or SnO_2 were held by spring pressure against optically flat and polished crystals of Type D.

b) Evaporated Ag electrodes were used with all other types of crystals. For dark current measurements electrodes were made also with conductive paint. Guard rings, also made of conductive paint, were used on all crystals to prevent leakage currents in the surface.

Excitation sources were used as follows: i) Hg-ultra high pressure lamp (Osram HBO 200); ii) Hg-high pressure lamp (Osram HQV 300); iii) Hg-low pressure lamp (UVS-11); iv) incandescent lights; v) arc lights with carbon electrodes.

Desired ranges of excitation wavelength were selected either with sets of filters or with a Zeiss single monochromator. In all experiments the incident radiation was directed perpendicularly at the absorbing surfaces of the crystals. Currents and voltages were measured with a vibrating reed electrometer (Cary 31) having switch selectable load resistors. Voltage measurements were set up as shown in Fig. 1a. Conductivity measurements (circuit in Fig. 1b) were performed as follows: α) Excitation at wavelength 365 or 250 nm was applied for 5 min at 4 °K while the electrodes were short-circuited and grounded; β) Voltage was applied to the crystal. Taking of data was delayed until the current had stabilized while keeping the temperature constant at 4,2 °K; γ) the crystal temperature was then raised linearly with time at constant applied voltage. The resulting conductivity glow-curve was registered on an ink-writing potentiometric recorder (H. & B. Linecomp). In some experiments, optical glow-curves were recorded simultaneously with the aid of an RCA 2020 photomultiplier.

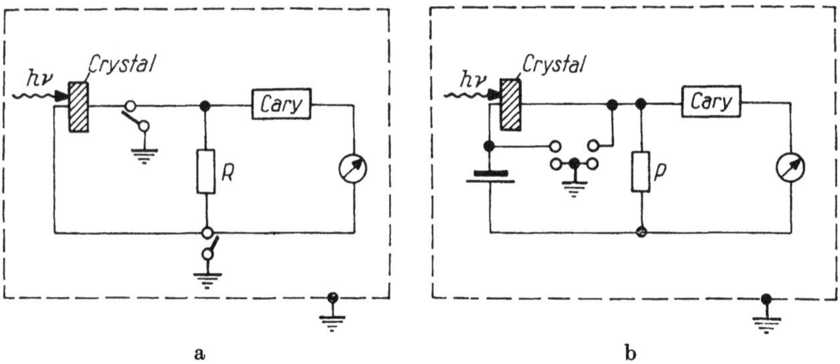

Fig. 1. Measuring circuits; a) voltage measurements, b) current measurements

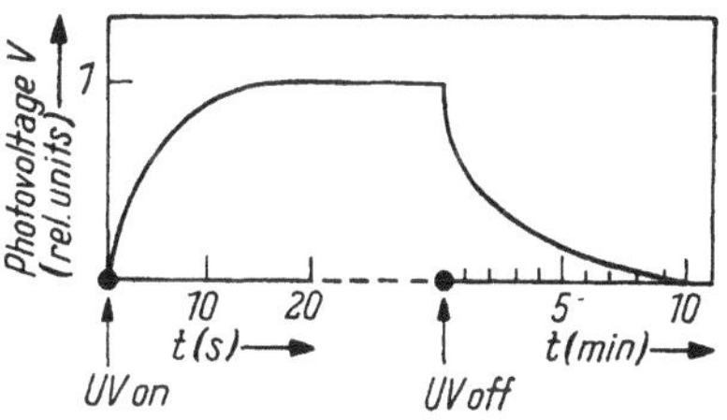

Fig. 2. Photovoltage of an anthracene single crystal as a function of time

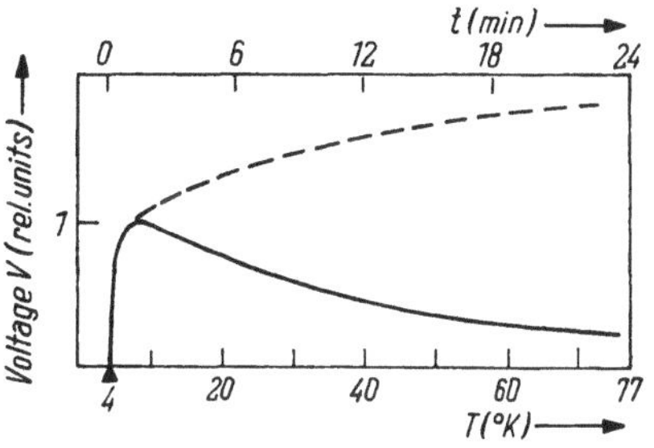

Fig. 3. Voltage glow-curve of an anthracene single crystal (after uv excitation and 5 min dark pause). Dashed curve: Calculated for infinite circuit resistance

4. Results

4.1 Photovoltages

Measurement of photovoltages was made possible by the rather high input impedance of the Cary electrometer. A typical graph of photovoltage vs. time is shown in Fig. 2 for a type D crystal at 4.2 °K. It was excited with uv light of a wavelength range of 350 to 370 nm peaking at 365 nm. The illuminated electrode became positive. The decay portion of the curve depends on the total circuit resistance (Fig. 1a). A total generated charge of 10^{-11} to 10^{-12} C was determined by numerical evaluation.

The steady release of trapped holes from a quasi-continuous hole trap distribution [5] generates a voltage glow-curve, shown in Fig. 3 for a type D crystal, which could be observed by the following procedure: After the excitation was turned off, the crystal was kept at a constant temperature of 4.2 °K until the (promt) voltage had decayed completely. (This could be speeded up by short—circuiting and grounding the electrodes). Then the crystal was heated linearly to 77 °K. Again, that electrode which had been illuminated became positive. This is quite plausible considering the small uv penetration and assuming that electrons at 4 °K show higher drift mobility than holes. This assumption is supported by other results discussed below. Fig. 3 shows further a curve computed for infinite load resistance (dashed curve). Previously reported small irregularities at the location of *optical* glow peaks were unobservable in this experiment regardless of the value of the load resistor. This experiment made directly observable the existence of a quasi-continuous hole trap distribution extending right down to the vicinity of the hole conducting band. A control experiment emphasizes this fact: After reaching a temperature of 20 °K during such a glow-curve experiment, heating was interrupted and the sample allowed to cool again to 4.2 °K. It was reheated after a waiting period of 10 min, during which the observed voltage disappeared. The voltage recorded previously at 20 °K was not reached until the temperature had risen again to about 20 °K.

4.2 Current-voltage characteristics

4.2.1 Dark currents

Dark current measurements were made on a number of newly grown crystals of type B, 10 to 100 μm thick, equipped with evaporated Ag-electrodes. Measurements were performed at 293°, 77°, and 4.2 °K with applied fields between 0.1

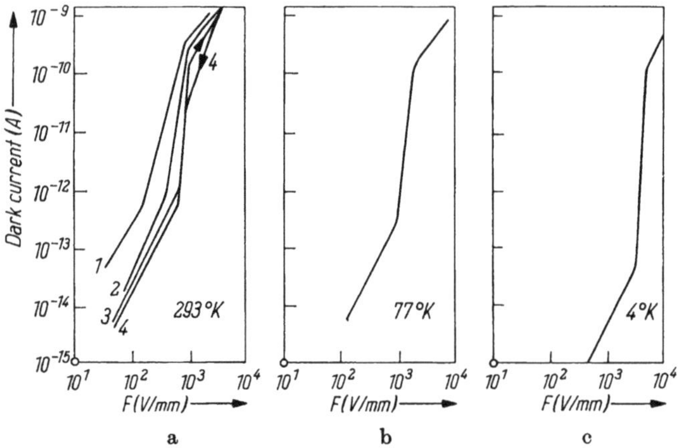

Fig. 4. Current-voltage characteristics (dark current) of an anthracene single crystal at various temperatures

and 8 kV/mm. After any change of applied field, measurements were delayed until complete stabilization had been reached. The results are shown in Fig. 4a to c. Data for curve 2, Fig. 4a, were taken immediately following those of curve 1; curves 3 and 4 were measured on the following day, curve 4 was taken immediately following curve 3 but in the direction of decreasing field strength. Three distinct regions are apparent at field strengths above 0.1 kV/mm. A linear, i.e. ohmic, region for lesser fields is not shown. In these plots of log i versus log F, the lower region shows slopes between 2 and 3, the intermediate region slopes of 5 to 10 and the upper region again values less than 3. The upper region does not reach saturation at any temperature. In the region of 5 to 7 kV/mm in some cases reversible breakthrough was observed. The effect was reversible in the sense that the crystal resistance broke down momentanously, then recovered and was subsequently able of withstanding higher voltage than that at breakdown. In the region of 6 to 8 kV/mm, destructive breakdown of the crystal resistance occurred. The breakdown field-strengths are very strongly dependent on defect concentrations.

At low and intermediate values of $i - F$ slope and field strength, the dark durrent decreases with decreasing temperature at constant applied fields, as shown in Fig. 4b and c. The dark currents were independent of temperature in the upper region (highest field strength) because of the increase of the $i - F$ slope with decreasing temperature. Each crystal shows a virgin curve characterized by somewhat lesser slopes in the first two regions, as indicated in Fig. 4a and supported by other experiments. This virgin curve, curve 1 of Fig. 4a, was found in similar and newly grown crystals. It was not reproducible (in the same sample) even after a delay of several days.

Curves 3 and 4 enclose, in the upper region, a type of hysteresis loop whose area is proportional to dielectric and other losses in the crystal. Dark current measurements on crystals of types C and D gave similar results. As was to be expected, lesser slopes were reached in the intermediate region.

4.2.2 Photocurrents

Photocurrents were measured on crystals of types B, C, and D as functions of excitation wavelength, excitation intensity, and applied field strength at

Fig. 5. Current-voltage characteristics (photocurrents) of 37 μm and 112 μm thick anthracene crystals at various temperatures. Exciting wavelength 365 nm.
a — hole current, b — electron currents; a', b' — currents at 77 °K; a'', b'' — currents at 4,2 °K

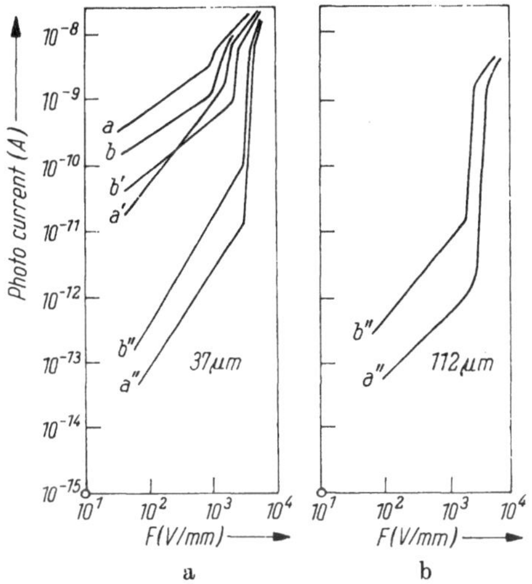

temperatures of 293, 77, and 4.2 °K. Current-field characteristics obtained with two samples of type B are shown in Fig. 5. The crystals were excited with a wavelength of 365 nm. The $i - F$ slopes are seen to increase with decreasing temperature. Above 4 kV/mm they approach the dark current characteristics. Neither reversible nor destructive resistance breakdown was affected by irradiation between 350 and 420 nm. When excited at 365 nm, the photocurrents were about proportional to the square of the excitation intensity J_0. The field dependence is evidenced only by a parallel displacement for lower field strengths. Fig. 5 is also typical for $i - F$ characteristics observed on crystals of types C and D equipped with evaporated Ag-electrodes. The dependence on the excitation intensity was again found to be $i \sim J^2$. The slopes of the photocurrenr vs. field characteristics were observed to decrease with crystal thickness in analogy with dark current measurements.

The wavelength dependence was measured with crystals of types C and D having evaporated electrodes. Results from type C crystals are shown in Fig. 6a and b.

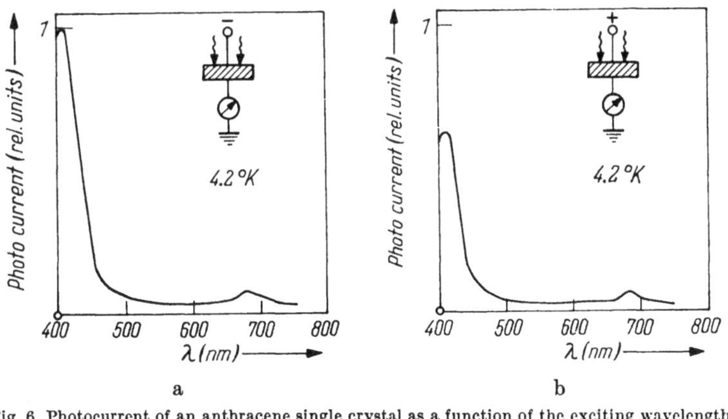

Fig. 6. Photocurrent of an anthracene single crystal as a function of the exciting wavelength; a) excited surface is negative, b) excited surface is positive

A remarkable difference is noted in Fig. 5, 6a and b: photocurrents excited with 365 nm at 293 °K were distinctly larger when the irradiated electrode was positive than with opposite polarity. The difference decreased markedly at 77 °K. At 4.2 °K the difference is negligible, reversing its sign for relatively thick crystals. To first approximation, the assumption seems justified by the strong uv absorption that with the illuminated electrode positive a hole current is observed and an electron current when the illuminated electrode is negative. These currents, outside the ohmic region below 0.1 kV/mm, are always space-charge limited or space-charge compensated since most of the generated carriers are trapped (for longer or shorter periods). Preliminary determination of drift mobility in anthracene at 4,2 °K shows distinctly higher values for the mobility of electrons compared to that of holes.

4.3 Conductivity glow-curves

Crystals of all four types were excited with uv light at 365 nm for 5 min at 4.2 °K. For these experiments all samples including those of type B had

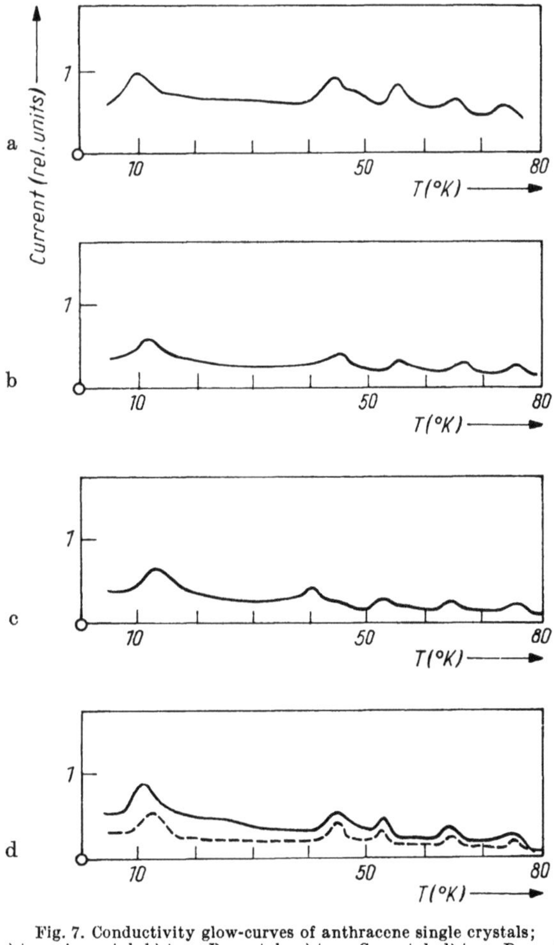

Fig. 7. Conductivity glow-curves of anthracene single crystals; a) type A crystal, b) type B crystal, c) type C crystal, d) type D crystal

evaporated Ag-electrodes. The method of measurement has already been described in Section 3. Typical conductivity glow-curves are shown in Fig. 7. For type D crystals, *optical* glow-curves were obtained together with the conductivity data and added to Fig. 7d. These weakly drooping curves are indicative of the steady release of holes from their quasi-continuously distributed traps. The low heating rate of 3 deg/min was selected to minimize the droop while a time constant of 1 s at the electrometer input was chosen to insure sufficient resolution. The amplitudes of the current peaks depend strongly on the excitation intensity. The samples were excited at 365 or 250 nm. The low pressure uv source for the shorter wavelengths had distinctly lower intensity than the high pressure lamps. The glow-curves with 250 nm excitation wavelength therefore show lower peak amplitudes without differing in other respects from the remaining measurements.

5. Discussion of Results

5.1 Photovoltages

To first approximation the potential jump at the metal anthracene interface is determined by the difference of the work functions. This difference cannot be studied with externally applied fields or carriers generated at random within the crystal volume because of the symmetry of the arrangement. But a photovoltage results if carriers are generated preferentially near an electrode interface by strongly absorbed uv irradiation. To calculate the photovoltage, we consider two charge exchange mechanisms: one is due to thermal agitation, the other is caused by the internal field in the region of the metal—anthracene interface. In order to compute the transport of carriers from metal to anthracene caused by thermal agitation, the electron velocity is needed. The fundamental equation of motion for electrons is

$$H\,\varphi_r = \left[-\frac{\hbar^2}{2\,m}\,\Delta + V(\boldsymbol{r})\right]\varphi_r = E_r\,\varphi(\boldsymbol{r})\,. \tag{1}$$

From the momentum operator

$$\boldsymbol{p} = \frac{\hbar}{i}\,\nabla$$

one finds

$$\boldsymbol{v} = \frac{1}{m}\,\boldsymbol{p} = \frac{1}{m}\,\frac{\hbar}{i}\int \varphi^*\,\nabla\varphi\,\mathrm{d}\boldsymbol{r} = \frac{1}{m}\,\frac{\hbar}{i}\,.$$

$$\int \varphi^*\,\mathrm{grad}\,\varphi\,\mathrm{d}\boldsymbol{r} = \frac{1}{\hbar}\,\mathrm{grad}_{\boldsymbol{k}}\,E_r(k)\,.$$

Obviously, only the x-component of the momentum contributes, thus we derive for the velocity

$$v = \frac{1}{\hbar}\,\frac{\partial E}{\partial \boldsymbol{k}} \quad \text{and} \quad v_x = \frac{1}{\hbar}\,\frac{\partial E}{\partial k_x}\,. \tag{2}$$

The density of the emission current is given by

$$i = \frac{2\,e}{8\,\pi^3}\int\limits_{\boldsymbol{k}} \frac{1}{\mathrm{e}^{E-\xi}+1}\,\frac{1}{\hbar}\,\frac{\partial E}{\partial k_x}\,\mathrm{d}k_x\,\mathrm{d}k_y\,\mathrm{d}k_z\,. \tag{3}$$

This equation is based on the assumption that the number of electrons is equal to that number of occupied levels which are coupled by the propagation vector.

Thus the number of electrons is determined by the level density and the probability of level occupation according to Fermi statistics.

Equation (3) was evaluated by the method given by Haug [6]. The result is a Richardson type law for the thermally generated "current" from metal to anthracene. The emission current density found in this manner at 4.2 °K is only e^{-3} A/cm^2 and thus would seem small enough to be equated to zero. It does, however, furnish a basis of comparison for the determination of the dynamic balance between thermally generated carriers emitted from the metal into the anthracene and those transported from anthracene into the metal due to the potential difference caused by light generated charge carriers. These charges increase the potential barrier at the interface. Their amount can be estimated from our measurements with sufficient accuracy.

Usually, photovoltages are derived from the assumption that the inner field is compensated by the external field of charges having passed the potential barrier, thus stopping a further flow of electrons. This view requires assumptions about the exact distribution of charges which are difficult to verify. The starting point of our discussion is the simple condition of dynamic equilibrium between field-dependent and thermally generated flow of charges when continued irradiation produces a stable voltage. An extension of (1) is used,

$$H' \varphi_r = \left[-\frac{\hbar^2}{2m} \Delta + V(\boldsymbol{r}) + F(\boldsymbol{r}) \right] \varphi_r = E_r' \varphi(\boldsymbol{r}) . \quad (4)$$

All energies are referred to the average crystal potential. $V(\boldsymbol{r})$ and $F(\boldsymbol{r})$ are combined into a larger term $V'(\boldsymbol{r})$. The difference between $V(\boldsymbol{r})$ and $V'(\boldsymbol{r})$ is given by $F(\boldsymbol{r})$ and determined by the difference between measured charge and the charge that would be required to generate a potential difference equal to the work function difference. We thus consider the stationary state when the x-component of current generated by the internal field across the potential barrier balances the thermally generated flow of charges. For electron motion in the x-direction we obtain

$$\frac{d^2\varphi}{dx^2} + \frac{2m}{\hbar^2} \cdot [E_x - V'(x)] \varphi = 0 . \quad (5)$$

The solution is found by the modified WKB method as given by Haug [6]. The details of the calculation will be given elsewhere [7]. The results are given for a fixed temperature of 4.2 °K,

$$i = \frac{e m}{2 \pi^2 \cdot \hbar^3} \int_0^{\xi_0} (\xi_0 - E_x) D(E_x) \, dE_x \quad (6)$$

with

$$\xi_0 = \frac{\hbar^2}{2m} (3 \pi n)^{2/3} , \qquad n = n_0 \exp \left[\frac{-(E_c + \xi)}{k T} \right] .$$

n is the charge carrier density, $D(E_x)$ the transparency of the potential barrier. The final result is (with e the elementary charge, $\varphi(y_0) = 1$, and w the work function)

$$i = \frac{\sqrt{2m}}{12} \cdot \frac{e^3 \mathcal{E}^2}{\pi^2 \cdot w \cdot \hbar^2} \cdot \xi \cdot \varphi(y_0) \cdot e^{-\frac{w^3}{e \mathcal{E}}} . \quad (7)$$

$\mathscr{E}$ is the additional field which depends on the light-generated charges and, therefore, on the intensity of irradiation. A photovoltage of 0.8 V is predicted for a type D crystal by comparing the currents (3) and (7). Experimentally, photovoltages between 0.2 and 5 V were found on different crystal samples of equal dimensions in reasonable agreement with the calculated value. The spread of experimental results may be due to differences of electrode transparency for the several samples.

Again, a voltage is observed when the exciting light source has been turned off, but the crystal is now heated *after* reaching a stationary state. Apparently, heating results in release of trapped charges as well as in diffusion of holes towards the previously illuminated electrode since it becomes positive.

The preferential reappearance of holes at the electrode appears plausible when it is remembered that the charges had been generated in the immediate vicinity of the electrode. Generated holes are trapped there with very high probabilities, while electrons have a chance to diffuse deeper into the crystal. Their higher mobility alone would support this point of view.

As discussed in the introduction, *conductivity* glow-curves exhibited peaks caused by electrons released from traps. As mentioned, corresponding maxima were not observed in *voltage* glow-curves. Their absence is caused by the large positive space charge in the vicinity of the electrode. The high impedance of the external circuit prevented charge removal, while the concentration of liberated electrons is too small to cause noticeable charge compensation. When conductivity glow-curves are measured, on the other hand, the external resistance is reduced to 10 to 10^{12} Ω, and an external field is applied. Then, the space charge can be rapidly removed, and the discrete electron trap distribution becomes observable.

5.2 Current field-strength characteristics

5.2.1 Dark currents

Dark current characteristics were measured for a field-strength range of 1 V/mm to breakdown. Ag-electrodes can inject charge carriers at all temperatures as is evident from the shapes of these characteristics. Their slopes increase, as was to be expected, with decreasing temperature and decreasing crystal thickness. When approaching dielectric breakdown, approximately equal currents are observed at all temperatures. An ohmic region of the dark current characteristic at field strengths below 0.1 kV/mm has been omitted in Fig. 4.

Three distinct regions are noticeable on these $\log i$ versus $\log F$ dark current diagrams. The first region is characterized by slopes between 2 and 3. It can be well described with equations derived by Rose [8] and Parmenter and Ruppel [9] for space charge limited currents in dielectrics, for instance

$$i = e\,\mu\,c\,F^2/d\,, \tag{8}$$

where F is the field strength, d the crystal thickness and c = const. The transition to the second region characterized by slopes of 5 to 10 occurs within a narrow region. The position of this transition depends on temperature and crystal thickness. Apparently, this second region is characterized by complete occupation of all avaiable traps. The number of such terms increases with decreasing temperature, explaining the rise of slope with decreasing temperature. The derivation of relations describing this region of extremely large slopes, requires assumptions about the detailed distribution of the injected space charge. When discussing the first region of relatively small slopes, the assumption of even distribution of charge in the crystal volume had been made. Such assumption is not warranted for region two. Here, we discuss the case of electron injection (the most probable mechanism of charge injection from metal electrodes at very high applied fields) into a crystal having its positive electrode grounded. This situation is governed by the following relations

$$i_n = e\,\mu \cdot n(x) \cdot F(x)\,, \tag{9}$$

$$\frac{\mathrm{d}i_n}{\mathrm{d}x} = e \cdot \mu \cdot \left[\frac{\mathrm{d}n(x)}{\mathrm{d}x} \cdot F(x) + n(x) \cdot \frac{\mathrm{d}F(x)}{\mathrm{d}x}\right], \tag{10}$$

$$\frac{\mathrm{d}F(x)}{\mathrm{d}x} = \frac{4\,\pi}{\varepsilon} \cdot e \cdot n(x)\,, \tag{11}$$

$$n(x) = c\,x^a + c' \tag{12}$$

(a is a whole number being different for each crystal)
with

$$n(x) = n_i(x) + n_0 - p_0 \tag{13}$$

and c and c' = const; e the elementary charge, x the distance from the injecting electrode, n_0 and p_0 are thermally generated charge carrier densities. They are always much smaller than the density $n_i(x)$ of injected electrons. From this system of equations results, for instance

$$i = e\,\mu\,k\,F^5\,, \qquad k = \text{const}\,. \tag{14}$$

The trap density is given by the formula of Lampert, Rose, and Smith [10],

$$V = e\,d^2\,N_t/2\,\varepsilon\,\varepsilon_0\,, \tag{15}$$

where V is the trap filled limit voltage, N_t the trap density, and d the crystal thickness, which yields a value of

$$N_t = 4.4 \times 10^{11}\ \text{cm}^{-3}\,.$$

Trap densities estimated from experimental observations on all our samples are compatible with this computed value. Therefore, it can be concluded that the actual distribution of injected charges is well approximated by this theory. The slope 7 found in (14) is approximately at the centre of the experimental values.

In the third and topmost region of the dark current versus field characteristics finally, injected free carriers by far exceed the total trapped space charge. The resulting slope corresponds to the case of untrapped charges described by Child's law. The shape of the darkcurrent characteristics is in close agreement with corresponding dark current characteristics of rather different substances, for instance ice [11] and CdS [10]. Therefore, it seems to be justified to conclude that this shape is a fundamental property of all insulators (with the exception of the reversible breakdowns). The nature of the observed reversible breakdown is not yet known. One may suspect a kind of punch-through confined to small crystal domains. The current voltage characteristic is terminated by the destructive breakdown as previously discussed.

5.2.2 Photocurrents

The slopes of photo-current vs. field characteristics as shown in Fig. 5 are generally smaller than those of dark currents. Again four regions are distinguishable. In the topmost region photocurrent and dark current merge. Neither reversible nor destructive breakdown are affected by any program of illumination. At high applied fields, field injection by far exceeds the rate of photo generation. Hole and electron currents can be separated by choosing the polarity of the irradiated electrode when uv excitation is used. Hole currents were found to exceed the electron currents at 293 °K while at 4.2 °K they were somewhat smaller. This was to be expected because at 4.2 °K the drift mobility of electrons appreciably exceeds that of holes. The spectral dependence of photocurrents is shown in Fig. 6. A weak increase of current in the region of triplet-absorption is found when these data are recalculated for constant quantum flux density. Apparently, charge carriers are formed in that region by bimolecular triplet-triplet annihilation.

We were able to support this view experimentally. Anthracene crystals at 4.2 °K were excited by large numbers of short flashes from a ruby laser without Q-switch. This form of operation generates relatively low intensities (compared to the Q-switched giant pulse) causing predominantly triplet absorption. Here, relatively large current pulses were observed.

5.3 Conductivity glow-curves

At equal heating rates, the peaks of *conductivity* glow-curves shown in Fig. 7 were observed to occupy the same position as peaks of *optical* glow-curves. We were able to exclude the possibility that in both cases the peaks were caused by annealing of a frozen-in polarization: The crystal was cooled with electric field applied and subsequently heated without having been uv excited. Alternately the crystal was cooled without field and again heated without having been illuminated. In both cases, no peaks in the conductivity curves were observable. At low heating rates, the steady release of holes from their quasi-continuous trap distribution contributes to the background current which decreases very slowly. Rapid heating, by contrast, causes a distinctly noticeable superposition of glow peaks on this background. No such peaks were observed in voltage glow-curves and this different behaviour is significant.

If the discrete traps, responsible for the peaks in the *conductivity* glow-curves, were hole traps then they should be well observable also with *voltage* glow curves. Their absence is thus a proof that the discrete traps are electron traps. We have shown in our previous paper [1] that this conclusion can be reached from other experiments. In the same paper we also tabulated the energetic depths of the traps below the conducting band. From the small energy separation of these traps, the conclusion can be drawn that in anthracene exists a quasi-continuous electron trap distribution as well as the quasi-continuous hole trap distribution discussed above. The electron trap distribution does, however, have a much lower density than the hole trap distribution. For this reason, it is not observable for temperatures above 77 °K with our present apparatus. It simulates the appearance of a discrete distribution at low temperatures, while in reality, it is symmetric to the hole trap distribution. The quasi-continuous electron trap distribution also affects the position of the quasi-Fermi level of the electrons. The traps have been identified with point defects associated with small lattice deformations capable of trapping electrons, excited states or both.

6. Conclusions

The existence of a quasi-continuous hole trap distribution in anthracene can be readily demonstrated with the method of voltage glow-curves. Evaporated Ag-electrodes and electrodes made from conductive paint can readily inject charge carriers at all temperatures. It is probable that these carriers are electrons (at very high field strengths). Dark current as well as photocurrent characteristics can be readily divided into four distinct regions when current is plotted versus field-strength in logarithmic scales. The bottom region, up to about 0.1 kV/mm is ohmic; very little charge is injected.

At higher applied fields, in the second region, current is proportional to the second to third power of field-strength. In this region, the amount of injected charge is of the same order as the charge set free by thermal agitation. The observed current is space charge limited. At still higher field strengths, and thus increased injection, all available traps are occupied. In this third region the injected space charge is no longer distributed evenly through the crystal volume. Rather, it is proportional to the second or third power of the distance from the electron injecting electrode and current is proportional to the 5th to 7th power of the applied field strength. In the fourth region at highest field strengths, injected charge dominates the total trapped space charge by far. In this region, the slope of the characteristic approaches again a value of 2 to 3. Here, reversible dielectric breakdown was observed but no current saturation was detectable. Destructive breakdown terminated the $i - F$ characteristic at about 6 to 8 kV/mm.

Conductivity glow-curves were obtained with circuit impedances small compared to those used for voltage glow-curves. The conductivity curves exhibit discrete peaks. Comparison of conductivity and voltage glow-curves as well as other measurements support the conclusion that there exist electron traps caused by lattice point defects. This electron trap distribution is quasi-continuous and mirror symmetric to the hole trap distribution. It has, however, much lower density compared to the latter.

Acknowledgement

The authors want to express their thanks to Mr. H. P. Glockmann for reading the manuscript.

References

[1] N. RIEHL and P. THOMA, phys. stat. sol., **16**, 159 (1966).
[2] W. HELFRICH and F. R. LIPSETT, J. chem. Phys., in the press.
[3] W. HELFRICH, N. RIEHL, and P. THOMA, Phys. Letters **10**, 31 (1964).
[4] P. THOMA and N. RIEHL, Organic Crystal Symposium, Chicago 1965 (p. 39).
[5] N. RIEHL, G. BECKER, and H. BÄSSLER, phys. stat. sol. **15**, 339 (1966).
[6] G. HAUG, Theoretische Festkörperphysik, Verlag F. Deuticke, Wien 1964 (p. 120).
[7] P. THOMA, to be published.
[8] A. ROSE, J. appl. Phys. **35**, 2664 (1964).
[9] P. H. PARMENTER and W. RUPPEL, J. appl. Phys. **30**, 1548 (1959).
[10] M. A. LAMPERT, A. ROSE, and R. W. SMITH, J. Phys. Chem. Solids **8**, 464 (1959).
[11] H. ENGELHARD and N. RIEHL, Phys. Letters **14**, 20 (1965).

(Received June 2, 1966)

phys. stat. sol. 16, 675 (1966)

Max-Planck-Institut für Metallforschung, Institut für Metallkunde, Stuttgart

On the Critical Resolved Shear Stress of Solid Solutions Containing Coherent Precipitates

By

V. GEROLD and H. HABERKORN

A quantitative estimate is made of the increase $\Delta\tau_0$ of the critical resolved shear stress for the case of solid solutions containing spherical and coherent precipitates. The interaction between edge dislocations and the coherency strain fields is believed to be the strengthening mechanism. The calculation is applied to experimental measurements on Al-Zn and Cu–Co alloys. For small particles (R/b $<$ 20) the dislocations tend to cut through the precipitates, whereas for larger particles the Orowan mechanism becomes predominant.

Es wird eine quantitative Abschätzung für die Erhöhung der kritischen Schubspannung $\Delta\tau_0$ von Mischkristallen mit kohärenten kugelförmigen Ausscheidungen gegeben. Dabei wird angenommen, daß die Wechselwirkung zwischen Stufenversetzungen und Kohärenzspannungen als der bestimmende Aushärtungsmechanismus anzusehen ist. Die Berechnungen werden mit den an Al-Zn- und Cu-Co-Legierungen vorliegenden experimentellen $\Delta\tau_0$-Werten verglichen. Kleine Teilchen ($R/b < 20$) werden von den Versetzungen durchschnitten, während bei größeren Ausscheidungen bevorzugt der Orowan-Mechanismus die Höhe von $\Delta\tau_0$ bestimmt.

1. Introduction

In numerous cases precipitation processes begin with the formation of coherent particles in the supersaturated solid solution. Because of the low nucleation barrier for such particles, they form with a high dispersion. The particles have sizes from 20 to 500 Å in diameter and their density is of the order of 10^{19} to 10^{16} cm^{-3}. The particles have the same lattice structure but a different lattice parameter as compared to the matrix. Because they are coherent with the matrix, elastic strain fields are introduced.

Coherent precipitation increases the strength of the alloy. In some cases the critical resolved shear stress (CRSS) is increased by more than a factor of 10. Among the explanations for this strength increase given by various investigators are the following:

(i) Coherency strains.
(ii) Differences in elastic moduli between the matrix and precipitate.
(iii) Differences in stacking fault energy between the matrix and precipitate.
(iv) Atomic order in the precipitates.

The last point has been discussed recently by Gleiter and Hornbogen [1]. Those authors gave a detailed quantitative analysis for the case where the particles exhibit long-range order. They then applied this theory to the precipitation of the ordered phase Ni_3Al in Ni-Al alloys.

Also recently, Hirsch and Kelly [2] discussed the influence on the CRSS of differences in the stacking fault energy of the precipitate. They applied their theory to the cases of precipitation in the alloys Al-Ag, Al-Zn and Cu-Co. It seems that this mechanism gives the main contribution to the strengthening of

Al-Ag alloys [3] where coherency strains are negligible [4]. However, for Al-Zn and Cu-Co the theory fails to explain the experimental data. Both alloys show coherency strains of the order of 1% [5, 6].

The differences in elastic moduli between solute and solved atoms seem to play an important role for the strengthening in solid solutions [7, 8]. In the case of precipitation this results in only a small contribution to the CRSS which cannot explain the experimental data.

There have been attempts [9] to calculate the influence of coherency strains on the CRSS, but they failed to explain the observed dependence on the size of the particles [10, 11, 12, 13]. The importance of the coherency strains has been demonstrated by Philips [14] by an indirect method. He found an anomalous temperature dependence of the yield stress in a copper alloy containing 3% cobalt, which decreased by 10% in changing from room temperature to 77 °K. This anomaly only can be explained by the decrease of the coherency strains with decreasing temperature. Such a decrease is to be expected because of the differences in the thermal expansion coefficients of copper and f.c.c. cobalt.

The present paper tries to give a quantitative analysis of the increase in the CRSS caused by coherency strains. In contrast to the well known theory by Mott and Nabarro [15], a model is assumed for this calculation where the interaction between dislocations and strain fields plays the dominant role. Similar calculations have been undertaken by Fleischer [7] for solid solution hardening. The following formula will be applied

$$\Delta\tau_0 = \frac{K}{b\,L}, \tag{1}$$

where K is the maximum repelling force of the strain field of an individual particle on a moving dislocation, L is the average distance between those force centers, b is the length of the Burgers vector of the dislocation, and $\Delta\tau_0$ is the difference between the CRSS of the alloy containing the precipitates and its value for the single-phase depleted matrix. The general problem is to find expressions for K and L. A rough estimate for the increase $\Delta\tau_0$ has already been given in a previous paper [16].

2. The Maximum Repelling Force K on a Dislocation

The dislocation is considered to move in its slip plane $z = z_0$ and to be parallel to the y-direction of a Cartesian coordinate system. The Burgers vector makes an angle α with the y-direction. The spherical precipitated particle is located at the origin of the coordinate system. It is surrounded by a spherical stress field. The shear stress of this stress field, which interacts with the dislocation, is written as

$$\left.\begin{aligned} \sigma(x, y, z) &= 6\,G\,\varepsilon\,\frac{(x \sin\alpha + y \cos\alpha)\,R^3\,z}{r^5} \quad &&\text{for} \quad r \geqq R\,, \\ &= 0 &&\text{for} \quad r < R\,, \end{aligned}\right\} \tag{2}$$

where $r^2 = x^2 + y^2 + z^2$ and R is the radius of the particle. Equation (2) assumes an isotropically elastic matrix with shear modulus G. The quantity ε is the linear misfit parameter, which describes the observed difference $\Delta a/a$ between the lattice parameters of the matrix and the particle. The stress field consists of two terms, the first interacting with the edge component of the dislocation, the second interacting with the screw component.

The interaction force F is defined by

$$F = b \int \sigma(x, y, z)\, \mathrm{d}s\,, \tag{3}$$

where the integral is taken along the dislocation line. Both the position and the shape of the dislocation line depend on the internal stress field σ and on the external field τ. Here only an approximation will be given for small interaction forces. In that case the shape of the dislocation line can be approximated as being straight in the neighbourhood of the particle, where the stress field σ is maximum. The straight line approximation results in

$$F \simeq b \int_{-\infty}^{+\infty} \sigma(x_0, y, z_0)\, \mathrm{d}y\,, \tag{4}$$

where $(x_0\, z_0)$ gives the position of the dislocation line. Combining equations (2) and (4) the result is

$$F = F_{\mathrm{e}} \sin\alpha + F_{\mathrm{s}} \cos\alpha\,.$$

The forces F_{e} and F_{s} are

$$F_{\mathrm{e}} = \frac{8\,G\,|\varepsilon|\,R^3\,x_0\,z_0\,b}{(x_0^2 + z_0^2)^2}\left\{1 - \frac{y_0\,(2\,R^2 + x_0^2 + z_0^2)}{2\,R^3}\right\}$$

with

$$\left.\begin{aligned} y_0^2 &= R^2 - x_0^2 - z_0^2 && \text{for} \quad x_0^2 + z_0^2 \leqq R^2\,,\\ &= 0 && \text{for} \quad x_0^2 + z_0^2 \geqq R^2\,, \end{aligned}\right\} \tag{5}$$

and

$$F_{\mathrm{s}} = 0\,.$$

In this approximation only the edge component of the dislocation gives a contribution to F. The reason is the symmetry of the strain field σ, which changes its sign with the sign of x_0 for an edge dislocation, but with the sign of y for a a screw dislocation.

For this reason the straight line approximation is a bad one for the screw component. Small changes in the shape of the dislocation line result in a considerable change of the interacting force, F_{s}. If one calculates the integral, equation (4), with the lower limit $y = 0$, the screw part results in

$$F_{\mathrm{s}}' \simeq 2\,G\,|\varepsilon|\,z_0\,b\,k \tag{6}$$

with

$$\begin{aligned} k &= 1 && \text{for} \quad x_0^2 + z_0^2 \leqq R^2\,,\\ &= \frac{R^3}{(x_0^2 + z_0^2)^{3/2}} && \text{for} \quad x_0^2 + z_0^2 \geqq R^2\,. \end{aligned}$$

This formula neglects the attractive part of the force and gives only the repulsive part.

The actual value of F_{s} is much smaller than given by equation (6), especially for small particles, where F_{s}' and therewith the curvature of the dislocation is small. The difference should decrease with increasing radius R. F_{s}' serves as an (much too large) upper limit for the interacting force, which is smaller than F_{e}.

By varying x_0 one gets a maximum force K for each slip plane $z = z_0$.

$$K_e(z_0/R) = 4\,G\,|\varepsilon|\,b\,R \cdot \varphi(z_0/R) \tag{7}$$

with

$$\varphi(s) = 2\,s\sqrt{1-s^2} \quad \text{for} \quad s^2 \leqq 3/4\,,$$
$$= \frac{3\sqrt{3}}{8\,s^2} \quad \text{for} \quad s^2 \geqq 3/4\,,$$

whereas the upper limit for the screw dislocation is

$$K_s'(z_0/R) = 4\,G\,|\varepsilon|\,b\,R \cdot \psi(z_0/R) \tag{8}$$

with

$$\psi(s) = \frac{s}{2} \quad \text{for} \quad s \leqq 1\,,$$
$$= \frac{1}{2\,s^2} \quad \text{for} \quad s \geqq 1\,.$$

Fig. 1 shows the magnitude of the interacting force K_e, which is given by the sizes of the positive or negative signs showing either repulsion (—) or attraction (+) by the stress field. The signs are given for a negative value of ε and for a sign of the Burgers vector as shown by the dislocation symbol. Fig. 2 demonstrates the dependence of K_e and K_s' with the distance z_0/R of the slip plane from the center of the particle.

The forces K lead to a bending of the dislocation line. The bending angle, $2\,\theta$, is given by

$$2 \sin\theta \simeq \frac{K}{2\,T}\,. \tag{9}$$

T is the line tension of the dislocation. A mean radius of curvature, ϱ, can be given which is defined by

$$\varrho \simeq \frac{2\,T}{K} \cdot R\,. \tag{10a}$$

Equations (6) to (8) are only valid if $\varrho \gtrsim 2\,R$ which gives the limit

$$K \lesssim T\,. \tag{10b}$$

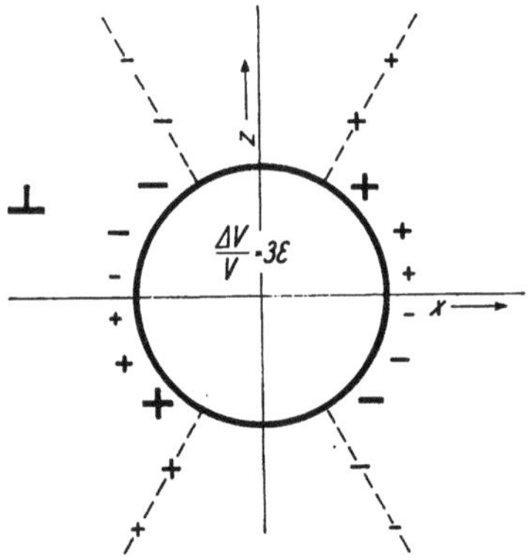

Fig. 1. Magnitude and sign of the interaction forces K_e for an edge dislocation lying on its lip plane $z = z_0$ and parallel to the y-direction

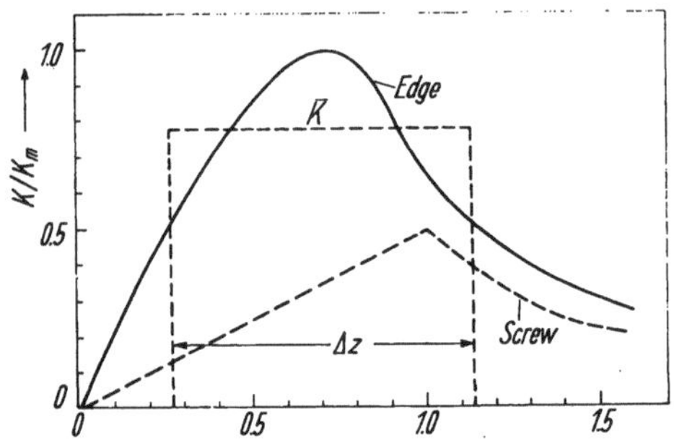

Fig. 2. Relative maximum interaction forces K/K_m as a function of the distance z_0/R between the slip plane z_0 and the center of the particle for both, an edge (K_e), and a screw dislocation (K_s'). $K_m = 4\,G\,\varepsilon\,b\,R$

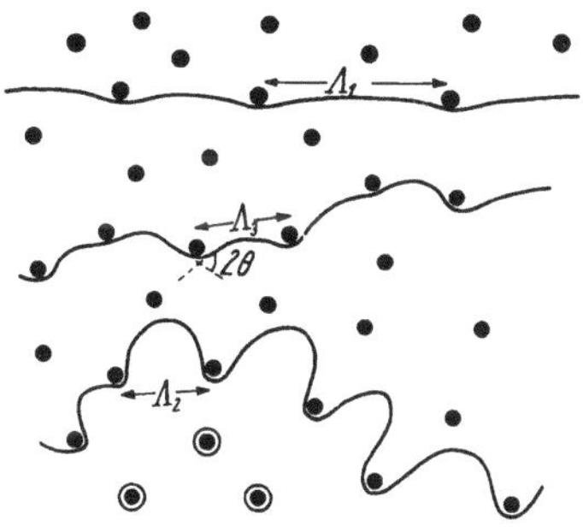

Fig. 3. Average distance Λ_i between particles for the case of a rigid (Λ_1), partially flexible (Λ_3), and totally flexible (Λ_2) dislocation line

3. The Average Distance L between Force Centers

The quantity L is the average distance between repulsive force centers along a dislocation line. This distance depends on the flexibility of the dislocation line as has been shown by Fleischer [7]. At first the formulae for the corresponding distances between particles will be given. If f is the volume fraction of the precipitation, $n = \frac{3f}{4\pi R^3}$ defines the density of the particles. The formulae for the average distance between two particles which are located at the dislocation line (Fig. 3 and 4) are for

a) a rigid dislocation line (Λ_1)

$$\pi R^2 \Lambda_1 n = 1$$

and

$$R/\Lambda_1 = \frac{3f}{4}; \tag{11}$$

b) a totally flexible dislocation line (Λ_2)

$$\Lambda_2^2 \, 2 R n = 1$$

and

$$R/\Lambda_2 = \left(\frac{3f}{2\pi}\right)^{1/2}; \tag{12}$$

c) a partially flexible dislocation line (Λ_3)

$$\frac{2}{3} \Lambda_3^2 \, \theta \, 2 R n = 1$$

and

$$R/\Lambda_3 = \left(\frac{\theta f}{\pi}\right)^{1/2}. \tag{13}$$

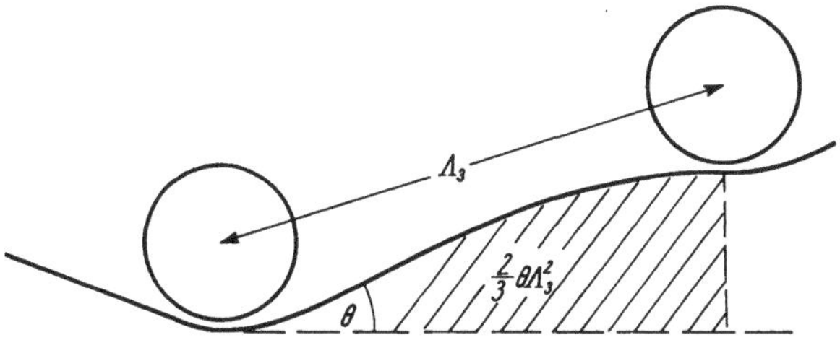

Fig. 4. Definition of the angle θ, to which the partially flexible dislocation line is bent due to the interaction force K

Equation (13) is only valid for bending angles

$$\frac{9\pi f}{16} < \theta < \frac{3}{2}. \tag{14}$$

For bending angles smaller than $\frac{9\pi f}{16}$ the dislocation must be treated as a rigid line, and equation (11) must be used. For bending angles near the upper limit of equation (14) the dislocation line is totally flexible, and equation (12) is applicable.

The problem of calculating the average distance L of repulsive force centers is much more difficult because these force centers have no defined volume. Therefore, the average range, Δz, of a force field along the z direction has to be defined. The resulting formulae are

$$R/L_1 = \frac{3f}{4} \cdot \frac{\Delta z}{R}, \tag{11a}$$

$$R/L_2 = \left(\frac{3f}{2\pi} \cdot \frac{\Delta z}{R}\right)^{1/2}, \tag{12a}$$

and

$$R/L_3 = \left(\frac{\theta f}{\pi} \cdot \frac{\Delta z}{R}\right)^{1/2} \quad \text{for} \quad \frac{9\pi f}{16} < \theta < \frac{3}{2}. \tag{13a}$$

4. The Critical Resolved Shear Stress

For relatively small values of K the dislocation cuts the particle after overcoming the force K by the applied shear stress τ.

In combining equation (1), (9), (10b), and (13a) one finds for the increase of the critical resolved shear stress

$$\Delta\tau_0 = \frac{K}{b\,R}\left(\frac{K f}{2\pi T} \cdot \frac{\Delta z}{R}\right)^{1/2} \quad \text{for} \quad \frac{9\pi f}{16} < \frac{K}{2\,T} < \frac{1}{2}. \tag{14a}$$

For an edge dislocation, $T \simeq G\,b^2/2$. With the use of equation (7) one finds a maximum value for the average $[(\Delta z/R)^{1/2} \cdot \varphi(\Delta z/R)^{3/2}]$, which is approx. 2/3. The final formula for the change in the CRSS is

$$\Delta\tau_0^e \simeq 3\,G|\varepsilon|^{3/2}\left(\frac{R f}{b}\right)^{1/2} \quad \text{for} \quad \frac{9\pi f}{16} < \frac{3\,|\varepsilon|\,R}{b} < \frac{1}{2}. \tag{15}$$

Similarly, for the screw dislocation one finds as an upper limit

$$\Delta\tau_0^s \simeq G\,|\varepsilon|^{3/2}\left(\frac{R f}{b}\right)^{1/2} \quad \text{for} \quad \frac{9\pi f}{16} < \frac{|\varepsilon|\,R}{b} < \frac{1}{2}. \tag{16}$$

For the case $K \gtrsim 2\,T$ the bending angle θ around the particle reaches 90°. In that case the dislocation bypasses the particle without cutting through it (Orowan mechanism). The CRSS is then given by

$$\Delta\tau_0 = \frac{2\,T}{b\,L_2}. \tag{17}$$

In the case of edge dislocations the CRSS becomes ($T \simeq G\,b^2/2$)

$$\Delta\tau_0^e = \frac{G\,b}{R}\left(\frac{3f}{2\pi} \cdot \frac{\Delta z}{R}\right)^{1/2} \quad \text{for} \quad \frac{K}{2\,T} > 1, \tag{18}$$

where Δz is the range in the z-direction where $K > 2\,T$. Since the value of K depends on the particle size, the value of Δz also depends on the particle size. Similarly, for a screw dislocation ($T \simeq G\,b^2$)

$$\Delta \tau_0^s = \frac{2\,G\,b}{R} \left(\frac{3\,f}{2\,\pi} \cdot \frac{\Delta z}{R}\right)^{1/2} \quad \text{for} \quad \frac{K}{2\,T} > 1\,. \tag{19}$$

A graph of equations (15) and (18) is shown in Fig. 7.

5. Comparison with Experimental Results

There are only a few experiments which can be compared with the theory. In the case of Cu-Co alloys Livingston [10] made extensive measurements on the dependence of the yield stress on particle size and volume fraction, but only on polycrystalline specimens.

Bonar [17] made some measurements on single crystals, but there is only one value reported in the literature by Kelly and Nicholson [9].

Very recently Haberkorn [11] investigated extensively the CRSS of single crystals of various Al-Zn alloys. In both cases an empirical relationship

$$\Delta \tau_0 = G\, f^{1/2}\, g(R/b) \tag{20}$$

can be established where g is a function depending only on the size of the particles. In the case of Cu-Co the yield stress [10] has to be converted into CRSS values. A conversion factor of 0.41 was chosen here as this gives good agreement with the reported single crystal value [9]. Fig. 5 shows the result of the analysis. The final curve taken from the experimental results of Haberkorn is shown in Fig. 6.

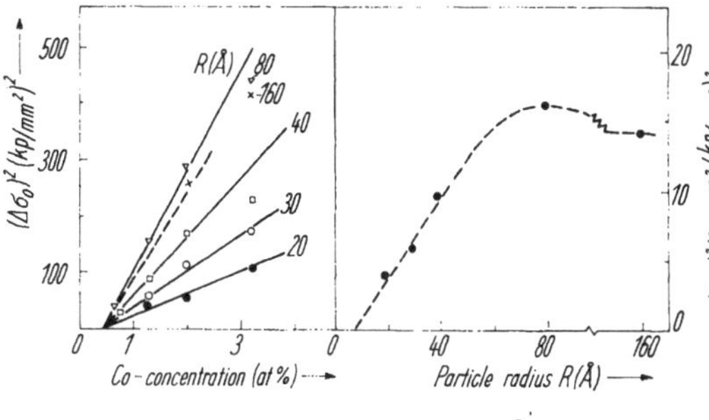

a b

Fig. 5. Increase of the yield stress $\Delta\sigma_0$ of zone hardened Cu-Co alloys from the experimental work of Livingston [10].
a) Square of the yield stress $\Delta\sigma_0$ vs. Co concentration
b) $\Delta\sigma_0^2/f$-values vs. particle radius R from the slopes of the lines in Fig. 5a

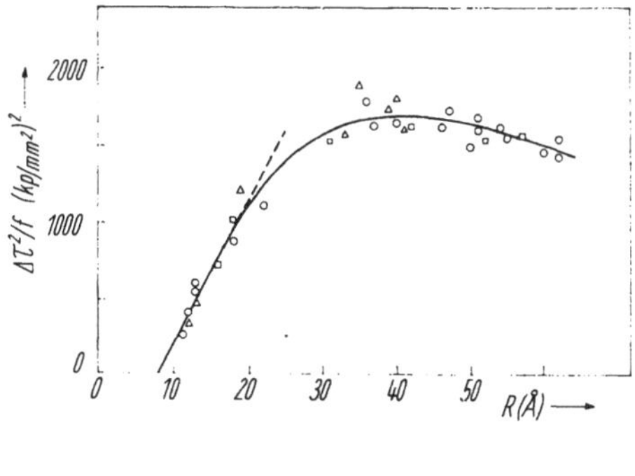

Fig. 6. $\Delta\tau^2/f$ as a function of the radius R of the Guinier-Preston zones in Al-Zn [11, 16]

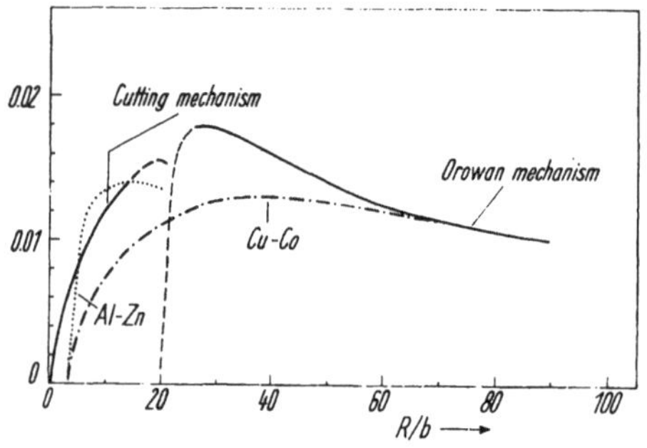

Fig. 7. Comparison of the theoretical quantity $\Delta\tau_0/G\sqrt{f}$ for edge dislocations with the experimentally observed values for the alloys Al–Zn [11, 16] and Cu–Co [10]. The curves are plotted against the normalized particle radius R/b

The comparison between theory and experiment is made in Fig. 7. For both alloys the distortion coefficient ε has nearly the same value of -1.2×10^{-2} [5, 6]. The following shear moduli have been used:

$$\text{Cu-Co} \quad G = 4100\ \text{kp/mm}^2,$$
$$\text{Al-Zn} \quad G = 2900\ \text{kp/mm}^2.$$

Two theoretical curves have been included in Fig. 7. The cutting process of an edge dislocation, equation (15), and the Orowan mechanism for an edge dislocation, resulting from equation (18). The cutting process of a screw dislocation, equation (16), shows a reduction in $\Delta\tau_0$ by more than a factor of three compared to the corresponding values for the edge dislocation. The Orowan mechanism for a screw dislocation comes into action for R/b-values larger than 80. The reason is the higher line tension of a screw dislocation and the lower value of the force K_s compared to K_e.

For small R/b-values only the cutting process is expected which increases with $(R/b)^{1/2}$ up to values of $R/b \approx 15$ according to equation (15). For larger particle radii there exists no explicit formula. When R/b reaches the value of 20 the Orowan mechanism comes into action for edge dislocations which determines the CRSS for R/b values larger than 25. The contribution to $\Delta\tau_0$ of the cutting process is reduced to nearly zero in this range.

From Fig. 7 it can be concluded that the edge dislocations determine the CRSS for both alloys. There are two interesting features to comment on. At first important differences exist between the curves for Cu-Co and Al-Zn, the latter being much steeper and reaching to higher values of $\Delta\tau_0/G\sqrt{f}$ for small particle radii. These differences are of the order of a factor of 0.5 and may arise from different widths of the extended dislocations in the different matrices. The width of an extended dislocation line is larger in the case of copper alloys where the stacking fault energy is lower. In this case the coherency strain field σ acts mainly on the leading partial dislocation. This reduces the repelling force K by a factor $v \gtrsim b_p/b = 0.58$. According to equation (14a), $\Delta\tau_0$ is proportional to $K^{3/2}$ which gives a reduction factor $v^{3/2} \gtrsim 0.44$ and is in agreement with the experimental factor of 0.5.

The second feature is the difference between theory and experiment for small particle sizes. The experimental $\Delta\tau_0$ is practically zero for particle radii smaller than 3 b. This has been found also by Jones [18] in his experiments with Al-Zn alloys. Small particle sizes are present only after short annealing times, where the precipitation process is in the beginning stage. There are probably three reasons for these discrepancies between theory and experiment.

First, the distortion parameter ε has been measured for larger particle sizes only where it was found to be independent of the particle radius. For very small particle sizes however this may no longer be true. The thickness of the interface now comes into play and this will reduce the effective value of ε responsible for the distortion field outside the particle. The second reason could be the inability of the theory to treat the case where the gradient of the coherent stress field is very large for small values of R. From equation (2) one can show that the gradient of σ is proportional to $1/R$ near the interface between the particle and the matrix. Because the interaction between the stress field and the dislocation is not restricted to the center of the dislocation line, a high gradient stress field may be overcome more easily by a dislocation than indicated by equation (4). It should also be expected that the temperature dependence of the CRSS increases with decreasing particle size, however, this was not observed [13]. Finally, there is a possibility that for very small particle sizes the volume fraction is smaller than the equilibrium volume fraction f used for the calculation. As observed experimentally, there exist differences in f in the earliest stages of the precipitation process. However, the f-values deduced from experiments were found to be only 20% below the equilibrium value at the beginning stage of the againg treatment [19, 20]. Because of the $\sqrt{f}$-dependence of $\Delta\tau_0$ there would be only an allowance for a 10% reduction compared to the theoretical curve in Fig. 7, which is far too small to explain the observed discrepancies.

6. Conclusions

It has been shown that the experimentally measured dependence of the CRSS on precipitate sizes and alloy concentration can be explained theoretically by the interaction of coherency strain fields and edge dislocations. This is valid for cases where the parameter ε given by the misfit between coherent spherical precipitates and matrix is of the order of 1%. For small particles, the dislocations cut through the precipitates, however, for larger particle sizes the Orowan mechanism will be observed. The approximate critical radius, at which the Orowan mechanism becomes operative is about $R \approx \frac{b}{3\,|\varepsilon|}$. The present theory is in better agreement with experiment than that developed by Hirsch and Kelly [2]. In this latter theory it is assumed that the different stacking fault energies for precipitate and matrix are the cause of the increased CRSS. Stacking fault hardening seems to be predominant in cases of coherent precipitates with ε being nearly zero [3].

Acknowledgements

The authors are grateful to Mr. H. Gleiter for valuable discussions on the interaction mechanisms between dislocations and coherent precipitates. The financial support of the research by the Deutsche Forschungsgemeinschaft is gratefully acknowledged. We also wish to thank Dr. G. R. Speich for reading and correcting the English manuscript.

References

[1] H. GLEITER and E. HORNBOGEN, phys. stat. sol. **12**, 251 (1965).
[2] P. B. HIRSCH and A. KELLY, Phil. Mag. **12**, 881 (1965).
[3] V. GEROLD and G. MAYER, to be published.
[4] V. GEROLD, Erg. exakt. Naturwiss. **33**, 105 (1961).
[5] M. F. ASHBY and L. M. BROWN, Phil. Mag. **8**, 1083 (1963).
[6] W. MERZ and V. GEROLD, Z. Metallk., in the press.
[7] R. L. FLEISCHER, in The Strengthening of Metals, Ed. D. Peckner, Reinhold Publ. Co., New York 1964.
[8] P. HAASEN, Z. Metallk. **55**, 55 (1964).
[9] A. KELLY and R. B. NICHOLSON, Progr. Mat. Sci. **10**, No. 3 (1963).
[10] J. D. LIVINGSTON, Trans. AIME **215**, 566 (1959).
[11] H. AUER, V. GEROLD, H. HABERKORN, and W. ZÜRN, phys. stat. sol. **6**, K 91 (1964).
[12] H. HREN and F. HINZNER, Z. Metallk. **56**, 305 (1965).
[13] H. HABERKORN, phys. stat. sol. **15**, 153 (1966).
[14] V. A. PHILLIPS, Phil. Mag. **11**, 775 (1965).
[15] N. F. MOTT and F. R. N. NABARRO, Proc. Phys. Soc. **52**, 86 (1940).
[16] H. HABERKORN and V. GEROLD, phys. stat. sol. **15**, 167 (1966).
[17] L. G. BONAR, Ph. D. Thesis, Cambridge University 1963.
[18] R. L. JONES, Acta metall. **14**, 81 (1966).
[19] V. GEROLD, phys. stat. sol. **1**, 37 (1961).
[20] J. J. BECKER, Trans. AIME **209**, 59 (1957); **212**, 138 (1958).

(Received June 2, 1966)

phys. stat. sol. **16**, 685 (1966)

Institut für Physik am Max-Planck-Institut für Metallforschung, Stuttgart, und Institut für theoretische und angewandte Physik der Technischen Hochschule Stuttgart

Electron Diffraction Contrast of Small Stacking Fault Tetrahedra with Applications to Quenched Gold

By

K.-P. CHIK

It is shown that the characteristic diffraction contrast of small stacking fault tetrahedra obtained by imaging with a {220} reflection can be used to distinguish small unresolved stacking fault tetrahedra from dislocation loops. Application of these results to quenched gold shows that vacancy clusters exist already in their early stage as stacking fault tetrahedra.

Es wird gezeigt, daß der an kleinen Stapelfehlertetraedern durch einen {220}-Reflex abgebildete charakteristische Beugungskontrast benutzt werden kann, um zwischen kleinen unaufgelösten Stapelfehlertetraedern und Versetzungsringen zu unterscheiden. Diese Ergebnisse werden zur Untersuchung der Leerstellenansammlung in abgeschrecktem Gold benutzt. Es zeigt sich, daß die Leerstellenagglomerate schon in ihrem Anfangsstadium als Stapelfehlertetraeder bestehen.

1. Introduction

In transmission electron microscopy, the resolving power of the electron diffraction contrast limits the details obtainable from small defect clusters. It has recently been demonstrated that valuable informations may be obtained by studying the 'strain contrast' of the defect clusters under dynamical diffraction condition. Ashby and Brown [1] showed that in this way quantitative informations about type of strain field around an elastic inclusion can be derived. Essmann and Wilkens [2] found that with low order reflections the 'black dots' in neutron irradiated copper exhibit the characteristic black-white contrast. The direction from the black side to the white side lies in $\langle 111 \rangle$-directions or their projections on the foil plane. Essmann and Wilkens explained this effect as diffraction contrast due to strain field of small dislocation loops of Frank type. This interpretation has proved to agree well with detailed theoretical calculations [3]. Thus by using the black-white contrast effect it is possible to identify small dislocation loops with diameter as small as 30 Å.

Another important type of defect clusters whose strain contrast has not yet been studied in details is the stacking fault tetrahedra (abbreviated as SFT), which can be produced for example in abundance in quenched gold [4]. Small SFT appear in electron micrographs as black dots similar to those due to dislocation loops. It is attempted here to exploit some special features of diffraction contrast from small SFT that may permit to distinguish them from the loops. The results are applied to study the initial vacancy clustering in quenched gold.

2. Diffraction Contrast of Stacking Fault Tetrahedra

2.1 Experimental methods

Well annealed thin gold foils of $\approx 60\ \mu m$ thickness were heated electrically in air, quenched into ice-brine and subsequently annealed. Tetrahedra of various sizes and densities can be obtained by controlling the quenching temperature and annealing conditions [5]. Crystal foils investigated had ⟨100⟩-, ⟨110⟩-, and ⟨112⟩-normals. In one special case, a foil with a ⟨111⟩-normal was obtained. Observations were made in a Siemens Elmiskop I with double tilting stage. Both bright field and dark field techniques were used.

Mainly investigated were stacking fault tetrahedra of moderate sizes (between 200 and 400 Å in edge lengths), since the details of such tetrahedra may still be resolved under the imaging conditions. It is expected that the results on the strain contrast can be applied to smaller tetrahedra, where the geometry cannot be directly visualized.

2.2 Results

2.2.1 Kinematical diffraction contrast

With low order diffraction vectors $\boldsymbol{g} = \{200\}$ and $\boldsymbol{g} = \{111\}$, all the four stacking fault faces of the tetrahedra are in contrast. Tetrahedra with edge lengths exceeding 200 Å appear then as squares in $\{100\}$-foils and as triangles in $\{110\}$-, $\{11\bar{2}\}$-, and $\{111\}$-foils. The characteristic fringes of overlapping stacking faults on non-parallel $\{111\}$-planes are also clearly visible [4]. Small SFT appear only as black dots.

With $\boldsymbol{g} = \{220\}$ the diffraction contrast depends on the foil normals. In $\{110\}$-foils a V-shaped contrast is obtained (Fig. 1). In $\{100\}$- and $\{11\bar{2}\}$-foils there is a "line of no contrast" which cuts the tetrahedra into two halfs (Fig. 2 and 3). This "line of no contrast" is always perpendicular to the direction of the operating $\boldsymbol{g} = \{220\}$. Truncated SFT appear only as two widely separated small triangles (A in Fig. 2). For very small SFT the contrast in $\{100\}$-foils then appears very similar to the "double arc" contrast of an inclined dislocation

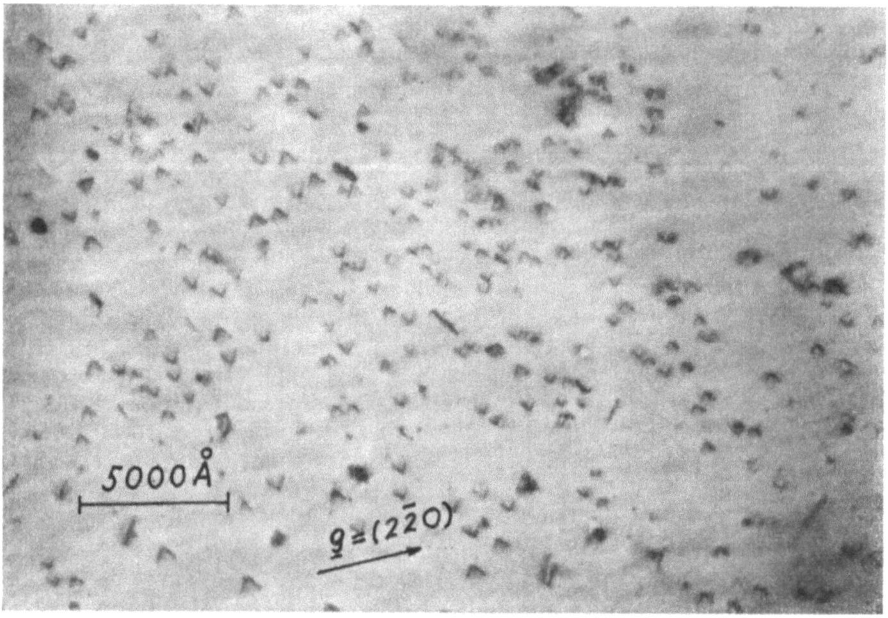

Fig. 1. Stacking fault tetrahedra imaged by $\boldsymbol{g} = (2\bar{2}0)$ in (110)-foil orientation ($s > 0$, bright field)

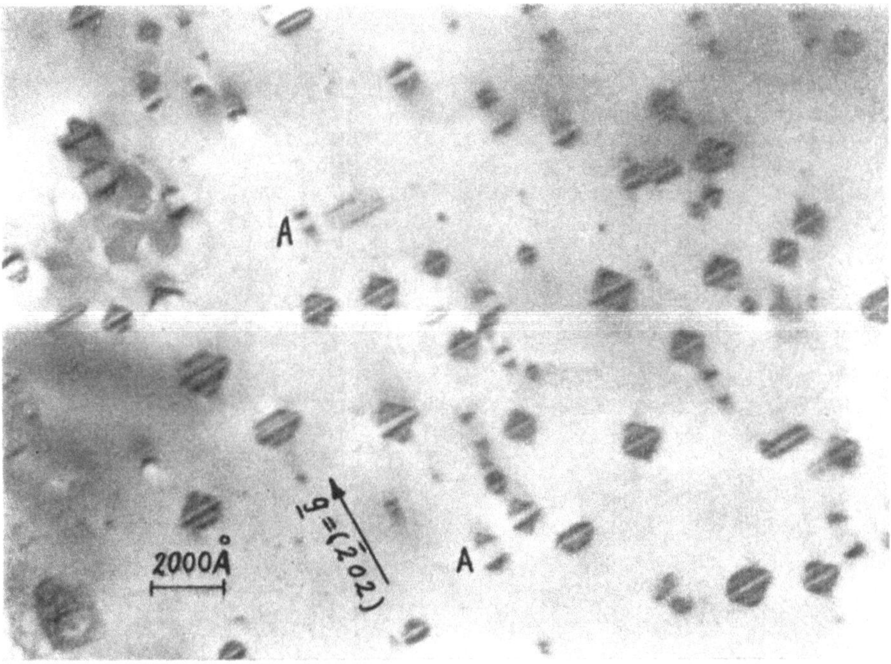

Fig. 2. Stacking fault tetrahedra imaged by $\boldsymbol{g} = (\bar{2}02)$ in (010)-foil orientation (bright field, $s > 0$). Truncated tetrahedra are denoted by A

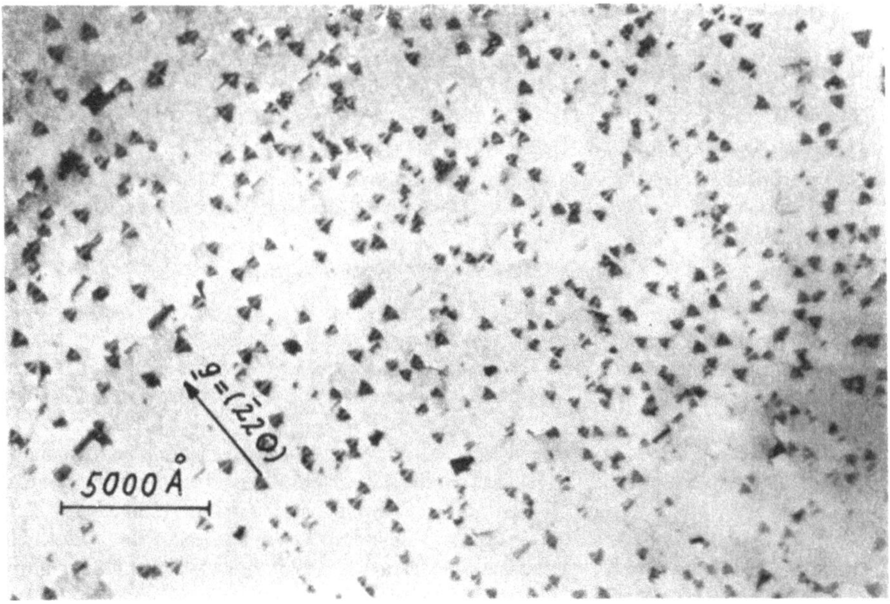

Fig. 3. Stacking fault tetrahedra imaged by $\boldsymbol{g} = (\bar{2}20)$ in $(11\bar{2})$-foil orientation (bright field)

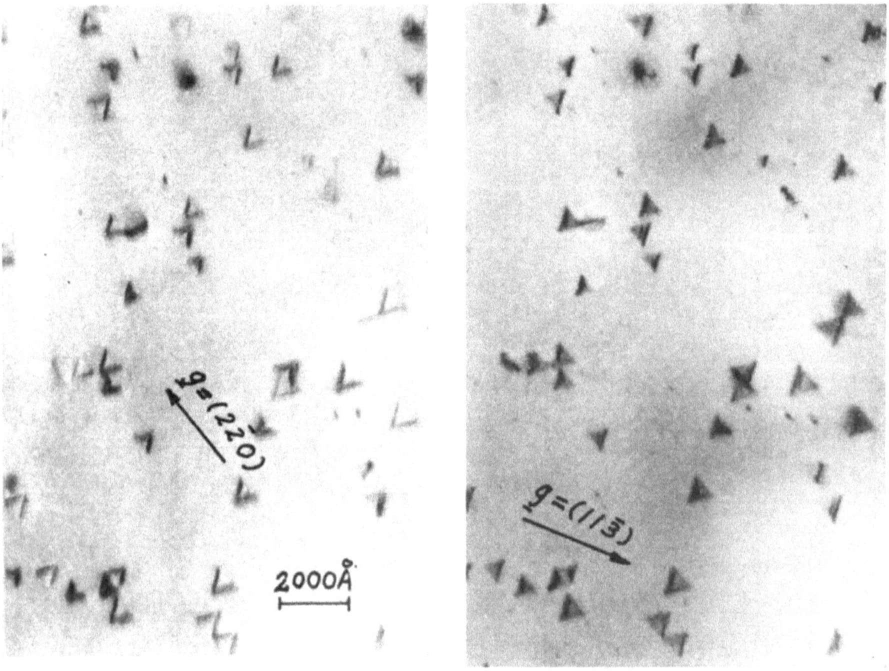

a b

Fig. 4. Diffraction contrasts of stacking fault tetrahedra imaged by two different reflections (bright field).
a) $\boldsymbol{g} = (2\bar{2}0)$ operating in $(11\bar{1})$-foil orientation.
b) The same area tilted to operate $(11\bar{3})$ reflection. The foil orientation is now near (332).
Most of the visible tetrahedra are cut by the foil surfaces

ring [6] or of an inclusion with elastic distortion in the matrix as described by Ashby and Brown [1]. It must be kept in mind that these three similar contrast effects all come from different origins. Fig. 4 shows the change of contrast in a (111)-foil by changing the operating $\boldsymbol{g} = (2\bar{2}0)$ into $\boldsymbol{g} = (11\bar{3})$. Note that the line of no contrast disappears in the latter case. The different contrast effects with different foil orientations result essentially from the three-dimensional nature of SFT.

2.2.2 Dynamical diffraction contrast

With low order reflections, i.e. $\boldsymbol{g} = \{111\}$ or $\boldsymbol{g} = \{200\}$, no appreciable 'strain contrast' in bright or dark field imaging is obtained irrespective of the size of tetrahedra observed. This result is to be expected on account of the weak strain field of the SFT and is in agreement with the predictions of Ashby and Brown [1].

By using $\boldsymbol{g} = \{220\}$, a definite black-white contrast effect analogous to that of dislocation loops [2, 3] is obtained. The fraction of tetrahedra showing black-white contrast, found by counting, is small and amounts to about ξ_g/t, where ξ_g is the extinction distance of the operating $\boldsymbol{g}$-vector and t the thickness of the foil. The truncated tetrahedra which are close to one of the surfaces show especially strong black-white contrast.

Due to the 3-dimensional nature of the SFT, the projection of the strain contrast in the foil plane appears differently in different foil orientations.

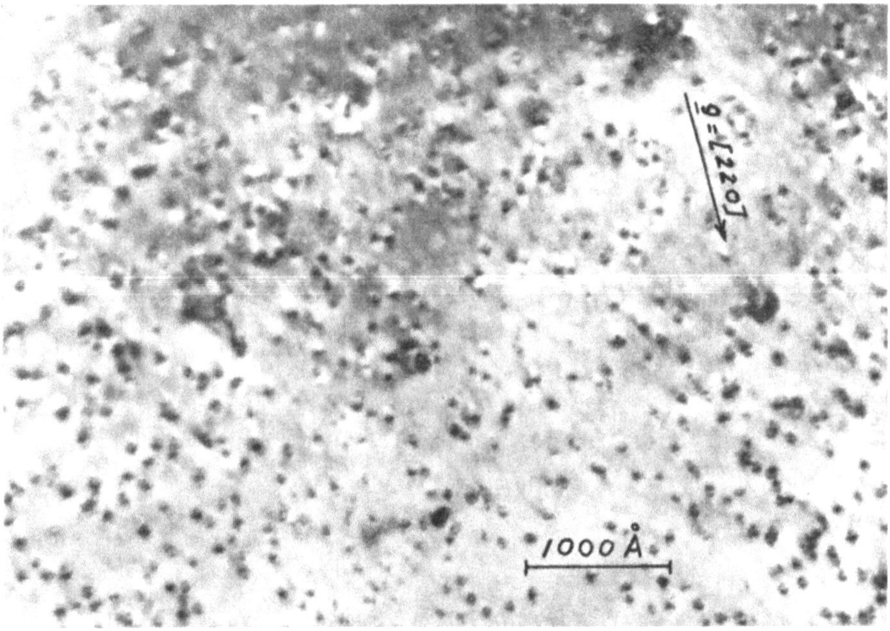

Fig. 5. Black-white contrasts of stacking fault tetrahedra in (010)-foil orientation (bright field)

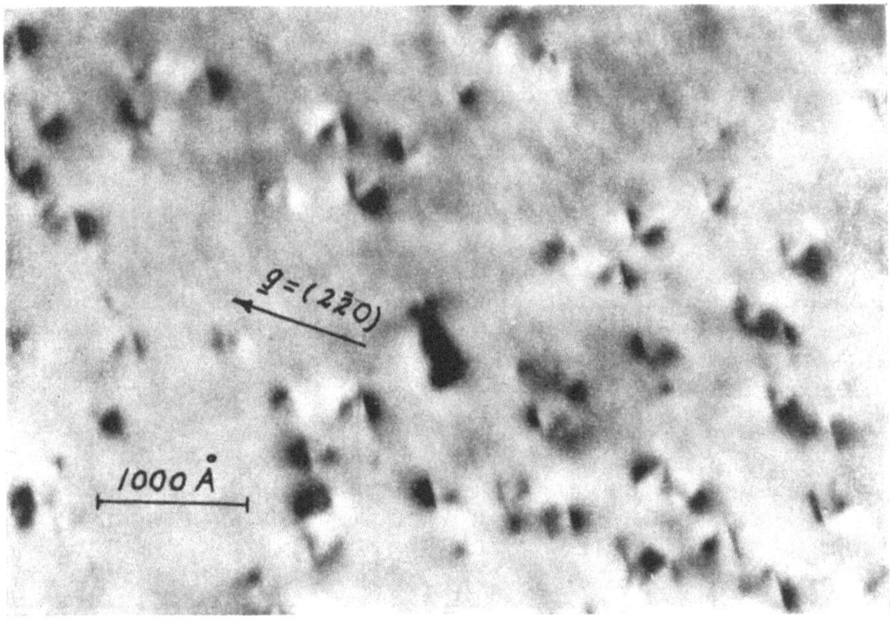

Fig. 6. Dynamical contrast effect of stacking fault tetrahedra in (110)-foil orientation with $\boldsymbol{g}$ = (2$\bar{2}$0) (bright field)

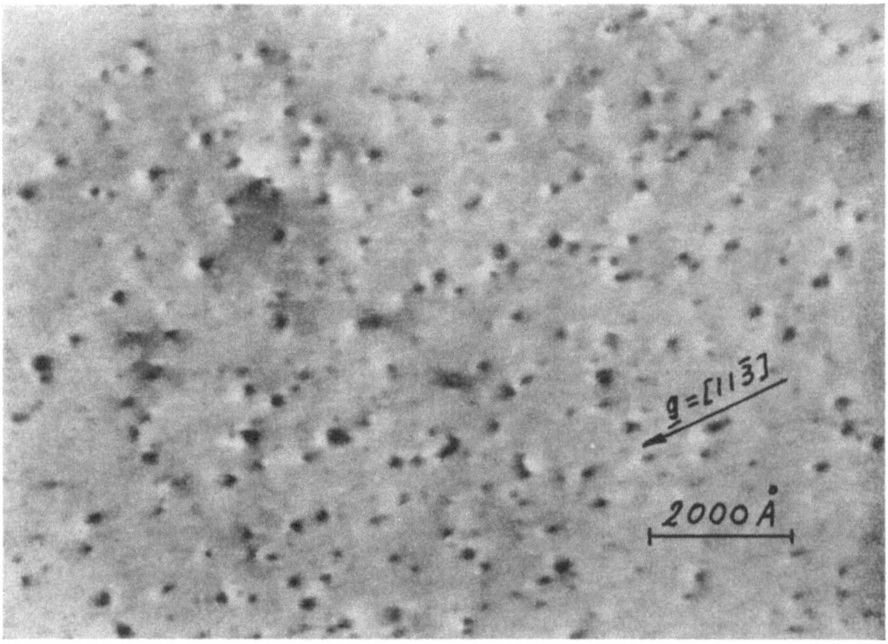

Fig. 7. Dynamical contrast effect with $\boldsymbol{g} = (11\bar{3})$ in (110)-foil orientation (bright field)

In {100}-foils, the SFT give a black-white contrast whose direction lies always parallel to the operating $\boldsymbol{g} = \{220\}$ (Fig. 5). The black as well as the white part has an elongated triangular shape reflecting the geometrical shape of the tetrahedron face (more apparent for large tetrahedra). Taking the direction from the black to the white side parallel to $\boldsymbol{g}$ as positive, approximately the same number of positive and negative direction is then observed in bright field images, while in dark field images the positive direction is predominant. In {110}- and {112}-foils the tetrahedra have a black-gray-white contrast (Fig. 6). Its properties are essentially the same as that of the black-white contrast.

A less defined black-white contrast is also found with $\boldsymbol{g} = \{11\bar{3}\}$ (Fig. 7). No dark field micrograph of good quality can be obtained with this reflection, because we have not succeeded to bring this diffracted beam to pass along the optical axis of the microscope.

2.3 Discussion

2.3.1 Kinematical case

In the case $s \gg 0$ the contrast effects can be explained by the kinematical theory of electron diffraction contrast [7], which gives the general criterion for disappearance of contrast of a crystal defect as

$$\boldsymbol{g} \cdot \boldsymbol{R} = 0 \,. \qquad (1)$$

$\boldsymbol{g}$ is the operating diffraction vector and $\boldsymbol{R}$ the displacement vector of the defect. For dislocations, $\boldsymbol{R}$ may to a first approximation be replaced by the

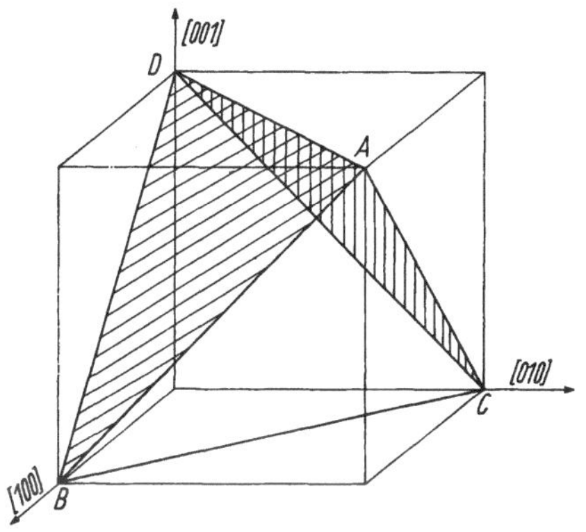

Fig. 8. Geometrical structure of a stacking fault tetrahedron. The stacking fault faces are ABC, ACD, ADB, and DBC

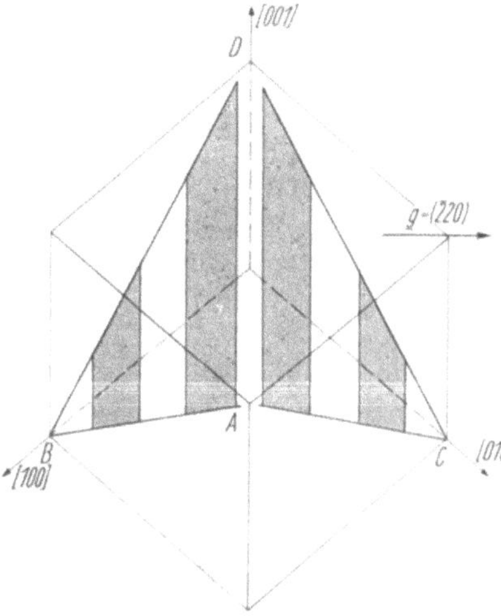

Fig. 9. Stacking fault tetrahedron viewed by $\boldsymbol{g} = (\bar{2}20)$ (schematic)

Burgers vector $\boldsymbol{b}$, i.e.

$$\boldsymbol{g} \cdot \boldsymbol{b} = 0\,. \qquad (2)$$

We discuss the contrast effects of the SFT with reference to Fig. 8. For $\boldsymbol{g} = (\bar{2}20)$, two of the four stacking fault faces, namely ABC and DBC satisfy equation (1) and one of the six stair-rods, BC, parallel to $\boldsymbol{g}$, satisfies (2). In addition to the line of no contrast AD, which will be discussed below, the above conditions give the 3-dimensional figure shown in Fig. 9. All the contrast effects for $\boldsymbol{g} = \{220\}$ discussed in 2.2.1 can be found by projecting this figure on the appropriate plane.

The stair-rod AD, which corresponds to the line of no contrast discussed in 2.2.1, however, does not satisfy (2) for $\boldsymbol{g} = (\bar{2}20)$. In this special case we have to consider directly the displacement of atoms in and near the core of the stair-rod, i.e. the displacement of atoms in the $(1\bar{1}0)$ symmetry plane at the intersection of the two inclined stacking fault faces and its neighbourhood. These atom displacements may be approximately visualized by considering the atoms as hard spheres. The bending of the crystal lattice planes in the neighbourhood of the stair-rod in (110) cross-section is shown schematically in Fig. 10. The atoms on the symmetry plane are only displaced in the plane itself. The neighbouring planes are

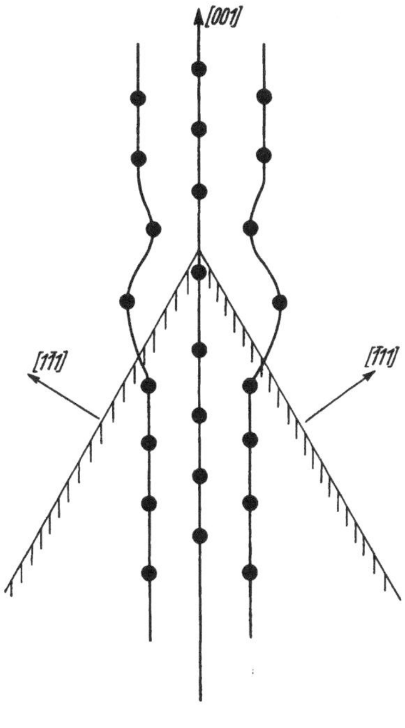

Fig. 10. Bending of crystal lattices around the edge of a stacking fault tetrahedron (schematic)

bent outwards below AD more strongly than the inward bending just above AD. Thus the electrons can pass through the symmetry plane undiffracted, but are diffracted near the stair-rod. In (100)-foils, the first fringe in the inclined stacking fault as counted from the line of no contrast should appear 'darker' than the rest. This can be seen in Fig. 2. The line of no contrast in SFT has also been discussed by Brown [8], who considered it to result directly from the symmetry of the dilatation of the defect. Some of the contrast effects described here are also discussed in the appendix of [9].

2.3.2 Dynamical case

The black-white contrast of SFT is characterized by the facts that it can only be imaged with $\boldsymbol{g} = \{220\}$ or $\boldsymbol{g} = \{113\}$, and that, due to the symmetry of the SFT, it changes direction together with the $\boldsymbol{g}$-vector. For example, the black-white contrast may remain parallel to the $\boldsymbol{g}$-vector, even though the $\boldsymbol{g}$-vector changes from, say, $(\bar{2}20)$ to (220). On the contrary, the black-white contrast of a Frank dislocation loop with Burgers vector $1/3\,[\bar{1}11]$ is formed by $\boldsymbol{g} = (\bar{2}20)$ but not by $\boldsymbol{g} = (220)$. Furthermore, the small fraction ξ_g/t of SFT showing the black-white contrast and the uni-direction of this contrast in dark field imaging indicate that only those tetrahedra near the surface of the foil can give an appreciable effect. This is supported by the strong black-white contrast exhibited by truncated SFT. The theoretical treatment of the dynamical contrast of small dislocation loops [3] shows that the black-white contrast changes its sign when the defect lies at a distance about $\xi_g/2$ from the surface. Assuming uniform distribution of defects in the specimen, we expect then an equal number of positive and negative directions of the black-white contrast in bright field *as well as* in dark field imaging. Since this is found not to be the case in SFT, we may assume that the tetrahedra on account of their smaller strain field are already ineffective in producing an asymmetric strain contrast at distance $\xi_g/2$ from the foil surface. If we examine carefully the micrographs of larger tetrahedra (edge length $\gtrsim 200$ Å) imaged by $\boldsymbol{g} = \{\bar{2}20\}$, we observe that some tetrahedra show an asymmetric (but not black-white) contrast. They appear on one side black and on the other side gray. This gray tone is sometimes lost in the background of the micrograph (Fig. 5). Such tetrahedra lie probably deeper in the foil where some dynamical strain contrast effect is still present.

The black-white contrast of SFT can be qualitatively explained analogous to the dislocation loops [3]. By imaging with a $\boldsymbol{g} = \{220\}$, two faces of the SFT are out of contrast (see 2.3.1). Under this condition the electrons 'see' only one half of the SFT. The remaining two stacking fault faces in contrast may then be considered approximately as two independent triangular stacking fault dislocation loops with a Burgers vector $\boldsymbol{b} = 1/12\,\langle 111\rangle$. When the dynamical imaging condition is fulfilled, each of the two faces will give a black-white contrast in the principal direction of the displacement field. However, the black-white contrast from these two faces has the same sign because the two triangular stacking fault loops are both of vacancy type. The strain field inside the SFT annihilates each other and should show only the background intensity. The white contrast is thus seen only on one side and the black contrast on the other side of the SFT. This explains the black-gray-white contrast observed with $\{110\}$- and $\{11\bar{2}\}$-foils, where the unstrained lattice inside the SFT is also projected.

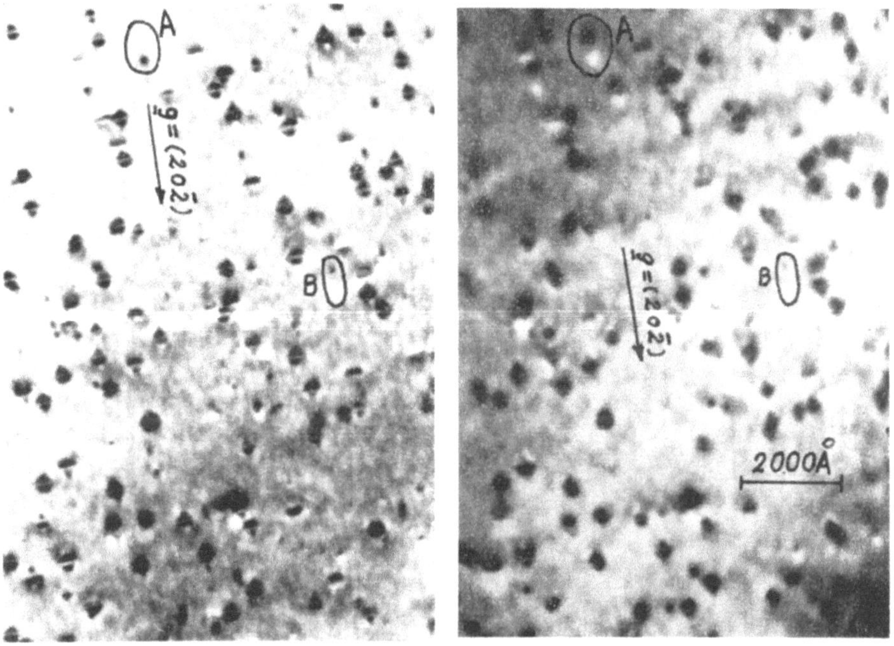

a b

Fig. 11. Comparison between bright (a) and dark (b) field micrographs. Operating $\boldsymbol{g}$-vector $(20\bar{2})$, foil orientation (010)

Fig. 11 shows the bright and dark field micrographs of the same area. The truncated SFT A reverses its sign from bright to dark field imaging, whereas the truncated SFT B has its sign unchanged. Since the truncated SFT must lie at one of the surfaces of the foil, the different behaviour of contrast between the two can only be explained by the dynamical theory of diffraction contrast [3]. According to the theory, the bright and dark field images are quasi-complementary at the bottom and approximately the same at the top due to anomalous absorption. Thus we may conclude that A lies at the bottom and B at the top of the foil.

Despite of the small fraction of tetrahedra giving black-white contrast, the characteristic diffraction contrast effects should be useful in identifying small unresolved SFT, and can be used to distinguish them from the dislocation loops. To do this, we need to study the same defects under different diffraction conditions, preferably using dark field technique. The following procedure is suggested. The defects are imaged with $\boldsymbol{g} = \{111\}$ or $\boldsymbol{g} = \{200\}$ under dynamical diffraction conditions. Then the SFT give *no* black-white contrasts, whereas the dislocation loops *do*. The same defects are now imaged with $\boldsymbol{g} = \{220\}$. Now, both the loop and SFT should give a black-white contrast. However, if the defects are loops, the total number of observable defects will be only approximately one half of that observed with $\boldsymbol{g} = \{111\}$ or $\boldsymbol{g} = \{200\}$. Furthermore by changing $\boldsymbol{g} = (\bar{2}20)$ into $\boldsymbol{g} = (220)$ which is perpendicular to the former reflection, these loops disappear. In the latter case, only those loops lying in (111) and $(11\bar{1})$ are imaged. The characteristic 'line of no contrast' in SFT can be observed *only* by $\boldsymbol{g} = \{220\}$ and is always perpendicular to the operating

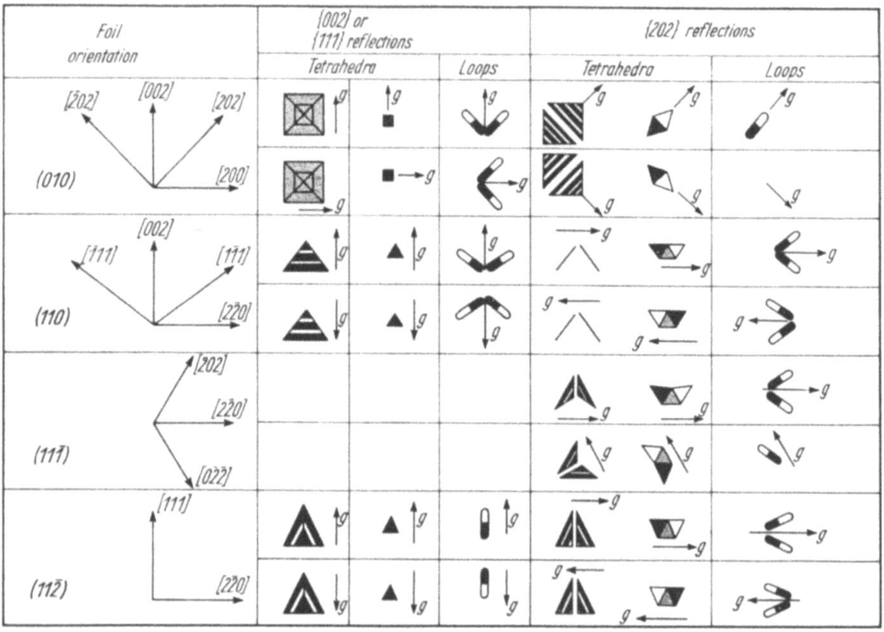

Fig. 12. A schematical representation of diffraction contrasts in stacking fault tetrahedra and dislocation loops. The left hand side of the column for tetrahedra gives the kinematical image. The rests are images under dynamical diffraction condition[1])

$\boldsymbol{g}$-vector. This line of no contrast is already recognizable for SFT with edge length $\gtrsim 100$ Å and can thus be used as a further identification of SFT. It differs from that of the elastic inclusions [1]. In the latter case, the line of no contrast may also be found with low-order reflections such as $\boldsymbol{g} = \{111\}$ or $\boldsymbol{g} = \{200\}$. The above informations are summarized in Fig. 12.

3. Application of the Contrast Effects of SFT

The Early Stage and Growth of Vacancy Clusters in Quenched Gold

The early stage of vacancy clustering in quenched gold is studied here with the help of the above described diffraction contrast of SFT. More detailed work on vacancy clustering in quenched gold has been described elsewhere [5].

Specimens were quenched from ≈ 900 °C into ice-brine at -10 °C. After quenching they were transferred quickly into liquid nitrogen. Each specimen was cut into three pieces. In order to study the vacancy clusters in their early stage, each piece was annealed for a definite time at 22 °C and then immediately thinned electrolytically at -25 °C. This low temperature prevents any further annealing of the quenched-in defects during the electropolishing. After thinning, foils transparent in the electron microscope have thicknesses between 500 and 2000 Å. This thickness is comparable with the average distance among vacancy clusters that have been formed during the annealing at 22 °C. The majority of the residual mobile vacancies in the foil after thinning will then go to the surface. We assume that the vacancy clusters have practically ceased to grow by vacancy absorption immediately after the thinning process has begun. The thinned specimens were studied in a Siemens Elmiskop I at 100 kV. The average quenched-in vacancy concentration in the specimens was estimated to be $\approx 1.5 \times 10^{-4}$.

[1]) The diffraction contrast effects shown in this figure have been simplified. The signs of the black-white contrast relative to $\boldsymbol{g}$-vektors should depend on the position of the defect in the foil and the imaging conditions (bright or dark field).

The results of microscopic observations are as follows:

1. *1 hour annealing at 22 °C.* No clusters were observable on the screen of the electron microscope. Micrographs of the specimens, however, showed a uniform distribution of small black dots with diameters between 30 and 70 Å. The cluster distribution remained unchanged after prolonged observation in electron microscope. Observations on well annealed specimens thinned under the same conditions showed that the black dots are not caused by etching or ion-irradiation in the microscope. According to [5], 1 hour annealing at 22 °C for a vacancy concentration of 1.2×10^{-4} corresponds to the final stage of the nucleation process.

2. *2 hours annealing at 22 °C.* Approximately the same total number of visible clusters as after treatment 1 are observed. Some of the clusters have grown to 200 Å in edge length. The larger clusters are already resolved as tetrahedra.

3. *Full annealing at 22 °C.* All visible clusters are now resolved as tetrahedra. The density is again approximately the same as above. No more black dots are present.

In order to study the nature of the small black dots, the procedure outlined in 2.3.2 is applied. It is found that none of the visible black dots give rise to black-white contrast with low-order reflections, i.e. $\boldsymbol{g} = \{111\}$ and $\boldsymbol{g} = \{200\}$, even when the dynamical diffraction condition is fulfilled. By applying $\boldsymbol{g} = \{220\}$, some black dots with diameter exceeding 50 Å *do* give the characteristic black-white contrast in dark field imaging (Fig. 13). The "lines of no contrast" are also clearly seen in most of the clusters. Thus the small black dots observed here can be assigned to SFT. The possibility of these clusters to

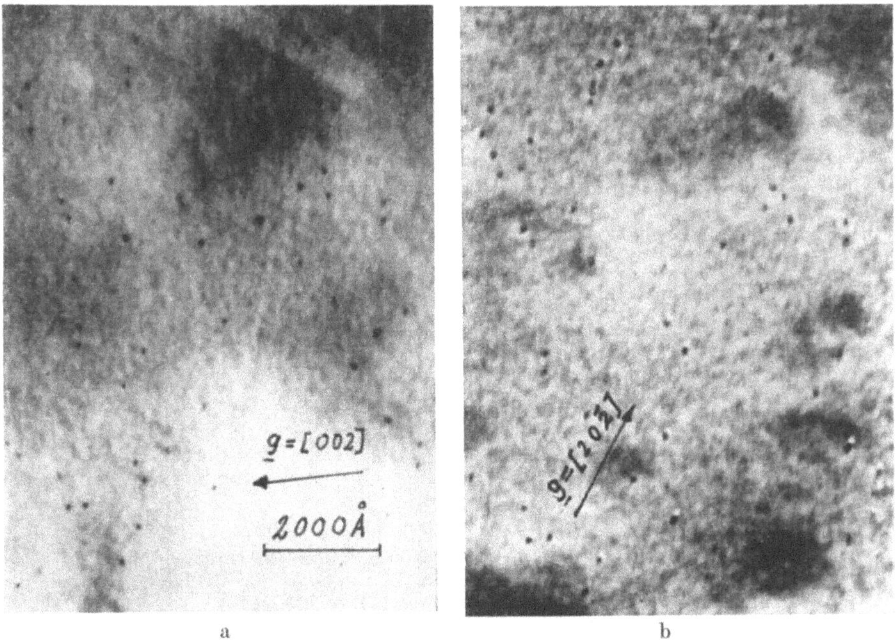

a b

Fig. 13. Dark field micrographs of small vacancy clusters in quenched gold, formed after annealing for about 1½ hours at 22 °C. Foil orientation is (001). Quenched-in vacancy concentration is $\approx 1.5 \times 10^{-4}$

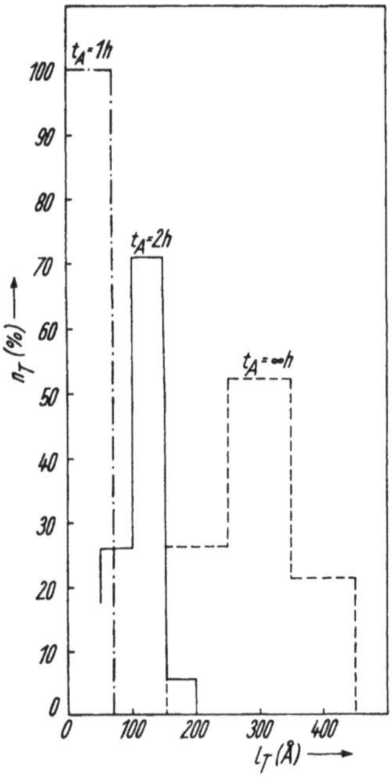

Fig. 14. Histograms of edge lengths of stacking fault tetrahedra after different annealing times t_A at 22 °C. Vacancy concentration $\approx 1.5 \times 10^{-4}$

be dislocation loops is excluded, since they do not give the strain contrast with low-order reflections.

The size distribution of the vacancy clusters for the three annealing times is given in Fig. 14. The average size as well as the maximum size of the clusters increases with the annealing time. This indicates that the SFT can grow by absorption of vacancies, in agreement with the ideas of Czjzek et al. [10] and with other experimental results [11, 12].

Comparison of electron microscopic observations with electrical resistivity measurements indicates that the nucleation period of vacancy clusters is only a few per cent of the total annealing period [13, 5]. The results of the present section show that the vacancy clusters formed at the end of the nucleation period exist already as SFT. Vacancy clusters existing early in the nucleation period (diameter < 50 Å) are too small to be imaged in electron microscope and could therefore not be studied with the present technique. From the present results, it may be concluded, however, that the SFT with size from 50 Å upwards grow into larger tetrahedra by absorption of vacancies. This is in agreement with the evidences from other experiments [5, 13] that the SFT probably grow directly from small three-dimensional vacancy aggregates such as hexavacancies.

Acknowledgements

The author wishes to express his sincere thanks to Professor Dr. A. Seeger and Dr. M. Wilkens for valuable discussions and to the Deutsche Forschungsgemeinschaft for the support of this work.

References

[1] M. F. Ashby and C. M. Brown, Phil. Mag. 8, 1083 (1963); 8, 1649 (1963).
[2] U. Essmann and M. Wilkens, phys. stat. sol. 4, K53 (1964).
[3] M. Rühle, M. Wilkens, and U. Essmann, phys. stat. sol. 11, 819 (1965).
[4] J. Silcox and P. B. Hirsch, Phil. Mag. 4, 72 (1959).
[5] K.-P. Chik, phys. stat. sol. 10, 659 (1965); 10, 675 (1965).
[6] P. B. Hirsch, J. Silcox, R. E. Smallman, and K. H. Westmacott, Phil. Mag. 3, 897 (1958).
[7] P. B. Hirsch, A. Howie, and M. J. Whelan, Phil. Trans. Roy. Soc. A 252, 499 (1960).
[8] C. M. Brown, Symposium on Electron Microscopy, Modena 1963, Ed. Paolo Buffa, CNR, Roma 1964.
[9] D. W. Pashley and A. E. B. Presland, Phil. Mag. 6, 1003 (1961).
[10] G. Czjzek, A. Seeger, and S. Mader, phys. stat. sol. 2, 558 (1962).
[11] W. Westdorp, H. Kimura, and R. Maddin, Acta metall. 12, 495 (1964).
[12] R. L. Segall and L. M. Clarebrough, Phil. Mag. 9, 865 (1964).
[13] M. Meshii, Lattice Defects in Quenched Metals, Academic Press 1965 (p. 387).

(Received June 2, 1966)

phys. stat. sol. **16**, 697 (1966)

Physics Department, University of Delaware, Newark, Delaware (a) and Allen-Bradley Co., Laboratory for Applied Electronic Research, Milwaukee, Wisconsin (b)

Properties of Recrystallized Evaporated CdS Layers[1])

By

K. W. Böer (a), J. W. Feitknecht (b), and D. G. Kannenberg (b)

Recrystallization of evaporated CdS layers in an environment containing N_2, HCl, O_2, Cu, and CdS at 630 °C are investigated using electron microscopic and electron probe analysis, and show an oxygen enhanced single crystal growth with preferably (001) parallel to the substrate surface. Photoconducting layers of unusual high frequency response for modulated light excitation (35% value at about 100 kHz), low electronic noise (above 60 Hz due to density fluctuation) and high photosensitivity, at 170 °K with a maximum gain factor of 10^8, and a "useful" light/dark current ratio (after 1 s decay) of 10^7 at 100 ft.-c are described. These layers have a trap density of $\lesssim 3 \times 10^{11}$ eV^{-1} cm^{-3} between 0.3 and 0.4 eV and show with 100 ft.-c bias illumination at room temperature electron lifetime (30 μs) determined relaxation.

Die Rekristallisation aufgedampfter CdS-Schichten bei 360 °C in einer Atmosphäre, die N_2, HCl, O_2, Cu und CdS enthält, wurde mit Hilfe elektronenmikroskopischer Analyse und mit einer Elektronen-Sonde untersucht, und zeigt ein durch Sauerstoff beschleunigtes Einkristallwachstum mit bevorzugter (001)-Orientierung parallel zur Unterlagenoberfläche. Photoleitende Schichten mit ungewöhnlich hoher Frequenzempfindlichkeit für Wechsellichtanregung (35%-Wert bei etwa 100 kHz), niedriges Elektronenrauschen (oberhalb 60 Hz verursacht durch Elektronendichteschwankungen) und hohe Photoempfindlichkeit, bei 170 °K mit einem maximalen Verstärkungsfaktor von 10^8 und einem „verwertbaren" Photo-Dunkelstrom-Verhältnis von 10^7 bei 100 ft.-c Anregung sind beschrieben. Diese Schichten haben eine Haftstellendichte von $\lesssim 3 \times 10^{11}$ eV^{-1} cm^{-3} zwischen 0.3 und 0.4 eV und zeigen mit 100 ft.-c Hintergrundbelichtung bei Zimmertemperatur eine durch die Elektronenlebensdauer bestimmte Zeitkonstante (30 μs).

1. Introduction

In a previous paper [1] it was shown that heat treatment of evaporated CdS layers in an atmosphere containing N_2, O_2, and HCl in a tube furnace on a bed of CdS and Cu powder at temperatures between 600 and 650 °C yield CdS with unusually high photosensitivity and speed of response. The electron mobility of these layers approaches the one known for single crystals. Measurements of the level distribution show a very low density (10^{13} eV^{-1} cm^{-3}) trap level in the 0.6 eV range, and it was concluded that the heat treatment reduces the density of some intrinsic defects, presumably of sulfur vacancies [2]. These defects are believed to cause the trap at about 0.6 eV, and usually render the speed of response of CdS at room temperature lower than some other known photoconductors, as e.g., CdSe. A favorable recrystallization in an atmosphere, supplying CdS, was suggested to be responsible for lowering the density of intrinsic defects within the crystallites, and for bridging the gaps between grains and thereby reducing electrical inhomogeneities.

[1]) Work done in Delaware in part supported by the National Aeronautics and Space Administration under Grant No. NsG-573.

In order to obtain further information about the recrystallization process, electron microscope and electron probe studies were performed and are reported in this paper. It was also observed that it is relatively easy to change electrical properties in a wide range by varying slightly the treatment parameters and thereby tailor CdS layers for desired electrical properties. As an example, some electrical properties of a new type of CdS layers are presented. It shows unusually fast response to modulated light with photocurrent modulation only 65% below the optical modulation at 100 kHz, very low noise, with the intrinsic density fluctuation spectrum already starting at 60 Hz, and a low trap density of $\lesssim 3 \times 10^{11}\ \mathrm{eV^{-1}\ cm^{-3}}$ in the range between 0.3 and 0.4 eV. Moreover, these layers exhibit maximum gain factors of 10^8 at 170 °K, yielding a useful[2]) ratio of photocurrent to dark-current of 10^7 at 100 ft.-c.

2. Experimental Procedure

The CdS layers were evaporated onto carefully cleaned[3]) glass slides to a thickness of about 1 μm and heat treated following the method described in the previous paper [1]. After treatment and intermediate storage in a dry atmosphere, Ti-Al contacts [3] were evaporated exposing a gap of 2.0 mm length L and 0.5 mm width W between the electrodes. For all electrical measurements 45 V were applied.

The electron microscope pictures were taken using one or two stage replica technique (see figure captions) and 45° shadowing with Pd-Pt. The electron probe measurements were done with an Applied Research Laboratory Model EMX probe.

Fig. 1 gives a schematic of the circuit used for investigation of the electron

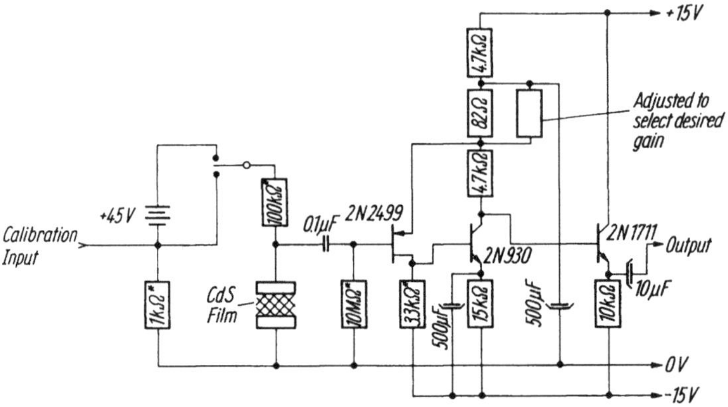

Fig. 1. Input circuit and preamplifier for measurements of electron noise and ac-current under modulated optical excitation. * indicates metal film resistor, 1% tolerance

[2]) One second after the optical excitation is switched off.

[3]) The cleaning consists of the following steps: 10 min immersion in an ultrasonically agitated solution of Alconox, followed by a deionized water rinse, 2 min rinse in ultrasonic cleaner with deionized water, rinse in running deionized water, drying in an isopropyl vapor dryer.

noise. A wide-band high impedance preamplifier (designed by P. Aemmer) with a bandwidth of 7 Hz to 100 kHz was used in conjunction with a Hewlett-Packard 302 A wave analyzer with 6 Hz bandwidth, with the output averaged electrically over 10 s using an external *RC* network. The CdS current measurements with modulated light excitations were done using the same input circuit and preamplifier in conjunction with a Tektronics 1A7 plug-in amplifier in a type 531 Oscilloscope and a Hewlett-Packard 3400 A rms Voltmeter. The light modulation was achieved using a Xenon ultrahigh pressure lamp (XBO 150 W/l), modulated with a maximum of 75 W ac power, and adjusted to 2% modulation up to 30 kHz. At higher frequencies the attainable degree of modulation dropped and disappeared in the noise of a vacuum photocell detector at about 250 kHz. The measured degree of modulation of the photocurrent above 30 kHz was corrected for 2% lamp modulation using a linear photocurrent vs. light intensity law.

The bias light intensity at the layer was 100 ft.-c. A Corning Cs-1-69 filter was used to reduce the IR-spectrum. The spectral response and conductivity glow curve were measured under N_2 ambient using a Bausch-Lomb grating monochromator, a Keithley 409 picoammeter and a Houston Inst. *x-y* recorder 96T. The light intensity was changed using a calibrated set of Wratten gray filters.

3. Experimental Results and Discussion

3.1 Recrystallization

A series of electron microscope pictures were taken before and after heat treatment under different conditions (see Fig. 2 and 3).

Fig. 2a shows a replica of a CdS surface directly after evaporation onto a glass substrate heated to a temperature of about 180 °C. The photograph shows crystallites of an average diameter of 100 nm, which are arbitrarily oriented in agreement to observations of Berger, Gutsche, and Kahle [4].

During heat treatment at 630 °C for 30 min on a bed of CdS powder containing about 10% by weight Cu-powder in an atmosphere of N_2, 7% of which passes through an aqueous solution of HCl (see [1]) and containing 0.1% O_2, a marked recrystallization takes place (see Fig. 2b). The average crystallite diameter has grown by a factor of about eight. There is a preference of crystal orientation with growth of the *c*-axis perpendicular to the surface, as can be verified by X-ray analysis.[4]) With increased oxygen concentration, an enhanced growth of crystallites can be observed also in agreement with [4] (Fig. 2c for 0.4% O_2 and Fig. 2d for 1.0% O_2). Three areas of rough appearance in Fig. 2d contain a higher than average amount of Cu, as observed by electron-probe measurements (see below).

Under optimized conditions, as described in [1], crystallite growth continues (see Fig. 3) again with preference of the (001) plane parallel to the substrate surface.

Electron-probe analysis reveals certain inhomogeneities for conductivity and density of backscattered electrons (Fig. 4a) due to regions which can be identified as crystallites rich in copper (Fig. 4b). This inhomogeneity can be essentially

[4]) Acknowledgment is given to J. Woyci for performing the X-ray analysis.

a

b

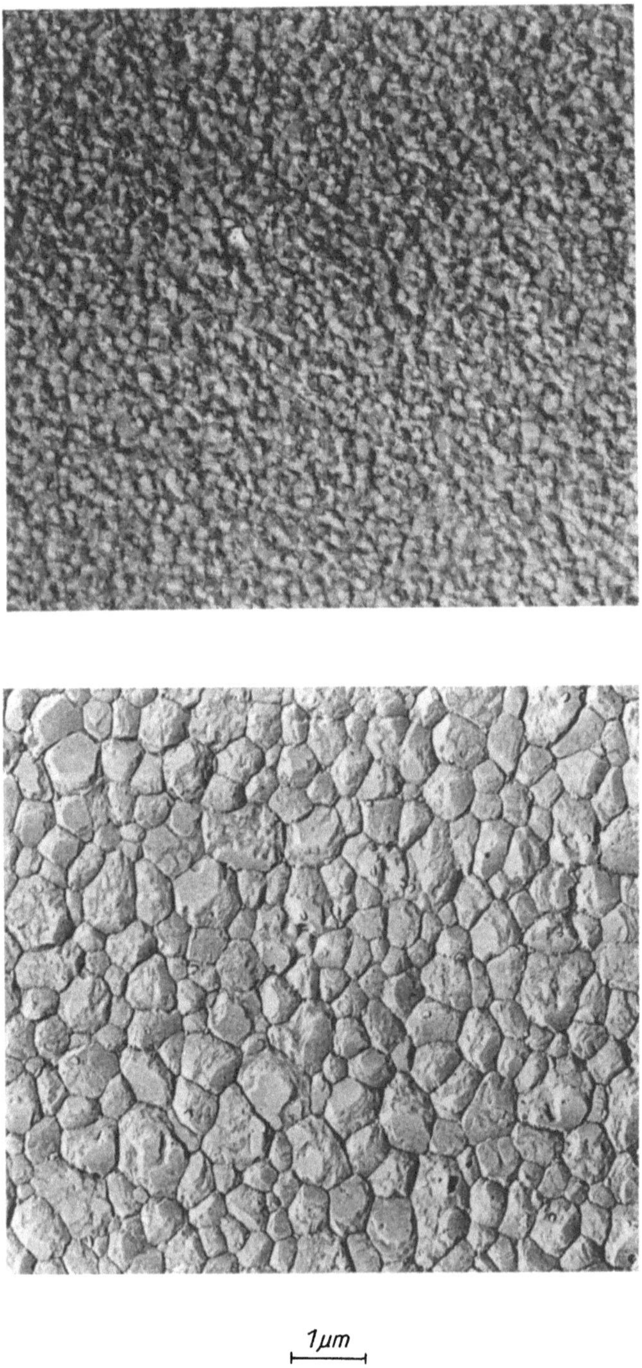

Fig. 2. One stage replica of CdS layers; electron microscdoic enlargement 1:10000, a) after evaporation, b) after heat treatment at 630 °C in an atmosphere containing 0.1% O_2, c) as b) but with 0.4% O_2, d) as b) but with 1% O_2

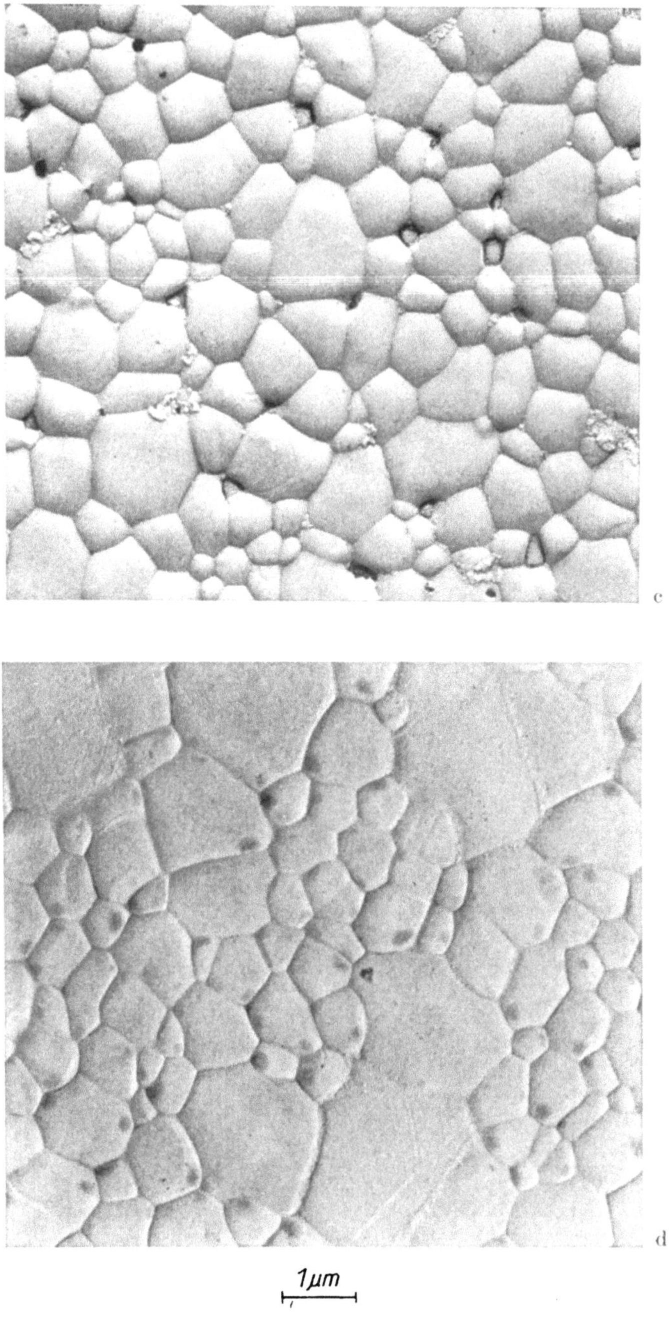

Fig. 2c and d

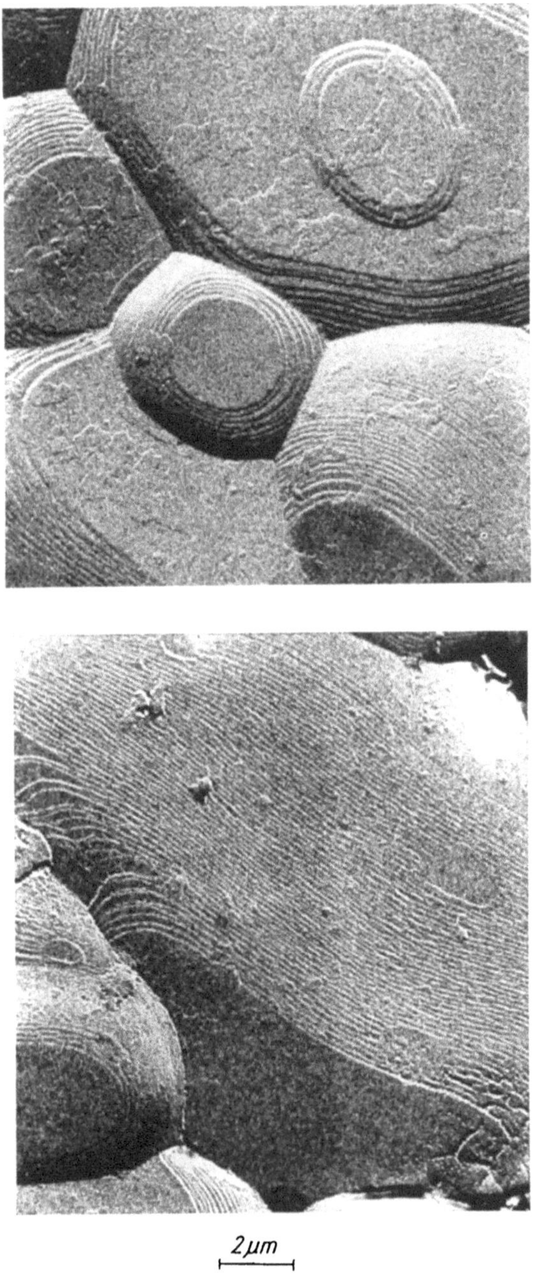

Fig. 3. Two stage replica of CdS layers, electron microscopic enlargement 1:20000, layers after heat treatment under optimized conditions as described in [1]

eliminated (see Fig. 4c and d) by a relative increase of the rate of CdS supplied through the vapor phase by the halogen transport mechanism and indicates a critical Cu doping density for homogeneous layers which lies well below 0.1% (limit of chemical analysis for the used portions of the layer). At higher Cu densities, although there is sufficient Cl for compensation offered during recrystallization, a new type of crystallite is observed, which probably incorporates all surplus copper and causes layer inhomogeneities to occur (see Fig. 2d).

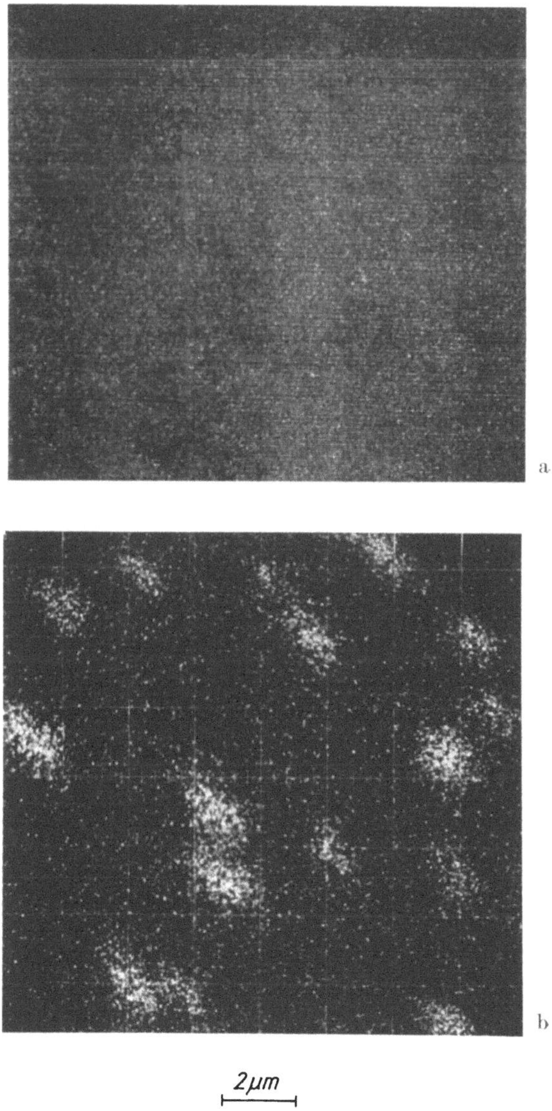

Fig. 4. Electron-probe analysis of recrystallized CdS, a) and c) back scattered electrons, b) and d) characteristic Cu emission (K_α)

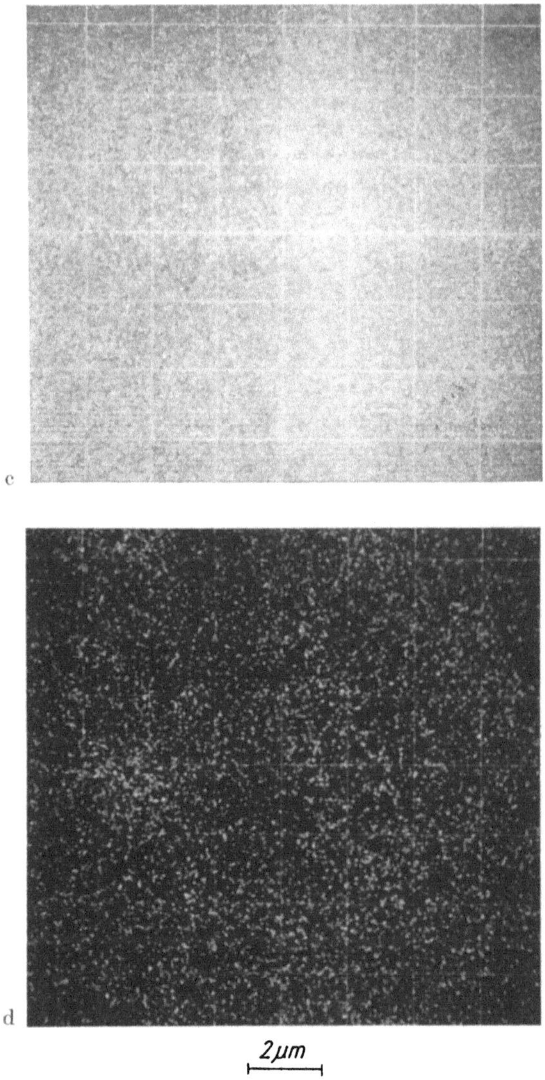

Fig. 4c and d

3.2 Electrical properties

By slight changes in recrystallization temperature, gas composition and arrangement of the CdS bed, one can change the electrical properties of the layer over a wide range. In the preceeding paper [1], layers of high photosensitivity and fast response time were described.

After reduction of the O_2 partial pressure and increase of the rate of supplied CdS through the vapor phase layers were obtained which are very sensitive for modulated light excitation up to frequencies of at least 250 kHz.[5])

[5]) Limit of detectable modulation of the lamp.

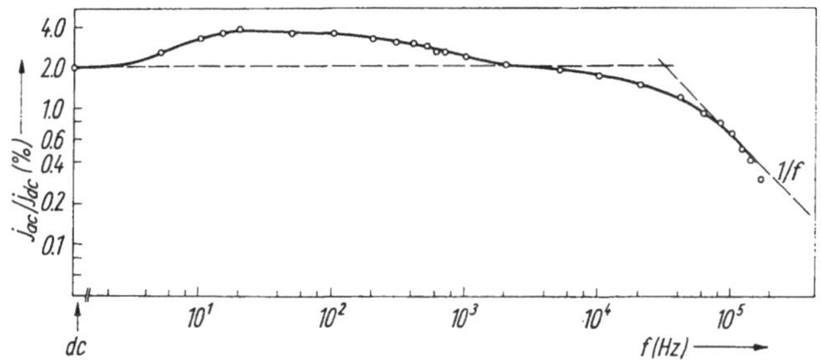

Fig. 5. Degree of modulation of the photocurrent under irradiation with 2% modulated light (5 Hz to 30 kHz) and corrected for 2% optical modulation above 30 kHz. Bias light 100 ft.-c, dc photocurrent 5×10^{-7} A

Fig. 5 shows a graph of the degree of photocurrent modulation vs. frequency of the light modulation. At 2% photocurrent modulation the ac part of the photocurrent (rms value) is 9.8×10^{-9} A. At values between 5 Hz and 1 kHz the photocurrent shows a higher degree of modulation (up to a factor of 1.9) than the exciting light. At frequencies above 80 kHz the photocurrent modulation decreases proportional to $1/f$ indicating an electron lifetime of about 30 μs. The steeper than linear decrease of the current modulation at 140 and 160 kHz is presumably due to increased errors, since the light modulation decreases to less than 0.3% and is comparable in magnitude to the lamp noise. Extrapolating the $1/f$ law to higher frequencies and using the same electron noise as observed at 10 kHz[6]), one should be able to use this CdS layer for modulation detection up into the 10 MHz range.

The increased (above 2%) modulation of the photocurrent in the intermediate frequency range is probably due to an overshoot of the photocurrent, as it is usually observed for superlinear photoconductors and shown in Fig. 6 for the investigated layer. Overshoot and overmodulation have the same magnitude as seen from Fig. 5 and 6 (factor 2) and also the frequency dependence in the 5 to 20 Hz range agrees reasonably well with the time dependence of the photo-"rise"-curve in the corresponding 50 to 200 ms range.

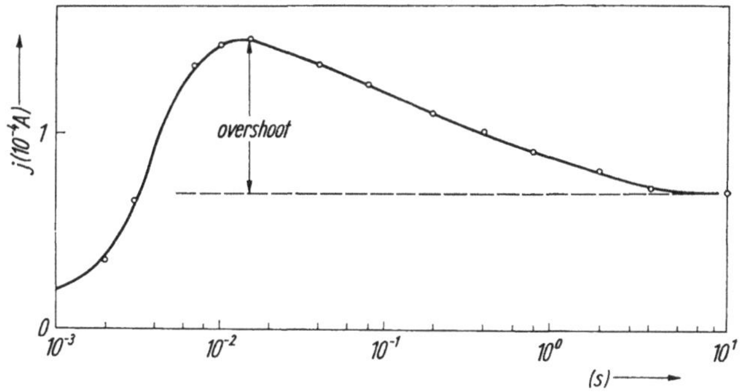

Fig. 6. Rise curve of photocurrent for 1500 ft.-c white (W-lamp, color temperature 2700 °K) light excitation

[6]) The intrinsic noise will certainly decrease at frequencies above $1/\tau$. However, the crystal RC will become a limiting factor, being the reason for the given somewhat conservative estimate.

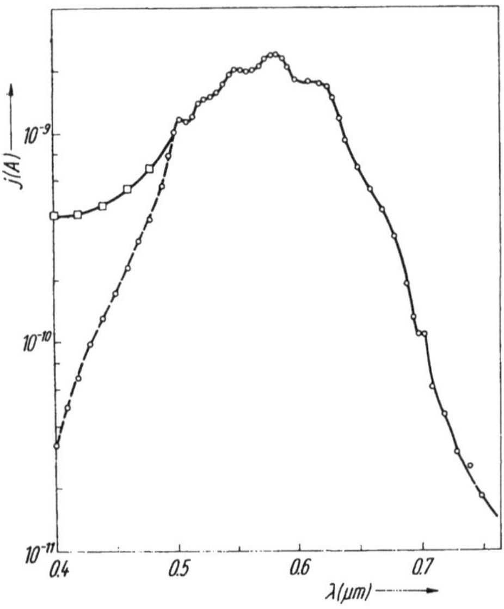

Fig. 7. Spectral distribution of the photocurrent at room temperature, solid curve corrected using superlinear slope 1.4 for equal photon flux of 4×10^{12} photons/cm² s, dashed part: measured curve

The "plateau" between 1 and 6 kHz indicates a limiting time constant of about 10^{-3} s for processes involving transitions over the valence band. These processes are responsible for the superlinear behaviour, which, in the investigated temperature and intensity range have a lifetime larger than the lifetime of electrons in the conduction band.

The spectral distribution of the photocurrent of this layer is given in Fig. 7 and shows a series of maxima at 505, 525, 550, 582, 620, and 705 nm, the first one due to close to band-edge excitation, and the following due to levels in the band gap at 2.36, 2.25, 2.13, 2.00, and 1.76 eV. From the light flux I_0 incident to the layer (4×10^{12} photons/cm² s) at the photocurrent maximum at 505 nm ($\eta = 50\%$ optical absorption) and the photocurrent j as given in Fig. 7 one obtains the lifetime τ of electrons in the conduction band,

$$\tau = \frac{j}{e \, \mu \, E \, L \, \eta \, I_0}, \tag{1}$$

where E is the field strength and μ is the electron mobility.

Using $\mu = 100$ cm²/Vs one obtains $\tau = 5 \times 10^{-7}$ s. At the photocurrent maximum at $\lambda = 582$ nm, the photocurrent is higher by a factor of 2, while the optical absorption is decreased at least by a factor of 5, thus for this excitation $\tau = 5$ μs. This value is a factor of six smaller than that obtained from the decay of the photocurrent modulation which was taken at considerably higher optical excitation and can be explained by the superlinear behaviour of the photocurrent vs. light intensity, which is observed and shown in Fig. 8 for the maximum of the photocurrent at room temperature.[7]) It is interesting to note that the slope does not depend on the photon energy of the exciting light between the band edge and 2.0 eV.

Fig. 9 shows a steep increase in photoconductivity with decreasing temperature. The exponential part of the curve indicates a thermal quenching with an activation energy of 0.53 eV. Between 170 °K and liquid nitrogen temperature the electron lifetime is increased to about 20 ms, yielding an unusually high photosensitivity. The maximum gain factor is

$$G_{\max} = \frac{\tau \, \mu \, E_b}{W} \approx 10^6$$

[7]) At 100 ft.-c illumination as used for measurements of Fig. 5, a photocurrent of 5×10^{-7} A was observed, which is 250 times higher than the photocurrent at the adsorption edge (Fig. 7). Using a slope of 1.4 for the entire photocurrent range one calculates an electron lifetime of 6 μs (at the low exitation intensity) which is in good agreement with the value reported above.

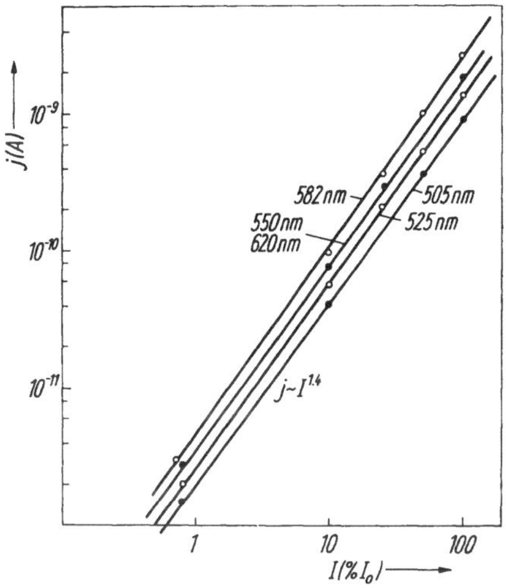

Fig. 8. Photocurrent vs. light intensity ($I_0 = 4 \times 10^{12}$ photons/cm² s) for monochromatic light of the wavelengths given in the graph at room temperature

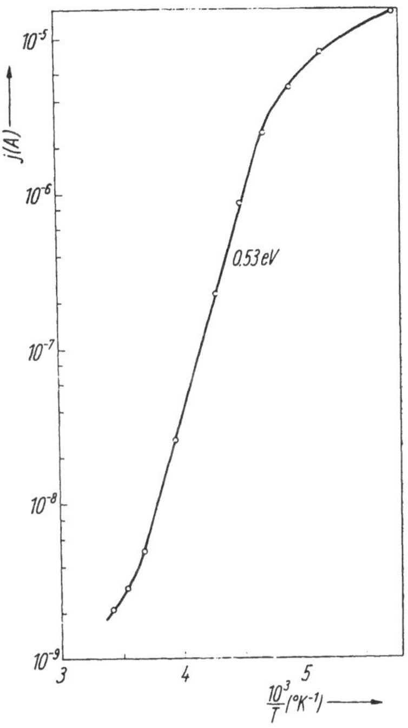

Fig. 9. Photocurrent vs. $1/T$ for monochromatic excitation at maximum photocurrent (at room temperature 582 nm and with decreasing temperatures shifted accordingly towards lower wavelength, at −180° at 573 nm)

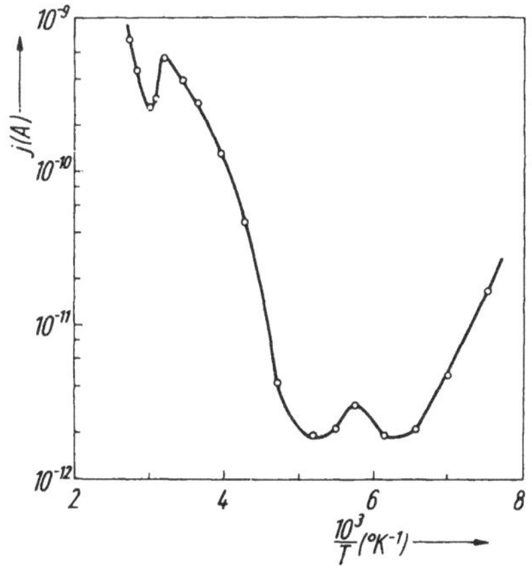

◀ Fig. 10. Conductivity glow curve, heating rate 0.2 deg/s

for the electrode configuration used and assuming 3×10^4 V/cm for the breakdown[8]) field strength E_b. With a narrower gap between the electrodes (25 μm gaps are easily achievable) the maximum gain factor lies at about 10^8.

The conductivity glow curve shows a very low peak at around 170 °K (Fig. 10), corresponding to a trap depth $E_c - E_t$ of about 0.35 eV, using the Randall and Wilkins [5] estimate of $E_c - E_t = 25\,kT_{max}$. With an estimated gain factor of 4×10^4 (using $\tau = 20$ ms and $\mu = 100$ cm²/Vs) the corresponding trap density becomes

$$N_t = \frac{\int_{t_1}^{t_2} j\,dt}{e\,G\,V_0} = 3 \times 10^{10}\ \text{cm}^{-3}\,, \tag{2}$$

[8]) Here determined by surface glide discharge.

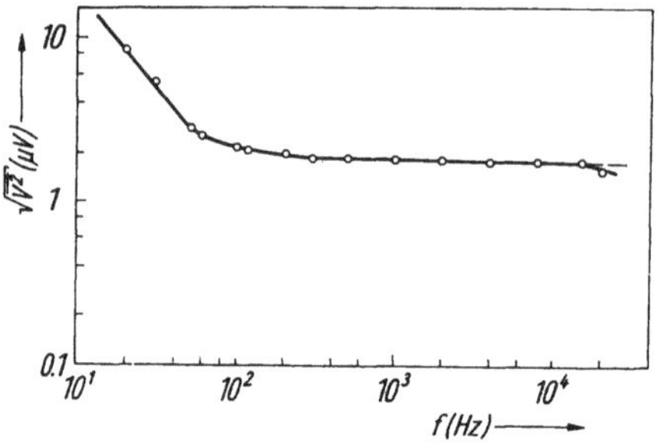

Fig. 11. Electron noise spectrum, 200 ft.-c illumination (W-lamp, 2700 °K)

where V_0 is the volume of the CdS layer between the electrodes ($\approx 10^{-6}$ cm^3). This low trap density permits even at low temperatures a fast decaying photocurrent, which, at 170 °K, shows a photocurrent–dark current ratio of more than 10^7, one second after excitation with white light (W-lamp 2700 °K) at 100 ft.-c.

The glow-curve exhibits an other peak at 310 °K, corresponding to 0.65 eV with a trap density[9]) of $\lesssim 4 \times 10^{17}$ cm^{-3}. It is interesting to note that in spite of the large trap density at 0.65 eV, the layer can be very fast for modulated light, provided the quasi-Fermi level is shifted into the range of low trap density. This is the case for the measured current modulation using a bias light of 100 ft.-c for which the quasi-Fermi level lies at 0.35 eV, i.e., in a range at which the trap density is less than 3×10^{11} eV^{-1} cm^{-3}.

Fig. 11 shows an electron noise spectrum, measured under white light excitation (tungsten lamp, color temperature 2700 °K) of 200 ft.-c. The $1/f$ range up to about 60 Hz is probably due to contact fluctuation and is orders of magnitudes less than usually observed at evaporated layers. The frequency independent part is due to electron density fluctuation [6],

$$\overline{V^2} = \frac{4\, e\, j\, V\, R^2\, \tau\, \mu\, \Delta f}{W^2\,(1 + \omega^2\, \tau^2)} \tag{3}$$

for $\omega = 2\pi f$, $\mu = 100$ cm^2/Vs, $R = 10^5\,\Omega$, $\Delta f = 6$ Hz and $j = 1.5 \times 10^{-6}$ A one obtains $\tau = 40$ µs in agreement with the value obtained from alternating light measurements.[10]) The slight decrease of the electron noise above 10 kHz also indicates a $\tau \approx 30$ µs. The low noise gives further evidence for good contacts at grain boundaries.

4. Summary

It has been shown that recrystallization of CdS layers to single crystals above 10 µm in diameter takes place at temperatures between 600 and 650 °C and there is evidence that oxygen in concentrations between 0.1 and 1% enhances the recrystallization markedly. The incorporation of copper can be rather inhomogeneous in such a way that a few percent of the crystallites are highly disordered and contain by far more copper than the majority of the single crystals. This kind of inhomogeneity can essentially be eliminated if an increased

[9]) This upper limit is obtained using the gain factor for the stationary photoconductivity at 310 °K. However, since for glow curves there is no hole supply to fast recombination centers due to optical excitation, the actual gain factor may be somewhat larger, and consequently the trap density reduced.

[10]) Taking superlinearity (slope 1.4) and increased optical excitation into consideration.

amount of CdS is supplied through the vapor phase, and suggests that copper doping in single crystals, although under highly compensated conditions, is not easily achievable above a critical value which, under given conditions, is well below 0.1%. The surplus copper diffuses to certain grains which grow in a highly disturbed fashion.

The electrical properties of the recrystallized layers can be varied over a wide range depending on slight changes of the recrystallization procedure, e.g., O_2 or HCl partial pressure, treatment temperature, contents of Cu in the bed. Certain types of layers can be produced with electrical properties surpassing the corresponding values known for good CdS single crystals grown at higher temperature from the vapor phase.

Layers with a very low trap density around 0.6 eV, having high photosensitivity and unusual short decay time of the photoconductivity at room temperature, were described in the preceeding paper [1]. Layers with extremely low trap density around 0.35 eV are discussed here which show an unusually high sensitivity for modulated light excitation, a very low noise spectrum, and at 170 °K a maximum gain factor of about 10^8 with a useful photo- to dark current ratio of at least 10^7 at 100 ft.-c illumination.

These layers have two electron trap levels, at 0.35 eV with a density of $\approx 3\times10^{11}$ eV^{-1} cm^{-3} and at 0.65 eV with a density of $\lesssim 4\times10^{17}$ cm^{-3} and five levels close to the valence band, at 2.36, 2.25, 2.13, 2.0, and 1.76 eV. Since the electron density in this layer for all investigated excitations is small compared with the trap density fillable up to room temperature, the recombination cross section is given by $q = (v_{th}\, \tau\, N_t)^{-1}$ and lies between 10^{-19} cm^2 (fast centers, thermally quenched region) and 10^{-23} cm^2 (slow centers, unquenched region), assuming for $N_t = 5\times10^{17}$ cm^{-3}.

The unusual low concentration of trap levels at 0.35 eV is presumably due to efficient annealing of defects during the recrystallization process at relatively (to single crystal growth) low temperatures.

Acknowledgement

We take the opportunity to extend our thanks to R. C. Allen for preparing the CdS layers and electrodes, to H. Köhler and G. H. Jaeger for making electron microscope pictures, to J. E. Bottkol and R. Nadolinski for running the electron probe analysis, to P. Voss for assisting during part of the electrical measurements, and to J. Abuya, T. Cyganek, R. Jackson, K. Platz, and M. Zosel for their assistance during several phases of the program. It is a pleasure to mention the encouragement and constant support the group has received from Dr. H. M. Schlicke. Finally, we gratefully acknowledge the permission of the Allen-Bradley Company to publish this paper.

References

[1] K. W. Böer, A. S. Esbitt, and W. M. Kaufmann, J. appl. Phys. **37**, 2664 (1966).
[2] R. H. Bube and L. A. Barton, RCA Rev. **20**, 564 (1959).
[3] K. W. Böer and R. B. Hall, J. appl. Phys., in preparation.
[4] H. Berger, E. Gutsche, and W. Kahle, phys. stat. sol. **7**, 674 (1964).
[5] R. Randall and M. Wilkins, Proc. Roy. Soc. **A 184**, 366 and 390 (1945).
[6] H. J. Gisolf, Physica **15**, 825 (1949).
K. W. Böer, Ann. Phys. (Germany) **14**, 87 (1954); **15**, 55 (1954).
K. M. van Vliet and J. Blok, Physica **22**, 231 and 525 (1956).

(Received June 2, 1966)

phys. stat. sol. **16**, 711 (1966)

Physikalisches Institut der Universität Freiburg i. Br.

Das Ultrarotspektrum von $NaNO_2$ zwischen 30 und 150 μm in der ferro- und paraelektrischen Phase

Von

H. VOGT und H. HAPP

Die folgende Arbeit befaßt sich mit dem Ultrarotspektrum des Ferroelektrikums $NaNO_2$, dessen Phasenübergang bei 163 °C ein Order-Disorder Übergang ist. Wir haben das Reflexionsspektrum im Bereich der äußeren UR-aktiven Gitterschwingungen (30 bis 150 μm,) gemessen, und zwar für Polarisationsrichtungen des elektrischen Feldes parallel zu den drei kristallographischen Achsen. In der ferroelektrischen Phase treten fünf Reflexionsbanden auf. Eine von diesen entspricht auf Grund der Gittersymmetrie einer reinen Translationsschwingung der Ionen, während die anderen zu Kombinationen von Translationsbewegungen und Torsionen der NO_2^--Ionen gehören. Bei Erhöhung der Kristalltemperatur von 30 bis 200 °C beobachteten wir eine starke Reduktion aller Reflexionsbanden.

The paper deals with the infrared spectrum of ferroelectric $NaNO_2$, which undergoes an order-disorder transition at 163 °C. The reflectivity in the external infrared-active lattice vibration range (30 to 150 μm) is measured for radiation polarized along the three crystallographic axes. In the ferroelectric phase five reflection bands appear. According to crystal symmetry one of these corresponds to a purely translational motion of the ions, while the others are belonged to combinations of translational motions and torsional motions of the NO_2^--ions. Increase of temperature from 30 to 200 °C yields a strong reduction of all reflection bands.

1. Einleitung

$NaNO_2$ ist ein ferroelektrischer Kristall des orthorhombischen Kristallsystems, dessen Curiepunkt bei 163 °C liegt [1]. In der ferroelektrischen Phase (Raumgruppe Im2m) sind die NO_2^--Dipole gleichsinnig längs der *b*-Achse ausgerichtet und die Na^+-Ionen aus der Mittellage zwischen den NO_2^--Ionen verschoben, wie Fig. 1 zeigt. Beim Phasenübergang wird die gleichsinnige Ausrichtung der NO_2^--Dipole aufgehoben: im paraelektrischen Zustand sind sie gleichwahrscheinlich sowohl in positiver als auch negativer *b*-Richtung orientiert [2]. Analog fluktuieren die Na^+-Ionen zwischen 2 Gleichgewichtslagen hin und her. Der Phasenübergang ist ein Order-Disorder Übergang, die Raumgruppe geht von Im2m nach Immm über.

Die Schwingungen des Kristallgitters lassen sich näherungsweise zerlegen in die inneren Schwingungen der NO_2^--Ionen und die äußeren Gitterschwingungen der NO_2^-- gegen die Na^+-Ionen.

Die inneren Schwingungen der NO_2^--Ionen wurden in mehreren Arbeiten untersucht. Sie erscheinen im Ultrarotspektrum bei Wellenlängen kleiner als 12 μm. Das Ramanspektrum [3, 4] zeigt neben den inneren Schwingungen äußere Schwingungen im Bereich zwischen 44 und 82 μm. Von den entsprechend der Gittersymmetrie zu erwartenden fünf ultrarotaktiven äußeren Gitterschwingungen wurden vier im Ramanspektrum beobachtet, zwei davon konnten eindeutig identifiziert werden.

Fig. 1. Einheitszelle von $NaNO_2$ im ferroelektrischen Zustand. Schraffierte Kreise Sauerstoff-, leere Kreise Natrium-, volle Kreise Stickstoffatome. Gitterkonstanten bei Raumtemperatur: $a = 3{,}56$ Å, $b = 5{,}56$ Å, $c = 5{,}38$ Å

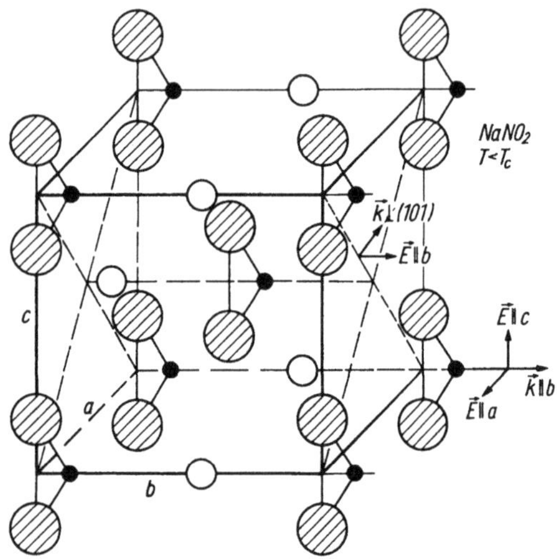

Das wesentliche Interesse der vorliegenden Arbeit besteht darin, den Einfluß eines ferroelektrischen Order-Disorder Überganges auf die ultrarot-aktiven Eigenschwingungen eines Systems mit übersichtlichem UR-Spektrum zu studieren. Wir haben daher das UR-Reflexionsspektrum von $NaNO_2$ im Bereich von 30 bis 150 μm mit polarisierter Strahlung für die drei Hauptrichtungen des Kristalles ausgemessen und in seiner Abhängigkeit von der Temperatur im Bereich zwischen 30 und 200 °C untersucht.

2. Experimentelles

Die in der vorliegenden Untersuchung benutzten $NaNO_2$-Kristalle wurden durch langsames Abkühlen einer Schmelze gewonnen [5]. Beim Unterschreiten des Schmelzpunktes (271 °C) bildet sich von der Tiegelmitte her ein auf der Oberfläche der Schmelze schwimmender Kristall. Die Wachstumsfläche des Kristalles parallel zur Oberfläche der Schmelze ist im allgemeinen eine (010)-Ebene. Auf ihr läßt sich die Domänen-Struktur im ferroelektrischen Zustand durch kurzzeitiges Ätzen mit Natronlauge sichtbar machen. Die Domänenstruktur gibt Auskunft über die Lage der a- und c-Achse des Kristalles [5].

Die Reflexionsmessungen wurden mit polarisierter Strahlung an polierten (010)-Wachstumsflächen und (101)-Spaltflächen der ungefähren Abmessung 15×15 mm² ausgeführt. Das elektrische Feld der einfallenden Strahlung schwingt näherungsweise bei Reflexion an der (010)-Fläche entweder parallel zur c- oder parallel zur a-Achse, bei Reflexion an der (101)-Fläche parallel zur b-Achse. Diese Polarisationsrichtungen sind zusammen mit den Einfallsrichtungen in Fig. 1 dargestellt. In der experimentellen Anordnung betrug der Einfallswinkel der Strahlung jeweils 10°, die Bündelapertur war 1:4,5.

Für jede Einfalls- und Polarisationsrichtung wurde das Reflexionsvermögen im Wellenlängenbereich zwischen 30 und 150 μm für mehrere Temperaturen zwischen 30° und 200° C mit einem Vakuum-Gitterspektrometer [6, 7] gemessen. Die Kristalle waren auf einer heizbaren Halterung montiert, deren Temperatur mit einer Konstanz von ±2 °C geregelt werden konnte.

Zur Polarisation der Strahlung dienten Drahtgitter mit Gitterkonstanten von 3,4 μm und 0,5 μm, die nach einem von M. Hass und M. O'Hara angegebenen Verfahren [8] auf Polystyrol- und Plexiglasfolien aufgedampft worden waren. Mit diesen in Transmission benutzten Polarisatoren ergaben sich im vorliegenden Spektralbereich Polarisationsgrade von über 96%.

Reflexionsmessungen an Kristallen aus verschiedenen Schmelzen ergaben unter den gleichen Meßbedingungen im Rahmen der Meßgenauigkeit übereinstimmende Resultate. Die Meßgenauigkeit für das Reflexionsvermögen betrug im Bereich maximaler Reflexionen etwa 2%, außerhalb dieses Bereiches etwa 5%.

3. Reflexionsspektrum bei Raumtemperatur

Fig. 2 zeigt das Reflexionsvermögen von $NaNO_2$ bei 30 °C für Polarisationsrichtungen des elektrischen Feldes E parallel zu den drei Hauptrichtungen des Kristalles. Für $E \| b$ existiert eine Reflexionsbande, für die beiden anderen Polarisationsrichtungen treten Doppelbanden auf. Im Bereich zwischen 34 und 42 μm konnte die kurzwellige Bande im Falle $E \| a$, sowie der Abfall der Banden für die beiden anderen Polarisationsrichtungen nicht ausgemessen werden, da ein geeignetes Beugungsgitter nicht zur Verfügung stand.

Das Reflexionsspektrum für die drei Hauptrichtungen in der ferroelektrischen Phase entspricht den auf Grund der Symmetrie des $NaNO_2$-Gitters zu erwartenden ultrarotaktiven Eigenschwingungen. Die primitive Einheitszelle, die z. B. durch die Basisvektoren

$$\boldsymbol{a}_1 = \frac{1}{2}(\boldsymbol{a} + \boldsymbol{b} + \boldsymbol{c})\,, \quad \boldsymbol{a}_2 = \boldsymbol{b}\,, \quad \boldsymbol{a}_3 = \frac{1}{2}(-\boldsymbol{a} + \boldsymbol{b} + \boldsymbol{c})$$

aufgespannt werden kann, enthält ein $NaNO_2$-Molekül, so daß 12 Gitterschwingungszweige vorliegen. Die 9 optischen Normalschwingungen für $\boldsymbol{q} = 0$ klassifizieren wir nach den Symmetrietypen der Punktgruppe m2m. Diese Beschränkung auf die Punktgruppe m2m ist möglich, da sich die Raumgruppe Im2m der ferroelektrischen Phase auffassen läßt als Produkt von m2m und der den erwähnten primitiven Basisvektoren entsprechenden Translationsgruppe, gegenüber der die Normalschwingungen im Bereich $\boldsymbol{q} = 0$ invariant sind.

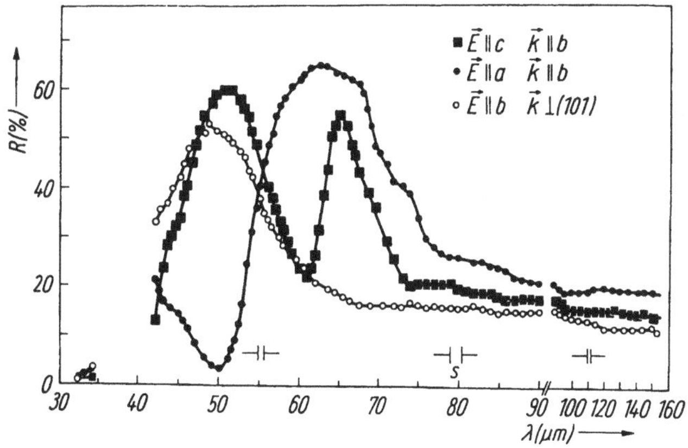

Fig. 2. Reflexionsspektrum von $NaNO_2$ bei 30 °C. s Wellenlängenäquivalent der Spaltbreite

Die Symmetrietypen und die jeweils dazu gehörige Anzahl von Normalschwingungen sind in Tabelle 1 angegeben. Hierbei unterscheiden wir zwischen

Tabelle 1

Symmetrietyp	A_1	A_2	B_1	B_2
Anzahl innerer Schwingungen	2	—	—	1
Anzahl äußerer Schwingungen	1	1	2	2

solchen Normalschwingungen, die wesentlich inneren Schwingungen des NO_2^--Ions entsprechen, und äußeren Schwingungen, bei denen NO_2^- als Ganzes schwingt.

Die inneren und äußeren Schwingungen betrachten wir wegen ihres relativ großen Frequenzunterschiedes (Frequenzverhältnis 1 : 5) als entkoppelt. Unter dieser Annahme sind die äußeren Schwingungen Kombinationen aus Drehschwingungen der NO_2^--Ionen und Translationsschwingungen der NO_2^--Schwerpunkte gegen die Na^+-Ionen. Für ihre Beschreibung geeignete Koordinaten in einem Koordinatensystem (x, y, z) parallel zu den Kristallachsen (a, b, c) sind: die Drehung $\boldsymbol{\varphi} = (\varphi_x, \varphi_y, \varphi_z)$ der NO_2^--Ionen um ihren Schwerpunkt und die Differenz $\boldsymbol{u} = (u_x, u_y, u_z)$ zwischen der Schwerpunktsauslenkung der NO_2^--Ionen und der Auslenkung der Na^+-Ionen.

Die so gewählten Koordinaten sind Symmetriekoordinaten und gehören zu den Symmetrietypen entsprechend Tabelle 2.

Tabelle 2

Koordinate	φ_x	φ_y	φ_z	u_x	u_y	u_z
Symmetrietyp	B_2	A_2	B_1	B_1	A_1	B_2

Die zu einem der 4 Symmetrietypen gehörigen Normalkoordinaten sind spezielle Linearkombinationen der in der Tabelle angegebenen zu diesem Symme-

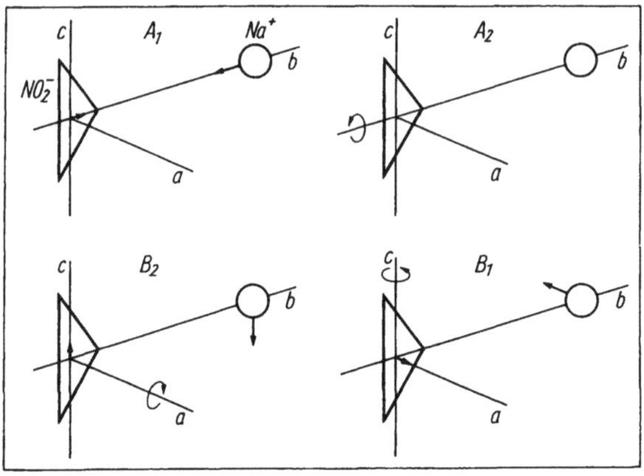

Fig. 3. Schwingungsbilder der äußeren Gitterschwingungen für $q = 0$

trietyp gehörigen Symmetriekoordinaten. Für die äußeren Gitterschwingungen bei $\boldsymbol{q} = 0$ ergeben sich die in Figur 3 dargestellten Schwingungsbilder. Als äußere Gitterschwingungen treten auf: Eine Translationsschwingung längs der b-Achse vom Typ A_1, eine Torsionsschwingung um die b-Achse vom Typ A_2, sowie je zwei Schwingungen der Typen B_1 und B_2, bei denen eine Torsion mit einer Translationsbewegung gekoppelt ist.

Die Torsionsschwingung des NO_2^- um die b-Achse ist ultrarotinaktiv, für die übrigen fünf Schwingungen dagegen ist Wechselwirkung mit UR-Strahlung möglich. Wie aus Fig. 3 hervorgeht, ist im Falle $E \parallel b$ eine von der Translationsschwingung längs der b-Achse herrührende Reflexionsbande, in den Fällen $E \parallel a$ und $E \parallel c$ dagegen je eine Doppelbande zu erwarten.

Diese äußeren Gitterschwingungen lassen sich phänomenologisch durch die folgenden Gleichungen beschreiben [9]:

$$\frac{\mu}{V_a} \ddot{\boldsymbol{u}} = -F\boldsymbol{u} - C\boldsymbol{\varphi} - G_1\boldsymbol{E},$$

$$\frac{\Theta}{V_a} \ddot{\boldsymbol{\varphi}} = -D\boldsymbol{\varphi} - \tilde{C}\boldsymbol{u} - \tilde{G}_2\boldsymbol{E},$$

$$\boldsymbol{P} = -G_1\boldsymbol{u} - G_2\boldsymbol{\varphi} - A\boldsymbol{E},$$

μ reduzierte Masse von Na^+ und NO_2^-, Θ Trägheitstensor von NO_2, V_a Volumen der primitiven Zelle, $\boldsymbol{P}$ Polarisation der Gitterschwingungen, $\boldsymbol{E}$ makroskopisches Feld. (Das lokale Feld, das der spontanen Polarisation entspricht, ist in den Kraftkonstanten berücksichtigt.)

In dem oben festgelegten Koordinatensystem sind F, D, G_1 und A diagonale Tensoren. Der Tensor C beschreibt Kopplungen zwischen Torsion und Translation und hat die Gestalt

$$C = \begin{pmatrix} 0 & 0 & C_{xz} \\ 0 & 0 & 0 \\ C_{zx} & 0 & 0 \end{pmatrix}.$$

Das gleiche Aussehen hat der Tensor G_2, der den von der Torsion der NO_2^--Ionen herrührenden Anteil der Polarisation P darstellt. Aus den obigen Gleichungen ergeben sich folgende Lösungen für transversal optische Gitterschwingungen ($E = 0$):

Typ A_1:

$$\omega_1^2 = \frac{F_{yy}}{\mu/V_a}, \quad \varphi = 0, \quad u_y \neq 0, \quad u_x = u_z = 0.$$

Typ B_1:

$$\omega_\pm^2(B_1) = \frac{1}{2}[\omega_r^2(c) + \omega_t^2(a)] \pm \sqrt{\frac{1}{4}[\omega_r^2(c) - \omega_t^2(a)]^2 + \frac{C_{xz}^2 V_a^2}{\mu\,\Theta_z}},$$

$$\begin{matrix} \varphi_z \neq 0, & \varphi_x = \varphi_y = 0, \\ u_x \neq 0, & u_y = u_z = 0, \end{matrix} \quad \frac{u_x}{\varphi_z} \begin{cases} < 0 \text{ für } \omega_+(B_1), \\ > 0 \text{ für } \omega_-(B_1). \end{cases}$$

Typ B_2:

$$\omega_{\pm}^2(B_2) = \frac{1}{2}[\omega_r^2(a) + \omega_t^2(c)] \pm \sqrt{\frac{1}{4}[\omega_r^2(a) - \omega_t^2(c)]^2 + \frac{C_{zx}^2 V_a^2}{\mu\,\Theta_x}},$$

$$\varphi_x \neq 0, \quad \varphi_y = \varphi_z = 0, \quad \frac{u_z}{\varphi_x}\begin{cases} < 0 \text{ für } \omega_+(B_2), \\ > 0 \text{ für } \omega_-(B_2). \end{cases}$$
$$u_z \neq 0, \quad u_x = u_y = 0,$$

Es ist

$$\omega_t^2(a) = \frac{F_{xx}}{\mu/V_a}, \quad \omega_t^2(c) = \frac{F_{zz}}{\mu/V_a}, \quad \omega_r^2(a) = \frac{D_{xx}}{\Theta_x/V_a}, \quad \omega_r^2(c) = \frac{D_{zz}}{\Theta_z/V_a},$$

ω_t bzw. ω_r wären die Schwingungsfrequenzen der Schwerpunktstranslation und der Torsion bei Entkopplung beider Bewegungen. Die Schwingungsrichtung bzw. die Drehachse ist jeweils im Argument angegeben.

Wie der Ausdruck für $\omega_{\pm}(B_1)$ bzw. $\omega_{\pm}(B_2)$ zeigt, bewirkt eine Kopplung von Torsion und Translation ein Auseinanderrücken der beiden Frequenzen, wobei der Mittelwert der Frequenzquadrate erhalten bleibt. Den Frequenzen $\omega_+(B_1)$ bzw. $\omega_+(B_2)$ entspricht eine Schwingung, bei der Torsion φ und zugehörige Translation u gegenphasig erfolgen. Bei den Frequenzen $\omega_-(B_1)$ bzw. $\omega_-(B_2)$ sind φ und u in Phase.

Die vorliegenden Messungen sind noch nicht vollständig genug, um aus einer Kramers-Kronig-Analyse ε' und ε'' und daraus die Dispersionsfrequenzen experimentell zu bestimmen. Durch Vergleich des Reflexionsspektrums mit dem von Tramer [3] untersuchten Ramanspektrum des $NaNO_2$ lassen sich jedoch den beiden Doppelbanden Dispersionsfrequenzen zuordnen, wie es in Tabelle 3 dargestellt ist.

Tabelle 3

Symmetrietyp	B_1		B_2		A_1
ω (cm^{-1})	$\omega_+(B_1)$ 226	$\omega_-(B_1)$ 148	$\omega_+(B_2)$ 185	$\omega_-(B_2)$ 151	ω_1 189
λ (µm)	44,2	67,5	54	66,2	53

Tramer konnte zwei bei 220 bzw. 160 cm^{-1} gelegene Linien des Ramanspektrums aus Polarisationsverhältnissen eindeutig den Schwingungstypen B_1 bzw. B_2 zuordnen. (Bei Tramer sind wegen einer anderen Wahl der Koordinatenachsen x und z die beiden Schwingungstypen B_1 und B_2 gerade umgekehrt bezeichnet wie hier.)

Der Ramanlinie bei 220 cm^{-1} (Typ B_1) entspricht nach Frequenzlage und Polarisation die kurzwellige Bande des Reflexionsspektrums für $E \parallel a$, während die Ramanlinie bei 160 cm^{-1} zur langwelligen Reflexionsbande im Falle $E \parallel c$ gehört. Die in Tabelle 3 angegebenen Frequenzen für diese beiden Ramanlinien beziehen sich auf Phononen, die sich senkrecht zur a- bzw. c-Achse ausbreiten. Die beiden Ramanlinien bei 148 cm^{-1} und 185 cm^{-1}, deren Symmetrietyp B_1 oder B_2 nicht eindeutig aus der Ramanuntersuchung erschlossen werden konnte, lassen sich zwanglos der langwelligen Reflexionsbande im Falle $E \parallel a$ (Typ B_1) bzw. der kurzwelligen Reflexionsbande im Falle $E \parallel c$ (Typ B_2) zuordnen. Eine weitere, in der Tabelle nicht angegebene Ramanlinie liegt bei 122 cm^{-1}. Sie gehört zum Typ A_2 und entspricht einer Torsionsschwingung der NO_2^--Ionen um die b-Achse, die nicht ultrarotaktiv ist.

Eine Ramanlinie vom Typ A_1 wurde nicht beobachtet. Die für A_1 angegebene Frequenz wurde aus der Anpassung einer klassischen Dispersionsformel an die gemessene Reflexionsbande $E \parallel b$ (Fig. 2) gewonnen. Hierbei wurden neben der Frequenz $\omega_1 = 189\ \mathrm{cm}^{-1}$ folgende Parameter benutzt: $\varepsilon_\infty = 2{,}8$, $\varepsilon_0 = 4{,}3$, $\gamma/\omega_1 = 0{,}12$. Die Abweichung der mit diesen Daten berechneten Reflexionskurve von der gemessenen beträgt im Bereich des Reflexionsmaximums zwischen 43 und 81 μm maximal 10%. Der aus dem langwelligen Ausläufer der Reflexionsbande bestimmte Wert $\varepsilon_0 = 4{,}3$ stimmt annähernd überein mit dem bei 3,3 GHz gemessenen Wert $\varepsilon_0 = 4{,}18$ [10].

4. Temperaturabhängigkeit des Reflexionsspektrums

Die Fig. 4, 5 und 6 zeigen die Veränderung der Reflexionsbanden mit der Temperatur und ihr Verhalten beim Übergang von der ferroelektrischen in die

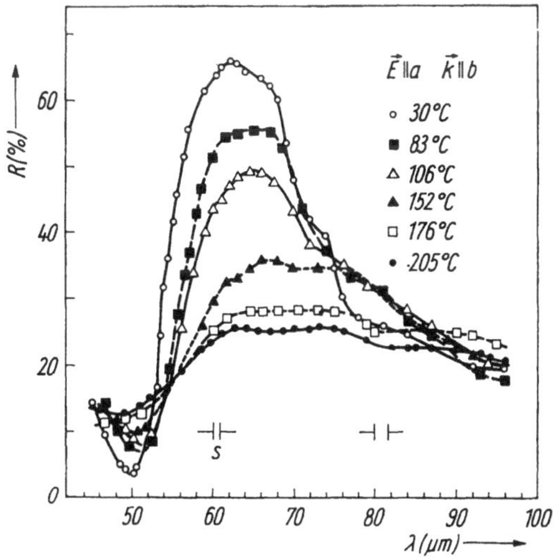

Fig. 4. Temperaturverhalten der Reflexion für $E \parallel a$

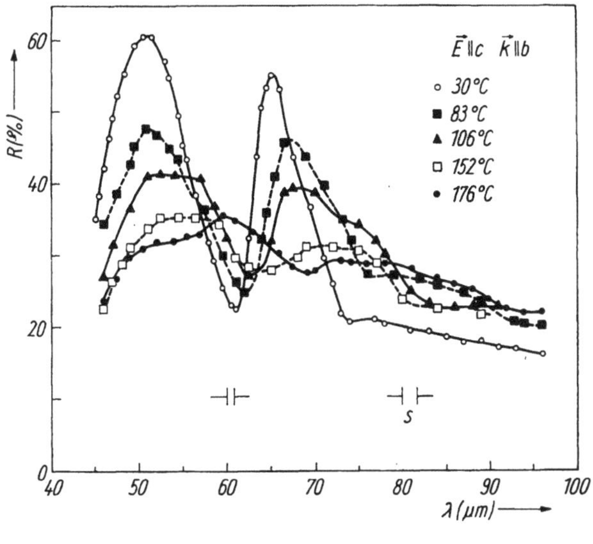

Fig. 5. Änderung der Doppelbandenstruktur mit der Temperatur für $E \parallel c$

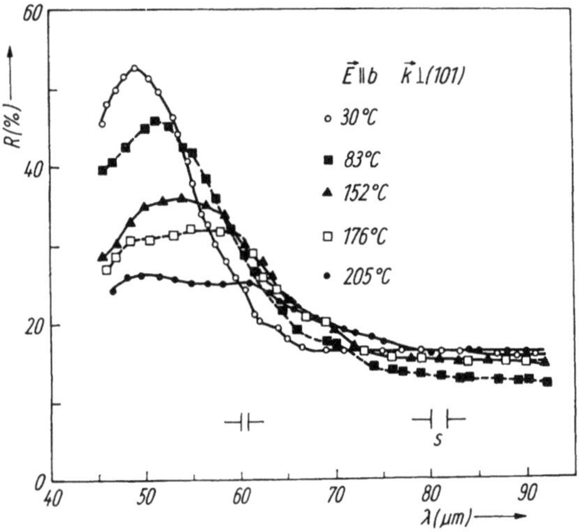

Fig. 6. Reduktion der Reflexionsbande für $E \parallel b$ mit wachsender Temperatur

paraelektrische Phase. Mit wachsender Temperatur verbreitern sich in allen drei Fällen die Banden, die Maxima der Reflexion verschieben sich in Richtung nach längeren Wellen. Gleichzeitig erniedrigt sich das maximale Reflexionsvermögen. Im Gebiet der langwelligen Ausläufer der Reflexionsbanden ist die Änderung des Reflexionsvermögens mit der Temperatur relativ gering und nicht monoton.

In den Fällen $E \parallel a$ und $E \parallel c$ wird beim Übergang in die paraelektrische Phase (Temperaturbereich um 160 °C) die Doppelbandenstruktur verwischt. In der paraelektrischen Phase verbleibt für alle drei Polarisationsrichtungen von den ursprünglichen ausgeprägten Reflexionsmaxima nur noch eine Andeutung in Form eines Gebietes flacher Reflexion, in dem das Reflexionsvermögen nur wenig größer ist als in den langwelligen Ausläufern.

Die starke Reduktion der Reflexionsbanden ist verknüpft mit dem Order-Disorder Übergang von $NaNO_2$ bei 163 °C. Die im ferroelektrischen Zustand längs der b-Achse gleichgerichteten NO_2^--Ionen sowie die gleichsinnig aus den Mittellagen zwischen den NO_2^--Ionen verschobenen Na^+-Ionen besitzen in der paraelektrischen Phase zwei Gleichgewichtslagen, die sie mit gleicher Wahrscheinlichkeit einnehmen [2]. Diesen Übergang kann man daher als einen Schmelzvorgang für die gleichsinnige Orientierung längs der ferroelektrischen Achse ansehen, der die Translationssymmetrie des Kristalles aufhebt. In einem solchen System, in dem die Bausteine zwischen zwei Gleichgewichtslagen statistisch hin und her fluktuieren, sind beim Durchgang elektromagnetischer Strahlung Resonanzen nicht mehr zu erwarten.

Im langwelligen, sich an das Gebiet der Reflexionsbanden anschließenden Bereich ist eine erhebliche Zunahme der Absorption mit der Temperatur festzustellen, während, wie oben bereits bemerkt, das Reflexionsvermögen nahezu konstant bleibt. Fig. 7 zeigt als Resultat einer vorläufigen Untersuchung die Transmission einer 0,45 mm dicken $NaNO_2$-Probe für Gesamtstrahlung oberhalb 200 μm. Bei 180 °C zeigen alle drei Transmissionskurven eine Anomalie. Entsprechend einer Untersuchung von Tanisaki tritt oberhalb 163 °C noch keine vollständige Disorder-Struktur auf, vielmehr existiert zunächst noch eine Nahordnung in Form von Mikrodomänen [11, 12]. Diese Nahordnung verschwindet

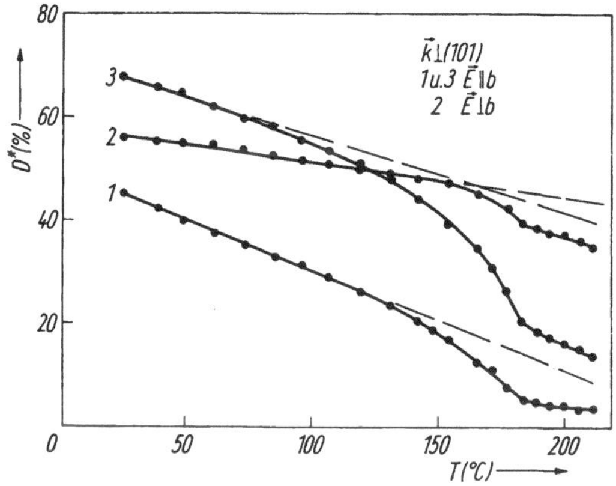

Fig. 7. Durchlässigkeit einer 0,45 mm dicken $NaNO_2$-Probe für auf der kurzwelligen Seite begrenzte Gesamtstrahlung einer Quarzquecksilberlampe.
Kurve 1: $\lambda > 200$ µm,
Kurve 2 und 3: $\lambda > 400$ µm

erst bei 178 °C, also bei einer Temperatur, die nahezu mit der Temperatur zusammenfällt, bei der die Anomalie in der Transmission auftritt.

Herrn Prof. Dr. L. Genzel danken wir für die Förderung dieser Arbeit und für anregende Diskussionen. Weiterhin möchten wir unseren Dank Herrn Dr. M. Hass sagen, der uns bei der Beschaffung der Polarisatoren behilflich war, sowie der Deutschen Forschungsgemeinschaft, die apparative Hilfsmittel zur Verfügung stellte.

Literatur

[1] S. Sawada, S. Nomura, S. Fujii und I. Yoshida, Phys. Rev. Letters **1**, 320 (1958).
[2] I. Shibuya, J. Phys. Soc. Japan **16**, 490 (1961).
[3] M. A. Tramer, C. R. Acad. Sci. (France) **248**, 3546 (1959).
[4] E. V. Chisler, Soviet Phys.-Solid State Phys. **7**, 1825 (1965).
[5] S. Nomura, Y. Asao und S. Sawada, J. Phys. Soc. Japan **16**, 917 (1961).
[6] L. Genzel und W. Eckhardt, Z. Phys. **139**, 578 (1954).
[7] R. Geick, Diplomarbeit, Frankfurt (Main) 1959.
[8] M. Hass und M. O'Hara, Appl. Optics **4**, 1027 (1965).
[9] M. Born und K. Huang, Dynamical Theory of Crystal Lattices, Clarendon Press, Oxford 1954 (S. 265 ff.).
[10] E. Nakamura, J. Phys. Soc. Japan **17**, 961 (1962).
[11] S. Tanisaki, J. Phys. Soc. Japan **16**, 579 (1961).
[12] S. Hoshino und I. Shibuya, J. Phys. Soc. Japan **16**, 1254 (1961).

(Received June 6, 1966)

phys. stat. sol. **16**, 721 (1966)

Centre d'Études d'Électronique des Solides, associé au C.N.R.S.,
Faculté des Sciences de Montpellier

Études de la durée de vie et du bruit de génération recombinaison controlés par les dislocations dans le germanium

Par

L. Gouskov, G. Lecoy, C. Llinares et M. Savelli

Une étude du bruit de génération-recombinaison et de la durée de vie mesurée par décroissance de la photoconductivité est effectuée sur une série d'échantillons des types p et n de germanium dopé en impuretés peu profondes. Les dislocations créées à la naissance du cristal, sont représentées par un niveau accepteur situé environ à 0,15 eV au dessous du bas de la bande de conduction, jouant le rôle de centre recombinant. La variance des porteurs et les durées de vie mesurées sont, dans le domaine 250 °K à 350 °K, dues aux processus du type Shockley-Read de transitions bande à bande, par l'intermédiaire de ces niveaux de dislocations.

Generation-recombination noise and lifetime are measured using the photoconductive decay, in n- and p-type Ge samples doped with shallow impurities. The dislocations formed during the crystal growth lead to acceptor levels, lying about 0.15 eV below the bottom of the conduction band which act as recombination centres. Carrier variance and lifetimes measured in the temperature range 250 to 350 °K, can be explained by a Shockley-Read type recombination process acting through these dislocationl evels.

1. Introduction

Le but de article est d'étudier dans le domaine de température 100 à 350 °K les processus dominants de génération et de recombinaison de porteurs libres dans des monocristaux de germanium des types (p) et (n) dopés en impuretés peu profondes.

Les résultats expérimentaux donnés ici sont relatifs aux mesures corrélatives de la durée de vie par la décroissance de la photoconductivité et d'analyse spectrale des fluctuations de la tension du bruit de génération-recombinaison.

Parmi la quinzaine d'échantillons étudiés, nous présenterons des résultats expérimentaux concernant cinq d'entre eux dont les caractéristiques essentielles sont données dans le tableau ci-dessous.

Tableau 1

N° de l'échantillon	Type	Dopage	Densité d'impuretés (cm^{-3})	Densité de dislocations (cm^{-2})	Résistivité à 300 °K (Ω cm)
1	n	As	$1{,}5 \times 10^{14}$		10,3
2	n	Sb	$1{,}15 \times 10^{14}$	$\approx 10^4$	15
3	n	P	4×10^{14}	1,8 à 2×10^3	4,5
4	p	Ga	$1{,}1 \times 10^{14}$	10^2 à 10^3	28
5	p	In	$2{,}7 \times 10^{14}$	3,2 à $3{,}3 \times 10^3$	12,5

Les dislocations dans ces échantillons n'ont pas été introduites volontairement, mais sont apparues lors de la fabrication des monocristaux.

Afin de n'avoir pas à tenir compte des phénomènes parasites de génération-recombinaison superficielle, nous avons travaillé sur des bâtonnets dont la section est toujours supérieure à 1 mm^2, soigneusement décapés et rincés à l'eau désionisée avant leur étude sous vide. Ceci a été vérifié en comparant les résultats de durée de vie avec ceux obtenus sur des échantillons de volume plus important.

L'appareillage d'analyse spectrale a une sélectivité de 30 Hz de 2 kHz à 50 kHz et de 500 Hz minimum de 50 kHz à 20 MHz.

Soient $P(t)$ et $N(t)$ le nombre total de trous et d'électrons libres à l'instant t fluctuant au gré des processus de transition et $I(t)$ l'intensité du courant instantané résultant en présence d'un champ électrique; on désigne par P_0, N_0 et I_0 les valeurs moyennes de ces quantités.

On a la relation suivante [10] entre les densités spectrales $\overline{i_\nu^2}$, $\overline{p_\nu^2}$, $\overline{n_\nu^2}$, $\overline{(p\,n)_\nu}$ respectivement transformées de Fourier de $\overline{\Delta I(0)\,\Delta I(t)}$, $\overline{\Delta P(0)\,\Delta P(t)}$, $\overline{\Delta N(0)\,\Delta N(t)}$, $\overline{\Delta P(0)\,\Delta N(t)}$:

$$\overline{i_\nu^2} = \left[\frac{I_0}{P_0 + b\,N_0}\right]^2 \{\overline{p_\nu^2} + 2\,b\overline{(p\,n)_\nu} + b^2\,\overline{n_\nu^2}\} . \tag{1}$$

On désigne par $\mathcal{N}_\nu$ le rapport de bruit, rapport de la puissance de bruit totale à la puissance de bruit thermique, dans la gamme de fréquences ν, $\nu + \Delta\nu$; on a

$$\mathcal{N}_\nu - 1 = \frac{R\,\overline{i_\nu^2}}{4\,k\,T} = \left[\frac{M_2 - M_1}{M_1}\right]_\nu . \tag{2}$$

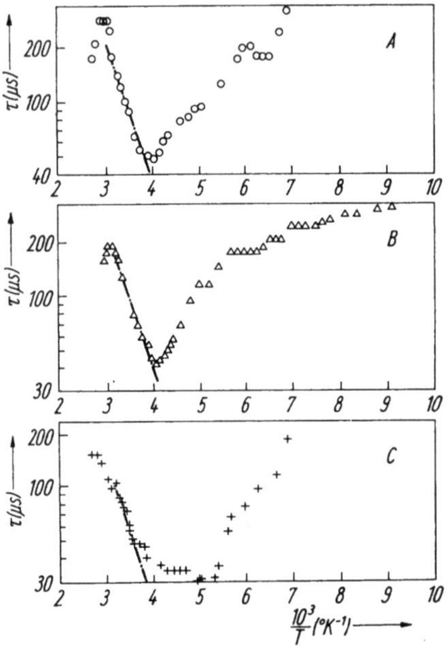

Fig. 1. Type (n): Variations expérimentales de la durée de vie en fonction de la température. (A): échantillon n° 1, (B): échantillon n° 2, (C): échantillon n° 3

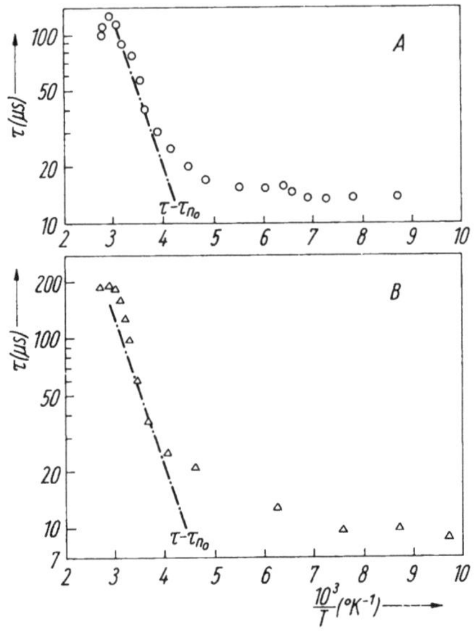

Fig. 2. Type (p): Variations expérimentales de la durée de vie en fonction de la température. (A): échantillon n° 4, (B): échantillon n° 5

M_2, M_1 étant respectivement les résultats des mesures d'analyse spectrale avec et sans champ électrique, R étant la résistance de l'échantillon.

2. Résultats expérimentaux

2.1 Résultats des mesures de durée de vie

Sur les Fig. 1 et 2 nous donnons les résultats expérimentaux des mesures de durée de vie en fonction de la température pour les trois échantillons de type n et les deux échantillons de type p. Nous pouvons noter la grande analogie entre les résultats obtenus sur les échantillons du même type.

2.2 Résultats des mesures de densités spectrales

Pour chacun de ces échantillons, dans la gamme de températures (270 à 350 °K) où le rapport de bruit est mesurable avec précision ($\mathcal{N} - 1 \geqq 0{,}1$), les mesures de $\mathcal{N} - 1$ en fonction de la fréquence ont été réalisées pour deux valeurs du courant I_0, au minimum. A titre d'exemple sur la Fig. 3, la variation de $\mathcal{N} - 1$ est donnée pour l'échantillon n° 4 en fonction de la fréquence et respectivement pour les températures 283 °K, 305 °K, 317 °K et 351 °K.

On peut faire les remarques suivantes:

$\mathcal{N} - 1$ est proportionnel à I_0^2 comme cela est prévisible à partir des relations (1) et (2).

L'ensemble de ces spectres obéit à une loi en $k\, I_0^2 \left(1 + \frac{\nu^2}{\nu_0^2}\right)^{-1}$ où les paramètres k et ν_0 dépendent de la température; par suite les spectres présentent un

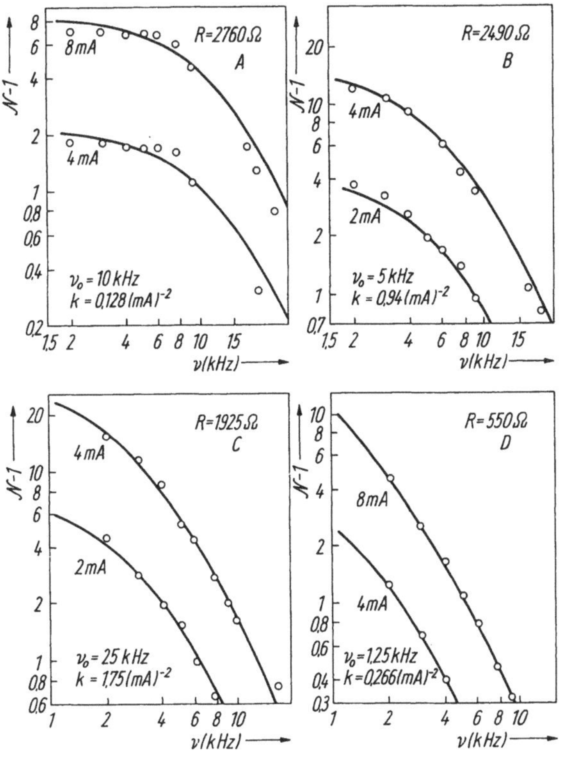

Fig. 3. Variation de $\mathcal{N} - 1$ en fonction de la fréquence.
(A): 283 °K,
(B): 305 °K,
(C): 317 °K,
(D): 351 °K pour l'échantillon n° 4.
○ points expérimentaux,
— courbe théorique de la forme
$k\, I_0^2 \left(1 + \frac{\nu^2}{\nu_0^2}\right)^{-1}$

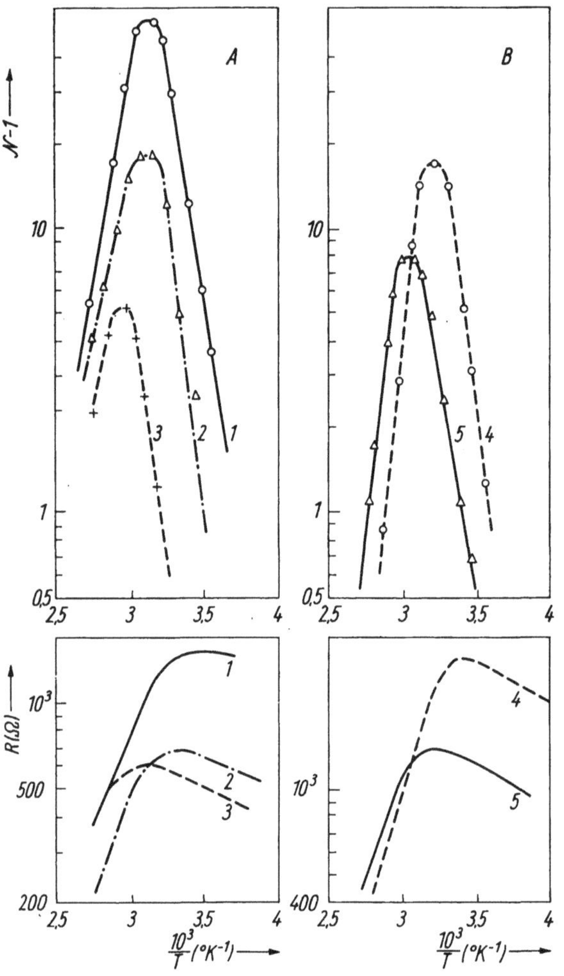

Fig. 4. Variation de $\mathcal{N} - 1$ à 2 kHz et de R en fonction de la température.
A type (n)
n° 1: $I_0 = 5$ mA,
n° 2: $I_0 = 10$ mA,
n° 3: $I_0 = 10$ mA,
B type (p)
n° 4: $I_0 = 4$ mA,
n° 5: $I_0 = 5$ mA

palier de valeur $k\,I_0^2$ dans le domaine $\nu \ll \nu_0$ et une décroissance en ν^{-2} dans le domaine $\nu \gg \nu_0$.

Sur la Fig. 3, on a tracé les courbes théoriques de la forme $k\,I_0^2\left(1 + \frac{\nu^2}{\nu_0^2}\right)^{-1}$ les plus compatibles avec les points expérimentaux, et précisé pour chaque température la valeur des paramètres k et ν_0.

De plus, on a mesuré pour $\nu = 2$ kHz la variation de $\mathcal{N} - 1$ en fonction de la température à courant I_0 constant. Sur la Fig. 4, on donne, en coordonnées semi-logarithmiques, pour les 5 échantillons cette variation ainsi que celle de la résistance R des échantillons (A type n, B type p). On peut voir que $\mathcal{N} - 1$ est maximum à l'amorce du régime intrinsèque.

3. Interprétation des résultats expérimentaux

On sait [4, 5] que les dislocations introduisent un niveau accepteur E_t, situé près du bas de la bande de conduction et pouvant exister dans les états de charge 0 et $-q$.

Fig. 5. Modèle des transitions électroniques à 3 niveaux

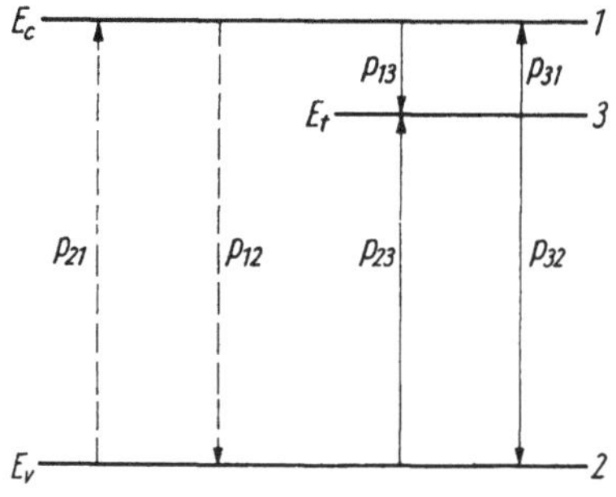

On considère donc les transitions électroniques entre les trois niveaux suivants: la bande de conduction (niveau 1), la bande de valence (niveau 2) et le niveau de dislocations (niveau 3). Sur la Fig. 5, on indique le flux p_{ij} de transition d'électrons du niveau i au niveau j, par unité de temps et unité de volume. Les flux de transitions intrinsèques du type p_{12} et p_{21} y figurent en pointillé, leur ordre de grandeur permet de les négliger devant les autres flux qui s'écrivent [8]

$$\left.\begin{aligned} p_{13} &= C_0^{\mathrm{n}}\, n\,(N_{\mathrm{t}} - N_{\mathrm{t}}^-)\,, & p_{31} &= C_0^{\mathrm{n}}\, n_1\, N_{\mathrm{t}}^-\,,\\ p_{23} &= C_-^{\mathrm{p}}\, p_1\,(N_{\mathrm{t}} - N_{\mathrm{t}}^-)\,, & p_{32} &= C_-^{\mathrm{p}}\, p\, N_{\mathrm{t}}^-\,. \end{aligned}\right\} \tag{3}$$

n, p et N_{t} sont les densités respectives d'électrons, de trous et des centres E_{t}, N_{t}^- la densité des dislocations ionisées, n_1 et p_1 respectivement les densités d'électrons et de trous si le niveau de Fermi était confondu avec le niveau 3. C_-^{p} et C_0^{n} désignent respectivement la probabilité par unité de temps de capture d'un trou ou d'un électron par un centre négatif ou neutre.

En étudiant le système de deux équations différentielles chiffrant la continuité des populations, on peut, comme le fait K. M. van Vliet [11], voir que:

En ce qui concerne la durée de vie, il apparait deux constantes de temps τ_1 et τ_2. Dans le cas, qui nous intéresse, où la densité des porteurs majoritaires est supérieure à N_{t}, on a (cf. formules (187) et (188) de [11])

$$\frac{1}{\lambda_1} = \tau_1 \approx \frac{\tau_{n_0}\tau_{p_0} N_{\mathrm{t}}}{\tau_2\,(n_0 + p_0)}\,, \tag{4}$$

$$\frac{1}{\lambda_2} = \tau_2 \approx \frac{\tau_{n_0}(p_0 + p_1) + \tau_{p_0}(n_0 + n_1)}{n_0 + p_0} \approx \tau_{\mathrm{SR}}\,. \tag{5}$$

Au dessus de 250 °K, on voit bien que la durée de vie τ mesurée par la photodécroissance de la conductivité est celle de Shockley-Read [8] et s'identifie avec τ_2. Dans cette gamme de température la variation expérimentale de τ est donc pour le type p en $\tau_{n_0}\left(1 + \frac{\tau_{p_0} n_1}{\tau_{n_0} p_0}\right)$ et pour le type n en $\tau_{p_0}\frac{n_1}{n_0 + p_0}$. La comparaison avec les variations expérimentales de τ permet de placer E_{t} à 0,15 eV en dessous du bas de la bande de conduction. Ce résultat confirme ceux obtenus par [4, 5, 6].

En ce qui concerne les densités spectrales (cf. formules (189) et (194) de [11]), on trouve

$$\left.\begin{aligned}\overline{n_\nu^2} &= 4\,V\left\{\frac{\lambda_1}{\lambda_1^2+\omega^2}\,\frac{\lambda_2-a_{11}}{(\lambda_2-\lambda_1)\,\lambda_1}\,\overline{p_{31}}-\frac{\lambda_2}{\lambda_2^2+\omega^2}\,\frac{\lambda_1-a_{11}}{(\lambda_2-\lambda_1)\,\lambda_2}\,\overline{p_{31}}\right\},\\ \overline{p_\nu^2} &= 4\,V\left\{\frac{\lambda_1}{\lambda_1^2+\omega^2}\,\frac{-1}{(\lambda_2-\lambda_1)\,\lambda_1}\,\overline{p_{23}}(\lambda_1-a_{11})+\frac{\lambda_2}{\lambda_2^2+\omega^2}\,\frac{1}{(\lambda_2-\lambda_1)\,\lambda_2}\,\overline{p_{23}}\,(\lambda_2-a_{11})\right\},\\ \overline{(p\,n)_\nu} &= 4\,V\left\{\frac{\lambda_1}{\lambda_1^2+\omega^2}\,\frac{1}{(\lambda_2-\lambda_1)\,\lambda_1}\,\overline{p_{31}}\,a_{21}-\frac{\lambda_2}{\lambda_2^2+\omega^2}\,\frac{1}{(\lambda_2-\lambda_1)\,\lambda_2}\,\overline{p_{31}}\,a_{21}\right\}.\end{aligned}\right\} \tag{6}$$

V étant le volume de l'échantillon et

$$\left.\begin{aligned}a_{11} &= C_0^{\mathrm{n}}\,\{n_0+n_1+N_{\mathrm{t}}-N_{\mathrm{t}}^-\}\,, & a_{12} &= C_0^{\mathrm{n}}\,\{n_0+n_1\}\,,\\ a_{21} &= C_-^{\mathrm{p}}\,\{p_0+p_1\}\,, & a_{22} &= C_-^{\mathrm{p}}\,\{p_0+p_1+N_{\mathrm{t}}^-\}\,.\end{aligned}\right\} \tag{7}$$

Les ordres de grandeurs nous conduisent à ne conserver dans les expressions (6) que les termes en $\tau_2\,(1+\omega^2\,\tau_2^2)^{-1}$; on a alors pour les types n et p au dessus de 250 °K

$$\overline{n_\nu^2} \approx \overline{p_\nu^2} \approx \overline{(n\,p)_\nu} \approx 4\,V\,\frac{n_0\,p_0}{n_0+p_0}\,\frac{\tau_2}{1+\omega^2\,\tau_2^2}\,. \tag{8}$$

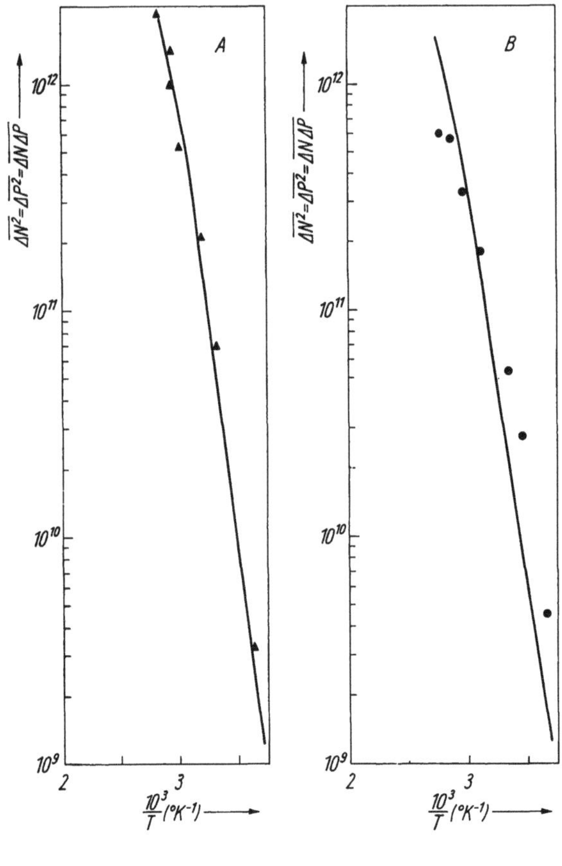

Fig. 6. Etude comparée, en fonction de la température, de la variance théorique (courbe) et de la variance déduite de mesures de bruit.
(A): échantillon n° 1 dopé As,
(B): échantillon n° 2 dopé Sb

Par suite, l'index théorique du bruit est (voir relations (1) et (2))

$$\mathcal{N}_\nu - 1 = \frac{R\, I_0^2\, (b+1)^2}{k\, T\, (P_0 + b\, N_0)^2} \frac{N_0\, P_0}{N_0 + P_0} \frac{\tau_2}{1 + \omega^2 \tau_2^2}. \tag{9}$$

On identifie donc τ_2 avec le paramètre expérimental $(2\,\pi\,\nu_0)^{-1}$.

Sur les Fig. 6 et 7, on a porté en fonction de la température la variance théorique obtenue en intégrant l'expression (8) sur toute la gamme de fréquence, soit:

$$\overline{\Delta N^2} \approx \overline{\Delta P^2} \approx \overline{\Delta N\, \Delta P} \approx \frac{N_0\, P_0}{N_0 + P_0}. \tag{10}$$

N_0 et P_0 étant déduits des mesures de résistivité et du volume de l'échantillon.

On a porté, sur les mêmes figures, quelques valeurs de la variance déduite de la mesure du rapport de bruit $\mathcal{N}_\nu$, dans la partie basse fréquence du spectre où $\omega\, \tau_2 \ll 1$ et des quantités expérimentales I_0, R, τ_2, P_0 et N_0, par la relation

$$\overline{\Delta N^2} = \overline{\Delta P^2} = \overline{\Delta N\, \Delta P} \approx (\mathcal{N}_\nu - 1) \frac{k\, T\, (P_0 + b\, N_0)^2}{R\, I_0^2\, (b+1)^2\, \tau_2}. \tag{11}$$

On constate un bon accord entre l'expérience et la théorie.

Remarque: Au dessous de 250 °K, pour les échantillons de type n, la forme de variation de τ en fonction de la température (voir Fig. 3) indique que le processus prépondérant dans l'élaboration de la durée de vie n'est pas un processus

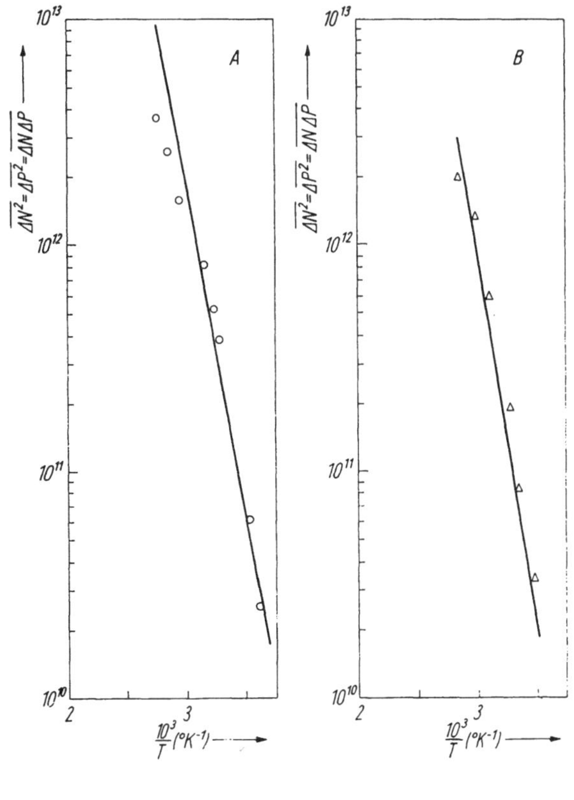

Fig. 7. Etude comparée, en fonction de la température, de la variance théorique (courbe) et de la variance déduite de mesures de bruit.
(A): échantillon n° 4 dopé Ga,
(B): échantillon n° 5 dopé In

de recombinaison du type Shockley-Read [8], prévoyant une augmentation de τ lorsque la température augmente. Il faut faire intervenir le piégeage des trous sur les dislocations pour rendre compte de cette variation.

4. Conclusion

L'étude comparative des mesures des durées de vie obtenues par la décroissance de la photoconductivité et de celles du bruit de génération-recombinaison, effectuées sur des échantillons de germanium dopés en impuretés peu profondes a été faite dans le domaine de températures 350 à 250 °K. En effet, au dessous de cette dernière température, la puissance de bruit est trop faible.

Dans cette gamme de températures, l'analyse des résultats de la durée de vie τ de photoconductivité et de l'index de bruit montre, en accord avec van Vliet [11] que la constante de temps τ_2 de corrélation s'identifie avec τ et avec la durée de vie prévue par les processus de génération-recombinaison du type Shockley-Read sur les niveaux E_t des dislocations situés à 0,15 eV au dessous du bas de la bande de conduction. Les variances des porteurs, égales à $\frac{N_0 P_0}{N_0 + P_0}$ ne sont pas affectées directement par la présence des centres recombinants, puisque leurs valeurs sont identiques à celles provenant des processus intrinsèques de création de paires électrons-trous.

Bibliographie

[1] A. Blanc-Lapierre et R. Fortet, Théorie des fonctions aléatoires, Masson & Cie, Paris 1953.
[2] R. E. Burgess, Proc. Phys. Soc. **B 68**, 541 (1955).
[3] R. E. Burgess, Proc. Phys. Soc. **B 69**, 1020 (1956).
[4] A. A. Gippius et V. S. Vavilov, Soviet Phys.-Solid State Phys. **4**, 1777 (1963).
[5] Yu. V. Gulyaev, Soviet Phys.-Solid State Phys. **3**, 796 (1961).
[6] M. I. Iglitsyn et L. I. Kolesnik, Fiz. tverd. Tela **2**, 1542 (1960).
[7] M. Lax, Rev. mod. Phys. **32**, 25 (1960).
[8] W. Shockley et W. T. Read, Phys. Rev. **87**, 835 (1952).
[9] K. M. van Vliet et J. Blok, Physica **22**, 231 (1956).
[10] K. M. van Vliet, Proc. IRE **46**, 1004 (1958).
[11] K. M. van Vliet et J. R. Fassett, Fluctuation Phenomena in Solids, Ed. R. E. Burgess, Academic Press, New York, London 1965 (p. 268 à 298).

(Received June 6, 1966)

phys. stat. sol. **16**, 729 (1966)

Institute of Metal Physics, Academy of Sciences of the USSR, Sverdlovsk

Effect of a DC Field on the Lattice Diffusion of Silver-110 in Copper and Gold

By

N. K. ARCHIPOVA, S. M. KLOTSMAN, A. N. TIMOFEEV, and I. SH. TRAKHTENBERG

A study is made of the effect of a direct current, with a density of about 100 to 150 A/mm², on the diffusion of silver-110 in gold (99.99%) and copper ("high purity" grade) at temperatures above 800 °C. Diffusion of silver-110 is measured by the residual activity method. The direction of the electromigration and the magnitude of the activated ion "effective charge" is clear evidence for electron drag. The "effective charge" decreases linearly with temperature increase. The temperature coefficient of the "effective charge" of silver in gold and copper is higher than the temperature coefficient of the matrix conductivity.

Der Einfluß eines Gleichstroms mit einer Dichte von 100 bis 150 A/mm² auf die Diffusion von Silber-110 in Gold (99,99%) und Kupfer („high purity") wird bei Temperaturen über 800 °C untersucht. Die Diffusion des Silber-110 wird mit der Restaktivitätsmethode untersucht. Die Richtung der Elektrowanderung und die Größe der aktivierten „effektiven Ionenladung" sind ein klarer Hinweis für Elektronendrag. Die „effektive Ladung" nimmt linear mit zunehmender Temperatur ab. Der Temperaturkoeffizient der „effektiven Ladung" von Silber in Gold und Kupfer ist größer als der Temperaturkoeffizient der Leitfähigkeit des Grundmaterials.

1. Introduction

Investigations of the influence of a dc field on the diffusion in solids are very interesting because the phenomena of electromigration are tightly connected with the properties of the ionic and electronic subsystems.

As was shown in [1 to 3], the force F acting on an activated ion in a metal is in the general case

$$F = F_1 + F_2 , \tag{1}$$

where F_1 is the force of direct action of the dc field on the charged particle (ion), and F_2 is the "drag force", e.g. the force with which electrons (or holes), scattered by the activated ion, act on these ions.

Since in metals usually $|F_2| > |F_1|$, the Einstein relation

$$q^* = \frac{k\,T\,v}{e \cdot E \cdot D} \tag{2}$$

permits only to determine some "effective charge" q^* of the diffusing ion. (k is the Boltzmann constant, e the electron charge, E the dc field strength, v the electromigration velocity, D the diffusion coefficient, and T the temperature.) It is obvious that $|q^*|$ should not be equal to the valency of the diffusing ion, and the sign of the effective charge may be opposite to the sign of the actual ion charge. Therefore, the drift in the dc field may be directed either to the anode, if the "electron drag" force is predominant, or to the cathode, if the force of the dc field or the drag caused by holes are predominant.

These conclusions are in agreement with the known experimental data. In metals with electronic conductivity (copper, silver, gold, aluminium) the electromigration is directed to the anode [4 to 9].

In this paper we investigate the influence of a dc field on the lattice diffusion of silver-110 in copper and gold at temperatures above 800 °C.

2. Materials and Specimen Preparation

Gold of 99.99% purity and vacuum melted electrolytic copper of "high purity" grade were used in the experiments. The total content of 11 impurities in copper was less than $2.5 \times 10^{-3}\%$.

The specimens were cut from rods into cubes with 3 to 4 mm edges, or into cylinders with diameters of 3 to 4 mm. On one specimen side a hole with 1 mm diameter was drilled for the thermocouple. The specimens were slightly etched and then annealed at 1000 °C for 10 h in vacuum. After annealing the ends were planed to give flat and parallel surfaces.

3. Experimental Procedure

Ag-110 was evaporated in vacuum on one face of the specimens. The thickness of the isotope layer (the diffusion source) was less than 10^{-5} cm. Two specimens with their active faces together were placed in a special device for electromigration [10]. The experiments were carried out in a vacuum of (10^{-4} to 10^{-5}) Torr for some hours, the direct current was (1 to 1.5)$\times 10^3$ A. The specimens were separated after annealing, their side faces removed to eliminate edge effects, and then they were sectioned. Layers about (5 to 10)$\times 10^{-4}$ cm thick were removed by etching of the specimen in a polishing etchant. The thickness of each section was determined with an error of $\pm 1\,\mu$m.

After removing the next layer the integral intensity of the remaining specimen was measured (the residual activity method [11]). The integral intensity of the γ-radiation of Ag-110 was measured with a 40×40 mm^2 NaJ (Tl) scintillation crystal and standard electronic devices. The initial γ-intensity of the specimens after annealing was about (5 to 6)$\times 10^3$ counts per second (the background intensity was about 20 counts/s).

In some experiments on silver electromigration in copper the residual β-intensity was measured with a G.-M. tube (window thickness 1.3 mg/cm^2). In these cases we measured the hard (with a 4 mm lucite filter) and the soft (as the difference between the intensities without and with filter) components of β-radiation. The initial β-intensity after annealing was about (2 to 5)$\times 10^3$ counts/min and (1 to 3)$\times 10^3$ counts/min for the hard and soft components, respectively (the backround was about 20 counts/min). An Ag-110 standard specimen was used for a control of the electronic device.

4. Methods of Calculation

The thickness of the diffusion source in our experiments was about 10^{-5} cm, the specimen thickness (3 to 4)$\times 10^{-1}$ cm, and the diffusion depth about (1 to 3) $\times 10^{-2}$ cm. Under these conditions (i.e. diffusion from an infinite thin source into a half-infinite body) the specific activity is [12]

$$i(X) = \frac{\mathrm{i}(0)}{2\sqrt{\pi D t}} \exp\left[-\frac{(X \pm v\,t)^2}{4\,D\,t}\right]. \tag{3}$$

Here $i(X)$ and $i(0)$ are the specific activities at distances X and zero from the diffusion source, t is the diffusion time. The negative sign in the expression $(X \pm v\,t)$ is related to a specimen in which the directions of electromigration and concentration gradient coincide.

The residual activity can be written

$$I(X) = \int_X^\infty i(\eta) \exp\left[-\mu\,(\eta - X)\right] d\eta$$

or

$$I(X) = \frac{i(0)}{2} \exp\left[\mu\,(X \pm v\,t) + \mu^2 D\,t\right] \cdot \operatorname{erfc}\left(\frac{X \pm v\,t}{2\sqrt{D\,t}} + \mu\sqrt{D\,t}\right), \qquad (4$$

where μ is the absorption coefficient and $\operatorname{erfc} Z = 1 - \frac{2}{\sqrt{\pi}} \int_0^Z \exp(-\zeta^2)\,d\zeta$.

From (4) it is easy to see that the curves $I(X)$ for anodic and cathodic specimens are displaced by $2\,v\,t$. This displacement was used for calculating v (Fig. 1).

With the known value of v the diffusion coefficient was then calculated for each specimen. The specific activity $i(X)$ was calculated from the relation

$$i(X) = I\left(X - \frac{\Delta X}{2}\right) - I\left(X + \frac{\Delta X}{2}\right) \cdot \exp(-\mu\,\Delta X)\,. \qquad (5)$$

Here ΔX is the thickness of the removed layer, and $I(X)$ the averaged residual activity resulting from a plot of I vs. X. For the hard component of the β-radiation and $(\Delta X)_{max} \approx 10^{-3}$ cm we have $\exp(-\mu\,\Delta X) \approx 1$; for the soft component we have $\exp(-\mu\,\Delta X) = 0.71$ for the same $(\Delta X)_{max}$ value.

For measurements of the residual γ-activity one can simplify expression (4) because of $\mu \ll 1$ for γ-radiation of Ag-110:

$$I_\gamma(X) = \frac{i(0)}{2} \operatorname{erfc}\left(\frac{X \pm v\,t}{2\sqrt{D\,t}}\right). \qquad (6)$$

It is obvious that in this case the value $2\,v\,t$ can be calculated from the displace-

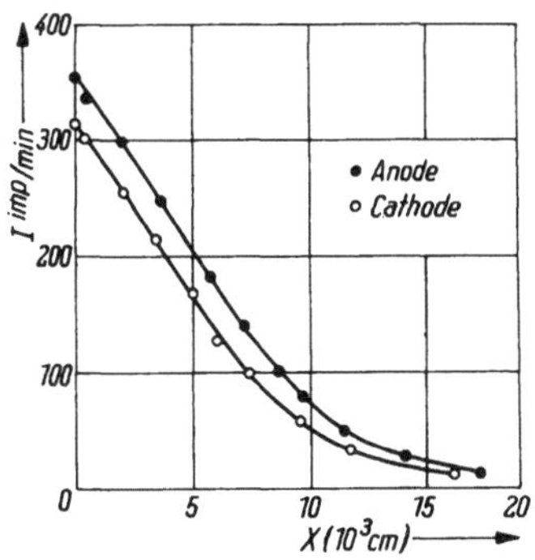

Fig. 1. Dependence of the residual β-activity I_β (hard component) vs. depth in the diffusion zone X for electromigration of silver in copper. Run No. 3 (Table 1)

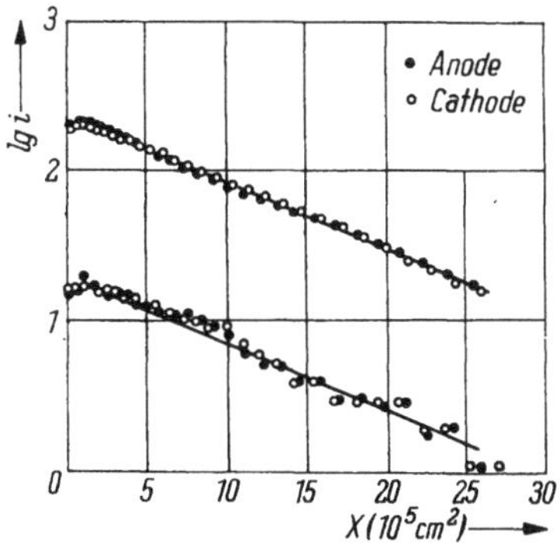

Fig. 2. Dependence of the logarithm of the specific β-activity vs. depth in the diffusion zone X for electromigration of silver in copper. The top line refers to the soft component and the bottom line to the hard component. Run No. 3 (Table 1)

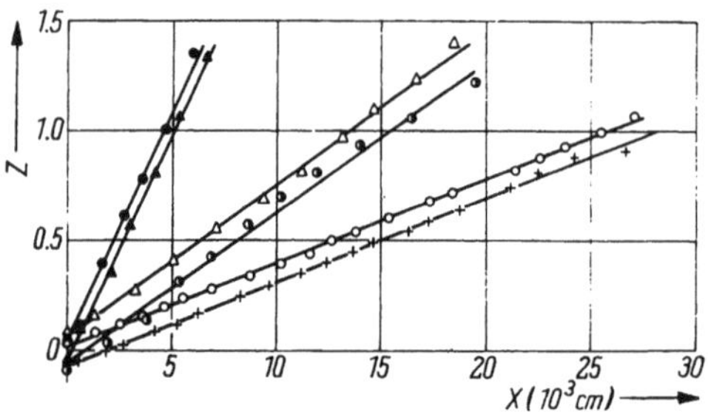

Fig. 3. Dependence of Z on X for electromigration of silver in copper (Table 1). Run No. 1: ▲ anod, ● cathod. Run No. 8: ◑ anod, Δ cathod. Run No. 9: + anod, ○ cathod

ment of the straight lines $Z = f\,(X \pm v\,t)$, where Z is the argument of

$$\operatorname{erfc} Z = \frac{2\,I_\gamma(X)}{I_\gamma^+(0) + I_\gamma^-(0)}\,. \tag{7}$$

Here $I_\gamma^{+-}(0)$ are the initial γ-activities of the anodic (+) and cathodic (−) specimens. The diffusion coefficient is calculated from the relation

$$D = \frac{1}{4\,t\left(\dfrac{\partial Z}{\partial X}\right)^2}\,. \tag{8}$$

This method of calculating D and v has two advantages. First, one can calculate both parameters D and v directly from the measured values of $I_\gamma(X)$ and v. Secondly, for calculating these parameters one uses straight line dependences which facilitates statistical calculations.

Fig. 3 and 4 show a plot of Z vs. X. The annealing temperature was corrected on the base of the calculated D values and the known dependence $D(T)$ for lattice diffusion of silver in copper [13] and gold [14]. The effective ion charge q^* was calculated from the relation

$$q^* = 8.62 \times 10^{-5} \cdot \frac{v\,T}{D\,j\,\varrho}\,, \tag{9}$$

where j is the current density in A/cm² and ϱ the specific resistance in Ωcm of the specimen at the annealing temperature (data are taken from [15]).

It must be pointed out that (9) does not contain the correlation factor f which connects the self-diffusion coefficient of the "normal" ions with the tracer diffusion coefficient. In the Einstein relation both values v and D belong to

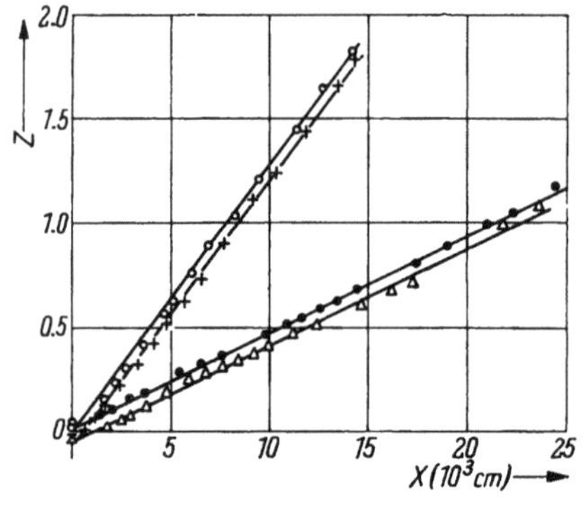

Fig. 4. Dependence of Z on X for electromigration of silver in gold (Table 2). Run No. 2: + anod, ○ cathod. Run No. 4: Δ anod, ● cathod

Table 1
Electromigration of silver in copper

No.	T (°C)	t (h)	Current density $\times 10^{-4}$ (A/cm²)	$\varrho \times 10^6$ (Ωcm)	Type of radiation	$D \times 10^9$ (cm²/s)	$v\,t \times 10^4$ (cm)	q^*	$\overline{q^*}$	Direction of electro-migration
1	789 ± 3	7	1.56	7.60	γ	0.210 ± 0.004	2.2 ± 0.3	32.3 ± 5.9	32.3 ± 5.3	
2	804 ± 4	13	1.56	7.63	soft β hard β	0.294 ± 0.005 0.300 ± 0.02	5.0 ± 0.4 4.7 ± 0.4	28.5 ± 3.3 26.2 ± 4.5	27.4 ± 3.9	
3	814 ± 2	19	1.14	7.73	soft β hard β	0.368 ± 0.007 0.350 ± 0.002	8.5 ± 0.7 6.9 ± 0.4	35.8 ± 4.4 30.6 ± 4.1	33.2 ± 4.2	To the anode
4	878 ± 7	7	1.36	8.20	γ	0.890 ± 0.030	4.6 ± 0.6	18.1 ± 3.4	18.1 ± 3.4	
5	906 ± 4	7	1.37	8.50	γ	1.300 ± 0.040	6.0 ± 1.1	15.9 ± 3.7	15.9 ± 3.7	
6	936 ± 4	7	1.49	8.85	γ	1.95 ± 0.12	9.0 ± 0.8	14.6 ± 2.5	14.6 ± 2.5	
7	943 ± 0.5	6	1.60	8.90	γ	2.16 ± 0.02	7.4 ± 0.1	11.6 ± 0.5	11.6 ± 0.5	
8	951 ± 3	9	1.44	8.95	γ	2.38 ± 0.08	11.0 ± 0.9	11.7 ± 1.6	11.7 ± 1.6	
9	1053 ± 0.5	6	1.42	9.90	γ	8.04 ± 0.21	11.8 ± 0.1	5.5 ± 0.3	5.5 ± 0.3	
10	1062 ± 2	4.3	1.22	10.0	soft β hard β	9.86 ± 0.20 9.44 ± 0.09	2.3 ± 0.8 4.0 ± 0.9	1.40 ± 0.54 2.56 ± 0.65	2.0 ± 0.6	

"normal" ions. If v and D are measured with the tracer method, the correlation factor must be excluded. If v is measured with mass migration methods (the marker method [6, 8], the weighting method [16]) and D with the tracer method, then it is necessary to introduce the correlation factor f and the Einstein relation becomes

$$q^* = 8.62 \times 10^{-5} \cdot \frac{v\,T\,f}{D^*\,j\,\varrho}, \tag{10}$$

where D^* is the tracer diffusion coefficient.

In the present experiments v and D are measured with the tracer method.

5. Results

The experimental conditions, the calculated values of the effective charge q^*, and the probable errors of all values in relation (9) are compiled in Tables 1 and 2. It is obvious that the principal error in the calculated q^* values is introduced by the error of the displacement values of the curves $I(X)$ and $Z(X)$.

It is clear from Fig. 1 to 4 that a few initial points lie lower than the straight line dependence calculated by the least-square method on the base of all points. Such activity reduction was also observed for the plots of $I(X)$, $Z(X)$, and $i(X)$ for lattice diffusion of silver in gold [14]. It is very likely that such activity reduction near the diffusion source is connected with evaporation of silver. Because the depth of this distorted region is very small in comparison to the total diffusion depth, this anomaly has no influence on the calculated v and D values. In fact, the dependence $D(T)$ for lattice diffusion of Ag-110 in gold [14] is unaffected by such activity reduction.

From Tables 1 and 2 one can see that the lattice electromigration of silver in copper and gold is directed to the anode. This shows that the predominant force, which acts on the activated ion, is the electron drag force. This result is in agreement with earlier results [5, 6, 8].

As one can see from the tables the effective charge of silver in copper and gold decreases with increasing temperature. The theory of this phenomena [1 to 3] shows that the effective charge of the diffusing ion must decrease with temperature increase, but the temperature coefficient of the effective charge must be the same as for the matrix conductivity. For comparison with the theoretical prediction about the temperature coefficient of the effective charge a straight line approximation was used for the dependences $1/\varrho(T)$ and $q^*(T)$ in the investigated temperature range (Fig. 5). The temperature dependence of $q^*(T)$ was

Table 2
Electromigration of silver in gold

No.	T (°C)	t (h)	Current density $\times 10^{-4}$ (A/cm²)	$\varrho \times 10^6$ (Ω cm)	$D \times 10^9$ (cm²/s)	$v\,t \times 10^4$ (cm)	q^*	Direction of electromigration
1	797 ± 7	15	1.53	9.8	0.52 ± 0.04	6.0 ± 1.2	13.1 ± 4.0	To
2	804 ± 2	7.75	1.61	9.9	0.52 ± 0.01	2.5 ± 0.2	9.9 ± 1.2	the
3	850 ± 3	7	1.35	10.5	1.26 ± 0.10	3.0 ± 0.5	6.5 ± 1.7	anode
4	919 ± 1	10	1.77	11.4	3.18 ± 0.09	7.4 ± 0.1	3.3 ± 0.2	
5	922 ± 9	9	1.57	11.45	3.33 ± 0.15	3.6 ± 0.9	1.9 ± 0.6	

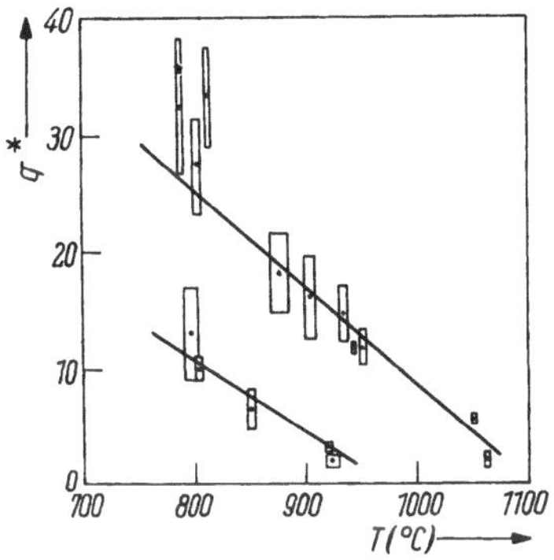

Fig. 5. Dependence of the effective charge q^* on T for electromigration of silver in copper (top line) and in gold (bottom line)

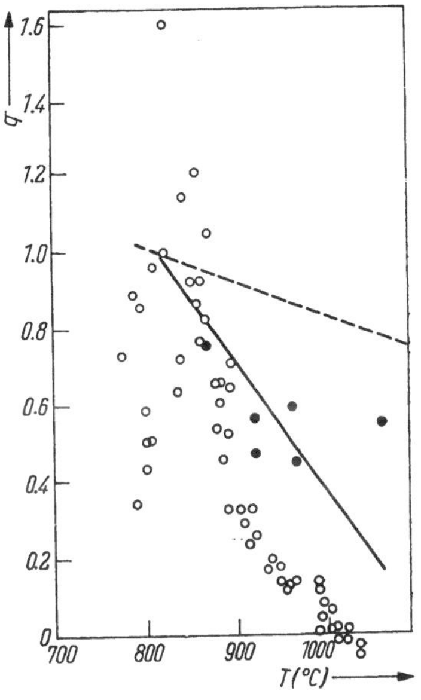

Fig. 6. Dependence of $q = q^*/q^*_{800\,°C}$ on T for electromigration of silver in copper. The solid line is the straight line approximation of the present data, the dashed line is the straight line approximation for the matrix conductivity in units $q^*_{800\,°C}\frac{\varrho_{800\,°C}}{\varrho}$. ○ data from [8], ● data from [9]

calculated by the least-square method with allowance for weights of each q^* value. The temperature coefficients $k = \mathrm{d}q^*/\mathrm{d}T$ and $k' = \mathrm{d}/\mathrm{d}T\ (q^*_{T_0}\varrho_{T_0}/\varrho)$, their probable errors σ, and the confidence interval boundaries (probability 95%) for k values are listed in Table 3. It is seen from this table that k' lies outside the confidence interval $k \pm \Delta k$. This means that $k > k'$ with 95% probability. Consequently, the effective charge of silver in copper and gold decreases with increasing temperature more rapidly than the matrix conductivity.

Such behaviour of the effective charge was earlier observed [6 to 8, 17]. The data from [5 to 8] and the present data (solid curves) in units $q = q^*/q^*_{T_0}$ are plotted for comparison in Fig. 6 and 7. The dashed lines are the straight line approximations of the dependence $q^*_{T_0}\varrho_{T_0}/\varrho$. T_0 was 800 °C for copper and 814 °C for gold.

One can see from Fig. 6 that only one q value for silver in copper (from data reported in [5]) does not agree with the present dependence (solid line). But all results of [8] are in full agreement with this dependence. The reliability of the effect $k > k'$ is confirmed by the sign reversal of the electromigration direction for self-diffusion in copper [8, 17].

Data on electromigration of silver in gold are plotted in Fig. 7. The great scatter of the experimental points in [6] does not permit a comparison of the

Table 3

System	$k \times 10^3$	$\sigma \times 10^3$	Confidence interval for $k \times 10^3$	$k' \times 10^3$
Ag–Cu	70	$\pm$ 8	36 to 104	22.9
Ag–Au	66	$\pm$11	20 to 102	12.8

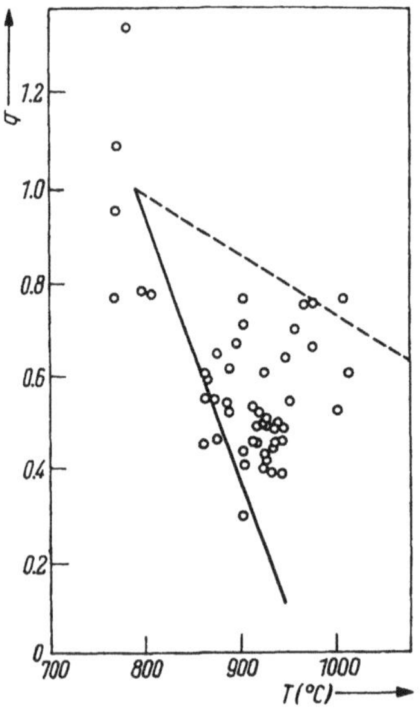

Fig. 7. Dependence of $q = q^*/q^*_{800\,°C}$ on T for electromigration of silver in gold. The solid line is the straight line approximation of the present data, the dashed line is the straight line approximation for the matrix conductivity in units. ○ data from [6]

data [6] with the present data (solid line). One can only say that all these data are in good qualitative agreement.

The present experiments on lattice electromigration of silver in copper and gold have shown that the electron drag force is predominant. The effective charge q^* decreases with increasing temperature faster than the matrix conductivity. The strong temperature dependence of the effective charge q^* may be connected with the temperature dependence of the activated ion scattering cross section for free charge carriers.

References

[1] V. B. Fiks, Fiz. tverd. Tela **1**, 1321 (1959).
[2] V. B. Fiks, Fiz. tverd. Tela **6**, 1589 (1964).
[3] L. Bosvieux and J. Friedel, J. Phys. Chem. Solids **23**, 123 (1962).
[4] P. P. Kuzmenko and E. I. Charkov, Ukr. fiz. Zh. **3**, 528 (1962).
[5] P. P. Kuzmenko and L. F. Ostrovskii, Fiz. tverd. Tela **4**, 1358 (1962).
[6] H. B. Huntington and A. R. Grone, J. Phys. Chem. Solids **20**, 76 (1961).
[7] R. V. Penney, J. Phys. Chem. Solids **25**, 335 (1964).
[8] A. R. Groun, J. Phys. Chem. Solids **20**, 88 (1961).
[9] P. P. Kuzmenko and L. F. Ostrovskii, Izv. Vyssh. Shkoli, Metallurgiya No. 11, 146 (1962).
[10] S. M. Klotsman, A. N. Timofeev, and I. Sh. Trakhtenberg, Fiz. tverd. Tela **5**, 3276 (1963).
[11] P. L. Grusin, Izv. Akad. Nauk SSSR 383 (1953).
[12] P. Ch. Malkovich, Fiz. tverd. Tela **2**, 2784 (1960).
[13] S. D. Gertsriken and A. L. Revo, Fiz. Metallov i Metallovedenie **9**, 518 (1960).
[14] S. M. Klotsman, N. K. Archipova, A. N. Timofeev, and I. Sh. Trakhtenberg, Fiz. Metallov i Metallovedenie **20**, 390 (1965).
[15] Handbook of Chemistry and Physics, Ed. Ch. D. Hodgman, Rubber Publ. Comp., Cleveland (Ohio) 1956 (p. 2355).
[16] P. P. Kuzmenko, Ukr. fiz. Zh. **7**, 117 (1962).
[17] H. Wever, Z. Elektrochem. **60**, 1170 (1956).

(Received April 18, 1966)

phys. stat. sol. 16, 737 (1966)

Institute of Metal Physics, Academy of Sciences of the USSR, Sverdlovsk

On the Magnetic Structure of Some Alloys of Transition Metals

By

S. K. SIDOROV and A. V. DOROSHENKO

A study is made of binary alloys, in which atoms of one of the components, and unlike atoms, interact ferromagnetically whereas the interaction between atoms of the other component is antiferromagnetic. Under these conditions a magnetic structure arises which involves a noncollinear alignement of the atomic magnetic moments. Expressions are given for calculating the average magnetic moment per atom and its dependence upon the concentration and the state of ordering. The results are compared with experimental data for Ni–Mn, Ni–Fe, and Pd–Fe alloys.

Рассмотрены бинарные сплавы, в которых разноименные атомы и атомы одной компоненты взаимодействуют между собой ферромагнитно, а атомы другой компоненты — антиферромагнитно. Показано, что при этом условии в сплавах должна возникать своеобразная магнитная структура с неколлинеарной ориентацией атомных магнитных моментов компонент. Даны формулы для вычисления среднего магнитного момента на атом сплава в зависимости от концентрации и состояния порядка. Сравнение с экспериментом проводится для сплавов Ni-Mn, Ni-Fe и Pd-Fe.

1. Basic Suggestions and Deduction of the Expression for the Average Magnetic Moment

It may be suggested that in binary alloys composed of two transition d-metals 1 and 2 the exchange interaction, which is responsible for magnetic order (of ferro- or antiferromagnetic type), is essentially due to direct coupling of localized 3 d shells and that this interaction is confined to the first coordination sphere. We may neglect therefore the interaction between the non-nearest atoms. We shall consider such alloys in which the energetic parameters (the exchange integrals) obey the condition

$$J_{11} > 0 \,, \quad J_{21} > 0 \,, \quad \text{and} \quad J_{22} < 0 \tag{1}$$

with an arbitrary relation between their absolute values.

Because of the well-known difficulties in the calculation of the exchange integrals [1, 2] the parameters (1) are regarded as determined experimentally, so that they are called exchange integrals only according to their physical implication.

Let us denote by $z P_{22}$ the average number of atoms 2, situated in the nearest positions to each atom of the same sort, where z is the first lattice coordination number. P_{22} may be considered as the probability for the nearest site to be occupied by an atom 2. This probability characterizes the nearest environment of atoms 2 and is defined in the general case of arbitrary state of order by the concentration, long-range order degree, and correlation parameters of the alloy.

The Nc atoms of sort 2 that are present in the alloy (N is the whole number of atoms, c the concentration of atoms 2) form $^1/_2\, NczP_{22}$ pairs of 2–2 type. The magnetic moments in each of them tend to align themselves antiparallel because of $J_{22} < 0$. However, each atom 2 is surrounded by zP_{21}, atoms 1, which tend to orientate the moments of atoms 2 parallel to the spontaneous magnetization ($J_{21} > 0$), because the magnetic moments of atoms 1 themselves are subjected to a similar orientating action from the neighbouring atoms 1 ($J_{11} > 0$). As a result of these opposite tendencies the magnetic moments of atoms 2 of the pairs 2–2 wouldn't be aligned antiparallel to each other, but will diverge to an angle less than 180°, whose magnitude is defined by the relation between $|J_{11}|$, $|J_{21}|$, and $|J_{22}|$ and by the real environment of each atom 2. The magnetic moments of the nearest atoms 1 must also deviate from mutual parallel orientation because of $J_{21} > 0$. The resulting magnetization of the alloy and the average magnetic moment per atom will be defined by the vectorial sum of all atomic moments. At a definite concentration c_0, whose value depends on P_{22} and the relation between $|J_{11}|$, $|J_{21}|$, and $|J_{22}|$, both these magnitudes may be equal to zero. The larger $|J_{22}|$ and P_{22} the smaller is c_0. Thus in an alloy with $c < c_0$ a magnetic state with spontaneous magnetization and noncollinear alignment of the magnetic moments occurs, while in an alloy with $c \geqq c_0$ a peculiar noncollinear antiferromagnetism must exist.

On the basis of these suggestions the following expression was obtained for the average magnetic moment per atom [3, 4]:

$$\bar{\mu}(c) = \mu_1 + (\mu_2 - \mu_1)\, c - z\, c\, P_{22}\, \lambda_2(P_{22})\, [z\, P_{21}\, \lambda_1\, (P_{22})\, \mu_1 + \mu_2]\,, \tag{2}$$

where μ_1 and μ_2 are the atomic magnetic moments of the alloy components and $P_{21} = 1 - P_{22}$.

It was also shown by the same authors that a) a mutual relation exists between the coefficients $\lambda_1(P_{22})$ and $\lambda_2(P_{22})$, so that they are expressed through a common parameter $\delta(P_{22})$, which generally may be positive, negative, or equal to zero, and when $c = c_0$ its value is

$$\delta(P_{22}^0) = -\frac{z\, c_0\, P_{22}^0 + c_0 - 1}{z\, c_0}\,. \tag{3}$$

$$\lambda_1(P_{22}) = \left[1 - \frac{\delta(P_{22})}{P_{21}}\right] \lambda_1\,, \qquad \lambda_2(P_{22}) = \left[1 + \frac{\delta(P_{22})}{P_{22}}\right] \lambda_2\,; \tag{4}$$

$$\lambda_1 = \frac{P_{22}^0 + \delta(P_{22}^0)}{P_{21}^0 - \delta(P_{22}^0)}\,, \qquad \lambda_2 = \frac{c_0}{1 - c_0}\,, \tag{5}$$

where P_{22}^0 is the value of P_{22} at $c = c_0$ and b) when $\delta(P_{22}) \leqq 0$

$$\delta(P_{22}) = -\frac{1}{z} \sum_{0}^{x-1} P_z(n, P_{22})\, n\,; \qquad x = 1, 2, \dots z\,, \tag{6}$$

where $P_z(n, P_{22})$ is the probability for an atom 2 to be surrounded by $n = 0, 1, 2, \dots z$ atoms 2, where z is the whole number of its nearest neighbour atoms. This probability in the absence of correlation in the alloy satisfies the binomial distribution

$$P_z(n, P_{22}) = \binom{z}{n} P_{22}^n\, P_{21}^{(z-n)}\,. \tag{7}$$

The integer parameter x in (6) characterizes the relation between ferromagnetic and antiferromagnetic interactions in the alloy. The smaller $|J_{22}|$ the larger is x.

The first two terms in (2) give the change in the average magnetic moment due to simple substitution of atoms 1 by atoms 2, when all atomic moments are aligned parallel. The third term, which is proportional to the number of 2–2 pairs in the alloy represents the diminution of the average moment which is due to the noncollinearity of the atomic magnetic moments.

2. Disordered Alloys

1. In disordered alloys the atoms of both components are randomly distributed over the crystal lattice sites (long-range order and correlation absent). It is obvious that $P_{22} = c$ and the formulae (2 to 7) are simplified in this case.

The solution of the equations (3) and (6) (when $c = c_0$) gives a set of z possible values of $c_0 - c_{01}, c_{02}, \ldots c_{0x}, \ldots c_{0z}$ which corresponds to different x. For a specified alloy the value of x (which is true for any ordered state) is defined from the condition that $c_{0x} = c_{0\,\exp}$, where $c_{0\,\exp}$ is the experimentally determined value of concentration at which $\bar{\mu}$ of the disordered alloys becomes equal to zero.

In the particular case of a f.c.c. lattice ($z = 12$) the graphical solution of equations (3) and (6) (see Fig. 1) gives the following set of possible c_0 values:

x	1	2	3	4	5	6	7	8	9	10	11	12
c_0 (at%)	25.0	25.4	27.0	29.6	32.8	36.8	41.5	46.9	52.4	58.7	65.6	74.4

2. There are reasons to suppose that in Ni–Mn [5] and Ni–Fe alloys [6, 7] the interactions between both Ni atoms and nonsimilar atoms are of ferromagnetic type while Mn and Fe atoms in the pairs Mn–Mn and Fe–Fe interact antiferromagnetically. Thus the condition (1) is fulfilled for these alloys if we denote the Ni atoms by 1 and the Mn and Fe atoms by 2.

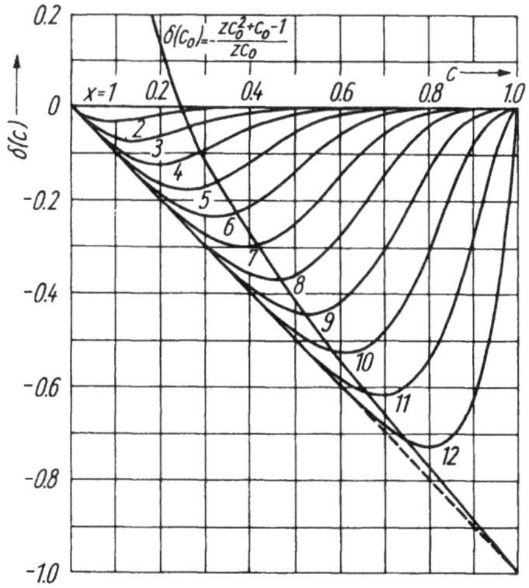

Fig. 1

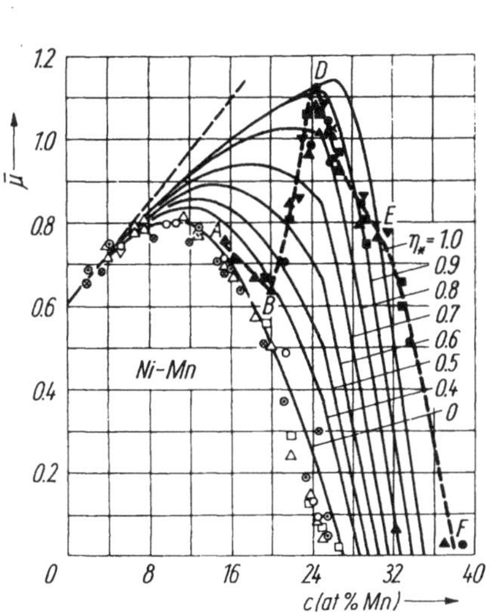

Fig. 2

Fig. 2. Concentration dependence of $\bar{\mu}$ for different η_*. Calculated curves and experimental data: ○ [3], × [9], ⊙ [16], ⊗ [17], △ [18], □ [19], ▽ [24]. Open symbols — disordered state, full symbols (and ×) — ordered state

The comparison of the results of $\bar{\mu}(c)$ calculation for Ni–Mn and Ni–Fe alloys with experimental data (for details of these calculations see [3, 8]) is illustrated by Fig. 2 and 3. The parameter x in these calculations was taken to be equal to 3 for Ni–Mn ($c_0 = 0.27$, $\delta(c_0) = -0.043$) and 12 for Ni–Fe ($c_0 = 0.744$, $\delta(c_0) = -0.715$), since $\bar{\mu}(c)$ in these alloys becomes zero according to the experimental data when the Mn concentration is ≈ 27 at% and the Fe concentration is ≈ 74 at%.

The atomic magnetic moment of Mn in Ni–Mn alloys is assumed to be constant and equal to 3.6 μ_B (this value is in better agreement with the magnetization saturation data of ordered Ni_3Mn than with the measured value 3.2 μ_B [9]). As to μ_{Ni} it is assumed to decrease from 0.6 μ_B for pure Ni to 0.3 μ_B for an alloy with 25 at% Mn [3]. In Ni–Fe alloys the magnetic moments of Ni and Fe are taken to be constant and equal to 0.6 μ_B for Ni and 2.8 μ_B for Fe [10, 11].

As it is seen from Fig. 2 (see the curve for $\eta_* = 0$) and 3 the calculated values of $\bar{\mu}(c)$ (solid line) are in good agreement with the experimental data. This agreement allows us to conclude:

a) In Ni–Mn and Ni–Fe alloys the mechanism of forming the resulting magnetization described above is apparently realized.

b) The atomic magnetic moments of Mn and Fe in Ni–Mn and Ni–Fe alloys seem to be almost constant (within the concentration limits considered). An analogous conclusion about μ_{Ni} cannot be stated, for it is too small in comparison with μ_{Mn} and μ_{Fe} so that even its considerable changes (especially to smaller values) can have only a weak influence on the concentration dependence of $\bar{\mu}(c)$.

In Pd–Fe alloys $J_{FeFe} < 0$ [12] and it may be that these alloys also obey the condition (1).

Therefore we have calculated $\bar{\mu}(c)$ for disordered Pd–Fe alloys using the equations (2 to 7). As in the case of Ni–Fe alloys the parameter $x = 12$ and the

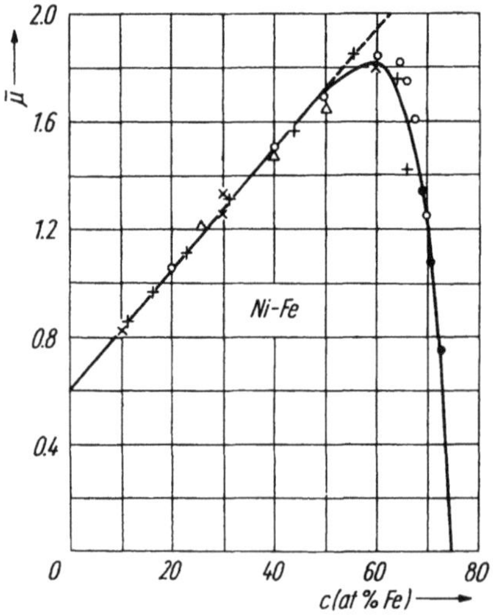

Fig. 3. Concentration dependence of $\bar{\mu}$ for Ni–Fe alloys. Calculated curve and experimental data: △ [9], × [11], + [20], ○ [21], ● [22]

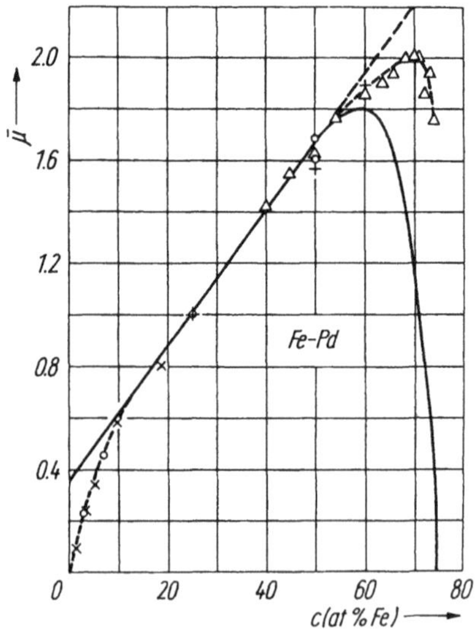

Fig. 4. Concentration dependence of $\bar{\mu}$ for Pd–Fe alloys. Calculated curve and experimental data: △ [12], + [13], × [14], ○ [15]

magnetic moments of Pd and Fe were assumed to be constant and equal to $\mu_{Pd} = 0.35\,\mu_B$ and $\mu_{Fe} = 3.0\,\mu_B$ according to [15].

The results of $\bar{\mu}(c)$ calculations are shown in Fig. 4 by the solid line. In the same figure the experimental data of many authors are also plotted [12 to 15]. It is seen that a deviation from a linear dependence in the region of 50 to 75 at% Fe may be qualitatively satisfactorily explained by the orientation mechanism of forming the average magnetic moment. But quantitatively there is a marked discrepancy between the calculated and experimental data. This may be explained possibly by the fact that an α-phase was present in the specimens investigated, since it is very difficult to obtain a pure γ-phase in these alloys.

3. The Influence of Ordering of the Alloy on the Concentration Dependence of $\bar{\mu}(c)$

1. When atoms in an alloy are ordered on short or long distances a tendency arises to surround each atom by atoms of the second component, and owing to that process the number of pairs of 2–2 type with antiferromagnetic interaction diminishes, so that P_{22} becomes smaller than c. Therefore, as it is seen from equation (2), the appearence of order in an alloy must be accompanied by an increase of $\bar{\mu}(c)$, if the third term of equation (2) plays a significant role in the concentration region, where the ordering process appears.

This perhaps accounts for the increase of the average magnetic moment in Ni–Mn alloys with concentrations ≈ 16 to 39 at% Mn as a result of annealing. In Ni–Fe and Pd–Fe alloys such an effect was not observed, because for Fe concentration less than 50 at%, the third term of equation (2) is negligible in comparison with the first two for these alloys owing to a weak antiferromagnetic interaction between the Fe atoms.

For a given crystal lattice and type of ordering P_{22} may be expressed through concentration, long-range order degree, and correlation parameters. In the special case of a f.c.c. lattice and ordering of $AuCu_3$-type it was shown [3] that

$$P_{22} = c - \frac{\gamma^2 \eta_*^2}{9\,c} + \frac{\varepsilon_{22}}{c}, \tag{8}$$

where η_* is the degree of long-range order, which is defined by the expression [23]

$$\eta_* = \frac{P_2^{(2)} - c}{\gamma}. \tag{9}$$

Here $P_2^{(2)}$ is the probability for site (2) (the regular lattice site for atoms 2) to be really occupied by an atom 2; $\gamma = c\,\frac{1-\nu}{\nu}$ for $c \leqq \nu$ and $\gamma = 1 - c$ for $c \geqq \nu$; $\nu = 1/4$ is the concentration of regular lattize sites for atoms 2 in the alloy; ε_{22} is the correlation parameter (for its expression see [3]).

2. For sake of comparison with experimental data we have calculated the dependence of $\bar{\mu}$ upon c for Ni–Mn alloys with different values of long-range order degree η_* assuming that correlation in the alloys is absent.

The calculations were carried out using the formulae (2 to 8) (for its details see [3]).

The results of the calculation of $\bar{\mu}(c)$ are represented in Fig. 2 for different values of η_*. The experimental values of $\bar{\mu}$ represented in Fig. 2 concern to annealed alloys in an ordered state of thermodynamical equilibrium, which corresponds to 350 to 400 °C.

Taking into account that the calculation of $\bar{\mu}(c)$ was carried out by assuming $\varepsilon_{22} = 0$, whereas in reality a correlation may exist in annealed alloys, the analysis of Fig. 2 leads to following conclusions:

a) Annealing of alloys of concentrations corresponding to the part AB of the curve leads probably to an appearance of short-range order but not to long-range order. In this case $P_{22} = c + \frac{\varepsilon_{22}}{c}$ is less than c and consequently ε_{22} must be negative. As a result the number of 2–2 pairs (e.g. Mn–Mn) in an ordered alloy will be smaller than in a disordered one. The disorientation of atomic magnetic moments in such alloys will also be weaker and $\bar{\mu}$ larger than for disordered alloys of the same composition, as it is known from experiments.

b) The part BD corresponds to the appearance and growth of long-range order (see the intersection points of the dashed line with the set of curves for different η_*). As a result of this process the ordering degree in the alloy of a composition near to Ni_3Mn (point D) increases up to unity.

c) In alloys, whose concentrations corresponds to part DE η_* decreases, passes through a minimum, and then increases up to unity at the point E. It is obvious that the value of ε_{22} at the points D and E is zero.

d) Within the concentration interval, corresponding to the points E and F we find $\eta_* = 1$ and $\varepsilon_{22} < 0$. This is seen from the fact that the experimental curve $\bar{\mu}$ vs. c in this plot lies above the calculated one (for $\eta_* = 1$ and $\varepsilon_{22} = 0$). Such behaviour may be possible only if a negative correlation arises in the alloys. As a result P_{22} (see equation (8)) decreases and $\bar{\mu}(c)$, as follows from (2), increases. At $\eta_* = 1$ and $\varepsilon_{22} < 0$ the sublattice of Mn atoms (one of the four simple cubic sublattices into which a f.c.c. lattice may be divided) is entirely occupied by Mn atoms. As to the other Mn atoms they distribute themselves in such a manner that the number of arising Mn–Mn pairs and consequently P_{22} would be smaller than for randomly distribution of the atoms over the sites of the other three sublattices. The probability P_{22} will be smallest if these atoms occupy the sites of only one of the three sublattices, and give thus a "supersuperstructure".

3. These conclusions permit an experimental proof. For this purpose we carried out a neutron diffraction investigation on ordered Ni–Mn alloys with 21 to 39 at% Mn [25].

The experimentally obtained dependence of the ratio of the superstructural and structural intensity maxima I_{110}/I_{111} upon the alloy composition is represented in Fig. 5.

From I_{110}/I_{111} the long-order degree η_* was calculated. The values η_* obtained for alloys with concentrations up to 26 at% Mn are shown in Fig. 5. Within the concentration interval 21 to 26 at% Mn η_* increases rapidly and reaches a value close to unity near the composition Ni_3Mn and then decreases and

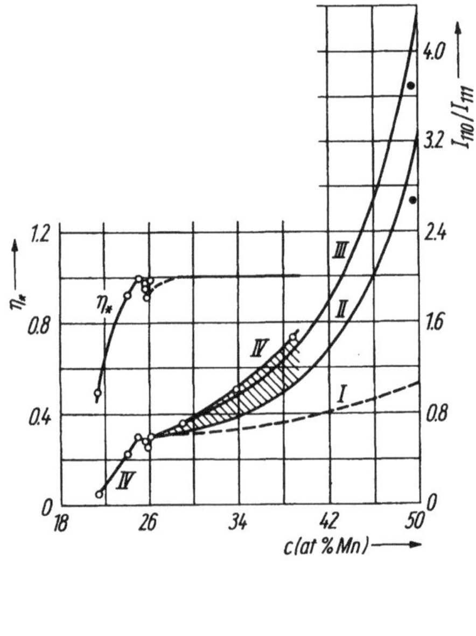

Fig. 5

passes a minimum. Such a behaviour is in good agreement with the conclusions b) and c) mentioned above.

For samples with 29.0, 33.7, and 38.8 at% Mn the values of η_* are larger than unity; this is of course meaningless and perhaps connected with deviations from the supposed distribution of Mn atoms over the lattice sites (prevailing in one of the sublattices and randomly in three others) beginning with concentrations less than 29 at% Mn.

The experimental results for I_{110}/I_{111} (within the interval 29 to 39 at% Mn) may be satisfactorily explained if one supposes that the lattice sites of one of the sublattice are entirely occupied by Mn atoms ($P_2^{(2)} = 1$ and therefore $\eta_* = 1$), and the excess atoms occupying only one of the other three superlattices.

The calculated results for I_{110}/I_{111} are given in Fig. 5. They were obtained by taking into account only nuclear scattering of neutrons for the case when all sites of one of the simple cubic sublattices are entirely occupied by Mn atoms and the excess atoms are distributed: a) with equal probability over the three other sublattices (curve I) and b) only in one of them (curve II). The comparison of curves I and II with the experimental one IV shows that the atoms are really distributed according to b), as may be concluded from the similarity of the curves II and IV. The difference between these curves may be explained by the presence of magnetic scattering of neutrons. It should also be noted that curve II is in good agreement with the experiment [26] (the low full circle in Fig. 5) for an alloy close to an equiatomic one.

4. The magnetic scattering of neutrons was taken into account supposing that within the concentration region 25 to 50 at% Mn a gradual transition takes place from a ferromagnetic state at 25 at% Mn (no Mn–Mn pairs exist) to an antiferromagnetic one at 50 at% Mn [26] through a set of noncollinear magnetic structures.

For the two extreme compositions 25 and 50 at% Mn the magnetic structure factors were calculated. In this calculation the square of the magnetic interaction vector was supposed to be equal to 2/3 for an ordered Ni_3Mn alloy. The calculated value was 0.76 for an equiatomic alloy in accordance with the magnetic lattice proposed in [26]. The atomic magnetic moments were accepted to be: for a composition with 25 at% Mn $\mu_{Mn} = 3.6\ \mu_B$, $\mu_{Ni} = 0.3\ \mu_B$; for that with 50 at% Mn $\mu_{Mn} = 4.0\ \mu_B$, $\mu_{Ni} = 0$. It was approximately assumed that within the interval 25 to 50 at% Mn the magnetic structure factors change linearly between the extreme values (for details see [25]).

Curve III in Fig. 5 shows the calculated ratio I_{110}/I_{111} vs. Mn concentration which has been obtained by taking into account both the nuclear (when the atoms are distributed according to b) and magnetic scattering. As it is seen, the curves III and IV are in good agreement with each other. This result confirms the idea that antiferromagnetic interaction in the Mn–Mn pairs causes in ordered alloys with 25 to 39 at% Mn the appearance of a noncollinear magnetic structure, where some features of ferro- and antiferromagnetism exist simultaneously.

It should be noted that the distribution of atoms according to b) leads to the highest possible increase of the nuclear part of I_{110}/I_{111} with Mn concentration. Any other distribution gives I_{110}/I_{111} values which lie between the curves I and II. The difference in the ordinates between the curves IV and II may be explained therefore only by the contribution of the magnetic scattering of neutrons to the intensity at the maxima.

5. From (8) at the values $\eta_* = 1$ and $c \geqq 1/4$, we get

$$P_{22} = c - \frac{(1-c)^2}{9c} + \frac{\varepsilon_{22}}{c}. \tag{10}$$

On the other hand when the Mn atoms are distributed according to b) (see (29) in [25])

$$P_{22} = \frac{4c-1}{6c}. \tag{11}$$

From (10) and (11) we obtain $\varepsilon_{22} = -\frac{16c^2 - 8c + 1}{18} < 0$ for $c > 1/4$.

Thus, an alloy with "supersuperstructure", which appears when the Mn atoms are distributed according to b) may be considered as a most ordered one (because $P_2^{(2)} = 1$ and $\eta_* = 1$) with a negative correlation. This confirms the conclusion expressed above that in well annealed alloys with Mn concentration above ≈ 30 at% $\eta_* = 1$ and $\varepsilon_{22} < 0$.

The authors are grateful to Professor S. V. Vonsovskii for useful discussions.

References

[1] R. STUART and W. MARSHALL, Phys. Rev. **120**, 353 (1960).
[2] A. J. FREEMAN and R. E. WATSON, Phys. Rev. **124**, 1439 (1961).
[3] S. K. SIDOROV and A. V. DOROSHENKO, Fiz. Metallov i Metallovedenie **20**, 44 (1965).
[4] S. K. SIDOROV and A. V. DOROSHENKO, Fiz. Metallov i Metallovedenie **18**, 811 (1964).
[5] M. J. MARCINOWSKI and N. BROWN, J. appl. Phys. **32**, 375 (1961).
[6] E. I. KONDORSKII, Zh. eksper. teor. Fiz. **37**, 1819 (1959).
[7] M. HATHERLY, K. HIRAKAWA, R. D. LOWDE, J. F. MALLETT, and M. F. STRINGFELLOW, Proc. Phys. Soc. **84**, 55 (1964).
[8] S. K. SIDOROV and A. V. DOROSHENKO, Fiz. Metallov i Metallovedenie **19**, 785 (1965).
[9] C. G. SHULL and M. K. WILKINSON, Phys. Rev. **97**, 304 (1955).
[10] M. F. COLLINS and J. B. FORSYTH, Phil Mag. **8**, 401 (1963).
[11] M. F. COLLINS, R. V. JONES, and R. D. LOWDE, J. Phys. Soc. Japan **17**, Suppl. B-III, 19 (1962).
[12] H. FUJIMORI and H. SAITO, J. Phys. Soc. Japan **20**, 293 (1965).
[13] M. FALLOT, Ann. Phys. (France) **10**, 291 (1938).
[14] J. GRANGLE, Phil. Mag. **5**, 335 (1960).
[15] J. W. CABLE, E. O. WOLLAN, and W. C. KOEHLER, Phys. Rev. **138**, A 755 (1965).
[16] S. KAYA and A. KUSSMANN, Z. Phys. **72**, 293 (1931).
[17] C. SADRON, Ann. Phys. (France) **17**, 371 (1932).
[18] G. R. PIERCY and E. R. MORGAN, Canad. J. Phys. **31**, 529 (1953).
[19] R. HAHN and E. KNELLER, Z. Metallk. **49**, 426 (1958).
[20] E. I. KONDORSKII and L. N. FEDOTOV, Izv. Akad. Nauk SSSR, Ser. fiz. **16**, 432 (1952).
[21] J. GRANGLE and G. C. HALLAM, Proc. Roy. Soc. **272**, 119 (1963).
[22] Y. BANDO, J. Phys. Soc. Japan **19**, 273 (1964).
[23] M. A. KRIVOGLAS and A. A. SMIRNOV, The Theory of Ordered Alloys, Fizmatgiz 1958 (in Russian).
[24] S. KAYA and M. NAKAYAMA, Proc. Physico-Math. Soc. Japan **22**, 126 (1940).
[25] S. K. SIDOROV and A. V. DOROSHENKO, Fiz. Metallov i Metallovedenie **20**, 850 (1965).
[26] J. S. KASPER and J. S. KOUVEL, J. Phys. Chem. Solids **11**, 231 (1959).

(Received April 18, 1966)

phys. stat. sol. 16, 745 (1966)

Institute of Metal Physics, Academy of Sciences of the USSR, Sverdlovsk

On Particularities of Magnetic Properties of Thin Films with Exchange Anisotropy[1])

By

A. A. Glazer, A. P. Potapov, R. I. Tagirov, and Ya. S. Shur

Investigations are made of the magnetic properties of multilayer magnetic films consisting of Fe–Ni–Mn and permalloy which have exchange anisotropy. It is found that a unidirectional anisotropy and an additional uniaxial anisotropy appear simultaneously. At the same time rotational hysteresis losses which do not vanish even in fields as high as 10 kOe, and a large magnetic after-effect are observed. The domain structure observations show that thin film remagnetization starts by means of a bidirectional rotation, and after the appearance of domains of opposite polarity the process of wall movement proceeds. The easy direction of the unidirectional anisotropy cannot be rotated at room temperature even in fields of 180 kOe.

Изучены магнитные свойства многослойных магнитных пленок с обменной анизотропией, состоящих из сплава Fe-Ni-Mn и пермаллоя. Было найдено, что вместе с появлением однонаправленной анизотропии возникают дополнительная одноосная анизотропия, потери на вращательный гистерезис, не изчезающие в полях до 10 кэ, большое магнитное последействие. Наблюдения доменной структуры показали, что при перемагничивании таких пленок вначале идет процесс двухстороннего вращения, а затем, после появления доменов с обратной ориентацией намагниченности — путем смещения границ. Легкое направление однонаправленной анизотропии не может быть повернуто при комнатной температуре магнитными полями до 180 кэ.

1. Introduction

Exchange or unidirectional anisotropy was first discovered ten years ago [1] in fine powders of cobalt coated with antiferromagnetic CoO. At the same time the explanation of the matter was found in the main as a result of the exchange interaction across the interface between the ferro- und antiferromagnetic phases. Two general features of the exchange anisotropy were pointed out — namely, a displaced hysteresis loop with a corresponding $\sin\theta$-torque curve and a nonvanishing rotational hysteresis at high fields $H > H_k$. However, this exchange interaction can lead to a number of unusual properties which are poorly studied. It is undoubtedly interesting to study these properties for the thorough understanding of the nature of the unidirectional anisotropy. The most suitable objects for such a study are double-layer films of manganese–permalloy where unidirectional anisotropy was found recently [2]. In such films one can change gradually the magnitude of the unidirectional anisotropy. All the features of the magnetic properties connected with the exchange anisotropy can be easily marked out because the magnetic properties of permalloy films are studied in detail. Besides, the unidirectional anisotropy in these films

[1]) Contribution to the International Colloquium on Magnetic Thin Films, 25th to 28th April 1966 in Jena.

exists at room temperature what is favourable to a study of their magnetic properties including their domain structure.

The purpose of the present work is a systematic investigation of the magnetic properties of composed manganese–permalloy films and an exposing of the features caused by the exchange coupling of ferromagnetic and antiferro· magnetic layers.

2. Experimental Results

1. The films were prepared by a consecutive evaporation of manganese and 82-permalloy in a magnetic field onto round glass substrates maintained at 250 °C. Just after the preparation the films had only a uniaxial anisotropy. The unidirectional anisotropy was appearing after an annealing at 350 °C in a magnetic field. In Fig. 1 the torque curves are shown for a film before and after annealing for 30 min, 3 and 6 h. The curves are obtained in a field of 1000 Oe. One can see that before annealing (Fig. 1a) the film has a sin 2 θ-torque curve and hence a uniaxial anisotropy. After annealing for 30 min at 350 °C in the field a great rotational hysteresis appears. The maximum of the torque curve corresponding to 220° is higher than the others, i.e., there exists some unidirectional anisotropy (Fig. 1b). After an annealing for 3 h this maximum becomes considerably higher, the amplitude of the torque curve and the rotational hysteresis increase appreciably (Fig. 1c). Finally, after an annealing for 6 h (Fig. 1d) the torque curve approaches to a sin θ-law, and the amplitude of this curve gets its maximum value. In Fig. 2 the constants K_{ud} of the unidirectional and K_u of the uniaxial anisotropies are shown as a function of annealing duration. (The values of K_{ud} and K_u were calculated by a Fourier analysis of the torque curves. The thickness of the film, used for calculation, was 800 Å, i.e. equal to the thickness of the permalloy layer.) The rotational hysteresis loss W_r, the coercive force H_c (i.e. the halfwidth of the hysteresis loop), and the shift ΔH (the distance from the origin of the coordinates to the

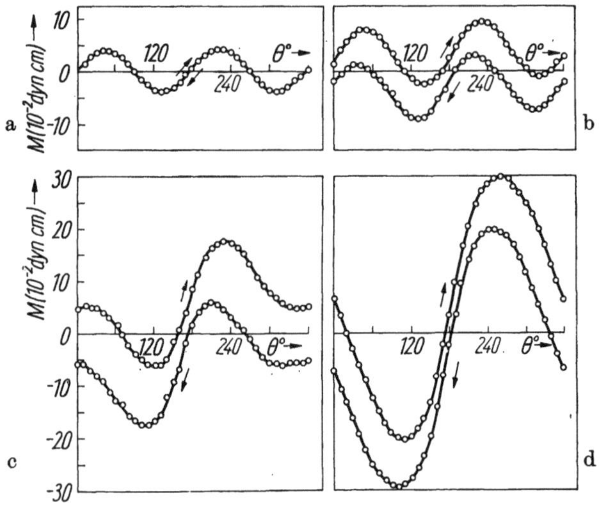

Fig. 1. Torque curves of a manganese-permalloy film at the original state (a) and after magnetic annealing at 350 °C for 30 min (b), 3 h (c) and 6 h (d)

Fig. 2. Various properties of a double-layer film as a function of time of magnetic annealing at 350 °C. 1 — H_c, 2 — ΔH, 3 — K_{ud}, 4 — K_u, 5 — W_r

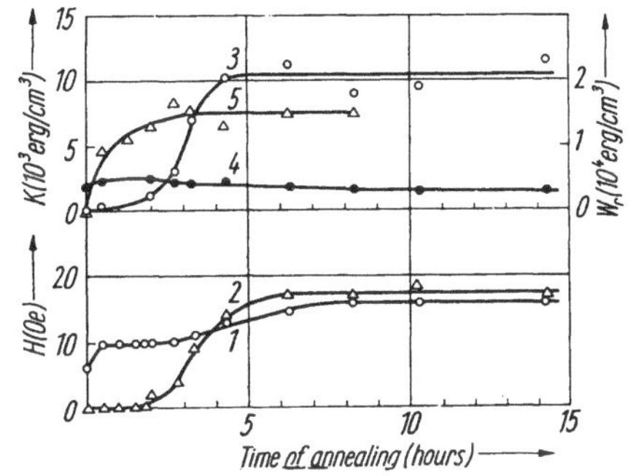

centre of the hysteresis loop) are also shown in Fig. 2. It is seen that K_{ud} and ΔH synchronously increase with the annealing duration, reach their maximum values in 5 h, and later on keep these values. This is connected, apparently, with an increasing of the thickness and of the crystalline anisotropy of the antiferromagnetic layer which is formed as a result of the mutual diffusion of Mn, Fe, and Ni. As long as this layer is not formed completely, the spins of the antiferromagnet are pulled from their easy directions due to exchange coupling with those of the ferromagnet during switching of the latter. This is a cause of the low values of ΔH and K_{ud} at the beginning of the annealing. During the annealing the thickness of the antiferromagnet grows and the deviations of the antiferromagnetic spins from their easy directions become smaller. At a definite thickness (which corresponds to an annealing for 5 h) these deviations vanish and the unidirectional anisotropy gets its maximum value. After that ΔH and K_{ud} do not increase even if the antiferromagnetic layer becomes thicker. A similar dependence of ΔH on the thickness of the antiferromagnet was observed by Bean [3] on an oxidized film of cobalt.

2. It is of essential importance for clarifying the nature of the unidirectional anisotropy to change the easy direction by a strong magnetic field. For this purpose strong magnetic fields were applied to a specimen having a shifted hysteresis loop. These fields were applied at 180 or 90° with respect to the easy direction. Permanent fields of 20 kOe or pulsed fields of 180 kOe were used. The duration of the latter was 1 ms (there was no opposite pulse). All attempts to change the easy direction failed. This result means that beyond of doubt the intermediate layer formed in the film is not a high coercive ferromagnet. Apparently, it is indeed an antiferromagnet with a threshold field of more than 180 kOe.

3. It is seen from Fig. 2 that the coercive force of films rises abruptly during the first 30 minutes of the annealing and then increases steadily with increasing of ΔH and K_{ud}. If the unidirectional anisotropy is uniform throughout the whole specimen, it will only shift the hysteresis loop and will not enlarge H_c. To find the cause of the increase of coercive force we studied changes of the domain structure of the films with a shifted hysteresis loop during switching parallel to the easy direction. The corresponding powder patterns are shown in Fig. 3. As one can see the film is switching by an incoherent rotation with subsequent locking of the rotation and appearing of domains magnetized parallel to the field. Such a behaviour of the film in the field means that it is inverted after an annealing at 350 °C, i.e. it is similar to the films described in [4]. Thus H_c increases. It is known that an inversion is caused by a large dispersion of easy

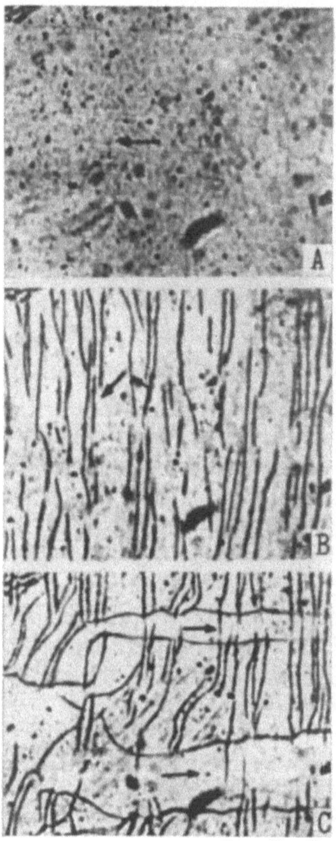

Fig. 3. Bitter patterns in a film with unidirectional anisotropy a various fields opposite to the easy direction. A — $H = 0$ Oe, B — $H = -40$ Oe, C — $H = -41$ Oe

axes. In our case a strong enlarging of the dispersion is connected with the formation of antiferromagnetic regions with the axes of antiferromagnetic anisotropy oriented randomly. The local deviations of the magnetization from the mean easy direction (i.e. the dispersion) arise due to the exchange interaction between the magnetization and the spins of the antiferromagnetic regions. This interaction becomes stronger during the annealing of the film and the dispersion (in angle and magnitude) increases to a definite value. The coercive force will increase correspondingly.

4. One can see from Fig. 1 and 2 that the rotational hysteresis increases during the first 2.5 hours of the annealing and then keeps constant. The rotational hysteresis losses do not vanish at high fields (we operated with fields up to 10 kOe). The nature of this hysteresis, apparently, is the same as that of oxidized cobalt particles [5]. Namely, these hysteresis losses are connected with irreversible rotations of spins in antiferromagnetic regions having a low anisotropy. Such regions exist at late stages of the annealing because the rotational hysteresis losses are high when K_{ud} reaches its saturation.
The antiferromagnetic regions with low anisotropy existing in the double-layer films after an annealing at 350 °C are, apparently, the cause of two further phenomena to be discussed below.

5. Fig. 2 shows that the *uniaxial* anisotropy K_u increases appreciably at the beginning of the annealing (as long as the unidirectional anisotropy is low), and then decreases to a constant value. Such a behaviour of K_u is unusual. Test experiments were done with a usual uniaxial 82-permalloy film evaporated onto the substrate at 250 °C. After the annealing of this film at 350 °C in a field parallel to the easy axis K_u loses 15% of its value, the whole change taking place in the first 30 minutes. Thus, the enlarging of K_u in the double-layer manganese-permalloy film during the annealing at 350 °C points out an additional uniaxial anisotropy which increases the total uniaxial anisotropy (although the M_s-induced anisotropy decreases). Vlasov and Mitsek [6] have predicted such an additional uniaxial anisotropy. Its nature is connected, apparently, with the exchange interaction of the ferromagnetic film with the antiferromagnetic regions having low anisotropy. When ferromagnetic spins are rotated by 180° by an external field, spins of weakly anisotropic antiferromagnetic regions rotate by 180° which is also due to exchange forces. The easy direction controlled by these regions is reversed. Such a small rotated unidirec-

tional anisotropy is similar to uniaxial anisotropy and its axis coincides with the direction of the exchange anisotropy.

6. A great magnetic after-effect was observed in films with unidirectional anisotropy. Their magnetization as a function of time was studied with a vibrating magnetometer in a permanent magnetic field at various points of the hysteresis loop. The results are shown in Fig. 4. The film was saturated in the hard direction, and then the field was lowered to get point 1 of the hysteresis loop shown schematically in Fig. 4. Just after the field became constant, the magnetization was measured as a function of time. The results are represented by curve 1. Similar measurements were done at other points of the hysteresis loop marked by the numbers 2 to 6 on the scheme. (The states corresponding to the points 4, 5, and 6 on the scheme were obtained after saturation of the specimen in the easy direction.) It is seen from Fig. 4 that the film has a large after-effect. The largest after-effect is observed on the steep part of that branch of the hysteresis loop, which is nearest to the coordinate axis (the curves corresponding to points 2 and 3). The after-effect on the other branch of the loop (at the points 5 and 6) is less noticeable. The feature of the after-effect at the points 2 and 3 is that the magnetization decreases in the positive external field, i.e., the magnetization changes *against* the action of the external field. The nature of this after-effect, apparently, is similar to the nature of the rotational hysteresis losses. Probably, when the external field reverses the magnetization of the ferromagnetic layer, spins of some antiferromagnetic regions are rotated by 180° irreversibly due to exchange forces. It lowers the internal field which is opposite to the external one. For this reason a ferromagnetic system of spins rotates by an additional angle what causes the rotation of spins of another antiferromagnetic regions, and so on. As a result of this process the magnetization is changing during some time in a fixed external field, i.e., the magnetic after-effect takes place.

7. Finally, we have studied the rotation of the easy direction by heating the film to a definite temperature and cooling it in a field normal to the original easy direction. The results are represented in Fig. 5. It follows from Fig. 5 that after heating to 250 °C and cooling in the field normal to the easy direction the latter rotates by 20° only. After heating to 310 °C with subsequent cooling in the field the easy direction rotates by 90°. If the temperature of heating changes from 230 to 310 °C the angle of rotation varies directly with tempera-

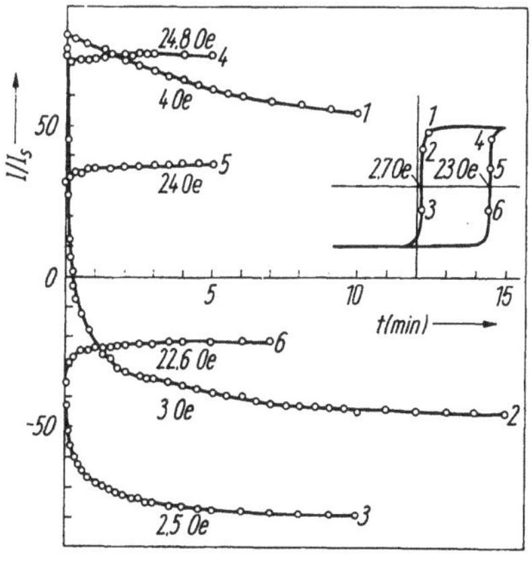

Fig. 4. Magnetic after-effect in a Mn–FeNi film with unidirectional anisotropy

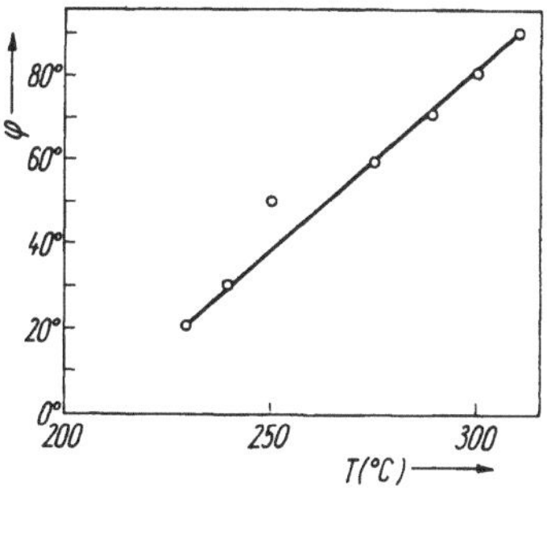

Fig. 5. Angle of rotation of the easy direction in a transverse field as a function of annealing temperature

ture. These results show that there is apparently a variety of Néel points the lowest of which is about 200 °C.

3. Conclusions

The results of the experiments described above show that the exchange interaction between ferro- and antiferromagnetic layers of double-layer manganese–permalloy films causes its following features:

1. The hysteresis loop is shifted and torque curve has sinθ-law. The easy direction of the unidirectional anisotropy cannot be rotated at room temperature by a field of 180 kOe at least.
2. The film is inverted.
3. Great rotational hysteresis nonvanishing in fields of 10 kOe at least.
4. The additional uniaxial anisotropy whose axis is parallel to the easy direction of the unidirectional anisotropy.
5. The presence of a large magnetic after-effect which has different values at various branches of the hysteresis loop.

It may be expected that the features found in the present work will occur also in other materials with unidirectional anisotropy.

References

[1] W. H. MEIKLEJOHN and C. P. BEAN, Phys. Rev. **102**, 1413 (1956).
[2] O. MASSENET and R. MONTMORY, C. R. Acad. Sci. (France) **258**, 1752 (1964).
[3] C. P. BEAN, Structure and Property of Thin Films, Wiley, New York 1959 (p. 331).
[4] E. FELDTKELLER, Elektron. Rechenanlagen **3**, 167 (1961).
[5] W. H. MEIKLEJOHN, J. appl. Phys. **33**, 1328 S (1962).
[6] K. B. VLASOV and A. I. MITSEK, Fiz. Metallov i Metallovedenie **14**, 487 (1962).

(Received May 9, 1966)

phys. stat. sol. **16**, 751 (1966)

Institute of Semiconductors, Academy of Sciences of the Ukrainian SSR, Kiev

Theory of the Photovoltaic Effect in Semiconductors

By

V. A. ZUEV and V. G. LITOVCHENKO

The detailed theory of "surface photovoltage" is developed for the case of small signals in Si and Ge. In addition to the usual factors considered (such as the spreading of space charge during illumination etc.) other important effects such as the trapping of photocarriers on the surface and the photoconductivity in the space charge region are taken into consideration. Δ_n involves the surface trapping parameters in addition to the electron–hole pair generation rate . A convenient set of approximate equations are chosen, from the usual relations, to describe the dependence of the photo e.m.f. on temperature, surface potential, and the parameters of the local centres. It is shown that under appropriate simplifying assumptions the expressions obtained reduce to those formerly described.

Eine ausführliche Theorie der „Oberflächenphotospannung" wird für kleine Signale in Si und Ge entwickelt. Zusätzlich zu den üblicherweise berücksichtigten (wie die Ausbreitung der Raumladung während der Belichtung) werden auch andere wichtige Einflüsse, wie das Anhaften der Photoleitungsträger an der Oberfläche und die Photoleitfähigkeit im Raumladungsgebiet, in Betracht gezogen. Δ_n schließt zusätzlich zu der Elektron–Lochpaar-Erzeugungsrate noch die Oberflächenanhaftparameter ein. Aus den gebräuchlichen Beziehungen wird ein geeignetes System von Näherungsgleichungen gewählt, um die Abhängigkeit der Photo-EMK von der Temperatur, dem Oberflächenpotential und den Parametern der lokalen Zentren zu beschreiben. Es wird gezeigt, daß unter passenden, vereinfachenden Annahmen die erhaltenen Ausdrücke sich auf früher beschriebene reduzieren.

1. Introduction

The theory of the "surface photo e.m.f." was developed in [1 to 6]. So the authors of [1] and [2] have considered semiconductors in which the bulk impurities ionised only to a small extent (such semiconductors as Cu_2O and CdS, extremely low temperatures). Therefore their relations are usually inapplicable for such semiconductors as Ge and Si. Complete ionisation of bulk impurities was supposed in [3, 4, and 6]; however, these authors did not consider in detail the electronic transitions between surface centres and band bending in the space charge region. The characteristics of the centres were included in the constant phenomenological parameter Δ_n (the generation rate of electron-hole pairs), and so there was no evident dependence of the photo e.m.f. V_{ph} on the parameters of the surface centres. Besides, as will be shown later on, this method which is useful for investigating the general behaviour of V_{ph} for semiconductors such as Ge, does not take into account some important particularities of the electronic transitions in the space charge region.

At last, the calculation [5] for a special model of electronic transitions was limited to the case of flat bands.

Moreover the works [5, 7, and 8] indicated that special features of the space charge, trapping of non-equilibrium photo-carriers at the surface, etc., are quite important for actual cases (silicon with extrinsic conductivity, semiconductors at low temperatures), especially in the presence of significant depletion or inversion layers at the surface.

Therefore the necessity has arised for a calculation of the photovoltage taking into account the following contributions:

1. the barrier voltage V_b,
2. the e.m.f. which arises from the accumulation of photo-carriers in surface levels ("trapping effect of photo-carriers"),
3. the e.m.f. which is due to the directed diffusion of electron-hole pairs (Dember voltage).

The aim of this work is to establish a connection of the surface photovoltage with the surface and volume parameters of the semiconductors.

2. Main Assumptions

The calculation is carried out under assumptions which may be realised experimentally:

1. The semiconductors exhibit extrinsic conductivity, the volume centres are fully ionised, and bipolar volume conductivity is realised.
2. A quasi Boltzmann distribution of carriers is present in the space charge region, and so the range of surface potential corresponding to this condition was taken into consideration (usually from $Y_{0s} \approx 2$ to inversion of surface conductivity). Criteria for this distribution are given, for instance, in [7].
3. The additional concentration of minority carriers (Δp for n-type) essentially exceeds the thermal concentration of minority carriers p_0 in the actual region but is considerably smaller than the thermal concentration of majority carriers n_0 ($n_0 \gg \Delta n, \Delta p$).
4. It was supposed that at the surface there is a system of discrete levels with concentration N_j, energetic position E_j (which is counted from the middle of the forbidden gap), capture cross-sections for electrons C_{nj} and holes C_{pj}.
5. Sinusoidally modulated light falls on the surface ($x = 0$) and creates g pairs of photo-carriers per unit volume and unit time. Then $g(x, t) = g_s e^{-kx+i\omega t}$, where k is the absorption coefficient, ω the frequency of modulation, and the subscript s corresponds to the illuminated surface.

3. Results and Discussion

It is necessary to solve the system of continuity equations and the Poisson equation, which may be written for germanium and silicon in the following form:

$$i\omega \Delta n + \frac{\Delta p}{\tau} + \frac{d}{dx} j_n(x) = g_s e^{-kx}, \tag{1}$$

$$\left(i\omega + \frac{1}{\tau}\right) \Delta p + \frac{d}{dx} j_p(x) = g_s e^{-kx}, \tag{2}$$

$$\frac{d^2 Y}{dx^2} = \frac{\varkappa^2}{n_v^0} (n - n_{0v} - p - p_{0v}), \tag{3}$$

$$\frac{d^2 \Delta Y}{dx^2} = \frac{\varkappa^2}{n_v^0} (\Delta n - \Delta p), \tag{3a}$$

where the subscripts n and p correspond to electrons and holes, respectively, the subscripts 0 and v refer to the thermal and bulk values respectively. Δ denotes the changes due to illumination. Y is the potential (in units $k\,T/e$) counted from the volume potential, $\varkappa^{-1}$ is the screening length, τ is the volume lifetime.

The first integral of the Poisson equation (3) under the assumptions 1. to 3. is

$$\frac{\mathrm{d}Y}{\mathrm{d}x} = \pm\sqrt{2}\left(\frac{n_s}{n_{0\,v}}\,\mathrm{e}^{-Y_s+Y} + \frac{\Delta p_s}{n_{0\,v}}\,\mathrm{e}^{Y_{0\,s}-Y_0} - Y + 1 + B\right), \tag{4}$$

where B is an integration constant.

The last expression is valid in the whole region of essential influence of space charge (region I). The limits of this region (from $x = 0$ to $x = l$) are defined in the following way: It is possible to neglect a) the net hole current j_p in comparison with the field hole current $\mu_p\,E_p$ in region I; b) the field current in comparison with the diffusion current in region II (from $x = l$ to $x = d$, where d is the thickness of the sample). Then the net hole current of region II must be equal to the field current of region I at the point $x = l$.

In the absence of illumination the first integral of the Poisson equation is

$$\frac{\mathrm{d}Y_0}{\mathrm{d}x} = \pm\,\varkappa\sqrt{2\,(\mathrm{e}^{Y_0} - Y_0 - 1)} = \pm\,\varkappa\,F(Y_0) \tag{5}$$

(up to inversion of surface conductivity). As follows from (5) the dependence $Y_0(x)$ has the following form near the point $x = l$ ($Y_0 \ll 1$):

$$Y_0 = Y_l\,\mathrm{e}^{-\varkappa(x-l)}\,. \tag{6}$$

Comparing the net hole current of region II and field hole current of region I at the point $x = l$ and using (6) we have

$$Y_{0\,l}\,\varkappa\,L_p = 1\,. \tag{7}$$

Then it follows from (5)

$$l = \pm\frac{1}{\sqrt{2}\,\varkappa}\int\limits_{(\varkappa L_p)^{-1}}^{Y_{0\,s}} \frac{\mathrm{d}Y_0}{\sqrt{\mathrm{e}^{Y_0} - Y_0 - 1}}\,. \tag{8}$$

For Si $l \approx$ (10 to 20) $\varkappa^{-1}$ is typical.

Using the expression for the constant value B (see Appendix A) and linearizing (4) we have

$$\frac{\mathrm{d}\Delta Y}{\mathrm{d}x} = \frac{\varkappa^2}{\left(\frac{\mathrm{d}Y_0}{\mathrm{d}x}\right)}\left[\frac{\Delta n_s}{n_{0\,v}}\,\mathrm{e}^{-Y_{0\,s}}(\mathrm{e}^{Y_0} - 1) + \frac{\Delta p_s}{n_{0\,v}}\,\mathrm{e}^{Y_{0\,s}}(\mathrm{e}^{-Y_0} - 1) - (\mathrm{e}^{Y_0} - 1)\,(\Delta Y - \Delta Y_s)\right]. \tag{9}$$

To determine ΔY_s it is necessary to integrate both parts of (9) and then to solve the integrated equation by transforming it into a difference equation.

The result is

$$\Delta Y - \Delta Y_s = (2\,n_{0\,v})^{-1}\Big\{\Delta n_s\,\mathrm{e}^{-Y_{0\,s}}\,f_1(Y_0) + \Delta p_s\,\mathrm{e}^{Y_{0\,s}}\,f_2(Y_0) + $$
$$+ F(Y_0)\int\limits_{Y_{0\,s}}^{Y_0} f_3(Y)\,F^{-1}(Y)\,[\Delta n_s\,\mathrm{e}^{-Y_{0\,s}}\,f_1 + \Delta p_s\,\mathrm{e}^{Y_{0\,s}}\,f_2]\,\mathrm{d}Y\Big\}, \tag{10}$$

where

$$\left.\begin{aligned} f_1 &= \int\limits_{Y_{0s}}^{Y_0} (e^Y - 1)\, F^{-2}(Y)\, dY\,, \qquad f_2 = \int\limits_{Y_{0s}}^{Y_0} (e^{-Y} - 1)\, F^{-2}(Y)\, dY\,, \\ f_3 &= \frac{e^{Y_0} - 1}{2}\, F^{-2}(Y_0)\,. \end{aligned}\right\} \tag{11}$$

For Ge and Si the following relation is typical (due to the small Dember voltage):

$$\Delta Y_s \gg \Delta Y_l\,.$$

From (10) we obtain

$$\Delta Y_s = \frac{\Delta n_s}{n_{0v}}\, e^{-Y_{0s}} - \frac{\Delta p_s}{n_{0v}}\, e^{Y_{0s}} [1 + f^*(Y_{0s}, Y_{0l})]\,, \tag{12}$$

where $f^* \ll 1$ for an accumulation layer and

$$f^* = -\frac{Y_{0l}}{1 + Y_{0s}}\, e^{-Y_{0s}} \left[\frac{e^{1/2}}{\sqrt{2}} - 1 - (-2 - 2\,Y_{0s})^{-1/2}\right] \approx 0.1\, \frac{e^{-Y_{0s}}}{Y_{0s}}\, (-\,Y_{0l})$$

for a depletion layer.

In order to estimate Y_{0l} let us use the relationship (7). The connection of ΔY_s with the parameters of surface levels is given by Δp_s and Δn_s (see Appendix B).

In the case of steady-state conditions, strongly absorbed light ($k^{-1} \ll l$), and "thick" sample ($d \gg L_p$) we have, using (12) and (B9, B10),

$$\Delta Y_s = \frac{\dfrac{g_s}{k}\, n_{0v}^{-1}\, e^{-Y_{0s}}}{\sum\limits_j C_{pj} N_j f_{nj} + \dfrac{D_p}{L_p}\, e^{Y_{0s}} (1 + L^-)} \times$$

$$\times \left[\frac{\sum\limits_j \dfrac{C_{pj}\, N_j f_{nj}^2}{C_{nj}\, n_{0s}} + \dfrac{1 - e^{Y_{0s}}}{E_{0s}}}{\sum\limits_j \dfrac{N_j}{n_{0s}} f_{pj} f_{nj} - \dfrac{1 - e^{-Y_{0s}}}{E_{0s}}} - e^{2Y_{0s}} (1 + f^*)\right]. \tag{13}$$

One can see that the second term in square brackets (for depletion and inversion layers at the surface) is negligible. This means that the influence of non-equilibrium minority carriers on ΔY_s is insignificant (in this region of potentials).

The relation (13) is especially simplified in some important cases. For example, it is known that only one level is effective in various separate regions of the gap.

At high temperatures (nearly $T > 250$ °K for Si) the terms $(e^{-Y_{0s}} - 1)\, E_{0s}^{-1}$ and $(e^{Y_{0s}} - 1)\, E_{0s}^{-1}$ are typically small in comparison with the neighbouring terms. Then

$$\Delta Y_s = \frac{g_s}{k\, n_{0v}}\, \frac{e^{-Y_{0s}}\, \dfrac{C_{pj} f_{nj}}{C_{nj} f_{pj}}}{C_{pj} N_j f_{nj} + \dfrac{D_p}{L_p}\, e^{Y_{0s}} (1 + L^-)} = V_{ph}\,. \tag{14}$$

In the case of strong recombination at the surface one has

$$\sum_j C_{pj} N_j f_{nj} \gg D_p L_p^{-1} e^{Y_{0s}} (1 + L^-) ,$$

thus

$$\Delta Y_s = \frac{g_s}{k n_{0v}} \frac{e^{-Y_{0s}}}{C_{nj} N_j f_{pj}} , \tag{15}$$

i.e., an exponential increase of ΔY_s with decreasing temperature is predicted:

$$\Delta Y_s \sim f_{pj}^{-1} \sim \exp (E_j - U_B) .$$

As the following inequalities hold at low temperatures:

$$\sum_j \frac{C_{pj} N_j f_{nj}^2}{C_{nj} n_{0s}} \gg (1 - e^{Y_{0s}}) E_{0s}^{-1} ,$$

$$\sum_j N_j n_{0s}^{-1} f_{pj} f_{nj} \ll \frac{1 - e^{Y_{0s}}}{E_{0s}} .$$

(13) gives

$$\Delta Y_s^* = \frac{g_s}{k n_{0v}} \frac{f_{nj}}{C_{nj} n_{0s}} E_{0s} = V_{ph}^* . \tag{16}$$

The last relation predicts the occurrence of saturation in the dependence of V_{ph} on temperature. When investigating the temperature dependence of V_{ph} (particularly when calculating V_{ph}^*/V_{ph}) the probability of actual changes of the surface levels should be taken into consideration. So if the value $\frac{kT}{e} Y_{0s}$ (the depth of the surface potential box) is constant, more shallow levels will become effective as the temperature decreases. If the value $\frac{kT}{e} (Y_{0s} + U_B) = \frac{kT}{e} U_s$ is constant, the same level will remain effective.

In the presence of strong accumulation-type band bending the expression for V_{ph} loses its dependence upon surface parameters (similarly as in [3]):

$$V_{ph} \approx - \frac{g_s}{k n_{0v}} \frac{L_p}{D_p} . \tag{17}$$

If there is no storage of photo-carriers in surface traps the following expression applies:

$$\sum_j C_{nj} N_j f_{pj} \Delta n_s = \sum_j C_{pj} N_j f_{nj} \Delta p_s . \tag{18}$$

Using (B9), (B10), and (18) we obtain

$$\frac{\sum_j r_{nj} g_{nj}^{-1}}{\sum_j r_{pj} g_{nj}} = e^{-Y_{0s}} .$$

Thus, from (13) we have

$$V_{ph,b} = \frac{g_s}{k n_{0v}} \frac{1 - (1 + f^*) e^{Y_{0s}}}{\sum_j C_{pj} N_j f_{nj} + D_p L_p^{-1} e^{Y_{0s}} (1 + L^-)} . \tag{19}$$

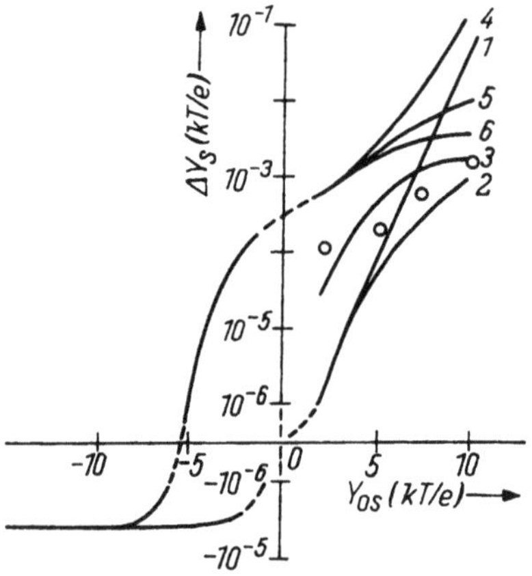

Fig. 1. Dependence of the linear photo e.m.f. V_{ph} on the surface potential Y_{0s} for parameter values which are typical of high-resistivity Si ($U_B = -7.9$, $L_n = 10^{-2}$ cm)
(1) Theory of barrier photo e.m.f. [3] assuming non-changing surface charge.
(2) Including effects connected with recombination in the surface charge region.
(3), (4), (5), and (6) Formula (13) with the parameters of surface centres [8], and the following capture cross-sections:
(3) $C_{nII} = 10^{-18}$ cm², $C_{pII} = 10^{-14}$ cm², $C_{pIII} = 10^{-15}$ cm²;
(4), (5), and (6) $C_{nII} = 10^{-18}$ cm², CpIII $= 10^{-17}$ cm², $C_{pII} = 10^{-16}$ cm², 10^{-15} cm², and 10^{-14} cm², respectively.
The circles represent the experimental data [5]. The generation rate g_s/k_s is taken to be 6.4×10^{10} cm^{-2} s^{-1}

Using the formula for photoconductivity [7] it can be shown that the generally used parameter "injection rate of pairs" (Δ_n) has the following form:

$$\Delta_n = \frac{g_s}{k\, n_{0v}} L_p\, \tau_{eff} = \frac{g_s}{k\, n_{0v}} \frac{1 + L^-}{e^{-Y_{0s}} \frac{L_p}{D_p} \sum C_{pj} N_j f_{nj} + (1 + L^-)} . \tag{20}$$

Comparing (19) and (20) we have

$$V_{ph,b} = \Delta_n \frac{1 - (1 + f^*)\, e^{Y_{0s}}}{1 + L^-} . \tag{21}$$

This formula corresponds to the relation (18) in [3].

The differences are the following:

1. In our work the generation rate is represented in terms of parameters of surface centres.

2. When we take into account the photoconductivity in the space charge region, we are led to the correction $V_{ph,b} = \frac{V^0_{ph,b}}{1 + L^-}$. Its influence is very strong for large band bending. In Fig. 1 we compare the theoretical curves $V^0_{ph,b}(Y_{0s})$ from [3] with curves $V_{ph,b}(Y_{0s})$ from the present work.

This figure also shows the dependence $V_{ph}(Y_{0s})$ for the case in which photocarriers are stored at the surface. The plotted curves were calculated from the general formula (13) using the parameters given in [8]. One can see that the value V_{ph} strongly increases with increasing Y_{0s}, which can also be concluded from (15). The experimental data [5] which we show by circles qualitatively follow the theoretical dependence (3).

In conclusion we note that the practical use of the formula obtained above is essentially facilitated when part of the surface parameters is determined from other experiments. In this case the photo e.m.f. data permit to obtain more exact information about the other parameters. For example, the value of C_{pII} was roughly estimated in [8] to be $\approx 10^{-17}$ to 10^{-14} cm². A comparison of the data [5] with formula (13) shows that the real value is $C_{pII} = 10^{-15}$ cm².

We express our gratitude to Prof. K.B. Tolpygo for a discussion of this paper.

Appendix A

Derivation of *B*

It is necessary to solve the system of linear equations which are obtained from (1) to (3) for region II:

$$\left.\begin{aligned} &D_{\rm n}\frac{\mathrm{d}}{\mathrm{d}x}\left(-\frac{\mathrm{d}\Delta n}{\mathrm{d}x}+n_0\frac{\mathrm{d}\Delta Y}{\mathrm{d}x}\right)+\frac{\Delta p}{\tau}+i\,\omega\,\Delta n=g_{\rm s}\,\mathrm{e}^{-k\,x}\,,\\ &-D_{\rm p}\frac{\mathrm{d}^2\Delta p}{\mathrm{d}x^2}+\left(i\,\omega+\frac{1}{\tau}\right)\Delta p=g_{\rm s}\,\mathrm{e}^{-k\,x}\,,\\ &\frac{\mathrm{d}^2\Delta Y}{\mathrm{d}x^2}=\varkappa^2\,n_{0\,\rm v}^{-1}\,(\Delta n-\Delta p)\,. \end{aligned}\right\} \tag{A1}$$

For a thick sample ($d \gg L_{\rm p}$) we have

$$\left.\begin{aligned} &\Delta n=C\,\mathrm{e}^{-\beta x}+\\ &+\frac{C_1\left(\varkappa^2-D_{\rm n}^{-1}\,\tau^{-1}\right)}{\beta^2-L_{\rm p}^{-2}}\,\mathrm{e}^{-x/L_p}+\frac{g_{\rm s}}{D_{\rm n}}\,\frac{1+\left(\varkappa^2-D_{\rm n}^{-1}\,\tau^{-1}\right)\left(L_{\rm p}^{-2}-k^2\right)}{\beta^2-k^2}\,\mathrm{e}^{-k\,x}\,,\\ &\Delta p=C_1\,\mathrm{e}^{-x/L_p}+g_{\rm s}\,D_{\rm p}^{-1}\,(k^2-L_{\rm p}^{-2})^{-1}\,\mathrm{e}^{-k\,x}\,, \end{aligned}\right\} \tag{A2}$$

where C, C_1 are integration constants, $\beta=\sqrt{\varkappa^2+\dfrac{i\,\omega}{D_{\rm n}}}$, $L_{\rm p}^{-1}=\sqrt{D_{\rm p}^{-1}\,(i\,\omega+\tau^{-1})}$, and D diffusion coefficient.
For $x = l$

$$\left(\frac{\mathrm{d}\Delta Y}{\mathrm{d}x}\right)_{x=l}=\frac{\varkappa^2}{\beta}\left[\frac{\Delta n_l}{n_{\rm v}^0}-\frac{\Delta p_l}{n_{\rm v}^0}\,\frac{\varkappa^2+D^{-1}\,\tau^{-1}+\beta\,L_{\rm p}^{-1}\left(D_{\rm p}\,D_{\rm n}^{-1}-1\right)}{\beta^2-L_{\rm p}^{-2}}\right]=\frac{\varkappa^2}{\beta}\,F\,. \tag{A3}$$

Linearization of the expression (4) for $x = l$ gives

$$\left(\frac{\mathrm{d}\Delta Y}{\mathrm{d}x}\right)_{x=l}=\frac{\varkappa^2}{\left(\dfrac{\mathrm{d}Y_0}{\mathrm{d}x}\right)_{x=l}}\left[\frac{\Delta n_l}{n_{0\,\rm v}}+\frac{\Delta p_l}{n_{0\,\rm v}}-\Delta Y_l+B\right]. \tag{A4}$$

Comparing (A4) and (A3) we define B as follows:

$$B=-\frac{\Delta p_l+\Delta n_l}{n_{\rm v}^0}+\Delta Y_l+\frac{\left(\dfrac{\mathrm{d}Y_0}{\mathrm{d}x}\right)_{x=l}}{\beta}\,F\,. \tag{A5}$$

The following relation is valid at surface potentials outside the inversion range:

$$E_0=-\frac{k\,T}{e}\,\frac{\mathrm{d}Y_0}{\mathrm{d}x}=\pm\frac{k\,T}{e}\,\varkappa\,[2\,(\mathrm{e}^{Y_0}-Y_0-1)]^{1/2}\,. \tag{A6}$$

Taking into account the condition $Y_{0l} \ll 1$, we obtain from (A6)

$$\left(\frac{\mathrm{d}Y_0}{\mathrm{d}x}\right)_{x=l}=-\varkappa\,Y_{0l}\,. \tag{A7}$$

Hence the last term in (A5) may be neglected in comparison with the first one. Using the assumptions 2. and 3. of this work, we find

$$B = -\frac{\Delta p_s e^{Y_{0s}} - \Delta n_s e^{-Y_{0s}}}{n_{0v}} - (\Delta Y_l - \Delta Y_s) e^{Y_{0l}} + \Delta Y_l . \quad (A8)$$

Appendix B

Derivation of Δn_s and Δp_s

We shall describe the exchange of carriers between the bands and the j-th surface level by the equation

$$\frac{dm_j}{dx} = r_{nj} \Delta n_s - g_{nj} m_j - r_{pj} \Delta p_s , \quad (B1)$$

where m_j is the change of the electron concentration on the j-th level due to illumination, r_{nj}, r_{pj} are the trapping rates of electrons and holes on the j-th level, ($r_{nj} = C_{nj} N_j f_{pj}$; $r_{pj} = C_{pj} N_j f_{nj}$), g_{nj} is the inverse lifetime of electrons and holes on the j-th level:

$$g_{nj} = C_{nj} n_{0s} f_{nj}^{-1} ; \quad f_{nj} = [1 + \exp(E_j - U_B - Y_{0s})]^{-1} ,$$

where U_B is the position of the Fermi level counted from the middle of the band gap in the bulk.

The expression (B1) is written for typical conditions (for Si and Ge), when the thermal excitation of minority carriers from traps can be neglected in comparison with the excitation of majority carriers.

It is possible to write the following boundary conditions for currents on the basis of B1):

$$- j_{ps} = \sum_j r_{pj} \Delta p_s , \quad (B2)$$

$$- j_{ns} = \sum_j \frac{g_{nj} r_{pj}}{g_{nj} + i\omega} \Delta p_s + i\omega \sum_j \frac{r_{nj}}{g_{nj} + i\omega} , \quad (B3)$$

and for the additional field

$$\left.\frac{d\Delta Y}{dx}\right|_{x=0} = \frac{\varkappa^2}{n_v^0} \left[\sum_j \frac{r_{nj}}{g_{nj} + i\omega} \Delta n_s - \sum_j \frac{r_{pj}}{g_{nj} + i\omega} \Delta p_s \right] . \quad (B4)$$

Integrating (1) from $x = 0$ to $x = l$ for the case of strong absorption ($k^{-1} \ll l$) we have

$$j_{ps} = j_{pl} + \int_0^l \frac{\Delta p}{\tau} dx - \frac{g_s}{k} , \quad (B5)$$

where Δp is defined on the basis of the assumptions 2. and 3. in the following form:

$$\Delta p_I = \Delta p_s e^{-Y_{0s} + Y_0} . \quad (B6)$$

The current $j_p(l)$ is calculated from (A2):

$$j_p(l) = -\frac{D_p}{L_p} C_1 e^{-l/L_p} . \quad (B7)$$

We determine C_1 by comparing the expressions for $\Delta p_{\rm I}(x)$ (B6) and $\Delta p_{\rm II}(x)$ (A2) at the point $x = l$:

$$C_1 = \mathrm{e}^{l/L_\mathrm{p}}\, \mathrm{e}^{Y_{0\mathrm{s}} - Y_{0l}}\, \Delta p_\mathrm{s}\,. \tag{B8}$$

Comparing (B5) and (B3) and using (B6, B8) we have

$$\Delta p_\mathrm{s} = \frac{\dfrac{g_\mathrm{s}}{k}}{\Sigma\, r_{\mathrm{p}j} + \dfrac{D_\mathrm{p}}{L_\mathrm{p}}\, \mathrm{e}^{Y_{0\mathrm{s}}} \left(1 + L^- \dfrac{L_\mathrm{p}^0}{L_\mathrm{p}}\right)}\,. \tag{B9}$$

Using the expressions (9), which is taken at the point $x = 0$, and (B4) one can derive Δn_s:

$$\Delta n_\mathrm{s} = \frac{\sum\limits_j \dfrac{r_{\mathrm{p}j}}{g_{\mathrm{n}j} + i\,\omega} + \dfrac{1 - \mathrm{e}^{Y_{0\mathrm{s}}}}{(\partial Y_0/\partial x)_{x=0}}}{\sum\limits_j \dfrac{r_{\mathrm{n}j}}{g_{\mathrm{n}j} + i\,\omega} - \dfrac{1 - \mathrm{e}^{-Y_{0\mathrm{s}}}}{(\partial Y_0/\partial x)_{x=0}}}\, \Delta p_\mathrm{s}\,. \tag{B10}$$

A more general approach to this problem is given in [7] and [9]. The following additional factors were taken into account: 1. thermal excitation of minority photo-carriers from traps into the V-band, 2. the finite thickness of the sample.

In the case of steady-state conditions and $k^{-1} \ll d$ the following expressions were obtained for Δp_s and Δn_s:

$$\Delta p_\mathrm{s} = \frac{g_\mathrm{s}}{k\, R\,(Y_{0\mathrm{s}})} \left\{ P(Y_{0\mathrm{s}})\, K(k) - \frac{k\, D_\mathrm{p}\, \mathrm{e}^{-kl}\, \mathrm{e}^{Y_{0\mathrm{s}}}}{\sinh^2 (d/L_\mathrm{p})\,(1 - k^2 L_\mathrm{p}^2)} \right\}, \tag{B11}$$

$$\Delta n_\mathrm{s} = \Delta p_\mathrm{s} \frac{\sum\limits_j \dfrac{r_{\mathrm{p}j}}{g_j} + \dfrac{1 - \mathrm{e}^{Y_{0\mathrm{s}}}}{E_{0\mathrm{s}}}}{\sum\limits_j \dfrac{r_{\mathrm{n}j}}{g_j} - \dfrac{1 - \mathrm{e}^{-Y_{0\mathrm{s}}}}{E_{0\mathrm{s}}}}\,, \tag{B12}$$

where

$$P(Y_{0\mathrm{s}}) = \frac{\mathrm{e}^{Y_{0\mathrm{s}}}}{\tau\, L_\mathrm{p}^{-1}} \left[L^- + \operatorname{ctgh} \frac{d}{L_\mathrm{p}} \right] + \sum_j \frac{g_{\mathrm{n}j}\, r_{\mathrm{p}j}}{g_j} +$$

$$+ \sum_j \frac{r_{\mathrm{n}j}\, g_{\mathrm{p}j}}{g_j} \frac{\sum\limits_j \dfrac{r_{\mathrm{p}j}}{g_j} - (1 - \mathrm{e}^{Y_{0\mathrm{s}}})\, E_{0\mathrm{s}}^{-1}}{\sum\limits_j \dfrac{r_{\mathrm{n}j}}{g_j} - (1 - \mathrm{e}^{-Y_{0\mathrm{s}}})\, E_{0\mathrm{s}}^{-1}}\,,$$

$$K(k) = 1 - 1 - k\, L_\mathrm{p} \operatorname{ctgh} \frac{d}{L_\mathrm{p}}\, \mathrm{e}^{-kl}\, (1 - k^2 L_\mathrm{p}^2)^{-1}\,,$$

$$R\,(Y_{0\mathrm{s}}) = P^2 - D_\mathrm{p}\, \mathrm{e}^{2 Y_{0\mathrm{s}}} \left[\tau \sinh^2 \frac{d}{L_\mathrm{p}} \right]^{-1},$$

$$E_{0\mathrm{s}} = \left(\frac{\mathrm{d} Y_0}{\mathrm{d}x} \right)_{x=0},$$

$$L^- = L_\mathrm{p}^{-1} \int\limits_0^l \mathrm{e}^{-Y_0}\, \mathrm{d}x\,,$$

$$g_j = g_{\mathrm{n}j} + g_{\mathrm{p}j} = C_{\mathrm{n}j}\, n_{0\mathrm{s}}\, f_{\mathrm{n}j}^{-1} + C_{\mathrm{p}j}\, p_{0\mathrm{s}}\, f_{\mathrm{p}j}^{-1}\,.$$

References

[1] V. E. Lashkarev, Zh. eksper. teor. Fiz. **18**, 917 (1948); Trudy Inst. Fiz. Akad. Nauk SSSR **3**, 3 (1952).
[2] K. B Tolpygo, Photoelectrical and Optical Phenomena in Semiconductors, Acad. Sci. Ukr. SSR, Kiev 1959 (p. 268) (in Russian).
[3] E. O. Johnson, Phys. Rev. **111**, 153 (1958).
[4] V. E. Primachenko and O. V. Snitko, Ukr. fiz. Zh. **5**, 488 (1960).
[5] V. M. Buimistrov, V. G. Litovchenko, and A. P. Gorban, Surface Sci. **3**, 445 (1965).
[6] F. Forlani and N. Minnaja, phys. stat. sol. **8**, 177 (1965).
[7] V. A. Zuev, A. Sachenko, and K. V. Tolpygo, Fiz. tverd. Tela **6**, 3174 (1964); Ukr. fiz. Zh. **10**, 275 (1965).
[8] V. G. Litovchenko, A. P. Gorban, and V. P. Kovbasyuk, Fiz. tverd. Tela **7**, 565 (1965).
[9] V. A. Zuev, A. V. Sachenko, and K. V. Tolpygo, Ukr. fiz. Zh. **10**, 1176 (1965).

(Received May 10, 1966)

phys. stat. sol. 16, 761 (1966)

Joint Institute for Nuclear Research, Dubna

On the Mechanism of Lithium Ion Drift in the Electrical Field of a p–n Junction in Silicon

By

A. S. ANTONOV

The mechanism of lithium ion drift in the electric field of a p-n junction is studied. In particular, the lithium concentration in the drift region and the change in the boundaries of the drift region with time are found. The distribution of the lithium concentration is calculated theoretically.

Der Driftmechanismus von Lithiumionen im elektrischen Feld des p-n Überganges von Silizium wird untersucht. Insbesondere werden die Konzentrationsverteilung des Lithiums im Driftgebiet sowie die Änderung der Grenzen des Driftgebietes in Abhängigkeit von der Zeit bestimmt. Die Konzentrationsverteilung des Lithiums wird außerdem theoretisch berechnet.

1. Introduction

The compensation of acceptors in p-silicon by lithium ion drift in the electrical field of a p-n junction was performed for the first time by Pell [1] in 1960. Since that time many publications on practical application of lithium drift have appeared. In general, this process is used for producing silicon p-i-n detectors which are widely used now in nuclear spectroscopy [2]. However, in spite of great success in this direction, the process of lithium ion drift has remained unexplained yet.

Pell [1] has found a law according to which the width of the lithium compensated region varies with respect to drift time. His results are in good agreement with experimental data on the p-n junction capacitance. However, Pell suggests that during the drift the position of the p-n junction is not varied.[1]) Besides, Pell does not determine the distribution of lithium ion concentration in the compensated region but assumes that it should be constant and equal to the acceptor concentration in the basic material.

Lehrer and Reiss [4] have given a general expression for the lithium ion concentration in the compensated region. However, to accomplish their calculation it is necessary to use experimental results, but the authors do not give them. Besides, the conclusions of Lehrer and Reiss are based on the assumption that the drift term in equation (3.26) of [4] is smaller compared to the diffusion term at the boundaries of the compensated region, and therefore the authors assume that the donor concentration in the drift term is equal to the acceptor concentration. Our investigations described below show, however, that this assumption is justified.

---

[1]) The development of junctions in silicion p-i-n detectors by electrochemical deposition of copper [3] shows, in particular, that the depth of the p-n junction is increased with increasing drift time.

This paper presents some experimental and theoretical investigations which shall clarify some basic laws of lithium ion drift in silicon.

2. Method

In reference [5] a method is described for determining the distribution of impurities in the region of the p–n junction in semiconductors. For the concentration $N = N_d - N_a$ (N_d is the donor concentration, N_a the acceptor concentration) the following expressions have been obtained:

$$|N(x_n)| = \left|\frac{dU}{dx_n}\right| \frac{\varepsilon}{4\pi e\,|x_p - x_n|}, \tag{1}$$

$$|N(x_p)| = \left|\frac{dU}{dx_p}\right| \frac{\varepsilon}{4\pi e\,|x_p - x_n|}, \tag{2}$$

where x_p and x_n are the coordinates of the boundaries of the negative and positive space charge regions (the origin of the coordinate system coincides with the boundary of the p–n junction). U is the reverse bias applied to the p–n junction, ε the dielectric permeability of the semiconductor, and e the electron charge.

From expressions (1) and (2) one can determine the distribution of the impurity concentration N in the region of the p–n junction, if it is known how the width of the regions of the positive and negative charges x_n and x_p vary with applied reverse bias. The dependence of the width of the negative space charge region upon reverse bias voltage $x_p = f(U)$ was found by means of a liquid electrophotographic developer [5]. The dependence $|x_p - x_n| = F(U)$ was found by measuring the capacitance of the p–n junction as a function of the reverse bias voltage (from the expression $C = \frac{\varepsilon S}{4\pi\,|x_p - x_n|}$ [6, 7], C is the capacitance and S the area of the p–n junction), hence, we find $x_n = g(U)$. Fig. 1 shows as an example the dependences $x_n = g(U)$ and $x_p = f(U)$, obtained for a silicon p–i–n detector.

The derivatives $\frac{dU}{dx_n}$ and $\frac{dU}{dx_p}$ can be obtained by graphical differentiation from the curves of the type shown in Fig. 1. Further, using (1) and (2) one can calculate the concentration N_d, and, consequently, the lithium concentration $N = N_d - N_a$ (since the acceptor concentration N_a in the basic material is known).

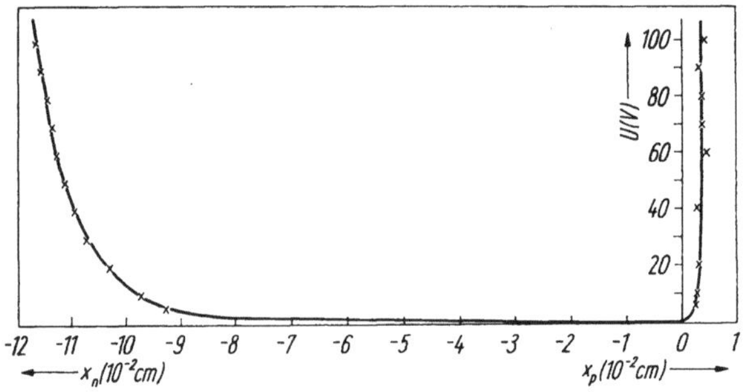

Fig. 1. The width of the regions of the positive and negative space charges x_n and x_p vs. reverse bias voltage U. Drift conditions: $T = 136\,°C$, $U = 100$ V, $t = 29$ h, boron concentration 4.9×10^{13} cm^{-3}

Simultaneously with the determination of the lithium concentration a study was made by electrochemical deposition of copper [3] how the position of the p–n junction and the boundary between the lithium compensated region and the highly lithium doped region varies with drift time. Such combined investigations allow to obtain all necessary values for describing lithium ion drift in silicon.

3. Experimental Results

Lithium diffusion and drift were performed by the method described in [1]. In particular, diffusion occured in argon at 180 to 560 °C for 3 to 30 min. The temperature of diffusion annealing was kept constant to an accuracy of ± 5 °C, whereas the diffusion time was measured with an error smaller than 6%.

Drift in samples occured at various temperatures and voltages in air of 1 Torr. The temperature during the drift was kept constant within ± 5 °C, whereas the drift voltage was kept constant to an accuracy of ± 10 V. The inital material was monocrystalline p-silicon containing boron $((1.8 \text{ to } 7) \times 10^{13}\ \text{cm}^{-3})$, the lifetime of the majority carriers being 85 to 145 s, the concentration of oxygen impurities being $< 10^{16}\ \text{cm}^{-3}$.

Fig. 2 shows how the depth of the p–n junction x_0 and the position of the boundary x' between the compensated region (the i-region) and the highly lithium doped region (the n-region) vary during the drift (x_0 and x' are counted from the sample surface). The curves 1 and 2 (1′ and 2′, respectively) refer to two samples in which diffusion took place under similar conditions ($T = 263$ °C, $t = 30$ min). The drift conditions are described in the captions of the figure. The error in the determination of the boundary position was not larger than 15 μm. The above results show that the boundary between the i- and n-regions is not practically changed during the drift and coincides with the initial position of the p–n junction after diffusion so that the compensated region is increased in the direction of drift in contrast to the assumptions made in [1] and [4].

In order to define the position of the right-hand side boundary x'' of the compensated region it is necessary to add x_0 to the width of the region of the negative space charge x_p obtained with a liquid electrochemical developer (see reference [5], and Fig. 1) at a voltage equal to the drift one. Thus, from the results on the electrochemical deposition of copper and on the development of the p–n junction

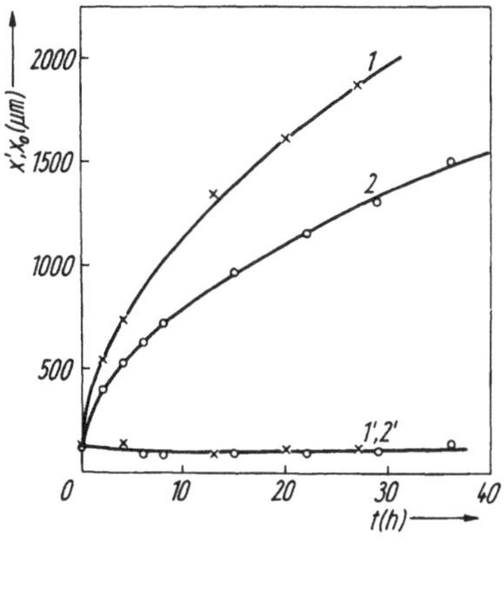

Fig. 2. The depths of the p–n junctions x_0 (curves 1 and 2) and the boundary between i- and n-regions x' (curves 1′ and 2′) vs. drift time. Drift conditions: $T = 136$ °C, $U = 200$ V, $N_a = 1.9 \times 10^{13}\ \text{cm}^{-3}$ (curves 1 and 1′) and $T = 136$ °C, $U = 100$ V, $N_a = 4.9 \times 10^{13}\ \text{cm}^{-3}$ (curves 2 and 2′)

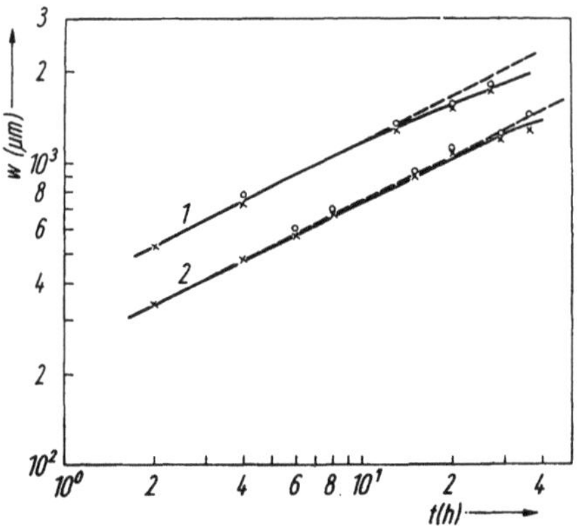

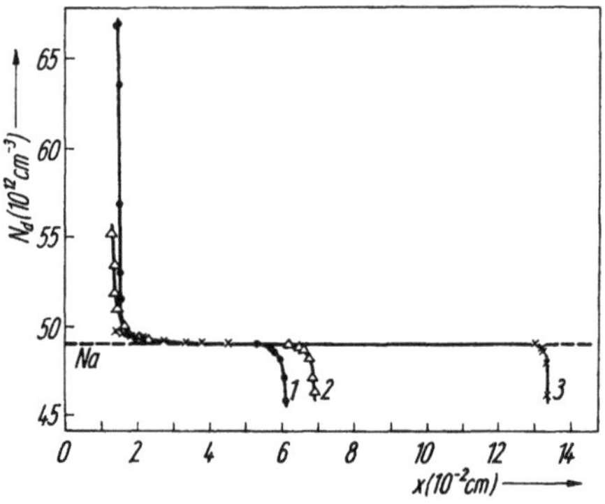

Fig. 3. Variation of the width of the lithium compensated region w vs. drift time. Values denoted by crosses have been obtained by measuring the p–n junction capacitance, whereas values denoted by circles have been obtained by developing the junctions by electrochemical deposition of copper and using a liquid electrophotographic developer. Continuous curves have been obtained experimentally, whereas dashed curves are straight lines with the angular coefficient 1/2. Drift conditions: $T = 136$ °C, $U = 220$ V, $N_a = 1.9 \times 10^{13}$ cm^{-3} (curve 1) and $T = 136$°C, $U = 100$ V, $N_a = 4.9 \times 10^{13}$ cm^{-3} (curve 2)

Fig. 4. Distribution of lithium concentration in the drift region. Drift conditions: $T = 136$ °C, $U = 100$ V, $N_a = 4.9 \times 10^{13}$ cm^{-3}. The drift time is 4 h (curve 1), 6 h (curve 2), and 29 h (curve 3)

with a liquid electrophotographic developer one can obtain the thickness of the lithium compensated region $w = x'' - x'$. It can also be obtained by measuring the capacitance of the p–n junction (see [1]) by using the expression $C = \frac{\varepsilon S}{4 \pi w}$. The error in the measurement of the compensated region w by both methods turned out to approximately the same and smaller than 6%.

Fig. 3 shows for two samples the variation of the width of the compensated region w found by the two above methods during drift. As is seen from the figure, both methods give identical results within the experimental error. The dashed line in the figure is the angle coefficient 1/2. It is seen that the width of the compensated region depends to a good approximation upon drift time according to the law $w = \text{const} \cdot t^{1/2}$. This dependence has been obtained by Pell [1]

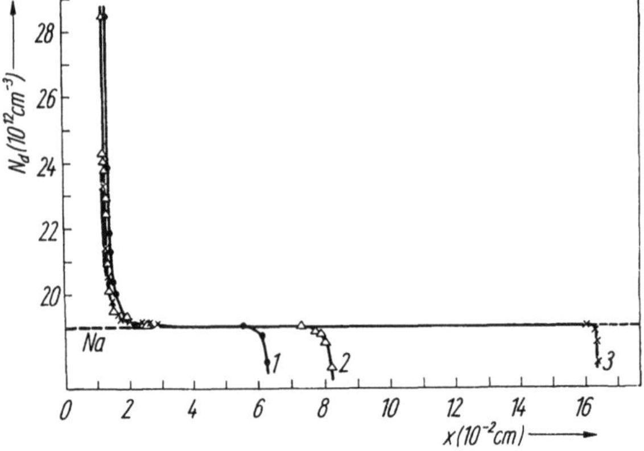

Fig. 5. Distribution of lithium concentration in the drift region. Drift conditions: $T = 136$ °C, $U = 200$ V, $N_a = 1.9 \times 10^{13}$ cm^{-3}. The drift time is 2 h (curve 1), 4 h (curve 2), and 20 h (curve 3)

for long drift times. Small deviations from this dependence according to our experimental data are observed at drift times longer than 12 h.

The variation of lithium concentration distribution during drift is shown in Fig. 4 and 5 for two samples. The measurements were made by the method described in Section 2. The distance x_0 from the crystal boundary surface to the p-n junction was found from the development of the p-n junction by the electrochemical deposition of copper [3]. In both samples lithium diffusion occurred under similar conditions ($T = 163\,°C$, $t = 30\,min$). Drift conditions are given in the captions.

The following conclusions can be drawn from the above results: 1. Within the experimental error (which in this case was 30 μm) the position of the boundary between the i- and n-regions is not varied during drift (as is the case with the electrochemical deposition of copper, see Fig. 2). 2. The lithium concentration in the direct vicinity of the above boundary is decreased during drift. 3. In the region of the drift front the lithium concentration sharply drops to zero. (It should be noted that the accuracy of the applied method for the determination of the impurity concentration in the p-n junction did not allow a determination of the lithium distribution over the whole drift front since at voltages close to the drift voltage the region of the negative space charge remains nearly the same — see Fig. 1.)

4. Theory of Lithium Ion Drift

4.1 Initial equations and assumptions

The process of lithium ion drift can be described by the following systems of equations:

$$\frac{\partial N_d}{\partial t} = D \frac{\partial^2 N_d}{\partial x^2} - \mu E \frac{\partial N_d}{\partial x} - \mu N_d \frac{\partial E}{\partial x}, \tag{3}$$

$$\frac{\partial E}{\partial x} = \frac{4 \pi e}{\varepsilon} (N_d - N_a), \text{ }^{2)} \tag{4}$$

and the boundary conditions

$$E(x') = E(x'') = 0, \tag{5}$$

$$\int_{x'}^{x''} (N_d - N_a)\, dx = 0, \tag{6}$$

$$\int_{x'}^{x''} E\, dx = U, \tag{7}$$

where E is the electrical field in the p-n junction, D the coefficient of lithium ion diffusion, and μ their mobility.

To simplify equation (3) its terms were evaluated by using the experimental results on lithium concentration in the drift region. The electrical field was calculated by a graphical integration of equation (4) when the boundary conditions (5) and (6) were used. The results of the evaluation of two various drift condi-

[2]) An approximate evalution has been made in agreement with [8] which showed that under real drift conditions one can neglect the contribution of free carriers in the Poisson equation.

Table 1

Evaluations of the terms of equation (3) under the following drift conditions: $T = 136\ °C$, $t = 4$ h, $U = 200$ V. The p-silicon monocrystal has a boron impurity concentration of $N_a = 1.9 \times 10^{13}\ cm^{-3}$ and on oxygen impurity concentration $< 10^{16}\ cm^{-3}$

$10^2 x$ (cm)	$10^{-12} \frac{D}{\mu} \frac{\partial^2 N_d}{\partial x^2}$ (CGS units)	$10^{-12} N_d \frac{\partial E}{\partial x}$ (CGS units)	$10^{-12} E \frac{\partial N_d}{\partial x}$ (CGS units)
1.16	864	147000	0
1.25	864	74000	−33700
1.4	82.2	4610	− 3300
1.5	8.32	2870	− 765
2	1.07	1196	− 204
2.5	0.34	561	− 24.1
4	<0.01	289	− 8.00
7.35	<0.01	0	− 8.34
7.75	−3.38	− 1250	- 471
8	−6.89	− 4110	1020
8.25	−136	−10540	− 3760

Table 2

Evaluation of the terms of equation (3) under the following drift conditions: $T = 136\ °C$, $t = 4$ h, $U = 100$ V. The p-silicon monocrystal has a boron impurity concentration of $N_a = 4.9 \times 10^{13}\ cm^{-3}$ and an oxygen impurity concentration $< 10^{16}\ cm^{-3}$

$10^2 x$ (cm)	$10^{-12} \frac{D}{\mu} \frac{\partial^2 N_d}{\partial x^2}$ (CGS units)	$10^{-12} N_d \frac{\partial E}{\partial x}$ (CGS units)	$10^{-12} E \frac{\partial N_d}{\partial x}$ (CGS units)
1.39	5890	840 000	0
1.44	5820	280 000	− 166000
1.49	2040	97 700	− 28000
1.54	820	38 900	− 5220
1.79	9.70	7 910	− 1140
2.29	2.90	3 050	− 58.3
3.79	<0.1	559	− 29.2
5.29	<0.1	0	− 29.8
5.54	−2.90	− 5180	− 710
5.79	−17.2	−11880	− 1560
6.04	−12.9	−31700	− 8130

tions are shown in Tables 1 and 2. The first two columns give the coordinate x (framed are the coordinates x' and x_0), the remaining columns give separate terms of equation (3). As it is seen from the tables one can neglect the diffusion term compared to the terms describing lithium ion drift in the electrical field. Then instead of equation (3) the following equation is obtained:

$$\frac{\partial N_d}{\partial t} = -\mu E \frac{\partial N_d}{\partial x} - \mu N_d \frac{\partial E}{\partial x} . \tag{3'}$$

4.2 On the quasi-stationary distribution of lithium concentration in the region of the coordinates (x', x_0)

As is seen from the curves shown in Fig. 4 and 5, the lithium concentration N_d remains nearly constant with time in the region of the coordinates (x', x_0) (from the left boundary of the drift region to the boundary of the p-n junction). Therefore, we can put in (3′) the derivative $\frac{\partial N_d}{\partial t} = 0$.[3]) Then, in the above region the equation

$$E \frac{\partial N_d}{\partial x} + N_d \frac{\partial E}{\partial x} = 0 \tag{3''}$$

will be valid instead of (3′).

By using the substitutions

$$\left.\begin{aligned} x &= \left(\frac{4 \pi e^2 N_a}{\varepsilon k T}\right)^{-1/2} \xi , \quad t = \frac{\varepsilon}{4 \pi \mu e N_a} \tau , \\ N_d &= N_a n , \quad E = \left(\frac{4 \pi N_a k T}{\varepsilon}\right)^{1/2} \mathscr{E} , \quad U = \frac{D}{\mu} \Psi , \end{aligned}\right\} \tag{8}$$

where k is the Boltzmann constant, the system of equations (3″) and (4) and the boundary conditions (5), (6), and (7) are reduced to the following dimensionless form:

$$\mathscr{E} \frac{dn}{d\xi} + n \frac{d\mathscr{E}}{d\xi} = 0 , \text{ [4])} \tag{9}$$

$$\frac{d\mathscr{E}}{d\xi} = n - 1 , \tag{10}$$

$$\mathscr{E}(\xi') = \mathscr{E}(\xi'') = 0 , \tag{11}$$

$$\int_{\xi'}^{\xi''} (n - 1) \, d\xi = 0 , \tag{12}$$

$$\int_{\xi'}^{\xi''} \mathscr{E} \, d\xi = \Psi , \tag{13}$$

where ξ' and ξ'' are found from the first substitution (8) with $x = x'$ and $x = x''$.

It follows from equation (9) that

$$\mathscr{E} n = \alpha , \tag{14}$$

---

[3]) This assumption (however, over the whole drift region) is made by the authors of [4].

[4]) The partial derivatives are replaced by ordinary ones since their dependence upon the time coordinate τ enters only under boundary conditions.

where α is a value depending only upon τ.

From equations (9), (10), and (14) one can obtain the following equation for n:

$$\alpha \frac{dn}{d\xi} = -(n-1)\, n^2\,, \tag{15}$$

to which we add the boundary condition

$$n(\xi') = n'\,. \tag{16}$$

The solution of equation (15) with the boundary condition of equation (16) is written in the following form:

$$-\alpha\left(\frac{1}{n} - \frac{1}{n'}\right) - \alpha \ln \frac{1 - \frac{1}{n}}{1 - \frac{1}{n'}} = \xi - \xi'\,. \tag{17}$$

Equation (17) together with equations (16) and (8) give the quasi-stationary distribution of lithium concentration in the (x', x_0) region.

Equation (17) is not solvable with respect to n. It follows from this equation that with $\xi \to \infty$, $n \to 1$. Since in our example n should be equal to unity for $\xi = \xi_0$[5]), then for the applicability of the quasi-stationary solution of (17) it is required that n differs sufficiently little from unity for $\xi = \xi_0$. Consider the quasi-stationary solution for $n \to 1$. In this case one can neglect on the left-hand side of (17) the first term compared to the secondary one and then one obtains

$$n = \frac{1}{1 - \frac{n'}{n'-1} \exp\left(-\frac{\xi - \xi'}{\alpha}\right)}\,.$$

Since usually $n' > 1$ (this follows from the experimental results shown in Fig. 4 and 5 with the substitution (8)) the condition for the applicability of the quasi-stationary solution is identical to the requirement $\frac{\xi_0 - \xi'}{\alpha} \gg 1$. However, $\xi_0 \approx \xi''$[6]), and therefore the previous inequality can be written as

$$\frac{\xi'' - \xi'}{\alpha} \gg 1\,. \tag{18}$$

Assume further that total compensation of impurities takes place over the whole drift region, i.e. $n = 1$. Then from (10), (13), and (14) it follows that

$$\alpha = \frac{\Psi}{\xi'' - \xi'}\,. \tag{19}$$

From (18), (19), and (8) we obtain that the condition for the applicability of the quasi-stationary solution is expressed by the inequality

$$w^2 \gg \frac{\varepsilon\, U}{4\pi\, e\, N_a}\,, \tag{20}$$

where w is the width of the compensated region.

---

[5]) ξ_0 is found from the first substitution (8) with $x = x_0$.

[6]) Usually, the relation $x_n \gg x_p$ holds (see Fig. 1). Since $x_n = x_0 - x'$, whereas $x_p = x'' - x_0$, it follows that $x_0 \gg x'' - x_0$ or $x_0 \approx x''$. Using the first substitution (8) we obtain $\xi_0 \approx \xi''$.

Under the assumption of total impurity compensation one can determine from the equation

$$w\,N_a = \int_0^t \mu\,E\,N_a\,dt \tag{21}$$

which expresses the dependence of lithium concentration in the compensated region $w\,N_a$ upon the flux of lithium ions $\mu\,E\,N_a$.

From (21) and (20) we obtain another form of the condition for the applicability of the quasi-stationary solution:

$$t \gg \frac{\varepsilon}{8\,\pi\,e\,\mu\,N_a}, \tag{22}$$

where t is the drift time.

With $T = 136$ °C and $N_a = 1.9\times10^{13}$ cm^{-3} (22) gives $t \gg 30$ s, whereas with $N_a = 4.9\times10^{13}$ cm^{-3} we obtain $t \gg 12$ s. Therefore, the quasi-stationary solution (17) is applicable to the results shown in Fig. 4 and 5.

The quasi-stationary solution (17) allows (with the given boundary ξ') a determination of the distribution of lithium concentration n as a function of ξ. The solution (17) depends upon two constants n' and α (with respect to the coordinate ξ). In Fig. 6 and 7 the theoretical distributions of lithium concentration for two drift cases are plotted in the coordinates. The constants n' and α are calculated from the experimental curves in Fig. 4 and 5. (More exactly the constant n' is determined from the value of the lithium concentration N_d at the boundary $x = x'$ when the third substitution (8) is used, whereas the constant α is calculated by the graphical solution of the Poisson equation (8) and with the help of the fourth substitution (8). For sake of comparison the experimen-

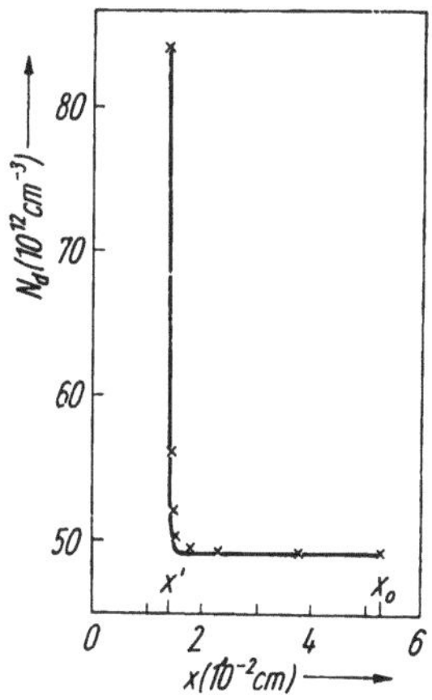

Fig. 6. Theoretical distribution of lithium concentration (continuous curve) in the coordinate region (x', x_0). Crosses denote experimental results. Drift conditions: $T = 136$ °C, $U = 100$ V, $N_a = 4.9\times10^{13}$ cm^{-3}, $t = 4$ h

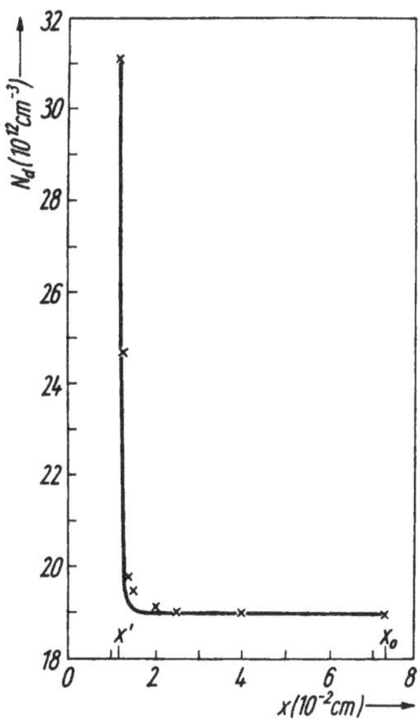

Fig. 7. Theoretical distribution of lithium concentration (continuous line). Drift conditions: $T = 136$ °C, $U = 200$ V, $N_a = 1.9\times10^{13}$ cm^{-3}, $t = 4$ h. Crosses denote experimental results

tally obtained values for the lithium concentration N_d taken from the curves 2 in Fig. 4 and 5 are denoted by crosses in Fig. 6 and 7. A small discrepancy between theory and experiment which is observed in the region of the concentration curve bending appears to be due to the neglection of the diffusion terms in the drift equation (3).

5. Conclusion

Experimental and theoretical investigations on the mechanism of ion lithium drift in p-type silicon are described. In particular, it is shown by electrochemical deposition of copper and also from measurements of the p–n junction capacitance that the lithium compensated region increases only in drift direction. By measuring the capacitance of the p–n junction and the development of the region of negative space charge with a liquid developer, the distribution of lithium concentration in the drift region has been obtained.

Separate terms of the drift equation have been evaluated, and it has been shown that over the larger part of the drift region one can neglect the diffusion term compared to terms describing lithium ion drift in the electrical field. Basing on this assumption and under the supposition that the contribution of free carriers can be neglected in the Poisson equation, the distribution of lithium concentration in the drift region from the coordinate x' to the coordinate x_0 has been found theoretically.

The theoretical calculations are compared to the experimental data obtained.

References

[1] E. M. PELL, J. appl. Phys. **31**, 291 (1960).
[2] N. A. BAILEY, J. W. MAYER, and R. J. GRAINGER, IRE Trans. No. 9, 91 (1962).
[3] A. S. ANTONOV, B. P. OSIPENKO, and L. G. YUSKESELIEVA, Soviet Phys. – J. phys. Chem. **39**, 2252 (1965).
[4] F. A. LEHRER and H. REISS, J. appl. Phys. **33**, 2353 (1962).
[5] A. S. ANTONOV and L. G. YUSKESELIEVA, Dokl. Akad. Nauk SSSR **164**, 1260 (1965).
[6] L. S. BERMAN, Nonlinear Semiconductor Capacitance, Fizmatgiz, Moscow 1963 (in Russian).
[7] A. S. ANTONOV, preprint 2029, Dubna 1965.
[8] S. P. MORGAN and F. M. SMITS, Bell Syst. tech. J. **39**, 1573 (1960).

(Received May 11, 1966)

phys. stat. sol. **16**, 771 (1966)

Institute of Physics of the R.S.R. Academy, Bucharest

Some Results Concerning the Co^{++} Site Symmetry in Alkali Halide Crystals

By

M. MUSA

Spectral studies are made of the Co^{++} ion site symmetry in LiCl and KBr host lattices. It is found that in LiCl and KBr the Co^{++} ion symmetry is octahedral and tetrahedral respectively. This conclusion is supported by a comparison with $CoCl_2$, Cs_3CoCl_5, and Cs_2CoBr_4 spectra. More detailed structures are found in the spectrum of a RbCl–Co(Zn)Cl_2 sample as compared with those exhibited by RbCl-Co^{++} single crystals grown from melt.

Es werden Untersuchungen über die Symmetrie des Co^{++}–Ionen-Platzes in LiCl- und KBr-Wirtsgittern durchgeführt. Es wird gefunden, daß in LiCl und KBr die Co^{++}–Ionen-Symmetrie oktaedrisch bzw. tetraedrisch ist. Diese Folgerung wird durch Vergleiche mit $CoCl_2$, Cs_3CoCl_5 und Cs_2CoBr_4 Spektren bestärkt. Detailliertere Strukturen werden im Spektrum von RbCl-Co(Zn)Cl_2 gefunden, im Vergleich zu RbCl-Co^{++} Einkristallen, die aus der Schmelze gezüchtet wurden.

1. Introduction

Recently (1963) Washimiya [1] has given theoretical support to the experimental work of Nasu und Kuwabara [2] indicating tetrahedral symmetry of the chlorine ions in the first coordination sphere of the Co^{++} ion in KCl single-crystals. Hills [3] has confirmed the theoretical [1] and experimental [2] results on KCl-Co^{++} doped crystals.

In a short note [4] we have concluded that the spectra of KCl-Co^{++} and RbCl-Co^{++} single-crystals are characteristic of tetrahedral symmetry around the Co^{++} ion; moreover, the diffuse aspect of the samples and the position of the maxima in their spectra have suggested to us that the small colored regions in the studied crystals are in fact double-salt precipitates such as M_2CoCl_4 or M_3CoCl_5, where M stands for K or Rb.

The aim of the present paper is to find out how the divalent Co^{++} ion enters other alkali halide crystals. The studied samples are LiCl-Co^{++}, RbCl-Co (Zn)Cl_2, Cs_2CoBr_4, and KBr-Co^{++}.

To support the tetrahedral type of symmetry we have compared the spectra with those obtained on Cs_3CoCl_5 single-crystals, which have this kind of arrangement around the Co^{++} ion, as is unanimously accepted [5 to 9]. To support the octahedral symmetry (in LiCl) we have made a comparison with the spectrum obtained on polycrystalline $CoCl_2$ layers in which the chlorine ions in the first coordination sphere have this symmetry (see [10] and also [11] and [12]).

Since in Cs_2CoBr_4 there is also tetrahedral symmetry around the Co^{++} ion [6] we have compared this spectrum with that of Cs_3CoCl_5, and, on the other hand, have used it to discuss the tetrahedral symmetry in alkali bromides, especially KBr-Co^{++}.

Our results obtained on single-crystals agree with those found by Tananaev et al. [13] on polycrystalline layers of alkali chlorides containing cobalt chlorides in known ratios.

2. Experimental

The single-crystal samples were obtained as follows:

Aqueous solutions of the salts taken in appropriate amounts to permit the growth of Cs_3CoCl_5, Cs_2CoBr_4, or RbCl-Co(Zn)Cl_2 single-crystals (with alkali halides in excess) were evaporated at room temperature, and single-crystals, homogeneously colored, of 5 to 10 mm^2 useful area were obtained. The chemical analysis of these compounds showed the following ratios: for Cs_3CoCl_5, Cs:Co:Br = 2.86:1:4.94, and for Cs_2CoBr_4, Cs:Co:Br = 2.45:1:4.46. Though Cs_3CoBr_5 seems to be formed too (about 45%) we shall designate the last sample as Cs_2CoBr_4. The RbCl-Co(Zn)Cl_2 single-crystals were prepared by adding a small amount of $ZnCl_2$ to a solution of RbCl and $CoCl_2$ (being in a proportion to form Rb_2CoCl_4); so hexagonal plates of about 8 mm^2 useful area were obtained. They were not as clear as the other crystals (Cs_3CoCl_5 or Cs_2CoBr_4) and after some time grew mat at the surface; their chemical structure was not established, so further we shall designate them as RbCl-Co(Zn)Cl_2 samples.

Before the measurements the samples were polished to suitable thickness.

The KCl-Co^{++} and KBr-Co^{++} single-crystals grown by the Czochralski method were obtained in argon atmosphere in our laboratory and contained $3^0/_{00}$ (weight) $CoCl_2$ or $CoBr_2$. The above mentioned crystals exhibit small deep colored regions which scatter light, so on these samples we have measured reflection spectra which show broadened and less structured bands.

The RbCl-Co^{++} and LiCl-Co^{++} samples (single-crystals grown by the Bridgman method) contained $3^0/_{00}$ $CoCl_2$. The crystals were obtained from the Institute of Solid State Physics in Prague, Czechoslovakia, and a detailed description of the growth method may be found in [14]. The LiCl-Co^{++} was transparent enough, without visible colored regions as exhibited by RbCl-Co^{++} samples grown under the same conditions.

The transmission and reflection spectra were measured at liquid nitrogen temperature in the spectral range 13,000 to 27,000 cm^{-1}. In the figures the transmission or reflection spectra of the samples are presented in the form: photocurrent response of the detector (in arbitrary units) versus wave number (in cm^{-1}).

The experimental technique has been described elsewhere [1, 10].

3. Results and Discussions

McClure [15] has shown that the crystal-field parameter $D\,q$ has to be multiplied by the factors 6 and 12 for the octahedral and tetrahedral symmetry, respectively, to determine the difference of the site energies for the $3d^7$ configuartion. Taking $D\,q_{\text{oct.}} = 690$ cm^{-1} and $D\,q_{\text{tetra.}} = 290$ cm^{-1} for chlorine ions as ligands of the Co^{++} ion from the papers of Ferguson et. al. [11, 17], we have estimated the preference energy of the octahedral site to be 1.84 kcal/mol. As this value is small in comparison with those found for other elements in the first transition group [15, 16] we may suppose that other facts could explain the behaviour of the Co^{++} ion in alkali halide single-crystals. For instance, we can analyse the parameters characterizing the host lattices. So from Table 1 it follows that the Co^{++} ion can easily substitute Li^+ or even Na^+ having nearly the same radii. Ferguson states that $D\,q_{\text{tetra.}}$ is roughly equal to $4/9\ D\,q_{\text{oct.}}$ for chlorine as ligand, i.e. the Co-Cl distance for an octahedral complex must be nearly the same

Table 1

Some parameters of the host lattices

Host lattice	Cation radius (Å)	Distance between nearest neighbours ($\frac{1}{2}$ (lattice constant)) (Å)	Co-Cl distance (Å) [2]
LiCl	0.78	2.56	1.81
NaCl	0.98	2.81	2.44
KCl	1.33	3.14	2.72
RbCl	1.48	3.27; 3.24	2.83; 2.81
CsCl	1.67	3.51; 3.56	2.91; 3.05
$[CoCl_6]^{4-}$ [1]	0.72		2.34
$[CoCl_4]^{2-}$ [1]	0.72		2.34

[1]) In $CoCl_2$ and $Cs_2(Co)ZnCl_4$ [11, 17].
[2]) If the Co^{++} ion occupies an interstitial site in the host lattice.

as that in Cs_3CoCl_5, where it is accepted to be 2.34 Å. So if the Co^{++} ion substitutes Li^+, the 2.56 Å distance between the nearest neighbours in the LiCl host lattice is even too large in comparison with the 2.34 Å existing in the $[CoCl_6]^{4-}$ complex. On the other hand, an interstitial position of the Co^{++} ion in LiCl would require a dilatation of the host lattice to keep the same Co-Cl distance.

Beginning with NaCl, the interstitial position seems to be more favourable than the substitutional one, and so passing from Na to Cs, a tetrahedral complex rather than an octahedral must be found.

Fig. 1 presents the spectrum of the LiCl-Co^{++} sample a) and that of the polycrystalline $CoCl_2$ layer. The general aspect and the maximum positions of the two spectra seem to be nearly identical. So we may conclude that in LiCl the site symmetry of the Co^{++} ion is octahedral, but the octahedron is deformed with respect to the pure $CoCl_2$ lattice: the chlorine ions in the first coordination sphere approach the Co^{++} to a distance which is nearly equal to that existing in polycrystalline $CoCl_2$ layers. The octahedron in LiCl must be more deformed than that

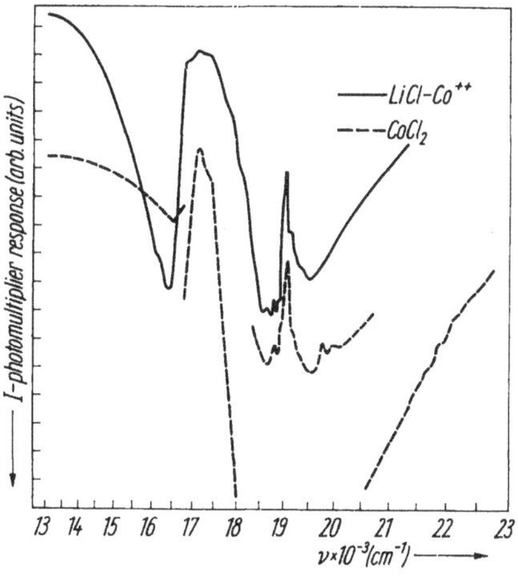

Fig. 1. Spectra of LiCl-Co^{++} and $CoCl_2$ samples

Table 2

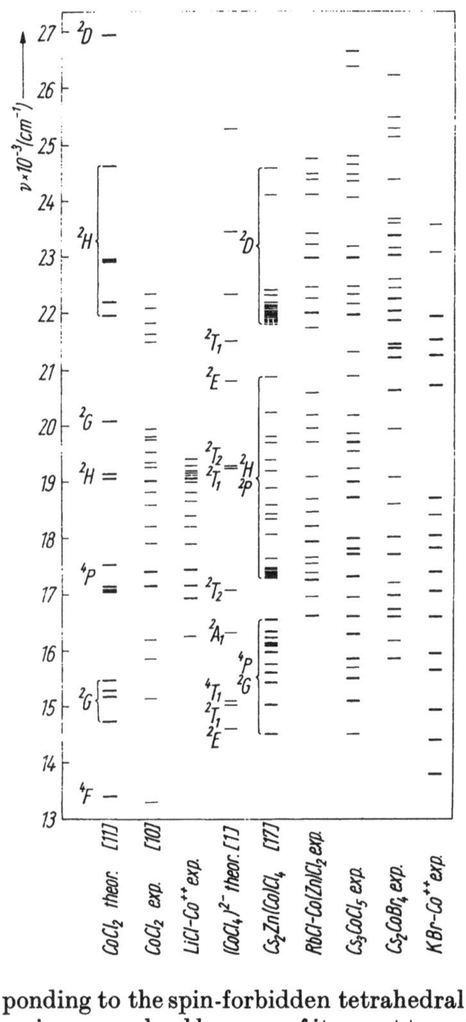

in $CoCl_2$ because the divalent Co^{++} requires a positive ion hole, probably in the second coordination sphere, to satisfy the neutrality condition. Perhaps the lower symmetry component of the crystal field favours the appearance of a new component in the principal band corresponding to the spin-allowed transition $T_1(^4F)-T_1(^4P)$ situated at about 17,000 cm^{-1}, and, on the other hand, bleaches the bands arising from the ^2G doublet (which is well resolved and situated at 19,000 cm^{-1} in the $CoCl_2$ spectrum) and the bands corresponding to the ^2H doublet (situated at about 22,000 cm^{-1} in $CoCl_2$, spectrum b).

Table 2 colletcts all these results and other data which are discussed in the following.

In [4] we have studied the reflection spectra of RbCl-Co^{++} single-crystals in comparison with the spectrum obtained on a Cs_3CoCl_5 single-crystal grown from the solution. Here we give the transmission spectrum of a RbCl-Co(Zn)Cl_2 sample in comparison with the reflection spectrum b) obtained on a RbCl-Co^{++} single-crystal grown from the melt [1]. It must be under lined that the spectrum a) has a clearer structure than b) in the range 17,000 to 27,000 cm^{-1} corresponding to the spin-forbidden tetrahedral transitions. The spin-allowed band remains unresolved because of its great transition probability and the low sensitivity of the photomultiplier in this spectral region (13,000 to 16,000 cm^{-1}). At the same time the spectra exhibit a structure which seems to have a vibrational nature. (A similar structure can be observed in the Cs_2CoBr_4 spectrum as well.) The identical maximum positions and the general aspect of the spectra a) and b) allow us to conclude that the small colored regions in the single-crystals grown from the melt are due probably to the instability of the solutions at the growth temperature, and that precipitates form in the host lattice.

From light scattering measurements made on KCl single-crystals doped with Ca, Sr, and Pb, Plint and Sibley [18] state that the divalent impurities precipitate in small spherical regions or concentrate along dislocations. From our spectral data we cannot say anything about the structure of the small light-scattering regions, but the resemblance of the spectra with those of samples having a defin-

Fig. 2. Spectra of RbCl–Co++ and RbCl–Co(Zn)Cl_3 samples

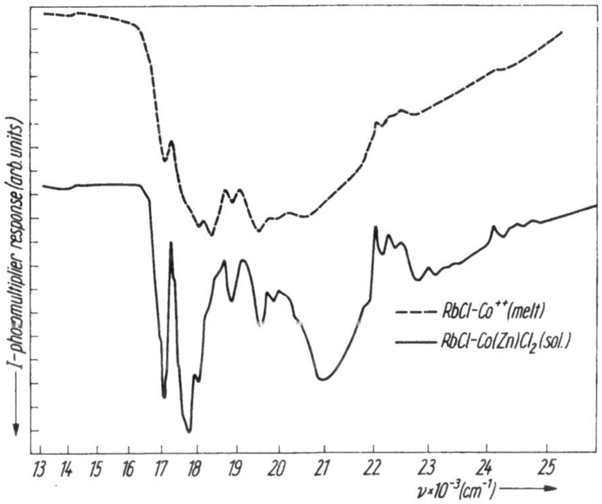

ite structure, such as Cs_3CoCl_5 or Cs_2CoBr_4, allow us to suppose that here we have to deal with unknown chemical compounds which retain, at least for Rb and K, tetrahedral symmetry of the first coordination sphere. Probably they are double salts such as M_2CoX_4 or M_3CoX_5, where M stands for K or Rb and X for Cl or Br, and the transition from the host structure to the new one (of the unknown compound) is accompanied by great distortions of the structures of the two pure compounds.

Fig. 3 shows the spectra of a Cs_3CoCl_5 sample a), of a $CsCoBr_4$ sample b), and of a KBr-Co^{++} single-crystal c). One may observe a correspondence between the band groups of a) and b) (see also Table 2) the bands being shifted, as aspected, towards longer wavelengths when passing from Cl to Br. The Cs_2CoBr_4 spectrum presents a more pronounced structure than that of the Cs_3CoCl_5 sample.

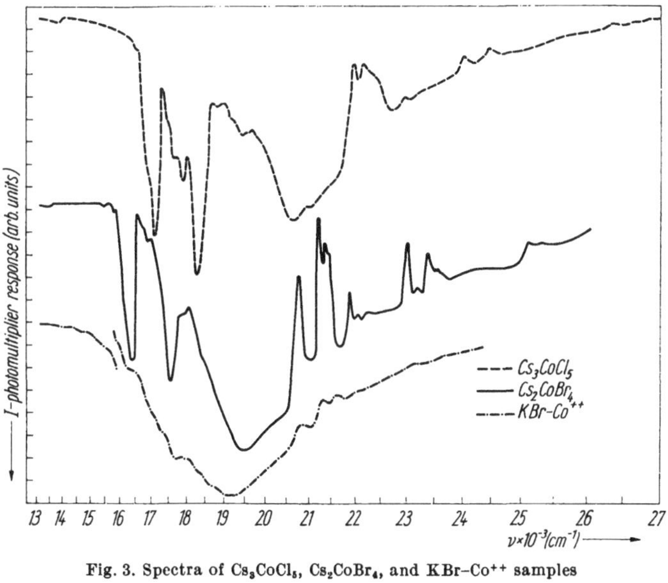

Fig. 3. Spectra of Cs_3CoCl_5, Cs_2CoBr_4, and KBr–Co^{++} samples

The reflection spectrum of the KBr-Co^{++} sample is less resolved (like that of the chlorine salt [4]), the band groups corresponding well to those found in the spectrum b). As we know the crystal structure of Cs_3CoCl_5, Cs_2CoCl_4, and Cs_2CoBr_4 [5 to 7] and may suppose that our salts have the same symmetry around the Co^{++} ion (slightly deformed tetrahedron), a comparison of the spectra b) and c) shows that the bands are due to the same complex species. Therefore we can conclude that the spectrum c) is due to a tetrahedral complex, too. Like the chlorine salt, the KBr-Co^{++} sample exhibits small regions which are blue-green colored.

Table 2 collects the data concerning the maximum positions of the discussed spectra and other available theoretical or experimental results which allow us to make a classification, at least of the band groups, on the basis of the energetic states of the free Co^{++} ion.

4. Conclusions

We have made spectral measurements on a LiCl-Co^{++} single-crystal sample, and found that the site symmetry in this host lattice is octahedral.

A clearer spectrum was obtained in the range of the tetrahedral forbidden transitions for a RbCl-Co(Zn)Cl_2 single-crystal grown from the solution, in comparison with the reflection spectrum of a RbCl-Co^{++} single-crystal sample grown from the melt which exhibits small colored regions. The identical position of the maxima supports the assumption that these small regions are precipitates which retain a tetrahedral symmetry around the Co^{++} ion.

Comparing the spectra of some Cs_3CoCl_5 and Cs_2CoBr_4 single-crystal samples grown from the solution, and of KBr-Co^{++} grown from the melt, we have shown that in the last sample the symmetry around the Co^{++} ion is tetrahedral, too.

Acknowledgements

I should like to thank Dr. A. Trutia for many useful discussions, Dr. M. Löbl for providing the LiCl-Co^{++} and RbCl-Co^{++} single-crystal samples and for helpful correspondence concerning some of the results, and Dr. G. Hlevka for performing a chemical analysis.

References

[1] S. WASHIMIYA, J. Phys. Soc. Japan **18**, 1719 (1963).
[2] NASU and KUWABARA, Nihom. Butzuri gakkai, No 4, 302 (1963).
[3] M. HILLS, J. Phys. Soc. Japan **19**, 760 (1964).
[4] M. MUSA, phys. stat. sol. **9**, K175 (1965).
[5] H. POWELL and A. WELLS, J. Chem. Soc. 359 (1935).
[6] M. PORAI-KOSHITZ, Trud. Inst. Krist. **100**, 117 (1954).
[7] G. TISCHENCO and Z. PINSKER, Dokl. Akad. Nauk SSR **100**, 913 (1955).
[8] L. KATZIN, J. Amer. Chem. Soc. **76**, 3089, (1954).
[9] N. ALLARD-PELLETIER, C.R. Acad. Sci. (France) **260**, 2170 (1965).
[10] A. TRUTIA and M. MUSA, phys. stat. sol. **8**, 663 (1965).
[11] J. FERGUSON, D. WOOD, and K. KNOX, J. chem. Phys. **39**, 881 (1963).
[12] J. SLIKER, Phys. Rev. **130**, 1749 (1963).
[13] I. TANANAEV and B. DJURINSKI, Dokl. Akad. Nauk SSR **139**, 120 (1961).
[14] M. LÖBL and I. TRNKA, Z. Physik **186**, 128 (1965).
[15] D. MCCLURE, J. Phys. Chem. Solids **3**, 311 (1957).
[16] I. DUNITZ and L. ORGEL, J. Phys. Chem. Solids **3**, 318 (1957).
[17] J. FERGUSON, J. chem. Phys. **39**, 116 (1963).
[18] C. PLINT and W. SIBLEY, J. chem. Phys. **42**, 1378 (1965).

(Received May 18, 1966)

phys. stat. sol. 16, 777 (1966)

Institut für Physikalische Chemie der Bulgarischen Akademie der Wissenschaften, Sofia

Elektronenmikroskopische Untersuchung der Wachstumsmorphologie von Zink- und Kadmiumeinkristallen

Von

CHR. NANEV

Es wird über eine nach dem Platin–Kohle-Abdruckverfahren durchgeführte elektronenmikroskopische Untersuchung der Oberflächenmorphologie von Zink- und Kadmiumeinkristallen berichtet. Die Kristalle werden unter bestimmten Bedingungen aus der Dampfphase gezüchtet. Die Untersuchung zeigt, daß die Oberflächenstruktur der Kristalle von den Wachstumsbedingungen abhängt. Die Temperaturabhängigkeit der Gleichgewichtsform der Kristalle wird geklärt und auf dieser Basis eine gute Übereinstimmung der experimentellen Daten mit früheren Ergebnissen über die Gleichgewichtsformen von Zink [3] gefunden, die durch direkte mikroskopische Beobachtung der Anfangswachstumsformen erhalten wurden.

The platinum-carbon replica technique is used to study the surface morphology of zinc and cadmium single-crystals grown under definite conditions from their vapours. It is shown that the surface relief depends on the growth conditions. Special attention is given to the temperature dependence of the equilibrium form of the crystals. In this connection an explanation is given of the good agreement between the experimental data and the results obtained previously when the equilibrium form was determined by direct microscopic observations of the initial growth forms [3].

1. Einleitung

Wichtige Schlußfolgerungen über die Gleichgewichtsformen von Kadmium- und Zinkeinkristallen wie auch über die energetischen Wechselwirkungen zwischen den Bausteinen des Kristallgitters sind von Stranski, Kaischew und Mitarbeitern [1, 2, 3] gezogen worden, indem sie die Anfangswachstumsformen der aus der Dampfphase gezüchteten Einkristalle mit Hilfe des Lichtmikroskops beobachtet haben. Für eine Untersuchung der feineren Kristallwachstumsmorphologie ist die Anwendung modernerer und auflösungsfähigerer Beobachtungsmittel, wie des Elektronenmikroskopes, notwendig.

Elektronenmikroskopische Untersuchungen der Wachstumsmorphologie von aus der Dampfphase gewachsenen Metalleinkristallen sind verhältnismäßig selten. Im Falle des Cadmiums wurden zwei solche Arbeiten — von Mahl und Stranski [4] und von Mutaftschiew [5] — publiziert. Die erstgenannte Veröffentlichung ist aber stark auf die Ursachen der sogenannten Blockstruktur der Kadmiumeinkristalle akzentuiert, und es wurden dort über die einzelnen Flächen und ihre Beschaffenheit keine näheren Daten angeführt. In dieser Hinsicht ist das Problem eingehender von Mutaftschiew behandelt worden. Die von ihm untersuchten Kristalle wurden jedoch unter nicht kontrollierbaren Bedingungen aus der Dampfphase gezüchtet. Es ist aber notwendig, um bestimmte Vergleiche mit der Theorie anstellen zu können, die untersuchten Kristalle unter definierten Bedingungen wachsen zu lassen. Im Falle des Zinks fehlen ähnliche elektronenmikroskopische Untersuchungen überhaupt. Es war auch von Interesse, den Einfluß eines in der Wachstumsampulle vorhandenen inerten Gases zu untersuchen.

2. Experimentelles

Die zu dieser Untersuchung benutzten Zink- und Kadmiumeinkristalle wurden in der schon an anderer Stelle beschriebenen Apparatur [3] aus der Dampfphase gezüchtet. Diese Apparatur soll noch einmal im Prinzip dargestellt werden (Fig. 1).

In einen kleinen Aluminiumblock B des Ofens A wurde eine Glasampulle eingesetzt, welche mit mehrmals unter Hochvakuum (bei 10^{-6} Torr) sublimiertem bzw. destilliertem Zink oder Kadmium beschickt war.[1]) An jenem Punkt der Glasampulle, den die Kühlspitze S berührte, wurde die notwendige Metallmenge übersublimiert (überdestilliert). Aus dieser Metallmasse wurde durch Schmelzen und nachfolgende langsame Abkühlung des erhaltenen Tropfens gewöhnlich ein Einkristall (K) gewonnen. Falls dies beim ersten Mal nicht gelang, wurde es durch nochmaliges teilweises Schmelzen erreicht [3]. Der Einkristall wuchs dann bei einer bestimmten Temperaturdifferenz weiter, welche durch seine eigene Temperatur und die Temperatur des den Dampf liefernden Metallvorrats gegeben war [3]. In einem bestimmten Wachstumsstadium wurde die Heizung des Ofens A ausgeschaltet.[2]) Infolge seiner großen Wärmeträgheit kühlte sich der Ofen jedoch zu langsam ab. Deshalb wurde während der Abkühlung des Ofens bis auf etwa 150 °C die Temperatur der Kühlspitze (vermittels des Temperaturreglers TR_2) so reguliert, daß eine Konstanthaltung der erwähnten Temperaturdifferenz gewährleistet wurde.

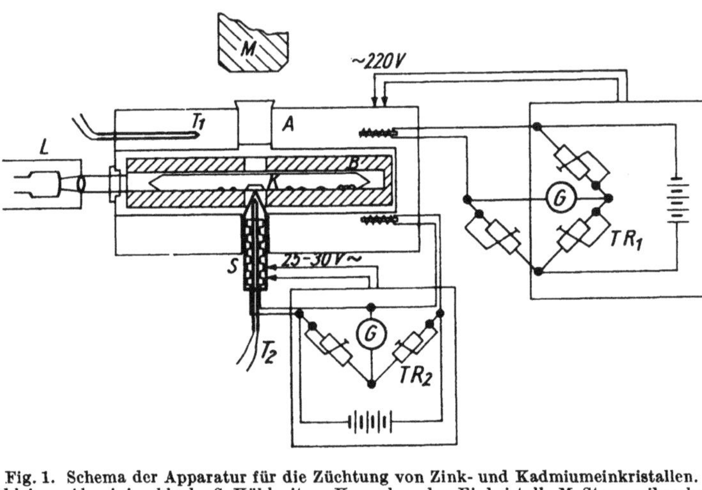

Fig. 1. Schema der Apparatur für die Züchtung von Zink- und Kadmiumeinkristallen.
A Ofen, B kleiner Aluminiumblock, S Kühlspitze, K wachsender Einkristall, M Stereomikroskop, L Lampe, T_1 und T_2 Thermoelemente, TR_1 und TR_2 Temperaturregler, G Nullgalvanometer

[1]) Es wurden spektralreine Metalle der Firma Johnson & Matthey benutzt.

[2]) Für unsere Zwecke eignen sich solche Kristalle besser, welche nur kurze Zeit aus der Dampfphase gewachsen sind.

Einige der untersuchten Kadmiumeinkristalle wurden in einer Wasserstoffatmosphäre gezüchtet. Dazu wurde in der evakuierten Ampulle ein Glaskölbchen zertrümmert, welche ein bekanntes Volumen Wasserstoffgas bei bekanntem Druck enthielt. Diese Kadmiumeinkristalle wurden in einer anderen Apparatur erzeugt, welche aber ähnliche Wachstumsbedingungen gewährleistete. Die Kristalle wurden bei Wasserstoffdrucken im Bereich von 20 bis 100 Torr und bei Temperaturdifferenzen von 50 bis 70 °C gezüchtet.

Der Einkristall wurde der Ampulle entnommen, indem sie genau über dem Ort des Kristalls angeritzt und dann zerbrochen wurde. Dann wurde der Kristall möglichst schnell in den Rezipienten einer Hochvakuum-Bedampfungsanlage hineingebracht und zur Präparation für das Elektronenmikroskop mit Platin-Kohle simultan nach dem Verfahren von Skatulla und Horn [6] bedampft. Der Kristall verblieb somit weniger als 30 s unter Atmosphärendruck. Der Abdruckfilm wurde vom Kristall elektrolytisch abgelöst. Als Elektrolyt diente eine 1 n Lösung des entsprechenden Sulfates. Die Stromdichte variierte in sehr weiten Grenzen (10 bis 50 mA/cm^2). Da der Platin-Kohle-Abdruckfilm vorher zerschnitten wurde, konnte dann bestimmt werden, von welchen Kristallbereichen die aufgefischten Abdruckteilbezirke stammen.

Bei der Präparation von Zinkeinkristallen muß als wichtige Besonderheit vermerkt werden, daß das erste Waschwasser angesäuert sein muß. Sonst sind die Abdrucke — wahrscheinlich infolge der teilweisen Hydrolyse von Zinksulfat — immer stark verunreinigt; sie sind mit vielen Salzkriställchen bedeckt.

Diese Präparationsmethode hat im Vergleich zu der von Mutaftschiew [5] angewendeten Auflösung in Säuren den Vorzug, daß sich bei der anodischen Auflösung des Kristalls kein Gas entwickelt. Außerdem sind unsere Bedingungen dadurch vereinfacht, daß die Kristalle an Luft gebracht werden können. Es zeigte sich beim Vergleich unserer elektronenmikroskopischen Bilder mit denen von Mutaftschiew, daß die von ihm benutzte Präparation unter völligem Luftabschluß nicht nötig ist. Unsere Bilder sind genügend klar, und man beobachtet keine Oxydationsspuren. Dies bedeutet entweder, daß die Oxydschicht dünner ist als das Auflösungsvermögen der Methode und daß sie die Beobachtung nicht stört, oder daß sie die ursprüngliche Oberflächenstruktur so gut wiedergibt, daß man die Oxydation nicht bemerkt.

3. Theoretische Betrachtungen

Vor der Darlegung der Ergebnisse sollen zwei theoretische Erläuterungen gegeben werden. Sie sind für die Besprechung der experimentellen Daten notwendig.

Die Untersuchung der Anfangswachstumsformen [3] zeigt, daß sich die Gleichgewichtsform von Zinkeinkristallen mit steigender Übersättigung vereinfacht. Bei einer Temperatur der Kristalle im Bereich von 370 bis 400 °C und $\Delta T > 7{,}5 \pm 0{,}5$ °C wird die Gleichgewichtsform des Zinks von den Flächen $\{0001\}$, $\{10\bar{1}0\}$, $\{10\bar{1}1\}$ begrenzt. Bei derselben Temperatur und ΔT zwischen $7{,}5 \pm 0{,}5$ °C und $2{,}1 \pm 0{,}3$ °C nehmen an der Gleichgewichtsform noch die Flächen $\{10\bar{1}2\}$ und $\{11\bar{2}0\}$ teil, und bei noch kleineren Temperaturdifferenzen wird die Gleichgewichtsform zusätzlich noch von den Flächen $\{11\bar{2}1\}$ und $\{11\bar{2}2\}$ begrenzt.

1. Die Gleichgewichtsform hängt nicht nur von ΔT, sondern auch von T ab. Hier soll die Temperaturabhängigkeit der Gleichgewichtsform untersucht werden.

Das Gleichgewicht Kristall–Dampf ist — wie bekannt — durch die Gleichung

$$\mu_a' - \mu_\infty' = k\,T \ln \frac{P_a}{P_\infty} = \mu_a - \mu_\infty = \Delta\mu \tag{1}$$

bestimmt. Hierin bedeuten μ_a' das chemische Potential der Dampfphase, die im Gleichgewicht mit einem Kristall der Größe a steht, μ_∞' das chemische Potential der Dampfphase, die im Gleichgewicht mit dem unendlich großen Kristall steht, k die Boltzmann-Konstante, T die absolute Temperatur und $\ln \frac{P_a}{P_\infty}$ die Übersättigung. μ_a und μ_∞, die chemischen Potentiale, entsprechen dem endlichen (der Größe a) und dem unendlich großen Kristall.

Berücksichtigt man noch die Clausius-Clapeyronsche Gleichung, so erhält man

$$\Delta\mu = k\,T \ln \frac{P_a}{P_\infty} = \Theta \frac{\Delta T}{T}, \tag{2}$$

wo $\Delta T = T_a - T_\infty$ und Θ die auf ein Atom bezogene spezifische Verdampfungswärme ist $\left(T = \sqrt{T_a \cdot T_\infty}\right)$.

Das Gleichgewicht wird durch den Unterschied zwischen den chemischen Potentialen $\Delta\mu$ bestimmt. Dieser hängt von ΔT und T und bei konstantem T nur von ΔT ab (Θ variiert sehr wenig mit T).

2. In konkaven Winkeln werden als Übergangswachstumsformen auch Flächen erscheinen, die nicht zur Gleichgewichtsform gehören [5].

4. Besprechung der Ergebnisse

Bemerkenswert ist die weitgehende Übereinstimmung zwischen der Makrostruktur der Kristalloberfläche — beobachtet mit dem Lichtmikroskop — und der Oberflächenmikrostruktur, aufgenommen mit dem Elektronenmikroskop. Es sei hervorgehoben, daß dieser Umstand die Deutung der elektronenmikroskopisch beobachteten Strukturen sehr erleichtert.

4.1 Im Vakuum gewachsene Zinkeinkristalle

Schon vor längerer Zeit hat Stranski den Gedanken ausgesprochen [7], daß „pyramiden- und treppenförmige Unebenheiten“ der Kristalloberfläche — die eine Vergröberung der Kristalle verursachen und die er als „Subindividuen“ bezeichnete — ebenfalls von Gleichgewichtsformflächen begrenzt sein sollten. Fig. 2a ist eine experimentelle Bestätigung dafür. Man sieht, daß die Subindividuen von den für mittlere Übersättigungen geforderten Gleichgewichtsformflächen $\{0001\}$, $\{10\bar{1}0\}$, $\{10\bar{1}1\}$, $\{10\bar{1}2\}$, $\{11\bar{2}0\}$ begrenzt werden. Auf der folgenden Aufnahme (Fig. 2b) ist nur der Bereich der Vertiefung zwischen den einzelnen Subindividuen wiedergegeben.

Die experimentelle Bestätigung des von Stranski geäußerten Gedankens, daß die Subindividuen auch von Gleichgewichtsformflächen begrenzt sind, ist wichtig, denn die Bestimmung der Gleichgewichtsform der Kristalle nach der Methode der Anfangswachstumsformen bei mit Vergröberungserscheinungen verbundenem Kristallwachstum beruht bekanntlich auf dieser Idee [1, 2, 3].

Die lichtmikroskopischen Beobachtungen zeigten, daß die Basisfläche $\{0001\}$ der aus der Dampfphase gewachsenen Zinkeinkristalle außerordentlich glatt ist

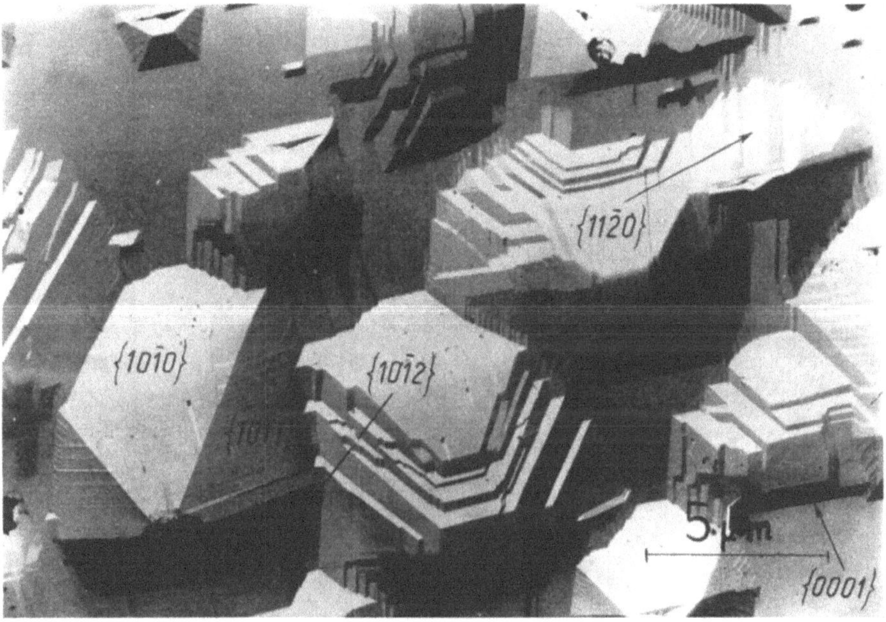

Fig. 2a. Elektronenmikroskopische Aufnahme eines bei $\Delta T = 2{,}5$ °C gewachsenen Zinkeinkristalls. Die Subindividuen sind von den Flächen {0001}, {10$\bar{1}$0}, {10$\bar{1}$1}, {10$\bar{1}$2}, {11$\bar{2}$0} begrenzt

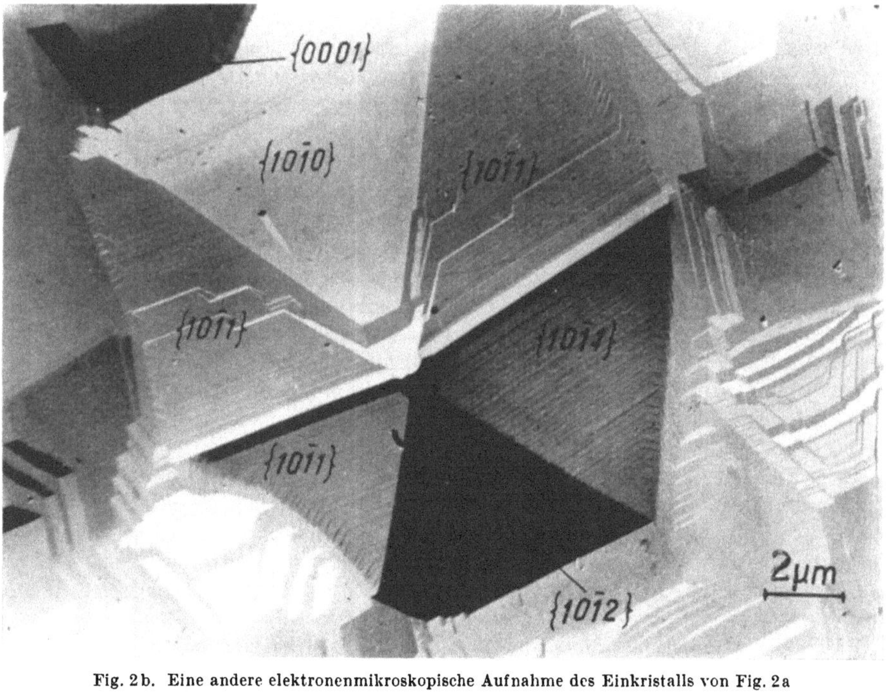

Fig. 2b. Eine andere elektronenmikroskopische Aufnahme des Einkristalls von Fig. 2a

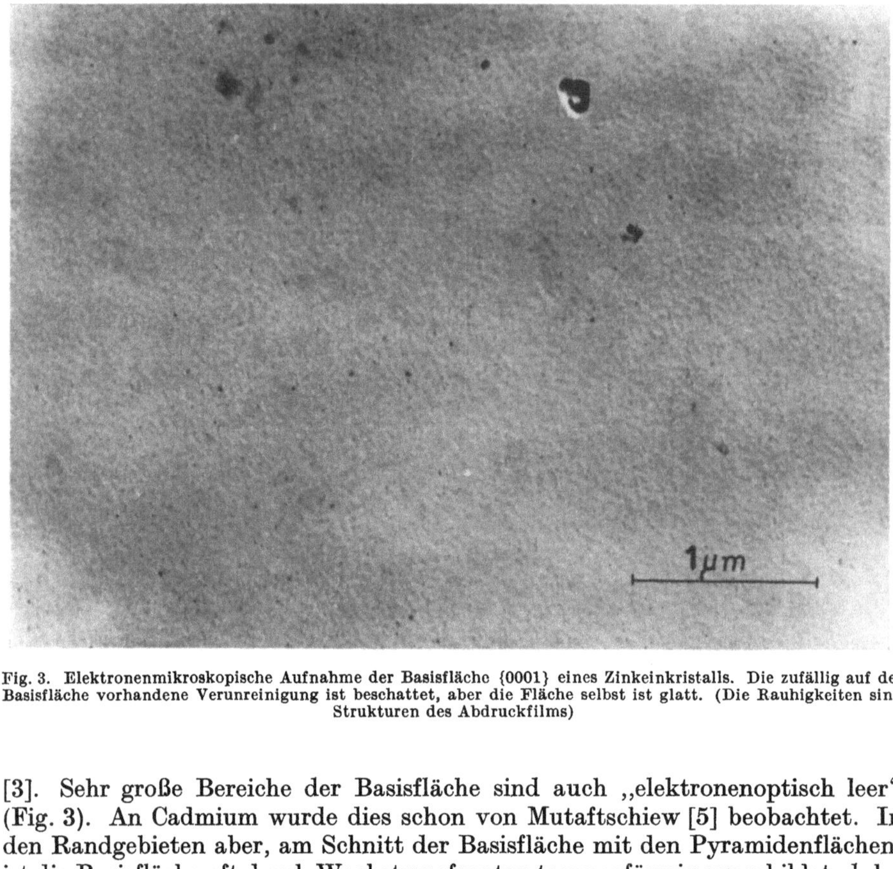

Fig. 3. Elektronenmikroskopische Aufnahme der Basisfläche {0001} eines Zinkeinkristalls. Die zufällig auf der Basisfläche vorhandene Verunreinigung ist beschattet, aber die Fläche selbst ist glatt. (Die Rauhigkeiten sind Strukturen des Abdruckfilms)

[3]. Sehr große Bereiche der Basisfläche sind auch „elektronenoptisch leer" (Fig. 3). An Cadmium wurde dies schon von Mutaftschiew [5] beobachtet. In den Randgebieten aber, am Schnitt der Basisfläche mit den Pyramidenflächen, ist die Basisfläche oft durch Wachstumsfronten treppenförmig ausgebildet, d. h., sie ist dort nicht völlig ausgebaut. Die vorwiegende Orientierung der Stufen auf {0001} zeigt die Existenz von Frontflächen erster Art [3]) (Fig. 4a). Dies folgt daraus, daß die Stufen untereinander meistens Winkel von 120° bzw. 60° einschließen. Unter diesen Winkeln müssen sich alle in die {0001}-Ebene projizierten Flächen erster Art untereinander schneiden. Es sind aber auch Fronten festzustellen, die Winkel von 150 bzw. 30° einschließen. Diese Winkel resultieren aus dem Schnitt von Flächen erster und zweiter Art. Solche Fronten sind seltener und kürzer, und sie gehen ohne Ausnahme von den konkaven Winkeln der Wachstumsfronten aus. Stärkere Vergrößerungen zeigen tatsächlich (Fig. 4b), daß solche Wachstumsstufen vom konkaven Winkel ausgehen und sich schnell ausbreiten. Es gibt deutliche Anzeichen dafür, daß diese Stufen verschwinden und nur Wachstumsstufen mit Frontflächen erster Art bleiben; denn auf {0001} werden im allgemeinen Wachstumsfronten mit Frontflächen zweiter Art seltener beobachtet, obwohl sie häufiger gebildet werden: In Fig. 4b kann man 7 solche in einem konkaven Winkel gebildete und weiterfortgeschrittene Fronten sehen. Abgesehen von den Flächen, die nicht zur Gleichgewichtsform gehören (manch-

[3]) Die Wachstumsstufen sind so niedrig, daß die Frontflächen nicht genau indiziert werden können.

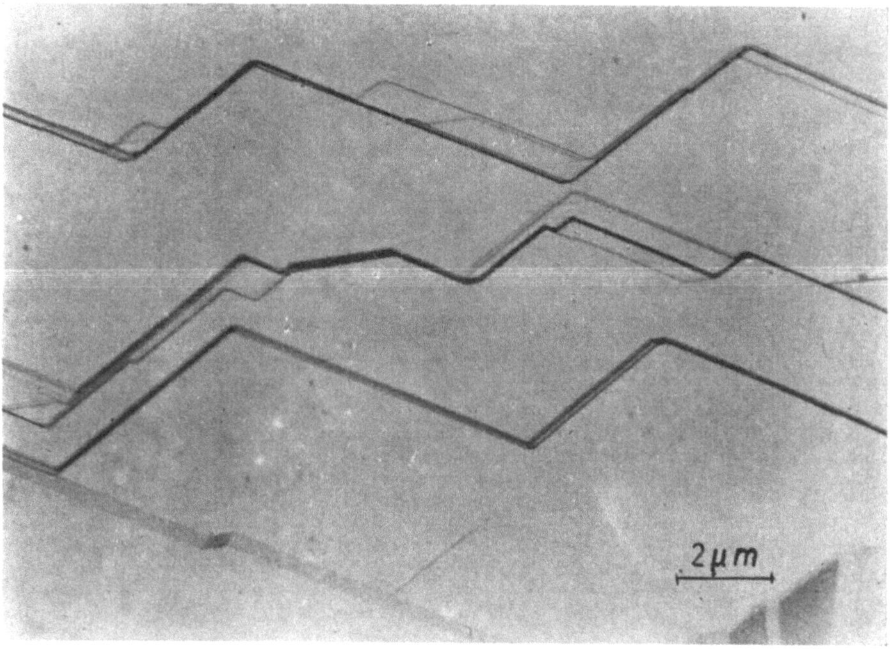

Fig. 4a. Aufnahme des Randgebiets der Basisfläche {0001} eines Zinkeinkristalls ($\Delta T = 8{,}0$ °C)

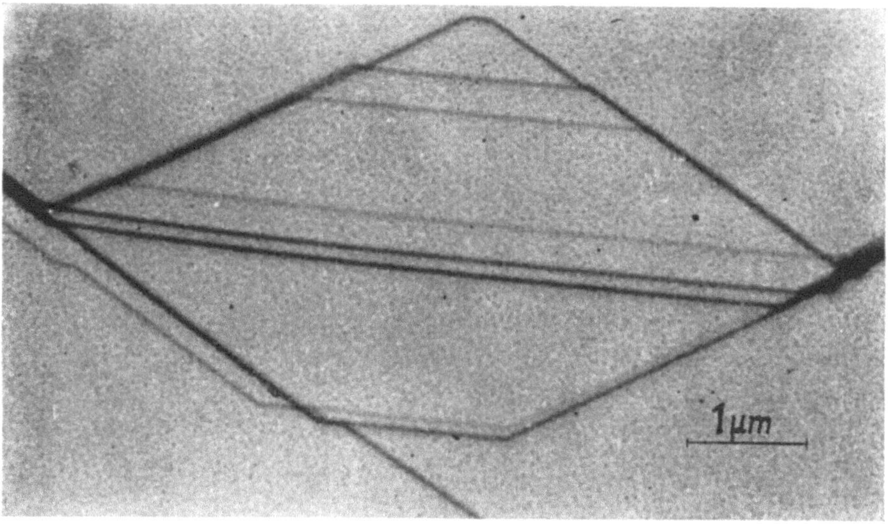

Fig. 4b. Einkristall von Fig. 4a. Die Wachstumsstufen breiten sich in Richtung $\langle 11\bar{2}0 \rangle$ aus

mal wird beobachtet, daß dies $\{11\bar{2}0\}$-Flächen sind) und die als Übergangswachstumsformen in den konkaven Winkeln erscheinen, werden bei $\Delta T = 8{,}0$ °C als Gleichgewichtsformflächen überhaupt nur die Flächen $\{0001\}$, $\{10\bar{1}0\}$, $\{10\bar{1}1\}$ festgestellt.

Wie schon erwähnt, wuchsen die Kristalle bei der Abkühlung von 400 °C bis auf etwa 150 °C bei einer konstanten Temperaturdifferenz weiter. Berücksichtigt man, daß 1. für die mit dem Elektronenmikroskop beobachtete Oberflächenstruktur die letzten Wachstumsstadien entscheidend sind und daß 2. die Kristalle ihr Wachstum bei ungefähr der gleichen Temperatur einstellen, sofern der Dampfdruck von Zink (unabhängig von ΔT) unter einen bestimmten kritischen Wert gefallen war, so folgt, daß ΔT die einzige variable Größe ist, die $\Delta\mu$ und damit die Gleichgewichtsform bestimmt. Tatsächlich kann diese Temperaturdifferenz nicht direkt mit ΔT_s [4]), wie sie in [3] eingeführt ist, verglichen werden, weil die Temperatur, bei der das Kristallwachstum aufhört, nicht genau bekannt ist. Die oben erwähnte Temperaturdifferenz kann man aber für eine halbquantitative Prüfung der früher gezogenen Schlußfolgerng über die Gleichgewichtsformen von Zinkeinkristallen benutzen. Formel (2) zeigt, daß sich die Gleichgewichtsform mit sinkender Temperatur und damit abnehmender Temperaturdifferenz vereinfachen muß, wenn $\Delta\mu_{\text{Übergang}}$ konstant bleiben soll. Daraus wird ersichtlich, daß $\Delta T = 8{,}0$ °C in das Gebiet der großen $\Delta\mu$ gehört, in welchem die einfachste Gleichgewichtsform ($\{0001\}$, $\{10\bar{1}0\}$, $\{10\bar{1}1\}$) vorliegt. Für mittlere $\Delta\mu$ muß eine Temperaturdifferenz innerhalb desjenigen Bedingungsgebietes gegeben sein, worin die Gleichgewichtsform von $\{0001\}$, $\{10\bar{1}0\}$, $\{10\bar{1}1\}$, $\{10\bar{1}2\}$, $\{11\bar{2}0\}$-Flächen begrenzt wird. Diese Bedingung wird nach Fig. 5 durch eine Temperaturdifferenz $\Delta T = 2{,}5$ °C erfüllt. Daraus geht hervor, daß die Gleichgewichtsform bei $\Delta T = 2{,}5$ °C erst bei ungefähr -50 °C einfacher werden sollte, während das Kristallwachstum bereits bei höheren Temperaturen aufhört.

Abweichend von der Basisfläche sind die Prismenflächen erster Art $\{10\bar{1}0\}$ und die Pyramidenflächen erster Art $\{10\bar{1}1\}$ charakteristisch parallel zu $\{0001\}$ gestreift. Auch in diesem Fall gibt das Elektronenmikroskop wesentlich feinere Einzelheiten wieder (Fig. 6).

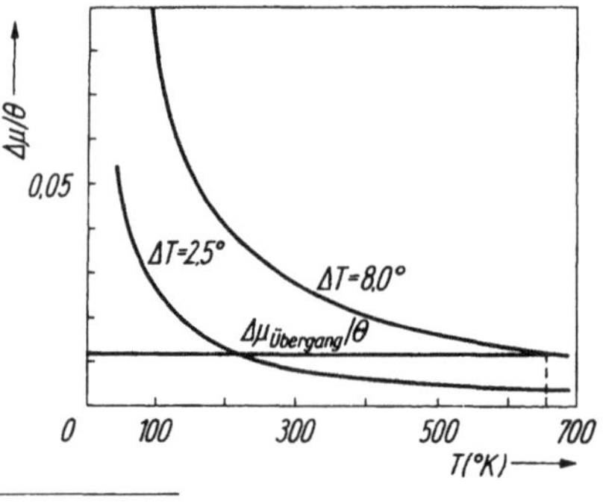

Fig. 5. $\Delta\mu/\Theta$ in Abhängigkeit von der absoluten Temperatur für $\Delta T = 8{,}0$ °C und $\Delta T = 2{,}5$ °C. $\Delta\mu_{\text{Übergang}}/\Theta$ ist durch $\Delta T = 7{,}5$ °C und $T = 384$ °C [3] bestimmt. Wenn Θ nicht von T abhängt, wird bei $\Delta T = 2{,}5$ °C $\Delta\mu_{\text{Übergang}}$ bei $T = -53$ °C erreicht

[4]) ΔT_s ist die kritische Temperaturdifferenz, oberhalb welcher die Gleichgewichtsform nur $\{0001\}$, $\{10\bar{1}0\}$, $\{10\bar{1}1\}$-Flächen enthält.

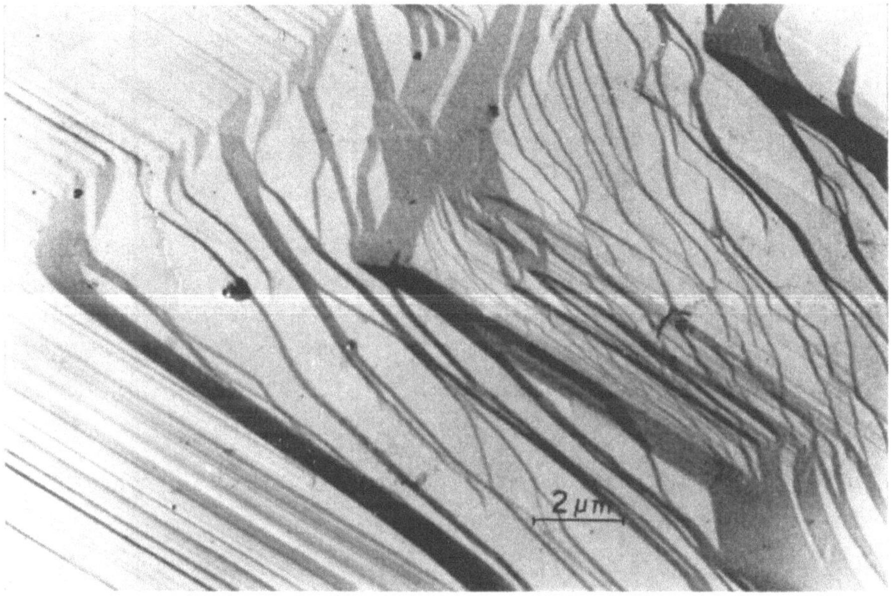

Fig. 6. Elektronenmikroskopische Aufnahme eines Zinkeinkristalls ($\Delta T = 2,5$ °C). Man erkennt gerade, parallele Wachstumsfronten, ähnlich den mit dem Lichtmikroskop beobachtbaren Streifungen, sowie gekrümmte Wachstumsfronten

4.2 Im Vakuum gewachsene Kadmiumeinkristalle [8]

Die Änderung der Gleichgewichtsformen beim Cadmium ist nicht untersucht worden, da sie bei sehr kleinen Temperaturdifferenzen stattfindet [2], die mit unserer Apparatur nicht verwirklicht werden konnten [3].

Es ist zu bemerken, daß die Kadmiumeinkristalle beim Wachstum aus der Dampfphase im allgemeinen leichter als die Zinkeinkristalle vergröbern. Manchmal beobachtet man auf der Basisfläche der Kadmiumeinkristalle eine charakteristische Oberflächenstruktur (Fig. 7a). Mit Hilfe des Lichtmikroskops kann man den Charakter der Spitze dieser flachen Wachstumspyramide nicht feststellen. Das elektronenmikroskopische Bild zeigt (Fig. 7b), daß in diesem Fall keine spiralförmige Wachstumsfront vorliegt. Daraus darf aber keinesfalls geschlossen werden, daß Wachstumsspiralen auf der Basisfläche der Kadmiumeinkristalle überhaupt fehlen. Bei den Metallen haben die Versetzungen einen Burgersvektor von einer (bis zwei) Gitterkonstante, und es ist also zu erwarten, daß die Stufenhöhe der spiralförmigen Wachstumsfronten unter dem Auflösungsvermögen der hier angewandten Methode liegt (höchstens 10 bis 20 Å Stufenhöhe). Daher kann nicht behauptet werden, daß die Oberfläche atomar glatt ist. Zur Entscheidung dieser Frage muß offensichtlich eine feinere Methodik entwickelt werden, die sich eventuell der Oberflächendekoration bedient.

4.3 In Anwesenheit von Wasserstoff gewachsene Kadmiumeinkristalle

Die elektronenmikroskopische Untersuchung zeigte in diesem Falle ebenfalls höher indizierte Flächen, die sich also von den Gleichgewichtsformflächen (beim

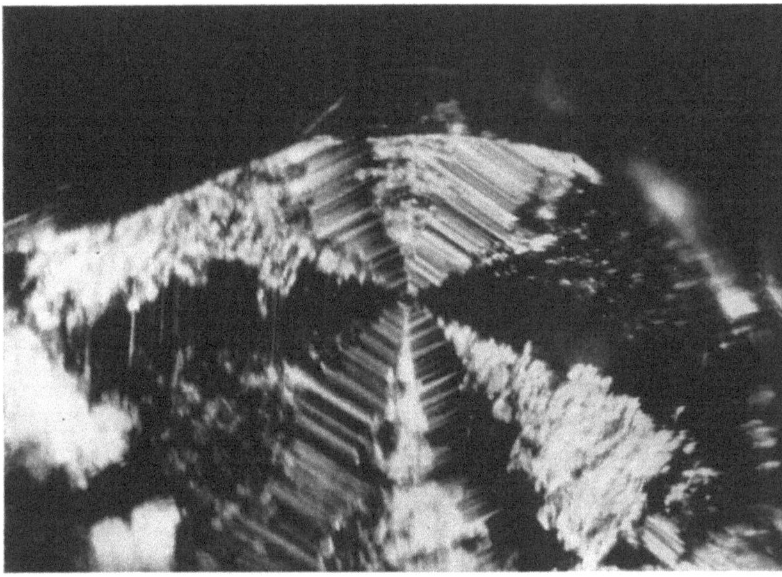

Fig. 7a. Lichtmikroskopische Aufnahme der Basisfläche {0001} eines Kadmiumeinkristalls (Vergrößerung 30 ×)

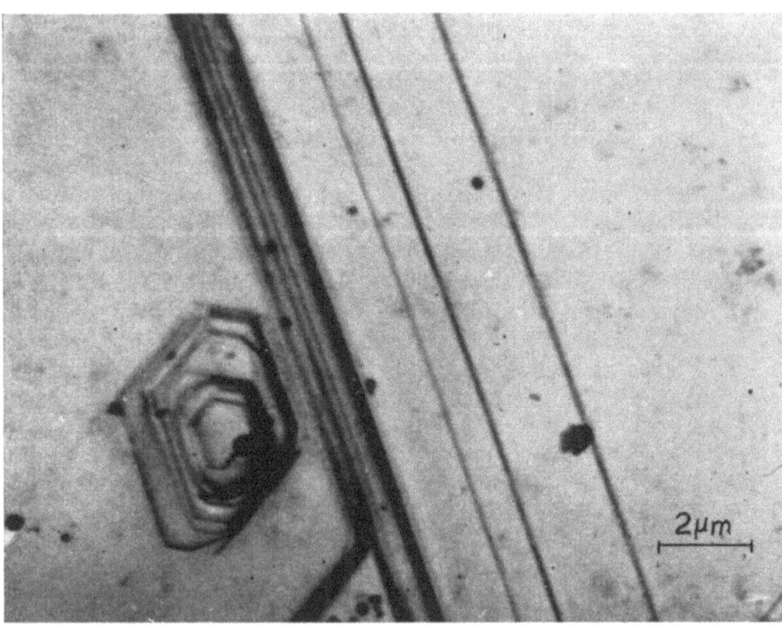

Fig. 7b. Die zu Fig. 7a entsprechende elektronenmikroskopische Aufnahme. (Ähnliche Bilder wurden auch von anderen Autoren aufgenommen [9])

Fig. 8. Elektronenmikroskopische Aufnahme eines Kadmiumeinkristalls, der in Wasserstoffatmosphäre gewachsen ist. Die dunkelsten Flächen sind {0001}-, die hellsten {22$\bar{4}$1}- und die übrigen {10$\bar{1}$0}-Flächen

Kristallwachstum im Vakuum) unterscheiden. Z. B. zeigt Fig. 8 eine interessante Struktur, die außer der Basisfläche {0001} noch Pyramidenflächen {22$\bar{4}$1} (die hellsten Flächen) und Prismenflächen {10$\bar{1}$0} enthält.

Herrn Prof. Dr. R. Kaischew möchte ich für die Anregung zu dieser Untersuchung und sein stets förderndes Interesse an dieser Arbeit herzlichst danken.

Literatur

[1] I. N. Stranski, Z. phys. Chem. **B 38**, 451 (1938).
[2] R. Kaischew, L. Keremidtschiew und I. N. Stranski, Z. Metallk. **34**, 201 (1942).
[3] R. Kaischew und Chr. Nanev, phys. stat. sol. **10**, 779 (1965).
[4] H. Mahl und I. N. Stranski, Z. Metallk. **35**, 147 (1943).
[5] B. Mutaftschiew, phys. stat. sol. **8**, 223 (1965).
[6] W. Skatulla und L. Horn, Exper. Tech. Phys. **8**, 1 (1960).
[7] Siehe z. B. I. N. Stranski, Ber. Dtsch. Chem. Ges. **A 72**, 141 (1939).
[8] Chr. Nanev, Diplomarbeit, Sofioter Universität, Sofia 1964.
[9] F. N. Tavadze und T. I. Kilitauri, Normalnyi i dislokatsionnyi rost kristallov nekotorykh tsvetnykh metallov, Nauka, Moskau 1965.

(Received May 18, 1966)

phys. stat. sol. **16**, 789 (1966)

Physikalisches Institut der Universität Münster (Westf.)

Recovery Processes in Nickel Thin Films between 20 and 300 °C[1])

By

R. DIETZ and W. HELLENTHAL

Evaporated nickel thin films about 500 Å thick are prepared by condensation at a rate of 5 Å/s onto glass substrates. Recovery processes are then studied by measuring the electrical resistance whilst the annealing temperature is increased linearly with time. This method gives the activation energy E and kinetic g of the recovery processes in individual samples. An investigation of the dependence of E on the pressure p during evaporation, controlled by emission of nitrogen gas, shows a reduction in E of about 1/4 as p is increased to around 10^{-4} Torr. By extrapolation of E to $p = 0$ a value of $E = (1.05 \pm 0.07)$ eV is obtained. The annealing kinetic appears to follow $g = 2$. These values apply to the temperature region between 60 and 110 °C. In the temperature range from about 230 to 270 °C another recovery process is observed with an activation energy E of approximately 1.4 eV.

An Nickel-Aufdampfschichten (niedergeschlagen auf Glasunterlage mit rund 5 Å/s, Dicke etwa 500 Å) wurden mit der Methode der zeitlinearen Aufheizung Erholungsuntersuchungen durch Messung des elektrischen Widerstandes ausgeführt. Hiermit ist es möglich, an einzelnen Proben Aktivierungsenergie E und Kinetik g von Ausheilprozessen zu bestimmen. Die Untersuchung der Abhängigkeit von E vom Gasdruck p während des Aufdampfens mittels geregelt eingelassenen Stickstoffs ergab eine Abnahme von E um etwa 1/4, wenn p bis etwa 10^{-4} Torr vergrößert wurde. Die Extrapolation auf verschwindendes p liefert $E = (1{,}05 \pm 0{,}07)$ eV. Die Kinetik dieses Prozesses läßt sich mit $g = 2$ beschreiben. Diese Werte wurden im Erholungsbereich zwischen 60 und 110 °C bestimmt. Im Temperaturintervall von 230 bis 270 °C konnte ein weiterer Erholungsvorgang beobachtet werden. Die aus Messungen an einigen Proben abgeschätzte Aktivierungsenergie betrug etwa 1,4 eV.

1. Introduction

Most of the published results on the dependence of the physical properties of thin films on their structure have been obtained by electron microscopy or diffraction. Only comparatively few systematic and quantitative investigations have been done on the influence of structural lattice defects and/or impurity atoms, which show an annealing behaviour corresponding to recovery processes.

Some papers dealing with these questions consider changes of M-induced anisotropy (cf. [1, 2, 3], further references are cited therein). Quantitative investigations of recovery in thin films for determining the activation energy E and kinetic g by measurements of the electrical resistivity on pairs of samples have recently been published [4]. In connection with the questions mentioned above we investigated the annealing behaviour of nickel thin films by observing changes in electrical conductivity during heating with linear temperature increase. This method has the advantage that measurements of electrical resistance at another (and mostly very low) reference temperature are not necessary and the evaluation of E and g requires only data from single samples.

[1]) Contribution to the International Colloquium on Magnetic Thin Films, 25th to 28th April 1966 in Jena.

With regard to the question of comparing results from films of different conditions of preparation, we have first varied the type of gas and its pressure during condensation of the samples.

2. Experimental

The nickel films were condensed onto glass substrates with contact leeds. For evaporation we used nickel wire of 99.97% purity. The condensation rate was automatically controlled to about 5 Å/s, using the electrical conductivity of films after suitable calibration by weight. The thickness of the samples lay in the range of 500 Å.

The double wall vacuum system with a tandem arrangement of diffusion pumps for the interior and the intermediate space enabled to work in a pressure range down to 10^{-9} Torr or to evacuate sufficiently far before immission of gases under controlled pressure for a systematic variation of the conditions of condensation.

Resistivity measurements on the films were made with a Wheatstone bridge by observing the deviation from balance with a compensation recorder. The sensitivity could be raised to 2% relative resistance variation for the full recording scale width (250 mm).

For the observation of recovery processes we have chosen the continuous linear increase of temperature with time (about 1 °C per minute). For the evaluation of E a good approximation to linearity is necessary.

3. Method of Measurement and Evaluation

While the temperature was increased linearly with time, nickel films which had been condensed at substrate temperatures between 20 and 40 °C showed an irreversible decrease of resistance ΔR_{irr} (cf. Fig. 1, curve (a)), which remained absent (Fig. 1, curve (b)) during repetition of the experiment with the same sample.

The temperature dependence of electrical resistance in the parts of the curves immediately outside the pronounced transition region appeared to be approximately equal. This is in accordance with the results of [5, 6] on the dependence

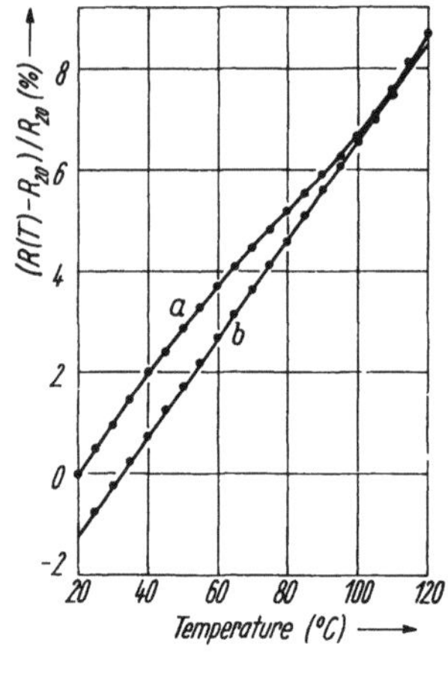

Fig. 1. Variation of resistance (related to the value of R_{20} at 20 °C) measured on a nickel film during heating with linear temperature increase with irreversible decrease of resistance in the first run (a), reversible behaviour in a succeeding run (b)

Fig. 2. Annealing of the irreversible decrease of resistance during heating with linear temperature increase

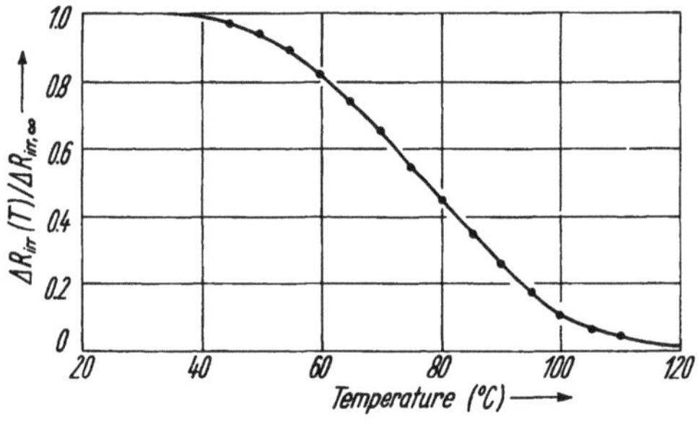

of $\mathrm{d}R/\mathrm{d}T$ on T for nickel bulk material. In the small temperature intervals used here, it was not necessary to consider a contribution from this effect.

Thus the variation of the resistance for a given rate of temperature increase may be written

$$R(T) = R_{\text{rev}}(T) + \Delta R_{\text{irr}}(T)\,. \tag{1}$$

In this representation the reversible part $R_{\text{rev}}(T)$ includes all contributions to resistivity, as far as they do not undergo irreversible changes in the temperature region of recovery which is investigated. They might show such a behaviour at sufficiently different temperatures. In the following $\Delta R_{\text{irr},\infty}$ represents the complete recovery of the process just investigated.

A more detailed study of such a recovery region becomes possible if it is assumed that there is a unique correlation between the irreversible decreasing part of the resistance $\Delta R_{\text{irr}}(T)$ (cf. Fig. 2) and the variation of defect concentration (or configuration) c. By assuming a linear dependence, i.e.

$$\Delta R_{\text{irr}}(T) \sim c(T) \tag{2}$$

and supposing that a single processes is responsible for the recovery in the temperature range investigated, one may use the method of Balarin and Zetzsche [7, 8] for the analysis of data obtained during a linear temperature increase with time. In the case of thermally activated recovery processes these authors use the equation

$$-\,\mathrm{d}c/c^{g} = K_0 \exp\{-\,E/kT\}\,\mathrm{d}t \tag{3}$$

for the variation of the concentration c with time (i.e. $\mathrm{d}c/\mathrm{d}t$). g determines the kinetic or reaction order of the process with activation energy E. K_0 is a frequency factor, k the Boltzmann constant and T the absolute temperature. For a linear variation of T with time t, i.e. a constant heating rate

$$\mu = \mathrm{d}T/\mathrm{d}t = \text{const} \tag{4}$$

a behaviour of the process involved is expected that is described by the differential equation

$$-\,\mathrm{d}c/c^{g} = (K_0/\mu)\exp\{-\,E/kT\}\,\mathrm{d}T\,. \tag{5}$$

The integration of this expression gives ([7, 8], see also [9])

$$\left(-\ln\frac{c(T)}{c_0}\right) = -\,\frac{E}{k\,T} - 2\ln\frac{E}{k\,T} + \Phi_1(\mu, K_0, E, T) \quad \text{for} \quad g = 1\,, \tag{6a}$$

$$\ln\left(\frac{1}{g-1}\left(\frac{1}{(c(T))^{g-1}} - \frac{1}{c_0^{g-1}}\right)\right) =$$

$$= -\,\frac{E}{k\,T} - 2\ln\frac{E}{k\,T} + \Phi_g(\mu, K_0, E, T) \quad \text{for} \quad g \neq 1\,. \tag{6b}$$

c_0 represents the initial defect concentration. Equation (6b) may be rewritten in terms of resistance values if (2) is used. By restricting to $g = 2$ which appears justified by the present results we obtain

$$\ln\left(\frac{\Delta R_{\text{irr},\infty} - \Delta R_{\text{irr}}(T)}{\Delta R_{\text{irr}}(T)}\right) = -\,\frac{E}{k\,T} - 2\ln\frac{E}{k\,T} + \Phi_2(\mu, K_0, E, T) \quad \text{for} \quad g = 2\,. \tag{6c}$$

The quantities Φ_1, Φ_g, and Φ_2 in equations (6a to 6c) are only weakly temperature dependent. For this reason the activation energy E may be evaluated by plotting the left sides of (6) versus $1/T$ as far as the measured values allow a linear interpolation. The present results are sufficiently well represented by equation (6c). The assumption $g = 1$ (and therefore the use of (6a)) gave a considerable deviation from a straight line.

For the evaluation of the activation energy E according to equation (6c) the relative irreversible decrease of resistance $\Delta R_{irr}(T)/\Delta R_{irr,\infty}$ which is directly obtained from experimental data or the relative concentration $c(T)/c_0$ are plotted versus temperature using special scales corresponding to (6c). This method has been quite useful for the evaluation of the activation energy E and the verification whether case $g = 2$ is sufficiently well approximated. Fig. 3 shows an example.

4. Results

4.1 Recovery in the temperature region 60 to 110 °C

At first we investigated the influence of dry nitrogen gas pressure during condensation of the films on the recovery in the temperature range between 60 and 110 °C. The relative irreversible part of the resistance $\Delta R_{irr,\infty}/R_{20}$ decreased monotonously with pressure during evaporation from nearly 5% at 10^{-4} Torr with better vacuum conditions adjusted for each preparation of a film. The activation energies E, evaluated as discussed in Section 3, varied with pressure p during condensation of the samples (cf. Fig. 4). An extrapolation to $p \to 0$ led to $E = (1.05 \pm 0.07)$ eV. This value agrees quite well with results of bulk nickel of similar purity measured after plastic deformation by Simson and Sizmann [10] ($E = 1.09$ eV for 99.8% Ni and $E = 0.92$ eV for 99.999% Ni). Seeger et al. [11] found $E = 1.06$ eV and Bell [12] evaluated 0.98 eV in the same temperature range. The papers mentioned give $g = 2$ for the order of reaction in the bulk material, which agrees with the linear dependence shown in Fig. 3.

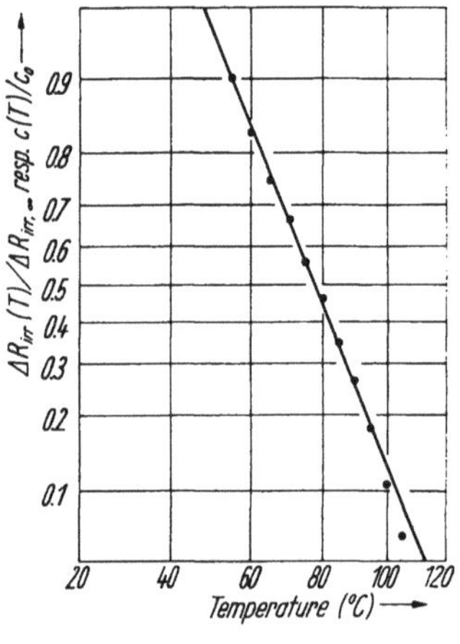

Fig. 3. Example for the evaluation of the activation energy following the method of Balarin and Zetzsche [7, 8] with the special diagram described in this paper

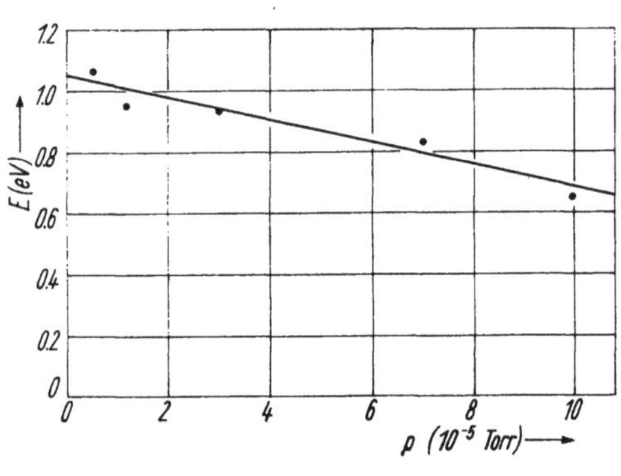

Fig. 4. Activation energies E measured in the recovery region between 60 and 110 °C as a function of the nitrogen gas pressure during condensation of the nickel films

Additionally some films were prepared in the presence of O_2, H_2, H_2O, He, or Ar. Oxygen gas has a much larger influence on the films than an equal pressure of nitrogen. It was necessary to improve the O_2 vacuum by about a factor of ten to get a similar height of the recovery process observed here by resistance measurements. Water vapour affected the films nearly as much as O_2. This demonstrates the necessity to control carefully contributions by a presence of these types of gases. It may be estimated that a reduction of the recovery process observed here to a relative height below about 1% necessitates a lowering of the partial pressure of O_2 or H_2O to $\lesssim 10^{-7}$ Torr for the condensation rate used.

The data for the rare gases He or Ar did not permit an evaluation for a uniform process but indicated two hardly separated activation energies which may influence the recovery of resistance. This has not yet been systematically investigated.

4.2 Recovery in the temperature region 200 to 300 °C

In this interval another process has been observed. With our present equipment only approximate data could be measured in this range. Up to date the following results were obtained:

a) A distinct process appears between 230 and 270 °C.

b) The relative irreversible decrease of resistance was considerably larger than for the same films in the other temperature region investigated (60 to 110 °C).

c) For the few samples measured an activation energy of about 1.4 eV was estimated.

5. Comment on the Results

The results from the annealing experiments reported here might be compared to the recovery regions III and IV of nickel bulk material. The values of the activation energies and the kinetic appear to be of the same order of magnitude as for the corresponding bulk materials. Although it appears to be possible to compare also the mechanisms attributed to recovery, it must be taken into consideration that the films investigated may be influenced by other impurities or impurity distributions and that a quite different relation of boundary area to the volume of the crystallites composing the samples exists in the films.

We would like to thank Prof. Dr. E. Kappler for his interest in this work.

References

[1] W. T. Siegle and W. R. Beam, J. appl. Phys. **36**, 1721 (1965).
[2] G. Kneer and W. Zinn, Proc. Internat. Symp. Clausthal-Göttingen 1965.
[3] D. O. Smith, G. P. Weiss, and K. J. Harte, J. appl. Phys. **37**, 1464 (1966).
[4] W. Sander and E. Strieder, Z. Phys. **188**, 99 (1965).
[5] W. Gerlach, H. Bittel, and S. Velayos, Sitz.-Ber. Bayer. Akad. Wiss., Math.-Nat. Kl., 1936 (S. 81).

[6] H. BITTEL and W. GERLACH, Ann. Phys. (Germany) **33**, 661 (1938).
[7] M. BALARIN and A. ZETZSCHE, phys. stat. sol. **2**, 1670 (1962).
[8] M. BALARIN and A. ZETZSCHE, phys. stat. sol. **3**, K 387 (1963).
[9] J. NIHOUL, phys. stat. stol. **3**, 2061 (1963).
[10] P. SIMSON and R. SIZMANN, Z. Naturf. **17a**, 596 (1962).
[11] D. SCHUHMACHER, W. SCHUELE, and A. SEEGER, Z. Naturf. **17a**, 228 (1962).
[12] F. BELL, Acta metall. **13**, **363** (1965).

(Received May 31, 1966)

phys. stat. sol. 16, 795 (1966)

Institut für Festkörperphysik der Tschechoslowakischen Akademie der Wissenschaften, Prag (a) und
Institut für Magnetische Werkstoffe der Deutschen Akademie der Wissenschaften zu Berlin, Jena (b)

Über die induzierte Anisotropieenergie manganhaltiger Ferrite

Von

R. GERBER (a), L. MICHALOWSKY (b), K. MOTZKE (b) und E. STEINBEISS (b)

Es wurde die Konstante K_u der induzierten einachsigen Anisotropieenergie von Manganferriten ($Mn_{1+x}Fe_{2-x}O_{4+\gamma}$) bei 88 °K in Abhängigkeit von x und γ gemessen. Dabei ergab sich, daß K_u im Bereich $0 < x \leqq 0{,}8$ linear von x abhängt und mit abnehmendem Sauerstoffgehalt der Proben zunimmt. Außerdem zeigte sich, daß Zusätze von Ni, Zn und Mg eine Abnahme von K_u bewirken, die der Konzentration dieser Ionen proportional und von der Ionensorte nahezu unabhängig ist. Eine einheitliche Deutung der Meßergebnisse konnte bisher nicht gefunden werden.

The constant, K_u, of the induced uniaxial anisotropy energy of manganese ferrites ($Mn_{1+x}F_{2-x}O_{4+\gamma}$) is measured at 88 °K as a function of x and γ. The measurements show that, for the range $0 < x \leqq 0.8$, K_u varies linearly with x and increases with decreasing oxygen content of the samples. Admixtures of Ni, Zn, and Mg, lead to a decrease of K_u which is proportional to the concentration of these ions and nearly independent of the type of ion. A unique interpretation of the experimental results could not be found.

1. Einleitung

Die Hysterese- und Impulseigenschaften manganhaltiger Ferrite werden bei tiefen Temperaturen (unterhalb 173 °K) wesentlich durch das Vorkommen einer induzierten Anisotropieenergie beeinflußt, die sich bei polykristallinen Proben in der Form

$$E = K_u \sin^2 \Theta \tag{1}$$

darstellen läßt [1] (Θ Winkel zwischen Vorzugsachse und der Magnetisierungsrichtung während der Messung).

Obwohl der Perminvareffekt dieser Ferrite schon mehrfach in der Literatur diskutiert wurde [2, 3], fehlten bisher quantitative Messungen über die Größe dieser Anisotropieenergie und die damit verbundenen Relaxationszeiten.

Während Giesecke [3] die Ursache der induzierten Anisotropieenergie in den besonderen Bindungsverhältnissen der Mn^{2+}- und Mn^{3+}-Ionen auf den Oktaederplätzen des Spinellgitters suchte, diskutierten Krupička und Mitarbeiter [4, 5] den Jahn-Teller-Effekt der Mn^{3+}-Ionen auf den Okktaederplätzen als Ursache von K_u. In jedem Falle sollte die Größe von K_u vom Mn-Gehalt der Proben abhängen.

Wir haben in der vorliegenden Arbeit eine Reihe von Manganferriten der Zusammensetzung $Mn_{1+x}Fe_{2-x}O_{4+\gamma}$ ($0 < x \leqq 0{,}8$) bezüglich der induzierten Anisotropieenergie näher untersucht. Darüber hinaus wurde noch der Einfluß von γ (Sauerstoffgehalt) und von Mg-, Zn- und Ni-Zusätzen auf die Relaxationszeit und Amplitude der induzierten einachsigen Anisotropieenergie untersucht, um zusätzlich Hinweise auf die Natur dieser Anisotropieenergie zu erhalten.

2. Meßverfahren und Probenherstellung

Die induzierte Anisotropieenergie wurde nach der Drehmomentenmethode an polykristallinen Scheiben und Kugeln gemessen. Die Meß- und Temperfeldstärken betrugen 5 kOe bei den Scheiben und 10 kOe bei den Kugeln. Die Scheiben hatten einen Durchmesser von 11 und eine Höhe von 3 mm, die Kugeln einen Durchmesser von 5 mm. Sie wurden bei Temperaturen von 1400 °C gesintert. Wie sich zeigte, hängen die induzierte Anisotropieenergie und die Relaxationszeit außer von der Zusammensetzung noch von der Glühtemperatur und vom Sauerstoffpartialdruck der Glühatmosphäre ab. Aus diesem Grunde wurden die Sinterkörper vor der Messung noch einer besonderen Glühbehandlung unterzogen. Diese Glühungen wurden generell bei 1200 °C und bei einem Luftdruck von 760 Torr bzw. 10^{-2} Torr durchgeführt. Die Glühdauer betrug ca. 10 Stunden. Durch die Glühung bei 760 Torr konnte erreicht werden, daß die Größe γ bei allen Proben in der Nähe von Null lag (vgl. Fig. 1, Kurve (a)). Nach der Anisotropiemessung wurden die Proben chemisch analysiert und die Zusammensetzung einschließlich der γ-Werte bestimmt. Die bei den Vakuumglühungen auftretenden γ-Änderungen wurden zusätzlich thermogravimetrisch gemessen. Die Ergebnisse stimmten in allen Fällen mit den chemisch ermittelten Werten überein.

Die Herstellungsbedingungen wurden so gewählt, daß nach dem von Bergstein und Kleinert [6] angegebenen Phasendiagramm die Einphasigkeit der Proben gewährleistet war. Als weitere Kriterien für die Einphasigkeit wurden Schliffbilder und Messungen der Hystereseschleifen an ringförmigen Proben herangezogen [7].

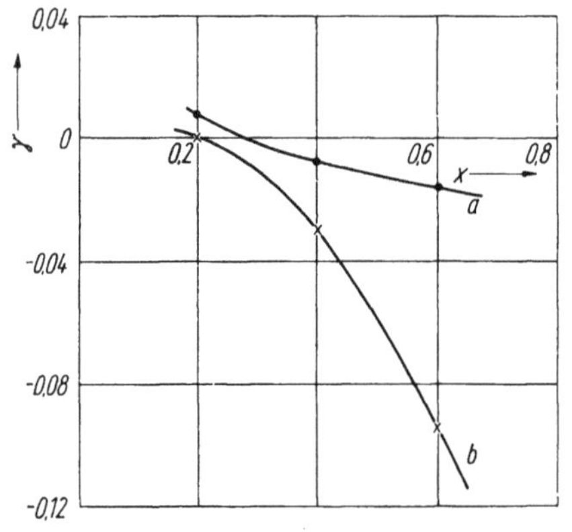

Fig. 1. Sauerstoffgehalt γ für verschiedene Manganferrite $(Mn_{1+x}Fe_{2-x}O_{4+\gamma})$.
Kurve (a): Glühung bei 1200 °C und 760 Torr,
Kurve (b): Glühung bei 1200 °C und 10^{-2} Torr

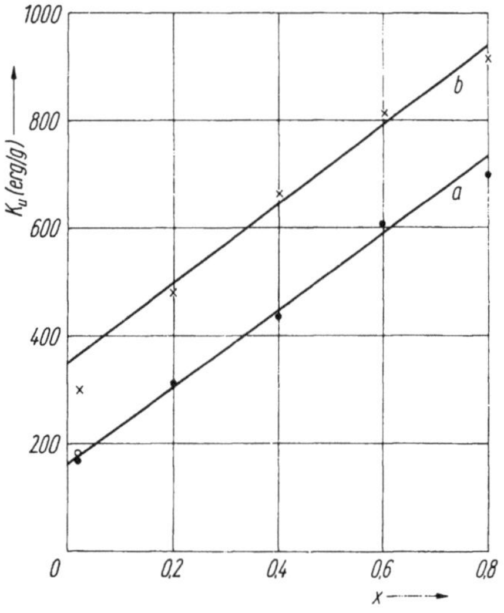

Fig. 2. Induzierte einachsige Anisotropieenergie K_u für verschiedene Manganferrite $(Mn_{1+x}Fe_{2-x}O_{4+\gamma})$ und Glühbehandlungen.
Kurve (a): Glühung bei 1200 °C und 760 Torr,
Kurve (b): Glühung bei 1200 °C und 10^{-2} Torr

3. Meßergebnisse

3.1 Mn-Ferrite

Für die Messungen der induzierten Anisotropieenergie wurde eine Temperatur von 88 °K gewählt, so daß die Relaxationszeiten eine Größe erreichten, bei der sowohl eine Magnetfeldtemperung als auch die Messung durchgeführt werden konnten. Lediglich die im Vakuum geglühte Probe mit der Zusammensetzung $Mn_{1,02}Fe_{1,98}O_{4+\gamma}$ mußte bei tieferen Temperaturen gemessen werden. Die Meßwerte wurden dann auf 88 °K extrapoliert. Dabei wurde für K_u die gleiche Temperaturabhängigkeit angenommen, die bei der Probe mit $\gamma = 0$ (Glühung bei 760 Torr) gemessen wurde. Diese Temperaturabhängigkeit zeigt unterhalb 20 °K noch einen zusätzlichen Anstieg von K_u, der bei 4,2 °K den Wert von 140 erg/g erreicht. Auch bei dynamischen Messungen fanden Závěta und Mitarbeiter [8] in diesem Temperaturgebiet Relaxationsprozesse, die vermutlich mit dem beobachteten Anstieg von K_u zusammenhängen. Für die Extrapolation wurde jedoch nur der Relaxationsprozeß oberhalb 20 °K zugrunde gelegt.

In Fig. 2 und 3 sind die Werte von K_u und $\tau_{1/2}$ (Halbwertszeit) für die beiden oben genannten Glühbehandlungen in Abhängigkeit vom Mn-Gehalt dargestellt.

Aus Fig. 2 ist zu entnehmen, daß K_u annähernd linear von x abhängt und bei $x = 0$ nicht verschwindet. Die Glühung bei einem Luftdruck von 10^{-2} Torr bewirkt eine beträchtliche Vergrößerung von K_u und eine Verkleinerung der Halbwertszeit (vgl. Kurven (b) in Fig. 2 und 3).

Obwohl die isothermen Glühungen bei unterschiedlichen O_2-Partialdrücken eine beträchtliche γ-Änderung hervorrufen, die vom Mangangehalt x abhängt, ist die Vergrößerung von K_u in einem weiten Bereich nahezu unabhängig von x.

Im Falle der Zusammensetzungen $Mn_{1,2}\,Fe_{1,8}O_4$ und $Mn_{1,4}\,Fe_{1,6}O_4$ wurde zusätzlich noch die induzierte Anisotropieenergie an Einkristallkugeln gemessen. Zur Beschreibung dieser Energie sind zwei Größen F und G erforderlich, die mit K_u (bei polykristallinem Material) durch die Beziehung

$$K_u = \frac{4}{10} F + \frac{3}{10} G \tag{2}$$

verknüpft sind. Die Messungen ergaben, daß die Größe G klein gegenüber F und

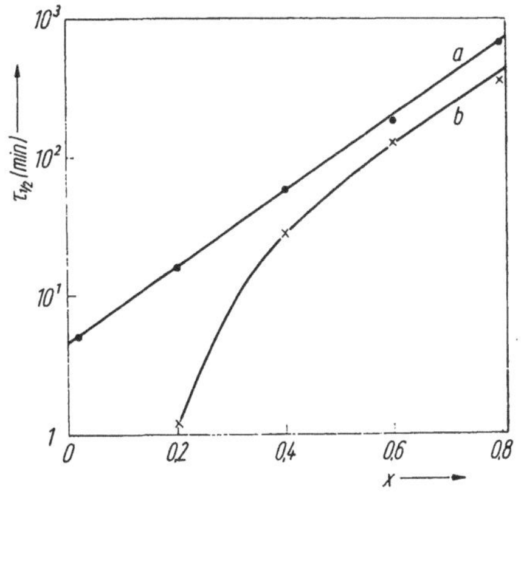

Fig. 3. Abhängigkeit der Halbwertszeit $\tau_{1/2}$ von der Zusammensetzung und Glühbehandlung für einige Manganferrite ($Mn_{1+x}Fe_{2-x}O_{4+\gamma}$).
Kurve (a): Glühung bei 1200 °C und 760 Torr,
Kurve (b): Glühung bei 1200 °C und 10^{-2} Torr

außerdem mit einer kleineren Relaxationszeit behaftet ist. Das bedeutet, daß bei einkristallinen Proben der größte Wert der einachsigen Anisotropieenergie durch Magnetfeldtemperung in einer [100]-Richtung erreicht wird und daß K_u bei Polykristallen in erster Näherung durch 4/10 F gegeben ist.

3.2 Manganferrit mit Zusätzen von Mg, Zn und Ni

Der Einfluß von Mg^{2+}-, Zn^{2+}- und Ni^{2+}-Ionen auf die induzierte Anisotropieenergie wurde an folgenden Zusammensetzungen untersucht:

$$Me_zMn_{1,4-z}Fe_{1,6}O_{4+\gamma} \quad \text{mit} \quad z = 0{,}22 \text{ und } 0{,}44.$$

Die Messungen sind in Fig. 4 und 5 dargestellt.

Alle drei Ionenarten bewirken eine Abnahme der induzierten Anisotropieenergie mit wachsender Konzentration dieser Ionen. Diese Abnahme ist in erster Näherung unabhängig von der Ionensorte, obwohl die Platzverteilung der verwendeten Ionen im Ferritgitter extrem unterschiedlich sein sollte.

Die Relaxationszeiten nehmen mit wachsender Konzentration der genannten Ionen ebenfalls ab (Fig. 5).

Durch Glühung der Proben bei 10^{-2} Torr ergibt sich wieder eine Vergrößerung von K_u, die nicht von der Zusammensetzung abhängt und die gleiche Größe wie bei den reinen Manganferriten erreicht.

Wie orientierende Messungen zeigten, nehmen die Relaxationszeiten nach Reduktionsglühungen ebenfalls ab.

4. Diskussion der Ergebnisse

Die Deutung der Meßergebnisse wird dadurch erschwert, daß weder der Mechanismus der induzierten Anisotropie noch die Verteilung der Ionen auf die Tetraeder- und Oktaederplätze des Spinellgitters genau bekannt sind. Für Mn-Ferrit sind in der Literatur im Zusammenhang mit Magnetisierungsmessungen und neutronographischen Untersuchungen verschiedene Formeln diskutiert worden.

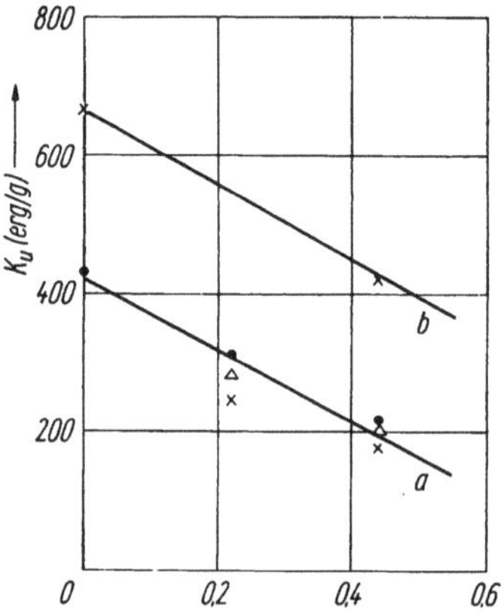

Fig. 4. Induzierte einachsige Anisotropieenergie K_u für einige Ferrite der Zusammensetzung $Me\cdot Mn_{1,4-z}Fe_{1,6}O_{4+\gamma}$ (●—●—●, Me = Zn, △—△—△ Me = Ni, ×—×—× Me = Mg). Kurve (a): Glühung bei 1200 °C und 760 Torr, Kurve (b): Glühung bei 1200 °C und 10^{-2} Torr

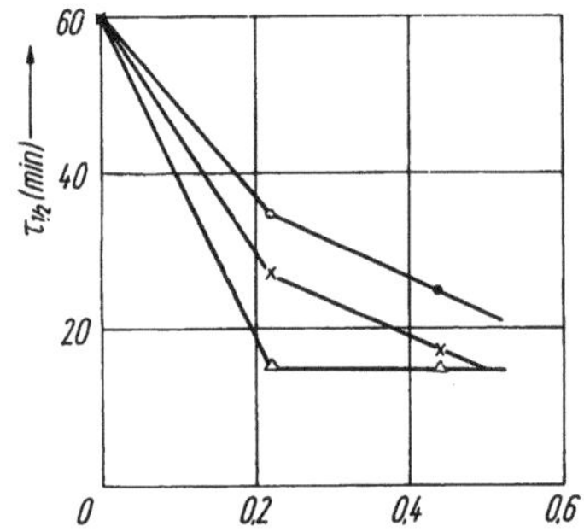

Fig. 5. Halbwertszeit $\tau_{1/2}$ für einige Ferrite der Zusammensetzung $Me_zMn_{1,4-z}Fe_{1,6}O_{4+\gamma}$ (Glühung bei 1200 °C und 760 Torr). ●—●—● Me = Zn, ×—×—× Me = Mg, △—△—△ Me = Ni

So erklärten Harrison und Kriessman [9] die Sättigungsmagnetisierung durch anomale magnetische Momente von Mn und Fe, Krupička, Brož und Závěta [10] nahmen Mn^{4+} und Fe^{2+} auf Oktaederplätzen an und Moruzzi [11] diskutierte gekantete Spinanordnungen. Beim stöchiometrischen Mn-Ferrit weisen neutronographische [12] und chemische Untersuchungen [13, 14] darauf hin, daß ein bestimmter Anteil des Mn auf Oktaederplätzen sitzt und dreiwertig ist und die gleiche Menge Fe^{2+} auf Oktaederplätzen vorkommt. Butler und Buessen [15] wiesen nach, daß dieser Anteil noch vom Sauerstoffgehalt der Proben abhängig ist.

Obwohl bisher die Existenz von Mn^{4+} bei höheren Mn-Gehalten nicht eindeutig ausgeschlossen werden konnte, neigt man zur Zeit mehr zu folgender Darstellung für Mn-Ferrite:

$$(Mn^{2+}_{0,8+\alpha}Fe^{3+}_{0,2-\alpha})[Mn^{3+}_{0,2+x-\alpha}Fe^{2+}_{0,2-\alpha}Fe^{3+}_{1,6-x+2\alpha}]O_{4+\gamma}, \quad (0 < \alpha < 0,2),$$

auf die wir uns im folgenden beschränken wollen.

Nach dieser Formel könnte man die lineare Abhängigkeit des K_u von x (Kurve (a) in Fig. 2) damit erklären, daß die induzierte Anisotropieenergie mit den Mn^{3+}-Ionen auf Oktaederplätzen zusammenhängt und daß die Größe α für $x = 0$ in der Nähe von Null liegt.

Von Krupička und Mitarbeitern [5] wurde z. B. der Jahn-Teller-Effekt der Mn^{3+}-Ionen auf Oktaederplätzen für die induzierte einachsige Anisotropieenergie verantwortlich gemacht. Nach den in [5] angegebenen Beziehungen erhält man für K_u in guter Näherung den Ausdruck

$$K_u = \frac{6\,N\,D^2}{5\,k\,T}\left[\frac{1-\frac{y}{2}-y^2-\frac{y^3}{2}+y^4}{1+y+y^2+y^3+y^4}\right]^2 \tag{3}$$

mit

$$y = \exp\left(-\frac{g\,\beta\,H_{ex}}{k\,T}\right).$$

In diesen Formeln bedeuten N die Anzahl der tetragonal verzerrten Oktaeder pro Gramm, H_{ex} ein effektives Austauschfeld, β das Bohrsche Magneton, g den Landé-Faktor, k die Boltzmannkonstante und D einen Parameter, der mit der Aufspaltung der Energieniveaus des Mn^{3+} zusammenhängt. Legt man die in [5] angegebenen Daten für D und $\beta\,H_{ex}$ zugrunde, dann erhält man nach Gleichung (3) die in Tabelle 1 zusammengestellten Werte für K_u in Abhängigkeit von x.

Tabelle 1

x	0,0	0,25	0,55	0,8
K_u (erg/g)	140×10^2	60×10^2	41×10^2	22×10^2

Ein Vergleich mit den Meßergebnissen in Fig. 2 zeigt, daß die diskutierte Theorie in dem untersuchten Temperaturbereich weder die Größe noch die gefundene Konzentrationsabhängigkeit von K_u erklären kann. Auch die Abhängigkeit des K_u vom Sauerstoffgehalt läßt sich mit dieser Theorie nicht deuten.

Nimmt man an, daß die induzierte Anisotropie statt durch Mn^{3+} durch andere Einzelionen hervorgerufen wird, so ergeben sich ebenfalls Abhängigkeiten, die den Meßergebnissen widersprechen.

Außer Einzelionen könnten auch Ionenpaare die induzierte Anisotropie verursachen. Von allen möglichen Ionenpaaren auf Oktaeder- bzw. Tetraederplätzen kämen nur die Mn^{3+}-Fe^{2+}-Paare auf Oktaederplätzen zur Erklärung der Konzentrationsabhängigkeit von K_u in Frage. Dazu muß jedoch vorausgesetzt werden, daß die beiden Partner statistisch verteilt sind, weil nur in diesem Falle die Anzahl der Paare dem Produkt der Konzentrationen von Mn^{3+} und Fe^{2+} proportional wäre.

Die Deutung der Leitfähigkeitsmessungen von Lotgering [16] durch Šimša [17] zeigt, daß diese Voraussetzung nicht erfüllt ist, da die Fe^{2+}-Ionen zu stark an Mn^{3+} gebunden sind.

Formal kann man weiterhin auch Paare von Mn^{2+}-Ionen auf Tetraederplätzen und Mn^{3+}-Ionen auf Oktaederplätzen in Betracht ziehen. Mit einem Produktansatz in der Form

$$K_u = A\,(0{,}8 - z) \cdot (0{,}2 - x) \tag{4}$$

lassen sich die Meßergebnisse (sowohl Kurve (a) in Fig. 2 als auch Kurve (a) in Fig. 4) in guter Näherung angleichen. Dabei ist für die reinen Manganferrite die oben angegebene Formel mit $\alpha = 0$, für MnZn-Ferrite die Darstellung

$$(Zn_z^{2+}Mn_{0,8-z}^{2+}Fe_{0,2}^{3+})\,[Mn_{0,2+x}^{3+}Fe_{0,2}^{2+}Fe_{1,6-x}^{3+}]O_4$$

mit $x = 0{,}4$ und für die untersuchten MnNi-Ferrite die Formel

$$(Mn_{0,8-z}^{2+}Fe_{0,2+z}^{3+})\,[Mn_{0,2+x}^{3+}Fe_{0,2}^{2+}Fe_{1,6-x-z}^{3+}Ni_z^{2+}]O_4$$

mit $x = 0{,}4$ zugrunde gelegt worden. (Bei MgMn-Ferriten liegt der Inversionsgrad zwischen diesen beiden Extremwerten.) Die Fe^{2+}-Konzentration wurde durchweg als konstant angenommen ($\alpha = 0$). Die möglicherweise auftretende geringe Abhängigkeit des α von x und z hat außerdem nur einen kleinen Einfluß auf K_u.

Der Anstieg der induzierten Anisotropieenergie nach einer Vakuumglühung der Proben läßt sich mit diesem Modell nicht erklären.

Da sich die Gesamtheit der Meßergebnisse durch keinen der genannten Einzelmechanismen deuten läßt, ist anzunehmen, daß die induzierte Anisotropieenergie bei den diskutierten Ferriten durch ein Zusammenwirken mehrerer Prozesse oder durch größere Ionenkomplexe hervorgerufen wird.

Herrn Dr. H. Schröder danken wir für die thermogravimetrischen Messungen, Herrn A. Funke für die Durchführung der chemischen Analysen und Herrn Dr. G. Elbinger für die Herstellung der Einkristalle. Den Herren Dr. R. Perthel und Dr. S. Krupička sind wir für anregende Diskussionen sehr zu Dank verpflichtet.

Literatur

[1] E. STEINBEISS, phys. stat. sol. **3**, K159 (1963).
[2] A. v. KIENLIN, Solid State Phys., Electron. Telecomn. **3**, 1 (1960).
[3] W. GIESEČKE, Z. angew. Phys. **11**, 91 (1959).
[4] S. KRUPIČKA und P. NOVÁK, phys. stat. sol. **4**, K117 (1964).

[5] S. Krupička, L. Červinka, P. Novák und K. Závěta, Proc. Internat. Conf. Magnetism, Nottingham 1964 (S. 650).
[6] A. Bergstein und P. Kleinert, Collect. Czechoslov. Chem. Commun. **29**, 2549 (1964).
[7] L. Michalowsky, Dissertation, Leipzig (1965).
[8] K. Závěta, E. I. Trinkler und F. Zounová, phys. stat. sol. **14**, K9 (1966).
[9] S. E. Harrison, C. J. Kriessman und S. R. Pollack, Phys. Rev. **110**, 844 (1958).
[10] J. Brož, S. Krupička und K. Závěta, Czech. J. Phys. **9**, 481 (1969).
[11] V. L. Moruzzi, J. appl. Phys., Suppl. **32**, 59S (1961).
[12] I. I. Yamzin, N. V. Belov und Yu. Z. Nozik, J. Phys. Soc. Japan **17**, Suppl. B III, 55 (1962).
[13] F. C. M. Driessens, Recueil Trav. chim. Pays-Bas **83**, 329 (1964).
[14] W. Bonsels und I. Linnemann, Z. analyt. Chem. **214**, 422 (1965).
[15] S. R. Butler und W. R. Buessem, J. appl. Phys. **34**, 1754 (1963).
[16] F. K. Lotgering, J. Phys. Chem. Solids **25**, 95 (1964).
[17] Z. Šimša, Czech. J. Phys. **B15**, 435 (1965).

(Received June 1, 1966)

Erratum

In the paper by G. Alefeld

Plasma-Type Oscillations of Kinked Dislocations

phys. stat. sol. **15**, 331 (1966)

the following corrections are to be applied:

On page 332 interchange superscript + and − on right-hand side of equations (4)and (5).

On page 336, line 4 read $\tau_0 = B\,L^2/E_L\,\pi^2$.

On page 335, line 25, after "... temperature." the following note added in proof is missing: A. D. Brailsford (J. appl. Phys. **36**, 3941 (1965)) recently arrived at essentially the same conclusion. Yet Brailsford apparently underestimated the size of this temperature dependence by using an equation which is valid only for non-interacting kinks. By employing a Debye-Hückel-type theory for interacting kinks, one can estimate that the kink density reaches saturation below room-temperature for Cu and Al.

See Short Note by M. Rühle and M. Wilkens

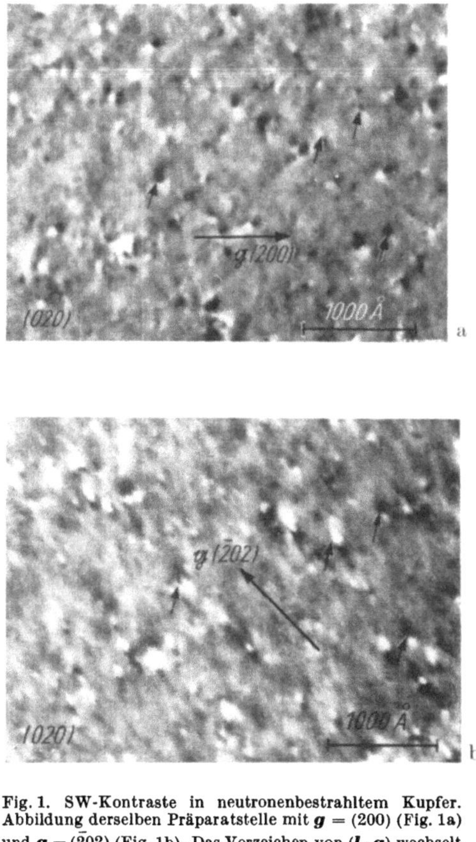

Fig. 1. SW-Kontraste in neutronenbestrahltem Kupfer. Abbildung derselben Präparatstelle mit $\boldsymbol{g} = (200)$ (Fig. 1a) und $\boldsymbol{g} = (\bar{2}02)$ (Fig. 1b). Das Vorzeichen von $(\boldsymbol{l} \cdot \boldsymbol{g})$ wechselt bei den mit ↑ gekennzeichneten SW-Kontrasten

See Short Note by H. Diepers and J. Diehl

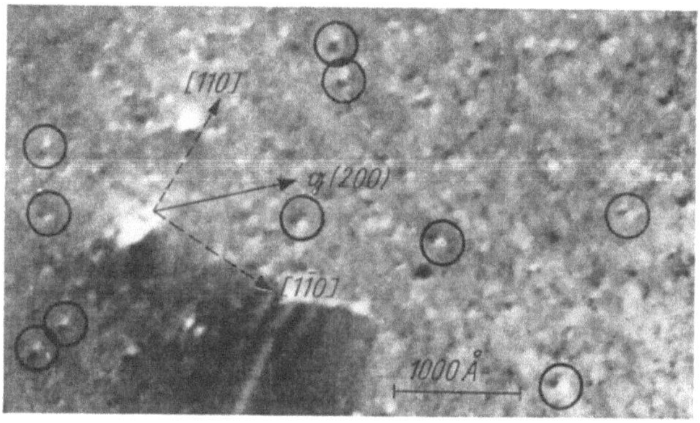

Fig. 1. Cu-Folien, untere Seite bestrahlt bei Raumtemperatur mit $0{,}8 \times 10^{15}$ Ar-Ionen/cm^2, Dunkelfeld

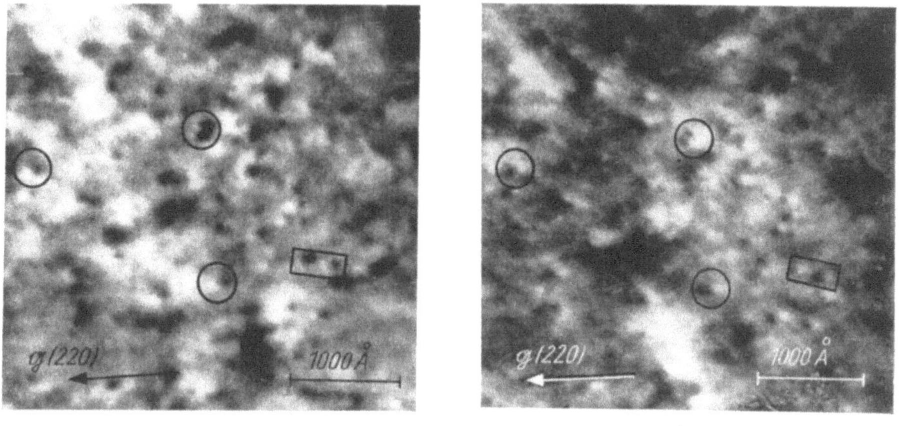

Fig. 3. Cu-Folie, untere Seite bestrahlt bei Raumtemperatur mit $1{,}5 \times 10^{15}$ Ar-Ionen/cm^2. a) Hellfeld, b) Dunkelfeld

See Short Note by G. B. ABDULLAEV et al.

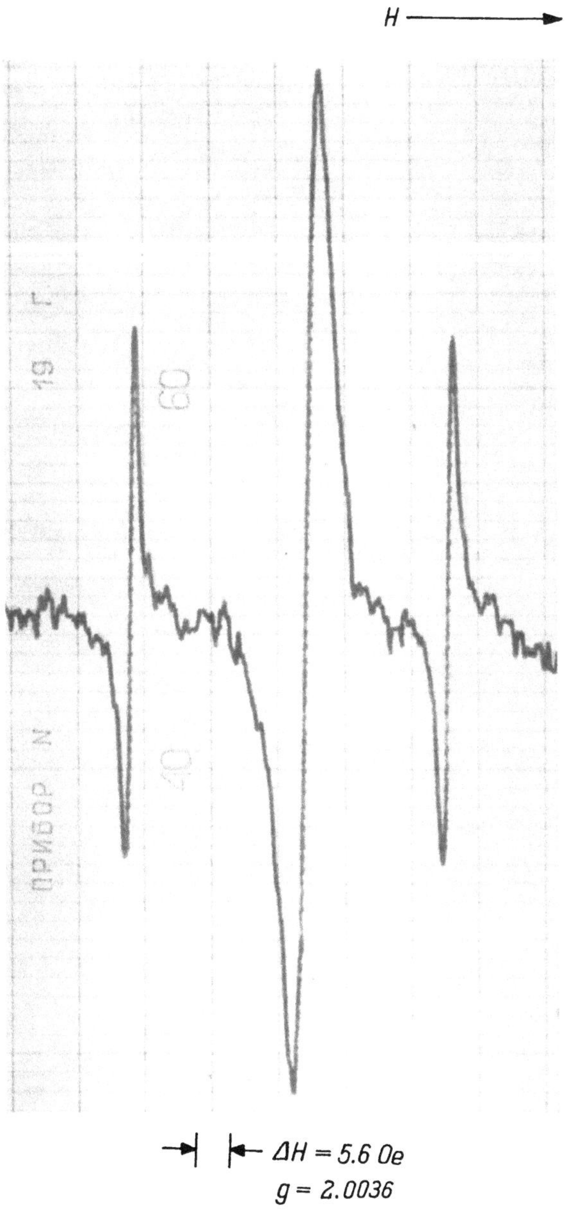

Fig. 1. ESR spectrum of selenium (central line). The lines on the left and on the right refer to Mn^{2+} ions in ZnS (natural sphalerite)

phys. stat. sol. 16, K105 (1966)

Institut für Physik am Max-Planck-Institut für Metallforschung, Stuttgart

Die Natur der elektronenmikroskopisch beobachtbaren Defekte in neutronenbestrahltem Kupfer

Von

M. RÜHLE und M. WILKENS

Makin und Mitarbeiter (1 bis 3) haben bei elektronenmikroskopischer Durchstrahlung von neutronenbestrahltem Kupfer schwarze Punkte beobachtet, deren Durchmesser kleiner als ≈ 200 Å waren. Essmann und Wilkens (4) haben gezeigt, daß bei Abbildung in dynamischer Zweistrahlbeugung ($w \approx 0$, w normierter Anregungsfehler) die schwarzen Punkte charakteristische Schwarz-Weiß-Kontraste (SW-Kontraste) zeigen. Aus den beobachteten kristallographischen Richtungen der (Quasi-)Symmetrieachsen durch die SW-Kontraste schlossen diese Autoren, daß die "schwarzen Punkte" von Frankschen Versetzungsringen auf {111}-Ebenen herrühren.

Die Berechnung des Beugungskontrastes von Frankschen Versetzungsringen durch Rühle, Wilkens und Essmann (5) ergab in Übereinstimmung mit der Annahme von (4), daß die Symmetrieachse durch die Projektion des Burgersvektors des Versetzungsringes auf die Folienebene verläuft. Führt man einen von "Schwarz" nach "Weiß" weisenden Vektor ℓ parallel zur Symmetrie achse ein, so folgt auf Grund der Rechnungen (5), daß das Vorzeichen des Produkts $(\mathfrak{g} \cdot \ell)$ ($\mathfrak{g}$ Beugungsvektor) periodisch mit dem Abstand des Defektes von der nächstbenachbarten Oberfläche wechselt, wobei die volle Periode dieser "Tiefenoszillationen" der Extinktionslänge ξ_g (genauer $\xi_g/\sqrt{1 + w^2}$) entspricht.

Wir haben die Kontrastrechnungen von (5) benutzt, um experimentell den Typ der Versetzungsringe (Leerstellen-Typ oder Zwischengitteratom-Typ) zu bestimmen. Nimmt man an, daß bei der Neutronenbestrahlung die Defekte gleichmäßig über den Folienquerschnitt verteilt sind und daß nur Defekte eines Typs auftreten, so sollten wegen der weiter oben erwähnten Tiefenoszillationen nicht nur im Hellfeld, sondern auch im Dunkelfeld etwa gleich viel SW-Kontraste mit $(\mathfrak{g}\cdot\ell) > 0$ und mit $(\mathfrak{g}\cdot\ell) < 0$ auftreten; die Regeln von Ashby und Brown (6) zur Bestimmung des Defekt-Typs sind demnach auf Versetzungsringe nicht anwendbar.

Um abzuzählen, innerhalb welcher Halbperiode der Tiefenoszillationen der Defekt liegt, benötigt man den Abstand des Gitterdefekts von der nächstbenachbarten Oberfläche. Diese Größe läßt sich direkt durch die quantitative Auswertung von Stereoaufnahmen ermitteln. Indirekt kann die Halbperiode durch Vergleich der Abbildung der gleichen Präparatstelle mit Beugungsvektoren verschiedener Extinktionslängen bestimmt werden. Wir haben beide Methoden angewandt.

Bei den im folgenden beschriebenen Experimenten wurde sorgfältig darauf geachtet, daß sowohl die von Rühle, Wilkens und Essmann (5) beschriebenen Polierdefekte als auch die durch Ionenbeschuß im Mikroskop entstehenden Defekte (siehe (7)) vermieden wurden. Bei den Untersuchungen von Essmann und Wilkens (4) waren den bei der Neutronenbestrahlung entstandenen Defekten die oben angegebenen störenden Defekte überlagert (siehe (5)). Die in (4) unter 4. und 5. beschriebenen Eigenschaften der SW-Kontraste waren durch diese zusätzlichen Defekte verfälscht.

Gut rekristallisierte Kupferfolien ($\approx$ 50 μm dick) wurden im Forschungsreaktor München bei $\approx 90\ ^{\circ}C$ mit 4×10^{17} n/cm^2 ($E_n > 0{,}1$ MeV) bestrahlt. Die dünnpolierten Folien wurden bei 150 kV im JEM-150-Mikroskop untersucht.

Die ersten drei der von Essmann und Wilkens (4) beschriebenen Eigenschaften der SW-Kontraste wurden bestätigt. Zusätzlich wurden folgende Experimente und Beobachtungen gemacht:

1. Hellfeld- und Dunkelfeldaufnahmen

Bei Hellfeldaufnahmen sind gleich viel SW-Kontraste mit "positivem" ($(\ell\cdot\mathfrak{g}) > 0$) als auch mit "negativem" ($(\ell\cdot\mathfrak{g}) < 0$) Vorzeichen vorhanden. Die in (5) unter 4a diskutierten Symmetriebeziehungen sind demnach erfüllt. - Bei Dunkelfeldaufnahmen sind ebenfalls beide Richtungen der SW-Kontraste ungefähr gleich häufig.

2. Stereoaufnahmen

Für die Stereoaufnahmen wurde der Anregungsfehler w so stark vergrößert, daß die SW-Kontraste zu schwarzen Pünktchen entarteten. Die Probe wurde zwischen den zwei (Hellfeld)-Aufnahmen um den Beugungsvektor $\mathfrak{g} = (200)$ um etwa 15° gekippt. Mit Hilfe des Zeiss-Stereoskops N 2 bestimmten wir quantitativ den Abstand der Defekte von den Oberflächen (siehe Diepers und Diehl (8)). Die Defekte, die bei der dynamischen Abbildung einen SW-Kontrast zeigen, sind im wesentlichen auf 4 Schichten in der Folie verteilt: In der obersten Schicht (sie reicht von der Ober-Fläche der Folie bis in eine Tiefe von ≈ 70 Å) ist stets $(\mathfrak{l} \cdot \mathfrak{g}) > 0$, in der 2. Schicht (von ~ 70 bis ≈ 150 Å) ist $(\mathfrak{l} \cdot \mathfrak{g}) < 0$. Nahe der Unter-Fläche der Folie sind ebenfalls 2 Schichten zu beobachten, jedoch hat $(\mathfrak{l} \cdot \mathfrak{g})$ jeweils das entgegengesetzte Vorzeichen wie in der entsprechenden Schicht nahe der Ober-Fläche.

3. Wechsel des Beugungsvektors

Wir haben dieselbe Präparatstelle ((020)-Folie) nacheinander mit $\mathfrak{g} = (\bar{2}02)$ und $\mathfrak{g} = (200)$ abgebildet. In einigen Fällen (siehe Fig. 1a und 1b) ändert der SW-Kontrast sein Vorzeichen, d.h. aus $(\mathfrak{l} \cdot \mathfrak{g}) > 0$ wird $(\mathfrak{l} \cdot \mathfrak{g}) < 0$ und umgekehrt. Da $\xi_{\bar{2}02} \approx 1,5\ \xi_{200}$ ist, liegen diese Defekte vermutlich bei $\mathfrak{g} = (\bar{2}02)$ gerade noch in der 1. und bei $\mathfrak{g} = (200)$ schon in der 2. Halbperiode der Tiefenoszillationen. Diese Aussage ist im Einklang mit den obenerwähnten Stereoaufnahmen. Damit sind die von der Theorie (5) erwarteten Tiefenoszillationen experimentell nachgewiesen.

Aus unseren Experimenten folgt: Die SW-Kontraste rühren zum überwiegenden Teil von Frankschen Versetzungsringen her, diese Aussage von (4) wird somit bestätigt. Die beobachteten Vorzeichen der Produkte $(\mathfrak{l} \cdot \mathfrak{g})$ und die obenerwähnten Experimente 2. und 3. in Verbindung mit den Kontrastrechnungen (5) zeigen, daß die Versetzungsringe vom Leerstellentyp sind.

Unter ca. 200 ausgewerteten SW-Kontrasten befand sich kein Zwischengitteratomring.

Der Deutschen Forschungsgemeinschaft und den Herren vom FR München sei für ihre Unterstützung gedankt.

Literatur

(1) M.J. MAKIN, A.D. WHAPHAM und F.J. MINTER, Phil. Mag. 6, 465 (1961).

(2) M.J. MAKIN, A.D. WHAPHAM und F.J. MINTER, Phil. Mag. 7, 285 (1962).

(3) M.J. MAKIN und S.A. MANTHORPE, Phil. Mag. 8, 1725 (1963).

(4) U. ESSMANN und M. WILKENS, phys. stat. sol. 4, K53 (1964).

(5) M. RÜHLE, M. WILKENS und U. ESSMANN, phys. stat. sol. 11, 819 (1965).

(6) M.F. ASHBY und L.M. BROWN, Phil. Mag. 8, 1083 (1963).

(7) M. RÜHLE und M. WILKENS, VI-th International Congress for Electron Microscopy, Kyoto, Japan, im Druck.

(8) H. DIEPERS und J. DIEHL, phys. stat. sol. 16, K109 (1966).

(Received June 8, 1966)

phys. stat. sol. 16, K109 (1966)

Institut für Sondermetalle am Max-Planck-Institut für Metallforschung, Stuttgart

Nachweis und Entstehung von Zwischengitteratom-Agglomeraten in ionenbestrahlten Kupferfolien

Von

H. DIEPERS und J. DIEHL

Bei elektronenmikroskopischer Durchstrahlung von Metallfolien, die mit Ionen beschossen worden waren, wurden von verschiedenen Autoren ähnlich wie nach Neutronenbestrahlung kleine schwarze Punkte beobachtet. Bei Bestrahlung mit schweren Edelgasionen niedriger Energie (einige 100 eV), die nur einige wenige Atomlagen tief in die Folie eindringen, sollen nach Brandon und Bowden (1, 2) diese Punktkontraste in Cu, Au und Ni von Zwischengitteratomen (ZG-Atomen) herrühren, die durch fokussierende Stoßfolgen (dynamische Crowdionen) tiefer ins Gitter eindringen und dort agglomerieren. (Bei der Bestimmung des Defekttyps - Agglomerate aus Leerstellen oder ZG-Atomen - wurden in (2) allerdings die Regeln von Ashby und Brown (3) verwendet, die im Falle von ebenen Punktfehleragglomeraten (Versetzungsringe) nach Rühle et al. (4, 5) nicht allgemein zutreffen. Im Gegensatz hierzu vermuten Castaing et al. (6), daß es sich nach Bestrahlung von Au mit 4-keV-Ar-Ionen bei den sichtbaren Defekten um Leerstellenagglomerate handelt. Zur näheren Aufklärung der Verhältnisse im keV-Bereich wurden epitaktisch gewachsene ca. 1000 Å dicke Cu-Folien (99,999%, Folienebene (100)), die mit 5-keV-Ar-Ionen bestrahlt worden waren, in einem Elmiskop I bei 100 keV Beschleunigungsspannung untersucht.

Dabei ergab sich folgendes:

1. Agglomeratkonfiguration

Bei Abbildung in dynamischer Zweistrahlbeugung zeigen die schwarzen Punkte die bekannten Schwarz-Weiß-Kontraste (SW-Kontraste) (siehe Fig. 1). Die (Quasi-)Symmetrieachsen (sie werden durch zu ihnen parallele, von "Schwarz" nach "Weiß" zeigende Vektoren ℓ charakterisiert) sind im allgemeinen nicht parallel zum Beugungsvektor g, wie man es für kugel-

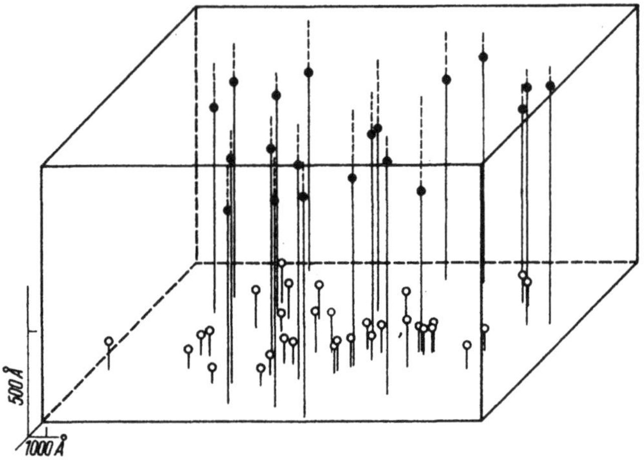

Fig. 2. Abstand der Defekte von der unteren Oberfläche in einer beidseitig mit $0{,}8 \times 10^{15}$ Ar-Ionen bestrahlten Cu-Folie. (Die eingezeichneten Abstände zur oberen Oberfläche geben wegen Dickenschwankungen der Folie die Tiefenlage der Defekte in der oberen Schicht nicht genau wieder)

förmige Agglomerate erwarten würde, sondern parallel zu einer ⟨110⟩-Richtung. Dies weist darauf hin, daß die Kontraste von Stapelfehlerscheibchen auf {111}-Ebenen herrühren, von denen man nach (4) erwartet, daß ℓ in der Projektion des Burgersvektors $\frac{a}{3}$⟨111⟩ der sie umrandenden Frankversetzung auf die Folienebene liegt.

2. Räumliche Anordnung der Defekte

Aus dem Vorzeichen des Produktes $(\mathfrak{g} \cdot \ell)$ kann man entnehmen, ob die Stapelfehlerscheibchen durch Agglomeration von Leerstellen oder von ZG-Atomen entstanden sind. Hierzu ist jedoch die Kenntnis der räumlichen Anordnung der Defekte erforderlich, da nach (4) das Vorzeichen des Produktes $(\mathfrak{g} \cdot \ell)$ mit dem Abstand des Agglomerates von der Oberfläche periodisch wechselt. Die Anordnung wurde Stereoaufnahmen entnommen (Kippwinkel $\approx 10^{\circ}$ um $\mathfrak{g}$). Fig. 2 zeigt das Ergebnis von Ausmessungen mit einem Zeiss-Stereoskop N 2 an einer Folie, die bei Raumtemperatur beidseitig bestrahlt wurde und deren Oberfläche zu ihrer Sichtbarmachung nach der Bestrahlung mit Au bedampft wurde. Die Agglomerate liegen ungefähr $t_D \approx 140$ Å unterhalb der bestrahlten Oberfläche innerhalb einer relativ dünnen Schicht von ca. 30 Å.

3. Defekttyp

Bei Abbildung mit $\mathfrak{g} = (200)$ entspricht t_D etwa $\xi_g/2$ (ξ_g Extinktionslänge; $\xi_{200} = 281$ Å). Dafür erwartet man nach (4) unter den experimentellen Bedingungen der Fig. 1 (Dunkelfeld, beschossene Oberfläche im Mikroskop unten) für Leerstellenagglomeration $(\mathfrak{g} \cdot \ell) < 0$ und für ZG-Atomagglomeration $(\mathfrak{g} \cdot \ell) > 0$. Für die SW-Kontraste in Fig. 1 gilt $(\mathfrak{g} \cdot \ell) > 0$. Daraus folgt, daß die Defekte aus ZG-Atomen aufgebaute Stapelfehlerscheibchen sind. Defekte, die einen Leerstellen entsprechenden SW-Kontrast aufweisen, wurden nicht gefunden.

Für Abbildung mit einem (220)-Reflex ist ξ_g um etwa 50% größer als für $\mathfrak{g} = (200)$, so daß die Tiefe t_D in der Nähe von $\xi_g/4$ liegt. In diesem Fall (Fig. 3) erhält man umgekehrten SW-Kontrast (Dunkelfeld $(\mathfrak{g} \cdot \ell) < 0$, Hellfeld $(\mathfrak{g} \cdot \ell) > 0$). Diese Kontrastumkehr entspricht der von der Theorie (4) geforderten Tiefenoszillation der SW-Kontraste (vgl. Fig. 1 in (4)). Sie bestätigt unseren Schluß über den Defekttyp und zeigt, daß man bei Anwendung der Regeln in (2) bei $\mathfrak{g} = (200)$ zu einer falschen Schlußfolgerung käme.

4. Defektentstehung und Crowdionreichweite

Die Defekte können nicht auf Ablagerung der eingeschossenen Ar-Ionen zurückgeführt werden, da deren normale Reichweite (R_{Ar}) nur etwa 50 Å beträgt. Ein mögliches tieferes Eindringen durch "Kanaleffekt" kann auf Grund von Versuchen mit verschiedenen Einschußrichtungen, die zu gleichartigen Ergebnissen führten, ausgeschlossen werden. Somit läßt sich die eingangs genannte Erklärung, wonach man es mit Agglomeraten von "Eigen"-Atomen zu tun hat, die durch dynamische Crowdionen in tieferen Schichten erzeugt werden, auch auf Ionenbeschuß im keV-Bereich anwenden. Zusätzliche Versuche unterhalb Raumtemperatur, bei denen die Folientemperatur stets < 200 °K war, d.h. in einem Bereich, in dem ZG-Atome thermisch nicht wandern können, führten zu keinen wesentlich verschiedenen Beobachtungen (gleiche SW-Kontraste, etwa gleiche Tiefenlage). Daraus läßt sich folgern, daß die Agglomerate großenteils athermisch durch gegenseitigen Einfang der ZG-Atome entstehen (bei Raumtemperatur kommt offenbar Keimbildung durch thermische Agglomeration hinzu, da die Anzahl der Defekte um bis zu einem Faktor 2 größer ist als bei $T < 200$ ^{o}K). Ferner ergibt sich, daß die Tiefenlage der Defekte mit einer mittleren Reichweite der Crowdionen (R_c) zu identifizieren ist. Aus dem in [100] gemessenen t_D erhält man nach Abzug von R_{Ar} für R_c in ⟨110⟩ ≈ 130 Å.

Die Mindestreichweite kann nicht sehr von R_c abweichen, da (auch bei Tieftemperaturversuchen) die Defekte immer in einer dünnen Schicht gefunden wurden. Im Gegensatz hierzu liegt bei den Rechnungen von Leibfried und Duesing (7) und den Auswertungen von Venables und Balluffi (8) die Annahme vor, daß keine Mindestreichweite auftritt, sondern Reichweiten bis praktisch auf den Wert Null herab möglich sind.

Literatur

(1) P. BOWDEN und D.G. BRANDON, Phil. Mag. 8, 935 (1963).
(2) P. BOWDEN und D.G. BRANDON, J. nuclear Mat. 9, 348 (1963).
(3) M.F. ASHBY und L.M. BROWN, Phil. Mag. 8, 1083 (1963).
(4) M. RÜHLE, M. WILKENS und U. ESSMANN, phys. stat. sol. 11, 819 (1965).
(5) M. RÜHLE und M. WILKENS, phys. stat. sol. 16, K105 (1966).
(6) R. CASTAING und B. JOUFFREY, J. Microscopie 1, 201 (1962); 2, 45 (1963).
(7) G. DUESING und G. LEIBFRIED, phys. stat. sol. 9, 463 (1965).
(8) J.A. VENABLES und R.W. BALLUFFI, Phil. Mag. 11, 1021 (1965).

(Received June 11, 1966)

phys. stat. sol. 16, K113 (1966)

Institute of Physics, Academy of Sciences of the Azerbaidzhan SSR, Baku

ESR Study of the Behaviour of Oxygen in Selenium

By

G.B. ABDULLAEV, N.I. IBRAGIMOV, SH.V. MAMEDOV, and T.CH. ZHUVARLY

The problem of oxygen in selenium is an interesting and almost inaccessible problem. At the same time the solution of this problem would lead, to some extent, to deeper understanding of many processes taking place in selenium. Probably, the disagreement between experimental results observed by different authors is due to oxygen.

Using selenium of high purity (99.9999%) we applied the ESR method to clarify the behaviour of oxygen in selenium. The microwave resonance equipment used was a standard radiospectrometer with double modulation of magnetic field (ν = 9304 MHz). As a standard DPPH and Mn^{2+} in ZnS (natural sphalerite) was used.

Previously (1) on amorphous, as well as on crystalline selenium which was not subjected to any special heat treatment, an almost symmetrical singlet without hyperfine structure was observed (see Fig. 1), with $g = 2.0036 \pm 0.0005$, peak-to-peak width $\Delta H = (5.6 \pm 0.5)$ Oe, and unpaired spin concentration of the order 10^{14} to 10^{15} cm^{-3}. In the range -196 to $+17$ ^{o}C the width and g-factor of the resonance line was independent of temperature, but an increase of intensity in agreement with Curie's law was observed. The most probable reason for the appearance of an ESR signal in selenium is the formation of donor-acceptor complexes with a charge transfer (C.Ch.T.) at the chain ends. As will be shown later, the action of such an active acceptor as oxygen on polymer molecules of selenium considerably increased the intensity of ESR signal. Evacuation to 10^{-5} Torr at room temperature did not lead to a considerable variation of the form and intensity of the signal.

In earlier experiments one did not give special attention to the prehistory and initial treatment of the samples, whereas subsequent experiments showed an appreciable influence of these factors on physical properties of selenium.

Selenium in the form of fine powder was filled in quartz ampoules which were evacuated to various pressures (10^{-1} to 10^{-5} Torr) and sealed. Then the ampoules were submitted to heat treatment at T = 350 °C during two hours, and at room temperature the ESR absorption of these samples was studied. It was found that selenium in ampoules evacuated to 10^{-5} Torr does not exhibit ESR absorption, and an ESR signal appears in the case of higher residual pressures (the parameters of the signal are the same as those mentioned above). The intensity of the signal decreased with increasing degree of evacuation. The maximum concentration was $\approx 10^{16}$ to 10^{17} unpaired spins per cm^3, and decreased nearly linearly with decreasing oxygen pressure. On the other hand, an increase of the temperature of the heat treatment (its length remaining constant) led to an increase of the signal intensity, during which the line parameters did not change. The upper limit of the heat treatment temperature at which the intensity increase is still observed was found to be $\approx$ 520 °C. Heat treatment at T > 520 °C leads to the disappearance of the ESR signal. The attempt to re-establish the ESR signal in the same samples by repeated heat treatment without contact with oxygen at T < 520 °C was unsuccessful. Probably, the signal disappears irreversibly. It must be noted that evacuation to 10^{-5} Torr at room temperature during a long time did not lead to a decrease of the ESR signal, which apparently indicates that oxygen is rather strongly bound to selenium. However, evacuation during heating to 400 °C, and recording at room temperature showed that the signal decreases with increasing evacuation time, and finally disappears.

If, as a result of heat treatment at T > 520 °C, or evacuation during heating, the ESR absorption has disappeared, a resonance absorption of the same nature is restored when the specimen is ground to fine powder in oxygen atmosphere and subjected to heat treatment at T $\leq$ 520 °C without evacuation. In this sense the process is completely reversible. This indicates that the surface plays an essential role in these processes. The attempt to detect an ESR signal in granulated selenium after different heat treatment in oxygen atmosphere led to a negative result. However, it was sufficient to enlarge the surface-volume ratio by pulverization (grain size $< 50\ \mu m$) in oxygen atmosphere and submit to heat treatment at $T \leq 520\ ^{o}C$, in order to make the stable ESR signal re-appear. This fact confirms the assumption of the fundamental role of the chain ends in the formation of paramagnetic centres. On the other hand, it appears that adsorbed oxygen is rather easily removed from selenium by evacuation, probably because it does not lead to the formation of C.Ch.T.; from this it follows that for the formation of C.Ch.T. thermal activation is required. A long storage of samples does not lead to a change of signal parameters.

It is known that the atoms in a selenium chain are bound covalently (saturated bonds), and the inter-chain binding is of the van der Waals type. Therefore one can regard selenium as a polymer. At the chain ends there are unsaturated bonds. Probably, oxygen bound physically at the chain ends forms C.Ch.T. during subsequent thermal activation. The high electronic polarizability and apparently relatively low ionization potential of the selenium chains as well as the positive electron affinity of oxygen molecules (+0.78 eV (2)) to a considerable degree contribute to this process.

The local paramagnetic centres formed in selenium containing oxygen, substantially behave as an admixture with respect to the basic structure of selenium. On the basis of this fact one may confirm that the conductivity of selenium submitted to heat treatment in the range 350 to 520 ^{o}C must be significantly larger than of selenium treated at $T > 520\ ^{o}C$. In our opinion, during high-temperature heat treatment ($T > 520\ ^{o}C$) a decomposition of C.Ch.T. occurs, and oxygen binds chemically with selenium. Because C.Ch.T. probably are suppliers of charge carriers, it is natural that the carrier concentration must sharply decrease with a decomposition of C.Ch.T.; as a result of this, the conductivity of selenium must decrease by three to four orders of magnitude. Apparently, the process of desoxygenization of selenium just involves a conversion of physically bound oxygen into chemically bound with subsequent evaporation of oxides from the bulk.

The crystallization of samples does not lead to any change of the ESR spectrum. Thus, in amorphous and crystalline selenium the number of chain ends per unit volume forming C.Ch.T. with oxygen is the same and does not exceed 10^{17} cm^{-3}. This means that the average number of atoms in a selenium chain amounts to 10^5 at room temperature. The maximum concentration of electrically active oxygen after the heat treatment procedure mentioned above does not exceed 10^{17} cm^{-3}.

Using corresponding activators one may significantly intensify the process of C.Ch.T. formation. Some control experiments showed that such a possibility exists.

References

(1) G.B. ABDULLAEV, N.I. IBRAGIMOV, SH.V. MAMEDOV, T.CH. ZHUVARLY, and G.M. ALIEV, Dokl. Akad. Nauk Azerb. SSR 20, 13 (1964).

(2) J.A. KAZARNOVSKI, Zh. fiz. Khim. 14, 320 (1940).

(Received June 16, 1966)

phys. stat. sol. 16, K117 (1966)

Department of Physics, School of Education, Gdańsk

Magnetic After-Effect of the Reversible Susceptibility of Cobalt

By

E. KLUGMANN

A decrease with time of the reversible susceptibility in polycrystalline cobalt was noted, both after demagnetization and after a quick change of a weak magnetic field. The effect of disaccommodation was observed by inductance measurements on a torus-shaped specimen in ac fields with frequencies of 800 and 8000 Hz and field strengths $H_{eff} < 100$ mOe, using a Maxwell-Wien bridge. The torus was made from electrically insulated cobalt wire (d = 0.5 mm) which had been heated in purified hydrogen at 1000 °C for 4 hours and gradually cooled to room temperature at a rate of about 5 degrees per minute. The cobalt wire (99.6%; impurities: 0.249% Ni, 0.016% Mn, 0.006% Cu, 0.004% Fe, 0.002% Cr, 0.048% Si) was obtained by a metallurgic powder method.

In specimens which had been annealed prior to measuring, for one hour at 300 °C, and then quickly cooled to -183 °C, an increase of the relaxation amplitude

$$A = \frac{\chi(t_1) - \chi(t_2)}{\chi(t_1)},$$

$t_1 = 30$ s, $t_2 = 10$ min, was noted to rise with temperature to 220 °C, this being followed by a monotonic fall with temperature to disappear at 350 °C.

Figs. 1, 2, and 3 show the relaxation amplitude as functions of temperature, ac measuring field, and magnetic field as well as the magnetic field dependence of reversible susceptibility.

As the magnetization of the specimen increases, the amplitude A falls considerably faster than the reversible susceptibility and disappears at field strengths where processes of reversible rotations of domain vectors J_s begin to appear.

The effect observed can be described by a logarithmic time dependence and occurs in the field range bringing about a mainly reversible movement of plane or cylindrically arching

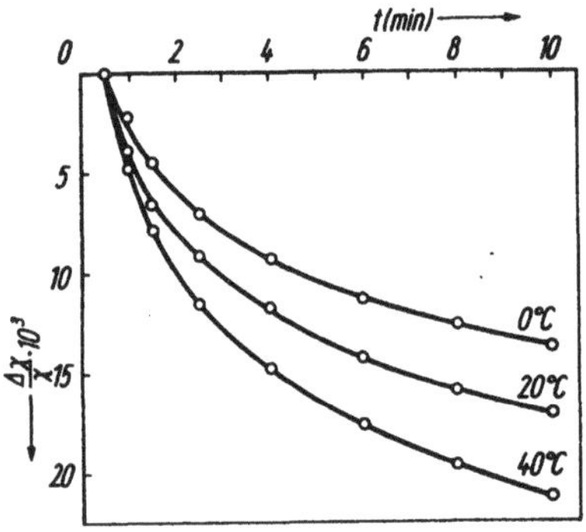

Fig. 1. Isotherms

$$\frac{\Delta\chi}{\chi} = \frac{\chi(t_1) - \chi(t)}{\chi(t_1)}$$

for a previously hardened specimen which was kept at different temperatures for 20 minutes. No decrease with time was noted at low temperatures (-183 °C)

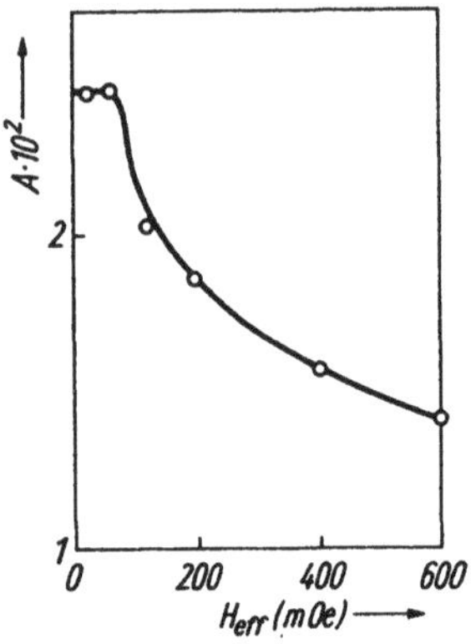

Fig. 2

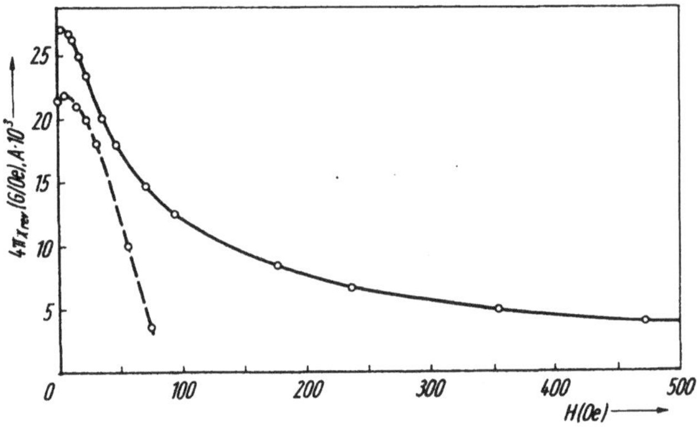

Fig. 3

Fig. 2. Relaxation amplitude at 100 °C as a function of an ac measuring field

Fig. 3. Magnetic field dependence of the relaxation amplitude (------) and reversible susceptibility (———) at 50 °C

mobile domain walls.

The average activation energy calculated from isotherms at 0 and 20 °C is found to be 0.1 eV, and this is near the value obtained from measurements of internal friction for a low-temperature cobalt phase (1). Among the lattice defects, which cause this delay, most probably a dominant role is played by the migration of divacancies or multiple vacancies, which are generated during deformation, become mobile during heating, and are frozen in during hardening. It should be mentioned that the migration energy of divacancies in Cu calculated by Johnson (2) amounts to 0.07 eV, and annealing of vacancies in Ni occurs in the so-called IV-th recovery stage at about 270 °C (3).

References

(1) R. KAMEL and K.Z. BOTROS, phys. stat. sol. 12, 399 (1965).
(2) R.A. JOHNSON, J. Phys. Chem. Solids 26, 75 (1965).
(3) P. SIMSON and R. SIZMANN, Z. Naturf. 17a, 596 (1962).

(Received June 2, 1966)

phys. stat. sol. 16, K121 (1966)

Physikalische Fakultät der Moskauer Staatlichen Universität, Moskau

Über die Ummagnetisierung dünner ferromagnetischer Schichten längs der Achse leichter Magnetisierung

Von

O.S. KOLOTOV, T.N. LETOVA, V.A. POGOZHEV und R.V. TELESNIN

In der Mehrzahl der Veröffentlichungen wird die Untersuchung von Prozessen der Impulsummagnetisierung dünner Schichten unter der Einwirkung magnetischer Felder mit einer maximalen Feldstärke H_S der Größenordnung 6 bis 10 Oe durchgeführt, d.h. solcher Felder, die die Grenzschwellenfelder für nichtkohärente Drehung unwesentlich überschreiten (1 bis 4). Offensichtlich ist es wünschenswert, den Feldstärkebereich der Ummagnetisierungsfelder zu erweitern, um detailliertere Information über die Eigenschaften des Mechanismus der nichtkohärenten Drehung zu erhalten.

In der vorliegenden Arbeit wurde das Ummagnetisierungsfeld H_S längs der leichten Achse der Schicht angelegt; dabei erreichte die Feldstärke H_S 36 Oe.

Die Erhöhung der Feldstärke des Ummagnetisierungsfeldes stellt besonders hohe Anforderungen an die Versuchsgeometrie: es ist unumgänglich, daß das Ummagnetisierungsfeld mit maximaler Genauigkeit in Richtung der mittleren leichten Achse der Schicht verläuft und daß die dazu senkrechte Feldkomponente gegen Null geht. Das konstante senkrechte Feld des Laboratoriums wurde mit einer Genauigkeit kompensiert, die nicht schlechter war als ±0,002 Oe. Um die Achse leichter Ummagnetisierung und das Feld H_S gleichzurichten, wurde von uns die Abhängigkeit der Form und Dauer des Ummagnetisierungssignals vom senkrechten Feld benutzt. Einzelne Elemente der Apparatur, die in unseren Untersuchungen benutzt wurde, sind in den Arbeiten (5 bis 7) behandelt.

Fig. 1 zeigt die Abhängigkeit der reziproken Ummagnetisierungszeit von der Feldstärke H_S für eine Schicht der Dicke 1100 Å, die durch thermisches Zerstäuben einer Legierung von 43% Fe + 57% Ni mit einer maximalen Winkeldispersion von 6 Grad

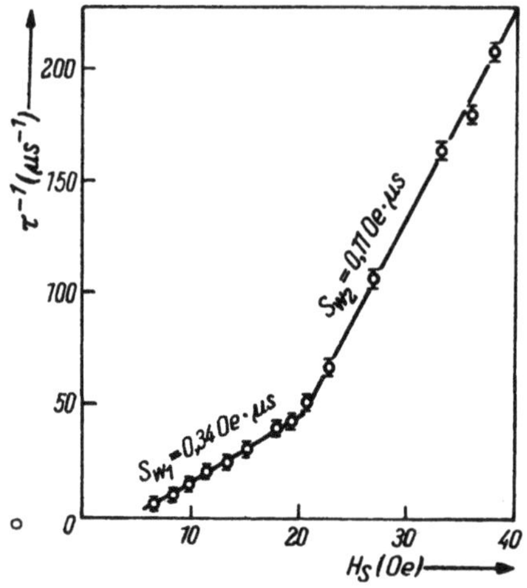

Fig. 1. Die Abhängigkeit $\tau^{-1} = f(H_S)$ für eine Schicht der Dicke 1100 Å

erhalten wurde. Man sieht, daß die Kurve $\tau^{-1} = f(H_S)$ bei einem Wert von $H_S' \approx 21$ Oe einen scharfen Knick besitzt, wobei die Werte des Schaltkoeffizienten auf den Kurvenabschnitten vor und nach dem Knick entsprechend 0,34 und 0,11 Oeμs betragen.

Die Autoren untersuchten 28 Schichten verschiedener Dicke (von 200 bis 3000 Å), die durch thermisches Zerstäuben folgender Legierungen erhalten wurden: 79% Ni + 4% Mo + 17% Fe; 80% Ni + 20% Fe; 43% Ni + 57% Fe und 83% Ni + 17% Fe. Alle Schichten besaßen einachsige Anisotropie. Der Knick der Kurve $\tau^{-1} = f(H_S)$ lag für die angeführten Schichten im Feldstärkebereich von 8 bis 27 Oe. Die erhaltenen experimentellen Ergebnisse erlauben noch nicht, eine eindeutige Abhängigkeit der Feldstärke H_S' von anderen Parametern der Schichten zu konstatieren. Geklärt ist auch nicht, ob der Knick in der Kurve $\tau^{-1} = f(H_S)$ einer einstelligen Veränderung des Ummagnetisierungsmechanismus entspricht. Die Ergebnisse lassen aber die Behauptung zu, daß hier kein Übergang zu einseitiger und um so mehr zu kohärenter Drehung vorliegt, da das senkrechte Signal sowohl vor als auch nach dem Knick einen Wert nahe Null besaß.

Im Verlauf der allgemeinen Diskussion über Impulseigenschaften von Schichten auf dem Internationalen Kolloquium für dünne magnetische Schichten (Jena, April 1966) erfuhren wir, daß ein derartiger Knick unabhängig von uns von Professor F.B. Humphry beobachtet wurde.

Literatur

(1) C.D. OLSON und A.V. POHM, J. appl. Phys. 29, 274 (1958).

(2) F.B. HUMPHRY, J. appl. Phys. 29, 284 (1958).

(3) F.B. HUMPHRY und E.M. GYORGY, J. appl. Phys. 30, 935 (1959).
(4) R.L. CONGER und F.C. ESSIG, Phys. Rev. 104, 915 (1956).
(5) O.S. KOLOTOV, Trudy IV. konferentsii po yadernoi radioelektronike, Atomizdat 1, 158 (1964).
(6) O.S. KOLOTOV und V.A. POGOZHEV, Izv. Akad. Nauk SSSR, ser. fiz. 29, 538 (1965).
(7) O.S. KOLOTOV und V.A. POGOZHEV, Prib. i Tekh. Eksper., Nr. 2 (1965).

(Received June 13, 1966)

phys. stat. sol. 16, K125 (1966)

A.F. Ioffe Physico-Technical Institute, Academy of Sciences
of the USSR, Leningrad

Cohesive Energy of Noble Metals

By
V.K. NIKULIN

1. The nature of the cohesive forces in noble metals is not yet understood. Fuchs (1) estimated the cohesive energy of copper in the Wigner-Seitz approximation. However, his result is not in agreement with the observed value. Both the quantum defect method (2) and the statistical method (3) have not proved capable of giving accurate results.

Friedel (4) and Mott and Stevens (5) suggested that the greater part of the cohesive energy is due to van der Waals forces between the ion cores.

Cohen (6) formulated the pseudopotential approach in the framework of the many-body problem.

The present paper reports an expression for the total energy which is slightly different from that given in (6). The value of the total cohesive energy in copper is calculated.

2. Let $\hat{H} = \hat{H}_o + \lambda\hat{V}$ be the Hamiltonian of the problem. $\hat{H}_o$ is the Hamiltonian in the Hartree-Fock (H-F) approximation, λ is the parameter of interaction, $\hat{V}$ represents both the interaction between the ions and between the electrons and ions:

$$\hat{V} = \int \psi^+(\mathfrak{r}) V_i(\mathfrak{r}) \psi(\mathfrak{r})\, d\mathfrak{r} + \frac{1}{2} \sum_{m,n}{}' \left[z^2 v(\mathfrak{R}_m - \mathfrak{R}_n) + W_{mn} \right],$$

where

$$V_i = \sum_m V_M(\mathfrak{r} - \mathfrak{R}_m) = -\int v(\mathfrak{r} - \mathfrak{r}') \varrho(\mathfrak{r}')\, d\mathfrak{r}'$$

is the superposition of the ion local pseudopotentials, $v(\mathfrak{r} - \mathfrak{r}') = \dfrac{e^2}{|\mathfrak{r} - \mathfrak{r}'|}$, and W_{mn} is the overlapping energy of the ion cores.

After making use of the relationship

$$\frac{\partial E}{\partial \lambda} = \left\langle \frac{\partial \hat{H}}{\partial \lambda} \right\rangle, \qquad (1)$$

the energy can be expressed as

$$E = E_0' + \int_0^1 d\lambda \int G_\lambda^<(\mathbf{r},\mathbf{r}) V_i(\mathbf{r})\, d\mathbf{r} + \frac{1}{2} \sum_{m,n}{}' \left[z^2 v(\mathbf{R}_m - \mathbf{R}_n) + W_{mn} \right], \quad (2)$$

where E_0' is the total H-F energy of the electron gas, $G_\lambda^<(\mathbf{r},\mathbf{r}) = \langle \lambda | \psi^+(\mathbf{r})\, \psi(\mathbf{r}) | \lambda \rangle$ is the correlation function for the perturbed electron gas.

The linear approximation for the correlation function is given by

$$G_\lambda^<(\mathbf{r},\mathbf{r}) = (1 - \lambda)\, G_0^<(\mathbf{r},\mathbf{r}) - \lambda \int \left[K(\mathbf{r}' - \mathbf{r}) - \delta(\mathbf{r}' - \mathbf{r}) \right] \varrho(\mathbf{r}')\, d\mathbf{r}', \quad (3)$$

where $(1 - \lambda)$ guarantees that the total charge equals zero, $K(\mathbf{r}' - \mathbf{r})$ is a dielectric response function for the interacting valence electrons.

After inserting (3) the integration brings (2) into the form

$$E = E_0 - \frac{3}{20} \frac{N}{r_s} \left(\frac{e^2}{a_0} \right) + \frac{1}{2} \sum_m \bar{V}_M + \sum_m U_m + \frac{1}{2} \sum{}' \left[U_{mn} + W_{mn} \right], \quad (4)$$

where E_0 is the H-F energy of the electron gas, the second term represents the interaction between the uniformly distributed electrons and point ions (N is the number of unit cells, r_s is the radius of the cell), $\bar{V}_M$ is the mean value of the deviation V_M from $v(\mathbf{r})$, the two last terms represent the self-energy of an ion and the interaction between the screened ions. Cohen's expression for the energy (6) is similar to equation (4) but his E_0 value is the exact energy of the electron gas, his second term is $\frac{1}{3} N E_F$ (E_F is the Fermi energy), and N stands instead of 1/2 in his third term.

E_0 has to be taken in the H-F approximation, as the correlation hole contributes to V_M (the correlation hole in the free atom and in the metal are supposed to be the same).

According to the general theorem ((7), p. 108) the perturbation in the linear Hartree approximation does not change locally the energy of the free electron gas. Cohen (6) treated the problem without regard to (7). However, taking into account (1) and (3) the theorem mentioned above can easily be extended to the exact energy obtained in the linear approximation.

3. The Hubbard approximation with screened exchange (8) for the static dielectric constant has been used here. Information from the spectroscopic data has been used to obtain the parameters of the model pseudopotential (9). In calculating W we followed (3) and employed data on the d-shell electron density listed in (10).

The results of our calculations of the cohesive energy ε_c and lattice constant L of Cu together with the observed values are summarized in the table.

	ε_c(eV)	L(Å)
calculated	3.677	3.520
observed	3.52	3.608

The author wishes to express his gratitude to Professor A.I. Gubanov for discussions and M.B. Trźaskovskaya for her assistance in programming the numerical calculation.

References

(1) K. FUCHS, Proc. Roy. Soc. A151, 585 (1935).
(2) K. KAMBE, Phys. Rev. 99, 419 (1955).
(3) P. GOMBÁS, Acta phys. Hung. I, 301 (1952).
K. LADÁNYI, Acta phys. Hung. V, 361 (1956).
(4) J. FRIEDEL, Proc. Phys. Soc. B65, 769 (1952).
(5) N.F. MOTT and K. STEVENS, Phil. Mag. 2, 1364 (1957).
(6) M.H. COHEN, J. Phys. Radium 23, 643 (1962).
(7) J.M. ZIMAN, Adv. Phys. 13, 89 (1964).
(8) T.M. RICE, Ann. Phys. 31, 100 (1965).
(9) V.K. NIKULIN, Fiz. tverd. Tela 7, 2708 (1965).
I.V. ABARENKOV and V. HEINE, Phil. Mag. 12, 529 (1965).
(10) F. HERMAN and S. SKILLMAN, Atomic Structure Calculations, New Jersey 1963.

(Received June 15, 1966)

phys. stat. sol. 16, K129 (1966)

Institute of Structural Research, Polish Academy of Sciences, Wrocław

Semiconducting Properties of CdP_4

By

W. ŻDANOWICZ and A. WOJAKOWSKI

During the course of investigations performed on $A^{II}B^{V}$ compounds, and in particular on those of cadmium with phosphorus (1, 2), cadmium tetraphosphide CdP_4 was obtained whose semiconducting properties are less known (3). According to data of Krebs et al. (4) CdP_4 crystallizes in the monoclinic system (a = 5.27 Å, b = 5.19 Å, c = 7.66 Å, β = 80°32'). After Pearson (5) CdP_4 can be included to the class of polyanionic valence compounds following the general valence rule:

$$\frac{(n_e + b_a - b_c)}{n_a} = 8$$

with $n_e = 22$, $n_a = 4$, $b_a = 10$, $b_c = 0$.

Electric and optical measurements have confirmed the semiconducting properties of this compound.

CdP_4 was obtained by saturating CdP_2 or Cd_3P_2 with phosphorus vapours under a pressure of 6 to 10 atm. The synthesis was carried out in evacuated silica tubes containing CdP_2 or Cd_3P_2 at one end (the temperature of this zone was about 700 °C) while red phosphorus was placed in the other end (the temperature of this zone was about 500 to 540 °C). Three times distilled phosphorus and cadmium purified by zone melting were used. By this procedure CdP_4 resulted in the bulk polycrystalline form or as small but well formed single crystals (3, 4). The lattice constants were in good agreement with results published previously (4).

Measurements of resistivity, Hall constant, and thermoelectric power for both poly- and monocrystalline samples of CdP_4 were performed within the temperature range 100 to 600 °K. From these measurements was deduced that CdP_4 is a p-type semiconductor, the resistivity of which for both poly- and monocrystalline samples at 300 °K is of the order of 2 to 8 Ωcm.

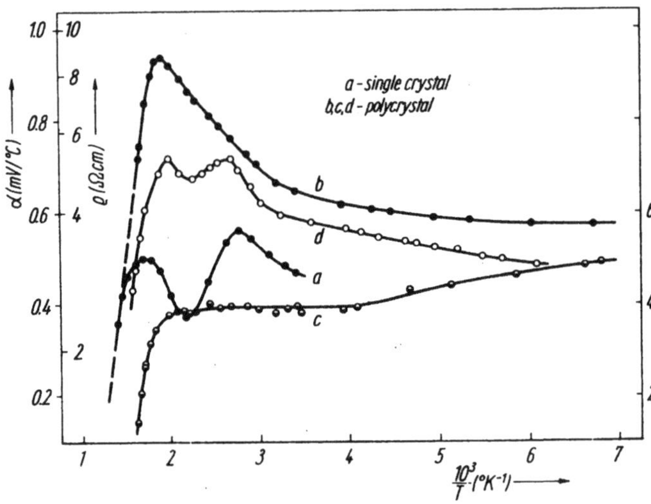

Fig. 1. Temperature dependence of the resistivity (a, b), Hall constant (c), and thermoelectric power (d) of CdP_4

The Hall constant for polycrystalline material at 300 °K is $R_H \approx 400\ cm^3/C$ ($p \approx 2 \times 10^{16}\ cm^{-3}$). The temperature dependence of the thermoelectric power (α) of polycrystalline CdP_4 shows two maxima (at 365 and 480 °K) at which α reaches a value of about 700 μV/°C, amounting at 300 °K about 600 μV/°C. Similar maxima appear also in the temperature dependence of resistivity of single crystals. In Fig. 1 the results of the temperature dependence of R_H, ϱ, and α are shown.

The optical measurements were carried out at room temperature using large polycrystalline CdP_4 samples 0.15 and 0.25 mm thick. The transmittance and reflectivity were measured from 0.5 to 3 μm. The maximum value of the absorption coefficient α is

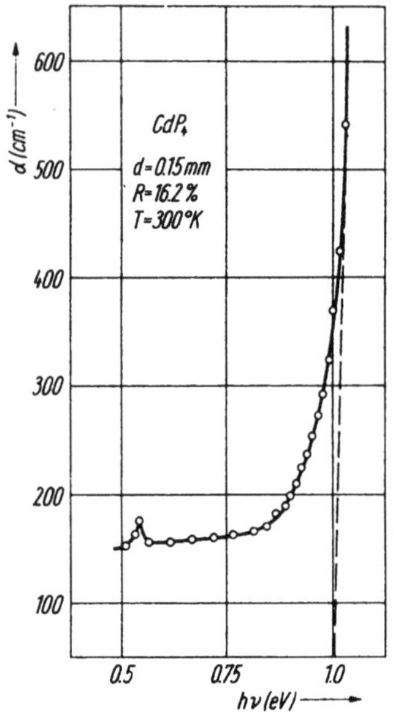

Fig. 2. The absorption coefficient α as a function of energy at 300 °K

equal to 600 cm^{-1} and drops beyond the absorption edge down to about 150 cm^{-1} (Fig. 2). The width of the forbidden band estimated from Fig. 2 is equal to $\Delta E_{opt} \approx 1.0$ eV. The above value is in satisfactory agreement with those estimated from measurements of the resistivity and Hall constant ($\Delta E_{el} \approx 1.15$ eV).

It seems, however, that the values ΔE obtained from optical and electric measurements for polycrystalline samples should be considered as some average ones. It is necessary to obtain larger single crystals of CdP_4 permitting a more precise estimation of ΔE and eventual an anisotropy of electric and optical properties.

The authors wish to express their sincere thanks for the help and interest received by Professor W. Trzebiatowski who supervised these investigations.

References

(1) W. ŻDANOWICZ and A. WOJAKOWSKI, phys. stat. sol. 8, 569 (1965).

(2) W. ŻDANOWICZ and A. WOJAKOWSKI, phys. stat. sol. 10, K93 (1965).

(3) I.S. KOVALEVA, A.V. DMITREV, and I.A. ŻIŻEJKO, Neorgan. Mat. 2, 403 (1966).

(4) H. KREBS, K.H. MULLER, and R. ZURN, Z. anorg. Chem. 285, 15 (1956).

(5) W.B. PEARSON, Acta cryst. 17, 1 (1964).

(Received June 21, 1966)

phys. stat. sol. 16, K133 (1966)

Institute of Physics, Polish Academy of Sciences, Warsaw

Kinetics of Photoconductivity in Boron

By

A.J. NADOLNY, J.W. OSTROWSKI, and M.Z. PRZEGALIŃSKA-MIESZKOWSKA

The kinetics of photoconductivity has been investigated in monocrystalline β-rhombohedral boron in the temperature range between -65 and +144 °C. The material was obtained by the method described by Niemyski and Olempska (1) and by Niemyski et al. (2).

The samples were illuminated by monochromatic or filtered light in the wavelength range between 0.6 and 1.0 μm. To avoid an uncontrolled effect of the atmosphere on the electrical conductance of boron samples all measurements were performed under the pressure of 0.1 Torr or in dry air (at lower temperatures).

The decay of photoconductivity may be presented as a sum of a fast and a slow component, each of them (except for its tail) is described by the formula

$$\Delta\sigma(t) = - C_\alpha \ \ln \frac{t + t_o}{\tau} , \qquad (1)$$

where $\Delta\sigma$ is the photoconductivity, t the time measured from the start of the decay, and C_α, t_o, τ are adjusted parameters for the fast and slow processes.

Fig. 1 shows the slow component on a $\ln(t + t_{02})$ time scale, where the parameter t_{02} has been so chosen as to make a straight line as long as possible. The fast component shown in Fig. 2 is obtained by subtracting the extrapolated part of the slow component (Fig. 1) from the total photoconductivity. The parameter t_{01} has been adjusted in the same way as t_{02}. The parameters τ correspond to the points where the straight lines of the respective components cross the abcissa. Thus we obtain $t_{01} = 1.7$ s, $\tau_1 = 57$ s and $t_{02} = 50$ s, $\tau_2 = 4\times10^4$ s for the fast and slow components, respectively.

The rise of photoconductivity can also be presented as a sum of two processes described by the formula

$$\Delta\sigma(t) = C_g \ \ln \frac{t + t_g}{\tau} , \qquad (2)$$

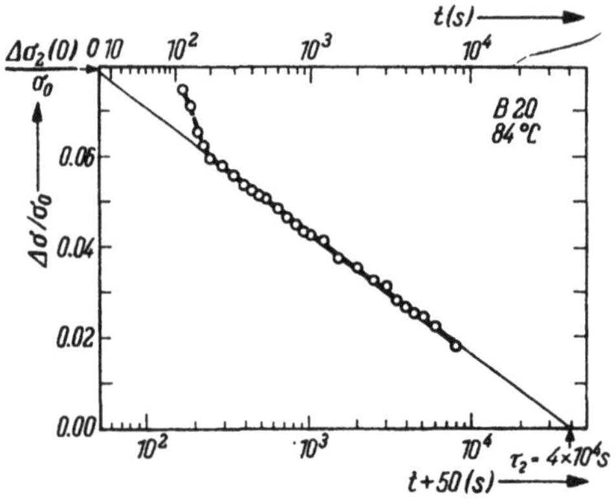

Fig. 1

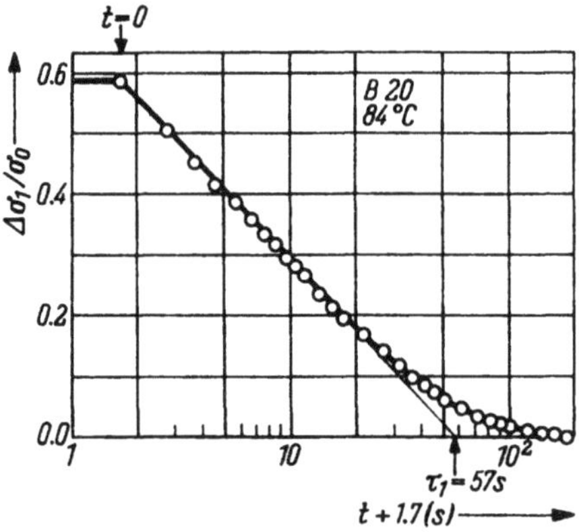

Fig. 2

Fig. 1. Slow component of photoconductivity decay in boron $\Delta\sigma_2$ (thick line) drawn according to the scale $\ln(t + t_o)$, where $t_{02} = 50$ s
$\Delta\sigma$ total photoconductivity,
σ_o equilibrium conductivity,
$\Delta\sigma_2(0)$ start level of the slow component at $t = 0$;
B 20 is the number of the sample

Fig. 2. Fast component of photoconductivity decay in boron $\Delta\sigma_1$ (thick line). The start level at $t = 0$ is shown

where C_q and t_q are parameters.

Neft and Seiler (3) and Gaulé et al. (4) have tried to present the relaxation of photoconductivity in β-rhombohedral boron as a sum of two exponential processes. If we plot the experimental data of the above mentioned authors according to our formula (1) (Fig. 3) we get a better fit than with the "exponential" formula. The values of the parameters t_o and τ corresponding to the experimental decay curve of Neft and Seiler measured at +35 °C are: $t_o = 0.7$ s, $\tau = 2.8\text{x}10^2$ s, and to that of Gaulé et al. (-77 °C): $t_o = 15$ s, $\tau = 8.8\text{x}10^3$ s. Both values of τ are similar to ours obtained at the respective temperatures (Fig. 4). From Fig. 3 it is seen that there is no

Fig. 3. Decay of photoconductivity in boron. Experimental dots obtained by Neft and Seiler (3) (Fig. 18) and Gaulé et al. (4) (Fig. 8) presented according to formula (1)

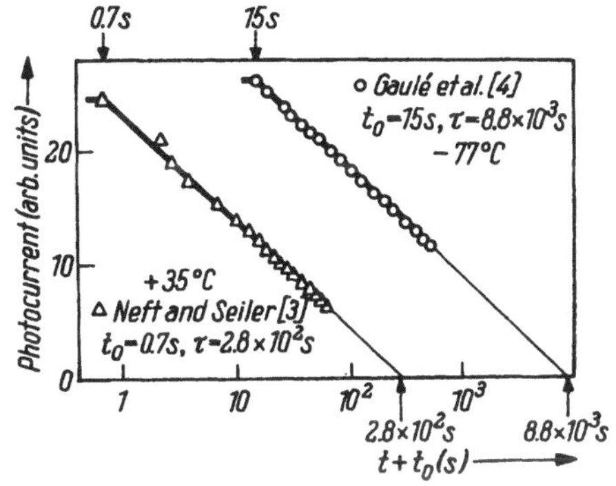

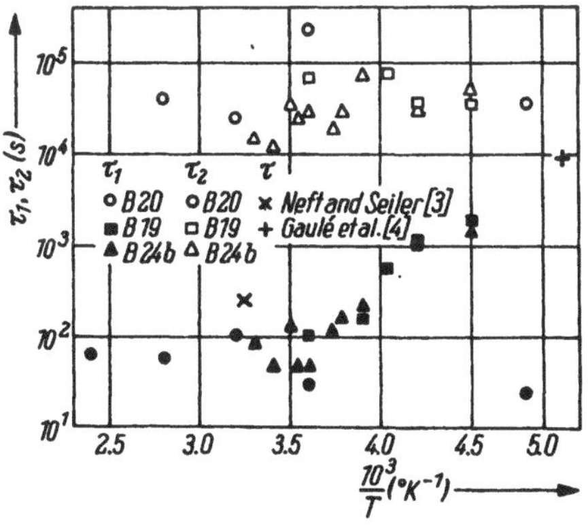

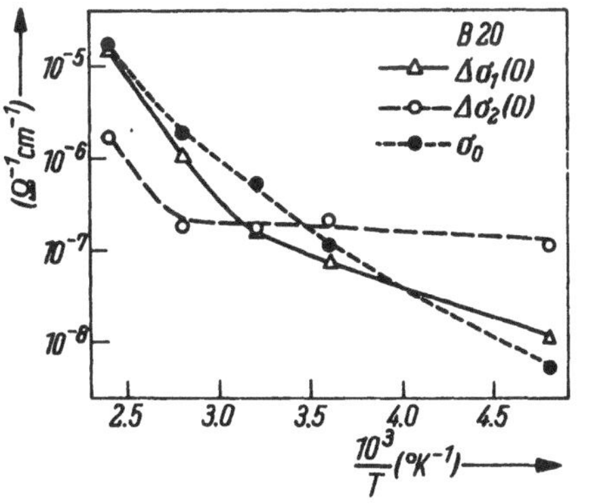

Fig. 4 Fig. 5

Fig. 4. Temperature dependences of the parameters τ_1 and τ_2 (see formula (1))

Fig. 5. Amplitudes of both components of photoconductivity relaxation and the dark conductivity as a function of temperature

slow component of the relaxation process. This can be due to the fact that the measuring time used by Neft and Seiler (60 s) and even by Gaulé et al. (500 s) was too short.

The obtained parameters τ_1 and τ_2 are independent of the excitation level (intensity of absorbed light), however, the parameters t_{01} and t_{02} increase rapidly when the excitation level decreases. Fig. 4 presents the parameters τ_1 and τ_2 as a function of temperature. Fig. 5 illustrates the temperature dependences of the amplitudes $\Delta\sigma_1(0)$ and $\Delta\sigma_2(0)$ of the fast and slow components of the relaxation process after 6000 s of a constant illumination as a function of temperature. The dark conductivity σ_0 is also shown in Fig. 5.

The decay of photoconductivity described by formula (1) has been anticipated by Figielski (5, 6) and observed by Jastrzębska and Figielski (7, 8) in plastically deformed germanium (with a high concentration of dislocations). Earlier Pratt and Kolm (9) have observed a similar "logarithmic" relaxation of work function in germanium when the surface was excited either by an electric field applied normally to it or by illumination. The results and considerations reported by Morrison (10) who measured the relaxation of the surface conductance may throw also some light on the problem of this kind of photoconductivity relaxation.

All above mentioned authors assume that the considered processes are determined by potential barriers whose height is a linear function of the number of carriers participating in the process.

It is too early to propose a model for boron, nevertheless it seems that potential barriers can play a leading role in the mechanism of photoconductivity in boron.

The authors wish to thank Professor L. Sosnowski for his kind interest, they are also greatly indebted to Dr. T. Figielski, Dr. J. Sochański, and Dr. A. Zaręba for their helpful discussions. Thanks are also due to Dr. T. Niemyski, Dr. Z. Olempska, Mrs. I. Pracka, and Mr. Z. Frukacz for providing us with boron samples.

References

(1) T. NIEMYSKI and Z. OLEMPSKA, J. less-common Metals 4, 235 (1962).

(2) T. NIEMYSKI, I. PRACKA, R. SZCZERBIŃSKI, and Z. FRUKACZ, Physics of Semiconductors, Braunschweig-Paris 1964 (p. 1295); Boron, Vol. 2, Preparation, Properties, and Applications, New York 1965 (p. 35).

(3) W. NEFT and K. SEILER, Boron, Vol. 2, New York 1965 (p. 143).
(4) G.K. GAULÉ, J.T. BRESLIN, and R.R. PATTY, Boron, Vol. 2, New York 1965 (p. 169).
(5) T. FIGIELSKI, phys. stat. sol. 6, 429 (1964).
(6) T. FIGIELSKI, phys. stat. sol. 9, 555 (1965).
(7) M. JASTRZĘBSKA and T. FIGIELSKI, phys. stat. sol. 7, K101 (1964).
(8) M. JASTRZĘBSKA and T. FIGIELSKI, phys. stat. sol. 14, 381 (1966).
(9) G.W. PRATT and H.H. KOLM, Semiconductor Surface Physics, Philadelphia 1957 (p. 297).
(10) S.R. MORRISON, Phys. Rev. 102, 1297 (1956).

(Received June 23, 1966)

phys. stat. sol. 16, K139 (1966)

Institute of Physics (a) and Institute of Semiconductors (b),
Academy of Sciences of the Ukrainian SSR, Kiev

Optical Properties and Energy Structure of CdS-CdTe Crystals

By

M.V. KURIK (a) and N.I. VITRIKHOVSKII (b)

Semiconducting solid solutions of the $A^{II}B^{VI}$ group represent a continuous set of crystals whose energy gap smoothly varies within a wide range. This creates great feasibility for a variety of technical uses of such semiconductors. In recent years some of these substances (CdS (1), CdS-CdSe (1, 2), ZnS-CdS (2)) have attracted increasing attention from the point of view of nonlinear processes in semiconductors. A successful technical use of semiconductors depends primarily upon the knowledge of its energy structure parameters, and therefore the study of this structure is of certain interest.

It is the purpose of this communication to describe the investigations of optical properties of CdS-CdTe anion solutions and give the calculations of their energy structure parameters: the width of the forbidden gap E_0, the magnitudes of spin-orbit splitting Δ_{so} and crystal field splitting Δ_c, and the values of electron and hole effective masses.

CdS crystals have wurtzite structure and CdTe zinc blende structure, this creates a limited solubility upon their mixing. It is comparatively easy to obtain mixed crystals with a low concentration of one of the components. The crystals under investigation were obtained by synthesis from the gas phase (3). It was shown (3) that this method is appropriate for single crystals containing 5% CdTe. However, crystals with 3.5% CdTe proved suitable for optical measurements. A recent work (4) reported about hexagonal CdS-CdTe crystals containing up to 20% CdTe obtained by a melting method in a quartz cell.

The absorption and reflection spectra were measured in polarized light at 300, 77, and 20.4 $^{\circ}$K. Fig. 1 shows results of absorption coefficient measurements in polarized light for CdS-CdTe mixed crystals of various composition at 77 $^{\circ}$K as

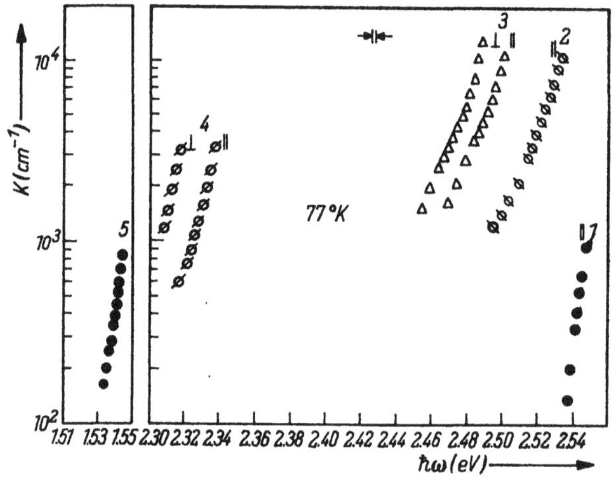

Fig. 1. Absorption spectra of CdS-CdTe crystals at 77 °K

1 - pure CdS, $e \parallel c$;

2 - CdS + 0.5% CdTe, $e \parallel c$;

3 - CdS + 1.5% CdTe, $e \perp, \parallel c$;

4 - CdS + 3.5% CdTe, $e \perp, \parallel c$;

5 - pure CdTe

well as for pure p-CdTe[1)] crystals. These results show that the absorption edge for pure CdS and CdTe crystals and mixed crystals depend exponentially on photon energy. Unlike pure crystals, the mixed crystals have a less steep absorption edge. This change in slope of the absorption edge of mixed crystals (Fig. 1) is probably associated with exciton broadening and a possible additional absorption. The latter is due to precipitation of atoms in the lattice upon mixed crystal formation. As the Te concentration increases, the absorption edge shifts towards longer wavelengths. The energy gap E_0 and its dependence on sample composition was determined from these data (Fig. 2).

The reflection spectra of CdS-CdTe crystals at 20.4 °K allow observations of the fine structure. The corresponding exciton bands A, B, and C (5) are sufficiently distinct at

---

1) p-CdTe single crystals were obtained by the Bridgman method.

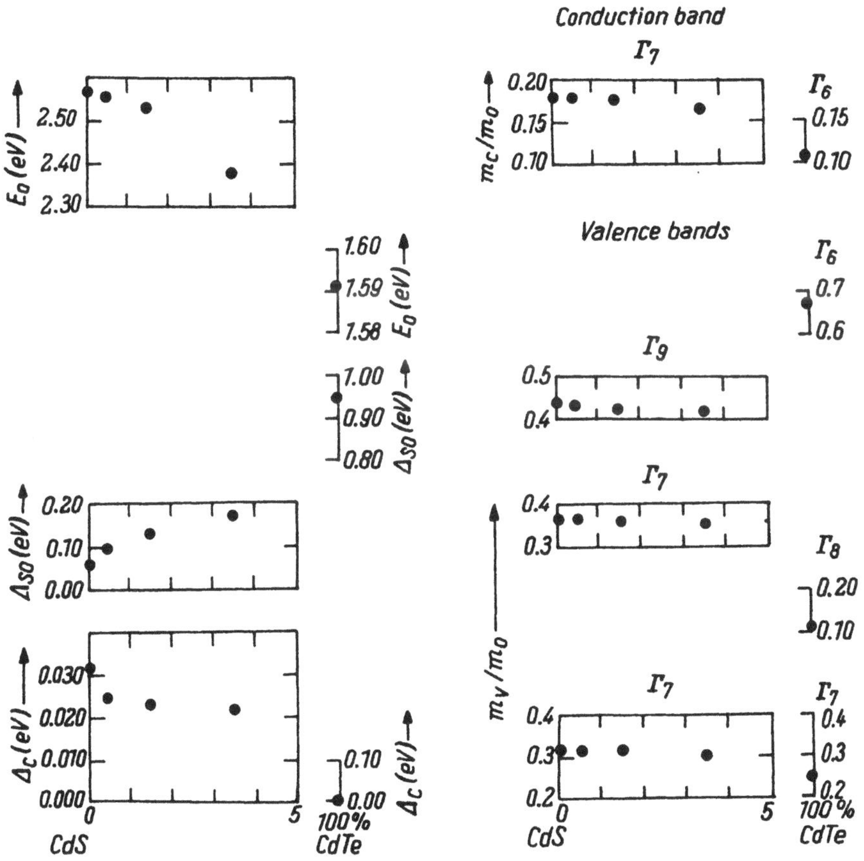

Fig. 2. The magnitudes of the energy gap E_0, spin-orbit splitting Δ_{so} and crystal field splitting Δ_c, and the electron and hole effective masses m_c and m_v, respectively, vs. sample composition for the CdS-CdTe system at 77 °K

low CdTe concentrations. These bands slightly differ in their relative position and shape from those of pure CdS crystals. With higher Te content these bands strongly broaden and at 3.5% they are practically not observable in reflection. With such concentrations the relative position of the exciton band is markedly changed. For cubic CdTe the reflection spectrum shows one exciton band A (6).

From measurements of exciton band positions the magnitudes of the valence band splittings Δ_{so} and Δ_c were calculated from Hopfield's theory (7). The splittings are given in Fig. 2. The figure shows that Δ_{so} and Δ_c vary in a complicated manner with composition. Δ_{so} increases from 0.063 eV for CdS to 0.95 eV for CdTe. Δ_{so} for CdTe is taken from (8). It was shown earlier (9) that in CdS-CdSe crystals, upon substitution of S atoms by Se atoms which have more inner electrons, Δ_{so} increased up to 0.41 eV for CdSe. When S atoms were substituted by Te atoms, which possess more inner electrons than Se, Δ_{so} increases up to 0.95 eV. Consequently, the spin-orbit splitting is substantially caused by inner electrons.

By using Cardona's method (10) the density-of-state effective masses of the valence and conduction bands were calculated and the results are presented in Fig. 2.

The authors are grateful to D.Sc. M.S. Brodin for his guidance and interest in this work.

References

(1) N.G. BASOV, A.S. GRASYUK, I.G. SUBAREV, and V.A. KATULIN, Fiz. tverd. Tela 7, 3639 (1965).
M.S. BRODIN, N.I. VITRIKHOVSKII, S.V. ZAKREVSKII, and V.YA. REZNICHENKO, Ukr. fiz. Zh. 11, 344 (1966).

(2) R.A. SORFT and H.W. MOOS, J. appl. Phys. 35, 2152 (1964).

(3) N.I. VITRIKHOVSKII and I.B. MIZETSKAYA, Fiz. tverd. Tela 1, 996 (1959).

(4) E. CRUCEANU and D. NICULESCU, C.R. Acad. Sci. 261, 935 (1965).

(5) D.J. THOMAS and J.J. HOPFIELD, Phys. Rev. 116, 573 (1959).

(6) D.J. THOMAS, J. appl. Phys. Suppl. 32, 2298 (1960).

(7) J.J. HOPFIELD, J. Phys. Chem. Solids 15, 97 (1960).

(8) R.E. DIETZ, Bull. Amer. Phys. Soc. II, 9, 619 (1964).

(9) M.S. BRODIN, M.V. KURIK, and S.P. YURTSENYUK, Fiz. tverd. Tela 7, 3112 (1965).

(10) M. CARDONA, J. Phys. Chem. Solids 24, 1543 (1963); 26, 1351 (1965).

(Received June 24, 1966)

phys. stat. sol. 16, K143 (1966)

Institute of Physics of the Academy of the R.P.R., Bucharest

Thermoelectric Power in Amorphous Germanium

By

R. GRIGOROVICI, N. CROITORU, and A. DÉVÉNYI

Amorphous germanium (a-Ge) layers deposited by evaporation on fused silica substrates and provided with ohmic silver contacts have been submitted to a temperature gradient when held at mean temperatures between 200 and 600 ^{o}K, the resulting thermoelectric power α being measured with the aid of a vibrating reed electrometer. Deposition and measurements have been performed in the same vacuum chamber ($p < 10^{-5}$ Torr).

The results are shown in Fig. 1 together with the corresponding log R vs. 1/T curves. The curves a refer to layers deposited or annealed at 300 ^{o}K; the curves b refer to layers annealed at more than 500 ^{o}K. The correlation between the temperature dependence of R and α is obvious.

It was shown in previous papers (1 to 3) that a-Ge is a p-type semiconductor with high acceptor densities (10^{17} to 10^{19} cm^{-3}). The three different activation energies of conductivity which are to be seen in the log R vs. 1/T plot have been attributed (2) to the following conduction mechanisms:

1. An electron hopping process between the acceptor levels at low temperatures; the electrons are provided by some compensated donors; the Fermi level lies somewhere between the acceptor energy level E_a and the valence band edge E_v.

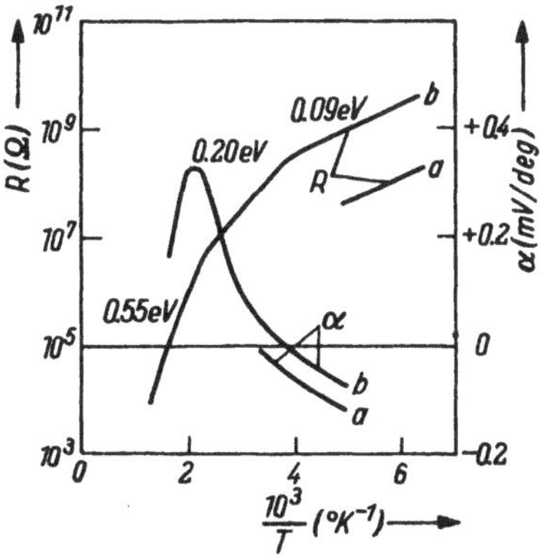

Fig. 1. Temperature dependence of the resistance R and the thermoelectric power α in amorphous Ge layers. Curves a: layers deposited or annealed at 300 ^{o}K; curves b: layers annealed at more than 500 ^{o}K

2. A hole conduction in the valence band at intermediate temperatures; the Fermi level rises in this temperature interval above the acceptor energy level E_a.

3. An intrinsic conduction by holes and electrons in the valence and conduction bands at temperatures higher than 450 $^{\circ}$K; the Fermi level lies near the centre of the energy gap.

Annealing is supposed to diminish the acceptor density and to increase therefore the resistivity at low and intermediate temperatures.

A comparison of the results of thermoelectric measurements with the proposed mechanisms is made by means of the general formula of the thermoelectric power

$$\alpha = - \frac{1}{eTi}\left[(E_a - E_F)i_a + (E_v - E_F)i_h + (E_c - E_F)i_e\right] ,$$

where i_a, i_h, and i_e are the electrical currents carried by electrons on acceptor levels, by holes in the valence band, and by electrons in the conduction band, respectively, and i is the total current. It can be shown that this formula is valid even if the effective mass approximation is not fulfilled (4).

At low temperatures only i_a is significant and $(E_a - E_F)$ is positive. Therefore α should be negative. Annealing must diminish the value of $(E_a - E_F)$ and α should become less negative.

At intermediate temperatures i_h is thought to be predominant; $(E_v - E_F)$ is negative and increases with temperature. Therefore α should become positive and rapidly increase with T.

At high temperatures the rising of the Fermi level is supposed to come nearly to a standstill, and i_e becomes progressively more important. As $(E_c - E_F)$ is positive, α should decrease again.

All these predictions are in qualitative agreement with our results (see also (5)).

It is much more difficult to understand from the point of view of carrier statistics how the intrinsic conduction range could be reached in such a highly doped semiconductor as a-Ge at temperatures as low as 450 ^{o}K. But this seems to be a general feature of amorphous semiconductors.

References

(1) R. GRIGOROVICI, N. CROITORU, A. DÉVÉNYI, and E. TELEMAN, Proc. Int. Conf. Semicond. Phys., Paris 1964 (p. 423).

(2) R. GRIGOROVICI, N. CROITORU, A. DÉVÉNYI, L. VESCAN, and A. BARNA, Rev. Roum. Phys. 10, 649 (1965).

(3) J. TAUC, R. GRIGOROVICI, and A. VANCU, phys. stat. sol., in the press.

(4) R. GRIGOROVICI, N. CROITORU, and A. DÉVÉNYI, Rev. Roum. Phys., in the press.

(5) B. LIDÉN and J. AULEYTNER, Ark. Fys. 22, 549 (1962).

(Received June 25, 1966)

phys. stat. sol. 16, K147 (1966)

Institut für Magnetische Werkstoffe der Deutschen Akademie der Wissenschaften zu Berlin, Jena

A New Nonlinear Mechanism for Establishing the Saturation Susceptibility during Spin Wave Parallel Pumping

By

TH. KLUPSCH and W. HAUBENREISSER

In the phenomenological theory (1 to 3), the spin waves are treated like classical damped oscillators which can be described in the presence of an rf field parallel to the dc field by the following linearized equation:

$$\frac{d}{dt} a_{k}(t) = i\left[\left(\omega(k) + i\eta^{(o)}(k)\right) a_{k}(t) - c(k)e^{i\omega t}a^{*}_{-k}(t)\right] \qquad (1)$$

with

$$c(k) = \frac{\pi\gamma^2 Mh}{\omega(k)} \sin^2 \vartheta_{k}\, e^{2i\varphi_k} \qquad (2)$$

(γ gyromagnetic ratio, t time variable, $k = k(|k|, \vartheta_k, \varphi_k)$). a_k, a_k^* denote the complex conjugated spin wave normal coordinates of wave vector k; $\omega(k)$, $\eta^{(o)}(k)$, ω, M, and h are the classical spin wave frequency, the linear damping constant, the microwave frequency, the magnetization, and the microwave field strength, respectively. The polar axis $\vartheta_k = 0$ is chosen in the direction of the dc field.

We can solve (1) by

$$a_k = a_{-k} = a_k^{(o)} \exp\left[\frac{i}{2}\omega - \eta^{(o)}(k) + s(k)\right]t, \qquad (3)$$

$$s(k) = \left[|c(k)|^2 - \left(\omega(k) - \frac{\omega}{2}\right)^2\right]^{1/2}. \qquad (4)$$

We see that in the case $\eta^{(o)}(k) = s(k)$ the behaviour of spin waves is similar to that of undamped oscillators, and that the spin system becomes unstable if h exceeds a critical value h_c which follows from $\min_{(k)} \mathrm{Re}\left[\eta^{(o)}(k) - s(k)\right] = 0$. To explain the steady state behaviour in the region $h > h_c$ it can be shown that the confluence of unstable spin waves gives rise to a nonlinearity of the damping constant $\eta(k)$ (4, 5) so that all the spin waves which have steady-state non-vanishing amplitudes obey the relation $\mathrm{Re}\left[\eta(k) - s(k)\right] = 0$. Also, the behaviour of such spin waves is like that of effectively undamped oscillators. It has been pointed out (5) that this theory cannot explain the magnitude of the saturation susceptibility by considering the confluence of two microwave excited spin waves as the process of highest probability.

To discuss the influence of thermal fluctuations of the spin system as a function of the time variable we have to add the expression of the thermal driving field $f_k(t)$ at the right hand side of (1). First, Schlömann (6) has derived the exact expression for f_k by the same method used in the Langevin theory of Brownian motion. He has found as a consequence of the fluctuation-dissipation theorem

$$\langle f_k(t) f_k^*(t') \rangle = 2\eta^{(o)}(k) n^{(o)}(k) \delta(t - t') , \qquad (5)$$

$$\langle f_k(t) f_{-k}(t') \rangle = \langle f_k(t) a_k(t') \rangle = \langle f_k(t) a_{-k}^*(t') \rangle = 0 . \qquad (6)$$

$n^{(o)}(k)$ denotes the thermal averaged squared k-spin wave amplitude which is equal to $\langle a_k^*(0) a_k(0) \rangle$. Assuming small deviations from the thermal equilibrium we find for the steady state value of $n(k)$ at presence of an rf field:

$$n(k) \equiv \langle a_k^*(t) a_k(t) \rangle = n^{(o)}(k) \left[1 + \frac{|c(k)|^2}{\left[\eta^{(o)}(k)\right]^2 - s^2(k)} \right] . \qquad (7)$$

The divergence of this expression for $\eta^{(o)}(k) = s(k)$ may be explained by the fact that an effective undamped oscillator can increase its amplitude by taking up energy from the thermal driving field, e.g. it becomes unstable. Therefore, the thermal driving field must cause an additional exaltation of the spin wave amplitude so that the nonlinear damping $\eta(k)$ exceeds the value $s(k)$. The same condition is valid in the region $h > h_c$, but the results of Gottlieb and Suhl (7) prove that it is insufficient to accept the thermal value of f_k because the calculated absorption becomes too small.

We have investigated this problem quantum-statistically using the method of non-equilibrium Green's functions given by

Kadanoff and Baym (8). From that we have found that the form of equation (1) including $f_{\mathbf{k}}$ and equations (5 to 7) remain invariant heuristically if we substitute $\eta^{(0)}(\mathbf{k})$, $n^{(0)}(\mathbf{k})$ by $\eta(\mathbf{k})$, $\tilde{n}(\mathbf{k})$ satisfying the following expressions for $n(\mathbf{k}) \gg 1$:

$$\eta(\mathbf{k}) = \frac{1}{2}\int \frac{d^3\mathbf{k}_1}{(2\pi)^2} K(\mathbf{k},\mathbf{k}_1)\Big[\frac{1}{2}\Big(n(\mathbf{k}-\mathbf{k}_1) + n(\mathbf{k}_1)\Big)\delta\Big(\hbar\omega(\mathbf{k}) - \hbar\omega(\mathbf{k}-\mathbf{k}_1) - \hbar\omega(\mathbf{k}_1)\Big) + \Big(n(\mathbf{k}_1-\mathbf{k}) - n(\mathbf{k}_1)\Big)\delta\Big(\hbar\omega(\mathbf{k}) + \hbar\omega(\mathbf{k}_1-\mathbf{k}) - \hbar\omega(\mathbf{k}_1)\Big)\Big], \quad (8)$$

$$\eta(\mathbf{k})\tilde{n}(\mathbf{k}) = \frac{1}{2}\int \frac{d^3\mathbf{k}_1}{(2\pi)^2} K(\mathbf{k},\mathbf{k}_1)\Big[\frac{1}{2} n(\mathbf{k}-\mathbf{k}_1) n(\mathbf{k}_1)\delta\Big(\hbar\omega(\mathbf{k}) - \hbar\omega(\mathbf{k}-\mathbf{k}_1) - \hbar\omega(\mathbf{k}_1)\Big) + n(\mathbf{k}_1-\mathbf{k}) n(\mathbf{k}_1)\delta\Big(\hbar\omega(\mathbf{k}) + \hbar\omega(\mathbf{k}_1-\mathbf{k}) - \hbar\omega(\mathbf{k}_1)\Big)\Big]; \quad (9)$$

$$K(\mathbf{k},\mathbf{k}_1) = \frac{16\pi^2\mu_B^3 M}{\hbar}\left|\sin 2\vartheta_{\mathbf{k}-\mathbf{k}_1} e^{-i\varphi_{\mathbf{k}-\mathbf{k}_1}} + \sin 2\vartheta_{\mathbf{k}_1} e^{-i\varphi_{\mathbf{k}_1}}\right|^2 \quad (10)$$

(μ_B Bohr's magneton),

considering the dipolar three-magnon process as the nonlinear process with the highest probability.

The resulting integral equations can be transformed into algebraic equations for sufficiently large h using

$$n(\mathbf{k}) \approx \pi|c(\mathbf{k})|^2\Big[\eta^2(\mathbf{k}) - |c(\mathbf{k})|^2\Big]^{-1/2}\tilde{n}(\mathbf{k})\delta(\omega(\mathbf{k}) - \tfrac{\omega}{2})$$

for $\omega(\mathbf{k}) \approx \frac{\omega}{2}$ and $n(\mathbf{k}) \approx \tilde{n}(\mathbf{k})$ for $\omega(\mathbf{k}) \approx \omega$.

It may be expected that our theory is valid in the region $H < H_{3M}$ of the static field strength H where the three-magnon confluence process of parametrically excited magnons is possible (9, 10).

Now, we find that the nonlinear damping constant agrees with that of Gottlieb and Suhl (7) while the quantity which corresponds to the thermal driving field in the limit $h \to 0$ increases by a factor higher than 10^4. This nonlinear effect has not yet been discussed. Finally, the resulting imaginary part of the susceptibility χ'' must be higher than the Gottlieb-Suhl value.

Also we have pointed out the characteristic behaviour of χ'' for occurrence of the three-magnon process of microwave excited magnons, i.e. the existence of a maximum saturation value in the limit $h \to \infty$. For experimental purposes the formula

$$\chi''_{max} \approx \frac{\gamma \mu_B M D \varkappa_o^2}{\hbar \omega^2} \tag{11}$$

can be used if $H_{3M} - H$ is sufficiently large. D and $\varkappa_o$ denote the dynamical exchange constant and the wave number of spin waves with frequency ω, respectively, which follows in the approximation of a spherical spin wave dispersion relation from

$$\varkappa_o^2 = \frac{1}{\gamma D}\left[\omega - \gamma H\right]. \tag{12}$$

For ultrapure YIG we find at X-band frequencies and room temperature $\chi''_{max} \approx 2 \times 10^{-2}$ in good agreement with experiments (9, 11)[1]. In the limit $H \to H_{3m}$ we have derived the numerical formula for the same material and frequencies:

$$\chi''_{max} \approx 1.7 (H_{3M} - H)^{-1}. \tag{13}$$

The expression seems acceptable for $H_{3M} - H \geq 50$ Oe but in the region $H_{3M} - H < 50$ Oe we have to take into account other non-linear interactions of spin waves and phonons to remove the singularity at $H = H_{3M}$, because the dipolar three-magnon interaction (see equation (10)) tends to zero in the region $\vartheta_\varkappa = \frac{\pi}{2}$.

We should like to express our appreciation to Professor G. Weber and to Dr. F. Voigt for valuable discussions.

References

(1) E. SCHLÖMANN, J.J. GREEN, and U. MILANO, J. appl. Phys. 31, 386S (1960).

(2) E. SCHLÖMANN and R.I. JOSEPH, J. appl. Phys. 32, 1006 (1961).

---

1) In (11) we have to correct χ'' by $4\pi\chi''$ in agreement with (9).

(3) F.R. MORGENTHALER, J. appl. Phys. 31, 95S (1960).
(4) E. SCHLÖMANN, J. appl. Phys. 33, 527 (1962).
(5) TH. KLUPSCH and W. HAUBENREISSER, Z. angew. Phys. 18, 230 (1964).
(6) E. SCHLÖMANN, Izv. Akad. Nauk SSSR 28, 454 (1964).
(7) P. GOTTLIEB and H. SUHL, J. appl. Phys. 33, 1508 (1962).
(8) L.P. KADANOFF and G. BAYM, Quantum Statistical Mechanics, New York 1962.
(9) J.J. GREEN and E. SCHLÖMANN, J. appl. Phys. 33, 535 (1962).
(10) W. HAUBENREISSER, phys. stat. sol. 3, K181 (1963).
(11) J.J. GREEN and E. SCHLÖMANN, J. appl. Phys. 33, 1358S (1962).

(Received June 25, 1966)

phys. stat. sol. 16, K153 (1966)

C.N.R.S., Laboratoire de Magnétisme, Bellevue

Effets à forte puissance dans les films minces ferromagnétiques, avec champ polarisant oblique [1)]

Par

H. PASCARD [2)]

Cette note présente les premiers résultats d'une étude des effets non linéaires à forte puissance lorsque la direction du champ polarisant varie de 0 à 90° par rapport à la normale au plan du film (1). L'étude des effets observés permet de préciser l'influence des inhomogénéités sur le début de la saturation du mode de résonance uniforme.

Nous avons étudié expérimentalement les variations, en fonction de la puissance, du rapport χ''/χ''_0 de la susceptibilité χ'' à la résonance et à fort niveau à la susceptibilité χ''_0 à très bas niveau. Les courbes χ''/χ''_0 en fonction du carré du champ hf, sont représentées pour différents angles de polarisation Φ et pour un film de permalloy (80 Ni, 20 Fe) de 1500 Å d'épaisseur, et en bande X (Fig. 1). Dans la région des niveaux de puissance modérés ($h < 15$ Oe), ce rapport varie linéairement avec h^2. Contrairement à ce que l'on observe

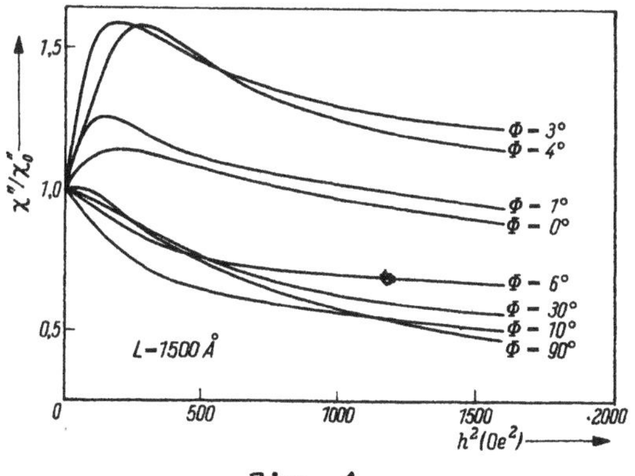

Fig. 1

1) Contribution au "Colloque International sur des films minces magnétiques", tenu à Jena, le 25ième au 28ième avril 1966.

2) Ce travail a été effectué en collaboration avec A.J. Berteaud.

classiquement pour de nombreux matériaux, la pente $\frac{d(\chi''/\chi''_0)}{d(h^2)}$ est positive dans cette région pour les directions de polarisation voisines de la normale au film.

Des effets analogues ont été quelquefois observés dans les ferrites par Schlömann, Green et Milano (2), qui ont montré théoriquement que la nature du mécanisme de dispersion de l'énergie du mode uniforme à des modes d'ondes de spin pouvait effectivement conduire à une pente $\frac{d(\chi''/\chi''_0)}{d(h^2)}$ positive. Ces auteurs ont établi une théorie microscopique concernant l'influence des inhomogénéités sur la saturation du mode de résonance uniforme. Ils ont montré que la loi de saturation à niveau de puissance modéré s'écrit sous la forme

$$\chi''/\chi''_0 = 1 - C\,(h/\Delta H)^2 + \text{(termes d'ordre plus élevé)}$$

et que la valeur du paramètre C est déterminée par la nature du mécanisme de dispersion : le transfert de l'énergie du mode uniforme à des ondes de spin de grandes longueurs d'onde conduit à des valeurs de C positives alors que le transfert de l'énergie à des ondes de spin de courtes longueurs d'onde tend à diminuer la valeur de C et peut conduire à des valeurs de C négatives.

En utilisant ces résultats théoriques nous avons inter-

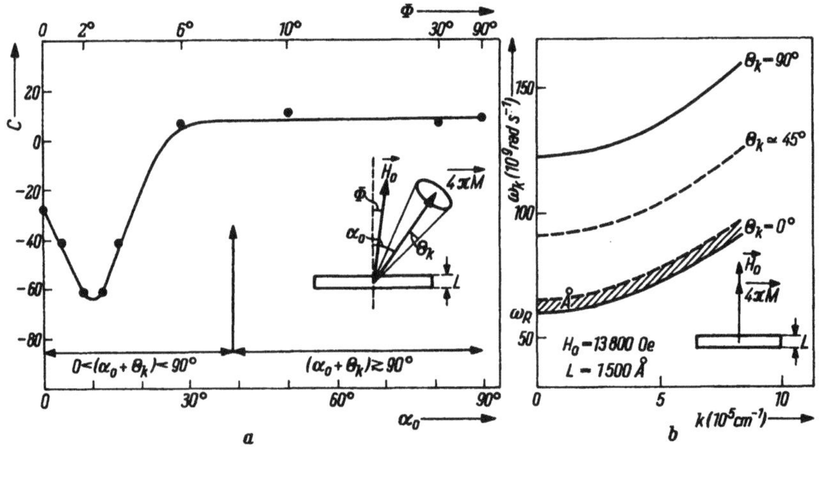

Fig. 2

prêté les variations de C en fonction de l'angle α_0 entre le vecteur aimantation et la normale au film (Fig. 2a). Dans le cas de polarisation normale ($\Phi = 0$) le mode uniforme est dégénéré avec le bas du spectre de dispersion où $\theta_k \approx 0$ (Fig. 2b). Par conséquent les seules ondes de spin dégénérées ayant la possibilité d'être excitées ont une direction de propagation voisine de la normale au film. Mais dans ce cas le domaine d'interaction entre le mode uniforme et ces ondes de spin est au plus égal à l'épaisseur du film et les longueurs d'ondes correspondantes sont donc nécessairement très courtes, ce qui expliquerait la valeur négative de C pour $\Phi = 0$. Lorsque Φ croît de quelques degrés l'échantillon ne peut plus être considéré comme un ellipsoïde aimanté suivant son axe de révolution, et le spectre de dispersion $\omega(k)$ est déformé : néanmoins le mode uniforme remonte légèrement par rapport au bas du spectre de dispersion (Fig. 2b) : θ_k est encore voisin de zéro et limite toujours le domaine d'interaction à l'épaisseur du film (donc faibles λ_k). Par contre, le nombre des ondes de spin dégénérées possibles s'est accru vers les plus courtes longueurs d'onde ce qui est en accord avec la décroissance de C observé. Dès que Φ atteint 3 à 4 degrés, l'angle α_0 atteint 10 à 15 degrés et θ_k croît également assez rapidement. Par conséquent, la direction limite des ondes de spin $\alpha_0 + \theta_k$ (Fig. 2a) atteint une valeur telle que le domaine d'interaction entre le mode uniforme et les ondes de spin ($\approx L\,[\cos(\alpha_0 + \theta_k)]^{-1}$) s'accroît rapidement et permet l'excitation d'ondes de spin de plus grandes longueurs d'onde : le paramètre C croît alors vers des valeurs positives pour des angles Φ encore assez faibles. Enfin lorsque $\alpha_0 + \theta_k$ atteint 90 degrés, le domaine d'interaction peut être infiniment grand. Toutes les ondes de spin dégénérées avec le mode uniforme peuvent être excitées, quelle que soit leur longueur d'onde. L'augmentation de Φ ne modifiant plus cette situation, on doit s'attendre à ce que le paramètre C reste constant, ce que confirme Fig. 2a.

Nous avons ainsi précisé la nature du mécanisme de dispersion de l'énergie du mode uniforme aux modes d'ondes de spin, dans la région des puissances modérées précédant la zone d'apparition des phénomènes de saturation et dans les couches minces ferromagnétiques.

Bibliographie

(1) H. PASCARD et A.J. BERTEAUD, Transactions on Magnetics of the I.E.E.E., September 1966.

(2) E. SCHLÖMANN, J.J. GREEN et U. MILANO, J. appl. Phys. 31, 386S (1960).

(Received June 28, 1966)

phys. stat. sol. 16, K157 (1966)

Physics Department, University of Delaware, Newark, Delaware

Gas Desorption from Virginal CdS Crystals[1)]

By

R. SCHUBERT

When a virginal CdS crystal platelet is placed in ultrahigh vacuo the dark- and photoconductance increases. After heating the crystal to temperatures up to 200 °C the conductance increases further; this cannot be completely accounted for by changes of the intrinsic defect structures (1, 2). Various authors (3 to 7) have shown that ambient gases at pressures as low as 10^{-5} Torr affect the electrical properties of the crystal through adsorbed or chemisorbed surface layers. For instance, chemisorbed oxygen creates acceptors, which can act as recombination centers at the surface and influence the dark- and photoconductance respectively. Desorption of oxygen has been observed at elevated temperatures.

It is the purpose of this paper to analyse the gas desorbed and to determine temperatures for maximum desorption. In the experiment to be described a CdS crystal was placed on a thermally-shielded copper crystal holder in an ultrahigh vacuum system which was pumped to the 10^{-10} Torr range using a triode ion and a Ti getter pump. The crystal temperature could be varied between -180 and +400 °C; crystal temperature and conductance were monitored continuously. A mass spectrometer, capable of measuring partial pressures down to 10^{-13} Torr, was mounted directly in front of the largest crystal surface and shielded against the crystal surroundings to allow desorption studies and to measure the partial pressure of ambient gases in the system.

Upon cooling the previously unheated crystal to liquid nitrogen temperature and then increasing the temperature time linearly, while monitoring the partial pressure of a specific mass number on the mass spectrometer one records a characteristic desorption curve, i.e. a plot of partial pressure versus

1) Supported by the Army Research Office, Durham, N.C. Contract No. DA-31-124-ARO (D)-173.

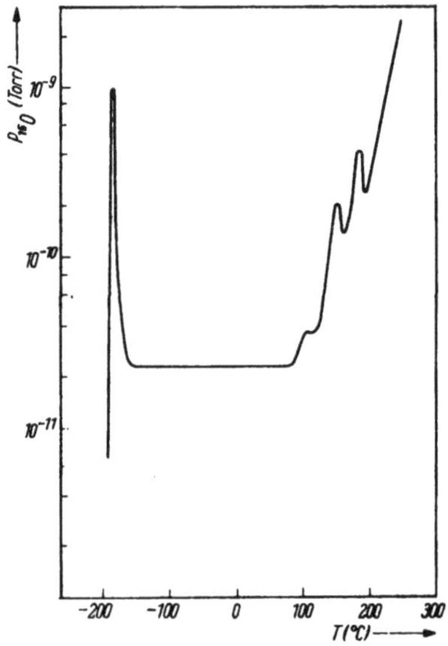

Fig. 1. Change in partial pressure of atomic mass 16 as a function of time linearly increasing CdS crystal temperature (0.4 deg/s)

increasing crystal temperature. A partial pressure maximum will indicate desorption with a certain activation energy.

Fig. 1 shows a typical desorption curve for atomic mass number 16 for a total initial pressure of 10^{-10} Torr. Four desorption maxima are seen. This indicates an adsorbed layer of very low binding energy which desorbs at about -180 °C and at least one[2)] chemisorbed layer (possibly three layers) which desorb between 110 and 185 °C. The observed desorption may have been enhanced by optical irradiation due to the electron source of the mass spectrometer.

Although a complete analysis of the mass spectrum of desorbed gases has not been carried out because of difficulties due to transient effects, it is probable that the main contribution to the mass 16 desorption spectrum is caused by oxygen rather than by other gas components of the same mass number (e.g. CH_4). After baking the CdS crystal at 300 °C in ultrahigh vacuo the desorption spectrum changes markedly and cannot be restored by oxygen adsorption only. This suggests that changes in the surface structure take place and also are responsible for the well known irreversable changes in electrical properties from the virginal to a later state.

The author thankfully acknowledges the proposal of this work and helpful discussions of Professor K.W. Böer.

2) There exists the possibility of desorption from three sites at the surface with different binding energies, leading to the observed three peaks in the measured case.

References

(1) K.W. BÖER and C.A. KENNEDY, to be published in phys. stat. sol.

(2) H.J. HÄNSCH, F. TUCHAN, and E.H. WEBER, phys. stat. sol. 3, 406 (1963).

(3) H. BERGER, K.W. BÖER, and E.H. WEBER, Z. Phys. 158, 501 (1960).

(4) W. MUSCHEID, Ann. Phys. (Germany) 13, 305 (1953).

(5) J. WOODS, J. Electronics 3, 225 (1957); 5, 417 (1958).
J. WOODS and D.A. WRIGHT, Solid State Physics in Electronics and Telecommunications, Proc. Int. Conf. Brussels (Academic Press) 2, 880 (1960).

(6) R. BUBE, Phys. Rev. 101, 1668 (1956).

(7) P. MARK, J. Phys. Chem. Solids 25, 911 (1964); 26, 959 (1965); Phys. Rev. 144, 751 (1966).

(Received June 28, 1966)

phys. stat. sol. 16, K161 (1966)

Institut für experimentelle Physik der Universität Halle

Umwandlung von U-Zentren in OH-Zentren bei KCl-Kristallen

Von

O. ADAMETZ[1)]

Vor einiger Zeit konnte gezeigt werden, daß sich durch Röntgenbestrahlung (1) und auch durch elektrolytische Verfärbung (2) von OH-haltigen KCl-Kristallen eine Umwandlung von "OH-Zentren" (OH-Ionen auf Anionenplätzen) in U-Zentren (H^--Ionen auf Anionenplätzen) erreichen läßt. Gleichzeitig wurden bei elektrolytischer Verfärbung durch Sauerstoff bedingte Absorptionsbanden gefunden. Es lag die Frage nahe, ob auch umgekehrt in Kristallen mit U-Zentren durch Temperung in Sauerstoff eine U → OH-Umwandlung stattfinden kann. Im folgenden wird über Experimente dazu berichtet.

Von größeren nach dem Kyropoulos-Verfahren aus analysenreinem Material an Luft bzw. unter Reinst-Argon (aus Pt-Tiegel) gezogenen KCl-Einkristallen wurden kleinere quaderförmige Proben abgespalten, additiv verfärbt und anschließend in einer Wasserstoffatmosphäre getempert, um definierte U-Zentren-Konzentrationen zu erreichen. Die Umwandlung von U-Zentren in OH-Zentren wurde mit Proben von 4x3x0,3 mm^3 durch eine Temperung in Sauerstoff bei 650 °C unter Normaldruck durchgeführt.

Der Einfluß der verschiedenen Behandlungen wurde in den Kristallen durch Absorptionsmessungen erfaßt. Die OH- und die U-Zentren besitzen glockenförmige Banden bei 204 bzw. 214 nm. Aus den Absorptionskurven lassen sich die Konzentrationen der zugehörigen OH- bzw. U-Zentren nach den Angaben von Anger, Fritz und Lüty (3) bzw. Gründig und Rühenbeck (4) errechnen.

Als Beispiel zeigt Fig. 1 in Kurve 1 zunächst die Absorption des unter 30 Torr Argon gezogenen Kristalles ohne zusätzliche thermische Behandlung ("as cleaved"-Probe). Man erkennt, daß auch unter diesen Bedingungen im Kristall eine schwache OH-Bande vorhanden ist, die dem KOH-Gehalt im Ausgangsmaterial zugeschrieben werden muß. Nach der Behandlung

1) Mitarbeiter am Institut für Einkristalle, Turnov, ČSSR.

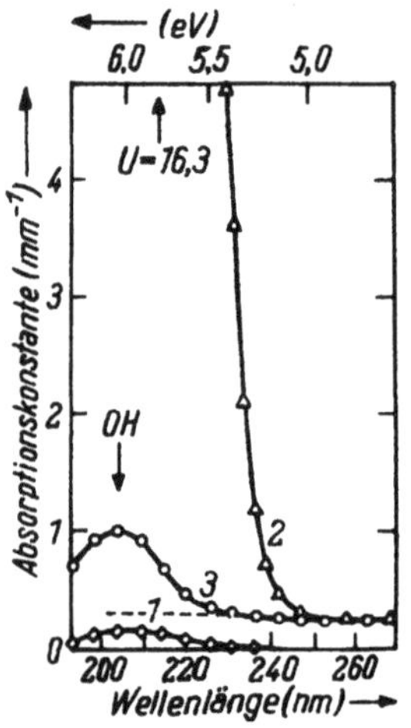

Fig. 1. Absorption eines KCl-Kristalles im UV
Kurve 1 im "as cleaved"-Zustand,
Kurve 2 nach Temperung in Wasserstoff,
Kurve 3 nach Temperung in Sauerstoff

im Wasserstoff zeigt sich eine starke U-Bande (Kurve 2). Wird eine solche Probe in Sauerstoffatmosphäre unter den umseitig angegebenen Bedingungen getempert, so verschwindet nach zwei Stunden die U-Bande vollkommen und es tritt statt dessen eine OH-Bande auf (Kurve 3). Andere Änderungen der Absorption konnten weder im ultravioletten noch im sichtbaren Spektralbereich nachgewiesen werden. Die neue OH-Bande ist größer als die ursprüngliche im "as cleaved"-Zustand, was darauf schließen läßt, daß die U-Zentren in OH-Zentren umgewandelt worden sind. Da die Oszillatorenstärke der OH-Bande wesentlich kleiner ist als die der U-Bande, ergibt sich aus der hohen U-Zentrenabsorption nur eine weit geringere OH-Zentrenabsorption.

Die Umwandlungen kann man auch quantitativ verfolgen. Unter der Annahme, daß ein U-Zentrum mit einem Sauerstoffatom zu einem OH-Zentrum reagiert, ergibt sich die folgende Gleichung (Schottky-Symbolik)

$$H_{\bullet}^{x}(Cl) + 1/2\ O_2 \longrightarrow OH_{\bullet}^{x}(Cl). \qquad (1)$$

Werden ferner die verschiedenen Konzentrationen wie folgt bezeichnet:

$[OH]_1$ Konzentration der OH-Zentren im "as cleaved"-Zustand (berechnet aus Kurve 1 in Fig. 1)

[U] Konzentration der U-Zentren nach der Behandlung in Wasserstoff (berechnet aus Kurve 2 in Fig. 1)

$[OH]_2$ Konzentration der OH-Zentren, die aus U-Zentren gemäß Gleichung (1) entstanden sind. Für diese Konzentration gilt $[OH]_2 = [U]$

$[OH]_3$ Konzentration der OH-Zentren nach Temperung in Sauerstoff (berechnet aus Kurve 3 in Fig. 1)

und verläuft die U→ OH-Reaktion quantitativ, so müßte sich für die Konzentration $[OH]_3$ ergeben:

$$[OH]_3 = [OH]_1 + [OH]_2 . \qquad (2)$$

In Tabelle 1 sind entsprechende Ergebnisse von drei Kristallen mit einer $[OH]_1$-Konzentration von $3{,}44 \times 10^{17}$ cm^{-3} (unter Argon gezogen) und einem Kristall mit einer $[OH]_1$-Konzentration von $5{,}86 \times 10^{17}$ cm^{-3} (an Luft gezogen) (Spalte 1) mit unterschiedlichen U-Zentren-Konzentrationen (Spalte 2) dargestellt. In Spalte 3 ist die Summe der Konzentrationen $[OH]_1 + [OH]_2$ dargestellt, in Spalte 4 die gemessene Konzentration $[OH]_3$.

Die Übereinstimmung der entsprechenden Werte in Spalte 3 und 4 darf als befriedigend angesehen werden, wenn man die systematischen Fehler (Genauigkeit der Formeln für die U- und OH-Zentren-Konzentrationen, Einfluß der thermischen Behandlung bei der additiven Verfärbung und Wasserstofftemperung) in Betracht zieht. Im Rahmen der vorliegenden Meßgenauigkeit ergibt sich damit, daß man durch Temperung in Sauerstoff eine Umwandlung der U-Zentren in OH-Zentren erreichen kann.

Die Eindiffusion von Sauerstoff in U-Zentren-haltige KCl-Kristalle läßt sich einfach nachweisen. Kristalle mit U-Zentren erhalten bekanntlich durch Röntgenbestrahlung eine intensive und gegenüber Lichteinstrahlung stabile Verfärbung durch F-Zentren. Es wurden deshalb in Sauerstoff getemperte größere Kristallproben mit Röntgenstrahlung verfärbt und danach

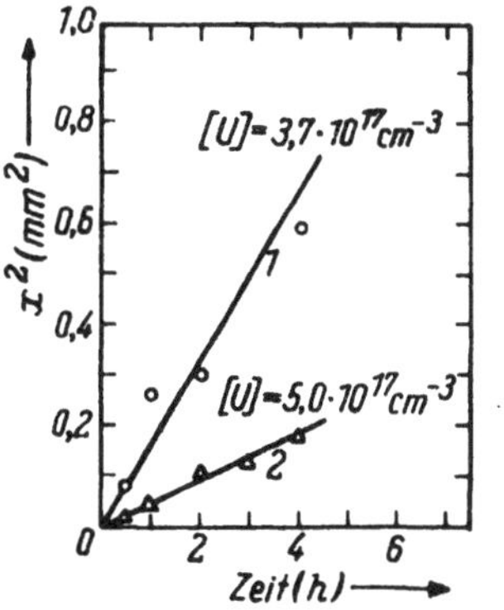

Fig. 2. x^2 als Funktion der Zeit, Temperatur 650 °C, Normaldruck. Kurve 1: Kristall an Luft gezüchtet, Kurve 2: Kristall unter 30 Torr Argon gezüchtet

Tabelle 1

[OH]$_1$ $\times 10^{-17}$ cm^{-3}	[U] = [OH]$_2$ $\times 10^{-17}$ cm^{-3}	[OH]$_1$ + [OH]$_2$ $\times 10^{-17}$ cm^{-3}	[OH]$_3$ $\times 10^{-17}$ cm^{-3}
3,44	3,37	6,81	9,85
3,44	4,20	7,64	9,96
3,44	5,13	8,57	7,20
5,86	4,72	10,58	12,07

mit weißem Licht bestrahlt, wobei nur die Randzone entfärbt wird, in der die U → OH-Umwandlung bereits stattgefunden hat. Für die Tiefe x der entfärbten Zone nach der Zeit t gilt die Beziehung für den "Anlaufvorgang" (5):

$$\frac{x^2}{t} = 2D\frac{c_r}{c_i}. \tag{3}$$

Dabei bedeuten D die Diffusionskonstante und c_r bzw. c_i die Randkonzentration des Sauerstoffes bzw. die Konzentration der als Indikator dienenden U-Zentren.

Fig. 2 zeigt an zwei Beispielen, daß die Meßwerte der Gleichung (3) relativ gut genügen. Die Streuung der Meßpunkte wird vor allem durch die nicht genügend homogene Verteilung der U-Zentren bedingt. Der Wert des Produktes Dc_r ist von den Züchtungsbedingungen der verwendeten Kristalle abhängig.

Herrn Prof. Dr. W. Messerschmidt danke ich für die freundliche Genehmigung, an seinem Institut arbeiten zu dürfen. Mein besonderer Dank gilt Herrn Dr. F. Fröhlich für die Anregung und für die Unterstützung bei den Experimenten und ihrer Diskussion.

Literatur

(1) F. LÜTY, J. Phys. Chem. Solids 23, 677 (1962).
(2) F. FRÖHLICH und O. ADAMETZ, erscheint demnächst in phys. stat. sol.
(3) J. ANGER, B. FRITZ und F. LÜTY, Z. Phys. 174, 240 (1963).
(4) H. GRÜNDIG und C. RÜHENBECK, Z. Phys. 183, 274 (1965).
(5) E. MOLLWO, Ann. Phys. (Germany) 29, 394 (1937).

(Received June 29, 1966)

phys. stat. sol. 16, K165 (1966)

Institute of Crystallography, Academy of Sciences of the USSR, Moscow

Nd^{3+} Optical Centres in $Y_3Al_5O_{12}$ Crystals

By

G.A. BOGOMOLOVA, A.A. KAMINSKII, and V.A. TIMOFEEVA

In recent years considerable interest has been shown to the study of the optical properties of garnet crystals doped with rare-earth elements (TR). This was stimulated by the application of these materials in optical quantum generators (lasers) (1).

It is known that TR^{3+} ions in garnet crystals form centres of different structure which are characterized by differences in the optical spectra (2, 3). The formation of any particular optical centre depends, among other factors, on the concentration of the activator.

In this communication we report preliminary results of an investigation of Nd^{3+} optical centres in $Y_3Al_5O_{12}$ crystals, grown in the Institute of Crystallography. The investigations were made with Nd^{3+} concentration from 0.2 to 12 wt% by the concentration series method (4 to 6) at 77 °K.

The luminescence and absorption spectra of $Y_3Al_5O_{12}$ single crystals activated with Nd^{3+} have been studied in great detail. The most complete data are given in (3, 7, 8).

The preliminary investigations of the absorption spectra were carried out at 300 °K in the region from 0.2 to 2.5 μm using a spectrophotometer type SP-700. The most convenient groups of lines were selected for further detailed investigation ($^4D_{3/2}$ - 3628 Å, $^2P_{3/2}$ - 3849 Å, $^2D_{5/2}$ - 4224 Å, $^2P_{1/2}$ - 4318 Å, $^4F_{3/2}$ - 8685; 8750; 8752; 8760; 8862 Å). The absorption in these groups was recorded at 77 °K using a diffraction spectrometer type DFS-12 with a resolution of 0.1 Å. Photomultipliers with antimony-cesium or oxygen-cesium photocathodes were used as optical detectors. The signal from the photomultiplier was amplified and then recorded with a potentiometer type EPP-09M3. Illumination was provided from an incandescent lamp supplied by stabilised voltage. The crystal

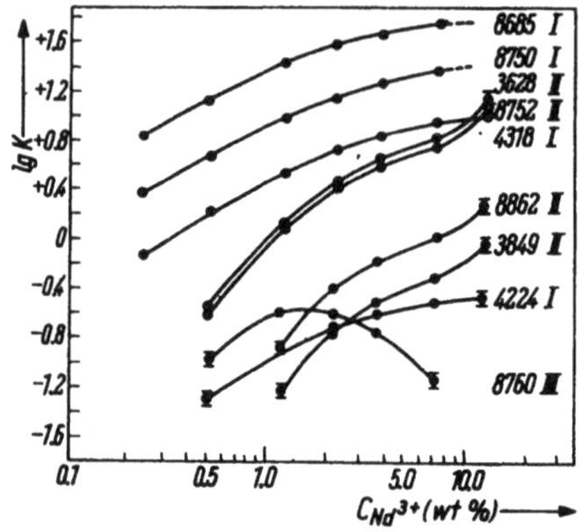

Fig. 1. Absorption coefficients at λ_{max} for lines of the $^4D_{3/2}$, $^2P_{3/2}$, $^2D_{5/2}$, $^2P_{1/2}$, and $^4F_{3/2}$ groups vs. concentration

was mounted between two quartz light guides and directly plunged into liquid nitrogen.

Fig. 1 represents an analysis of the $Y_3Al_5O_{12}$-Nd^{3+} absorption spectra. The absorption coefficients at the maximum of the individual lines have been plotted in a double-logarithmic scale as functions of the concentration. This was done for only a few of the lines in order to avoid overcrowding of the figure.

A comparison of the curves shows clearly that they can be divided into three systems designated by I, II, and III in the figure. Each of these systems is a family of parallel curves. These systems of lines obviously belong to centres having different structures.

References

(1) A.A. KAMINSKII and V.V. OSIKO, Neorgan. Mat. 1, 2049 (1965).

(2) W.P. WOLF, M. BELL, M.T. HUTHINS, M.J.M. LEASK, and A.F.G. WYATT, J. Phys. Soc. Japan 17, Suppl. BI, 443 (1962).

(3) A.A. KAMINSKII, Zh. eksper. teor. Fiz. 51, 50 (1966).

(4) YU.K. VORONKO, A.A. KAMINSKII, and V.V. OSIKO, Zh. eksper. teor. Fiz. 49, 420 (1965).

(5) YU.K. VORONKO, A.A. KAMINSKII, and V.V. OSIKO, Zh. eksper. teor. Fiz. 49, 724 (1965).

(6) YU.K. VORONKO, A.A. KAMINSKII, and V.V. OSIKO, Zh. eksper. teor. Fiz. 50, 15 (1966).

(7) J.A. KONINGSTEIN and J.E. GEUSIC, Phys. Rev. 136, A711 (1964

(8) P.P. FEOFILOV, V.A. TIMOFEEVA, M.N. TOLSTOI, and L.M. BELYEV, Optika i Spektroskopiya 19, 817 (1965).

(Received July 6, 1966)

phys. stat. sol. 16, K167 (1966)

Research Institute for Technical Physics of the Hungarian Academy of Sciences, Budapest

Electron Mobility in Plastically Deformed Germanium

By
B. PÖDÖR

A plastic deformation influences strongly the electrical properties of semiconductor single crystals. If the dislocations, generated by plastic deformation, have an edge component, they introduce acceptor centres along the dislocation line which capture electrons from the conduction band of an n-type semiconductor (1). The dislocation line becomes negatively charged, and a space charge is formed around it. The resulting potential field scatters the conduction electrons and so reduces the electron mobility. The effect of dislocations on the electron mobility in n-type germanium was examined experimentally in (2 to 4), and theoretically in (5) and (6).

The screened potential of a negatively charged dislocation line at a great distance from the line is according to (7)

$$U(r) = - \frac{ef}{2\pi\varepsilon a} K_o\left(\frac{r}{\lambda_D}\right), \qquad (1)$$

where a is the distance between the acceptor centres along the dislocation line, f is the occupation rate of these acceptor sites, and λ_D the Debye screening length:

$$\lambda_D = \left(\frac{\varepsilon kT}{e^2 n}\right)^{1/2} .$$

Knowing the scattering potential we determined with the Born approximation the scattering cross section. Thus the relaxation time perpendicular to the system of parallel dislocations has been deduced as follows:

$$\tau_{disl} = \frac{8\varepsilon^2 m^2 \left(\frac{\hbar^2}{4m^2\lambda_D^2} + v_\perp^2\right)^{3/2}}{N_{disl}\, e^4 f^2\, \lambda_D} ,$$

where $v_\perp$ is the component of electron velocity perpendicular to the dislocations. Averaging with the equilibrium distribution function, the mobility is

$$\mu_{disl} = \frac{30(2\pi)^{1/2}\varepsilon^2 a^2 (kT)^{3/2}}{N_{disl}\ e^3 f^2\ \lambda_D\ m^{1/2}} . \qquad (2)$$

Equation (2) is slightly different from the corresponding expression given in (6), based on a potential function different from (1). Equation (2) indicates a stronger scattering than the corresponding expression in (6). At about 100 $^{\circ}$K it gives a mobility which is more than one order of magnitude smaller than that obtained from the expression in (8). This high mobility is due to the scattering effect of the mechanical deformation field around the dislocation line. Since the reciprocal values of the relaxation times resulting from different physical mechanisms are additive, the scattering caused by the charge of dislocations in n-type semiconductors gives the dominant effect below room temperature.

We measured the electron mobility on plastically deformed germanium single crystals between 80 and 300 $^{\circ}$K. n-type high-purity germanium single crystals with a room temperature resistivity of 25 Ωcm and a dislocation density of 4×10^3 cm^{-2} were bent about the [112] axis at 730 $^{\circ}$C in a H_2 stream by graphite shapes with a known curvature. The samples were plated with tin in order to prevent contaminations during the heat treatment (9). According to our measurements the heat treatment alone did not produce acceptor centres, the electron concentration and mobility remained unchanged.

The electron mobility was determined from the conductivity and the Hall coefficient. The ratio of Hall mobility to conduction mobility was taken to be $3\pi/8$. The magnetic field was normal to the axis of bending i.e. to the direction of the dislocations, and the current perpendicular to both. Typical results of the measurements on non-deformed and deformed samples are plotted in Fig. 1.

The mobility measured on deformed samples is the compo-

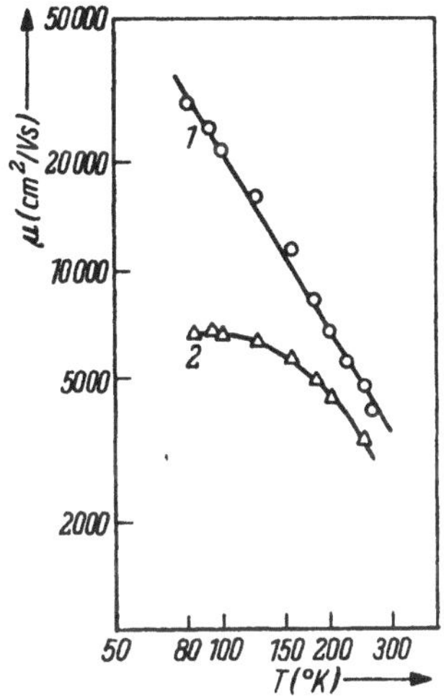

Fig. 1. Electron mobility vs. temperature in non-deformed and plastically deformed germanium single crystals
Curve 1: unbent crystal;
curve 2: bent crystal, bending axis [112], bending radius 50 mm

sition of the scattering effects of electrons by acoustic phonons and by dislocations. The resulting mobility in the usual approximation is

$$\mu_{def}^{-1} = \mu_{disl}^{-1} + \mu_{lattice}^{-1} .$$

From the results of the measurements at 100 °K we find μ_{disl} = 9600 cm^2/Vs. At the same time we obtain μ_{disl} = 15000 cm^2/Vs from equation (2) with $\varepsilon/\varepsilon_0 = 16$, $m = 0.3\ m_{el}$, $a = 3.46\times10^{-8}$ cm, $n = 2.1\times10^{13}$ cm^{-3} (from Hall measurements on deformed crystals), $N_d = 6.53\times10^{13}$ cm^{-3} (from Hall measurements on non-deformed crystals), and N_{disl} = $= 5\times10^6$ cm^{-2} (from the bending radius).

Thus the agreement between the experimental and theoretical results is acceptable.

I wish to express my acknowledgement to Dr. Z. Bodó for valuable discussions.

References

(1) W.T. READ, Phil. Mag. 45, 775 (1954).
(2) G.L. PEARSON, W.T. READ, and F.J. MORIN, Phys. Rev. 93, 666 (1954).
(3) R.A. LOGAN, G.L. PEARSON, and D.A. KLEINMANN, J. appl. Phys. 30, 885 (1959).
(4) L.I. KOLESNIK, Fiz. tverd. Tela 6, 1253 (1964).
(5) W.T. READ, Phil. Mag. 46, 111 (1954).
(6) E.M. KUZNETSOVA, Fiz. tverd. Tela 3, 1987 (1961).
(7) V.L. BONCH-BRUEVICH and V.B. GLASKO, Fiz. tverd. Tela 3,

36 (1961).
(8) D.L. DEXTER and F. SEITZ, Phys. Rev. 86, 964 (1952).
(9) T.R. FIGIELSKI and A.D. BELYAEV, Fiz. tverd. Tela 6, 2146 (1964).

(Received July 6, 1966)

phys. stat. sol. 16, K171 (1966)

Pädagogische Hochschule Katowice

Weitere Untersuchungen über Relaxationserscheinungen in Fe-Si-C(N)-Legierungen

Von

J.W. MOROŃ und B. PILCH

Außer der Snoek-Relaxation (I1) wurden in α-Fe-Si-C und α-Fe-Si-N zwei weitere Maxima (I4, I5) gefunden, die auch durch den in Lösung befindlichen Kohlenstoff bzw. Stickstoff verursacht werden (1 bis 8). Allgemein nimmt man an, daß die erste dieser Erscheinungen (I4) durch elementare Sprünge von C- und N-Atomen auf Zwischengitterplatz, die in der Nähe von gelösten Siliziumatomen verbleiben, hervorgerufen wird. In (9, 10) wurde gezeigt, daß sich I5 aus einer ganzen Reihe von Relaxationseffekten zusammensetzt, von denen jeder ein kleines Maximum bildet. Auch die Analyse einer bei 21,3 °C gemessenen Desakkommodationskurve zeigte, daß ihr Verlauf durch zwei Relaxationen beschrieben werden kann (11); bei dieser Temperatur wird der zeitliche Abfall der Permeabilität durch I5 verursacht (1, 10).

In der vorliegenden Arbeit wurde die Temperaturabhängigkeit der magnetischen Verluste an einer weiteren Eisen-Silizium-Probe gemessen. Die Probe enthielt 2,73 Gew.% Si, 0,023% C und kleine Zusätze von anderen Elementen (Ni: 0,06%, Cr: 0,03%, Mo: 0,02%, Mn: $<$0,01%, Ti: $<$0,01%, Co: $<$0,002%, P: 0,009%, S: 0,002%, Spuren von Cu und Pb). Der Gehalt an Stickstoff wurde nicht bestimmt. Die Probe enthielt jedoch eine kleine Menge dieses Elements, was an dem Verlauf der experimentellen und theoretischen $\text{tg}\delta = f(\frac{1}{T})$-Kurven zu sehen war.

Die ringförmige Probe, die aus einem Band von 0,1 mm Dicke und 10,1 mm Breite gewickelt wurde, hat einen Innendurchmesser von 25 mm und einen Außendurchmesser von 33 mm gehabt. Sie wurde mehreren Glühungen bei 800 bis 900 °C in einem Vakuumofen unterworfen. Danach wurde sie langsam abgekühlt und gewöhnlich bei 180 °C gealtert.

Die Messungen wurden mit einer Maxwell-Wien-Brücke durchgeführt, die in (9) beschrieben wurde. Das Magnetisierungsfeld betrug 7,9 mOe, seine Frequenz 33,3 bis 166,7 Hz. Die Tempera-

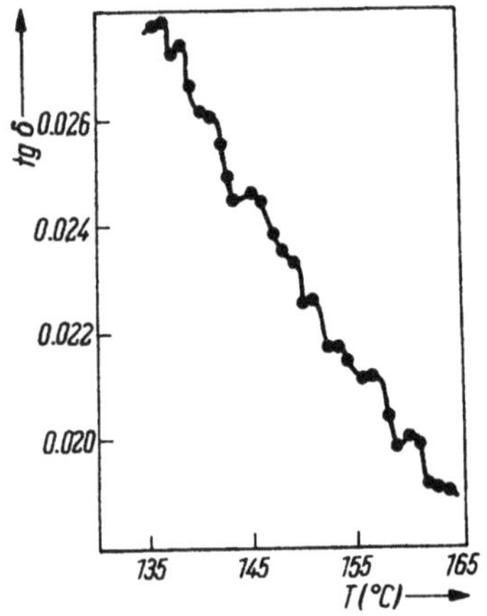

Fig. 1. tgδ = f(T) nach einer sorgfältigen Entspannungsglühung; f = 33,3 Hz

turschwankungen waren kleiner als ±0,1 °C. Hysterese und Wirbelstromverluste wurden im allgemeinen nicht abgezogen. Die Temperaturen der Maxima wurden aus den graphischen tgδ = f(T)-Darstellungen abgelesen.

Es wurden mehrere Messungen im Bereich von 100 bis 300 °C durchgeführt. Fig. 1 zeigt beispielsweise eine Kurve, die nach einer sehr sorgfältigen Entspannungsglühung bei 33 Hz erhalten wurde. Sie enthält eine Anzahl von kleinen Maxima und mehrere sehr deutliche Schultern, die zweifellos weiteren Relaxationen entsprechen. Auch in allen anderen Fällen wurden ähnliche Resultate erhalten: die Maxima traten bei allen Meßfrequenzen und in allen Meßreihen auf. Die Fehlerrechnung zeigte, daß die Fehler von tgδ viel kleiner sind als die Abweichungen von einem eventuellen monotonen tgδ = f(T)-Verlauf.

Für einige größere Maxima ist es aus Messungen bei fünf Frequenzen gelungen, die Aktivierungsenergie (Q) und den Vorfaktor (Θ_o) des Arrhenius-Gesetzes zu ermitteln. In Fig. 2 ist lnω als Funktion von $\frac{1}{T_p}$ für einige untersuchte Fälle dargestellt (ω Kreisfrequenz des Magnetisierungsstromes, T_p Temperatur in Grad Kelvin). Die beigefügte Tabelle faßt die Ergebnisse zusammen. Spalte 2 enthält die Temperaturen der Maxima T_p bei 33,3 Hz.

Die Messungen zeigten auch, daß bei höheren Temperaturen weitere Maxima auftreten (bei 66,7 Hz bis 260 °C). Es wurde auch festgestellt, daß sich ähnlich wie in (9) keine I5-Relaxation von den anderen Prozessen derselben Art hervorhebt.

Die in (9, 10) und in der vorliegenden Arbeit gefundenen Maxima werden sicher nicht durch die in Lösung befindlichen

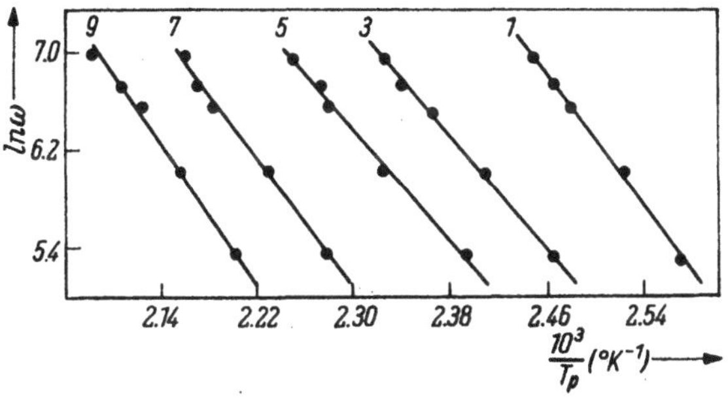

Fig. 2. $\ln\omega = f\left(\frac{1}{T_p}\right)$ für die Relaxationen 1, 3, 5, 7, 9

anderen Zusatzelemente hervorgerufen. Sie wurden auch in einer Probe, die nur Si und C(N) enthielt (Probe RB-RBN in (10)), gefunden. In letzter Zeit wurden ähnliche Relaxationen ebenfalls in mehreren α-Fe-C(N)-Proben entdeckt (12), und es scheint so, als ob I5 auch nicht mit dem in ungeordneter Lösung befindlichen Silizium verbunden wären.

Die hier erörterten Erscheinungen können durch Paare und größere Anhäufungen kleiner Zwischengitteratome hervorgerufen werden. Sie können auch durch Sprünge einzelner, sich infolge Segregationen in einer Art von Guinier-Preston Zone befindenden C(N)-Atome verursacht werden.

Tabelle

Die Q- und θ_o-Werte einiger I5-Relaxationen

Relaxation	T_p (°C)	Q (cal/g-Atom)	θ_o (s)
1	115,5	25500 ± 800	$1,8\times10^{-17}$
2	123,0	26400 ± 1300	$1,4\times10^{-17}$
3	132,2	24600 ± 1400	$2,6\times10^{-16}$
4	140,5	21900 ± 1100	$1,4\times10^{-14}$
5	145,5	21700 ± 300	$1,5\times10^{-14}$
6	154,3	27700 ± 1000	$3,3\times10^{-17}$
7	165,7	26000 ± 1000	$2,1\times10^{-16}$
8	174,5	26600 ± 800	$4,5\times10^{-16}$
9	179,7	27100 ± 1200	$1,1\times10^{-17}$

Literatur

(1) A. FERRO und G. MONTALENTI, Ricarca Scientifica 25, 3069 (1955).
(2) D.A. LEAK, W.R. THOMAS und G.M. LEAK, Acta metall. 3, 501 (1955).
(3) D.A. LEAK und G.M. LEAK, J. Iron Steel Inst. 189, 256 (1958).
(4) A.J. BOSMAN, Dissertation, Amsterdam 1960.
(5) R. RAWLINGS und P.M. ROBINSON, J. Iron Steel Inst. 197, 211 (1961).
(6) H. BORCHERS, Arch. Hüttenwesen 34, 453 (1963).
(7) P.E. BROMMER, private Mitteilung (1965).
(8) M.A. KRISHTAL, Fiz. Met. Metallov. 19, 111 (1965).
(9) J.W. MOROŃ, phys. stat. sol. 5, K77 (1964).
(10) J.W. MOROŃ, Acta phys. Pol. 26, 1117 (1964).
(11) W. HAMPE und D. WIDMANN, Z. angew. Phys. 15, 360 (1963).
(12) A. DOBRZAŃSKA, J.W. MOROŃ und J. RASEK, in Vorbereitung.

(Received July 7, 1966)

phys. stat. sol. 16, K175 (1966)

Department of Physical and Inorganic Chemistry,
University of Adelaide

Misfit and Hysteresis at Structure Transformations in Rubidium and Cesium Nitrates; and Thermal Expansions

By

S.W. KENNEDY, G.F. TAYLOR, and (in part) J.H. PATTERSON

$CsNO_3$ and $RbNO_3$ are suitable substances in which to study symmetry-controlled transformation in single crystals, especially since they are isomorphous with the higher temperature forms of thallous nitrate in which several different mechanisms were seen (1). This transformation is also of intrinsic interest in relation to theories of ferro- and antiferroelectricity (2, 3). Trigonal rubidium nitrate is considered to be antiferroelectric (4), and it has been suggested that the cubic form anomalously contracts on heating.

In the present work it is found that the transformation is discontinuous, the expansion coefficient of the cubic form is normal and positive, that the small discontinuity produces measurable distortion in the product phase and the annealing of this can be followed, that supercooling can occur until the driving force becomes about 160 cal/mol, and that the hysteresis in propagation in single crystals is about one-tenth of that measured by dilatometry using bulk samples, but increases if the crystals are recycled and aged.

The room temperature form, $RbNO_3$ IV and $CsNO_3$ II, is trigonal. At higher temperatures the structure is cubic, space group Pa3 (5). (Only an older result Pm3m, based on powder work, has been quoted in several recent papers.) Both structures are based on sub-cells having a cation at each corner and a nitrate ion within (1). It is clear that in the transformation any three-fold axis of the cubic structure becomes the trigonal axis of the room temperature form (6). The unit cell may be polarized in either sense along the trigonal axis.

The equilibrium temperatures and the supercoolings were measured using vacuum dried small single crystals of high purity sealed in thin-walled hard glass tubes immersed in a large thermostatted oil bath and observed between crossed polars. As the interface was quite mobile it was possible to note precisely the temperature at which the phases co-existed. Thermal expansions were measured mainly by single crystal X-ray diffractometry. The expansion coefficients for trigonal phases were derived by a least squares treatment. To measure any spacing changes across the phase interface without extrapolation the crystal was kept partly transformed in a temperature gradient of 0.4 °C/mm, and the interface was set within the fine X-ray beam. The (90.0) and (00.9) reflections were measured. The strains and imperfections introduced by the transformation are indicated by the mosaic spread which was measured using only the $K\alpha_1$ component of the Cu radiation.

The expansion of cubic $RbNO_3$ was normal, the approximate expansion coefficient being $\alpha = 85 \times 10^{-6}$ $^{\circ}C^{-1}$. The expansion coefficients of trigonal $RbNO_3$ and $CsNO_3$ were linear. Table 1 shows the expansion coefficients α_i for the direction of the trigonal form parallel to the axis indicated by the subscript, the percentage change of lattice spacing at the interface in the change to cubic, mosaic spreads (quoted as half-width at half-height), and the maximum supercooling ΔT. The supercooling was not systematically altered by recycling followed by annealing. Generally only one nucleus grew.

At the observed supercooling of 17 °C the free energy of transformation calculated from the specific heat data, is

Table 1

	Equil. temp. (°C)	$10^6\alpha_c$ (°C^{-1})	$10^6\alpha_a$ (°C^{-1})	T_c (°C)	Δc (%)	Δa (%)	Mosaic spread (degrees arc)		ΔT (°C)
							trig. before	cubic after	
$RbNO_3$	164	101.5	97.1	164	−0.70	+0.78	0.03	0.135	17
$CsNO_3$	153.7	71.4	68.8	153.8	0	+0.59			12

about 160 cal/mol. If embryos are present then this would be the driving force needed to overcome the interfacial and compressive or tensile energies.

In contrast to the delay in nucleation, the hysteresis in growth (movement of the phase interface) was 0.1 °C in fresh single crystals which had been transformed once or twice. (This was measured with the micro-thermostat and polarised light.) However when $RbNO_3$ crystals had been maintained at about 164 °C for over a week and had been recycled several times, the maximum hysteresis increased to 1 °C. It was then much less reproducible and apparently was greater the longer the interface had been stationary.

On annealing transformed fresh crystals for about 20 hours at 140 °C, polygonization occurred. Annealing of the cubic form for 5 hours produced only a slight reduction in the mosaic spread. It has previously been shown that after the similar transformation in $TlNO_3$ corresponding splitting of the X-ray reflections was accompanied by formation of a network of subgrain boundaries (1, 7). Preliminary measurements on single crystals (Chaplin and Kennedy) show that the changes in X-ray reflections on annealing are paralleled by changes of 20 to 40% (in different crystals) in the ionic conductivity over 20 hours.

It can be concluded that since this is a rapidly propagating symmetry-controlled transformation, misfits (for given elastic constants) which would demand a reconstructive or martensitic mechanism would be greater than those involved here. This correlates with more extensive studies of $TlNO_3$, where invariant planes are considered.

The measurements showing the production and annealing of strains while an orientation relation is maintained are consistent with the hypothesis (1) that the temperature independent time dependent nucleation rates occurring in the trigonal phase of $TlNO_3$ were due to competition between nucleation and the annealing out of nucleation sites: they are consistent too with the further hypothesis that the critical stage of nucleation in solids may be precipitated by rather sudden release of interfacial and compressive stresses by a process involving the migration of dislocations and the loss of previously complete coherence. The increase of growth hysteresis with time and recycling suggests that interface movement is held up, possibly through dislocation pinning, by segregation of the products of slight decomposition. The observation that the growth hysteresis in fresh single crystals is much less than that found by dilatometry suggests that the latter values were not entirely due to Clausius-Clapeyron effects as supposed, and points to the difficulty in dilatometry of distinguishing between kinetic effects and pre-transformation expansion anomalies, especially since delays intended to allow the system to reach a steady state may themselves increase the kinetic hysteresis. Alternative precise single crystal techniques are being developed.

References

(1) S.W. KENNEDY and J.H. PATTERSON, Proc. Roy. Soc. A283, 498 (1964).

(2) A.S. SONIN and I.S. ZHELUDEV, Kristallografiya 4, 487 (1959).

(3) K. AIZU, J. Phys. Soc. Japan 19, 918 (1964).

(4) A.YA. DANTSIGER and E.G. FESENKO, Kristallografiya 8, 894 (1963); Soviet Phys. - Cryst. 8, 717 (1964).

(5) U. KORHONEN, Ann. Acad. Sci. Fenn. A VI 102 (1951); 150 (1953).

(6) J.H. PATTERSON, Thesis; 1959 Ph.D. Thesis, Adelaide University, 1962.

(7) S.W. KENNEDY and J.H. PATTERSON, Nature 194, 1072 (1962).

(Received July 11, 1966)

phys. stat. sol. 16, K179 (1966)

Physikalisch-Technisches Institut der Deutschen Akademie der Wissenschaften zu Berlin, Bereich elektrischer Durchschlag, und IV. Physikalisches Institut der Humboldt-Universität zu Berlin

Einfluß flacher Haftstellen auf die Strom-Spannungs-Charakteristik bei Doppelinjektion

Von

H.J. ROHDE

Doppelinjektion in Isolatoren mit ungleichen Lebensdauern für Löcher und Elektronen wurde von Lampert (1) an einem einfachen Modellisolator untersucht. Dieser enthält ein Rekombinationszentrum, das im thermischen Gleichgewicht gefüllt ist und für das $\sigma_n \ll \sigma_p$ gilt. (σ_n, σ_p Einfangquerschnitte dieses Zentrums für Elektronen im unbesetzten bzw. Löcher im besetzten Zustand). Die Theorie ergibt eine Grenzspannung V_{th}, bei der der Doppelinjektionsstrom vertikal einsetzt. Daran schließt sich ein Bereich negativen differentiellen Widerstandes an. Am Ende dieses Bereiches haben Elektronen und Löcher gleiche Lebensdauern, und die Elektronen aus den Rekombinationszentren befinden sich stationär im Leitungsband. Bei noch höheren Strömen liegt Doppelinjektion in einen Halbleiter mit gleichen Lebensdauern für Löcher und Elektronen vor, und der Strom ist proportional dem Quadrat der Spannung (2).

Lampert vernachlässigt in seiner Theorie die Raumladung, nimmt also an, daß die injizierten Löcher- und Elektronenladungen sich gegenseitig weitgehend neutralisieren. Dies ist der Fall, wenn die Elektronenlebensdauer im thermischen Gleichgewicht unbegrenzt groß ist (vollbesetzte Rekombinationszentren), so daß freie Elektronen immer an den Ort der Löcherladung gelangen können, um diese zu neutralisieren. Fig. 1, Kurve 1 zeigt die für das Lampertsche Modell gültige Kennlinie.

Die von Lampert entwickelte Theorie berücksichtigt nicht die in Photoleitern vom Typ CdS vorhandenen flachen Haftstellen. Diese können den Einträgerinjektionsstrom wesentlich modifizieren (3, 4), und es ist wünschenswert, auch ihren Einfluß auf den Doppelinjektionsstrom zu untersuchen.

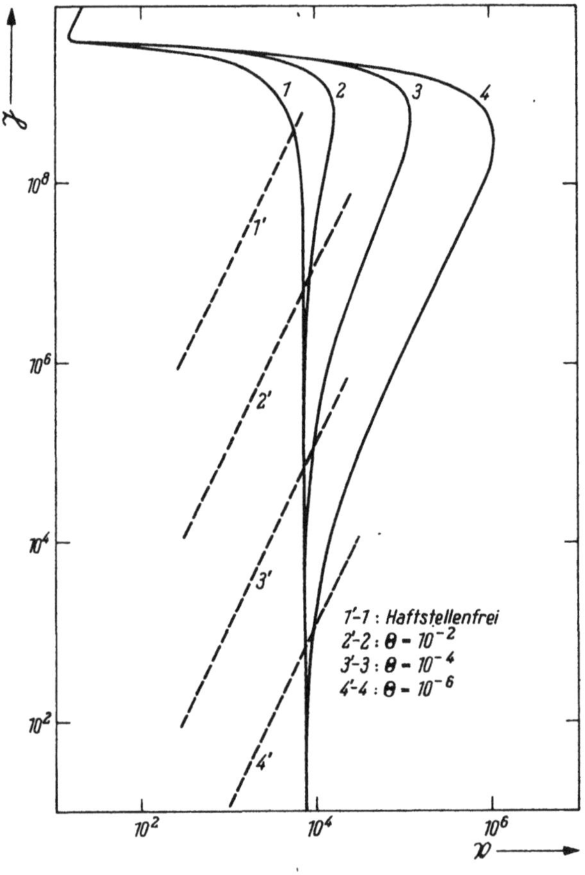

Fig. 1. Berechnete Strom-Spannungs-Kennlinien bei Doppelinjektion
Kurve 1: haftstellenfreier Isolator;
Kurven 2 bis 4: Isolator mit Haftstellen;
Kurven 1' bis 4': die zu den Kurven 1 bis 4 gehörenden Einträgerinjektions-Charakteristiken

Der von Lampert gegebene Formalismus zur Berechnung der Doppelinjektion läßt sich auf das Modell eines Photoleiters mit einem gefüllten Rekombinationszentrum und einer diskreten Haftstelle übertragen. Es gelten folgende Gleichungen:

Stromflußgleichung

$$j = j_n + j_p = e(\mu_n n + \mu_p p)E, \qquad (1)$$

Kontinuitätsgleichung

$$\frac{1}{e}\mathrm{div}\, j_n = v_t \sigma_n p_R = -\frac{1}{e}\mathrm{div}\, j_p, \qquad (2)$$

Neutralitätsgleichung

$$p + p_R - n - h = 0 \qquad (3)$$

und die reaktionskinetischen Ausdrücke

$$v_t \sigma_n n p_R = v_t \sigma_p p(N_R - p_R), \qquad (4)$$

$$h = \frac{nH}{n + \Theta H} . \qquad (5)$$

Dabei bedeuten

μ_n, μ_p Elektronen- bzw. Löcherbeweglichkeit,

v_t thermische Geschwindigkeit freier Träger,

h, p_R Konzentration von Elektronen in den Haftstellen bzw. von Löchern in den Rekombinationszentren,

H, N_R Konzentration flacher Haftstellen bzw. Rekombinationszentren,

Θ Trappingfaktor. Es gilt

$$\Theta = \frac{N_c}{H} e^{-\frac{\Delta E_H}{kT}} \qquad (6)$$

mit

N_c effektive Zustandsdichte im Leitungsband,

ΔE_H energetischer Abstand flacher Haftstellen vom Leitungsband.

Aus (1) bis (5) gewinnt man ähnlich wie in (1) die Strom-Spannungs-Charakteristik in Parameterform. Eine numerische Auswertung für verschiedene Werte von Θ liefern die in Fig. 1 gezeigten Kennlinien (Kurven 2 bis 4). Es wurden für CdS übliche Werte gewählt, d.h. $\frac{H}{N_R} = 10^{-2}$, $\frac{\sigma_n}{\sigma_p} = 10^{-3}$ und $\frac{\mu_n}{\mu_p} = 10$.

Man sieht, daß die Grenzspannung $\mathcal{v}_{th}$ für $j \rightarrow 0$ mit dem von Lampert gefundenen Wert übereinstimmt (j, $\mathcal{v}$ dimensionsloser Strom bzw. Spannung). Bei höheren Strömen verlaufen die Kennlinien steil, jedoch nicht mehr vertikal und biegen nach Erreichen einer Spannung $\mathcal{v}_{max}$ in den Bereich negativen differentiellen Widerstandes ein. Die Steilheit des Stromanstieges im Kennlinienteil vor der Spannung $\mathcal{v}_{max}$ ist um so geringer, je kleiner der Trappingfaktor θ ist. Die auf der Grundlage der Neutralitätsgleichung (3) berechneten Kennlinien gelten nicht für beliebig kleine Ströme, weil der Transport dann raumladungsbegrenzt wird. Die Gesamtkennlinie für ein bestimmtes θ wird daher bei kleinen Strömen durch raumladungsbegrenzte Einträgerinjektion bestimmt (in Fig. 1 die gestrichelten Geraden) und mündet bei $\mathcal{v}_{th}$ in die steile Doppelinjektions-Kennlinie ein.

Weitere theoretische und experimentelle Ergebnisse zur Doppelinjektion in CdS-Einkristallen werden demnächst veröffentlicht.

Literatur

(1) M.A. LAMPERT, Phys. Rev. 125, 126 (1962).
(2) M.A. LAMPERT und A. ROSE, Phys. Rev. 121, 26 (1961).
(3) A. ROSE, Phys. Rev. 97, 1538 (1955).
(4) M.A. LAMPERT, Phys. Rev. 103, 1648 (1956).

(Received July 11, 1966)

Pre-printed Titles and Abstracts

Papers to be published in "physica status solidi"
Vol. 17, No. 1

Original Papers

A.J. MORTON, C.S.I.R.O. Division of Tribophysics, University of Melbourne and C.M. WAYMAN, Department of Mining, Metallurgy, and Petroleum Engineering, University of Illinois, Urbana, On the Feasibility of a "Composite" Martensite Model for the $(225)_f$ Austenite-Martensite Transformation

A.F. KRAVCHENKO and W.S. SARDARYAN, Institute of Semiconductors, Academy of Sciences of the USSR, Novosibirsk, On the Structure of the Bottom of the Conduction Band in GaAs
The magnetoresistance and the Hall effect of oriented n-type GaAs specimens with carrier concentrations of 5×10^{15} to 1×10^{18} cm^{-3} are studied at temperatures between 78 and 800 °K. Almost all samples showed an anisotropy of the transverse magnetoresistance and the non-vanishing longitudinal magnetoresistance. The energetic position of the minima and the effective mass are estimated from the temperature dependence of the Hall coefficient: ΔW = 0.12 to 0.36 eV, $m_2^* = 1.2\ m_o$, $m_\parallel^* = 1.98\ m_o$, $m_\perp^* = 0.37\ m_o$. A model for the band structure in the vicinity of the conduction band edge is discussed and shown to be consistent with the experimental data when anisotropy of electron scattering by weakly oriented dipoles is taken into account.

A.A. ABDURAKHMANOVA and M.I. ALIEV, Institute of Physics, Academy of Sciences of the Azerbaidzhan SSR, Baku, Electrical Properties of GaSb-Ga_2Te_3 and GaSb-GaTe Solid Solutions
Measurements are made of the Hall coefficient, electrical conductivity, and thermoelectric power of GaSb-GaTe and GaSb-Ga_2Te_3 solid solutions as functions of composition and temperature in the range 100 to 550 °K. It is shown that the maximum electron concentration and electrical conductivity are less in the first case due to considerable distortion of the lattice

periodicity by the GaTe. A decrease of Hall mobility is observed as the composition is varied and this is attributed to electron scattering by the disordered alloy structure. The mobility varies with temperature approximately as T^{-1} at high temperatures. This law fails for $T < 290\ ^{o}K$ due to the electron gas becoming degenerate. The electrons appear to be dominantly scattered by the acoustic lattice vibrations over the temperature range considered.

T.J. KOPPENAAL, Metallurgy Division, Argonne National Laboratory, Argonne, Illinois, and R.J. ARSENAULT, Metallurgy Division, Oak Ridge National Laboratory, Oak Ridge, Tennessee, Some Notes on the Strengthening Mechanism in Neutron Irradiated Copper Single Crystals Discussions are presented in some of the controversial areas regarding the interpretation of the strengthening mechanism in neutron irradiated copper single crystals. The specific topics that are discussed include the following: the correlation of cross-head speed changes to strain rate changes, the strain rate dependence of the flow stress, the temperature dependence of the pre-exponential factor, the free energy and the enthalpy of activation, the defect and obstacle size spectrum, and the nature of the force-distance behavior for the interaction of mobile dislocations with irradiation produced obstacles. It is shown that many of the recent arguments made against our previously published explanation of the strengthening mechanism can be readily explained without changing our interpretation.

CH. SCHWINK and H. NEUHÄUSER, II. Physikalisches Institut der Universität München, On the Strengthening Mechanism in Neutron Irradiated Copper Single Crystals - Critical Comments to the Preceding Paper by Koppenaal and Arsenault (I) Measurements of strain rate changes for analysing the flow stress of n-irradiated copper crystals led us to the following conclusions: a) The "active crystal length", i.e. the number of simultaneously active slip planes, increases with the strain rate $\dot{\gamma}$, Consequently $\dot{\gamma}$ is not proportional to $\dot{l}$ (cross-head speed). - b) The critical shear stress (css) varies with $\dot{l}$ more rapidly than is expected by theory. In the preceding paper arguments are advanced against these statements. All of these arguments will be invalidated. The proportionality $\dot{\gamma} \sim \dot{i}$ is shown to represent a rather special case from an atomic point of view, by far more general is a non-proportionality $\dot{\gamma} \nsim \dot{i}$. Taking this into account one succeeds in describing straightforwardly the results of b). New experiments supporting our former statements are reported. All conclusions drawn previously on the strengthening mechanism in copper and based on a) and b) can be fully maintained.

J. DIEHL and G.P. SEIDEL, Institut für Sondermetalle am Max-Planck-Institut für Metallforschung, Stuttgart, On the Strengthening Mechanism in Neutron Irradiated Copper Single Crystals - Critical Comments to the Preceding Paper by Koppenaal and Arsenault (II) Koppenaal's and Arsenault's analysis of the thermally activated glide processes in neutron irradiated copper is discussed and compared with a similar investigation of the present authors. It is shown that Koppenaal and Arsenault use a number of very restrictive and not very realistic assumptions, e.g. regarding the behaviour of the dislocations and the nature of the dislocation obstacles, which are not necessary in our investigations. Koppenaal's and Arsenault's arguments against our analysis can easily be refuted.

V.V. KOBELEV, Institute of Exact Mechanics and Computing Technique, Academy of Sciences of the USSR, Moscow, Ferromagnetic Films in Non-Uniform Fields An analysis of the magnetization processes in magnetic films must take into account the fields arising from the non-uniformity of the film magnetization. Equations are obtained involving the fields of magnetic poles distributed over the film surface. These equations are applicable to continuous films as well as to separate magnetic thin film strips. The magnetization distribution is determined for a hard direction field by a numerical solution of the basic integro-differential equations.

M.G. ASCH, M.M.C. FELIX et M.R. ONGARO, Laboratoire d'Électronique et de Physique du Solide, Faculté des Sciences de Lyon, Électrisation du polyéthylene et du polystyrène après irradiation

On montre qu'après avoir été exposés au rayonnement γ, le polyéthylène et le polystyrène peuvent acquérir une électrisation superficielle stable sous l'influence d'un champ électrique. La densité superficielle de charge est calculée en admettant que l'électrisation est due au déplacement d'électrons qui, initialement piégés, sont progressivement libérés par excitation thermique. L'expression de la densité de charge ainsi obtenue est conforme aux résultats expérimentaux.

V.A. ROMANOV, I.P. ZHADKO, and I.I. BOIKO, Institute of Semiconductors, Academy of Sciences of the Ukrainian SSR, Kiev, Photoconductivity of Anisotropically Deformed Intrinsic Germanium Theoretical and experimental investigations are made of the photoconductivity in anisotropically deformed germanium in conditions favourable to the occurrence for electrical pinch. A saturation of the photocurrent for sufficiently high electric fields is predicted and observed. The effect of the electric field on the spatial distribution of the carriers in an illuminated crystal is investigated using the infrared spot method. Similar values of the surface recombination velocities are obtained from the current-voltage and relaxation characteristics.

H. HORA and R. KANTLEHNER, Physikdepartment der Technischen Hochschule München, The Anomalous Vectorial Photoemission from a Cesium-Antimony Layer The photoemission from a transparent cesium-antimony layer is measured. Here the special arrangement of the cell allows a 75° angle of incidence of the light. Moreover, the reflected and transmitted intensities can thus be determined simultaneously. With an E-vector perpendicular or parallel to the plane of incidence, the quantum yields Q in relation to the absorbed light are measured. In this way it is possible for the first time to determine the absolute ratios of the polarization yield $K_{abs} = Q_{\perp}/Q_{\parallel}$ for large angles of incidence. With glass-side illumination the vector effect is anomalous in agreement with previous results, having maximum K_{abs} values of 1.7 in the region of photon energies of 2.2 eV. By contrast, the K_{abs} values are within 10% of the value 1 with vacuum-side illumination. Besides the virtual absence of the vector effect, there is a small systematic variation around the value 1. This shows that the excitation process of the electrons is almost isotropic counter to the flight direction of the photons. Only in the forward direction can anisotropy with a preferred direction perpendicular to the E-vector be deduced. This is analogous to the Compton effect and may be explained as a predominance of indirect transitions. The ratio of the possibility of indirect to direct transitions is estimated, and from this the band structure beyond the minima may be determined.

D.V. EDDOLLS, Allen Clark Research Centre, The Plessey Company Limited, Caswell, Towcester, Northants, Electrical Properties of n-Type Epitaxial Gallium Arsenide Measurements have been carried out to determine the temperature dependence of the Hall coefficient and resistivity of n-type epitaxial gallium arsenide in the range 4.2 to 300 °K. Such material has in general a lower impurity concentration and a higher mobility than melt grown material. The temperature dependence of the carrier concentration is described by the expression

$$n(N_A + n)/(N_D - N_A - n) = N_c \Big/ \sum_{r=1}^{p} g_r \exp(E_D - E_r)/kT$$

where g_r is the degeneracy factor and E_r the excitation energy of the r-th excited state of the donor atom, the number of excited states depends upon the donor concentration. The ionisation energy varies from 1.6×10^{-3} eV to 4.7×10^{-3} eV when the donor concentration changes from 5×10^{15} cm^{-3} to 3×10^{14} cm^{-3}. Qualitative comparison between the measured temperature variation of the Hall mobility and published theory shows reasonable agreement. The highest mobility measured was 103,000 cm^2/Vs at 52 °K.

K.N. PATHAK and B. DEO, Department of Physics, Indian Institute of Technology, Kanpur, Thermal Expansion of a Linear Lattice An expression for the coefficient of linear expansion of a linear lattice is obtained by using the thermodynamic Green's function method.

T.A. THOMPSON, Carnegie Institute of Technology, Pittsburgh Pennsylvania and H. CHANG, IBM Watson Research Center, Yorktown Heights, New York, Nanosecond Microscopic Measurements of Magnetic Film Switching by Kerr Magneto-Optic Apparatus With the use of a laser as the stable high-intensity light source and a nanosecond-response photomultiplier tube as sensing device nanosecond switching of 25 μm diameter spots in a Ni-Fe film has been measured on a Kerr magneto-optic apparatus. A sampling oscilloscope is used for examining the switching waveform obtained from the photomultiplier. Because the photon shot noise dominates the photomultiplier output, filtering techniques must be applied to the sampled output. The transient response of the whole system is limited by the anode pulse rise time of the photomultiplier which is a few nanoseconds. Examples of useful data obtained with this apparatus are given.

M. BALARIN und M. SCHENK, Zentralinstitut für Kernforschung der Deutschen Akademie der Wissenschaften zu Berlin, Rossendorf, Zur Verschiebung des Umwandlungspunktes zwischen festen Phasen durch Gitterdefekte Es wird eine thermodynamische Behandlung der Verschiebung von Phasenumwandlungspunkten polymorpher Festkörper infolge geringer Konzentrationen strahleninduzierter Gitterdefekte durchgeführt. Bei hohen Defektkonzentrationen kann es zu strahlenbedingten Phasenumwandlungen kommen. Die im Fall des Bariumtitanats untersuchte, durch Neutronenbestrahlung hervorgerufene Curiepunktverschiebung und schließliche Phasentransformation tetragonal → kubisch wird mit den theoretischen Schlußfolgerungen verglichen. Die experimentellen Ergebnisse stimmen mit der Theorie überein.

N. SINGH and M.D. KARKHANAVALA, Atomic Energy Establishment Trombay, Chemistry Division, Bombay, Studies on the Electrical Properties of Uranium Oxides-II, Non-Stoichiometric U_3O_8 The electrical conductivity σ and the Hall coefficient R_H have been measured in the range 250 to 400 °K on pellets of U_3O_8 sintered for different period (upto 120 h) at 900 °C resulting in a progressive decrease of the O/U ratio from 2.666 to 2.63 and a simultaneous increase in the density from 5.62 ("green" density) to 8.08 g/cm^3. The values of electrical conductivity were corrected for porosity. With decreasing O/U ratio, the conductivity and Hall mobility increased, while the activation energy of conduction decreased. The results for the highly sintered (96.5% theoretical density) material of composition $UO_{2.63}$ have been interpreted on the basis of "hopping" of "small" polarons.

G. KLUGE, Theoretisch-Physikalisches Institut der Friedrich-Schiller-Universität Jena, Akustische Aktivität Mit Hilfe der Methode der langen Wellenlängen werden die akustischen Zweige bis zur 3. Näherung untersucht. Breiten sich die Schallwellen in einer Richtung aus, in der sich der Kristall akustisch isotrop verhält, so erhält man für nichtprimitive Gitter neben der longitudinalen Welle eine rechts- und eine linkszirkular polarisierte Welle, die bei gleicher Frequenz verschiedene Wellenlängen haben. Das heißt, die Polarisationsebene einer transversalen Schallwelle wird gedreht, der Kristall verhält sich akustisch aktiv. Es wird weiter ein Experiment vorgeschlagen, mit dem man die akustische Aktivität nachweisen könnte.

R.T. DELVES, Chelsea College of Science and Technology, London, The Theory of the Stability of the Solid-Liquid Interface under Constitutional Supercooling (II) The theory of the stability of the solid-liquid interface during crystal growth given by the author previously is extended to include thermal effects, to first order in the amplitude of the initial disturbance. For dilute solutions the stability conditions are the same as those derived by Mullins and Sekerka (1) using a simpler approach with time independent diffusion equations. The original criterion of Tiller et al. for the occurrence of constitutional supercooling does not predict instability correctly for some special non-ideal solutions. The response of the interface to oscillatory variations of temperature is also calculated.

(1) W.W. MULLINS and R.F. SEKERKA, J. appl. Phys. 35, 444 (1964).

L.V. KIRENSKII, V.G. PYNKO, G.P. PYNKO, G.I. RUSOV, I.S. EDELMAN, Institut für Physik der Sibirischen Abteilung der Akademie der Wissenschaften der UdSSR, Krasnoyarsk,

R.V. SUKHANOVA und N.I. SIVKOV, Krasnoyarsker Pädagogisches Institut, Magnetische Eigenschaften von Einkristallschichten aus Eisen, Nickel, Kobalt und deren Legierungen It is shown that the energy of anisotropy of crystallites in epitaxial films is additive (for example in Co films). The field applied during the evaporation does not noticably influence the anisotropy of monocrystalline Fe films, but produces a strong uniaxial component in Ni and β-Co films. A domain fine structure in monocrystalline Fe-Ni and Co-Ni films with weak crystal anisotropy is found. The width of the walls in monocrystalline Fe, Ni, and β-Co films is measured electron microscopically. It is shown that in films with a strong biaxial anisotropy the coercivity changes with the change of direction of the switching field is satisfactorily described by

$$H_c = \frac{p}{J_s}(\cos\theta_1 - \cos\theta_2)^{-1}$$

where p is the pressure on the wall due to the field, θ_1 and θ_2 are the angles between the field and the magnetization vectors on both sides of the wall, and J_s is the saturation magnetization.

L.V. KIRENSKII, V.G. PYNKO, G.P. PYNKO, Institut für Physik der Sibirischen Abteilung der Akademie der Wissenschaften der UdSSR, Krasnoyarsk, N.I. SIVKOV, R.V. SUKHANOVA, Krasnoyarsker Pädagogisches Institut und M.A. OVSYANNIKOV, Sibirisches Technologisches Institut, Krasnoyarsk, Züchtung von Einkristallschichten aus Eisen, Nickel, Kobalt und deren Legierungen A study is made of epitaxial film growth of ferromagnetic metals and their alloys on NaCl, KCl, and LiF crystals which have been cleaved in vacuum or heat-treated in a vacuum of 10^{-4} Torr. It is found that in the latter case especially favourable conditions for epitaxy exist at a definite composition of the residual gases.

O.N. EFIMOV and E.K. KOVYEV, Institute of Semiconductors, Academy of Sciences of the USSR, Leningrad, Integral Characteristics of Anomalous Transmission of X-Rays Integral characteristics of anomalous transmission of X-rays are discussed on the basis of an experimental and theoretical analysis of the integrated intensity at the example of the symmetrical Laue-diffraction case, (220) germanium reflection, and CuK_{α_1} radiation. Experimental and calculated data are in good agreement. The physical meaning of these characteristics and the method of their determination are found. These characteristics are valid for the region of approximation to a "thick crystal", which is defined by the condition that only one wave field reaches the exit surface with anomalously low absorption (pure Borrmann effect). They are connected with the real and imaginary parts of the structure factor, $\mu_i \sim F_{ih}$ and $y_i \sim F_{rh}$ respectively. The necessity for the use of both integral characteristics is emphasized by the analysis of a general degree of perfection of the crystals with different types of defects and especially for cases where a "kinematic component" in scattering can be essential.

Y.T. CHOU, Edgar C. Bain Laboratory For Fundamental Research, United States Steel Corporation, Research Center, Monroeville, Pennsylvania, Screw Dislocations in and near Lamellar Inclusions Using the method of images, the stress field of a screw dislocation in and near a lamellar inclusion is determined. The general solution holds for many simple cases involving straight boundaries, such as metals with surface film, thin plates, or composite materials of laminated structure. The image force on the dislocation is derived and discussed in terms of the thickness and rigidity of the inclusion.

Y. KAZUMATA, T. KAMIKAWA, and K. OZAWA, Japan Atomic Energy Research Institute, Tokai-mura, Zero-Phonon Lines of Color Centers in LiF and NaF Optical absorptions of LiF and NaF irradiated by electrons or neutrons are measured at LHeT (liquid He temperature) and above. In these crystals zero-phonon lines are observed on the F-aggregate centers. From the temperature dependence of relative intensities between zero-phonon lines and broad bands, Debye temperatures are deduced using the following two approximations: (i) the Debye approximation, and (ii) the semi-local mode approximation based on the Debye model. Debye temperatures obtained by the above method are different in each F-aggregate center. The coupling between the trapped electrons in the lattice defects and the lattice waves is discussed and interpreted in terms of both the local and normal modes.

L. ELSTNER, VEB Halbleiterwerk Frankfurt/Oder, Werkteil Stahnsdorf, Die Temperaturabhängigkeit der Defektelektronenbeweglichkeit in Siliziumeinkristallen Es wurde die Temperaturabhängigkeit der Hallbeweglichkeit an hochgereinigten und dotierten p-leitenden Siliziumeinkristallen im Temperaturbereich von 20 bis 300 °K gemessen. Unterhalb von 100 °K können die Ergebnisse durch eine kombinierte Streuung der Defektelektronen an akustischen Gitterschwingungen und an ionisierten Störstellen erklärt werden. Bei alleiniger Streuung an akustischen Phononen gilt auch im p-Silizium das $T^{-3/2}$-Gesetz. Zwischen 100 und 300 °K wird der Einfluß einer zusätzlichen Streuung an optischen Phononen diskutiert.

R. PERTHEL, G. ELBINGER und W. KEILIG, Institut für Magnetische Werkstoffe der Deutschen Akademie der Wissenschaften zu Berlin, Jena, Über die Anisotropie des magnetischen Momentes und die Kristallanisotropie bei Kobaltferrit The anisotropic magnetic moment of the Co^{++} ion is studied by measuring the magnetic moment of a Co ferrite at room temperature in the [100], [110], and [111] directions. The values are 88.74, 87.87, and 87.70 emu g^{-1}, respectively. Using the relations for the magnetic moment and crystal anisotropy obtained by Slonczewski and Tachiki the magnetization curves are calculated for the [110] and [111] directions. Comparison between the calculated and experimental curves gives the magnetic crystal anisotropy constant as $K_1 = +(2.7 \pm 0.1) \times 10^6$ erg cm^{-3} and the constants of the anisotropy of the magnetization as $k_1 = 3.9 \times 10^{-2}$ and $k_2 = -3.7 \times 10^{-2}$. Measurements of the torque curves in the (100) plane show that $K_1 = +2.6 \times 10^6$ erg cm^{-3} and $k_1 = 3 \times 10^{-2}$.

J. PRZYSTAWA, Institute of Theoretical Physics, University of Wrocław, On the Magnetization Directions in the Heisenberg Antiferromagnet The isotropic Heisenberg antiferromagnet with two interpenetrating, translationally invariant, ferromagnetic sublattices is investigated by Green's functions technique in the random-phase approximation. From the requirement that Green's functions must not contain poles at E = 0, the equations for the directions of the magnetization vector are obtained. These equations are the same as derived in a previous paper (1) with the help of all other methods which are available up to now.

(1) J. PRZYSTAWA, Acta phys. Polon. 29, No. 5 (1966).

P. BAUDUIN et F. BUCHY, Laboratoire Central des Industries Électriques, Fontenay-aux-Roses (Seine), Oscillations de courant dans le sulfure de cadmium photoconducteur De nombreux travaux rapportent l'existence d'oscillations de courant dans un barreau de sulfure de cadmium quand le champ électrique appliqué atteint une certaine valeur critique. Dans le présent travail, où la conductivité de l'échantillon est controlée par un flux lumineux, les oscillations apparaissent, dans les conditions d'amplification ultrasonore, quand l'anode est masquée. Trois types d'oscillations de différentes fréquences sont alors observés dans un échantillon à géométrie particulière; deux d'entre eux sont étudiés avec une sonde flottante.

F. VAN DE WIELE and G. JACOBS, Laboratorium voor Kristallografie en Studie van Vaste Stoffen, Rijksuniversiteit Gent, Optical and Thermal Properties of Color Centers in CsBr Single Crystals Irradiated at Liquid-Helium Temperature Absorption bands at 435 and 437 nm, induced in CsBr single crystals by X-irradiation at liquid-helium temperature, are interpreted as being due to H-centers. Underlying bands are observed in the region of the F-absorption band. Optical and thermal interactions between the F- and V-type and α-bands are shown to be present and mechanisms for these interactions are proposed. The formation of a new absorption band at 320 nm during the thermal bleaching is observed. Some optical and thermal properties of this band are examined.

P. WEBER, Hauptabteilung Mikroelektronik des VEB Keramische Werke Hermsdorf und W. RUSKE, Institut für Magnetische Werkstoffe der Deutschen Akademie der Wissenschaften zu Berlin, Jena, Temperature Dependence of Oblique-Incidence Anisotropy in Permalloy Films The temperature dependence of the oblique-incidence anisotropy of approximately non-magnetostrictive permalloy films evaporated onto unheated substrates is investigated. Characteristic experimental results are as follows: (i) For isothermal annealing the time dependence of the constant of uniaxial anisotropy, K_u, is a logarithmic one. (ii) During annealing K_u changes its sign. (iii) The reversible changes of K_u as a function of temperature corresponds to that of M_s^2 (M_s saturation magnetization). A tentative explanation of the results is given on the basis of known structural properties of oblique-incidence films.

S.A. ALIEV, F.P. KESAMANLY, T.S. LAGUNOVA, and D.N. NASLEDOV, A.F. Ioffe Physico-Technical Institute, Academy of Sciences of the USSR, Leningrad, and Institute of Physics, Academy of Sciences of the Azerbaidzhan SSR, Baku, Hall Effect and Magnetoresistance of n-InP Crystals at Low Temperatures A study was made of the temperature dependence of the Hall constant R, the electrical conductivity σ, and the magnetoresistance $\Delta\varrho/\varrho$ between 1.7 and 300 $^{\circ}$K in specimens with electron concentrations from 2×10^{16} to 10^{18} cm^{-3}. A maximum in R(T) has been observed in the temperature range 20 to 100 $^{\circ}$K, $\Delta\varrho/\varrho$ is negative in all specimens below the maximum temperature of R(T). The results are explained by the participation of impurity band conduction.

W. ANDRÄ and K. STEENBECK, Institut für Magnetische Werkstoffe der Deutschen Akademie der Wissenschaften zu Berlin, Jena, Magnetic Anisotropy Induced in Thin Films at Low Temperatures In Einkristallschichten der Zusammensetzung 96%Ni-4%Co wurde die magnetische Ausrichtung von Wasserstoffatomen untersucht, über die von Adler (1 bis 4) berichtet worden ist. Die Orientierungsabhängigkeit, die dieser Autor vorausgesagt hat, konnte bestätigt werden. Es wurde gefunden, daß eine zusätzliche Anisotropie, die auch in reinem Ni ohne Wasserstoffgehalt existiert, bei Temperaturen unter Zimmertemperatur induziert werden kann. Diese Anisotropie kann möglicherweise mit der Wirkung von Nickel-Zwischengitteratomen in Hantellage oder von Doppelleerstellen erklärt werden. Es wurde außerdem festgestellt, daß eine Anisotropie durch Magnetfeldtemperung bei

$T = 80\ ^{\circ}K$ innerhalb von etwa 1 min induziert werden kann.

(1) E. ADLER, Z. angew. Phys. 15, 250 (1963).
(2) E. ADLER, Z. Metallk. 56, 249 und 294 (1965).
(3) E. ADLER and C. RADELOFF, Z. angew. Phys. 18, 482 (1965).
(4) E. ADLER, Z. Metallk. 56, 361 (1965).

H.J. FISCHBECK, Physikalisch-Technisches Institut der Deutschen Akademie der Wissenschaften zu Berlin, Über die Pippardsche Theorie des magnetischen Durchbruchs in Metallen Die Pippardsche halbphänomenologische Theorie (1) des magnetischen Durchbruchs in Metallen wird durch Transformation und WKB-Behandlung der Schrödingergleichung quantenmechanisch begründet. Dadurch kann der Pippardsche Durchbruchparameter unmittelbar definiert werden. Aus seiner Magnetfeldabhängigkeit ergibt sich die Blountsche Bedingung $(\Delta E)^2 \ll \hbar\omega_c \varepsilon_F$ für das Auftreten von magnetischem Durchbruch.

(1) A.B. PIPPARD, Proc. Roy. Soc. A256, 317 (1964).

E. KRÖNER und D. KESSEL, Institut für Theoretische Physik der Bergakademie Clausthal, Technische Hochschule, Clausthal-Zellerfeld, Zur nichtlokalen klassischen Theorie der Dielektrika Dielektrische Stoffe, in denen Ladungen über endliche Strecken verschoben werden, müssen durch eine nichtlokale Theorie beschrieben werden. Die Form der Theorie kann aus der Elektrostatik des Vakuums und der Vorstellung der endlichen Ladungsverschiebung allein abgeleitet werden. Die Grundgleichungen sind, abgesehen von der Interpretation, mit den von Rukhadze und Silin (1) diskutierten Gleichungen identisch. Eine Verschiebungsreichweite R wird definiert und das Kraftgesetz zwischen zwei Punktladungen für zwei Modellsubstanzen berechnet, das bei Abständen der Ladungen $< R$ erheblich vom Coulombschen Gesetz abweicht (Abschnitt 5). Abschnitt 6 bringt u.a. einiges über Anwendungen und eine kurze Studie über die Genauigkeit der Formel von Clausius und Mosotti.

(1) A.A. RUKHADZE und V.P. SILIN, Uspekhi fiz. Nauk 4, 459 (1961).

P. DELAVIGNETTE, R. TRIVEDI, R. GEVERS, and S. AMELINCKX, S.C.K.-C.E.N., Mol, Electron Microscopic Images of Dislocation Ribbons The image profiles of dislocation ribbons, as observed in diffraction electron microscopy, are calculated numerically and compared with observations on graphite. It is found that parallel dislocations with the same Burgers vector may exhibit quite different image profiles under identical diffraction conditions. The effect results from the overlap of the strain-fields due to the two parallel dislocations; it is particularly large in graphite as a consequence of the large anisotropy of this material. Analytical calculations using the kinematical theory are made for threefold ribbons; the contrast of the three dislocations is found to increase or decrease systematically; the sense of the variation depends on the sign of s.

F. VAN DER WOUDE, Laboratory for Solid State Physics, University of Groningen, Mössbauer Effect in α-Fe_2O_3 Experimental data obtained for the antiferromagnetic compound α-Fe_2O_3 with the Mössbauer technique, and their interpretation are given. α-Fe_2O_3 exhibits interesting properties, viz. the quantization-axis of the spins of the iron ions changes direction by 90° at T_t = 260 °K, and between 260 °K and the Néel temperature of 956 °K it shows weak ferromagnetism. The changes of the internal fields at the transition temperature are studied in detail. The effective magnetic field H_{eff} drops abruptly by 8 kOe on heating α-Fe_2O_3 through the transition temperature. Furthermore, the H_{eff} vs. T-curve shows temperature-hysteresis around T_t. An explanation for the discontinuity in H_{eff} at T_t can be given on the basis of a small asymmetry of the charge distribution of the 3d-electrons of the Fe^{3+} ions; this produces a contribution to the effective magnetic field of the orbital magnetic moment and the spin magnetic moment of these electrons in addition to the contact field. The temperature dependence of H_{eff} has also been determined and cannot be described in terms of a Brillouin curve for S = 5/2.

K. ISEBECK, F. RAU, W. SCHILLING, K. SONNENBERG, P. TISCHER, and H. WENZL, Physik-Department der Technischen Hochschule München, Stored Energy, Volume, and Resistivity Change in Neutron Irradiated Aluminium After irradiation of pure aluminium samples at 4 °K in the core of the Munich Research Reactor we have measured the liberation of the stored energy ΔQ as well as the changes of the length Δl and of the electrical resistivity $\Delta \varrho$ during annealing between 4 and 300 ^{o}K. For the first annealing stage, observed between 4 and 55 ^{o}K, these measurements yield $\Delta Q / \Delta \varrho$ = (38.3 ± 1.3) Ws/g$\mu\Omega$cm, $\frac{\Delta l/l}{\Delta \varrho}$ = (0.12 ± 0.005) %/$\mu\Omega$cm. The corresponding ratios for the two other stages observed have the same values. The experimental results suggest the following values for specific properties of defects in aluminium: E_F = 3.4 eV; E_I = 2.6 eV; E_V = 0.8 eV; V_F = 1.2 atomvol.; ϱ_F = 340 $\mu\Omega$cm; ϱ_V = 220 $\mu\Omega$cm; ϱ_I = 120 $\mu\Omega$cm. (E_F, E_I, E_V energy per Frenkel defect, per interstitial, per vacancy, respectively; V_F volume change per Frenkel defect; ϱ_F, ϱ_I, ϱ_V resistivity change per unit concentration of Frenkel defects, interstitials, vacancies, respectively).

D. AST und J. DIEHL, Institut für Sondermetalle am Max-Planck-Institut für Metallforschung, Stuttgart, Untersuchung des Zerfalls der Leerstellenagglomerate in neutronenbestrahltem Nickel mit Hilfe der kritischen Schubspannung. An Ni-Einkristallen, die bei 80 ^{o}C mit schnellen Neutronen bestrahlt worden waren, wurde die Erholung der kritischen Schubspannung untersucht. Dabei ergab sich, daß die für die kritische Schubspannung maßgebenden Leerstellenagglomerate ("verdünnte Zonen") bei ca. 400 ^{o}C nach einer Reaktion erster Ordnung mit einer Aktivierungsenergie von 2,28 eV zerfallen. Zur Deutung der Ergebnisse wird angenommen, daß die Agglomerate durch Abspaltung von einzelnen Leerstellen oder Mehrfachleerstellen zerfallen, wobei die Abspaltung der ersten Leerstelle der geschwindigkeitsbestimmende Schritt ist. Berechnungen der Stabilität möglicher Agglomeratkonfigurationen machen es wahrscheinlich, daß Versetzungsschleifen vom Frankschen oder R-Typus vorliegen, die ca. 10 Leerstellen enthalten. Der Haupterholungsstufe ist ein breites Erholungsband vorgelagert, das auf die Rekombination von an Verunreinigungen gebundenen Zwischengitteratomen mit den Leerstellenagglomeraten zurückgeführt wird.

D.N. SEIDMANN and R.W. BALLUFFI, Department of Materials Science and Engineering and Materials Science Center, Cornell University, Ithaca, New York, On the Efficiency of Dislocation

Climb in Gold The climb rate of the three-dimensional dislocation network in gold is investigated experimentally under the following non-equilibrium vacancy conditions: (1) supersaturation at elevated temperatures during rapid down-quenching; (2) supersaturation at lower temperatures following a rapid down-quench from an elevated temperature; (3) subsaturation during up-pulsing and holding at an elevated temperature. It is found that the network climbs with a quite high efficiency under all the above conditions; that is, at a rate which is within a factor of one-tenth to one of the maximum possible rate calculated on the assumption that vacancy equilibrium is maintained at the dislocation cores and that the climb rate is therefore vacancy diffusion limited.

W. METZDORF and H.E. WIEHL, Wernerwerk für Bauelemente der Siemens und Halske AG, München, Negative Oblique-Incidence Anisotropy in Magnetostriction-Free Permalloy Films An oblique-incidence anisotropy of the order of some oersteds, lying parallel to the plane of incidence exists also in magnetostriction-free permalloy films which were evaporated at small and medium angles of incidence. This is true especially for those, which were evaporated onto hot substrates. It could be shown, that a columnar grain growth, resulting in slightly anisotropic surface roughness of the continuous film is responsible for this small oblique-incidence anisotropy parallel to the plane of incidence. This negative oblique-incidence anisotropy can be used to compensate for the normal, positive one at medium substrate temperatures, in order to get low easy axis skew. The anisotropy of permalloy films which were obliquely deposited at room temperature could not be inverted by annealing at higher temperatures, whilst the negative oblique-incidence anisotropy could be enlarged by annealing.

J. NIHOUL and L. STALS, Solid State Physics Department, S.C.K.-C.E.N., Mol, On the Reaction Order of Bimolecular Recovery Processes in Solids A general expression is derived for calculating the reaction order of recovery processes when the rate equation is known. This expression is independent of the rate constant but entirely determined by the relative concentration of the defects being considered. On the basis of this expression the reaction orders of some bimolecular recovery processes are discussed as a function of defect concentrations, the ratio of the initial concentrations and the capture radius. It is shown that, unless rather severe conditions are fulfilled, the reaction order may vary in a wide range, even in cases where the application of some widely-used methods points to an apparently constant reaction order.

K.W. BÖER and P.L. QUINN, Physics Department, University of Delaware, Newark, Delaware, Inhomogeneous Field Distribution in Homogeneous Semiconductors Having an N-Shaped Negative Differential Conductivity Layer-like field inhomogeneities in homogeneous semiconductors which have been shown to occur whenever the conductivity decreases faster than linearly with increasing field strength, are discussed using Poisson and transport equations including drift and diffusion terms only. Stationary solutions are obtained for differential negative conductivity due to field-dependent carrier concentration or to field-dependent mobility in an approximation neglecting the diffusion term in the transport equation. When the diffusion is included, the method of characteristics is used to compare solutions for the case of field-dependent carrier concentration with the case of field-dependent mobility. A linear analysis of the behavior of solutions near singular points is conducted for more detailed information about the solution near boundaries.

E.V. CHISLER and M.S. SHUR, A.F. Ioffe Physico-Technical Institute, Academy of Sciences of the USSR, Leningrad, The Raman Spectrum of $NaNO_2$ in the Ferroelectric Phase The Raman spectrum of $NaNO_2$ single crystals in the ferroelectric phase is obtained and the polarization of the lines investigated. The forms of normal vibrations are considered on the basis of symmetry coordinates and line intensities. Formulae for translational vibrations are derived and numerical calculations are performed taking into account short-range repulsive forces between the nearest neighbours and long-range Coulomb forces.

E.V. CHISLER and M.S. SHUR, A.F. Ioffe Physico-Technical Institute, Academy of Sciences of the USSR, Leningrad, Raman

Spectrum and Phase Transition in Ferroelectric $NaNO_2$ Crystals The temperature dependence of the frequencies of the bands at 119, 153, and 220 cm^{-1} in the Raman spectrum of ferroelectric $NaNO_2$ crystals is investigated in the temperature region 40 to 220 ^oC. Anomalous decrease of the frequency of the dipole vibration at 153 cm^{-1} is discussed in connection with the Ginzburg-Anderson-Cochran theory (1 to 3). On the basis of the experimental data it is concluded that the phase transformation is due to reorientation of the NO_2^- group around an axis perpendicular to the anion plane.

(1) V.L. GINZBURG, Zh. eksper. teor. Fiz. 19, 36 (1949).
(2) P.W. ANDERSON, Physics of Dielectrics, Izd. Akad. Nauk SSSR, Moscow 1960.
(3) W. COCHRAN, Adv. Phys. 9, 387 (1960).

P. VIGIER, C. MINIER-CASSAYRE, P. MOSER et V. HIVERT, Centre d'Études Nucléaires de Grenoble, Étude de défauts ponctuels créés par bombardement neutronique dans le fer contenant certaines impuretés Les auteurs étudient les défauts ponctuels crées à 77 ^oK par irradiation aux neutrons dans du fer pollué. Utilisant simultanément des méthodes sensibles soit à leur réorientation, soit à leur concentration globale, ils mettent en évidence un processus d'agglomération des défauts élémentaires au niveau des stades I et II.

Z. KAMBERSKA and V. KAMBERSKY, Institute of Physics, Czechoslovak Academy of Sciences, Prague, Oblique-Incidence Anisotropy in Cobalt Films

G. KNEER and W. ZINN, Siemens und Halske AG, Forschungslaboratorium, München, On the Origin of M-Induced Anisotropy in Nickel and Permalloy Films In thin nickel and 81/19 nickel-iron films the anisotropy field strength H_k was measured immediately after deposition and during a subsequent systematic annealing in external magnetic fields. These measurements were performed with the films kept in ultra high vacuum using a planar Hall effect method. An analysis of the H_k-isotherms and of the temperature dependence of H_k revealed the spectrum of H_k-contributions of a given film with their relaxation times and activation energies. In non-magnetostrictive permalloy the main contribution of 60 to 85% to the anisotropy was attributed to directional ordering of the iron pairs. In nickel a complicated concurrence of several anisotropy contributions was found. The contribution of the magnetoelastic energy dominates; it can be understood in terms of the constraint model.

R. FISCHER, F. STÖCKMANN und J. STUKE, Institut für angewandte Physik der Technischen Hochschule Karlsruhe, Raumladungsbegrenzte Ströme in inhomogen belichteten CdS-Einkristallen Raumladungsbegrenzte Ströme wurden an photoleitenden CdS-Kristallen untersucht, bei denen ein schmaler Streifen zwischen den Elektroden schwächer belichtet wurde als der übrige Kristall. Bei diesen inhomogen belichteten Kristallen wird der Strom in dem schwach belichteten Steifen bereits bei kleinen Spannungen durch Raumladungen begrenzt, wobei das angrenzende stärker belichtete Gebiet auf der Seite der Kathode als injizierender Kontakt wirkt. In dieser Anordnung war es möglich, alle für raumladungsbegrenzte Ströme charakteristischen Abhängigkeiten von der Spannung, dem Elektrodenabstand und der Dunkelkonzentration der Ladungsträger an ein und derselben Probe zu messen. Alle Meßergebnisse stimmen sehr gut mit den Gesetzmäßigkeiten überein, die für Kristalle mit flachen Elektronen-Haftstellen gelten.

J. STUKE und G. WEISER, Institut für angewandte Physik der Technischen Hochschule Karlsruhe, Franz-Keldysh-Effekt an trigonalem und amorphem Selen An dünnen trigonalen Se-Einkristallen und an aufgedampften Schichten von anorphem Se wird die Verschiebung ΔE_g der Absorptionskante durch ein elektrisches Feld untersucht. Dabei ergibt sich, daß ΔE_g recht genau proportional zum Quadrat der Feldstärke F und zum Quadrat der Steigung S der exponentiell verlaufenden Kante ist. Die spezifische Kantenverschiebung $\Delta E_g/F^2$ hängt bei Kristallen stark von der Orientierung des Feldes und der Polarisation des Lichtes ab. Daraus ist zu schließen, daß die reduzierte effektive Masse in Richtung der c-Achse um etwa den Faktor 6,5 kleiner ist als senkrecht dazu. Bei amorphem Se stimmt der Franz-Keldysh-Effekt qualitativ mit dem von Se-Kristallen überein.

R.B. LAIBOWITZ and H.S. SACK, Department of Engineering Physics, Cornell University, Ithaca, New York, Luminescence and Photoconductivity in Silver Chloride with Laser Excitation

In zone refined AgCl, luminescence can be excited by ruby laser light through two photon processes, while the photoconductivity is a result of ionization of donors. The luminescence spectrum shows some phonon structure not observed with steady uv excitation.

R. VON JAN, Theoretical Physics Division, Atomic Energy Research Establishment, Harwell, Berkshire, The Interpretation of Recovery Stages in Irradiated Metals

Recent experiments suggest that in stages I and II bound Frenkel pairs only anneal. Such an interpretation leads to the following model: a vacancy and an interstitial form a bound Frenkel pair when their separation is less than about 10 lattice constants; the pairs which anneal above stage I are produced by multiple displacements or focused replacement sequences, and all pairs arising from defocused single displacements anneal in stage I, secondary pairs (with partners from different damage events) produce effects similar to those expected from a freely migrating interstitial. A consistent interpretation of the recovery in Au, Cu, and Al can be given in this way.

Original Papers published in "Fizika Tverdogo Tela"
Vol. 8, No. 6

L.E. GUREVICH and I.V. IOFFE, Leningrad, Instability of Non-Uniform Current or Field Distribution A criterion is deduced for the occurrence of a discontinuity in the uniform distribution of a current or field. This criterion is applied to cases in which the external fields or the mean free path are inhomogeneous. The case of conductors with one type of carrier, the effect of alternating external fields, and the peculiarities of the effects in anisotropic crystals are discussed.

S.K. TURKOV and T.D. SHERMERGOR, Voronezh, Effect of Screw Dislocations on the Internal Friction of Paraelastic Solids The internal friction due to stopping oscillating screw dislocations is calculated for the relaxation of the strain fields of the dislocations in a standard linear medium. The dependence of this type of internal friction on the defect of the modulus of the medium and on the magnitude of the applied stress is studied. It is shown that the ratio of the height of the dislocation peak to the height of the peak of a dislocation-free solid decreases with increasing defect of the modulus as well as with increasing applied stress.

A.E. NIKIFOROV, Sverdlovsk, Effect of Electric Field on the Optical Spectrum of Exchange-Coupled Pairs of $3d^3$ Ions in Corundum The Stark effect for spectral lines due to exchange-coupled pairs of $3d^3$ impurity ions in corundum is studied theoretically. In this case the combined charge-dipole interaction of the ions of the pair and the external electric field must be taken into account. The splitting values and the structural parameters of the pairs are expressed in terms of the splitting values of the R-lines in diluted samples. As distinct from the Stark effect for R-lines in diluted samples, a splitting can also be observed for the electric field perpendicular to the crystal axis.

B.YA. LYUBOV and V.A. SOLOVEV, Moscow, Kinetics of the Decay of Dislocation Cracks on Polygonal Walls of Edge Dislocations The decay of deformation-induced metastable dislocation cracks on polygonal walls of edge dislocations produced by means of diffusion to the crack surface is investigated theoretically. Estimates are made of the distribution of the dislocations in such walls and of the time constants characterizing the process by assuming the crack sides to be planar. The possible role of this process in brittle fracture and mechanical-thermal treatment is discussed.

T.V. MASHOVETS and R.YU. KHANSEVAROV, Leningrad, Low Temperature γ-Irradiation and Annealing of InSb Investigations are made of the effect of Co^{60} γ-irradiation at 77 °K on the electrical properties of InSb. The effect of isochronous defect annealing in the temperature range 77 to 300 °K is also considered. The dose-dependence of the defect production rate, and the boundary positions of the Fermi levels in n- and p-type material are determined. The results are explained in terms of a complex (monopolar) defect annealing process operating during the irradiation with a particular dependence of the annealing rate on the charge state of the defect. Information is obtained on the energy spectrum of the γ-radiation-induced defects.

O.V. VAKULENKO, M.P. LISITSA, and YA.F. KONONETS, Kiev, Infrared Absorption by Free Carriers in PbS The absorption spectrum of PbS is investigated in the wavelength range 3 to 15 μm at 293 and 100 °K. When the background which is apparently connected with mechanical defects is eliminated, the absorption coefficient for free carriers is found to be proportional to $\lambda^{2.8 \pm 0.2}$ (λ wavelength). Possible scattering mechanisms for the free carriers are discussed.

E.P. GUENOK and A.YU. KUDZIN, Dnepropetrovsk, Effect of Moisture on the Polarization of $BaTiO_3$-Ta_2O_5 Single Crystals It is found that an increase in the moisture of the surrounding atmosphere leads to a large increase in the low-field dielectric constant of a $BaTiO_3$-Ta_2O_5 single crystal and also facilitates the polarization of the crystal in strong fields. It is shown that the observed phenomena are due to the properties of the surface layer of the crystal.

V.M. GUSEV, V.V. ZADDE, A.P. LANDSMAN, and V.V. TITOV, Some Characteristics of Ion-Bombardment-Produced p-n Photoelements Some p-n photoelements are obtained by bombarding p-type silicon with 30 keV-phosphorus ions. The threshold irradiation dose for which a p-n junction is formed in 4 Ωcm silicon, and the depth of the junction are determined. After the irradiated crystals are annealed, the p-n junction appears at a depth of ≈ 1 μm with average donor concentration of $\approx 10^{19}$ cm^{-3} in the doped layer. The efficiency of these photoelements is 6 to 8%. The photoelectric properties of the photoelements are measured.

S.L. GINZBURG, Leningrad, Conductivity Oscillations in Bismuth due to the Electron-Phonon Interaction A theoretical study is made of the oscillations of the transverse magnetoresistance in bismuth due to inelastic scattering of electrons by optical phonons. It is shown that the investigation of such oscillations can give additional information on the band structure.

L.YA. SHEKUN, Kazan, Theory of Optical Spectra and EPR of Nd^{3+} in $PbMoO_4$ Single Crystals The crystalline Stark effect of Nd^{3+} in $PbMoO_4$ is calculated. The sub-levels of the ^4I_J and $^4F_{3/2}$ terms observed experimentally are described by a tetragonal potential with parameters $B_2^0 = +260$, $B_4^0 = -83$, $B_6^0 = -10$, $B_4^4 = -750$, and $B_6^4 = -500$ cm^{-1} yielding average deviations from 4 to 10 cm^{-1}. The wavefunctions are found taking into account J-mixing. Using these wavefunctions the parameters of the spin Hamiltonian are calculated and are found to agree with EPR data. A comparison of the theoretical interpretation of the polarization of the reflection spectra with the experimental results indicates a S_4- or D_{2d}-type symmetry of the crystal field.

N.S. RYTOVA, Moscow, Chemical Potential and Density of States of a Non-Ideal Electron Gas in a Doped Thin Film The density of states and the chemical potential of the electron gas in a semiconducting thin film are calculated by the Green's function method taking the inter-electronic interaction and the random distribution of impurities into account. For this calculation it is assumed that the motion of the electrons perpendicular to the plane of the film is quantized and only one quantum level is filled for which the electron gas is two-fold degenerate.

P.P. KUZMENKO, N.N. NOVIKOV, N.YA. GORIDKO, and L.I. FEDORENKO, Kiev, Photomechanical Effect in Germanium Doped with Weakly Soluble Elements The photomechanical effect is measured in germanium with small concentrations of antimony, indium, and gallium; also in antimony with small concentrations of germanium. In all cases the decrease in the hardness of the illuminated surface depends strongly on the impurity concentration. The photomechanical effect vanishes when the impurity concentration is near the solubility limit. The nature of the impurity appears (within the present experimental error) to have no effect on the characteristics of the photomechanical effect. The nature of the impurity does, however, slightly affect the value of the "thermal" micro-hardness of the sample. It is concluded that the mechanical properties of germanium depend essentially on the quantity rather than the nature of the impurity.

S.V. KARYAGIN, Moscow, Intensities in the Hyperfine Structure of Mössbauer Spectra The intensities in the hyperfine structure (hfs) for multi-field emission of the mixed E-M-type are calculated for the case in which the local field responsible for the hfs is asymmetric. The local field is considered as a superposition of the magnetic and quadrupole electric field. The effect of the anisotropy of the Mössbauer factor f' on the hfs intensities is studied for powders. The case of single crystals with some sublattices is discussed in the context of a transition from simple single crystals to polycrystals. Tables are given of the functions calculating the angular distributions in crystals under an asymmetric local field. Typical results for the anisotropy of f' calculated by a numerical method are presented. The relations deduced give detailed information about the crystals and nuclei if used in conjunction with very accurate measurements of the hfs intensities.

V.L. BONCH-BRUEVICH, Moscow, Movement of Electric Domains in Semiconductors with Hot Electrons The field and charge density distribution in a semiconductor with hot electrons is treated theoretically for the one-dimensional nonlinear case. It is shown that under specified conditions propagation occurs of stationary waves (of fixed amplitude) of the space charge and field. This leads to oscillations being experimentally observable during domain movement. The vibration period is related to the sample length and to the phase velocity of the wave. The order of magnitude of this velocity is estimated.

A.V. GOLUBKOV, E.D. DEVYATKOVA, V.P. ZHUZE, V.M. SERGEEVA, and I.A. SMIRNOV, Leningrad, Thermal Conductivity of Lanthanum and Its Monochalcogenides The thermal conductivities of La, LaTe, LaSe, and LaS are measured over the temperature range 80 to 480 ^{o}K, and the electronic and lattice components are separated. It is shown that the lattice component plays an important role in the thermal conductivity of these substances. The temperature dependence of the lattice thermal conductivity is explained by the phonon-phonon and phonon-electron scattering.

YU.N. OBRAZTSOV, Leningrad, Quantum Oscillations of the Thermo E.M.F. in Semiconductors in a Transverse Magnetic Field The thermo e.m.f. of a semiconductor, with a simple band structure, in a transverse magnetic field is calculated for the range of quantum oscillations which are connected with the filling of levels with small Landau quantum number ($\hbar\Omega \sim \xi_F$; Ω cyclotron frequency, ξ_F Fermi level).

M.S. BRESLER, R.V. PARFENEV, and S.S. SHALYT, Leningrad, Quantum Oscillations of the Thermo E.M.F. in n-InSb The quantum oscillations of the transverse and longitudinal magneto-thermo e.m.f. on n-InSb are measured at liquid-helium temperatures and compared with the oscillations of the transverse and longitudinal magnetoresistance and the Hall coefficient. A spin splitting of the Landau energy spectrum is observed which allows the g-factor of the spectroscopic splitting to be estimated. A phase shift of the oscillating longitudinal and transverse magneto-thermo e.m.f. coefficients is found and the relations governing the increase of these coefficients are obtained in the quantum limit. Comparing the experimental results with theory for the transverse magneto-thermo e.m.f. demonstrates the importance of the effect of an entrainment of the electrons by phonons.

A.P. LUKIRSKII, O.A. ERSHOV, T.M. ZIMKINA, and E.P. SAVINOV, Leningrad, Spectral Distribution of the Coefficients of Absorption, Reflection, and Photoemission of LiF in the Range 60 to 120 eV The coefficients of absorption, reflection, and quantum yield in LiF are measured in the range 60 to 120 eV. Over a range of about 15 eV near the K-absorption edge of the Li ion pronounced fluctuations in the absorption spectrum are observed and these are completely reproduced in the spectral distribution of the quantum yield. The high values of the quantum yield and its correlation with the absorption coefficient indicate the fundamental role of Auger electrons in the photoemission near the absorption edge. A fine structure is observed in the reflection spectrum near the K-absorption edge of the Li ion which agrees qualitatively with the fine structure of the absorption spectrum.

P.V. PAVLOV, E.I. ZORIN, and D.I. TETELBAUM, Gorkii, Inversion Layers in n-Germanium Formed by Boron- and Aluminium-Ion-Bombardment The conductivity of inversion layers in Ge formed by B- and Al-ion-bombardment with subsequent annealing is investigated for a wide range of irradiation doses. The results are compared with those of analogous experiments using C, Ne, and Ar ions. The significance of the type of bombarding ion used is demonstrated. The distribution of the bombardment-introduced impurities with depth is studied.

R.A. ZHITNIKOV and A.P. PAUGURT, Leningrad, Paramagnetic Resonance of Free Silver Atoms Stabilized on Solid Surfaces Silica-gel with adsorbed silver ions is irradiated with X-rays at 77 ^{o}K. As a result of this, free silver atoms are formed which are stabilized on the silica-gel surfaces. The stabilized silver atoms are studied by paramagnetic resonance measurements. The formation mechanism of the stabilized silver atoms and their interaction with the surface are discussed.

M. EREZHEPOV, Tashkent, Theory of Hall Effect and Magnetoresistance of Semiconductors in Strong Magnetic Fields Including the Field of Charged Impurity Centres The effects of the screened Coulomb field of charged impurity centres on the Hall field and magnetoresistance are calculated for the case of non-quantizing magnetic fields.

V.M. GAVRILYUK and V.K. MEDVEDEV, Kiev, Field Emission Microscopic Study of the Adsorption of Lithium on the Surface of a Tungsten Single Crystal The adsorption of Li on the faces (110), (111), (112), and partly (100) of a W single crystal is investigated using a field emission microscope. The dependence of the work function φ on the concentration n of the adsorbed atoms is measured. The Li concentration at the minimum of the curve $\varphi(n)$ for the face (110) is found to be $(4.7 \pm 0.3)\times 10^{14}$ cm^{-2}. Irreversible changes in $\varphi(n)$ are observed between 112 and 167 ^{o}K. The phenomena observed can be explained by assuming the existence of two processes: chemisorption for $n < 4\times 10^{14}$ cm^{-2} and the formation of a metallic film which is superimposed on the first process at high n. The irreversible temperature changes seem to be connected with changes in the structure of the film.

G.F. KVENTSEL, Kiev, Two-Exciton Optical Absorption Calculations are made of the probability of optical absorption by the molecules of a molecular crystal by the simultaneous excitation of two excitons. Detailed consideration is given to the case in which one of the transitions from the ground state is forbidden. It is shown that the integral intensity of this absorption is principally determined by the energy of the resonance interaction of the neighbouring molecules.

A.M. VASILEV and L.I. TRUSOV, Moscow, Contribution of Umklapp Processes and Inter-Valley Transitions to the Free Electron Absorption in n-Si A theoretical investigation is made of the contribution of inter-valley transitions and Umklapp processes to the free electron absorption in n-Si. The contribution of these processes is estimated and found to be commensurable with the contribution of intra-valley electron-phonon scattering.

A.A. PREDVODITELEV, N.K. RAKOVA, and E.S. CHEBOTAREVA, Moscow, Dislocation Movement during Creep in NaCl Crystals The movement of dislocations during creep is studied in NaCl crystals. It is found that under these conditions the dislocations are not multiplied and only ≈ 40% of the mobile dislocations take part in the process. The magnitude of the creep is calculated from the movement parameters measured.

M.I. KURKIN and N.G. PARFENOVA, Moscow, Effect of the Interaction of Nuclear Spin Waves on the Quadrupole Splitting of NMR Lines A theoretical study is made of the NMR spectrum in antiferromagnetics, with the antiferromagnetic axis lying in the basal plane, taking into account quadrupole interaction and the interaction of nuclear spins with spin waves.

SH.F. MURTAZIN, Kazan, Effect of "Phonon Bottle Neck" in Nuclear Magnetic Relaxation Using the results of Holstein (1) an expression is derived for the spin-lattice relaxation time of crystals due to the interaction of nuclear and electron spins in the presence of a "phonon bottle neck". Gaussian and Lorentz forms are considered for the EPR lines of the paramagnetic centres. It is shown that the width of the spectral distribution of excited phonons with a frequency equal to the sum of the electronic and nuclear frequencies, and also the width of the excitation of phonon "holes" with a different frequency, can considerably exceed the linewidth of the electronic resonance. The theoretical results agree with experiment (2).

(1) T. HOLSTEIN, Phys. Rev. 72, 1212 (1947); 83, 1159 (1951).
(2) S.M. DAY, E. OTSUKA, and B. JOSEPHSON, Phys. Rev. 137, A108 (1965).

T.A. POLYANSKAYA, G.A. SIKHARULIDZE, V.M. TUCHKEVICH, and YU.V. SHMARTSEV, Leningrad, Galvanomagnetic Effects in $CdSnAs_2$ The galvanomagnetic effects in n- and p-type $CdSnAs_2$ single crystals are investigated in the temperature range 1.3 to 450 ^{o}K. It can be shown that the experimental results do not contradict the theoretical prediction of the structure of the conduction band. It is assumed that the predominant scattering mechanism for $T > 300$ ^{o}K is the interaction of electrons with optical phonons. The hole effective mass is found to be approximately 0.1 m_o, and the ratio of electron to hole mobility is 25 (at T ≈ 300 °K). It is suggested that in p-type samples at low temperatures fast holes give an anomalous contribution to the electrical conductivity. This assumption is confirmed by the complicated dependence of the magnetoresistance on the magnetic field.

A.V. KALYUSH, Chernovtsy, Theory of Galvanomagnetic Effects with Anisotropic Electron Energy Spectrum A theoretical study is made of the effect of an anisotropic electron energy spectrum on the spectrum of the acoustic branch of excited systems of electrons and ions in a homogeneous magnetic field. The fluctuations which are connected with the thermal motion of the ions are calculated with the inclusion of the collective Coulomb interaction between electrons and ions. The concentration dependence of the anisotropy parameter and the Hall mobility for T = 300 °K are found.

O.V. KONSTANTINOV and G.V. TSARENKOV, Leningrad, Recombination Waves in a Finite Sample The conditions for the formation of recombination waves in a sample with ohmic contacts are studied theoretically. The recombination waves appear when the drift velocity of the carriers in a constant electric field exceeds a critical value. In this case in a sample with contacts of low injection ability current oscillations are generated. When the field is slightly lower than the critical value the sample impedance shows resonance peculiarities due to the existence of damped recombination waves.

N.V. STAROSTIN, Theory of Frenkel Excitons in Cu_2O A variant of the theory of Frenkel excitons is developed for cubic crystals with a complicated elementary cell of Cu_2O-type. A detailed investigation is made of the properties of a separate band of Frenkel excitons at wavevector $\vec{k} = 0$ which belong to the irreducible tensor representation of the cubic group Γ_5. The wavefunctions deduced in this paper are used for the calculation of the dielectric tensor $\varepsilon_{i,ij}(\omega, \vec{k})$ in the range of quadrupole absorption (ω frequency).

L.S. DUBINSKAYA and I.I. FARBSHTEIN, Leningrad, Anisotropy of Scattering by Ionized Impurities in Tellurium The concentration dependence of the anisotropy of the relaxation time in tellurium is calculated for scattering by ionized impurities at 4.2 °K. As a result the observed anisotropy of the electrical conductivity is found to agree with the anisotropy of the effective mass known from cyclotron resonance experiments. A qualitative agreement is found between the calculated and experimental concentration dependences of the anisotropy of the electrical conductivity.

A.G. LUNDIN and S.P. GABUDA, Krasnoyarsk, Anisotropy of the Magnetic Screening of F^{19} Nuclei in LaF_3 Single Crystals The NMR of F^{19} in LaF_3 single crystal is studied in the temperature range -100 °C to room temperature and at frequencies of 15 and 37 MHz. The magnetic screening constants of the two non-equivalent groups in the structure of F atoms are determined. With these constants some parameters of the electronic structure of the crystals are calculated. It is shown that the diffusion constants of the two non-equivalent groups of F ions are different.

N.V. VOLKENSHTEIN and G.V. FEDOROV, Sverdlovsk, Hall Effect in Holmium The Hall effect of holmium (99.9% purity, resistivity ratio $\varrho_{300\ °K}/\varrho_{4.2\ °K} = 12$) is measured in the temperature range 4.2 to 300 °K. The field and temperature dependence of the Hall effect is compared with the dependences of two other rare-earth metals.

A.F. KRAVCHENKO, V.S. SARDARYAN, and L.I. MAGARILL, Novosibirsk, Phenomenological Theory of the Longitudinal Hall Effect A phenomenological theory of the longitudinal Hall effect in cubic crystals with anisotropic energy bands, two-band conductivity and a tensor relaxation time is developed.

M.L. BELLE, Leningrad, Effect of Temperature and Doping on the Conduction Band Edge of Some n-Type Group $A^{III}B^{V}$-Semiconductors The reflection spectra of GaAs and InAs with different concentrations of n-type impurities are measured at 100 and 290 °K in the range of the reflection doublet $\Lambda_3 \rightarrow \Lambda_1$. With increasing impurity concentration the doublet is shifted to lower energies indicating a reduction of the energy gap. The asymmetric form of the reflection band at T = 100 °K can be explained by the existence of exciton states near the L-minimum of the conduction band. The disappearance of the asymmetry with increasing impurity concentration allows the binding energy of the L-exciton to be estimated.

V.M. FRIDKIN, G. GULYAMOV, V.A. LYAKHOVITSKAYA, V.N. NOSOV, and N.A. TIKHOMIROVA, Moscow, Anomaly of Optical Properties of the Ferroelectric SbSI in the Phase Transition Region The dependence of the energy gap E_g on hydrostatic pressure p and temperature T is measured in the ferroelectric phase transition region of SbSI. It is shown that in the vicinity of the Curie point dE_g/dT and $(\partial E_g/\partial p)_T$ have anomalously large values. Away from the phase transition point the change of E_g with temperature is determined, as usual, by direct electron-phonon interaction $dE_g/dT \approx (\partial E_g/\partial T)_V$. In the phase transition region, however, the value of dE_g/dT is determined by the thermal expansion of the crystal and $dE_g/dT \gg (\partial E_g/\partial T)_V$. During the transition from the ferroelectric to the paraelectric state the electron-phonon interaction term is reduced by a factor of two.

A.G. BOGANOV, I.I. CHEREMISIN, and V.S. RUDENKO, Leningrad, Direct Calculation of the Electrostatic Energy of Ionic Lattices A summation method for the calculation of the electrostatic energy of ionic lattices is developed which is applicable to complicated cubic (PrO_2, C-Pr_2O_3) and hexagonal (A-Pr_2O_3, α-Al_2O_3) lattices.

YU.E. ZAKIEV, Makhach-Kala, Transverse Magnetoresistance of n-InAs in the Ultra-Quantum Limit The transverse magnetoresistance of n-InAs is measured in the temperature range 20.4 to 600 °K in magnetic fields up to 420 kOe. The measured dependences of the magnetoresistance on magnetic field and temperature are compared with theory (1, 2).

(1) E.N. ADAMS and T.S. HALSTEIN, J. Phys. Chem. Solids 10, 254 (1959).

(2) V.L. GUREVICH and YU.K. FIRSOV, Zh. teor. Fiz. 40, 1 (1961).

S.V. STARODUBTSEV, D. KAIPNAZAROV, L.P. KHIZNICHENKO, and P.F. KROMER, Tashkent, Low-Temperature Internal Friction in Silicon The dislocation relaxation in silicon single crystals is investigated by low-temperature internal friction measurements. The experimental results are analyzed on the basis of the theory of Seeger, Donth, and Pfaff (1). It is shown that the ratio of the Peierls strain to the shear modulus is 1.5×10^{-5}, i.e. one order of magnitude smaller than in metals.

(1) A. SEEGER, H. DONTH, and F. PFAFF, Disc. Faraday Soc. 23, 19 (1957).

Short Notes published in "Fizika Tverdogo Tela" Vol. 8, No. 6

A.P. LUKIRSKII, T.M. ZIMKINA, and S.A. GRIBOVSKII, Leningrad
Photoionization of d-Electrons in Te, Sn, Pb, PbTe, and SnTe

YU.A. BYKOVSKII and V.F. ELESIN, Moscow
Possibility of a Photomagnetic Effect for "Semi-Light" Holes in Germanium

V.K. MEDVEDEV, Kiev
Migration of Lithium on the Surface of an Adsorbed Lithium Film

A.F. KRAVCHENKO and V.S. SARDARYAN, Novosibirsk
Theory of the Electron Energy Spectrum of Doped Semiconductors

E.V. CHISLER, Leningrad
Opalescence and Combination Scattering Spectrum during the Phase Transition KNO_3 II - KNO_3 I

N.B. GORNYI, Leningrad
Plasmon Photoeffect and Secondary Electron Emission

I.V. KRYUKOVA, L.I. PADUCHIKH, and V.G. KARNAUKHOV
Recombination Radiation from p-n Junctions in GaSb

A.P. KULSRESHTA and V.A. GORYUNOV, Moscow
Calculation of Thermally Stimulated Currents

B.I. BERKIN and K.L. DUDKO, Kharkov
Origin of the Photomagnetic Anomaly in Germanium

YU.V. POLIKANOV, G.F. LYMAR, and L.M. ZHUKOVA
Radiative Recombination in a Volume Charge Layer of a p-n Junction

R.E. ANDREICHIN, G.K. GETOV, and P.A. SIMIDCHIEVA, Sofia
Effect of Silver Impurity on the Position of the Fundamental Absorption Edge in Vitreous As_2S_3

A.Z. GRASYUK, V.F. EFIMKOV, I.G. ZUBAREV, V.A. KATULIN, and A.N. MENTSER, Moscow
CdSe Laser with Two-Photon Excitation

L.N. PELIKH, Kharkov
Quantum Oscillations of the Fermi Level in Sb

A.V. RZHANOV, K.K. SVITASHEV, and V.M. SHEPEL, Novosibirsk
Effect of Trapping by Surface Defects on the Intrinsic Photoconductivity Spectrum of Thin Germanium Samples

S.A. KIZHAEV and V.A. BOKOV, Leningrad
Magnetic Properties of $PbCo_{0.5}W_{0.5}O_3$ and $BaNi_{0.5}W_{0.5}O_3$

S.M. FAINSHTEIN and V.N. NEKHODTSEV
Effect of a Damaged Layer on the Surface Recombination Velocity in Germanium

I.P. GRIGAS and A.S. KARPUS, Vilnyus
Peculiarities of the Interaction of a Microwave Field with Sb_2S_3 Single Crystals

V.S. VAVILOV, M.I. GUSEVA, E.A. KORONOVA, V.V. KRASNOPEVTSEV, V.F. SERGIENKO, and V.V. TITOV, Moscow
Ion-Bombardment-Produced Semiconducting Diamonds

YA.A. KRAFTMAKHER and T.YU. ROMASHINA, Novosibirsk
Heat Capacity of Cobalt near the Curie Point

B.YU. LOTOTSKII and L.K. CHIRKIN, Leningrad
Negative Differential Resistance due to Micro-Heating

E.L. FRANKEVICH, E.I. BALABANOV, and G.V. VSELYUBSKAYA, Moscow
Study of the Change of the Photoconductivity of Organic Semiconductors in a Magnetic Field

B. ORAZGULYEV, Makhach-Kala
Dependence of the Transverse Hall Effect in p-Si on the Magnetic Field Strength

YU.E. ZAKIEV, Makhach-Kala
Oscillation of the Longitudinal Magnetoresistance in Heavily Doped n-InAs

List of Short Notes

List of Short Notes - Continued

Contents — Continued

Short Notes

Pre-printed Titles and Abstracts

of papers to be published in this or in the Soviet journal „Физика Твердого Тела" (Fizika Tverdogo Tela).

Manuscripts and letters should be sent to

physica status solidi, 102 Berlin, Neue Schönhauser Str. 20
or Internationaler Buchversand GmbH, 1 Berlin 19, Eschenallee 21.

The object of this journal is to publish original work in the general field of solid state physics. Each issue will be published on the first day of the month. Items published will fall into the following categories:

1. Review articles on topical subjects in the field of solid state physics written by arrangement with the editors.
2. Original papers on new results of research, which have not been submitted for publication elsewhere, after being reviewed. (The latest date for acceptance of contributions is the 8th day of each month, 50 days before the delivery of the journal.)
3. Short notes on topical results of research. These will be published in the photographically reproduced part of the journal and are the sole responsibility of the author. (The latest date for acceptance of manuscripts for this part is the 12th day of each month, 16 days before the delivery of the journal.)
4. Pre-printed titles and abstracts of papers to be published in this or in the Soviet journal „Физика Твердого Тела" (Fizika Tverdogo Tela), the latter translated into English.

Editorial Information

Manuscripts may be submitted in English, German, French, and Russian. Authors are asked to observe the following rules:

1. Manuscripts should be typewritten on one side only, with at least double spacing. For both review articles and original papers two copies should be sent. Manuscripts of original papers and short notes should not exceed 20 and 3 pages, respectively.
2. The author's name, initials and institution should be given.
3. An abstract of about ten lines, and, if possible, a translation of it into one of the above languages is required.
4. Drawings should be enclosed on separate sheets of paper; photographs should be submitted, on white paper, only for review articles and original papers.
5. Captions for the figures should be submitted on separate sheets of paper, and special instructions with regard to the size of illustrations should be given on these.
6. References should be numbered in order of appearance and should be given, on a separate sheet, in the form used in this journal.
7. Units, symbols, etc. should be used in accordance with the recommendations given by the IUPAP.

The authors will be sent two galley proofs of review articles and original papers. Proof reading of short notes and pre-printed titles and abstracts will be done the editorial staff.

For each original paper, and short note, a total of 75 reprints is available free of charge. The number of reprints free of charge per review article is 25. In addition review articles will be remunerated.

Bei Fragen zur Produktsicherheit wenden Sie sich bitte an:
If you have any questions regarding product safety,
please contact:

Walter de Gruyter GmbH
Genthiner Straße 13
10785 Berlin
productsafety@degruyterbrill.com